Current Approaches and Applications in Natural Language Processing

Current Approaches and Applications in Natural Language Processing

Editors

Arturo Montejo-Ráez
Salud María Jiménez-Zafra

MDPI • Basel • Beijing • Wuhan • Barcelona • Belgrade • Manchester • Tokyo • Cluj • Tianjin

Editors

Arturo Montejo-Ráez
Universidad de Jaén
Jaén, Spain

Salud María Jiménez-Zafra
Universidad de Jaén
Jaén, Spain

Editorial Office
MDPI
St. Alban-Anlage 66
4052 Basel, Switzerland

This is a reprint of articles from the Special Issue published online in the open access journal *Applied Sciences* (ISSN 2076-3417) (available at: https://www.mdpi.com/journal/applsci/special_issues/Language_Processing).

For citation purposes, cite each article independently as indicated on the article page online and as indicated below:

LastName, A.A.; LastName, B.B.; LastName, C.C. Article Title. *Journal Name* **Year**, *Volume Number*, Page Range.

ISBN 978-3-0365-4439-7 (Hbk)
ISBN 978-3-0365-4440-3 (PDF)

Contents

About the Editors

Arturo Montejo-Ráez

Arturo Montejo-Ráez is an Assistant Professor at Universidad de Jaén (Spain). He holds a European PhD in Computer Science from the University of Granada. He started his research career at the European Laboratory for Particle Physics in Geneva (Switzerland), where he worked from 2000 to 2004. He is a member of the SINAI research group, part of the scientific committee of the Center for Advanced Studies in ICT (CEATIC), member of the Spanish Society for Natural Language Processing and founder and Chief Technological Officer of the spin-off Yottacode S.L., which develops NLP-related solutions for people with disabilities. His scientific activity focuses on human language technologies, with special attention to machine learning techniques. Current research topics are the analysis of stereotype bias in language models, user profiling in social networks and explainability in foundational language models. With 20 JCR articles and more than 90 research contributions. He received the ISCA Award for the best paper published in Computer Speech & Language, 2012-2016, among other mentions and recognitions for his work in international cooperation for development projects in Africa.

Salud María Jiménez-Zafra

Salud María Jiménez-Zafra is a post-doctoral researcher at Universidad de Jaén (Spain). She holds an International PhD in Computer Science, a Specialist in Internet Information Processing and a Diploma in Statistics from Universidad de Jaén. She is a member of the research group SINAI (TIC 209) and belongs to the Spanish Society for Natural Language Processing, the PLN.net network and the DiverTLes community. She has been awarded, on a national competitive basis, several research contracts. Her scientific interests are focused on artificial intelligence, specifically in the field of Natural Language Processing, with her specialty being negation processing in Spanish and sentiment analysis. The research carried out so far has resulted in scientific contributions that have been cited in more than 2,350 works. She is also the author, together with Arturo Montejo-Ráez, of the book "Curso de Programación Python" published by Anaya. Throughout her career she has been awarded 7 research prizes, one of which is the award for the best doctoral thesis in Natural Language Processing granted by the Spanish Society for Natural Language Processing in the XIX Edition of the SEPLN Award.

Preface to "Current Approaches and Applications in Natural Language Processing"

Current approaches to Natural Language Processing (NLP) have shown impressive improvements in many important tasks: machine translation, language modeling, text generation, sentiment/emotion analysis, natural language understanding, and question answering, among others. The advent of new methods and techniques, such as graph-based approaches, reinforcement learning, or deep learning, have boosted many NLP tasks to a human-level performance (and even beyond). This has attracted the interest of many companies, so new products and solutions can benefit from advances in this relevant area within the artificial intelligence domain.

This Special Issue, focusing on emerging techniques and trendy applications of NLP methods, reports on some of these achievements, establishing a useful reference for industry and researchers on cutting-edge human language technologies.

Arturo Montejo-Ráez and Salud María Jiménez-Zafra

Editors

applied sciences

MDPI

Editorial

Current Approaches and Applications in Natural Language Processing

Arturo Montejo-Ráez * and Salud María Jiménez-Zafra *

Departamento de Informática, SINAI Research Group, CEATIC, Universidad de Jaén, Campus Las Lagunillas s/n, 23071 Jaén, Spain
* Correspondence: amontejo@ujaen.es (A.M.-R.); sjzafra@ujaen.es (S.M.J.-Z.)

Citation: Montejo-Ráez, A.; Jiménez-Zafra, S.M. Current Approaches and Applications in Natural Language Processing. *Appl. Sci.* **2022**, *12*, 4859. https://doi.org/10.3390/app12104859

Received: 9 May 2022
Accepted: 10 May 2022
Published: 11 May 2022

1. Introduction

Artificial Intelligence has gained a lot of popularity in recent years thanks to the advent of, mainly, Deep Learning techniques. These algorithms have broken many of the barriers in difficult computer based tasks such as computer vision, decision making or machine translation, among others. Nevertheless, many of the applications and problems overcome were already attempted with traditional algorithms in machine learning, heuristic approaches or knowledge-based systems. The big difference from previous approaches is that the current proposals are data-driven: they are able to learn from large amounts of data and build models to perform different tasks with a level of success never reached previously by other solutions.

This shift has been especially dramatic for Natural Language Processing (NLP). Linguistic-based methods have been surpassed by end-to-end architectures, where no prior knowledge on language is needed, although only when a massive amount of data is available. During the last two years we have witnessed the birth of amazing language models with impressive results in many different tasks, defining the new state-of-the-art in all of them. These models do not include, explicitly, traditional language processing tasks such as morpho-syntactic tokenization, lemmatization, stop-words removal, syntactic parsing, part of speech labeling, and other linguistic treatments on the text. New models seem to learn all of this linguistic information just from data.

Thus, NLP research has shown impressive improvements in many major tasks: machine translation, language modeling, text generation, sentiment/emotion analysis, natural language understanding, and question answering, among others. The advent of new methods and techniques such as graph-based approaches and reinforcement learning over deep learning architectures have boosted many of the tasks in NLP to reach human-level (and even further) performance. This has attracted the interest of many companies, so new products and solutions can profit from the advances of this relevant area within the artificial intelligence domain.

However, intensive research is still being conducted using deep learning approaches. Many new relevant features are being proposed, mainly related to stylometry, personality, or psicolinguistics. All of them are ad hoc features computed from texts that try to capture profile information, which, as we will see, can be used together with traditional machine learning algorithms to overcome user-centered tasks.

This Special Issue focuses on emerging techniques and trendy applications of NLP methods as an opportunity to report on all these achievements, establishing a useful reference for industry and researchers on cutting edge human language technologies. The contributions included in this issue propose new NLP algorithms and applications of current and novel NLP tasks. In addition, some trends, potential future research areas and new commercial products have been identified.

2. Review of Issue Contents

The contributions collected in this Special Issue tackle diverse tasks in NLP: text classification, text summarization, question and answering, machine translation, etc. We have organized these papers according to these topics.

2.1. Text Classification

Text classification is still a major concern in NLP research. Several contributions are related to this topic. For example, ref. [1] predict whether a patient had been diagnosed with a mental disorder and, if so, the specific mental disorder type. LIWC, spaCy, fastText, and RobBERT were used to analyze Dutch psychiatric interview transcriptions. LIWC, in combination with the random forest classification algorithm, performed the best in predicting whether a person had a mental disorder or not. SpaCy, in combination with random forest, best predicted which particular mental disorder a patient had been diagnosed with. When studying the results obtained with RobBERT and fastText, it was found, by applying LIME analysis, that the difference between mental disorder and no disorder was more prevalent in the manner of speaking than in the topics or the semantic content. Again, classical ML techniques such as Random Forest are still very useful.

Multimodal approaches are also present. In [2], a novel approach to fuse textual and visual features using a scaled dot-product attention mechanism is proposed. This is used in a multimodal classification system applied in fake news detection. The attention mechanism allows fine-grained combination of both visual embeddings and word embeddings taken from the image and text found in posts. The system achieves competitive results on the Weibo dataset.

Another paper studies the automatic detection of misogyny in web content by building an annotated corpus from several sources and then training a system for classifying texts [3]. The system is based on BERT embeddings and a final linear regression classifier. The results are good, although not comparable to other systems. A major contribution is the way in which the corpus is generated, which can allow for augmenting training datasets on misogyny detection.

Text classification can also be used to classify types of texts at high-level semantics. For instance, ref. [4] explores different machine learning and neural network techniques for the classification of strings as problems, non-problems, solutions, and non-solutions. The algorithm that provided the best results was a convolutional neural network.

To addresses the detection of fake news, the authors of [5] present a solution based on three steps: stance detection, author credibility verification, and machine learning classification. Stance detection verifies the relevance between the title and paragraphs of a news item; if there is a match, the next module checks whether the author is authentic to determine whether the news item should be believed or not. Finally, machine learning algorithms are used to classify the news item.

Text classification can also be applied to user profiling. A proposal for personality recognition relying on the dominance, influence, steadiness, and compliance (DISC) model together with a Bag-of-Words model of language is presented in [6]. Classical machine learning algorithms such as AdaBoost and Random Forests achieved good performance.

Topic detection is still stimulating research. Ref. [7] applies BERT word embeddings and a classical clustering algorithm (spherical k-means) to assign documents to topics. The proposal encodes documents as a linear combination of word embeddings and word frequencies in the document. Topics have been previously identified using the spherical k-means algorithm over all word embeddings in the corpus. Finally, documents are associated with topics using cosine distance. This method outperforms other approaches such as PLSA (Probabilistic Latent Semantic Analysis) and do not need to fine-tune the deep learning model.

In addition to systems and methods, this Special Issue includes some overviews. Related to this topic, ref. [8] provides a review on corpora related to deception detection on several approaches to the study of deception and on previous research into its linguistic detection. Moreover, the author explores the linguistic cues of deception in the Spanish language.

One last contribution to text classification is the creation of a new multi-modal Wikimedia Commons dataset based on concrete/abstract words [9], along with a novel multi-modal pre-training approach based on curricular learning. The authors use the curricular learning method to train the model on the concepts through images and their corresponding captions to achieve multimodal language modeling. BERT and Resnet-152 models are employed in each modality and combined using attentional pooling to perform pre-training on the dataset.

2.2. Name Entity Recognition

Among major natural language understanding tasks, information extraction is still attracting much of the research. Named Entity Recognition (NER) is a central problem here. This Special Issue covers some novel approaches to NER in different languages. For instance, ref. [10] proposes an approach to entity linking (associate mentions in documents to existing entities in a knowledge graph) that profits from structural information of the graph, so correlation information between entities is enriched. No deep learning is used here, nor machine learning. It is a fully distance-based approach.

Nevertheless, transformers are the most prominent approach to NER. In [11], the task of a nested named entity recognition over two and four levels of annotation is accomplished by fine-tuning a BERT model. The results outperform state-of-the-art approaches such as Bi-LSTM-CRF. Thus, this approach is easier to generalize as it does not need specific feature extraction methods.

Another contribution to fine-grained NER is [12]. This work proposes a system for using character-level embeddings over LSTM networks multi-stacked for feature fusion. The unbalance problem usually found in fine-grained NER is solved by means of contextual information of coarse-grained named entities. The system is able to outperform other state-of-the art NER systems.

To close the papers related to NER, an interesting overview is also included, but it is focused on the clinical domain [13]. The paper summarizes the current status of named entity recognition techniques and clinical relationship extraction in the clinical domain, discussing the existing models for the two tasks and their performances, the current challenges and future directions.

2.3. Question and Answering

Staying in natural language understanding tasks, Question and Answering (Q & A) systems still emerge as a continuous topic of research. In this regard, the paper by [14] proposes an attention model to solve question difficulty estimation in Question–Answering tasks. The method first relates question and information components using dual multi-head co-attention. Then, a self-attention model is applied over these relationships. This approach sets a new state-of-the-art in question difficulty estimation.

Expanding the number of question-answer pairs of Thai Question Answering corpora using Multilingual Text-to-Text Transfer Transformer (mT5) is the approach proposed by [15]. In addition, the authors propose a new syllable-level evaluation metric, which they consider more suitable for the Thai language because there is no ambiguity in syllable tokenization.

One last contribution to the Q & A topic is the paper by [16]. This paper introduces a privacy-preserving machine reading comprehension system capable of working with private data at a large scale and that is language independent.

2.4. Machine Translation

The problem of out-of-vocabulary (OOV) or rarely occurring words that limit the performance of neural machine translation models is known in automatic machine translation. The authors of [17] present a post-processing method for correcting machine translation results using a named entity recognition (NER) model to overcome this problem and conduct experiments on Chinese to Korean translation.

Another relevant issue is the estimation of the quality of a translation system. In [18], a pure performance comparison between several multilingual pretrained linguistic models (mPLM) is performed. As a result of the experiments, the authors confirm that the XLM-TLM model performs better and that the induced learning of cross-language alignment during pre-training had a positive impact. Furthermore, they perform experiments using mBART, and its additional noise schemes had a positive effect.

Bilingual embeddings are the subject of [19]. To train English–Welsh bilingual embeddings, the authors combine a Welsh corpus of approximately 145 million words with an English Wikipedia corpus. To learn the monolingual embeddings, they use word2vec and fastText. In addition, they explore three cross-language alignment strategies: cosine similarity, inverted softmax, and cross-domain similarity local scaling (CSLS). Different combinations of these approaches were evaluated on two tasks, bilingual dictionary induction and cross-lingual sentiment analysis. The best results were obtained using fastText monolingual embeddings and the CSLS metric.

2.5. Dialogue Systems

Conversational agents and chatbots are leveraging the research in dialogue systems. Two papers are included in this Special Issue with two totally different approaches. One is based on classical algorithms, and another uses a large language model. The former paper [20] proposes a system architecture for conversational agents that performs language understanding by intent detection and slot filling. The answering mechanism is based on a text retrieval engine (BM25). A classical CRF model is applied to perform the filling task, and the SVM algorithm was used for intent classification. No deep learning models were needed. The second approach is a novel task-oriented Arabic dialogue dataset (Arabic-TOD) and proposes an end-to-end generative dialogue system based on the multilingual mT5 [21]. The experiments show a performance comparable to high-resourced languages, such as English, and that a joint-training strategy with English and Chinese leads to better results.

2.6. Other Tasks

Explainability is a matter of study which is gaining deserved interest in recent years, in order to guarantee trustworthy systems. The work presented in [22] proposes a system to represent multilingual sentences using a natural machine language. The paper generates related universal concepts that are intuitive, according to human evaluation. Also related to explainability, the aim of the work presented in [23] is to provide people with an understandable representation of the complications of a disease. The authors present an approach to extract disease causal pathways, through cause–effect relation extraction, from documents on diabetes, kidney disease, heart disease, and arterial disease posted on Thai hospital web boards.

Related to Information Retrieval systems, we can find a proposal for query expansion [24]. This paper proposes a query expansion technique for Information Retrieval systems. A supervised expansion technique using the Naïve Bayes Multinomial Naïve Bayes algorithm is presented to extract relevant terms from the first documents retrieved by the initial query. In the evaluation of the proposed method, more accurate results are obtained compared to those achieved by the systems which participated in the TREC2017 Precision Medicine Track.

As an additional contribution, this time related to text summarization, is the work presented in [25]. It is a monolingual approach for abstractive summarization in Catalan and Spanish. The approach is based on a Transformer encoder–decoder pretrained and fine-tuned specifically for the language under studied. The performance of the monolingual models is compared with two of the most widely used multilingual models in text summarization, mBART, and mT5. Moreover, the authors present a new metric, content reordering, intended to help quantify the reordering of original content within an abstractive summary.

3. Conclusions

This Special Issue covers some of the most trending tasks in natural language processing: text classification, machine translation, information extraction, explainability, question and answering, or dialogue systems, among other topics. Many of the contributions in this Special Issue set a new state-of-the-art in targeted tasks.

It is remarkable how multilinguality is fostering research to cover what are considered "low-resourced" languages (i.e., those different from English or Chinese). In addition, we can confirm from the set of contributions that deep learning models (LSTM, BERT, mT5, among others) have irrupted the NLP arena to move approaches from computational linguistics to end-to-end solutions. Still, classical machine learning algorithms such as CRF, SVM, or Random Forest, just to cite few, are valid choices in many scenarios and are integrated in some competitive systems.

As a last remark, we find interesting the advent of hybrid approaches, such as those based in the combination of multiple features (word embeddings, char embeddings, BoW, etc.). In this ensemble of methods and techniques, graph-based and knowledge-based approaches deserve the focus of a growing number of studies.

As a main conclusion, this Special Issue offers a wide and varied insight into current NLP research, a domain of research which has already been considered as the main frontier in artificial intelligence.

Funding: This work was supported by Project LIVING-LANG (RTI2018-094653-B-C21) funded by MCIN/AEI/10.13039/501100011033 and by ERDF A way of making Europe, Fondo Social Europeo and Administration of the Junta de Andalucía (DOC_01073), Grant P20_00956 (PAIDI 2020) and grant 1380939 (FEDER Andalucía 2014-2020) from the Andalusian Regional Government.

Conflicts of Interest: The authors declare no conflict of interest.

References

1. Spruit, M.; Verkleij, S.; de Schepper, K.; Scheepers, F. Exploring Language Markers of Mental Health in Psychiatric Stories. *Appl. Sci.* **2022**, *12*, 2179. [CrossRef]
2. Wang, J.; Mao, H.; Li, H. FMFN: Fine-Grained Multimodal Fusion Networks for Fake News Detection. *Appl. Sci.* **2022**, *12*, 1093. [CrossRef]
3. Aldana-Bobadilla, E.; Molina-Villegas, A.; Montelongo-Padilla, Y.; Lopez-Arevalo, I.; Sordia, O.S. A Language Model for Misogyny Detection in Latin American Spanish Driven by Multisource Feature Extraction and Transformers. *Appl. Sci.* **2021**, *11*, 10467. [CrossRef]
4. Mishra, R.B.; Jiang, H. Classification of Problem and Solution Strings in Scientific Texts: Evaluation of the Effectiveness of Machine Learning Classifiers and Deep Neural Networks. *Appl. Sci.* **2021**, *11*, 9997. [CrossRef]
5. Islam, N.; Shaikh, A.; Qaiser, A.; Asiri, Y.; Almakdi, S.; Sulaiman, A.; Moazzam, V.; Babar, S.A. Ternion: An Autonomous Model for Fake News Detection. *Appl. Sci.* **2021**, *11*, 9292. [CrossRef]
6. Hernández, Y.; Martínez, A.; Estrada, H.; Ortiz, J.; Acevedo, C. Machine Learning Approach for Personality Recognition in Spanish Texts. *Appl. Sci.* **2022**, *12*, 2985. [CrossRef]
7. Cheng, Q.; Zhu, Y.; Song, J.; Zeng, H.; Wang, S.; Sun, K.; Zhang, J. Bert-Based Latent Semantic Analysis (Bert-LSA): A Case Study on Geospatial Data Technology and Application Trend Analysis. *Appl. Sci.* **2021**, *11*, 11897. [CrossRef]
8. Almela, Á. A Corpus-Based Study of Linguistic Deception in Spanish. *Appl. Sci.* **2021**, *11*, 8817. [CrossRef]
9. Sezerer, E.; Tekir, S. Incorporating Concreteness in Multi-Modal Language Models with Curriculum Learning. *Appl. Sci.* **2021**, *11*, 8241. [CrossRef]
10. Li, Q.; Li, F.; Li, S.; Li, X.; Liu, K.; Liu, Q.; Dong, P. Improving Entity Linking by Introducing Knowledge Graph Structure Information. *Appl. Sci.* **2022**, *12*, 2702. [CrossRef]
11. Agrawal, A.; Tripathi, S.; Vardhan, M.; Sihag, V.; Choudhary, G.; Dragoni, N. BERT-Based Transfer-Learning Approach for Nested Named-Entity Recognition Using Joint Labeling. *Appl. Sci.* **2022**, *12*, 976. [CrossRef]
12. Kim, H.; Kim, H. Fine-Grained Named Entity Recognition Using a Multi-Stacked Feature Fusion and Dual-Stacked Output in Korean. *Appl. Sci.* **2021**, *11*, 10795. [CrossRef]
13. Bose, P.; Srinivasan, S.; Sleeman, W.C.; Palta, J.; Kapoor, R.; Ghosh, P. A Survey on Recent Named Entity Recognition and Relationship Extraction Techniques on Clinical Texts. *Appl. Sci.* **2021**, *11*, 8319. [CrossRef]
14. Song, H.J.; Yoon, S.H.; Park, S.B. Question Difficulty Estimation Based on Attention Model for Question Answering. *Appl. Sci.* **2021**, *11*, 12023. [CrossRef]

15. Phakmongkol, P.; Vateekul, P. Enhance Text-to-Text Transfer Transformer with Generated Questions for Thai Question Answering. *Appl. Sci.* **2021**, *11*, 10267. [CrossRef]
16. Ait-Mlouk, A.; Alawadi, S.A.; Toor, S.; Hellander, A. FedQAS: Privacy-Aware Machine Reading Comprehension with Federated Learning. *Appl. Sci.* **2022**, *12*, 3130. [CrossRef]
17. Lee, J.; Lee, J.; Lee, M.; Jang, G.J. Named Entity Correction in Neural Machine Translation Using the Attention Alignment Map. *Appl. Sci.* **2021**, *11*, 7026. [CrossRef]
18. Eo, S.; Park, C.; Moon, H.; Seo, J.; Lim, H. Comparative Analysis of Current Approaches to Quality Estimation for Neural Machine Translation. *Appl. Sci.* **2021**, *11*, 6584. [CrossRef]
19. Espinosa-Anke, L.; Palmer, G.; Corcoran, P.; Filimonov, M.; Spasić, I.; Knight, D. English–Welsh Cross-Lingual Embeddings. *Appl. Sci.* **2021**, *11*, 6541. [CrossRef]
20. Chuang, H.M.; Cheng, D.W. Conversational AI over Military Scenarios Using Intent Detection and Response Generation. *Appl. Sci.* **2022**, *12*, 2494. [CrossRef]
21. Fuad, A.; Al-Yahya, M. AraConv: Developing an Arabic Task-Oriented Dialogue System Using Multi-Lingual Transformer Model mT5. *Appl. Sci.* **2022**, *12*, 1881. [CrossRef]
22. Qin, P.; Tan, W.; Guo, J.; Shen, B.; Tang, Q. Achieving Semantic Consistency for Multilingual Sentence Representation Using an Explainable Machine Natural Language Parser (MParser). *Appl. Sci.* **2021**, *11*, 11699. [CrossRef]
23. Pechsiri, C.; Piriyakul, R. Causal Pathway Extraction from Web-Board Documents. *Appl. Sci.* **2021**, *11*, 10342. [CrossRef]
24. Silva, S.; Seara Vieira, A.; Celard, P.; Iglesias, E.L.; Borrajo, L. A Query Expansion Method Using Multinomial Naive Bayes. *Appl. Sci.* **2021**, *11*, 10284. [CrossRef]
25. Ahuir, V.; Hurtado, L.F.; González, J.Á.; Segarra, E. NASca and NASes: Two Monolingual Pre-Trained Models for Abstractive Summarization in Catalan and Spanish. *Appl. Sci.* **2021**, *11*, 9872. [CrossRef]

applied sciences

MDPI

Article

Exploring Language Markers of Mental Health in Psychiatric Stories

Marco Spruit [1,2,*], Stephanie Verkleij [3], Kees de Schepper [4] and Floortje Scheepers [4]

1 Leiden University Medical Center (LUMC), Campus The Hague, Leiden University, Turfmarkt 99, 2511 DC The Hague, The Netherlands
2 Leiden Institute of Advanced Computer Science (LIACS), Leiden University, Niels Bohrweg 1, 2333 CA Leiden, The Netherlands
3 Department of Information and Computing Sciences, Utrecht University, Princetonplein 5, 3584 CC Utrecht, The Netherlands; stephanieverkleij@hotmail.com
4 University Medical Center Utrecht (UMCU), Utrecht University, Heidelberglaan 100, 3584 CX Utrecht, The Netherlands; c.w.m.deschepper@umcutrecht.nl (K.d.S.); f.e.scheepers-2@umcutrecht.nl (F.S.)
* Correspondence: m.r.spruit@lumc.nl

Abstract: Diagnosing mental disorders is complex due to the genetic, environmental and psychological contributors and the individual risk factors. Language markers for mental disorders can help to diagnose a person. Research thus far on language markers and the associated mental disorders has been done mainly with the Linguistic Inquiry and Word Count (LIWC) program. In order to improve on this research, we employed a range of Natural Language Processing (NLP) techniques using LIWC, spaCy, fastText and RobBERT to analyse Dutch psychiatric interview transcriptions with both rule-based and vector-based approaches. Our primary objective was to predict whether a patient had been diagnosed with a mental disorder, and if so, the specific mental disorder type. Furthermore, the second goal of this research was to find out which words are language markers for which mental disorder. LIWC in combination with the random forest classification algorithm performed best in predicting whether a person had a mental disorder or not (accuracy: 0.952; Cohen's kappa: 0.889). SpaCy in combination with random forest predicted best which particular mental disorder a patient had been diagnosed with (accuracy: 0.429; Cohen's kappa: 0.304).

Keywords: language marker; mental disorder; deep learning; LIWC; spaCy; RobBERT; fastText; LIME

Citation: Spruit, M.; Verkleij, S.; de Schepper, K.; Scheepers, F. Exploring Language Markers of Mental Health in Psychiatric Stories. *Appl. Sci.* **2022**, *12*, 2179. https://doi.org/10.3390/app12042179

Academic Editors: Arturo Montejo-Ráez and Salud María Jiménez-Zafra

Received: 21 September 2021
Accepted: 15 February 2022
Published: 19 February 2022

Publisher's Note: MDPI stays neutral with regard to jurisdictional claims in published maps and institutional affiliations.

1. Introduction

Mental disorders make up a major portion of the global burden of disease [1], and in 2017, 10.7% of the global population reported having or having had a mental disorder [2]. This prevalence is not staying stead, but is rising mainly in developing countries [1]. Furthermore, mental disorders have a substantial long term impact on individuals, caregivers and society [3]. The challenge of diagnosing a mental disorder is the complexity of multiple genetic, environmental and psychological contributors and individual risk factors [4].

Research has shown that people with mental health difficulties use distinctive language patterns [5]. Until now, the Language Inquiry and Word Count (LIWC) toolkit has been the main focus for identifying language markers [6]. This toolkit of Natural Language Processing (NLP) techniques calculates the number of words of certain categories that are used in a text based on a dictionary [7]. LIWC is a traditional programme in the sense that it analyses texts with symbolic (i.e., deterministic and rule-based) techniques, predominantly at the word level. LIWC does not use subsymbolic (i.e., probabilistic and vector-based) NLP techniques such as word vector representations within neural networks.

The objective of our research was to compare the performance of LIWC with the performances of other NLP techniques in the quest to provide useful insights into Dutch psychiatric stories. In this paper, we compare the performances of LIWC [6], spaCy [8], fastText [9] and RobBERT [10] when applied to psychiatric interview transcriptions. SpaCy provides, among other things, a dependency grammar parser to syntactically process texts. This NLP technique can provide insights by unravelling the grammatical structure of each sentence, and it will provide information about the grammatical relationships between words [11]. By using this technique, we aimed to uncover the different uses of grammar by patients with different mental illnesses. This provides further insights into the stylistic differences between people with and without mental disorders. fastText and RobBERT were selected because both techniques employ deep learning models. Deep learning exploits layers of non-linear information processing for both supervised and unsupervised tasks [12]. We hypothesise that deep learning techniques can provide more insights than other methods into these complex mental health disorders.

2. Related Work

This research is not the first to attempt to identify language markers associated with mental disorders. Several researchers already compared mental disorders using the LIWC tool [5,13]. We introduce and compare several state-of-the-art alternative NLP approaches to identifying language markers' associations with mental health disorders.

2.1. Language Markers for Mental Health Disorders

A literature study was performed to review earlier work related to language markers for mental health disorders. The snowballing method was used to find the relevant literature. Both backward snowballing and forward snowballing were employed [14]. A curated set of recent papers on language markers in mental healthcare was used as the starting point [5,6,13,15,16]. Then, one or two levels deep were snowballed back and forth. The number of levels snowballed depended on whether new relevant literature was found. Whenever a dead end was reached, the snowballing procedure was stopped. We selected Google Scholar (with a proxy from Utrecht University) to execute the following search queries:

- "Language marker" "mental health" "LIWC"
- "Language marker" "mental health" "language use"
- "Mental health" "deep learning"
- "Dutch" "parser" "NLP"
- "BERT" "mental health" "classification"
- "Alpino" "dependency parser"
- "spaCy" "lemma" "dependency parser"
- "Language" in conjunction with the words below:
  - ADHD
  - Autism
  - Bipolar Disorder
  - Borderline personality disorder
  - Eating disorder
  - Generalised anxiety disorder
  - Major depressive disorder
  - OCD
  - PTSD
  - Schizophrenia

Table 1 summarises our findings related to ten different mental disorders, highlighting their uses of language. These include mainly characteristic use of pronouns (Pron), the degree ([n]ormal/[i]mpaired) of semantic coherence (SC) and usage of topical words. We only list the disorders that appear in our dataset as the main diagnosis; the N column shows the number of patients.

We found that people with attention deficit hyperactivity disorder (ADHD) use more third-person plural (3pl) pronouns, less words of relativity [13] and more sentences, but less clauses per sentence [17] than normal. Autism is strongly linked to motion, home, religion and death features [18]. Furthermore, people with autism are more self-focused, because they use more first-person singular (1sg) pronouns [18]. People who are bipolar are also more self-focused and use more words related to death [19]. The use of more swear words, words related to death and third-person singular (3sg) pronouns, and less use of cognitive emotive words are associated with borderline personality disorder (BPD) [5]. Eating disorders, consisting of bulimia, anorexia and eating disorders not otherwise specified, are associated with the use of the words related to the body, negative emotive words, self-focused words and cognitive process words [13]. People with generalised anxiety disorder (GAD) produce more sentences which lack semantic coherence [20]. Furthermore, they use more tentative words and impersonal pronouns, and they use more words related to death and health [13]. Major depressive disorder (MDD) has a strong appearance of being more self-focused, involving more past tense and repetitive words and producing short, detached and arid sentences [21]. Obsessive compulsive disorder (OCD) is associated with words related to anxiety and cognitive words. Researchers do not yet agree on the language cues associated with post-traumatic stress disorder (PTSD). One study showed that there were no cues [13], yet another study showed that people with PTSD use more singular pronouns and words related to death and less cognitive words [22]. Finally, research shows that a lack of semantic cohesion [23], usage of words related to religion and hearing voices and sounds are associated with schizophrenia [5]. Further details are available in [24].

Table 1. Overview of associated language markers for ten mental health disorders.

Disorder	Pron	SC	Word Use	More	N
ADHD	3pl	-	-	Relativity, more sentences, less clauses	4
Autism	1sg	-	Motion, home, religion and death	-	5
Bipolar	1sg	-	Death	-	7
BPD	3sg	n	Death	Swearing, less cognitive emotion words	5
Eating	1sg	-	Body	Negative emotion words	10
GAD	imprs	i	Death and health	Tentative words	4
MDD	1sg	i	-	Inverse word-order and repetitions	11
OCD	1sg	-	Anxiety	More cognitive words	4
PTSD	sg	-	Death	Less cognitive words	6
Schizophrenia	3pl	i	Religion	Hearing voices and sounds	16

2.2. NLP Techniques for Identifying Language Markers

We investigated the following four basic approaches in NLP for identification of language markers: lexical processing from a lexical semantics perspective, dependency parsing from a compositional semantics viewpoint, shallow neural networks in a stochastic paradigm and deep neural networks employing a transformer-based architecture.

2.2.1. Lexical Processing

Research so far on exploring language markers in mental health has been done mainly with Linguistic Inquiry and Word Count (LIWC) [6]. LIWC is a computerised text-analysis tool and has two central features: a processing component and dictionaries [15]. The processing feature is the program which analyses text files and goes through them word by word. Each word is compared with the dictionaries and then put in the right categories. For example, the word "had" can be put in the categories verbs, auxiliary verbs and past tense verbs. Next, the program calculates the percentage for each category in the text; for example, 17% of the words may be verbs. A disadvantage of the LIWC program is that it ignores context, idioms, sarcasm and irony. Furthermore, the 89 different categories are based on language research. However, this does not guarantee that these categories represent reality, because categories could be missing.

2.2.2. Dependency Parsing

The syntactic processing of texts is called dependency parsing [25]. This processing is valuable because it forms transparent lexicalised representations and it is robust [25]. Furthermore, it also gives insights into the compositional semantics, i.e., the meanings of a sentence's individual words or phrases [26]. Small changes in the syntactic structure of a sentence can change the whole meaning of the sentence. For example, *John hit Mary* and *Mary hit John* contain the same words, but have different meanings. It is said that compositionality is linked to our ability to interpret and produce new remarks, because once one has mastered the syntax of a language, its lexical meanings and its modes of composition, one can interpret new combinations of words [27]. Compositionality is the semantic relationship combined with a syntactic structure [28]. Compositional semantics is driven by syntactic dependencies, and each dependency forms, from the contextualised sense of the two related lemmas, two new compositional vectors [29]. Therefore, the technique required for extracting the compositional semantics needs to contain a dependency parser and a lemmatizer. Choi et al. [25] compared the ten leading dependency parsers based on the speed/accuracy trade-off. Although Mate [30], RBG [31] and ClearNLP [32] perform best in unlabeled attachment score (UAS), none of them includes a Dutch dictionary, which was needed for this research. However, spaCy does include a Dutch dictionary. Other Dutch dependency parsers are Frog [33] and Alpino [34]. Both Frog (https://github.com/LanguageMachines/frog/releases/, accessed on 17 October 2021) and spaCy (https://spaCy.io/models/nl, accessed on 17 October 2021) include the Dutch dictionary corpus of Alpino, but due to equipment constraints, we selected spaCy for the dependency parsing task.

2.2.3. Shallow Neural Networks

Features made for traditional NLP systems are frequently handcrafted, time consuming and incomplete [35]. Neural networks, however, can automatically learn multilevel features and give better results based on dense vector representations [16]. The trend toward neural networks has been caused by the success of deep learning applications and the concept of word embeddings [16]. Word embeddings, such as the skip-gram model and the continuous bag-of-words (CBOW) model [36], distribute high-quality vector representations and are often used in deep learning models as the first data processing layer [16]. The word2vec algorithm uses neural networks to learn vector representations [37]. It can use the skip-gram model or the CBOW model, and it works for both small and large datasets [37]. However, out-of-vocabulary (OOV) words, also referred to as unknown words, are a common issue for languages with large vocabularies [16]. The fastText model overcomes this problem by handling each word as a bag-of-character n-gram. This is achieved by using the skip-gram model from word2vec as an extension. These n-grams are used to represent the sums of the n-gram vectors [9]. Finally, it is worth noting that both Word2vec and fastText are said to employ a shallow neural network architecture; i.e., their neural networks only define one hidden layer, which explains why these models are known to be many orders of magnitude faster in training and evaluation than other deep learning classifiers, while often performing as well as those classifiers in terms of accuracy [38].

2.2.4. Deep Neural Networks

In 2017 the transformer neural network architecture was introduced [39], which much improved NLP tasks such as text classification and language understanding [40]. Bidirectional encoder representations from transformers (BERT) is an immensely popular transformer-based language representation model designed to pretrain, from unlabelled text, deep bidirectional representations [41]. The multilingual version of BERT is simply called mBERT. A more recent and improved version of BERT is RoBERTa, which stands for robustly optimised BERT approach [42]. The main changes are that RoBERTa trains for longer, on more data, with bigger batches and on longer sequences [42].

2.2.5. Neural Networks for Dutch

In Table 2 an overview of the different neural networks can be seen. The choice of best fit is limited, because of the small and Dutch dataset. Two neural networks were chosen for this research, one based on words and one based on sentences. Furthermore, the neural networks had to have a Dutch model. Thus, the choice was between word2vec and fastText at the word-level and between BERT, mBERT and RoBERTa at the sentence level. Other models, such as ClinicalBERT, could also be used in combination with a transfer learning model such as the Cross-lingual Language Model (XLM) to tackle the Dutch data. However, these models have not yet been used extensively in the medical domain [43]. This could be because the interpretability and performance of a model are equally important in the medical domain. Even though deep learning models can perform better than the more traditional models, they are hard to explain or understand [44]. Hence, this approach was not used for this research. Furthermore, fastText has proven that it results in better performance in comparison to Word2vec [45] and it is able to handle OOV words as well, because of the n-grams.

Table 2. Overview of neural network models under consideration for identifying language markers in Dutch.

Model	Dutch	Architecture	Input Level	Selected
Word2Vec	Yes	CBOW & Skip-gram	Word	No
fastText	Yes	RNN	Word	Yes
ELMo	Yes	(Bi)LSTM	Sentence	No
ULMFit	Yes	Transformer	Sentence	No
GPT	No	Transformer	Sentence	No
GPT-2	No	Transformer	Sentence	No
GPT-3	No	Transformer	Sentence	No
BERT	Yes	Transformer	Sentence	No
RoBERTa/RobBERT	Yes	Transformer	Sentence	Yes
ClinicalBERT	No	Transformer	Sentence	No
XLnet	No	Transformer-XL	Sentence	No
StructBERT	No	Transformer	Sentence	No
ALBERT	No	Transformer	Sentence	No
T5	No	Transformer	Sentence	No

The Dutch version of BERT is called BERTje [46], the Dutch version of RoBERTa is called RobBERT [10] and mBERT is the multilingual BERT with support for more than 100 languages, including Dutch [41]. A choice between the three BERTs was made by looking at their performances with respect to the classification task, because that was the focus of this research. The research of Delobelle et al. [10] shows that RobBERT (ACC = 95.1%) performs best on classification tasks compared to mBERT (ACC = 84.0%) and BERTje (ACC = 93.0%) with a full dataset. Therefore, the neural networks selected for this research were fastText and RobBERT.

3. Methodology

3.1. Dataset and Preprocessing

The dataset used for this research was obtained from the Verhalenbank ("Storybank") of the University Medical Centre Utrecht (UMCU) in The Netherlands. Its psychiatry department has been collecting stories about mental illness of people who have or had psychiatric issues or were in contact with people with psychiatric issues. Interviews were conducted with (ex-)patients, caregivers and medical employees to gain new leads which could benefit the recovery of patients. The interviews were then transcribed into anonymous stories and put on the website of the Verhalenbank (https://psychiatrieverhalenbank.nl/, accessed on 17 October 2021). The dataset consists of 108 interviews with 11 diagnostic labels; 36 are without mental disorder labels. The diagnoses were assigned by multiple doctors and based on other material than the interviews. The interviews were all between 60 and 90 min long, and the corresponding transcripts are between 6782 and 9531 words in length. The split used for this research was 80% training and 20% testing. There were not enough data to have a validation set. Source code for the data analysis is available at: https://github.com/StephanieVx/ExploringLinguisticMarkers, accessed on 17 October 2021.

3.2. Data Analysis

This exploratory study compares the classification performances of different NLP techniques and looks at which language cues could predict if a person has a mental disorder, and if so, which kind of mental disorder. The four different techniques were applied to the two tests. The first test consisted of deciding between mental disorder and no mental disorder; and the second one consisted of deciding between the different mental disorders. After applying the techniques, predictions were made. For LIWC and spaCy, the classification algorithms decision tree, random forest and support vector machine (SVM) were used by means of the default configurations of the R packages *rpart*, *randomForest* and *e1071*, respectively. The deep learning techniques used their default prediction models without incorporating a transfer learning step [47]. Next, the techniques and predictions were applied again after removing the stop words, as listed in the Dutch portion of the NLTK Python package [48], after which the interviews and the predictions were compared. Furthermore, to gain further insight into the predictions of fastText and RobBERT, LIME (Local Interpretable Model Agnostic Explanation) was applied [49].

4. Results

4.1. Descriptive Statistics

An overview of the number of people per mental disorder in our dataset is shown in Figure 1. The group with dissociation (a disconnection between a person's memories, feelings, perceptions and sense of self) contains the least number of people in this dataset; the group with psychosis is the largest. Furthermore, there are two labels about personality. Personality includes obsessive-compulsive personality disorder, avoidant personality disorder, dependent personality disorder and unspecified personality disorders. Personality+ in this research only includes borderline personality disorder (BPD). Figure 2 shows a boxplot of the number of words per mental disorder, which indicates that people with eating disorders use less words than people without eating disorders.

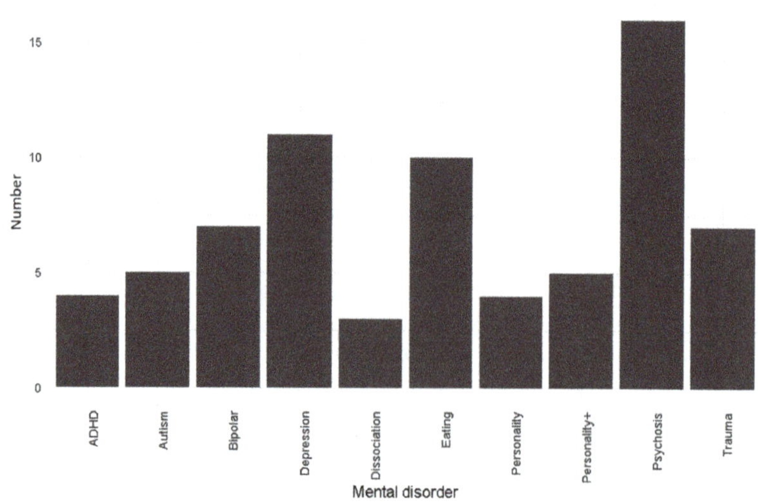

Figure 1. Columnchart of number of people per mental disorder in the dataset.

4.2. Predictions

Table 3 shows the accuracies in the two tests and Cohen's Kappa per prediction. The best performing classifiers are highlighted in bold text. The LIWC program in combination with the random forest algorithm achieved the highest accuracy when comparing mental disorder to no mental disorder (accuracy: 0.952). SpaCy reached the highest accuracy when comparing the different kinds of mental disorder (accuracy: 0.429).

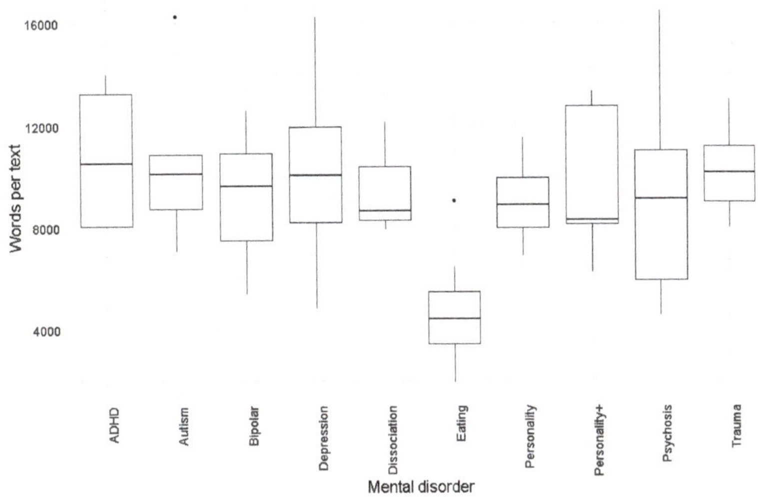

Figure 2. Boxplot of number of words per mental disorder in the dataset.

Cohen's kappa was used to assess the inter-classifier agreement [50]. This metric takes the probability that the 10 different labels (in this case) agree by chance into consideration when quantifying how much they agree. Cohen's kappa was calculated for each model and prediction algorithm. If the coefficient is below 0.4, there is a slight correlation between the models (and with a negative kappa it is even below chance level). A kappa of above 0.6 means that the classifiers have a substantial agreement; for example, see the LIWC-output with the SVM model in the MD (mental disorder) vs. control group comparison. When the kappa is between 0.8 and 1.0, this indicates that the classifiers have almost perfect agreement. This applies to the LIWC-output with the random forest model in the second comparison with a kappa of 0.889. Care should be taken when interpreting Cohen's kappa [51], but the fact that the item with the highest kappa also has the highest accuracy is reassuring. The low accuracy of the second comparison can be explained due to a dataset having only 72 interviews from people with mental disorders and 10 different kinds of mental disorders.

What also can be seen in Table 3 in the sixth and seventh columns is that without stop words spaCy performed less accurately, while LIWC, fastText and RobBERT performed almost the same in both comparisons.

Table 3. Accuracy and Cohen's Kappa for the model predictions (with and without stop words).

Comparison	Input	Model	Accuracy	Kappa	Accuracy No Stopwords	Kappa No Stopwords
Mental Disorder vs. No Mental Disorder	LIWC-output	decision tree	0.857	0.667	0.857	0.674
	LIWC-output	random-Forest	**0.952**	**0.889**	**0.952**	**0.877**
	LIWC-output	SVM	0.857	0.64	0.905	0.738
	spaCy	decision tree	0.810	0.391	0.444	−0.309
	spaCy	random-Forest	0.762	0.173	0.389	−0.370
	spaCy	SVM	0.714	0.115	0.528	−0.275
	raw data	fastText	0.643	0.172	0.607	0.072
	raw data	RobBERT	0.607	0.000	0.607	0.000
Mental Disorder multiclass	LIWC-output	decision tree	0.286	0.157	**0.286**	**0.177**
	LIWC-output	random-Forest	0.214	0.120	0.214	0.144
	LIWC-output	SVM	0.286	0.114	0.143	0.0718
	spaCy	decision tree	0.143	−0.0120	0.071	−0.052
	spaCy	random-Forest	**0.429**	**0.304**	0.214	0.078
	spaCy	SVM	0.357	0.067	0.143	0.091
	raw data	fastText	0.286	0.000	0.200	0.000
	raw data	RobBERT	0.200	0.000	0.267	0.120

4.3. Interpretation

In this section, we elaborate on our findings regarding the performances of the LIWC, SpaCy, fastText and RobBERT approaches to NLP for language marker identification.

4.3.1. Lexical Processing with LIWC

Figure 3 shows the decision tree for the LIWC-output. If an interview transcription consisted of more than 5.4% of the first-person singular pronoun, than it was classified as being of a person with a mental disorder. If not and if less than 8.5% of the words were related to social concepts, then the interview was classified as being of a person with no mental disorder. Furthermore, the decision tree categories of the LIWC tool were visualised in a stripchart (jitter) plot, a fragment of which is shown in Figure 4. In particular, this plot effectively illustrates the potential to identify people with and without a mental disorder based on the empirical frequencies of hypothesised LIWC category occurrences, such as first-person singular pronoun (1sg), further strengthening the rationale behind this feature being the root decision of the LIWC decision tree shown in Figure 3.

Furthermore, we investigated the LIWC's feature importance using a random forest classifier to determine which variables added the most value to our binary predictions. Figure 5 shows the top 10 variables that impacted the classification.

4.3.2. Dependency Parsing with SpaCy

Similarly, we investigated the SpaCy feature importance using a random forest classifier to determine which n-grams added the most value to our binary predictions. Figure 6 shows the top 10 variables that impact the classification. In addition, we present the mean, standard deviation (sd) and standard error (se) for each n-gram in Figure 7. A Mann–Whitney U test revealed no significant difference between people with and without mental disorders in their usage of the following four spaCy variables: denken_denken_ROOT, gaan_gaan_ROOT, ja_ja_ROOT and zijn_zijn_ROOT. Finally, we provide example sentences for each of the identified SpaCy language markers in Table 4.

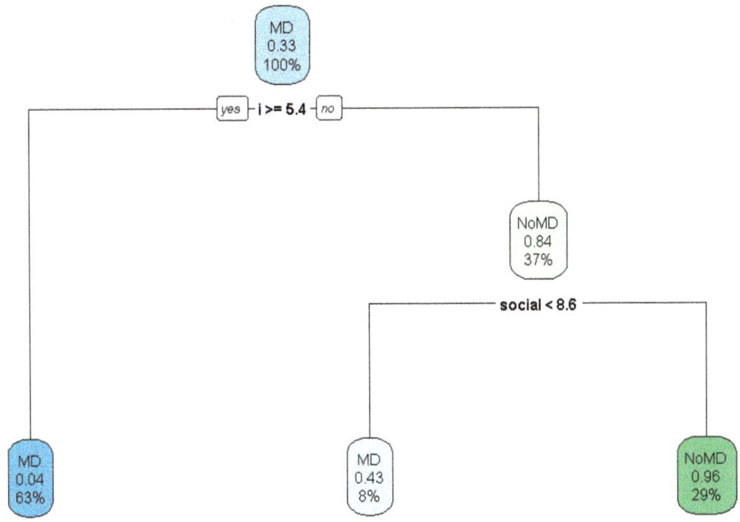

Figure 3. Example decision tree with two LIWC parameters (parameter *i* means the percentage of first-person pronouns and parameter *social* the percentage of words referring to others, such as *they*; each box lists the choice between mental disorder or not, the chance of the class being no mental disorder and the percentage of the data that fall in this box).

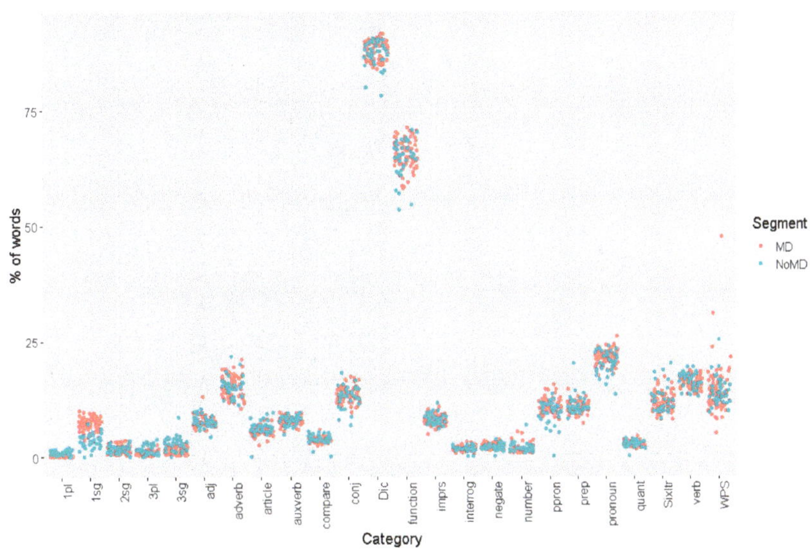

Figure 4. This stripchart plot illustrates the potential to identify people with and without a mental disorders based on the empirical frequencies of hypothesised LIWC category occurrences, e.g., first-person singular pronoun (1sg).

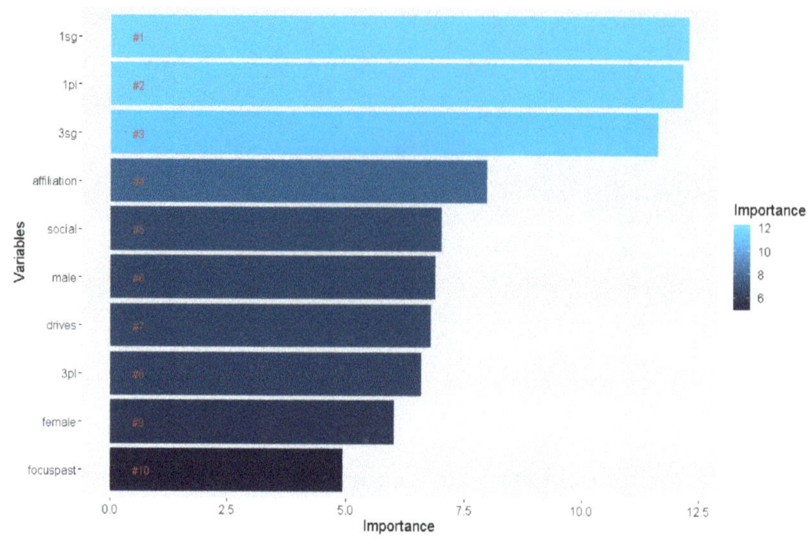

Figure 5. Top 10 LIWC features by importance in binary classification.

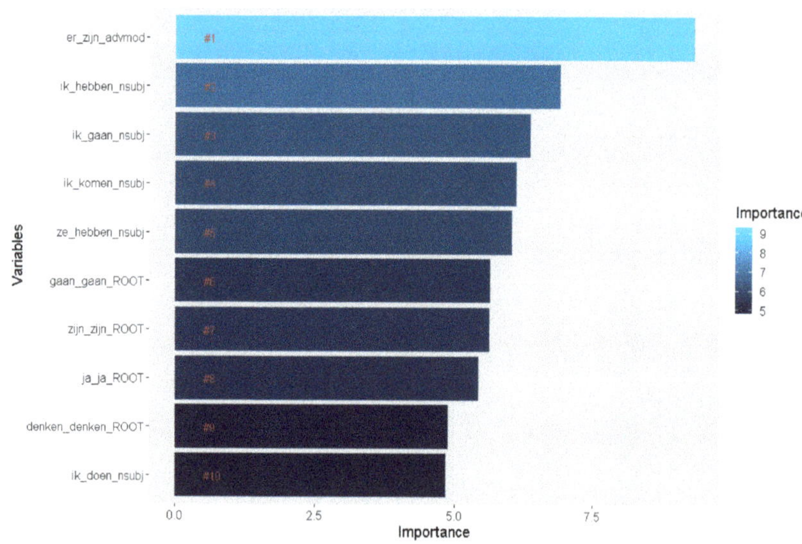

Figure 6. Top 10 SpaCy features by importance in binary classification.

4.3.3. Neural Networks with fastText and RobBERT

LIME was applied to both fastText and RobBERT to gain further insight into the black-box neural network models. LIME is a well-known and well-understood surrogate model-based approach to help explain model predictions by learning surrogate models using an operation called input perturbation [49]. For each sentence, subsamples of words were generated and fed to the model, so for each word the predictions for subsamples with and without this word could be compared, and subsequently the contribution of this word could be assessed. For example, quote 1 was from someone who had been diagnosed with schizophrenia, and the text was labelled by RobBERT as mental disorder. The word "eh" has been highlighted because it explains according to LIME why it was labelled as mental disorder (class = 0). Note that the original quote is in Dutch, but for convenience we provide English translations here. In addition, "[silence]" means a pause that was judged as meaningful by the transcriber of the interview. In Figure 8, the ten words with the highest usage can be seen. Some words appear multiple times in the figure. This is because LIME looks locally at a text and every word appears in a different context. This also means that sometimes a word will be an explanation for a mental disorder and other times not, especially for context sensitive algorithms like RobBERT.

ngram	Disorder	n	mean	sd	se
denken_denken_ROOT	MD	72	19.375	17.03057524	2.00707254
	NonMD	36	20.22222222	13.26889189	2.211481982
er_zijn_advmod	MD	72	13	12.07255064	1.422763737
	NonMD	36	18.80555556	9.080023425	1.513337237
gaan_gaan_ROOT	MD	72	30.30555556	22.68749043	2.673746389
	NonMD	36	36.77777778	27.19675518	4.532792529
ik_doen_nsubj	MD	72	11.47222222	10.76898549	1.269137111
	NonMD	36	5.75	8.083051051	1.347175175
ik_gaan_nsubj	MD	72	17.29166667	14.12463386	1.664604064
	NonMD	36	9.638888889	11.58854799	1.931424664
ik_hebben_nsubj	MD	72	49.34722222	38.89355634	4.583649572
	NonMD	36	25.61111111	20.6013561	3.433559349
ik_komen_nsubj	MD	72	6.75	7.084500042	0.834916337
	NonMD	36	1.777777778	4.427905736	0.737984289
ja_ja_ROOT	MD	72	27.33333333	25.80206346	3.04080234
	NonMD	36	30.58333333	24.41940095	4.069900159
ze_hebben_nsubj	MD	72	2.902777778	5.39994711	0.63638987
	NonMD	36	11.58333333	15.08144555	2.513574259
zijn_zijn_ROOT	MD	72	16.90277778	13.56690227	1.598874766
	NonMD	36	20.72222222	13.15753149	2.192921915

Figure 7. Top 10 SpaCy n-gram features in binary classification.

Table 4. Example sentences containing the top 6 spaCy variables.

spaCy Variable	Example Sentence
ik_doen_nsubj I_do_nsubj	Ik doe normaal, haal mijn studie en gebruik geen drugs en ben niet irritant 'I do normal, get my degree and do not use drugs and am not irritating'
ik_gaan_nsubj I_go_nsubj	ik ben meer waard dan dit, ik ga voor mezelf opkomen. 'I am worth more than this, I'm going to stand up for myself'
ik_hebben_nsubj I_have_nsubj	Ik heb ook behandelingen gehad, of een behandeling gehad 'I have also gotten treatments, or got a treatment'
ik_komen_nsubj I_come_nsubj	Ja, ik kwam in de bijstand 'Yes, I came into welfare'
er_zijn_advmod there_are_advmod	Er zijn zo veel vrouwelijke sociotherapeuten in heel [naam][centrum] die opgeroepen kunnen worden 'There are so many female sociotherapists in [name][centre] who can be called'
ze_hebben_nsubj they_have_nsubj	Al een tijdje maar ze hebben nooit wat aan mij verteld 'For some time, but they have never told me anything'

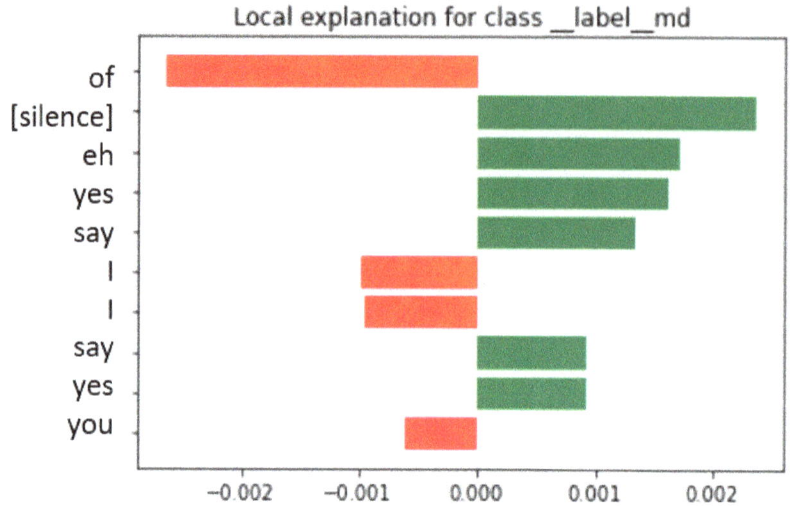

Figure 8. LIME explanation for quote 1 (top 10 words and how much they approximately contribute to the classification decision).

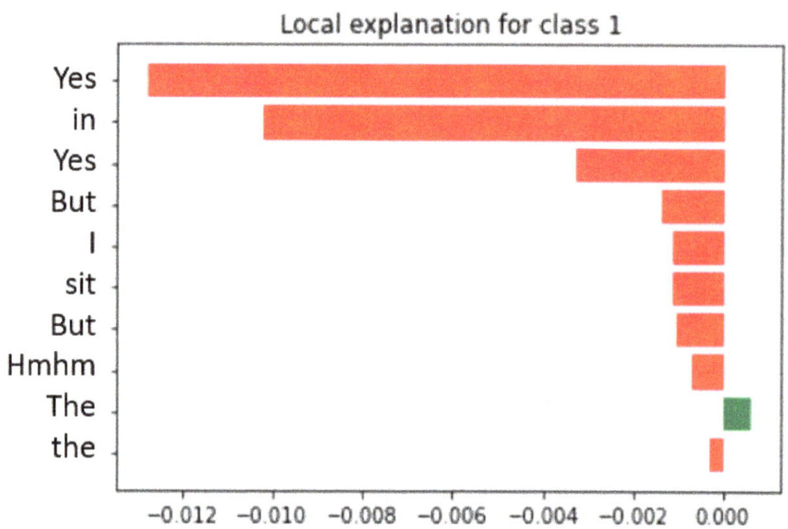

Figure 9. LIME explanation for quote 2 (top 10 words and how much they approximately contribute to the classification decision).

> Quote 1: "I ehm, [silence] the most poignant I will you. Yes, the most poignant what I can tell you is that, I have weekend leave on the weekend and then [name_of_wife] and I lay together in bed. Furthermore, nothing happens there. As I do not need that, haha. However, I cannot even feel that I love her. I know it, that I love her. Furthermore, I know that my wife is and I, and I. However, that is all in here eh, but I do not feel it. Furthermore, that is the biggest measure which you can set··· Yes. Furthermore, I talked about it with her."

Quote 2 is from someone with an eating disorder and was analysed with fastText. The word "eh" was highlighted because it explained why the transcription was labelled as coming from a patient with a mental disorder (class = _label_md). Figure 9 shows the ten words with the highest probabilities from that transcription.

> Quote 2: "Yes it gives kind of a kick or something to go against it and to see that people you really eh yes I don't know. That your that your eating disorder is strong and people find that then. Then, you think oh I am good at something. Then, yes I don't know. Then you want there that you want to be doing something you are good at··· Eh I am able to walk again since two months. Before I eh stayed in bed and in a wheelchair around half a year, because I eh could not walk myself. Furthermore, I was just to weak to do it. and eh yes I still cannot do quite a lot of things. I am really happy that I can walk again by myself."

Other text also heavily featured conversational words such as "eh," "well," and "yes" in the LIME analyses. This suggests that perhaps for these interviews the difference between mental disorder and no disorder was more prevalent in the manner of speaking than in the topics they addressed.

Table 5 shows samples of eight interviews whose words resulted in the assignment of the mental disorder (MD) label or the no mental disorder (noMD) label. The first four interviews were analysed with stop words, and as can be seen, most of the words are stop words or "generally not meaningful" words. They could, however, be related to insightful words, which are also showen in the quotes. This could be supposedly because RobBERT looks both left and right in the context of a word in all layers of the transcription and then conditions it. Apparently, some words appear both in the mental disorder column and in the no mental disorder column, simply because these words appear in different contexts. Such words can contribute to a mental disorder classification in some language contexts, whereas in another context they do not. To further investigate, we removed all stop words from the last four interviews to determine whether LIME found more meaningful words. For example, in interview 7 with the fastText model, LIME found the words "psychiatrics" and "performance" as markers for a mental disorder, whereas in interview 8 LIME found the words "healing" and "job". In conclusion, without stop words we tended to find moderately more insightful words than with stop words. However, the words found by LIME are different for almost every interview and thus not yet applicable to support analyses of other interviews.

Table 5. LIME output of fastText and RobBERT for a sample of eight interviews.

ID	MD	SW	RobBERT	fastText	Words MD BERT	Words noMD BERT	Words MD fastText	Words noMD fastText
1	Y	Y	0.68	0.77	everyone, too, because, Yes, For example, too , Yes, I, did	-	yes, with, is, · · · , common, me	from, common, common, eh
2	Y	Y	0.55	0.69	feel, allowed, I, really, eh, angry, they, You	[name], there	together, am, well, well.	am, I, me, my
3	N	Y	0.39	0.45	happy, the, looking back, Well, belongs, eh, always, no, well, think	-	say, come, yes, and, causing	not, that, [place name], week, say
4	N	Y	0.37	0.23	could, can, Furthermore, That, sat, be, chats, and, whole	walked	protected, to, is, do, bad, have, is, physical, am	walks
5	Y	N	0.68	0.77	ehm, one, bill, yes, distraction, recovery	sat, eh, real, goes	yes, well, that, yes, well, rest	if, but, better, care
6	Y	N	0.58	0.65	eh	hospital, Furthermore, whole, whole, she, one, also, eh, again	whole, completely, · · · , further, times	stood, sick, selfish, and, ehm
7	N	N	0.41	0.46	eh, nineteen ninety seven, of, notices of objection, say, team	car, ehm, team, through, However,	psychiatrics, performance, one, he	that, en route, exciting, we, go, and
8	N	N	0.49	0.43	married, common, a, sit, heaven, times, and, The	ehm, ehm	sewn, healing, and, but, job	huh, hear, term, ready, busy

4.4. Summary of Findings: Language Markers

Table 6 shows an overview of the uncovered language markers for LIWC and spaCy. The 1SG LIWC pronoun notably came out as a language marker for a person with a mental disorder. In spaCy, 1SG was also the basis for labelling a mental disorder. The **W; $p < 0.05$** caption of the rightmost column refers to the Mann–Whitney two-tailed U tests that were performed to determine whether the means of the two groups per variable were equal to each other.

Unfortunately, we did not uncover clear patterns in the LIME results of the RobBERT and fastText neural network-based models, as different words were found for every interview to indicate either a mental disorder or no mental disorder.

Table 6. Summary of language markers uncovered by LIWC and spaCy.

	Language Marker	Mental Disorder	W; $p < 0.05$
LIWC	1sg	Yes	2487
	focuspast	Yes	1856
	affiliation	No	380
	drives	No	568
	female	No	937
	male	No	767
	3sg	No	454
	social	No	281
	3pl	No	882
	1pl	No	217.5
spaCy	ik_doen_nsubj	Yes	1700.5
	ik_gaan_nsubj	Yes	1726
	ik_hebben_nsubj	Yes	1796.5
	ik_komen_nsubj	Yes	1852.5
	er_zijn_advmod	No	849
	ze_hebben_nsubj	No	768.5

4.5. Focus Group

Furthermore, the results of the different models were discussed in a qualitative focus group session with UMCU data scientists, researchers and psychologists to better understand the outcomes. We discussed three key observations. First, the data scientists noted that the data used for this research are static data—i.e., somebody told their story and that was it. No new data from this particular person were added at a later time. The group hypothesised that following a person in their healing process, including their language usage, over a longer period of time, would result in additional relevant datapoints, and therefore could reveal additional interesting outcomes.

Second, the language markers found by LIWC and spaCy were discussed. The data originated from both people with mental disorders who told their own personal stories and from medical employees and family members who talked about people with mental disorders. This dual data origin situation likely influenced the outcome of this research. When an individual tells his own personal story, he will probably use more 1sg pronouns. Furthermore, when a health professional discusses an experience with a patient, he will likely use more 3sg and 3pl pronouns. Finally, people with mental disorders also shared their personal stories when they were not in an acute phase, and then, they could talk more about a completed story in their past. Therefore, the uncovered language markers actually make a lot of sense, according to the experts.

Third, rigid classifications are being abandoned in psychiatry, because they do not really help a person, according to some psychologists. However, if the current outcome classification will be changed depending on how far someone is in their healing process, one could find additional interesting results. The models discussed in this research could be applied for this new direction. To exemplify this, it was hypothesised that a person who is further into his healing process will tell a more integrated story about his past than a person who is less far. In other words, "focuspast" could be a marker for someone being further into the healing process. Another proposition was that this research could be used to look at symptoms instead of being used for diagnostic assistance: what kind of treatment will help a person based on how he speaks? Another idea is to look at suicidality or aggression: what can a text tell us about that? Put differently, find out what a person is not explicitly saying, by analysing the deeper layers to find possible patterns or symptoms. One domain expert concluded: "The strength of this research lays not in the exact results, but in the application of the different models and the potential questions which could be answered by these models."

5. Discussions and Conclusions

We have explored language markers in Dutch psychiatric interview transcriptions. We particularly focused on comparing the performances of traditional machine learning algorithms trained on LIWC and spaCy inputs with neural network approaches such as fastText and RobBERT, in predicting mental disorders. We found that the best performing technique in terms of determining whether a person has a mental disorder based on their word choices was LIWC in combination with random forest as the classification algorithm, which reached an accuracy of 0.952 and a Cohen's kappa of 0.889. Our hypothesis that the neural network approaches of fastText and RobBERT would perform best was not borne out. Several reasons may be posited. First, the pretrained language models of fastText and RobBERT did not for the most part consist of (transcribed) interview data. Second, the dataset was rather small (108 interviews) and the concept under consideration (mental illness) is not immediately apparent from a text. This suggests that for similar tasks with small datasets it may be best to use a dedicated algorithm such as LIWC, as it uses only a small selection of curated variables.

With regard to differentiating between mental illnesses, spaCy in combination with random forest predicted best which mental disorder each person had with an accuracy-score of 0.429 and a Cohen's kappa of 0.304. This moderate accuracy score can be explained due to the fact that the dataset of people with mental disorders only included 72 interview transcriptions and yet 10 mental disorder labels.

Finally, stop words did not appear to have that much influence on the performance of the classifiers except when employed using spaCy. We presume that is ws due to spaCy analysing the text from a grammatical point of view. When stop words are missing, spaCy cannot deduce the correct syntactic dependencies. Further work will focus on exploring additional model explainability techniques with differing explainability mechanisms and visualisation techniques in comparison to LIME, and investigating alternative NLP models in combination with an expanded data collection.

Ultimately, we argue that better understanding of a person's language use through the identification of language markers will result in better diagnosis of that person's mental health state, similar to the identification of a person's biomarkers. The impressive recent advancements within the field of Natural Language Processing are now allowing us to recalibrate our ambitions regarding language marker identification in informal patient narratives.

Author Contributions: Conceptualization, M.S. and F.S.; Data curation, K.d.S.; Formal analysis, S.V.; Funding acquisition, M.S.; Investigation, S.V., M.S. and K.d.S.; Methodology, M.S. and S.V.; Project administration, S.V.; Resources, M.S., K.d.S. and F.S.; Software, S.V.; Supervision, M.S., K.d.S. and F.S.; Validation, F.S.; Writing—original draft, S.V.; Writing—review and editing, M.S. All authors have read and agreed to the published version of the manuscript.

Funding: This research was supported by the Computing Visits Data (COVIDA) research programme of the Strategic Alliance Fund of Utrecht University, University Medical Center Utrecht and Technical University Eindhoven (round 2019).

Institutional Review Board Statement: The UMC Utrecht Medical Research Ethics Committee on 5 October 2016 confirmed with reference number WAG/mb/16/030724 that the Medical Research Involving Human Subjects Act (WMO) does not apply in the context of the Psychiatrieverhalenbank project with reference 16/626.

Informed Consent Statement: Informed consent was obtained from all subjects involved in the Verhalenbank study.

Data Availability Statement: See Section 3.1 for more information on the Verhalenbank ("Storybank") dataset which is available at https://psychiatrieverhalenbank.nl/, accessed on 17 October 2021.

Conflicts of Interest: The authors declare no conflict of interest.

References

1. Whiteford, H.A.; Degenhardt, L.; Rehm, J.; Baxter, A.J.; Ferrari, A.J.; Erskine, H.E.; Charlson, F.J.; Norman, R.E.; Flaxman, A.D.; Johns, N.; et al. Global burden of disease attributable to mental and substance use disorders: Findings from the Global Burden of Disease Study 2010. *Lancet* **2013**, *382*, 1575–1586. [CrossRef]
2. Ritchie, H.; Roser, M. Mental Health. In *Our World in Data*; 2020. Available online: https://ourworldindata.org/mental-health (accessed on 17 October 2021).
3. McIntosh, A.M.; Stewart, R.; John, A.; Smith, D.J.; Davis, K.; Sudlow, C.; Corvin, A.; Nicodemus, K.K.; Kingdon, D.; Hassan, L.; et al. Data science for mental health: A UK perspective on a global challenge. *Lancet Psychiatry* **2016**, *3*, 993–998. [CrossRef]
4. Russ, T.C.; Woelbert, E.; Davis, K.A.; Hafferty, J.D.; Ibrahim, Z.; Inkster, B.; John, A.; Lee, W.; Maxwell, M.; McIntosh, A.M.; et al. How data science can advance mental health research. *Nat. Hum. Behav.* **2019**, *3*, 24–32. [CrossRef] [PubMed]
5. Lyons, M.; Aksayli, N.D.; Brewer, G. Mental distress and language use: Linguistic analysis of discussion forum posts. *Comput. Hum. Behav.* **2018**, *87*, 207–211. [CrossRef]
6. Calvo, R.A.; Milne, D.N.; Hussain, M.S.; Christensen, H. Natural language processing in mental health applications using non-clinical texts. *Nat. Lang. Eng.* **2017**, *23*, 649–685. [CrossRef]
7. Pennebaker, J.W.; Francis, M.E.; Booth, R.J. Linguistic inquiry and word count: LIWC 2001. *Mahway Lawrence Erlbaum Assoc.* **2001**, *71*, 2001.
8. Honnibal, M.; Johnson, M. An Improved Non-monotonic Transition System for Dependency Parsing. In Proceedings of the 2015 Conference on Empirical Methods in Natural Language Processing, Lisbon, Portugal, 17–21 September 2015; pp. 1373–1378.
9. Bojanowski, P.; Grave, E.; Joulin, A.; Mikolov, T. Enriching word vectors with subword information. *Trans. Assoc. Comput. Linguist.* **2017**, *5*, 135–146. [CrossRef]
10. Delobelle, P.; Winters, T.; Berendt, B. RobBERT: A dutch RoBERTa-based language model. *arXiv* **2020**, arXiv:2001.06286.

11. Davcheva, E. Text Mining Mental Health Forums—Learning from User Experiences. In Proceedings of the 26th European Conference on Information Systems: Beyond Digitization—Facets of Socio-Technical Change, ECIS 2018, Portsmouth, UK, 23–28 June 2018; Bednar, P.M., Frank, U., Kautz, K., Eds.; AIS eLibrary: Atlanta, GA, USA, 2018; p. 91.
12. Deng, L.; Yu, D. Deep learning: Methods and applications. *Found. Trends® Signal Process.* **2014**, *7*, 197–387. [CrossRef]
13. Coppersmith, G.; Dredze, M.; Harman, C.; Hollingshead, K. From ADHD to SAD: Analyzing the language of mental health on Twitter through self-reported diagnoses. In Proceedings of the 2nd Workshop on Computational Linguistics and Clinical Psychology: From Linguistic Signal to Clinical Reality, Denver, CO, USA, 5 June 2015; pp. 1–10.
14. Webster, J.; Watson, R.T. Analyzing the past to prepare for the future: Writing a literature review. *MIS Q.* **2002**, *26*, xiii–xxiii.
15. Tausczik, Y.R.; Pennebaker, J.W. The psychological meaning of words: LIWC and computerized text analysis methods. *J. Lang. Soc. Psychol.* **2010**, *29*, 24–54. [CrossRef]
16. Young, T.; Hazarika, D.; Poria, S.; Cambria, E. Recent trends in deep learning based natural language processing. *IEEE Comput. IntelligenCe Mag.* **2018**, *13*, 55–75. [CrossRef]
17. Kim, K.; Lee, S.; Lee, C. College students with ADHD traits and their language styles. *J. Atten. Disord.* **2015**, *19*, 687–693. [CrossRef] [PubMed]
18. Nguyen, T.; Phung, D.; Venkatesh, S. Analysis of psycholinguistic processes and topics in online autism communities. In Proceedings of the 2013 IEEE International Conference on Multimedia and Expo (ICME), San Jose, CA, USA, 15–19 July 2013; pp. 1–6.
19. Forgeard, M. Linguistic styles of eminent writers suffering from unipolar and bipolar mood disorder. *Creat. Res. J.* **2008**, *20*, 81–92. [CrossRef]
20. Remmers, C.; Zander, T. Why you don't see the forest for the trees when you are anxious: Anxiety impairs intuitive decision making. *Clin. Psychol. Sci.* **2018**, *6*, 48–62. [CrossRef]
21. Trifu, R.N.; Nemeş, B.; Bodea-Hategan, C.; Cozman, D. Linguistic indicators of language in major depressive disorder (MDD). An evidence based research. *J. Evid.-Based Psychother.* **2017**, *17*, 105–128. [CrossRef]
22. Papini, S.; Yoon, P.; Rubin, M.; Lopez-Castro, T.; Hien, D.A. Linguistic characteristics in a non-trauma-related narrative task are associated with PTSD diagnosis and symptom severity. *Psychol. Trauma Theory Res. Pract. Policy* **2015**, *7*, 295. [CrossRef] [PubMed]
23. Corcoran, C.M.; Cecchi, G. Using language processing and speech analysis for the identification of psychosis and other disorders. *Biol. Psychiatry Cogn. Neurosci. Neuroimaging* **2020**, *5*, 770–779. [CrossRef] [PubMed]
24. Verkleij, S. Deep and Dutch NLP: Exploring Linguistic Markers for Patient Narratives Analysis. Master's Thesis, Department of Information and Computing Sciences, Utrecht University, Utrecht, The Netherlands, 2021.
25. Choi, J.D.; Tetreault, J.; Stent, A. It depends: Dependency parser comparison using a web-based evaluation tool. In Proceedings of the 53rd Annual Meeting of the Association for Computational Linguistics and the 7th International Joint Conference on Natural Language Processing (Volume 1: Long Papers), Beijing, China, 15 July 2015; pp. 387–396.
26. Hermann, K.M. Distributed representations for compositional semantics. *arXiv* **2014**, arXiv:1411.3146.
27. Liang, P.; Potts, C. Bringing machine learning and compositional semantics together. *Annu. Rev. Linguist.* **2015**, *1*, 355–376. [CrossRef]
28. Guevara, E.R. A regression model of adjective-noun compositionality in distributional semantics. In Proceedings of the 2010 Workshop on GEometrical Models of Natural Language Semantics, Uppsala, Sweden, 16 July 2010; pp. 33–37.
29. Gamallo, P. Sense Contextualization in a Dependency-Based Compositional Distributional Model. In Proceedings of the 2nd Workshop on Representation Learning for NLP, Vancouver, BC, Canada, 3 August 2017; pp. 1–9.
30. Bohnet, B. Top accuracy and fast dependency parsing is not a contradiction. In Proceedings of the 23rd International Conference on Computational Linguistics (Coling 2010), Beijing, China, 23–27 August 2010; pp. 89–97.
31. Lei, T.; Xin, Y.; Zhang, Y.; Barzilay, R.; Jaakkola, T. Low-rank tensors for scoring dependency structures. In Proceedings of the 52nd Annual Meeting of the Association for Computational Linguistics (Volume 1: Long Papers), Baltimore, MD, USA, 14 June 2014; pp. 1381–1391.
32. Choi, J.D.; McCallum, A. Transition-based dependency parsing with selectional branching. In Proceedings of the 51st Annual Meeting of the Association for Computational Linguistics (Volume 1: Long Papers), Sofia, Bulgaria, 13 August 2013; pp. 1052–1062.
33. Van den Bosch, A.; Busser, B.; Canisius, S.; Daelemans, W. An efficient memory-based morphosyntactic tagger and parser for Dutch. *LOT Occas. Ser.* **2007**, *7*, 191–206.
34. Van der Beek, L.; Bouma, G.; Malouf, R.; Van Noord, G. The Alpino dependency treebank. In *Computational Linguistics in The Netherlands 2001*; Brill Rodopi: Amsterdam, The Netherlands, 2002; pp. 8–22.
35. Otter, D.W.; Medina, J.R.; Kalita, J.K. A survey of the usages of deep learning for natural language processing. *IEEE Trans. Neural Netw. Learn. Syst.* **2020**, *32*, 604–624. [CrossRef] [PubMed]
36. Mikolov, T.; Sutskever, I.; Chen, K.; Corrado, G.S.; Dean, J. Distributed representations of words and phrases and their compositionality. *Adv. Neural Inf. Process. Syst.* **2013**, *2*, 3111–3119.
37. Mikolov, T.; Chen, K.; Corrado, G.; Dean, J. Efficient estimation of word representations in vector space. *arXiv* **2013**, arXiv:1301.3781.
38. Joulin, A.; Grave, E.; Mikolov, P.B.T. Bag of Tricks for Efficient Text Classification. *EACL* **2017**, *2017*, 427.
39. Vaswani, A.; Shazeer, N.; Parmar, N.; Uszkoreit, J.; Jones, L.; Gomez, A.N.; Kaiser, Ł.; Polosukhin, I. Attention is all you need. *Adv. Neural Inf. Process. Syst.* **2017**, *30*, 5998–6008.

40. Wolf, T.; Chaumond, J.; Debut, L.; Sanh, V.; Delangue, C.; Moi, A.; Cistac, P.; Funtowicz, M.; Davison, J.; Shleifer, S.; et al. Transformers: State-of-the-art natural language processing. In Proceedings of the 2020 Conference on Empirical Methods in Natural Language Processing: System Demonstrations, 16–20 November 2020; pp. 38–45.
41. Devlin, J.; Chang, M.W.; Lee, K.; Toutanova, K. Bert: Pre-training of deep bidirectional transformers for language understanding. *arXiv* **2018**, arXiv:1810.04805.
42. Liu, Y.; Ott, M.; Goyal, N.; Du, J.; Joshi, M.; Chen, D.; Levy, O.; Lewis, M.; Zettlemoyer, L.; Stoyanov, V. Roberta: A robustly optimized bert pretraining approach. *arXiv* **2019**, arXiv:1907.11692.
43. Khattak, F.K.; Jeblee, S.; Pou-Prom, C.; Abdalla, M.; Meaney, C.; Rudzicz, F. A survey of word embeddings for clinical text. *J. Biomed. Inform.* **2019**, *4*, 100057. [CrossRef]
44. Miotto, R.; Wang, F.; Wang, S.; Jiang, X.; Dudley, J.T. Deep learning for healthcare: Review, opportunities and challenges. *Briefings Bioinform.* **2018**, *19*, 1236–1246. [CrossRef]
45. Le, N.Q.K.; Yapp, E.K.Y.; Nagasundaram, N.; Yeh, H.Y. Classifying promoters by interpreting the hidden information of DNA sequences via deep learning and combination of continuous fasttext N-grams. *Front. Bioeng. Biotechnol.* **2019**, *7*, 305. [CrossRef] [PubMed]
46. de Vries, W.; van Cranenburgh, A.; Bisazza, A.; Caselli, T.; van Noord, G.; Nissim, M. Bertje: A dutch bert model. *arXiv* **2019**, arXiv:1912.09582.
47. Sarhan, I.; Spruit, M. Can we survive without labelled data in NLP? Transfer learning for open information extraction. *Appl. Sci.* **2020**, *10*, 5758. [CrossRef]
48. Loper, E.; Bird, S. NLTK: The Natural Language Toolkit. *arXiv* **2002**, arXiv:cs/0205028.
49. Ribeiro, M.T.; Singh, S.; Guestrin, C. "Why should i trust you?" Explaining the predictions of any classifier. In Proceedings of the 22nd ACM SIGKDD International Conference on Knowledge Discovery and Data Mining, New York, NY, USA, 13–17 August 2016; pp. 1135–1144.
50. Cohen, J. A coefficient of agreement for nominal scales. *Educ. Psychol. Meas.* **1960**, *20*, 37–46. [CrossRef]
51. McHugh, M.L. Interrater reliability: The kappa statistic. *Biochem. Medica* **2012**, *22*, 276–282. [CrossRef]

applied sciences

MDPI

Article

FMFN: Fine-Grained Multimodal Fusion Networks for Fake News Detection

Jingzi Wang, Hongyan Mao * and Hongwei Li *

Shanghai Key Laboratory of Trustworthy Computing, East China Normal University, Shanghai 200062, China; jingziwang@163.com
* Correspondence: hymao@sei.ecnu.edu.cn (H.M.); 51194501008@stu.ecnu.edu.cn (H.L.)

Abstract: As one of the most popular social media platforms, microblogs are ideal places for news propagation. In microblogs, tweets with both text and images are more likely to attract attention than text-only tweets. This advantage is exploited by fake news producers to publish fake news, which has a devasting impact on individuals and society. Thus, multimodal fake news detection has attracted the attention of many researchers. For news with text and image, multimodal fake news detection utilizes both text and image information to determine the authenticity of news. Most of the existing methods for multimodal fake news detection obtain a joint representation by simply concatenating a vector representation of the text and a visual representation of the image, which ignores the dependencies between them. Although there are a small number of approaches that use the attention mechanism to fuse them, they are not fine-grained enough in feature fusion. The reason is that, for a given image, there are multiple visual features and certain correlations between these features. They do not use multiple feature vectors representing different visual features to fuse with textual features, and ignore the correlations, resulting in inadequate fusion of textual features and visual features. In this paper, we propose a novel fine-grained multimodal fusion network (FMFN) to fully fuse textual features and visual features for fake news detection. Scaled dot-product attention is utilized to fuse word embeddings of words in the text and multiple feature vectors representing different features of the image, which not only considers the correlations between different visual features but also better captures the dependencies between textual features and visual features. We conduct extensive experiments on a public Weibo dataset. Our approach achieves competitive results compared with other methods for fusing visual representation and text representation, which demonstrates that the joint representation learned by the FMFN (which fuses multiple visual features and multiple textual features) is better than the joint representation obtained by fusing a visual representation and a text representation in determining fake news.

Keywords: fake news detection; feature fusion; attention mechanism; social media

Citation: Wang, J.; Mao, H.; Li, H. FMFN: Fine-Grained Multimodal Fusion Networks for Fake News Detection. *Appl. Sci.* **2022**, *12*, 1093. https://doi.org/10.3390/app12031093

Academic Editors: Arturo Montejo-Ráez and Salud María Jiménez-Zafra

Received: 13 December 2021
Accepted: 17 January 2022
Published: 21 January 2022

1. Introduction

With the rapid development of social networks, social media platforms have become ideal places for news propagation [1]. Due to its convenience, people are increasingly seeking out and consuming news through social media. However, the convenience also facilitates the rapid spread and proliferation of fake news [2], which has a devasting impact on individuals and society [3].

As one of the most popular social media platforms, microblogs, such as Twitter and Weibo, allow people to share and forward tweets, where the tweets with both text and images are more likely to attract attention than the text-only tweets. This advantage is also exploited by fake news producers, who post tweets about fake news on microblogs by manipulating text and forging images. If these tweets are not verified, they may seriously jeopardize the credibility of microblogs [4]. Therefore, it is crucial to detect fake news on microblogs.

In recent years, methods for fake news detection have gradually evolved from unimodal to multimodal approaches. The question concerning how to learn a joint representation that contains multimodal information has attracted much research attention. Jin et al. [4] use local attention mechanism to refine the visual representation, but the refined visual representation cannot reflect the similarity between the visual representation and the joint representation of text and social context. Wang et al. [5] propose a model based on adversarial networks to learn an event-invariant feature. Khattar et al. [6] propose a model based on variational autoencoder (VAE) to learn a shared representation. However, these models view the concatenation of unimodal features as a joint representation, which cannot discover dependencies between modalities. Song et al. [7] leverage an attention mechanism to fuse a number of word embeddings and one image embedding to obtain fused features, and further extract key features from the fuse features as a joint representation. Although the joint representation captures the dependencies, the fusion is not fine-grained enough. This is due to the fact that they do not use multiple feature vectors representing different visual features to fuse with textual features, and ignore correlations between different visual features.

To overcome the limitations of the aforementioned methods, the fine-grained multimodal fusion networks (FMFN) is proposed for fake news detection. Our approach includes the following three steps. First, we use deep convolutional neural networks (CNNs) to extract multiple visual features of a given image and RoBERTa [8] to obtain deep contextualized word embeddings of words, each of which can be considered as a textual feature. Then, the scaled dot-product attention [9] is employed to enhance the visual features as well as the textual features, and fuse them. Finally, the fused feature is fed into a binary classifier for the detection.

The contributions can be summarized as follows:

1. To effectively detect fake news with text and image, we propose a novel model for fine-grained fusion of textual features and visual features.
2. The proposed model utilizes attention mechanism to enhance the visual features as well as the textual features, and fuse the enhanced visual features and the enhanced textual features, which not only considers the correlations between different visual features but also captures the dependencies between textual features and visual features.
3. We conduct extensive experiments on the real-word dataset. The results demonstrate the effectiveness of the proposed model.

This paper is organized as follows. In the next section, we review related work on fake news detection and scaled dot-product attention. Section 3 provides details of the proposed model. Section 4 presents the experiments. Section 5 gives the ablation analysis. In Section 6, we conclude the paper with a summary and give an outlook on future work.

2. Related Work

Fake news is defined as the news that is deliberately fabricated and is verifiable false [10,11]. Existing work on fake news detection can be divided into two categories: unimodal and multimodal. Scaled-dot product attention has been applied to the fields of natural language processing (NLP) and computer vision (CV). In NLP and CV, the extraction of corresponding features, such as textual features and visual features, is a fundamental task, and it is also a key step in fake news detection. In this section, we review the related work on unimodal fake news detection, multimodal fake news detection, and the scaled dot-product attention.

2.1. Unimodal Fake News Detection

Only one modality of content is utilized for unimodal fake news detection, such as text content, visual content, and social context. The text content of news plays an important role in determining the authenticity of the news. Ma et al. [12] use RNN to learn text representations from text content. Yu et al. [13] propose a CNN-based method to extract local-global significant features of text content. The two methods concentrate on detecting fake news at the event level, and thus require event labels, which increases the cost of the detection. To learn a stronger indicative representation of rumors, a GAN-style model is proposed by Ma et al. [14]. Besides text content, image is also crucial, which has a great influence on news propagation [15,16]. Qi et al. [17] use RNN and CNN-RNN to extract visual features in the frequency domain and the pixel domain, respectively. The visual features in different domains are then fused using an attention mechanism. In addition to textual features and visual features, social context features are also widely used for fake news detection on social media. To capture propagation patterns of news, Wu et al. [18] develop an SVM classifier based on kernel methods, which combine some social context features. For early detection of fake news, Liu et al. [19] extract user characteristics from user profiles to judge the authenticity of the news.

2.2. Multimodal Fake News Detection

Multimodal fake news detection relies on multimodal information, rather than information from one modality of content. The process involves feature extraction and feature fusion. In feature extraction, textual feature extractors can be implemented using Bi-LSTM [20,21], textCNN [22,23], or BERT [24], and visual features are typically extracted by CNNs. In feature fusion, there are several typical methods as follows. Jin et al. [4] exploit text content, image, and social context to produce a joint representation. An attention mechanism is leveraged to refine the visual representation. However, the refined visual representation cannot reflect the similarity between the visual representation and the social-textual representation, since the attention values are only calculated from the social-textual representation. Wang et al. [5] are inspired by the idea of adversarial networks and thus propose an event adversarial neural network (EANN), which contains an event discriminator used to identify the event label of news, in addition to the feature extractors and the detector. To learn a more general joint representation, a minimax game is set up between the event discriminator and feature extractors. Khattar et al. [6] proposed a multimodal variational autoencoder (MVAE) for fake news detection, which is composed of an encoder, a decoder, and a fake news detector. The encoder first extracts textual features and visual features, which are converted to a sampled multimodal representation. Then, the decoder reconstructs the textual features and visual features from the sampled multimodal representation. Finally, the encoder, the decoder, and the detector are jointly trained to learn a shared representation of multimodal information. Nevertheless, the above three methods [4–6] obtain a joint representation by simply concatenating unimodal features without considering the dependencies between modalities. Song et al. [7] leverage an attention mechanism to fuse a number of word embeddings and one image embedding to obtain fused features, and further extract key features from the fuse features as a joint representation. Although the fusion considers inter-modality relations, it is not fine-grained enough.

2.3. Scaled-Dot Product Attention

The scaled dot-product attention first appears in transformer [9], which is originally used for machine translation tasks. The scaled dot-product attention enables the transformer to capture global dependencies between input and output, which represent text content in two different languages, respectively.

For NLP, Transformer architecture based on the scaled dot-product attention has become the de-facto standard [25]. Some pretrained language models, such as BERT [24], XLNET [26], and GPT-3 [27], have achieved state-of-the-art results on different NLP tasks. Inspired by NLP success, there are multiple works [28,29] that combine CNNs and the scaled dot-product attention in CV. For capturing global information, the scaled dot-product attention has some advantages over repeated convolutional operations, leading to application of the scaled dot-product attention in CV. Thus, some works [25,30] interpret an image as a sequence of words and process them by the Transformer's encoder solely based on the scaled dot-product attention.

Considering the power of the scaled dot-product attention, we propose to fuse textual features and visual features with the scaled dot-product attention. Like the transformer, the feature fusion in our method is entirely based on the scaled dot-product attention, and the proposed method is expected to improve the performance of fake news detection.

3. Model

3.1. Model Overview

Given news with text and image, the proposed model aims to determine whether the news is real or fake. The architecture of the model is shown in Figure 1, which consists of three parts. The first part is composed of a textual feature extractor and a visual feature extractor, which extract textual features and visual features, respectively. This is followed by the feature fusion, where scaled dot-product attention is used for fine-grained fusion of the textual features and the visual features. The last part is a fake news detector that exploits the fused feature to judge the truth of the news.

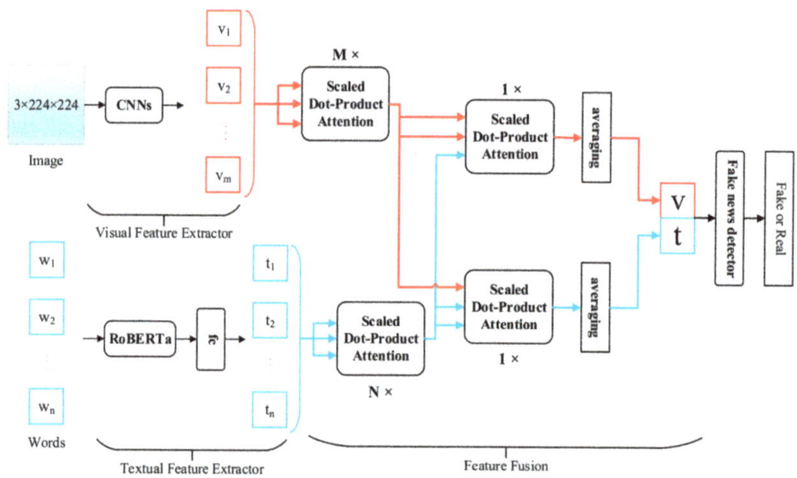

Figure 1. The architecture of our FMFN model.

3.2. Visual Feature Extraction

CNNs have achieved great success in CV. In CNNs, multiple feature maps are obtained by applying convolutional operations of different convolution kernels over an image and can be considered as visual features of the image.

Instead of a visual representation that represents the image, we exploit multiple visual features of the image to fully fuse with textual features, where each visual feature is represented by a feature vector. To learn different features of the image, the VGG-19 [31] is employed, which contains 16 convolutional layers, and 3 feed-forward layers. For an image, the VGG-19 network outputs one vector containing different features, which is not conducive to fine-grained fusion with textual features. Thus, the last three fully-connected layers are removed, and several additional convolutional layers are added behind the 16 convolutional layers of the VGG-19. In this way, the visual feature extractor is composed entirely of convolutional layers and yields a specified number of feature maps $P = [p_1, p_2, \ldots, p_m]$, where m is determined by the number of convolution kernels in the last convolutional layer and each feature map p_i is a $h \times w$ dimensional vector. By collapsing the spatial dimensions of each feature map p_i, we obtain the visual features $R_V = [v_1, v_2, \ldots, v_m]$, each of which is a $hw \times 1$ dimensional vector.

3.3. Textual Feature Extraction

The text content is tokenized into a sequence of tokens denoted as $W = [w_1, w_2, \ldots, w_n]$, where n is the number of tokens. For fine-grained fusion, we obtain the word embedding of each token, rather than a vector representation that represents the text content.

In the NLP field, pretrained language models have achieved state-of-the-art results on different NLP tasks. In particular, the BERT and its variants are widely used due to the ability to utilize both left-to-right and right-to-left contextual information. RoBERTa [8], an improved pretraining procedure for BERT, performs better than BERT on some benchmarks, which removes the next sentence prediction task and adopts the dynamic masking scheme. Thus, RoBERTa is employed to extract word embeddings of the tokens, which is denoted as $E = [e_1, e_2, \ldots, e_n]$.

Compared with other methods of learning word representations, such as word2vec [32], GloVe [33], and fastText [34], word representations generated by the RoBERTa contain contextual information, which means that each word embedding e_i contains information about the entire text content, and therefore can be considered as a textual feature. To adjust the dimensionality of each textual feature, a fully connected layer with ReLU activation function (denoted as "fc" in Figure 1) transforms $E = [e_1, e_2, \ldots, e_n]$ to $R_T = [t_1, t_2, \ldots, t_n]$, where each textual feature t_i is a $d \times 1$ dimensional vector.

3.4. Feature Fusion

Transformer is originally used for machine translation tasks. For a task to translate from English to French, the transformer draws dependencies between English sentences and French sentences thanks to the scaled dot-product attention. We apply the scaled dot-product attention to multimodal fusion so as to capture dependencies between textual features and visual features. In addition, the scaled dot-product attention also can be used to capture global information between these visual features since we extract multiple visual features instead of a visual representation.

Motivated by the above observations, scaled dot-product attention (See Figure 2) is used for fine-grained fusion of textual features and visual features. The scaled dot-product attention block is defined as $ScaledDotProductAttn(Queries,\ Keys,\ Values)$, where $Queries$, $Keys$ and $Values$ are mapped into three representations Q, K, and V with three linear layers, then the scaled dot-product attention is computed on Q, K, and V.

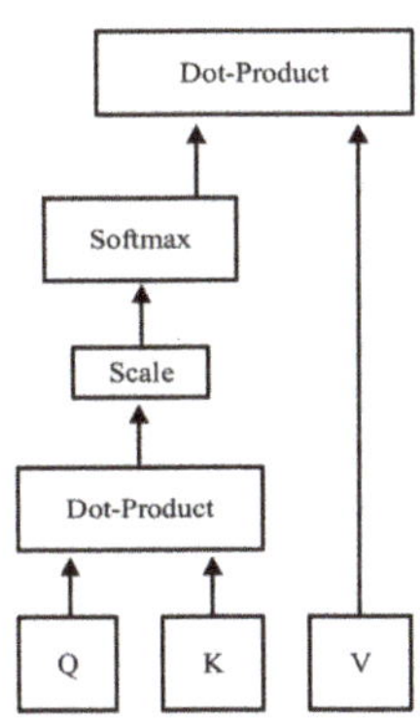

Figure 2. The scaled dot-product attention.

We first enhance the visual features and the textual features using scaled dot-product attention blocks, which can capture global information. For visual features, it enables these features to be further correlated, although global features are obtained by deep CNNs. The process is as follows.

$$R_V^1 = ScaledDotProductAttn(R_V,\ R_V,\ R_V) \quad (1)$$

$$R_V^2 = ScaledDotProductAttn\left(R_V^1,\ R_V^1,\ R_V^1\right) \quad (2)$$

$$R_V^M = ScaledDotProductAttn\left(R_V^{M-1},\ R_V^{M-1},\ R_V^{M-1}\right) \quad (3)$$

where M is the number of the scaled dot-product attention blocks and $R_V^M = \left[v_1^M,\ v_2^M,\ \ldots,\ v_m^M\right]$ represents a number of enhanced visual features. Several scaled dot-product attention blocks (The number of the blocks is N) are also applied to the textual features R_T to obtain $R_T^N = \left[t_1^N,\ t_2^N,\ \ldots,\ t_n^N\right]$ in the same way.

Then, two scaled dot-product attention blocks are utilized to refine the enhanced visual features R_V^M and the enhanced textual features R_T^N, respectively. The process to refine the visual features R_V^M is as follows.

$$R_V' = ScaledDotProductAttn\left(R_T^N,\ R_V^M,\ R_V^M\right) \quad (4)$$

The $R_V' = \left[v_1',\ v_2',\ \ldots,\ v_m'\right]$ are the refined visual features representing the fine-grained fusion with the textual features R_T^N. Note that the queries come from the enhanced textual features, and the keys and the values come from the enhanced visual features. Therefore, it can capture the dependencies between visual features and textual features. The R_T' is also obtained by computing the scaled dot-product attention, where queries come from the enhanced visual features, and the keys and the values come from the enhanced textual features.

Finally, the refined features R_V' and R_T' are transformed to two vectors v and t by the averaging. The process of averaging the refined features R_V' to produce the vector v is as follows.

$$v = \frac{v_1' \oplus v_2' \oplus \ldots \oplus v_m'}{m} \quad (5)$$

where $\oplus$ denotes element-wise sum. The two vectors v and t are concatenated into a vector r as the joint representation, which not only considers the correlations between different visual features but also reflects the dependencies between textual features and visual features.

3.5. Fake News Detector and Model Learning

The fake news detector is a fully connected layer with SoftMax function, which takes the joint representation r as input to make the prediction as follows.

$$\hat{y} = softmax(W \times r + b) \quad (6)$$

where W is parameters of the fully connected layer and b is the bias term.

To configure the model for training, the loss function is set to cross entropy as follows.

$$L(\theta) = -ylog(\hat{y}) - (1-y)log(1-\hat{y}) \quad (7)$$

where θ represents all of the learnable parameters of the proposed model, and $y \in \{0,\ 1\}$ denotes the ground-truth label.

4. Experiments

4.1. Dataset

We evaluate the effectiveness of the proposed model on the dataset collected by Jin et al. [4], on which the real news is collected from an authoritative news source, Xinhua News Agency, and the fake news is verified by Weibo's official rumor debunking system.

For the dataset, we only focus on tweets with text and images in order to fuse textual features and visual features. Thus, tweets without text or images are removed. The data split scheme is the same as the benchmark scheme, and the data are preprocessed in a similar way to the work [4]. The detailed statistics of the dataset are listed in Table 1.

Table 1. The Weibo dataset.

	Training Set	**Test Set**
fake news	3345	862
real news	2807	835
images	6152	1697

4.2. Settings

The optimizer used is Adam [35] with a learning rate of 0.001, $\beta1 = 0.9$ and $\beta2 = 0.999$.

For the textual feature extraction, the Chinese BERT with whole word masking [36,37] is used, and the max length of text is set to 160. For efficient training, the feature-based approach is adopted on the pretrained language model, which means that the parameters of the pretrained language model are fixed. Only the fully connected layer with ReLU activation function (denoted as "fc" in Figure 1) is trained, and its hidden size is 100.

For the visual feature extraction, the first 16 convolutional layers and the first four max-pooling layers of VGG19 are adopted, which means that we remove the last three fully-connected layers, and the last max-pooling layer of VGG19. The parameters of the 16 convolutional layers are frozen. Two additional convolutional layers with ReLU activation function, the first with 256 convolution kernels and the second with 160 convolution kernels, are added behind these layers and trained. For these convolution kernels, the receptive field is 3×3, and the convolution stride is 1. Thus, 160 visual features are produced by the visual extractor, each of which is a 100×1 dimensional vector.

As above, the number of visual features m is equal to the number of textual features n, and the dimensionality of each visual feature and each text feature are also equal, which facilitates the computation of the Scale-Dot Product Attention.

For the M and N, they are set to 3 and 1, respectively, which achieves the best performance.

4.3. Baselines

For comparison with other methods, two unimodal models and six multimodal models are chosen as baselines, which are listed as follows:

- Textual: All scaled dot-product attention blocks and the visual feature extractor are removed from the proposed model FMFN. The textual features R_T obtained by the textual feature extractor are transformed to a vector by the averaging, and the vector is fed into a binary classifier to train a model. For a fair comparison, the parameters of the RoBERTa in the textual feature extractor are frozen.
- Visual: Similar to textual, the visual feature extractor, and a binary classifier are jointly trained for fake news detection. For a fair comparison, the parameters of the first 16 convolutional layers in the visual feature extractor are fixed.
- VQA [38]: The objective of visual question answering is to answer questions concerning certain images. The multi-class classifier in the VQA model is replaced with a binary classifier, and one-layer LSTM is used for a fair comparison.
- NeuralTalk [39]: The model aims to produce captions for given images. The joint representation is obtained by averaging the outputs of RNN at each time step.
- att-RNN [4]: A novel RNN with an attention mechanism is utilized to fuse multimodal features for effective rumor detection. For a fair comparison, we do not consider the social context, and only fuse textual features and visual features.
- EANN [5]: The model is based on adversarial networks, which can learn event-invariant features containing multimodal information.

- MVAE [6]: By jointly training the VAE and a classifier, the model is able to learn a shared representation of multimodal information.
- CARMN [7]: An attention mechanism is used to fuse word embeddings and one image embedding to obtain fused features. From the fuse features, key features are extracted as a joint representation.

4.4. Comparison with Baselines

Table 2 shows the results of different methods on Weibo dataset. We can observe that our proposed model achieves competitive results.

Table 2. The results of different methods on Weibo dataset.

Method	Accuracy	Fake News			Real News		
		Precision	Recall	F_1	Precision	Recall	F_1
Textual	0.725	0.763	0.661	0.708	0.677	0.774	0.722
Visual	0.657	0.682	0.617	0.648	0.622	0.68	0.65
VQA	0.736	0.797	0.634	0.706	0.695	0.838	0.76
NeuralTalk	0.726	0.794	0.613	0.692	0.684	0.84	0.754
att-RNN	0.772	0.854	0.656	0.742	0.72	0.889	0.795
EANN	0.782	0.827	0.697	0.756	0.752	0.863	0.804
MVAE	0.824	0.854	0.769	0.809	0.802	0.875	0.837
CARMN	0.853	**0.891**	0.814	0.851	0.818	0.894	0.854
FMFN	**0.885**	0.878	**0.851**	**0.864**	**0.874**	**0.896**	**0.885**

Specifically, the proposed model FMFN achieves an accuracy of 88.5% on the dataset and outperforms all of the baseline models except the precision of fake news. In these baseline systems, CARMN performs best, which can be attributed to the attention mechanism. The attention mechanism in CARMN can capture the dependencies between textual features and visual features, but other multimodal methods, which simply concatenate unimodal features, cannot learn the dependencies. The dependencies include consistency between text content and image content. The news with inconsistent text and image is generally fake. It is difficult to identify if the dependencies between textual features and visual features cannot be captured. Compared with CARMN, our model boosts accuracy by about 3%. It is the fined-grained fusion of word embeddings and multiple visual features that achieves significant improvements, whereas CARMN only fuses word embeddings and one image embedding. It illustrates the importance of the fine-grained fusion, which facilitates a better capture of such dependencies.

5. Ablation Analysis

5.1. Component Analysis

To verify the impact of each component of FMFN, three baselines are constructed as follows.

- FMFN(CONCAT): The last two scaled dot-product attention blocks are removed from the proposed model FMFN. By the averaging, the R_V^M and R_T^N are transformed to two vectors, respectively. The concatenation of the two vector is fed into the fake news detector. Therefore, it cannot capture the dependencies between textual features and visual features.
- FMFN(TEXT): We do not use the refined visual features R'_V and only use the refined textual features R'_T. The refined textual features R'_T are transformed to a vector by the averaging, and the vector is fed into the fake news detector.
- FMFN (M = 0): The number of scaled dot-product attention blocks M is set to 0, which means that we do not consider the correlations between different visual features.

From Table 3, we can see that our proposed method FMFN outperforms all baselines. If we remove one of the components from the model, both the accuracy and F_1 scores will drop. The results show that all components of the model are indispensable.

Table 3. The results of FMFN (CONCAT), FMFN (TEXT), FMFN (M = 0), and FMFN.

Method	Accuracy	Fake News F_1	Real News F_1
FMFN (CONCAT)	0.867	0.839	0.872
FMFN (TEXT)	0.874	0.845	0.876
FMFN (M = 0)	0.877	0.851	0.880
FMFN	**0.885**	**0.864**	**0.885**

Compared with FMFN (CONCAT), the accuracy of FMFN increases from 86.7% to 88.5%. It shows that the scaled dot-product attention blocks used to capture the dependencies between visual features and textual features are critical for performance improvement. For FMFN (CONCAT), simply concatenating multiple visual features and textual features can yield relatively good results (an accuracy of 86.7%) without using attention, which shows the importance of representing different features of an image with multiple feature vectors. If we only use the refined textual features, the accuracy will drop about 1%, which indicates that both the refined textual features and the refined visual features are important. For the hyper-parameter M, there will be a performance loss as well if we set it to 0. This indicates that it is useful to use attention to make multiple visual features correlated.

5.2. Visualization of the Joint Representation

To further illustrate the impact of the feature fusion, the joint representation r learned by FMFN and the joint representation learned by FMFN(CONCAT) are visualized with t-SNE [40]. As depicted in Figure 3, two colors represent fake news and real news, respectively.

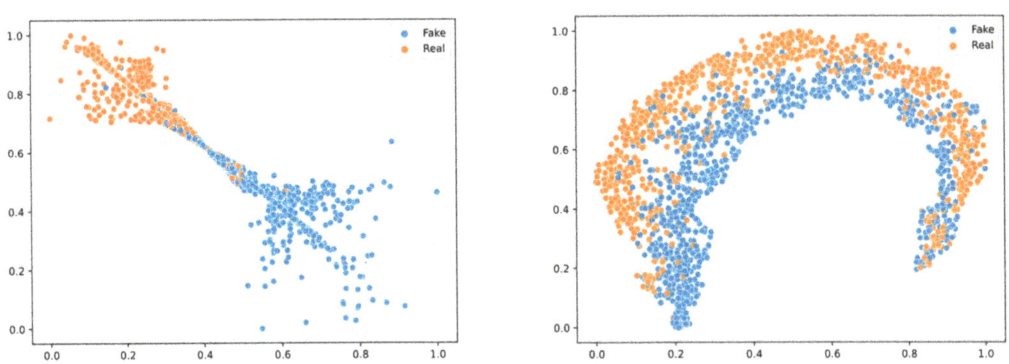

(a) (b)

Figure 3. Visualization of the joint representation: (**a**) FMFN; (**b**) FMFN (CONCAT).

From Figure 3, we can see that FMFN can learn more discriminable representations compared with FMFN (CONCAT). As is shown in Figure 3a, the representations of the different categories are in the upper left and lower right regions of the image. In addition, the representations of the same category are more easily aggregated, which makes the number of points in Figure 3a look small. For FMFN (CONCAT), it basically distinguishes between two types of representations. However, there are many representations that are difficult to distinguish. The visualization illustrates the effectiveness of the feature fusion.

6. Conclusions and Future Work

We propose a novel fine-grained multimodal fusion network (FMFN) to fully fuse textual features and visual features for fake news detection. For a tweet with text and image, multiple different visual features of the image are obtained by deep CNNs and word embeddings of words in the text are extracted by a pretrained language model, each of which can be considered as a textual feature. The scaled dot-product attention is employed to enhance the visual features as well as the textual features and fuse them. This is a fine-grained and adequate fusion, which not only considers the correlations between different visual features but also captures the dependencies between textual features and visual features. Experiments conducted on a public Weibo dataset demonstrate the effectiveness of FMFN. In comparison with other methods for fusing the visual representation and the text representation, FMFN achieves competitive results. It shows that the joint representation learned by the FMFN, which fuses multiple visual features and multiple textual features, is better than the joint representation obtained by fusing a visual representation and a text representation in determining fake news.

In the future, we plan to fuse social context features in addition to textual features and visual features. Moreover, the visual features in the frequency domain [17] are considered to further improve the performance of fake news detection.

Author Contributions: Conceptualization, H.L. and H.M.; investigation, J.W.; methodology, H.L., H.M. and J.W.; software, H.L.; project administration, J.W.; validation, H.L., H.M. and J.W.; writing—original draft preparation, H.L.; writing—review and editing, H.L., H.M. and J.W.; visualization, J.W., H.L. and H.M.; funding acquisition, J.W. and H.M. All authors have read and agreed to the published version of the manuscript.

Funding: This research was funded by the National Natural Science Foundation of China (No. 61872145, 62032024), MOE International Joint Lab of Trustworthy Software, East China Normal University (No. 2021-5) and the Key Laboratory of Embedded System and Service Computing, Ministry of Education, Tongji University (No. ESSCKF2021-03).

Institutional Review Board Statement: Not applicable.

Informed Consent Statement: Not applicable.

Data Availability Statement: Not applicable.

Conflicts of Interest: The authors declare no conflict of interest.

References

1. Czeglédi, C.; Valentinyi, K.V.; Borsos, E.; Járási, É.; Szira, Z.; Varga, E. News Consuming Habits of Young Social Media Users in the Era of Fake News. *WSEAS Trans. Comput.* **2019**, *18*, 264–273.
2. Helmstetter, S.; Paulheim, H. Collecting a Large Scale Dataset for Classifying Fake News Tweets Using Weak Supervision. *Future Internet* **2021**, *13*, 114. [CrossRef]
3. Zakharchenko, A.; Peráček, T.; Fedushko, S.; Syerov, Y.; Trach, O. When Fact-Checking and 'BBC Standards' Are Helpless: 'Fake Newsworthy Event' Manipulation and the Reaction of the 'High-Quality Media' on It. *Sustainability* **2021**, *13*, 573. [CrossRef]
4. Jin, Z.; Cao, J.; Guo, H.; Zhang, Y.; Luo, J. Multimodal Fusion with Recurrent Neural Networks for Rumor Detection on Microblogs. In Proceedings of the 25th ACM International Conference on Multimedia, Mountain View, CA, USA, 23–27 October 2017; pp. 795–816.
5. Wang, Y.; Ma, F.; Jin, Z.; Yuan, Y.; Xun, G.; Jha, K.; Su, L.; Gao, J. EANN: Event Adversarial Neural Networks for Multi-Modal Fake News Detection. In Proceedings of the 24th ACM SIGKDD International Conference on Knowledge Discovery & Data Mining, London, UK, 19–23 August 2018; pp. 849–857.
6. Khattar, D.; Goud, J.S.; Gupta, M.; Varma, V. MVAE: Multimodal Variational Autoencoder for Fake News Detection. In Proceedings of the World Wide Web Conference, San Francisco, CA, USA, 13–17 May 2019; pp. 2915–2921.
7. Song, C.; Ning, N.; Zhang, Y.; Wu, B. A multimodal fake news detection model based on crossmodal attention residual and multichannel convolutional neural networks. *Inf. Process. Manag.* **2021**, *58*, 102437. [CrossRef]
8. Liu, Y.; Ott, M.; Goyal, N.; Du, J.; Joshi, M.; Chen, D.; Levy, O.; Lewis, M.; Zettlemoyer, L.; Stoyanov, V. Roberta: A Robustly Optimized BERT Pretraining Approach. *arXiv* **2019**, arXiv:1907.11692.

9. Vaswani, A.; Shazeer, N.; Parmar, N.; Uszkoreit, J.; Jones, L.; Gomez, A.N.; Kaiser, Ł.; Polosukhin, I. Attention is all you need. In Proceedings of the 31st International Conference on Neural Information Processing Systems, Long Beach, CA, USA, 4–9 December 2017; pp. 6000–6010.
10. Islam, N.; Shaikh, A.; Qaiser, A.; Asiri, Y.; Almakdi, S.; Sulaiman, A.; Moazzam, V.; Babar, S.A. Ternion: An Autonomous Model for Fake News Detection. *Appl. Sci.* **2021**, *11*, 9292. [CrossRef]
11. Alonso, M.A.; Vilares, D.; Gómez-Rodríguez, C.; Vilares, J. Sentiment Analysis for Fake News Detection. *Electronics* **2021**, *10*, 1348. [CrossRef]
12. Ma, J.; Gao, W.; Mitra, P.; Kwon, S.; Jansen, B.J.; Wong, K.-F.; Cha, M. Detecting rumors from microblogs with recurrent neural networks. In Proceedings of the Twenty-Fifth International Joint Conference on Artificial Intelligence, New York, NY, USA, 9–15 July 2016; pp. 3818–3824.
13. Yu, F.; Liu, Q.; Wu, S.; Wang, L.; Tan, T. A convolutional approach for misinformation identification. In Proceedings of the 26th International Joint Conference on Artificial Intelligence, Melbourne, Australia, 19–25 August 2017; pp. 3901–3907.
14. Ma, J.; Gao, W.; Wong, K.-F. Detect Rumors on Twitter by Promoting Information Campaigns with Generative Adversarial Learning. In Proceedings of the World Wide Web Conference, San Francisco, CA, USA, 13–17 May 2019; pp. 3049–3055.
15. Jin, Z.; Cao, J.; Zhang, Y.; Zhou, J.; Tian, Q. Novel Visual and Statistical Image Features for Microblogs News Verification. *IEEE Trans. Multimed.* **2017**, *19*, 598–608. [CrossRef]
16. Shu, K.; Sliva, A.; Wang, S.; Tang, J.; Liu, H. Fake News Detection on Social Media: A Data Mining Perspective. *ACM SIGKDD Explor. Newsl.* **2017**, *19*, 22–36. [CrossRef]
17. Qi, P.; Cao, J.; Yang, T.; Guo, J.; Li, J. Exploiting Multi-domain Visual Information for Fake News Detection. In Proceedings of the 2019 IEEE International Conference on Data Mining (ICDM), Beijing, China, 8–11 November 2019; pp. 518–527.
18. Wu, K.; Yang, S.; Zhu, K.Q. False rumors detection on Sina Weibo by propagation structures. In Proceedings of the 2015 IEEE 31st International Conference on Data Engineering, Seoul, Korea, 13–17 April 2015; pp. 651–662.
19. Liu, Y.; Wu, Y.-F.B. Early Detection of Fake News on Social Media Through Propagation Path Classification with Recurrent and Convolutional Networks. In Proceedings of the 32th AAAI Conference on Artificial Intelligence (AAAI-18), New Orleans, LA, USA, 2–7 February 2018.
20. Schuster, M.; Paliwal, K.K. Bidirectional recurrent neural networks. *IEEE Trans. Signal. Process.* **1997**, *45*, 2673–2681. [CrossRef]
21. Hochreiter, S.; Schmidhuber, J. Long Short-Term Memory. *Neural Comput.* **1997**, *9*, 1735–1780. [CrossRef]
22. Kim, Y. Convolutional Neural Networks for Sentence Classification. In Proceedings of the 2014 Conference on Empirical Methods in Natural Language Processing, Doha, Qatar, 25–29 October 2014. [CrossRef]
23. Zhang, Y.; Wallace, B. A Sensitivity Analysis of (and Practitioners' Guide to) Convolutional Neural Networks for Sentence Classification. *arXiv* **2015**, arXiv:1510.03820.
24. Devlin, J.; Chang, M.-W.; Lee, K.; Toutanova, K. BERT: Pre-training of Deep Bidirectional Transformers for Language Understanding. *arXiv* **2018**, arXiv:1810.04805.
25. Dosovitskiy, A.; Beyer, L.; Kolesnikov, A.; Weissenborn, D.; Zhai, X.; Unterthiner, T.; Dehghani, M.; Minderer, M.; Heigold, G.; Gelly, S.; et al. An Image is Worth 16 × 16 Words: Transformers for Image Recognition at Scale. *arXiv* **2020**, arXiv:2010.11929.
26. Yang, Z.; Dai, Z.; Yang, Y.; Carbonell, J.G.; Salakhutdinov, R.; Le, Q.V. XLNet: Generalized Autoregressive Pretraining for Language Understanding. In Proceedings of the 32nd Annual Conference on Neural Information Processing Systems (NeurIPS 2019), Vancouver, BC, Canada, 8–14 December 2019; Volume 32, pp. 5753–5763.
27. Brown, T.B.; Mann, B.; Ryder, N.; Subbiah, M.; Kaplan, J.; Dhariwal, P.; Neelakantan, A.; Shyam, P.; Sastry, G.; Askell, A.; et al. Language Models are Few-Shot Learners. *arXiv* **2020**, arXiv:2005.14165.
28. Wang, X.; Girshick, R.; Gupta, A.; He, K. Non-local Neural Networks. In Proceedings of the 2018 IEEE/CVF Conference on Computer Vision and Pattern Recognition, Salt Lake City, UT, USA, 18–23 June 2018; pp. 7794–7803.
29. Carion, N.; Massa, F.; Synnaeve, G.; Usunier, N.; Kirillov, A.; Zagoruyko, S. *End-to-End Object Detection with Transformers*; Springer: Cham, Switzerland, 2020; pp. 213–229.
30. He, K.; Chen, X.; Xie, S.; Li, Y.; Dollár, P.; Girshick, R.B. Masked Autoencoders Are Scalable Vision Learners. *arXiv* **2021**, arXiv:2111.06377.
31. Simonyan, K.; Zisserman, A. Very Deep Convolutional Networks for Large-Scale Image Recognition. *arXiv* **2014**, arXiv:1409.1556.
32. Mikolov, T.; Sutskever, I.; Chen, K.; Corrado, G.; Dean, J. Distributed Representations of Words and Phrases and their Compositionality. In Proceedings of the Advances in Neural Information Processing Systems, Harrahs and Harveys, Lake Tahoe, NV, USA, 5–10 December 2013.
33. Pennington, J.; Socher, R.; Manning, C. Glove: Global Vectors for Word Representation. In Proceedings of the Conference on Empirical Methods in Natural Language Processing, Doha, Qatar, 25–29 October 2014; pp. 1532–1543.
34. Joulin, A.; Grave, E.; Bojanowski, P.; Mikolov, T. Bag of Tricks for Efficient Text Classification. *arXiv* **2016**, arXiv:1607.01759.
35. Kingma, D.P.; Ba, J. Adam: A Method for Stochastic Optimization. *arXiv* **2015**, arXiv:1412.6980.
36. Cui, Y.; Che, W.; Liu, T.; Qin, B.; Yang, Z.; Wang, S.; Hu, G. Pre-Training with Whole Word Masking for Chinese BERT. *arXiv* **2019**, arXiv:1906.08101. [CrossRef]
37. Cui, Y.; Che, W.; Liu, T.; Qin, B.; Wang, S.; Hu, G. Revisiting Pre-Trained Models for Chinese Natural Language Processing. *arXiv* **2020**, arXiv:2004.13922.

38. Antol, S.; Agrawal, A.; Lu, J.; Mitchell, M.; Batra, D.; Zitnick, C.L.; Parikh, D. VQA: Visual Question Answering. In Proceedings of the 2015 IEEE International Conference on Computer Vision (ICCV), Santiago, Chile, 7–13 December 2015; pp. 2425–2433.
39. Vinyals, O.; Toshev, A.; Bengio, S.; Erhan, D. Show and tell: A neural image caption generator. In Proceedings of the 2015 IEEE Conference on Computer Vision and Pattern Recognition (CVPR), Boston, MA, USA, 7–12 June 2015; pp. 3156–3164.
40. Van der Maaten, L.; Hinton, G. Viualizing data using t-SNE. *J. Mach. Learn. Res.* **2008**, *9*, 2579–2605.

applied sciences

MDPI

Article

A Language Model for Misogyny Detection in Latin American Spanish Driven by Multisource Feature Extraction and Transformers

Edwin Aldana-Bobadilla [1,2], Alejandro Molina-Villegas [1,3,*], Yuridia Montelongo-Padilla [2], Ivan Lopez-Arevalo [2] and Oscar S. Sordia [3]

1 CONACYT, Mexico City 03940, Mexico; edwyn.aldana@cinvestav.mx
2 Centro de Investigación y de Estudios Avanzados del I.P.N., Unidad Tamaulipas, Victoria 87130, Mexico; yuridia.montelongo@cinvestav.mx (Y.M.-P.); ilopez@cinvestav.mx (I.L.-A.)
3 Centro de Investigación en Ciencias de Información Geoespacial, Merida 97302, Mexico; osanchez@centrogeo.edu.mx
* Correspondence: amolina@centrogeo.edu.mx

Abstract: Creating effective mechanisms to detect misogyny online automatically represents significant scientific and technological challenges. The complexity of recognizing misogyny through computer models lies in the fact that it is a subtle type of violence, it is not always explicitly aggressive, and it can even hide behind seemingly flattering words, jokes, parodies, and other expressions. Currently, it is even difficult to have an exact figure for the rate of misogynistic comments online because, unlike other types of violence, such as physical violence, these events are not registered by any statistical systems. This research contributes to the development of models for the automatic detection of misogynistic texts in Latin American Spanish and contributes to the design of data augmentation methodologies since the amount of data required for deep learning models is considerable.

Keywords: automatic hate speech detection; multisource feature extraction; Latin American Spanish language models; natural language processing

Citation: Aldana-Bobadilla, E.; Molina-Villegas, A.; Montelongo-Padilla, Y.; Lopez-Arevalo, I. and S. Sordia, O. A Language Model for Misogyny Detection in Latin American Spanish Driven by Multisource Feature Extraction and Transformers. *Appl. Sci.* **2021**, *11*, 10467. https://doi.org/10.3390/app112110467

Academic Editor: Arturo Montejo-Ráez

Received: 21 September 2021
Accepted: 2 November 2021
Published: 8 November 2021

1. Introduction

According to a recent report released by the World Health Organization, *"Physical or sexual violence is a public health problem that affects more than one third of all women globally"* [1]. Nevertheless, the problem seems even more prominent in Latin America when looking at the regional data. For instance, the regional prevalence rate for sexual violence among all women older than 15 years is 36.1% for the Americas region and 27.2% for Europe [2]. The Commission on the Status of Women (CSW), one of America's leading promoters of women's human rights, have been covering issues related to women's social and economic rights and, very recently, online violence. The sixty-fifth session of the CSW revolved around the theme of *"The participation of women in public life and the elimination of violence"* [3] in response to an increasingly online, gender-based abuse, cyberbullying, and sexual harassment. Out of all the recommendations to prevent and eliminate violence against women in public life (https://undocs.org/E/CN.6/2021/3, accessed on 2 November 2021, Par. 65), we can highlight the following three given their relationship with online violence against women (the original labels are used):

(i) reform legal frameworks to criminalize violence against women in political and public life, both online and offline, and to end impunity;
(o) set standards on what constitutes online violence against women in public life so that the media and companies running social media platforms can be held accountable for such content; and
(p) increase the capacity of national statistical systems to collect data regularly and systematically (both online and offline) on violence against women in public life.

It is of great importance that online violence against women is included in the recommendations above since many authors have considered that subtle violence can stratify to more severe violence. Johan Galtung, a renowned Norwegian pacifist and sociologist, assures that "*Cultural violence makes direct and structural violence appear, and even perceived, as charged with reason—or at least not bad*" [4]. Thus, for Galtung, hate speech, such as misogyny, precisely represents expressions of cultural violence because, through language, the misogynistic expressions legitimize and naturalize rejection and contempt towards women. Similarly, for Michel Foucault [5], discursive practices produce effects on the world, so that hate speech not only involves violence in itself but also implies the risk of generating direct violence on disadvantaged groups in addition to the fact that makes structural violence invisible.

However, creating effective mechanisms to detect misogyny online automatically represents significant scientific and technological challenges. The complexity of recognizing misogyny through computer models lies in the fact that it is a subtle type of violence, it is not always explicitly aggressive, and it can even hide behind seemingly flattering words, jokes, parodies, and other expressions (see Reference [6]). Currently, it is even difficult to have an exact figure for the rate of misogynistic comments online because, unlike other types of violence, —such as physical violence—, these events are not registered by any statistical systems.

Given this scenario, recent efforts to quantify and visualize the incidence of hate speech in digital media have recently been made mainly by the Natural Language Processing community, as is described in the Related Work Section 2.

Our research contributes to the development of models for the automatic detection of hate speech, particularly misogynistic texts, and to the design of Spanish data augmentation methodologies (since the amount of data required for deep learning models is considerable). However, in addition to the scientific contribution, we have the goal of doing science with social relevance. We seek raise awareness about the proliferation of misogyny in social networks in Latin America.

2. Related Work

Several recent studies evidence the growing interest of the scientific community on automatic detection of hate speech, mainly for English [7–12]. This research area has grown mainly thanks to the competitions organized at SemEval [13] (e.g., HatEval, OffensEval, and Toxic Spans Detection) and other venues, such as TRAC [14] and HASOC [15]. These competitions are essential since they provided participants with widely used benchmark datasets (e.g., OLID [16]). Regarding aggressiveness detection for Latin American Spanish, the most relevant competition is MEX-A3T track at IberLEF 2019 [17], where the organizers considered two tasks focused on the authorship and aggressiveness in Mexican tweets, and IberEval 2018 [18], with the first shared task on Automatic Misogyny Identification.

Two very notorious aspects emerge from the state of the art on detection of hate speech: the target language defines the degree of maturity of the existing models and the target group to which the hate speech is directed defines the specific challenges. Regarding the first aspect, there are several research in different languages, most of them including data compilation: German [19,20]; French [21]; Danish[22]; Greek [23]; Italian [24]; Hindi [20]; Arabic [21,25]; Indonesian [26]; Polish [27]; Turkish [28]; and Spanish [29]. However, the lack of language-specific corpora for all possible languages and variants have being created an important gap between the research maturity and results in English face to other languages but also had motivated innovation in research to deal with this challenge. To cope with data scarcity, researchers have explored different solutions, such as feature engineering, data augmentation, and multilingual models [30,31].

The other aspect is that, in hate speech, there are few specific groups to which the attacks are systematically directed (women and immigrants, for instance [32]). This is why talking about hate speech is still generally in the context of the state of the art on automatic misogyny detection. In this sense, below, we will pay greater attention to the state of the art in the specific task of detecting and/or classifying misogynistic language.

The authors of Reference [33] present an experimental analysis using different NLP features and ML models to detect misogynous tweets in English labeled from different perspectives and an exploratory investigation using NLP features and ML models to detect and classify misogynistic language. Several ML models from scikit-learn were used: Linear Support Vector Machine (SVM), Random Forest (RF), Naïve Bayes (NB), and Multi-layer Perceptron Neural Network (MPNN). The best reported model, SVM, arises an accuracy of 0.7995. In addition, for automatic identification of misogynistic language in English, in Reference [34], the authors propose a Long Short-Term Memory (LSTM) classifier using a pretrained LSTM-based Language Model to build an accurate classification model with a small training set. A "Bayesian interpretation" of Transfer Learning is presented as a regularization technique to estimate the uncertainty in the pre-training. The method is relevant since misogynistic tweet is a highly unbalanced class against general tweets, so the regularization proposal avoids overfitting. The best model arises the following scores: accuracy 0.846, precision 0.806, and F1-score 0.781. In Reference [35], the authors present an exploratory work detecting Misogyny for English and Spanish using the IberEval 2018 data [18]. They test different ML classifiers obtaining the best result using an ensemble technique (majority voting) to combine the predictions of SVM, Random Forest, and Gradient Boosting classifiers. The best model arises accuracy 87.05 for English and 81.35 for Spanish. To the best of our knowledge, the work presented in Reference [36] is the only that compiled messages harassing women in Spanish from Latin America. The proposal created the MisoCorpus-2020, a balanced corpus regarding misogyny in Spanish. The authors also present models combining word embeddings and linguistic features for three ML classifiers: Random Forest (RF), a decision tree classifier, Sequential Minimal Optimization (SMO), and a Support Vector Machine achieving the best accuracy of 85.17%.

In a deeper research, Fulper et al. [37] explored whether social media can be used as an indicator of sexual violence in the U.S., by tracking misogynistic tweets. Using the FBI Uniform Crime Reports provides rape statistics in the U.S. at the state level and a 10% sample of the Twitter stream produced during 2012. The authors manually compiled a list of 90 terms that are commonly used as misogynistic insults. With such filters, they obtained georeferenced tweets that contain misogynistic language and location (either latitude/longitude or a free-form location string) and mapped them to states, using state boundary data from the U.S. Census Bureau. The final dataset contains roughly 170 million georeferenced tweets, of which 1.2 million contain misogynistic language. As a result, the authors found a significant association between tweets that contain misogynistic language and rape crime statistics for each state in the U.S. A similar project is presented in [38], where the author delves into the relationship between the rate of misogynistic tweets and the rate of femicides in Mexico. Data consisted of femicide reports from the Executive Secretariat of the National Public Security System of México (SESNSP) and about twelve million georeferenced tweets in 2017–2018. Some regions were found to have particularly high rates of both misogyny and femicides. Furthermore, the Spearman correlation coefficient between both variables is 0.2515 with a significance level of 0.16; in other words, there is an interdependence, although very low, between both indicators, but the risk of concluding that there is a correlation, when, in reality, there is not, is only 16%.

3. Misogyny Detection Approach

The first challenge in recognizing written misogyny is obtaining a representative dataset containing a wide set of examples of what may or may not be a manifestation of violence. Another challenge that arises is the complexity of extracting the convenient features of the text from which is possible to create computational models able to recognize manifestations of written violence. As mentioned before, the complexity lies in the fact that the written violence is subtle; it is not always explicitly aggressive. Guided by these challenges, we have designed an integrated proposal consisting of several techniques, from gathering data to training a model capable of recognizing misogynistic manifestations.

Our proposal includes three main stages: *Gathering*, *Feature Extraction*, and *Modeling*, illustrated in Figures 1 and 2 described in the subsequent sections.

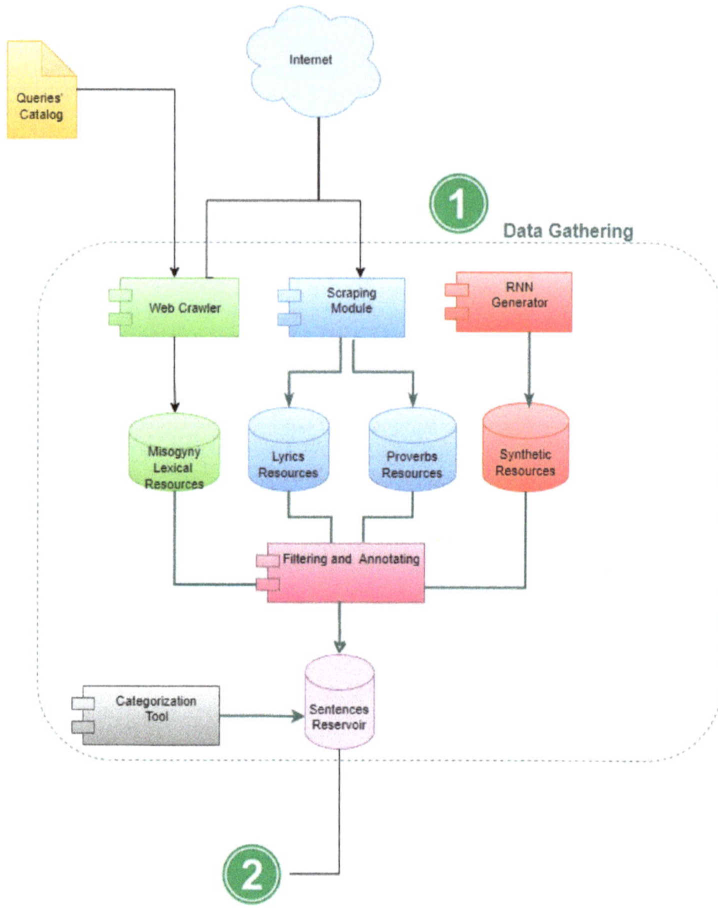

Figure 1. Overview of data gathering stage. As result of this stage, there is a corpus containing annotated sentences which will be the input of the subsequent stage (Feature Extraction).

3.1. Data Gathering

This stage comprises a set of components that allow us to obtain an appropriate set of documents for our purpose. The components of Gathering, illustrated inside the upper box in Figure 1, are:

- Web Crawler: It allows us to seek and obtain documents (in HTML and PDF format) containing misogynistic expressions. The search of documents is guided by a set of what we have called *Queries Catalog*. This catalog contains 64 sentences in Spanish made up of key words allusive to misogyny, that intend to focus the search on those documents containing misogynistic elements. For example, sentences of the form: *comportamiento misógino (misogynistic behavior)*, *discriminación y violencia contra la mujer (discrimination and violence against women)*, *chistes misóginos (misogynistic jokes)*, *misoginia en la política (misogyny in politics)*, and so on. Additionally, the catalog also contains a set of n-grams that frequently appeared in text with misogynistic bias, according to a preliminary study reported in [38]. For example, the n-gram *eres una puta (you are a whore)* and *malditas feminazis (fucking feminazis)*, among others. From the described catalog, the web crawler could find an initial set of 991 documents of different length containing text in Spanish with a high probability of having misogynistic expressions.

- Scraping Module: Unlike the web crawling, scraping is a process that allows us to obtain the content of prior identified resources. Relying on the assumption that the text of several songs could be a suitable resource to find misogynistic expressions (an interesting study is reported in Reference [39]), we focus on identifying lyrics in Spanish singled out as resources with sexist content and violence against women, under the perception of some group of people. In this regard, we use a set of keywords on the search engines of Google, Facebook, Twitter, and YouTube to identify those songs that are commonly associated with misogynistic content. From the results of the searches, we build a catalog of 163 titles of songs with a high probability of including valuable sentences for our purposes; this resource is available at http://shorturl.at/lptzT (accessed on 4 November 2021). Relying on the catalog's titles, the Scraping Module is executed in order to obtain their corresponding lyrics, from the website https://www.letras.com (accessed on 4 November 2021). We also configured this module to scrap and filter documents available at https://proverbia.net (accessed on 4 November 2021), which are short documents (proverbs) just containing expressions that people often quote for giving advice or some philosophical reflection. Those proverbs with misogynistic content were not considered. It is worth mentioning that all the texts, including the proverbs, were manually revised to avoid including misogynistic phrases.
- RNN Generator: Although the above datasets allowed us to obtain an large number of documents, it was insufficient to encompass the study phenomenon. For this reason, our proposal includes a strategy to overcome the lack of data, based on a Recurrent Neural Network (RNN) capable of learning the intrinsic semantic of misogynistic expressions contained within the collected lyrics and generating documents that contain synthetic text. The length of the generated text is determined by a parameter corresponding to the number of words desired. For purposes of our work, we set this parameter to a constant value of 300. Since the quality of the generated text is much lower than that of the lyrics text, a lot of generated documents could not contain valuable sentences to be considered in our corpus. Despite this, we achieved to obtain a valuable set of sentences by exhaustive manual inspection (see Table 1).
- Other components: At this point, we take the collected documents as input to execute the module that we have called *Filtering and Annotating*: Filtering is the process by which expressions and sentences (in general) are extracted from the text of each document; this process was manual and did depend on the criteria from who executes it to define what is a sentence. From this, a set of 7191 "raw sentences" was obtained. Relying on these sentences, we executed an annotation process in which each sentence was annotated by 2 independent annotators judging the presence or absence of misogynistic content (binary decision). Of the 7191 sentences, we obtained 6747 agreements (93.84%) and 3624.8 agreements expected by chance (50.41% of observations) resulting in a kappa value of 0.87 indicating a suitable agreement. For those sentences where the annotators did not agree, there was a third annotator to judge and make a final decision based on the mode of the three annotations. We try to keep the annotation guideline as simple as possible, so that annotators could easily make a decision. The guideline included only three rules regarding a misogynistic sentence:
  - Unigrams sentences containing rudeness that could be directed at a woman. For instance, *puta (whore)*, *fea (ugly)*, *gorda (fat)*, *tonta (silly)*.
  - Any sentence containing one or more rudeness and violent language explicitly directed at a woman or a group of them. For example, *me engañaste pinche*

alcohólica (you cheated on me, fucking alcoholic), te odio pinche zorra (I hate you fucking bitch), etc.

- Any phrase that, in the judgment of the annotator, indicates explicitly or implicitly submission, inferiority or violence, but it does not necessarily include rudeness. For example, *ellas también tiene que respetarse, se visten así y luego se quejan cuando les pasa algo (they have to respect themselves, they dress that way and then complain when something happens to them), vete a la cocina y prepararme un sándwich (go to the kitchen and make me a sandwich)*.

In Table 1, a summary that illustrates the number of documents and annotated sentences gathered via the described components is shown. The documents are of different length; the longest documents were those obtained via the web crawler with a length in the range of 476 to 2492 words. The length of the lyrics is in the range of 80 to 300 words, while the length of the proverbs is in the range of 3 to 50 words.

Table 1. Documents and annotated sentences via data gathering components.

Component	Documents	Sentences
Web Crawler	991	3310
Scraping (Lyrics)	200	733
Scraping (Proverbia)	2196	2196
RNN Generator	1000	952
	Total sentences	7191

The synthetic documents obtained via RNN Generator have a constant length of 300 words. Since the quality of the generated text is much lower than that of the lyrics text, a lot of generated documents could not contain valuable sentences to be considered in the corpus. On the other hand, a single generated document could contain more than one valuable sentence. The above means that some groups of the sentences found come from the same document.

At this point, we have a corpus containing annotated sentences that will be use by the remaining stages, which are illustrated in Figure 2, and described in subsequent sections.

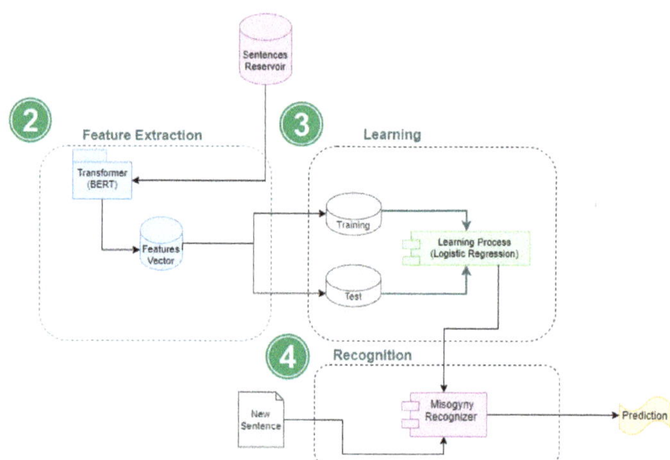

Figure 2. Overview of Feature Extraction, Learning, and Recognition stages. The first two make up the pipeline by the means of which a model of language is obtained on the basis of the inherent knowledge and experience conveyed in the corpus.

3.2. Feature Extraction

Having overcome the lack of data on Latin American Spanish, now, the main challenge is to encode such information into a model able to recognize that misogyny could be a subtle type of violence, and it can even hide behind seemingly flattering words, jokes, parodies, and other expressions.

We resort to recent approaches known as *transformers* [40] which incorporate the so-called *attention mechanisms* to identify these relations. In general, transformers provides thousands of pre-trained models to perform tasks on texts, such as classification, information extraction, question answering, summarization, translation, and text generation, in a lot of languages. They have been trained on large amounts of raw text in a self-supervised fashion (the objective is automatically computed from the inputs of the model). For practical purposes, we can build our models on top of already trained models, reducing the overall compute cost, and this process is usually known as *transfer learning*. We focus on a state-of-the art transformer known as BERT (Bidirectional Encoder Representations from Transformers) [41]. BERT is described as "bidirectional" because, unlike methods, such as Word2Vec, it can read a text or a sequence of words all at once, with no specific direction. Thanks to its bidirectionality, this model can understand the meaning of each word based on context both to the right and to the left of the word.

In general, any model based on a *transformer* architecture involves high computational resources to process and store a huge amount of training data and parameters. This complexity is latent to the resulting pre-trained language models that keep getting larger and heavier to new problems. Under this drawback, we decide to focus on distillation technique [42,43] that allows us to compress a large model, called *the teacher*, into a smaller model, called *the student*. Specifically, we use a "distilled" version of BERT known as DistilBERT, reported in Reference [44] as a suitable approach comparable to the performance of state-of-the-art transformers. The process of transfer learning via BERT allows us to extract the features of the sentences gathered in previous stage and denoted in what follows as $\mathbb{D}$; such a process involves the following steps:

1. Obtaining of an instance of a BERT model pre-trained on a large unlabeled dataset in Spanish.
2. Fine-tuning [45], where the model is initialized with the pre-trained parameters and all of them are fine-tuned using a labeled data regarding sentences previously categorized as misogynistic and non-misogynistic.

So far, $\mathbb{D}$ is a set of sentences that require be prepared in the form that BERT expects. For this aim, we use a tokenizer provided by BERT which we have called *BertTokenizer*, obtaining a set $\mathbb{T}$ containing encoded sentences in the form of multidimensional arrays. For each $\vec{t}$ in $\mathbb{T}$, we pad it with zeros until its length is equal to the longest array in $\mathbb{T}$. As mentioned, one of the main characteristics of BERT is the transformer structure, where its encoder pay attention to the sentence as a whole and not dividing it into tokens. We need to tell BERT which part of the whole tokenized and padded array $\vec{t} \in \mathbb{T}$ contains useful information. So, for each $\vec{t}$, we create a binary vector $\vec{m}$ indicating those positions in which there is a tokenized or padded value (encoded as 1 and 0, respectively). At this point, we have a set of binary vectors known as *attention mask* and denoted as $\mathbb{M}$. Finally, both $\mathbb{T}$ and $\mathbb{M}$ are propagated through BERT model (fine tuning). As a result, we obtain a high-dimensional vector (embedding) for each sentence in $\mathbb{D}$ representing its feature vector; the set of feature vector is denoted as $\mathbb{F}$. The above process is summarized in Algorithm 1 and illustrated in Figure 3.

Algorithm 1: Feature Extraction

Data:
$\mathbb{D}$: Sentences reservoir,
Result: Set of feature vector from $\mathbb{D}$

```
/* Creating instances of Bert Model and Bert Tokenizer */
BertModel ← get_BertModel();
BertTokenizer ← get_BertTokenizer();
/* Tokenizing sentences */
𝕋 ← BertTokenizer.encode(𝔻);
/* Padding Sentences and Attention Mask */
maxLen ← max_length(𝕋);
𝕄 ← ∅;
foreach t ∈ 𝕋 do
    t ← padding(t, maxLen);
    m ← get_mask(t);
    𝕄 ← 𝕄 ∪ {m};
end
𝔽 ← BertModel.transform(𝕋,𝕄);
return 𝔽
```

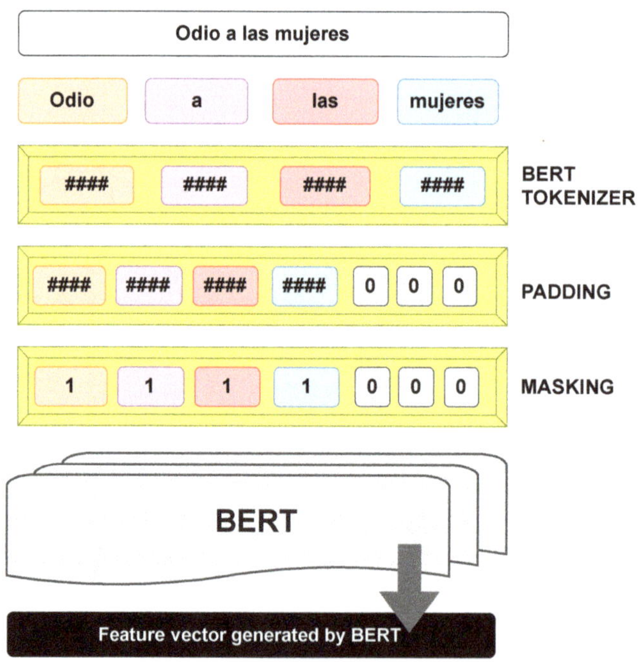

Figure 3. Illustration of feature extraction via BERT for the sentence in Spanish "Odio a las mujeres" (*"I hate women"*).

3.3. Learning Process

Having determined the semantic features of whole sentences in $\mathbb{D}$, we aim to find a model that is able to recognize the presence or absence of misogynistic patterns in the encoded sentences. At this point, we resort to Logistic Regression because it is fast, easily understandable, and appropriate for a dichotomous dependent variable as it is our case. So, from the encoded sentences $\mathbb{F}$ and their corresponding labels $\mathbb{L}$, we now define the so-called training and test datasets denoted as $\mathbb{X}_{train}$, $\mathbb{T}_{train}$ and $\mathbb{X}_{test}$, $\mathbb{Y}_{test}$. We execute an exhaustive search over an specified set of hyper-parameter values $\mathbb{P}$ for obtaining an appropriate model. This search involved cross-validation with $k = 10$ folds on the train set in order to obtain the best parameter values that attain the most reliable model. The set $\mathbb{P}$ included 27 tuples of the form $[optimizer, penalty, C]$ corresponding to optimization algorithm (lbfgs, sag, saga), norm used for penalization (l_1,l_2), and the inverse of regularization strength (1, 5, and 10 in our case), respectively. The above is illustrated in Algorithm 2, when the optimal $p^* \in \mathbb{P}$ is found, we create an instance of Logistic Regression using p^* and fit it to $\mathbb{X}_{train}$. Since this instance has a high degree of certainty of getting the best prediction score on $\mathbb{X}_{test}$, we consider it the best model for our purposes. The score was defined in terms of accuracy, precision, recall, and F1-score; the results regarding these metrics are discussed more fully in Section 4. Finally, the best hyper-parameter values obtained after executing an exhaustive search using cross-validation are: $C = 10$, lbfgs as optimizer and l_2 regularization.

Algorithm 2: Learning Process

Data:
$\mathbb{F}$: Encoded Sentences
$\mathbb{L}$: Labels of Sentences
$\mathbb{P}$: Set of tuples of hyper-parameters for estimator (Logistic Regression)
Result: Misogyny Recognizer Model

/* Defining train and test datasets */
$\mathbb{X}_{train}, \mathbb{Y}_{train}, \mathbb{X}_{test}, \mathbb{Y}_{test} \leftarrow$ ***split_data***$(\mathbb{F}, \mathbb{L})$;
/* Applying a stratified sampling with k-folds on training data */
$i \leftarrow 0, k \leftarrow 10$;
$\mathbb{X}_{folds}, \mathbb{Y}_{folds} \leftarrow$ ***get_folds***$(\mathbb{X}_{train}, \mathbb{Y}_{train}, k)$;
$\mathbb{S} \leftarrow \emptyset$;
$score \leftarrow 0$;
foreach $p \in \mathbb{P}$ **do**
  while $i < k$ **do**
    /* Creating an instance of the model */
    $model \leftarrow LogisticRegression.$***getInstance***$(p)$;
    /* Validation set */
    $\mathbb{X}_{val}, \mathbb{Y}_{val} \leftarrow \mathbb{X}_{folds}[i], \mathbb{Y}_{folds}[i]$;
    /* Training set */
    $\mathbb{X}_{train}, \mathbb{Y}_{train} \leftarrow (\mathbb{X}_{folds} - \mathbb{X}_{folds}[i]), (\mathbb{Y}_{folds} - \mathbb{Y}_{folds}[i])$;
    /* Training Model */
    $model.$***fit***$(\mathbb{X}_{train}, \mathbb{Y}_{train})$;
    /* Validation Model */
    $\mathbb{Y}_{predicted} \leftarrow model.$***predict***$(\mathbb{X}_{val})$;
    $score \leftarrow score +$ ***get_score***$(\mathbb{Y}_{val}, \mathbb{Y}_{predicted})$;
    $i \leftarrow i + 1$;
  end
  $avg\_score \leftarrow score / k$;
  $\mathbb{S} \leftarrow \mathbb{S} \cup \{avg\_score\}$;
end
$p^* \leftarrow$ ***get_best_params***$(\mathbb{S}, \mathbb{P})$;
$BestModel \leftarrow LogisticRegression.$***getInstance***$(p^*)$;
/* Training Model */
$BestModel.$***fit***$(\mathbb{X}_{train}, \mathbb{Y}_{train})$;
/* Testing Model */
$\mathbb{Y}_{predicted} \leftarrow BestModel.$***predict***$(\mathbb{X}_{test})$;
$score \leftarrow$ ***get_score***$(\mathbb{Y}_{test}, \mathbb{Y}_{predicted})$;
return $BestModel$

3.4. Recognition Process

Having obtained a reliable model, now, we aim to estimate the degree of misogyny in a sentence s that the model has not seen before. In order to make s amenable to the model, we again use the previous instances of BERT model and BERT tokenizer. At this point, we have an encoded sentence in the form of a high dimensional vector denoted as f which is given as input to the model in order to obtain the prediction probabilities associated with the presence or absence of misogynistic elements in s. The above process is summarized in Algorithm 3.

Algorithm 3: Misogyny Recognition

```
Data:
s: Sentence to be analyzed,
Result: Prediction vector
/* Loading previous models                                */
BertModel ← get_BertModel();
BertTokenizer ← get_BertTokenizer();
MisogynyRecognizer ← BestModel;
/* Tokenizing sentence to be analyzed                     */
t ← BertTokenizer.encode(s);
/* Padding Sentences and Attention Mask                   */
maxLen ← BertModel.get_max_length();
t ← padding(t, maxLen);
m ← get_mask(t);
f ← BertModel.transform(t, m);
p ← MisogynyRecognizer.predict(f);
return p
```

4. Experiments and Results

The experiment consisted of assessing the performance of our approach from two aspects: (1) the learning ability of the model in terms of the well-known evaluation metrics (accuracy, precision, recall, and F1-score) to predict the misogyny degree on a test dataset consisting of sentences belonging to the corpus gathered via our proposal, and (2) the recognition ability to determine the presence or absence of misogynistic patterns in a real world dataset that includes sentences which the model has not seen during the learning process.

4.1. Learning Ability Assessment

As pointed out in Algorithm 2, the learning process is performed on the sets $\mathbb{F}$ and $\mathbb{L}$ corresponding to the set of encoded sentences in the form of high dimensional vector and their corresponding labels, respectively. From these datasets, we determine training and test datasets. On the training dataset, we performed a sampling strategy known as k-folds (with $k = 10$) in order to find the most reliable model that minimizes the overfitting. At this point, we assessed the yield of the resulting model based on the test dataset. This assessment is defined in terms of metrics, such as *accuracy*, *precision*, *recall*, and *F1-score*. Since a single iteration of the above process does not guarantee the overall performance of the model, we repeated it 100 times by changing random seed values, obtaining a statistical approximation to the real performance values, as it is illustrated in Figure 4.

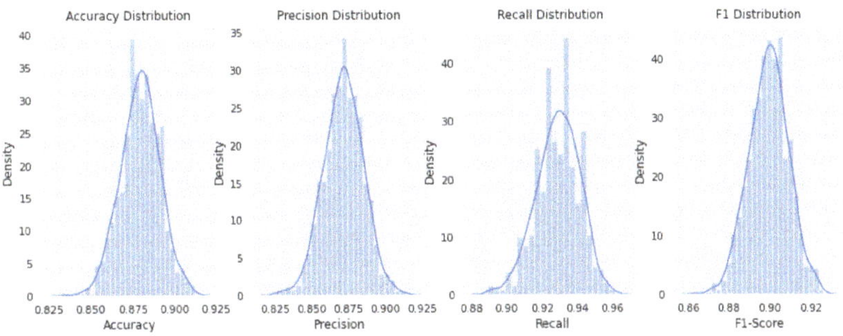

Figure 4. Approximation to the density function of the values corresponding to accuracy, precision, recall, and F1-Score exhibited by the model during the assessment of learning ability.

We analyzed the variability in the performance of our method using quartile summary statistics illustrated in Table 2. From the Interquartile Range (IQR), we can see that there are not large differences in the experiments for a particular metric, and Confidence Interval (CI) allows us to quantify how we expect the average performance to be. In general, the performance of our model was mostly kept inside acceptable range in all metrics.

Table 2. Quartile analysis for the assessment of *learning ability* of the model.

Statistic	Accuracy	Precision	Recall	F1-Score
Q_1	0.87	0.86	0.92	0.89
Q_2	0.88	0.87	0.93	0.89
Q_3	0.88	0.88	0.93	0.90
IQR	0.01	0.01	0.01	0.01
CI	[0.84, 0.91]	[0.83, 0.90]	[0.89, 0.96]	[0.87, 0.92]

4.2. Assessment for Real World Data

Finally, we conducted a set of experiments on a real-world dataset reported in Reference [13]. This work shows the results of different working teams grouped by task. We focused on the work regarding *Task 5* regarding the analysis of hate speech against women and immigrants on a corpus of labeled tweets in Spanish and English. The labels were encoded as three binary values described as follows:

- **HS**—a binary value indicating the presence or absence of hate speech against one of the given target people (women or immigrants).
- **TR**—a binary value indicating if the target of the hate speech is a generic group of people (0) or a specific individual (1).
- **AG**—a binary value indicating if the hate speech present in the tweet is aggressive (1) or not (0).

For our purposes, we filtered the data to retain only those tweets in Spanish and labeled with HS = 1. Then, we selected manually those tweets containing expressions focused on women (most labeled with HS = 1,TR = 1). As result a set of instances containing misogynistic tweets were obtained. On the other hand, instances containing non-misogynistic tweets were obtained simply by filtering tweets tagged with HS = 0. In total, a set of 1774 tweets (with a comparable number of classes) was obtained. On this data, we executed 100 times our model attempting to recognize the presence or absence of misogynistic manifestations. We resort again to the metrics of accuracy, precision, recall, and F1-score to quantify the recognition ability. In Figure 5, the approximation to the density function for the executed experiments is shown.

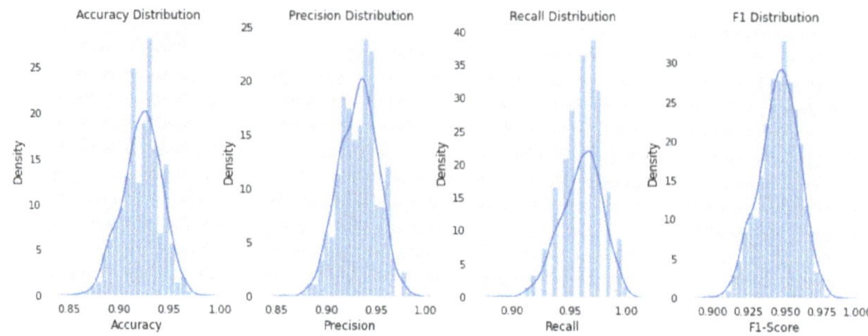

Figure 5. Approximation to the density function of the values corresponding to accuracy, precision, recall, and F1-Score exhibited by the model during the assessment of *recognition ability*.

As in the above assessment, the model exhibited a successful performance. The quartile analysis in Table 3 shows a close variability to that of learning assessment. It means that the recognition ability of our method (on unknown data) is statistically comparable to its ability exhibited during the learning process. To formalize this assumption, we finally conducted a hypothesis test that allowed us to show that there is not a significant difference between what we have called learning ability and recognition ability (p-value > 0.05).

Table 3. Quartile analysis for the assessment for the proposed Latin American Spanish misogyny *recognition* model. The metrics reported are Accuracy, Precision, Recall, and F1-Score.

Statistic	Accuracy	Precision	Recall	F1-Score
Q_1	0.90	0.91	0.93	0.93
Q_2	0.92	0.93	0.94	0.94
Q_3	0.93	0.94	0.95	0.95
IQR	0.02	0.02	0.01	0.01
CI	[0.86, 0.98]	[0.87, 0.98]	[0.90, 0.98]	[0.90, 0.98]

5. Discussion

We have been able to generate a Language Model based on previous knowledge and using extracted data from a pipeline that we designed with the purpose of covering up as much sources as we could where we could find a misogynistic attitude. We must emphasize certain points we think should be considered on this work and in the reported results. First, setting boundaries was very important from the beginning because we are aware of the challenges that NLP tasks have, as well as the subjectivity in opinions while working with topics, such as misogyny. So, we kept following this path knowing we could not get rid of any kind of subjectivity, domain, social and geographic context, polarization, or controversy. On the other hand, we selected Spanish for the lack of particular works on misogyny detection in this language and the importance of pointing out this behavior through social media and its social impact in the Latin America context. On the other hand, we also need to clarify that, even though it is pretty clear we were the ones who selected the data directly, tagging each sentence as misogyny or not misogyny, the process of retrieving has its own subjectivity since what we got is what the internet is indexing as misogyny. So, putting this in mind, we can ensure that the more data we add to the current dataset, the bigger the context will be around this particular type of hate speech. It is also true that our model is capturing exactly what it learned. It is identifying what we taught it was misogyny, considering the full sentence. We know this by comparing the probability that our model generates for each tag in certain sentences. In Table 4, we can see the difference between passive misogyny sentences that do not have any negative word that can be considered as aggressive or insulting, active misogyny that actually uses an insulting word, and a non-misogynistic sentence. The results the model gave us in these sentences show us how it finds more probability of a sentence being misogyny where it is talking directly to women or about women, as it should, but it still does find differences between nice words inside a sentence that itself is misogynistic and one that could be aggressive but that does not specify a gender. We must say the results can improve at using our pipeline for gathering data more than one time, collecting more data from the web that can be used as misogyny sentences, but we are also concerned that our model still lacks in the negative class since we have only trained with proverbs and some other sentences we obtained from the documents we collected. However, at this point, we know the path we should keep working on in order to make our model more robust.

Table 4. Examples of instances used in the Misogyny Recognizer. We can note the difference between passive misogyny sentences, active misogyny that actually use an insulting word and a non-misogynistic sentence.

	Original Tag	Not Misogyny	Misogyny
Y tu eres mi esposa ve a la cocina zorra	1	0.0886	0.9113
Solo sirves para lavar los trastes	0	0.8533	0.1466
Me encanta mi novia porque hace las tareas doméstica s	1	0.3099	0.6900
Ella también tiene que respetarse, se visten así y luego se quejan cuando les pasa algo	1	0.0167	0.9832

6. Conclusions

Throughout this article, we have discussed the relevance of having mechanisms for detecting hate speech online, particularly for detecting misogyny. Thanks to the recent scientific advances, it will be possible, in the short term, to use automatic detection to build indicators of violence against women. We have also discussed the challenges of bringing this type of technology to under-resourced languages through Feature Extraction methods, and we have taken Latin American Spanish as a case study.

As part of the technical aspects presented, we consider it most important to highlight the following. We have proposed a pipeline to collect, filter, tag, and generate documents' features for training a recognition model, stating a path where we can be sure we would get a fair quantity of data to use, and it promises to be helpful in future research. The misogynistic corpus we generated is a labeled resource for future investigations, and we plan to expand it for future work. This resource is also available at http://shorturl.at/lptzT (accessed on 4 November 2021).

Our model accurately discerns misogynistic comments based on the data we collected, and, for this, it seems to identify contextual cues of a sentence to generate the probabilities. The model reacts to subtleties in the language. The trained model can be tested at http://contralamisoginia.org/ (accessed on 4 November 2021). We are sure that the continuation of this project and other similar projects will transcend the creation of awareness around misogyny in Latin America.

Author Contributions: Conceptualization, E.A.-B., A.M.-V.; Methodology, E.A.-B., A.M.-V., Y.M.-P.; Software, E.A.-B. and Y.M.-P.; Validation, E.A.-B., Y.M.-P.; Investigation, I.L.-A., E.A.-B., A.M.-V., Y.M.-P.; Resources, I.L.-A., E.A.-B. and O.S.S.; Data curation, E.A.-B. and Y.M.-P.; Writing—original draft preparation, E.A.-B., A.M.-V., Y.M.-P.; Writing—review and editing, I.L.-A., E.A.-B., A.M.-V., Y.M.-P. All authors have read and agreed to the published version of the manuscript.

Funding: This research received no external funding.

Informed Consent Statement: Not applicable.

Data Availability Statement: Publicly available datasets were analyzed in this study. This data can be found here: http://shorturl.at/lptzT (accessed on 4 November 2021).

Conflicts of Interest: The authors declare no conflict of interest.

References

1. WHO. Violence against women: A global health problem of epidemic proportions. In *WHO News Release*; WHO: Geneva, Switzerland, 2013.
2. WHO. *Global and Regional Estimates of Violence against Women: Prevalence and Health Effects of Intimate Partner Violence and Non-Partner Sexual Violence*; World Health Organization: Geneva, Switzerland, 2013.
3. CSW. Report of the Secretary-General of the Commission on the Status of Women, United Nations, Sixty-Fifth Session. 2021. Available online: https://undocs.org/E/CN.6/2021/3 (accessed on 5 August 2021).
4. Galtung, J. Cultural violence. *J. Peace Res.* **1990**, *27*, 291–305. [CrossRef]
5. Foucault, M. *The Order of Discourse (L'ordre du Discours)*; Galimart: Paris, France, 1971. (In French)
6. Hewitt, S.; Tiropanis, T.; Bokhove, C. The problem of identifying misogynist language on Twitter (and other online social spaces). In Proceedings of the 8th ACM Conference on Web Science, Hannover, Germany, 22–25 May 2016; pp. 333–335.
7. Hardaker, C.; McGlashan, M. "Real men don't hate women": Twitter rape threats and group identity. *J. Pragmat.* **2016**, *91*, 80–93. [CrossRef]
8. Waseem, Z.; Hovy, D. Hateful symbols or hateful people? predictive features for hate speech detection on twitter. In Proceedings of the NAACL Student Research Workshop, San Diego, CA, USA, 12–17 June 2016; pp. 88–93.
9. Davidson, T.; Warmsley, D.; Macy, M.; Weber, I. Automated hate speech detection and the problem of offensive language. In Proceedings of the International AAAI Conference on Web and Social Media, Montréal, QC, Canada, 15–18 May 2017; Volume 11.
10. Yao, M.; Chelmis, C.; Zois, D.S. Cyberbullying ends here: Towards robust detection of cyberbullying in social media. In Proceedings of the World Wide Web Conference, San Francisco, CA, USA, 13–17 May 2019; pp. 3427–3433.
11. Ridenhour, M.; Bagavathi, A.; Raisi, E.; Krishnan, S. Detecting Online Hate Speech: Approaches Using Weak Supervision and Network Embedding Models. In *International Conference on Social Computing, Behavioral-Cultural Modeling and Prediction and Behavior Representation in Modeling and Simulation*; Springer: Berlin, Germany, 2020; pp. 202–212.
12. Lynn, T.; Endo, P.T.; Rosati, P.; Silva, I.; Santos, G.L.; Ging, D. A comparison of machine learning approaches for detecting misogynistic speech in urban dictionary. In Proceedings of the 2019 International Conference on Cyber Situational Awareness, Data Analytics and Assessment (Cyber SA), Oxford, UK, 3–4 June 2019; pp. 1–8.
13. Basile, V.; Bosco, C.; Fersini, E.; Nozza, D.; Patti, V.; Rangel Pardo, F.; Rosso, P.; Sanguinetti, M. SemEval-2019 Task 5: Multilingual Detection of Hate Speech Against Immigrants and Women in Twitter. In Proceedings of the 13th International Workshop on Semantic Evaluation, Minneapolis, MN, USA, 6–7 June 2019; pp. 54–63. [CrossRef]
14. Kumar, R.; Ojha, A.K.; Lahiri, B.; Zampieri, M.; Malmasi, S.; Murdock, V.; Kadar, D. Proceedings of the Second Workshop on Trolling, Aggression and Cyberbullying. In Proceedings of the Second Workshop on Trolling, Aggression and Cyberbullying, Marseille, France, 11–16 May 2020.
15. Mandl, T.; Modha, S.; Majumder, P.; Patel, D.; Dave, M.; Mandlia, C.; Patel, A. Overview of the hasoc track at fire 2019: Hate speech and offensive content identification in indo-european languages. In Proceedings of the 11th Forum for Information Retrieval Evaluation, Kolkata, India, 12–15 December 2019; pp. 14–17.
16. Zampieri, M.; Malmasi, S.; Nakov, P.; Rosenthal, S.; Farra, N.; Kumar, R. Predicting the type and target of offensive posts in social media. *arXiv* **2019**, arXiv:1902.09666.
17. Aragon, M.; Carmona, M.A.; Montes, M.; Escalante, H.J.; Villaseñor-Pineda, L.; Moctezuma, D. Overview of MEX-A3T at IberLEF 2019: Authorship and aggressiveness analysis in Mexican Spanish tweets. In Proceedings of the 1st SEPLN Workshop on Iberian Languages Evaluation Forum (IberLEF), Bilbao, Spain, 24 September 2019.
18. Fersini, E.; Rosso, P.; Anzovino, M. Overview of the Task on Automatic Misogyny Identification at IberEval 2018. *IberEval@ SEPLN* **2018**, *2150*, 214–228.
19. Bretschneider, U.; Peters, R. Detecting offensive statements towards foreigners in social media. In Proceedings of the 50th Hawaii International Conference on System Sciences, Hilton Waikoloa Village, HI, USA, 4–7 January 2017.
20. Kovács, G.; Alonso, P.; Saini, R. Challenges of Hate Speech Detection in Social Media. *SN Comput. Sci.* **2021**, *2*, 9. [CrossRef]
21. Ousidhoum, N.; Lin, Z.; Zhang, H.; Song, Y.; Yeung, D.Y. Multilingual and multi-aspect hate speech analysis. *arXiv* **2019**, arXiv:1908.11049.
22. Sigurbergsson, G.I.; Derczynski, L. Offensive language and hate speech detection for Danish. *arXiv* **2019**, arXiv:1908.04531.
23. Pitenis, Z.; Zampieri, M.; Ranasinghe, T. Offensive language identification in Greek. *arXiv* **2020**, arXiv:2003.07459.
24. Bosco, C.; Felice, D.; Poletto, F.; Sanguinetti, M.; Maurizio, T. Overview of the evalita 2018 hate speech detection task. In Proceedings of the EVALITA 2018 Sixth Evaluation Campaign of Natural Language Processing and Speech Tools for Italian, Turin, Italy, 12–13 December 2018; Volume 2263, pp. 1–9.
25. Albadi, N.; Kurdi, M.; Mishra, S. Are they our brothers? analysis and detection of religious hate speech in the arabic twittersphere. In Proceedings of the 2018 IEEE/ACM International Conference on Advances in Social Networks Analysis and Mining (ASONAM), Barcelona, Spain, 28–31 August 2018; pp. 69–76.
26. Ibrohim, M.O.; Budi, I. Multi-label hate speech and abusive language detection in Indonesian twitter. In Proceedings of the Third Workshop on Abusive Language Online, Florence, Italy, 1–2 August 2019; pp. 46–57.

27. Ptaszynski, M.; Pieciukiewicz, A.; Dybała, P. Results of the Poleval 2019 Shared Task 6: First Dataset and Open Shared Task for Automatic Cyberbullying Detection in Polish Twitter. 2019. Available online: https://ruj.uj.edu.pl/xmlui/bitstream/handle/item/152265/ptaszynski_pieciukiewicz_dybala_results_of_the_poleval_2019.pdf?sequence=1&isAllowed=y (accessed on 4 November 2021).
28. Hussein, O.; Sfar, H.; Mitrović, J.; Granitzer, M. NLP_Passau at SemEval-2020 Task 12: Multilingual Neural Network for Offensive Language Detection in English, Danish and Turkish. In Proceedings of the Fourteenth Workshop on Semantic Evaluation, Barcelona, Spain, 13–14 September 2020; pp. 2090–2097.
29. Pereira-Kohatsu, J.C.; Quijano-Sánchez, L.; Liberatore, F.; Camacho-Collados, M. Detecting and monitoring hate speech in Twitter. *Sensors* **2019**, *19*, 4654. [CrossRef] [PubMed]
30. Corazza, M.; Menini, S.; Cabrio, E.; Tonelli, S.; Villata, S. A multilingual evaluation for online hate speech detection. *ACM Trans. Internet Technol. TOIT* **2020**, *20*, 1–22. [CrossRef]
31. Ranasinghe, T.; Zampieri, M. Multilingual offensive language identification with cross-lingual embeddings. *arXiv* **2020**, arXiv:2010.05324.
32. Pamungkas, E.W.; Basile, V.; Patti, V. Misogyny detection in twitter: a multilingual and cross-domain study. *Inf. Process. Manag.* **2020**, *57*, 102360. [CrossRef]
33. Anzovino, M.; Fersini, E.; Rosso, P. Automatic Identification and Classification of Misogynistic Language on Twitter. In *Natural Language Processing and Information Systems*; Silberztein, M., Atigui, F., Kornyshova, E., Métais, E., Meziane, F., Eds.; Springer International Publishing: Cham, Switzerland, 2018; pp. 57–64.
34. Bashar, M.A.; Nayak, R.; Suzor, N. Regularising LSTM classifier by transfer learning for detecting misogynistic tweets with small training set. *Knowl. Inf. Syst.* **2020**, *62*, 4029–4054. [CrossRef]
35. Frenda, S.; Bilal, G. Exploration of Misogyny in Spanish and English tweets. In Proceedings of the Third Workshop on Evaluation of Human Language Technologies for Iberian Languages (IberEval 2018), Sevilla, Spain, 18 September 2018; Volume 2150, pp. 260–267.
36. García-Díaz, J.A.; Cánovas-García, M.; Colomo-Palacios, R.; Valencia-García, R. Detecting misogyny in Spanish tweets. An approach based on linguistics features and word embeddings. *Future Gener. Comput. Syst.* **2021**, *114*, 506–518. [CrossRef]
37. Fulper, R.; Ciampaglia, G.L.; Ferrara, E.; Ahn, Y.; Flammini, A.; Menczer, F.; Lewis, B.; Rowe, K. Misogynistic language on Twitter and sexual violence. In Proceedings of the ACM Web Science Workshop on Computational Approaches to Social Modeling (ChASM), Bloomington, IN, USA, 23–26 June 2014; pp. 57–64.
38. Molina-Villegas, A. La incidencia de las voces misóginas sobre el espacio digital en México. In *Jóvenes, Plataformas Digitales y Lenguajes: Diversidad Lingüística, Discursos e Identidades*; Pérez-Barajas, A.E., Arellano-Ceballos, A.C., Eds.; Elementum: Pachuca, Mexico, 2021; in press.
39. Cundiff, G. The influence of rap and hip-hop music: An analysis on audience perceptions of misogynistic lyrics. *Elon J. Undergrad. Res. Commun.* **2013**, *4*.
40. Vaswani, A.; Shazeer, N.; Parmar, N.; Uszkoreit, J.; Jones, L.; Gomez, A.N.; Kaiser, Ł.; Polosukhin, I. Attention is all you need. In *Advances in Neural Information Processing Systems*; Curran Associates, Inc.: Red Hook, NY, USA, 2017; pp. 5998–6008.
41. McCann, B.; Bradbury, J.; Xiong, C.; Socher, R. Learned in Translation: Contextualized Word Vectors. In *Advances in Neural Information Processing Systems*; Curran Associates, Inc.: Red Hook, NY, USA, 2017.
42. Buciluǎ, C.; Caruana, R.; Niculescu-Mizil, A. Model compression. In Proceedings of the 12th ACM SIGKDD International Conference on Knowledge Discovery and Data Mining, Philadelphia, PA, USA, 20–23 August 2006; pp. 535–541.
43. Hinton, G.; Vinyals, O.; Dean, J. Distilling the knowledge in a neural network. *arXiv* **2015**, arXiv:1503.02531.
44. Sanh, V.; Debut, L.; Chaumond, J.; Wolf, T. DistilBERT, a distilled version of BERT: smaller, faster, cheaper and lighter. *arXiv* **2019**, arXiv:1910.01108.
45. Devlin, J.; Chang, M.; Lee, K.; Toutanova, K. BERT: Pre-training of Deep Bidirectional Transformers for Language Understanding. In Proceedings of the 2019 Conference of the North American Chapter of the Association for Computational Linguistics: Human Language Technologies, NAACL-HLT 2019, Minneapolis, MN, USA, 2–7 June 2019; Volume 1 (Long and Short Papers); Burstein, J., Doran, C., Solorio, T., Eds.; Association for Computational Linguistics: Cambridge, MA, USA, 2019; pp. 4171–4186. [CrossRef]

applied sciences

MDPI

Article

Classification of Problem and Solution Strings in Scientific Texts: Evaluation of the Effectiveness of Machine Learning Classifiers and Deep Neural Networks

Rohit Bhuvaneshwar Mishra * and Hongbing Jiang

School of Management Engineering, Zhengzhou University, Zhengzhou 450001, China; jhbymx@foxmail.com
* Correspondence: rohit.bnmishra123@gmail.com

Abstract: One of the central aspects of science is systematic problem-solving. Therefore, problem and solution statements are an integral component of the scientific discourse. The scientific analysis would be more successful if the problem–solution claims in scientific texts were automatically classified. It would help in knowledge mining, idea generation, and information classification from scientific texts. It would also help to compare scientific papers and automatically generate review articles in a given field. However, computational research on problem–solution patterns has been scarce. The linguistic analysis, instructional-design research, theory, and empirical methods have not paid enough attention to the study of problem–solution patterns. This paper tries to solve this issue by applying the computational techniques of machine learning classifiers and neural networks to a set of features to intelligently classify a problem phrase from a non-problem phrase and a solution phrase from a non-solution phrase. Our analysis shows that deep learning networks outperform machine learning classifiers. Our best model was able to classify a problem phrase from a non-problem phrase with an accuracy of 90.0% and a solution phrase from a non-solution phrase with an accuracy of 86.0%.

Keywords: discourse analysis; problem–solution pattern; automatic classification; machine learning classifiers; deep neural networks

Citation: Mishra, R.B.; Jiang, H. Classification of Problem and Solution Strings in Scientific Texts: Evaluation of the Effectiveness of Machine Learning Classifiers and Deep Neural Networks. *Appl. Sci.* **2021**, *11*, 9997. https://doi.org/10.3390/app11219997

Academic Editor: Arturo Montejo-Ráez

Received: 20 September 2021
Accepted: 20 October 2021
Published: 26 October 2021

1. Introduction

Problem-solving is not a standardized exercise. Problems are different in domain, content, type, and linguistic properties [1–7]. In the most general sense, a problem is an unknown that arises from any situation where a person aims to satisfy a need or accomplish a goal. A problem comes into consideration when there is a "felt desire" to seek a solution to eliminate the problem or to find ways to solve the differences [8]. Problems traditionally have a problem area or domain, a problem category, a problem-solving approach, and a solution. The area or domain defines the problem constructs, laws, and fundamentals [9]. The problem category defines the type or nature of the problem [10,11]. The problem-solving strategy is then determined, and finally, we present the solution.

Mayer and Wittrock [10] categorized problem types as "poorly-defined", "well-defined", "routine," and "non-routine." Jonassen [11] classified well-structured problems from ill-structured problems by identifying individual variations in cognitive functioning. Smith [12] identified external variables from internal problem-solver characteristics, including domain and complexity. According to the researchers, there is increasing consensus that problems differ in content, structure, and method [13]. Problems also differ concerning their form, sophistication, and abstractness (domain specificity). While these three factors are similar, they are neither independent nor identical. Among these variables, there is enough independence to merit separate consideration.

Problem-solving is widely recognized as the most significant cognitive task [14–16]. However, the exploration of problem-solving techniques is severely limited in academic papers. According to studies, to decipher an unknown phenomenon we apply problem-solving methods. Previous knowledge, imaginative guesswork, and logical inference are the tools to solve problems in day-to-day life [17–19]. However, systematic problem-solving can be understood only after formulating a problem that we want to solve. Therefore, discovering the problem to be solved is the first component in problem-solving practice [20,21]. After identifying the problem, a search for a suitable or ideal solution starts in the problem-solving process. In the final step, we apply algorithmic analysis.

In the past, various problem-solving techniques were developed. One of the most well-known problem-solving models, the IDEAL model [22], describes problem-solving as a "structured process of identifying potential problems, defining the problem, representing the problem, exploring potential solutions, implementing strategies to find the solution to the problem, and then reflecting on and evaluating the activities' outcomes." Although the IDEAL model recommends applying these approaches to various problems in different ways, there are no specific guidelines for how to do so. Another model by Gick [2], as shown in Figure 1, describes a problem-solving model that must involve the following three processes: creating a problem representation, looking for solutions, and implementing and tracking solutions. Similarly, Smith [12] tried to offer a uniform problem-solving theory, but it was not entirely successful.

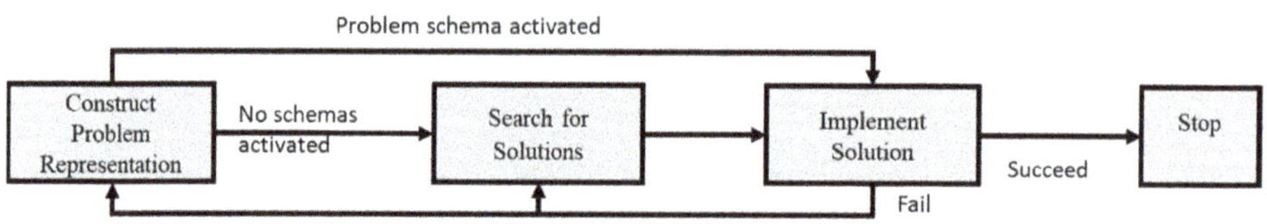

Figure 1. Problem-Solving Model (Gick, 1986).

Problem-solving methods develop knowledge. However, with the increase in content from journals, social media, business press releases, and scientific articles and discourse, knowledge generation will be highly challenging in the future [23–28]. Digitization initiatives in nearly all sectors would only increase the volume and variety of unstructured data. To apply outdated problem-solving models on this ever-increasing unstructured data is already out of scope. Hence, we need innovative and automatic techniques for the data and problems of the twenty-first century.

In the backdrop of all these developments, it is essential to look for automated techniques of problem–solution differentiation for information generation. With this paper, we aim to improve the linguistic and educational aspects of studying the problem–solution patterns. Furthermore, we seek to increase our understanding and associated assessments of problem–solution patterns. We look at real-world examples in published scientific literature to understand the complex problem–solution patterns. This analysis uses a collection of sentence features to apply various machine learning classifiers and deep learning models [29,30] in order to intelligently identify a string as a problem string or a solution string. We examine the parsed dependencies of our test word (problem, solution, or their synonyms) in the subject position in a sentence structure, then select its syntactic argument as a test phrase for our automatic classification. By extracting the problem and solution strings from scientific articles, our method can significantly improve knowledge mining. It will aid us in learning the gist of the papers and significantly improve information processing and visualization. Our methods can aid in the collection of scientific information and improve the efficiency of scientific searches. It will also aid in comparing related papers and, in the long run, lead to the automated production of field-specific review papers. In comparison with the previous studies, the work presented in this paper makes the following innovative and distinguishable key findings:

- Based on our review of the relevant literature, the proposed technique is the first to compare Machine Learning Classifiers and Deep Neural Networks for problem and solution string classification.
- Our approach is unique in applying both data iteration and cross-validation approach in assessing the effectiveness of Machine Learning Classifiers and Deep Neural Networks.
- Additionally, we perform parameter tuning to enhance the accuracy of our models.

2. Literature Review

In academic literature, the problem–solution pattern is pervasive [13,31,32]. Our writings, according to Jordan [33], represent our problem-solving, thought-action process. The research focused on linguistic and educational studies to develop a complete view of the complex problem–solution mechanism and to explain how we communicate these systems in the literature. The structure, domain specificity (abstractness), and complexity of the problems were defined by Jonassen [34]. He specified the continuum of outcomes for problem-solving learning. He also differentiated between well-structured problems and ill-structured problems in terms of the instructional design criteria.

Flowerdew [35] studied how particular keywords can be commonly utilized to discover specific aspects of the discourse structure. The keyword study was conducted on two corpus forms: the technical corpus and the student corpus. The phraseology of the keywords in both corpora was examined, and the examination revealed that the students' writing lacked various grammatical and lexical patterns used in expressing the problem–solution patterns and its elements. He discussed the pedagogical implications of these learning issues as well as the data-driven learning concepts.

In their analysis, the researchers [36] established adverbials that belong to the semantic category of "Result and Inference." Upton and Connor [37] used the corpus method to propose a text–linguistic approach that considers the unique characteristics of the genre-specific corpora. Charles [38] analyzed the problem–solution trend using discourse markers instead of a keyword-based approach. He analyzed adverbials such as "therefore, hence, and then" in two different corpora. He analyzed about 190,000 words from politics and about 300,000 words from materials science to check how they signal a pattern of problem-solving. According to the findings, combining corpus methods with discourse analysis would provide richer insights into academic discourse.

Technical texts have a four-part structure, according to Winter [39], which includes a situation, a problem, a solution, and an evaluation. This pattern is similar to Van Dijk's [40] pattern of "Introduction-Theory, Problem-Experiment-Comment, and Conclusion." SPRE, one of the most commonly used problem-solving patterns, is introduced by Hoey [41,42]. S stands for the situation; P for the query, purpose, problem, or the knowledge required; R for the reply, answer, response, the methods applied, and so on (depending on the case); and E for evaluation, in which a successful evaluation (the sequence comes to an end), or an unfavorable evaluation is given (the sequence is recycled).

In academic research texts, the assumption is that the problem identification or formulation comes before the solution [1,5,6,31,32]. In most scientific texts, the problem's condition is set at the start of the solution and remains unchanged. However, in some cases [43–46], the problem's initial specification is eventually reformulated or re-specified as the problem is solved.

The CARS ('Create a Research Space') model is one of the most well-known models of research article introductions [47]. The model's different movements cover similar ground to that of the SPRE model. The first move, 'Establishing a Territory', is similar to the Situation from SPRE; the second move, 'Establishing a Niche', is similar to the Problem in SPRE, and it identifies a knowledge void or an issue in the research field; and the third move, 'Occupying the Niche', enables the researcher to fill the void by announcing their findings, thus forming a Response step.

Based on the extensive literature review, in this work we address the challenge of defining patterns for problem-solving in the scientific text [48,49]. We chose the problem-solving model defined by Hoey [42] for our study. We aimed to classify strings of problems and solutions through various machine learning classifiers and deep learning networks [30,50,51]. We restricted ourselves to ML classifiers and deep neural networks, and we did not include non-standard techniques, such as morphological neural networks with dendritic processing and spiking neural networks for our study [52–57]. We were more concerned with evaluating sentence features in various syntactic variations [48,49] than trying all available classifiers. We theorized the multiple explications for our results and tried to ascertain the source of the failure/success of our models in increasing the accuracy.

3. Research Methodology

The methodology section consists of five sub-sections. In Section 3.1, we describe the SPRE problem–solution model with an example. In Section 3.2, we talk about the corpora and dataset preparation. Section 3.3 describes the wide range of syntactic variations, which we consider for the problem and solution strings. In Section 3.4, we discuss the training data preparation. Finally, in Section 3.5, we discuss our algorithm and model development.

3.1. Problem–Solution Pattern

A sentence's various features aid in the identification of problem–solution patterns. We use the SPRE model [42] for our study, which consists of Situation, Problem, Response, and Evaluation. We analyzed several sentences and found that the situation is an optional element in many sentences. As a result, we agreed to focus our model-building efforts on Problem, Response, and Evaluation. It is an excellent place to start describing the pattern with the problem because it is unusual for an author to present a problem and then leave the problem without any answer. There are exceptional cases, as when authors present a problem and then leave it for future research; however, in principle, the problem variable should have enough information for our model to work. The second knowledge parameter we are searching for is Response and Evaluation in the same sentence. We consider providing sufficient information to help our automatic classifier recognize the pattern externally for Response and Evaluation. Let us go through each part of the SPRE model with a real-life illustration. It will assist us in effectively understanding the functionality of the SPRE model.

Situation: Background information on situations; details about the persons, topic, case, or location involved in the debate. Example: John was doing experiments in the school laboratory.

Problem: An aspect of a situation that requires a solution; a need, a dilemma, a puzzle, or an obstacle that is being discussed; flaws in the current situation. Example: John discovered an error in the experiment code.

Response: Problem–solution(s); discussion of a way(s) to deal with or solve the problem. Example: John modified and executed the code.

Evaluation: A determination of the efficacy of the proposed solution(s); if several solutions are available, which one is the best. Example: The code ran successfully, and John got the desired output.

The flowchart explaining the flow of processes in the SPRE model is shown in Figure 2 below. As shown, the process ends if the evaluation is positive, or else the whole process is recycled again.

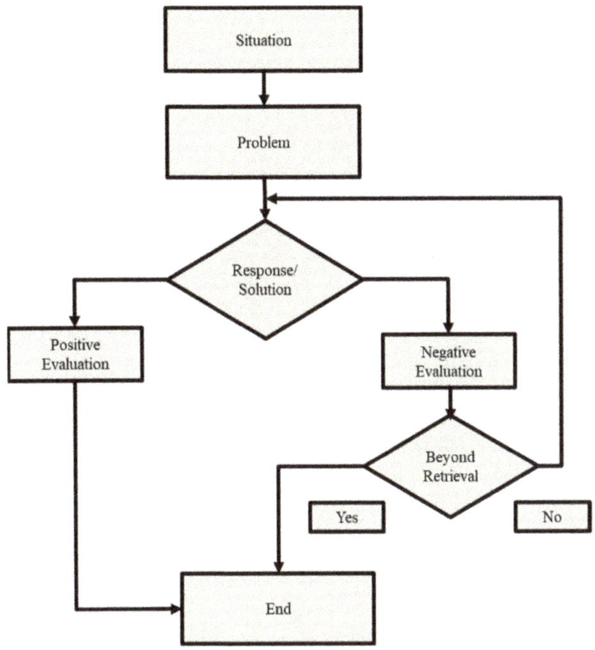

Figure 2. SPRE Model given by Hoey.

The word "problem" has various meanings. For example, the word "problem" can mean that something needs to be accomplished (a task), and another meaning is that something is troublesome, negative, and needs a solution. We are restricted to the use of the problem only for the second scenario. The use of the problem-defining task is beyond the scope of this research.

3.2. *Corpora*

There are numerous public datasets available for citation and text analysis. The datasets with citation data are used for cluster analysis with network and edge information, analyzing influence in the citation network, recognizing the most impactful papers, and for topic modelling analysis [58–60]. Text analysis datasets on the other hand are used to implement different techniques for practical problems in artificial intelligence, computational science, scientific data analysis, and other fields [61–64].

We use the dataset provided by [48]. This dataset is prepared from the March 2016 ACL anthology corpus [65]. The data provided by [48] checks the parsed dependencies and the searches for "problem/solution" or their synonyms in the subject position. The data include synonyms of problem and solution words in order to maximize the search. The dataset manually selects 28 synonyms for the "problem" word and 19 synonyms for the "solution" word. The synonyms are manually selected from the semantically closest words, trained using Word2Vec on PubMed articles [66,67]. The words chosen for the problem candidate phrase extraction, as from [48], are "*Bottleneck, Caveat, Challenge, Complication, Conundrum, Difficulty, Dilemma, Disadvantage, Drawback, Fault, Flaw, Impediment, Issue, Limitation, Mistake, Obstacle, Pitfall, Problem, Quandary, Riddle, Shortcoming, Struggle, Subproblem, Threat, Tragedy, Trouble, Uncertainty*, and *Weakness*." Similarly, the words chosen for the solution candidate extraction are "*Alternative, Answer, Answers, Approach, Approaches, Idea, Method, Methodology, Proposal, Remedy, Solution, Suggestion, Scheme, Step, Strategy, Technique, Task, Way*, and *Workaround.*"

We chose exact and straightforward strings in suggesting a positive or negative condition for the problem/solution argument. A problem string denotes an unexplained event, a research query, or an item that has failed to meet its specified requirements. A solution argument, on the other hand, is a successful answer to the evaluation. However, we chose the phrase so that its status as a problem or solution phrase was not revealed lexically. This additional check was used to reject inputs to the classifier that were overly evident. For example, we rejected the sentence if it contained the words "troublesome" or "shortcomings" because it would be far too easy for the classifier to pick up on such signals. For both the problems and the solutions, there are corresponding negative examples (non-problem and non-solution). The negative strings were chosen to imitate as closely as possible the apparent characteristics of the positive examples, hence offering no extra information for differentiation on the surface. Negative examples were selected from a collection of sentences which had a similar length and a similar POS (part of speech) to that of the test problem and the test solution strings. Negative examples were carefully sampled to match the positive examples' POS pattern and sentence length. Sentences that lacked words for problems/solutions were flagged, and one syntactic subtree within them was chosen at random to create negative samples. Finally, 500 problem strings, 500 non-problem strings, 500 solution strings, and 500 non-solution strings made up the data sample. These 2000 strings were used to train and test our machine learning and deep learning models.

3.3. Selection of Problem and Solution Strings

Our model aimed to identify strings of problems and solutions that appear in a wide range of syntactic variations. Furthermore, we wanted to test the problem and solution scenarios even when the problem or solution status was not explicitly specified. Moreover, in certain instances, two or more sentences are used to explain the problem or the solution. These kinds of sentences are beyond the research scope of the present study. Here, we just looked at one-sentence-long problem and solution phrases.

3.4. Creating the Training Sample

The dataset was limited, so we applied two methods for preparing the training and testing data. In the first method, we randomly divided the test and train data and performed various iterations. In each iteration, the train dataset and the test dataset were different. However, in each iteration, the percentage of train data was 67%, and the test data was 33% [68–70]. This method achieved a better estimate of accuracy (average accuracy), even on this limited dataset. In the second method, we applied multifold cross-validation to train and test the data [71,72]. We applied 5-fold and 10-fold validation, and we realized that the accuracy changed. We have mentioned the accuracies obtained from the best scenarios in this paper.

3.5. Method and Model Development

The dataset [48] listed a collection of features without considering the context of the phrase. We used traditional machine learning classifiers, and deep learning networks to test features, such as BOW (bag of words), transitivity, modality, polarity, syntax, doc2vec, word2vec, and word2vec smoothed, as discussed by Heffernan and Teufel [48]. We improved the findings with a comparison with those from [48]. We theorized different explanations for our findings and attempted to determine the cause of failure/success in terms of the classification-accuracy improvement.

The semantic disambiguation ability of the problem/solution definition was tested using traditional machine learning classifiers and deep learning networks. The problem/solution keywords enabled template-based search. The syntactic complement of the keyword in the subject position was used. To ensure that there were no "give-away" phrases (phrases that provide additional information) inside the phrases, we modeled only the keyword from the sentence and excluded the rest of the sentence. This was conducted to aid in the generalization of our models for real-world problem/solution differentiation tasks. The following is the computational algorithm:

1. Obtain the features from the dataset. Start with baseline feature (bag of words).
2. Method 1: Divide the data into 67% training and 33% testing. Obtain the accuracy for differentiating the problem strings from the non-problem strings. In the next iteration, different train and test data with the same proportion are selected, and the classification is performed again. The absolute accuracy is the average of the accuracies in all the iterations. We apply the same steps for differentiating the solution strings from the non-solution strings.
3. Method 2: Perform multifold cross-validation of data. Obtain the accuracy for differentiating the problem strings from the non-problem strings. We apply the same steps for differentiating the solution strings from the non-solution strings.
4. Add one feature extra on top of the BOW. Apply method 1/method 2. Repeat.
5. Compare the results after applying the machine learning classifier/deep learning model on all the available features.
6. Improve the best model by hyperparameter tuning.

4. Evaluation

We evaluated the machine learning classifiers and the deep neural network models separately for methods one and two. The results varied for the different methods. For example, the accuracy values were high for both the machine learning classifiers and the deep neural networks when we used the data iteration method, while accuracy decreased when we used the multifold validation.

4.1. Evaluation of Machine Learning Classifiers

Machine learning is all about discovering patterns and applying those to new datasets. To evaluate an algorithm, we can divide the dataset into two parts: train and test. In the first method, we divided the data into 67% training and 33% testing. We trained the data in 5–10 iterations and applied ML classifiers, such as logistic regression (LR), multilayer perceptron (MLP), support vector classifier (SVC), random forest (RF), Naive Bayes classifier (NV), AdaBoost (AB), and gradient boosting (GB). We used the scikit-learn machine learning library for our experiments [73]. It is a free, open-source, and trendy machine learning package for Python [74].

We used a bag of words (BOW) for the first feature, which is also our baseline feature. We applied the ML classifiers and checked the accuracy. In the next step, we added the transitivity feature on top of the baseline feature. Similarly, we kept adding extra features in the given order: bag of words (BOW), transitivity, modality, polarity, syntax, doc2vec, word2vec, and word2vec smoothed [48]. We calculated the disambiguation capacity of the problem/solution through these ML classifiers. The accuracy of each model for both scenarios is shown in Tables 1 and 2 below.

Table 1. Classification of problems from non-problems using ML classifiers by training data iteration.

Feature Set	LR	MLP	SVC	RF	NB	AB	GB
Baseline	0.69	0.70	0.66	0.68	0.66	0.68	0.67
+transitivity	0.69	0.69	0.64	0.68	0.65	0.68	0.66
+modality	0.69	0.68	0.66	0.68	0.65	0.67	0.67
+polarity	0.70	0.70	0.65	0.69	0.65	0.67	0.67
+syntax	0.76	0.75	0.75	0.77	0.65	0.73	0.73
+doc2vec	0.77	0.77	0.76	0.76	0.66	0.75	0.68
+word2vec	0.73	0.69	0.61	0.70	0.66	0.72	0.72
+word2vecSmoothed	0.76	0.74	0.75	0.77	0.65	0.73	0.72

Table 2. Classification of solutions from non-solutions using ML classifiers by training data iteration.

Feature Set	LR	MLP	SVC	RF	NB	AB	GB
Baseline	0.71	0.70	0.65	0.65	0.67	0.67	0.66
+transitivity	0.70	0.70	0.68	0.69	0.68	0.66	0.68
+modality	0.70	0.70	0.68	0.67	0.67	0.66	0.68
+polarity	0.71	0.70	0.68	0.69	0.68	0.67	0.69
+syntax	0.74	0.73	0.74	0.76	0.67	0.71	0.74
+doc2vec	0.76	0.75	0.75	0.75	0.68	0.76	0.64
+word2vec	0.76	0.73	0.71	0.79	0.68	0.80	0.73
+word2vecSmoothed	0.77	0.72	0.70	0.78	0.66	0.80	0.71

For method one, the three most effective ML classifiers for the classification of the problems from the non-problems by training data iteration were **logistic regression (LR), multilayer perceptron (MLP),** and **random forest (RF),** giving an accuracy of 77%.

For the classification of the solutions from the non-solutions, the most effective ML model in method one was **AdaBoost**, giving an accuracy of **80%**.

In the second method, we applied a 5–10-fold cross-validation for the same ML models. In a similar way to method 1, we started with the BOW, added the other features in the same order as before, and calculated the accuracy. We used the same scikit-learn machine learning library for method two as well. The results show that for most cases, the accuracy in method 2 decreases when compared to that of method 1. Moreover, the accuracy of the models increases from 5-fold validation to 10-fold validation. Hence, we present the results from the 10-fold validation. The accuracy of each model for method two is shown in Tables 3 and 4 below.

Table 3. Classification of problems from non-problems using ML classifiers by 10-fold cross validation.

Feature Set	LR	MLP	SVC	RF	NB	AB	GB
Baseline	0.71	0.71	0.68	0.71	0.66	0.70	0.71
+transitivity	0.70	0.71	0.67	0.70	0.66	0.70	0.70
+modality	0.70	0.71	0.67	0.69	0.66	0.69	0.67
+polarity	0.71	0.71	0.67	0.70	0.66	0.71	0.66
+syntax	0.73	0.72	0.69	0.71	0.66	0.70	0.68
+doc2vec	0.74	0.73	0.70	0.76	0.66	0.73	0.73
+word2vec	0.74	0.67	0.61	0.67	0.66	0.70	0.67
+word2vecSmoothed	0.73	0.74	0.69	0.71	0.66	0.70	0.67

Table 4. Classification of solutions from non-solutions using ML classifiers by 10-fold cross validation.

Feature Set	LR	MLP	SVC	RF	NB	AB	GB
Baseline	0.72	0.71	0.68	0.69	0.69	0.68	0.70
+transitivity	0.72	0.71	0.68	0.70	0.69	0.68	0.70
+modality	0.72	0.69	0.69	0.70	0.69	0.69	0.70
+polarity	0.72	0.70	0.69	0.72	0.69	0.69	0.70
+syntax	0.71	0.71	0.68	0.70	0.69	0.69	0.68
+doc2vec	0.73	0.72	0.69	0.75	0.69	0.75	0.67
+word2vec	0.75	0.71	0.72	0.79	0.69	0.79	0.75
+word2vecSmoothed	0.75	0.70	0.72	0.79	0.69	0.81	0.75

For method two, the most effective ML model for classifying problems from non-problems was **random forest (RF)**, giving an accuracy of **76%**.

For the classification of solutions from non-solutions, the most effective ML model for method two was **AdaBoost**, giving an accuracy of **81%**.

4.2. Evaluation of Deep Learning Models

Although the simple machine classifiers performed well as per our experimental settings, they still can be improved. If an AI algorithm returned an incorrect prediction for a few scenarios, we had to interfere. A deep learning model can efficiently improve our results because a deep learning model may use neural networks to figure out on its own whether a prediction is correct or not [75,76]. Hence, we compared the classifications between the ML classifiers and the deep learning models. A deep learning model's reasoning structure is close to how a person can conclude. This is achieved using a layered system of algorithms called an artificial neural network. An artificial neural network's architecture was inspired by the human brain's neural network [77,78], contributing to a learning mechanism that is far more capable than the machine learning classifiers discussed earlier.

For the first method, we applied three deep learning models to train the dataset. We divided the data into 67% training and 33% testing. We trained the data in 5–10 iterations and applied the deep learning models: Long short-term memory (LSTM), neural network (NN), and convolutional neural network (CNN) [79,80]. As with the ML classifiers, we started with BOW, added the other features in the same order as before, and calculated the accuracy. We used Keras, an open-source library, for the application of the deep learning models [74,81]. It offers a python interface for artificial neural networks and also serves as a frontend for TensorFlow. The accuracy of each model for method 1 (data iteration) is shown in Tables 5 and 6 below.

Table 5. Classification of problems from non-problems using deep learning networks by training data iteration.

Feature Set	LSTM	NN	CNN
Baseline	0.49	0.70	0.82
+transitivity	0.49	0.69	0.82
+modality	0.50	0.71	0.83
+polarity	0.50	0.69	0.83
+syntax	0.50	0.75	0.86
+doc2vec	0.47	0.76	0.84
+word2vec	0.50	0.71	0.82
+word2vecSmoothed	0.50	0.75	0.82

Table 6. Classification of solutions from non-solutions using deep learning networks by training data iteration.

Feature Set	LSTM	NN	CNN
Baseline	0.50	0.72	0.77
+transitivity	0.48	0.71	0.85
+modality	0.49	0.70	0.82
+polarity	0.48	0.71	0.78
+syntax	0.52	0.75	0.81
+doc2vec	0.50	0.77	0.83
+word2vec	0.54	0.71	0.83
+word2vecSmoothed	0.57	0.76	0.83

For method one, the most effective deep learning model for the classification of the problems from the non-problems by training data iteration is **CNN**, giving an accuracy of **86%**.

For the classification of solutions from non-solutions, the most effective deep learning model in method one is again **CNN**, giving an accuracy of **85%**.

For the second method, we applied 5–10-fold validation cross-validation for the same deep learning models. The results show that for most cases, the accuracy in method 2 decreases when compared to method 1. Moreover, the accuracy of the models decreases from 5-fold validation to 10-fold validation. Hence, we present the results from the 5-fold validation. In a similar way to method 1, we started with BOW, added the other features in the same order as before, and calculated the accuracy. For method two, we also used the Keras library for the application of the artificial neural networks.

The accuracy of each model for method two is shown in Tables 7 and 8 below.

Table 7. Classification of problems from non-problems using deep learning networks by 5-fold cross validation.

Feature Set	LSTM	NN	CNN
Baseline	0.53	0.63	0.61
+transitivity	0.53	0.65	0.61
+modality	0.47	0.65	0.65
+polarity	0.55	0.66	0.57
+syntax	0.50	0.67	0.57
+doc2vec	0.50	0.70	0.55
+word2vec	0.48	0.61	0.59
+word2vecSmoothed	0.52	0.67	0.59

Table 8. Classification of solutions from non-solutions using deep learning networks by 5-fold cross-validation.

Feature Set	LSTM	NN	CNN
Baseline	0.50	0.59	0.64
+transitivity	0.51	0.56	0.63
+modality	0.38	0.59	0.64
+polarity	0.51	0.58	0.65
+syntax	0.37	0.57	0.61
+doc2vec	0.50	0.62	0.56
+word2vec	0.63	0.64	0.63
+word2vecSmoothed	0.58	0.64	0.68

For method two, the most effective deep learning model for the classification of the problems from the non-problems is **NN**, with an accuracy of **70%**.

For the classification of the solutions from the non-solutions, the most effective model in method two is **CNN**, with an accuracy of **68%**.

4.3. Hyperparameter Tuning

To improve the accuracy further, we performed hyperparameter tuning for the best neural network model. Hyperparameter tuning methods include scanning through the space of possible hyperparameter values to find possible model architecture candidates [82,83]. Finding the optimal values is also referred to as "searching" the hyperparameter space. If the learning rate is too low for a model, the model will miss important data patterns. If the model is too heavy, it can have collisions leading to decreased accuracy. Hence, choosing suitable hyperparameters is key to the success of our neural network architecture. Even for small models, the number of hyperparameters may be significant. It is quite an intricate task to tune the hyperparameters but doing so improves the efficiency of our model significantly. Here, in Table 9, we can see that our model improves on performing the hyperparameter tuning. We applied hyperparameter tuning to CNN networks in method one as it was the best performing model and method.

Table 9. Hyperparameter tuning the CNN model for the data iteration method.

Feature Set	Problem Phrase	Solution Phrase
Baseline	0.85	0.84
+transitivity	0.90	0.85
+modality	0.87	0.81
+polarity	0.86	0.85
+syntax	0.88	0.86
+doc2vec	0.85	0.84
+word2vec	0.85	0.85
+word2vecSmoothed	0.83	0.84

As shown in the above results, hyperparameter tuning impacts the classification results immensely. The accuracies improved for most of the feature list when we tuned our models. For our analysis, the accuracy improved to **90%** for classifying the problems from the non-problems and **86%** for classifying the solutions from the non-solutions.

To get an idea of the impact of hyperparameter tuning, we show the results for the top classifier (after hyperparameter tuning) in the figures below. First, we present the best CNN model results for the classification of the problem from the non-problem string, as shown below in Figure 3. The top 10 values are in Table 10, and the graph is in Figure 3 to help us understand the effect of tuning on a model.

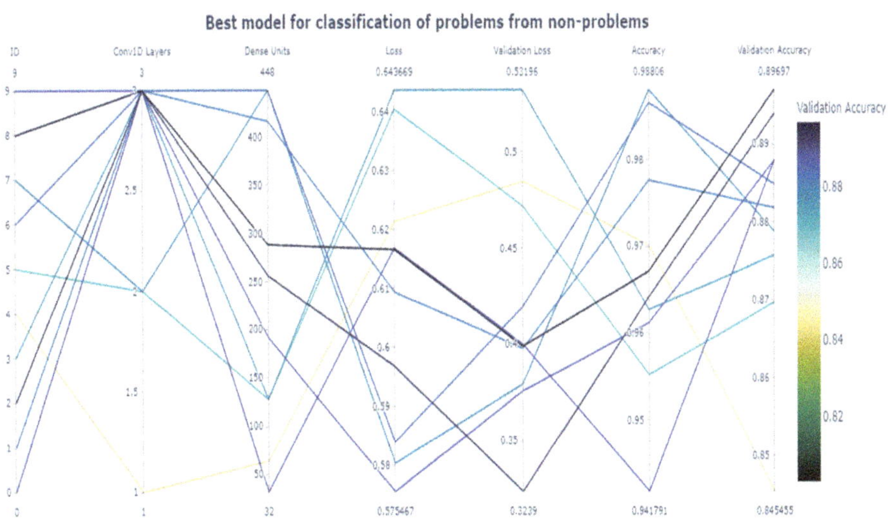

Figure 3. Hyperparameter tuning results for the classification of problems from non-problems.

Table 10. Hyperparameter tuning results for the classification of problems from non-problems.

Uid	Num_Conv1D	Units	Accuracy
0	3	192	0.89
1	3	416	0.88
2	3	256	0.89
3	3	128	0.88
4	1	64	0.85
5	2	128	0.87
6	3	448	0.88
7	2	448	0.88
8	3	288	0.90
9	3	32	0.89

The best accuracy of **90%** occurs for **three one-dimensional convolutional networks** and **288 units**. The number of units is an essential hyperparameter. The neural networks are universal function approximators, and they need enough 'power' to learn for the prediction task. The number of units is the critical indicator of the learning ability of the model. The simple role can require fewer units. The greater the number of units, the more complex the role of the parameter tuning. The number of units used for the best model indicates that our model is highly complex. Hence, it achieves good accuracy.

In a similar way to the tuning of the problem phrase classifier, we present the hyperparameter tuning results for the top 10 values of the best CNN model for the classification of the solutions from the non-solutions in Table 11 and Figure 4.

Table 11. Hyperparameter tuning results for the classification of solutions from non-solutions.

Uid	Num_Conv1D	Units	Accuracy
0	2	192	0.86
1	1	512	0.83
2	2	64	0.82
3	1	480	0.81
4	2	416	0.81
5	1	256	0.77
6	2	320	0.82
7	1	192	0.78
8	4	288	0.82
9	3	480	0.82

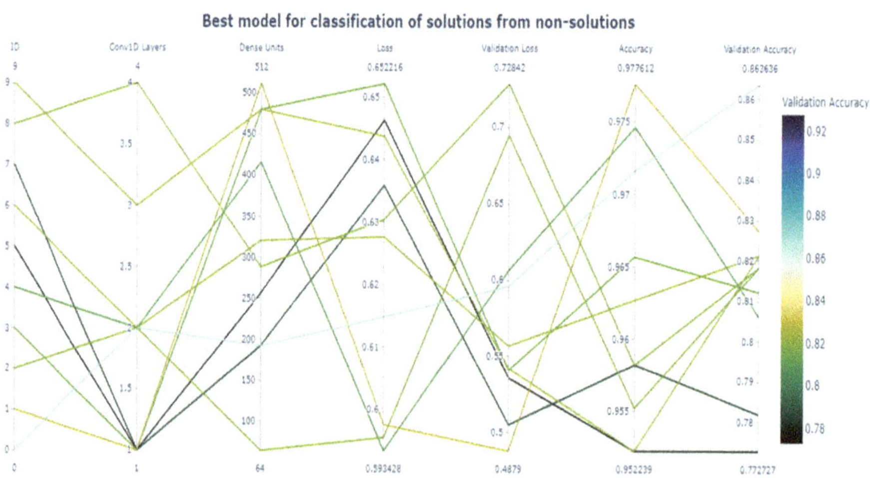

Figure 4. Hyperparameter tuning results for classification of solutions from non-solutions.

The best accuracy of **86%** is achieved for **two one-dimensional convolutional networks** and **192 units**.

4.4. Final Results

We report the results for the best model, method, and hyperparameter tuning parameters in Table 12 below.

Table 12. Best performing CNN model.

Strings	Best Accuracy	Model	Method	Hyperparameters
Problem Strings	90.0%	CNN	Method 1	Number of Conv1D: 3 Units: 288
Solution Strings	86.0%	CNN	Method 1	Number of Conv1D: 2 Units: 192

5. Discussion

The research discussed in this paper focuses on comparing machine learning classifiers with deep learning models that consist of an empirical assessment. In our paper, the theory, arguments, and solution methods proposed were empirically tested. Even though the dataset size was small, the high accuracy of the models indicates that our models were highly efficient. We addressed our views on how to improve the accuracy of classification using different methods. We performed hyperparameter tuning to improve the accuracy for the used models. We provided an overview of some of the potential ways to increase accuracy. We did our best to minimize the challenges to the problems of validity. We extracted various characteristics from the sentence, but we did not conduct any relation or meta-data analysis that could be regarded as external variables.

We used a dataset that is available in the public domain. The dataset was annotated and checked by more than one person [48], ensuring that the annotation of the dataset was of good quality and that no errors in the annotation and calculation were present. Moreover, we carried out the experiments more than once to ensure that there were no mistakes when performing the experiments and that our findings were replicable.

We would also like to highlight the challenges we faced while performing the classifications. The first hurdle was locating the correct dataset for our study. Then, it was challenging to apply various ML and neural networks to the feature list present in the dataset. The next issue we faced was identifying various permutations and combinations of the models and the data training approach discussed in the paper. To ensure that our algorithms were robust, we tried various approaches and models, but mentioned only the significant ones in this paper. This was the most challenging part of our analysis.

Next, we would talk about the future scope of our experiments. In future research, we would like to have a few valuable additions. Firstly, our corpus consisted entirely of scientific articles from computer linguistic research, constituting a particular subset of textual information. We want to test our model on scientific articles from different countries and domains of science. We used an existing dataset; hence, we could not change the existing parameters, such as the feature set, the synonyms of problem words, the synonyms of solution words and the number of sentences in the corpus. In the future, we would like to prepare our corpus. It would offer us more flexibility in the way we approach our research question. We would like to make a corpus that contains articles from different domains. It would help us to identify different characteristics of problem–solution patterns specific to a research domain. We would like to apply the model to a corpus containing research articles from multiple research domains and to try to find a generalized model for problem–solution classification. Once we achieve a generalized model for problem–solution classification, the next stage of our research would be linking the problem string with its corresponding solution string. This would help in knowledge mining from research articles which would eventually lead to idea discovery from the scientific texts. This would improve the existing mechanisms of text summarization and text abstraction. It would also help review articles and help in the classification of research domains based on the texts from the research papers.

6. Conclusions

We presented a technique focused on machine learning classifiers, such as logistic regression (LR), multilayer perceptron (MLP), support vector classifier (SVC), random forest (RF), Naive Bayes classifier (NV), AdaBoost (AB), gradient boosting (GB), and deep learning neural networks, such as long short-term memory (LSTM), neural network (NN), and convolutional neural network (CNN). We constructed various classification models in which we used sentence features to distinguish the problem from the non-problem strings and the solution strings from the non-solution strings. For the distinction of the solution from the non-solution, our best model was able to achieve an 86% accuracy. Likewise, an accuracy of 90% was obtained by the best model for differentiating the problems from the non-problems. In both cases, CNN worked the best, and we performed hyper-parameter tuning to achieve the best CNN outcome.

Author Contributions: Conceptualization, R.B.M. and H.J.; formal analysis and writing—review and editing, R.B.M.; data curation and writing—original draft preparation, R.B.M.; supervision, H.J. All authors have read and agreed to the published version of the manuscript.

Funding: This research was funded by National Natural Science Foundation of China, Project no. 71801195.

Institutional Review Board Statement: Not applicable.

Informed Consent Statement: Not applicable.

Data Availability Statement: Not applicable.

Acknowledgments: The authors would like to thank the three anonymous reviewers for providing their constructive comments.

Conflicts of Interest: The authors declare no conflict of interest.

References

1. Albay, E.M. Analyzing the Effects of the Problem Solving Approach to the Performance and Attitude of First Year University Students. *Soc. Sci. Humanit. Open* **2019**, *1*, 100006. [CrossRef]
2. Gick, M.L. Problem-Solving Strategies. *Educ. Psychol.* **1986**, *21*, 99–120. [CrossRef]
3. Hembree, R. Experiments and Relational Studies in Problem Solving: A Meta-Analysis. *J. Res. Math. Educ.* **1992**, *23*, 242–273. [CrossRef]
4. Hidayati, N.; Permana, D. Assessment of Problem Solving Abilities and Student Learning Activities Based on Learning Tools: The Basis of Problem Based Learning Development. *Int. J. Sci. Technol. Res.* **2019**, *8*, 453–456.
5. Molnár, G.; Csapó, B. The Efficacy and Development of Students' Problem-Solving Strategies during Compulsory Schooling: Logfile Analyses. *Front. Psychol.* **2018**, *9*, 302. [CrossRef] [PubMed]
6. Priemer, B.; Eilerts, K.; Filler, A.; Pinkwart, N.; Rösken-Winter, B.; Tiemann, R.; Belzen, A.U.Z. A Framework to Foster Problem-Solving in STEM and Computing Education. *Res. Sci. Technol. Educ.* **2020**, *38*, 105–130. [CrossRef]
7. Rausch, A.; Schley, T.; Warwas, J. Problem Solving in Everyday Office Work—A Diary Study on Differences between Experts and Novices. *Int. J. Lifelong Educ.* **2015**, *34*, 448–467. [CrossRef]
8. Sinnott, J.D. *Everyday Problem Solving: Theory and Applications*; Praeger: New York, NY, USA, 1989; ISBN 978-0-275-92691-5.
9. Kim, D.-K.; El Khawand, C. An Approach to Precisely Specifying the Problem Domain of Design Patterns. *J. Vis. Lang. Comput.* **2007**, *18*, 560–591. [CrossRef]
10. Mayer, R.E.; Wittrock, M.C. Problem-solving transfer. In *Handbook of Educational Psychology*; Prentice Hall International: London, UK, 1996; pp. 47–62, ISBN 978-0-02-897089-9.
11. Jonassen, D.H. Instructional Design Models for Well-Structured and III-Structured Problem-Solving Learning Outcomes. *ETRD* **1997**, *45*, 65–94. [CrossRef]
12. Smith, M.U. *Toward a Unified Theory of Problem Solving: Views from the Content Domains*; Erlbaum: Hillsdale, MI, USA, 1991; ISBN 0-8058-0510-9.
13. Hoey, M. Problem-Solution Patterns. *Encycl. Lang. Linguist.* **2006**, *1*, 112–115. [CrossRef]
14. Delahunty, T.; Seery, N.; Lynch, R. Exploring Problem Conceptualization and Performance in STEM Problem Solving Contexts. *Instr. Sci.* **2020**, *48*, 395–425. [CrossRef]
15. Greiff, S.; Holt, D.; Funke, J. Perspectives on Problem Solving in Cognitive Research and Educational Assessment: Analytical, Interactive, and Collaborative Problem Solving. *J. Probl. Solving* **2013**, *5*, 71–91. [CrossRef]
16. Huitt, W.G. Problem Solving and Decision Making: Consideration of Individual Differences Using the Myers-Briggs Type Indicator. *J. Psychol. Type* **1992**, *24*, 33–44.
17. Bronkhorst, H.; Roorda, G.; Suhre, C.; Goedhart, M. Logical Reasoning in Formal and Everyday Reasoning Tasks. *Int. J. Sci. Math. Educ.* **2020**, *18*, 1673–1694. [CrossRef]
18. Galotti, K.M. Approaches to Studying Formal and Everyday Reasoning. *Psychol. Bull.* **1989**, *105*, 331–351. [CrossRef]
19. Hintikka, J. Is Logic the Key to All Good Reasoning? *Argumentation* **2001**, *15*, 35–57. [CrossRef]

20. Christ, T.J.; Christ, T.J. *Best Practices in Problem Analysis*; National Association of School Psychologists: Bethesda, MD, USA, 2008.
21. Narula, S.C. Systematic Ways to Identify Research Problems in Statistics. *Int. Stat. Rev. Rev. Int. De Stat.* **1974**, *42*, 205–209. [CrossRef]
22. Bransford, J.; Stein, B.S. *The Ideal Problem Solver. A Guide for Improving Thinking, Learning, and Creativity*; Series of books in psychology; W. H. Freeman and Company: New York, NY, USA, 1984; ISBN 978-0-7167-1669-3.
23. Farrington, T.; Alizadeh, A. On the Impact of Digitalization on R&D: R&D Practitioners Reflect on the Range and Type of Digitalization's Likely Effects on R&D Management. *Res. Technol. Manag.* **2017**, *60*, 24–30. [CrossRef]
24. Hausberg, J.P.; Liere-Netheler, K.; Packmohr, S.; Pakura, S.; Vogelsang, K. Research Streams on Digital Transformation from a Holistic Business Perspective: A Systematic Literature Review and Citation Network Analysis. *J. Bus. Econ.* **2019**, *89*, 931–963. [CrossRef]
25. Nadkarni, S.; Prügl, R. Digital Transformation: A Review, Synthesis and Opportunities for Future Research. *Manag Rev. Q* **2021**, *71*, 233–341. [CrossRef]
26. Nelson, G.; Ellis, S. The History and Impact of Digitization and Digital Data Mobilization on Biodiversity Research. *Philos. Trans. R. Soc. B Biol. Sci.* **2019**, *374*, 20170391. [CrossRef]
27. Reis, J.; Amorim, M.; Melão, N.; Cohen, Y.; Rodrigues, M. Digitalization: A Literature Review and Research Agenda. In *Proceedings on 25th International Joint Conference on Industrial Engineering and Operations Management—IJCIEOM*; Anisic, Z., Lalic, B., Gracanin, D., Eds.; Springer International Publishing: Cham, Switzerland, 2020; pp. 443–456.
28. Reis, J.; Amorim, M.; Melão, N.; Matos, P. Digital Transformation: A Literature Review and Guidelines for Future Research. In *Trends and Advances in Information Systems and Technologies*; Rocha, Á., Adeli, H., Reis, L.P., Costanzo, S., Eds.; Springer International Publishing: Cham, Switzerland, 2018; pp. 411–421.
29. Bui, D.T.; Tsangaratos, P.; Nguyen, V.-T.; Liem, N.V.; Trinh, P.T. Comparing the Prediction Performance of a Deep Learning Neural Network Model with Conventional Machine Learning Models in Landslide Susceptibility Assessment. *CATENA* **2020**, *188*, 104426. [CrossRef]
30. Nanehkaran, Y.A.; Zhang, D.; Salimi, S.; Chen, J.; Tian, Y.; Al-Nabhan, N. Analysis and Comparison of Machine Learning Classifiers and Deep Neural Networks Techniques for Recognition of Farsi Handwritten Digits. *J. Supercomput.* **2021**, *77*, 3193–3222. [CrossRef]
31. Khaw, L.L. Problem-Solution Patterns in the Introductions of Chemical Engineering Research Articles: Pedagogical Insights. In Proceedings of the 2020 IEEE Global Engineering Education Conference (EDUCON), Porto, Portugal, 27–30 April 2020; pp. 78–84.
32. Khaw, L.L.; Tan, W.W. Creating Contexts in Engineering Research Writing Using a Problem-Solution-Based Writing Model: Experience of Ph.D. Students. *IEEE Trans. Prof. Commun.* **2020**, *63*, 155–171. [CrossRef]
33. Jordan, M.P. Short Texts to Explain Problem–Solution Structures–and Vice Versa. *Instr. Sci.* **1980**, *9*, 221–252. [CrossRef]
34. Jonassen, D.H. Toward a Design Theory of Problem Solving. *ETRD* **2000**, *48*, 63–85. [CrossRef]
35. Flowerdew, L. *Corpus-Based Analyses of the Problem–Solution Pattern*; John Benjamins Publishing Company: Amsterdam, The Netherlands, 2008; ISBN 978-90-272-2303-6.
36. Biber, P.D.; Finegan, E.; Johansson, S.; Conrad, D.S.; Leech, G. *Longman Grammar Spoken & Written English Cased*; Longman: Harlow, UK, 1999; ISBN 978-0-582-23725-4.
37. Upton, T.; Connor, U. Using Computerized Corpus Analysis to Investigate the Textlinguistic Discourse Moves of a Genre. *Engl. Specif. Purp.* **2001**, *20*, 313–329. [CrossRef]
38. Charles, M. Adverbials of Result: Phraseology and Functions in the Problem–Solution Pattern. *J. Engl. Acad. Purp.* **2011**, *10*, 47–60. [CrossRef]
39. Winter, E.O. A Clause-Relational Approach to English Texts: A Study of Some Predictive Lexical Items in Written Discourse. *Instr. Sci.* **1977**, *6*, 1–92. [CrossRef]
40. van Dijk, T.A. *Text and Context: Explorations in the Semantics and Pragmatics of Discourse*; Longman: London, UK, 1977; ISBN 978-0-582-55085-8.
41. Hoey, M. On the Surface of Discourse. *Language* **1983**, *61*, 734–735. [CrossRef]
42. Hoey, M. *Textual Interaction: An. Introduction to Written Discourse Analysis*, 1st ed.; Routledge: London, UK; New York, NY, USA, 2000; ISBN 978-0-415-23169-5.
43. Kurup, U.; Bignoli, P.G.; Scally, J.R.; Cassimatis, N.L. An Architectural Framework for Complex Cognition. *Cogn. Syst. Res.* **2011**, *12*, 281–292. [CrossRef]
44. Schön, D.A. *The Reflective Practitioner: How Professionals Think in Action*; Routledge: London, UK, 1992; ISBN 978-1-85742-319-8.
45. Smith, R.P.; Eppinger, S.D. Identifying Controlling Features of Engineering Design Iteration. *Manag. Sci.* **1997**, *43*, 276–293. [CrossRef]
46. Thomke, S.; Fujimoto, T. The Effect of "Front-Loading" Problem-Solving on Product Development Performance. *J. Prod. Innov. Manag.* **2000**, *17*, 128–142. [CrossRef]
47. Swales, J. *Genre Analysis: English in Academic and Research Settings*, 1st ed.; Cambridge University Press: Cambridge, UK, 2014; ISBN 978-0-521-33813-4.
48. Heffernan, K.; Teufel, S. Identifying Problems and Solutions in Scientific Text. *Scientometrics* **2018**, *116*, 1367–1382. [CrossRef]
49. Heffernan, K.; Teufel, S. *Identifying Problem Statements in Scientific Text*; University of Potsdam: Potsdam, Germany, 2016. [CrossRef]

50. Haq, A.U.; Li, J.; Memon, M.; Khan, J.; Din, S.U.; AHAD, I.; Sun, R.; Lai, Z. Comparative Analysis of the Classification Performance of Machine Learning Classifiers and Deep Neural Network Classifier for Prediction of Parkinson Disease. In Proceedings of the 2018 15th International Computer Conference on Wavelet Active Media Technology and Information Processing (ICCWAMTIP), Chengdu, China, 14–16 December 2018.
51. Apruzzese, G.; Colajanni, M.; Ferretti, L.; Guido, A.; Marchetti, M. On the Effectiveness of Machine and Deep Learning for Cyber Security. In Proceedings of the 2018 10th International Conference on Cyber Conflict (CyCon), Tallinn, Estonia, 29 May–1 June 2018. [CrossRef]
52. Franchi, G.; Fehri, A.; Yao, A. Deep Morphological Networks. *Pattern Recogn.* **2020**, *102*, 107246. [CrossRef]
53. Zamora, E.; Sossa, H. Dendrite Morphological Neurons Trained by Stochastic Gradient Descent. *Neurocomputing* **2016**, *260*, 420–431.
54. Arce, F.; Zamora, E.; Azuela, J.H.S.; Barrón, R. Differential Evolution Training Algorithm for Dendrite Morphological Neural Networks. *Appl. Soft Comput.* **2018**, *68*, 303–313. [CrossRef]
55. Sossa, H.; Guevara, E. Efficient Training for Dendrite Morphological Neural Networks. *Neurocomputing* **2014**, *131*, 132–142. [CrossRef]
56. Sussner, P.; Campiotti, I. Extreme Learning Machine for a New Hybrid Morphological/Linear Perceptron. *Neural Netw.* **2020**, *123*, 288–298. [CrossRef] [PubMed]
57. Jenkinson, G.; Khezeli, K.; Oliver, G.R.; Kalantari, J.; Klee, E.W. Universally Rank Consistent Ordinal Regression in Neural Networks. *arXiv* **2021**, arXiv:2110.07470.
58. Peroni, S.; Shotton, D. OpenCitations, an Infrastructure Organization for Open Scholarship. *Quant. Sci. Stud.* **2020**, *1*, 428–444. [CrossRef]
59. Sinha, A.; Shen, Z.; Song, Y.; Ma, H.; Eide, D.; Hsu, B.-J. (Paul); Wang, K. An Overview of Microsoft Academic Service (MAS) and Applications. In Proceedings of the 24th International Conference on World Wide Web, Florence, Italy, 18–22 May 2015; Association for Computing Machinery: New York, NY, USA, 2015; pp. 243–246.
60. Tang, J.; Zhang, J.; Yao, L.; Li, J.; Zhang, L.; Su, Z. ArnetMiner: Extraction and Mining of Academic Social Networks. In Proceedings of the 14th ACM SIGKDD International Conference on Knowledge Discovery and Data Mining, Las Vegas, NV, USA, 24–27 August 2008; Association for Computing Machinery: New York, NY, USA, 2008; pp. 990–998.
61. NCBI Resource Coordinators Database Resources of the National Center for Biotechnology Information. *Nucleic Acids Res.* **2018**, *46*, D8–D13. [CrossRef]
62. Europe PMC. Over 15,300 Full Text COVID-19 Now Available in Europe PMC. Available online: http://blog.europepmc.org/2021/02/full-text-covid19-preprints.html (accessed on 15 October 2021).
63. Lo, K.; Wang, L.L.; Neumann, M.; Kinney, R.; Weld, D. S2ORC: The Semantic Scholar Open Research Corpus. In Proceedings of the 58th Annual Meeting of the Association for Computational Linguistics, Online, 5–10 July 2020; pp. 4969–4983.
64. Lu Wang, L.; Lo, K.; Chandrasekhar, Y.; Reas, R.; Yang, J.; Eide, D.; Funk, K.; Kinney, R.; Liu, Z.; Merrill, W.; et al. CORD-19: The Covid-19 Open Research Dataset. *arXiv* **2020**, arXiv:2004.10706.
65. ACL Anthology. Available online: https://www.aclweb.org/anthology/ (accessed on 12 May 2021).
66. McKeown, K.; Daume, H.; Chaturvedi, S.; Paparrizos, J.; Thadani, K.; Barrio, P.; Biran, O.; Bothe, S.; Collins, M.; Fleischmann, K.R.; et al. Predicting the Impact of Scientific Concepts Using Full-Text Features. *J. Assoc. Inf. Sci. Technol.* **2016**, *67*, 2684–2696. [CrossRef]
67. Mikolov, T.; Chen, K.; Corrado, G.; Dean, J. Efficient Estimation of Word Representations in Vector Space. *arXiv* **2013**, arXiv:1301.3781.
68. Breiman, L.; Spector, P. Submodel Selection and Evaluation in Regression—The X-Random Case. *Int. Stat. Rev.* **1991**, *60*, 291–319. [CrossRef]
69. Stone, M. Cross-Validatory Choice and Assessment of Statistical Predictions. *J. R. Stat. Society. Ser. B* **1974**, *36*, 111–147. [CrossRef]
70. Xu, Y.; Goodacre, R. On Splitting Training and Validation Set: A Comparative Study of Cross-Validation, Bootstrap and Systematic Sampling for Estimating the Generalization Performance of Supervised Learning. *J. Anal. Test.* **2018**, *2*, 249–262. [CrossRef]
71. Jung, Y. Multiple Predicting K -Fold Cross-Validation for Model Selection. *J. Nonparametric Stat.* **2017**, *30*, 1–19. [CrossRef]
72. Kohavi, R. A Study of Cross-Validation and Bootstrap for Accuracy Estimation and Model Selection. In Proceedings of the Fourteenth International Joint Conference on Artificial Intelligence (II), Montreal, QC, Canada, 20–25 August 1995; Volume 2, pp. 1137–1143. [CrossRef]
73. Pedregosa, F.; Varoquaux, G.; Gramfort, A.; Michel, V.; Thirion, B.; Grisel, O.; Blondel, M.; Prettenhofer, P.; Weiss, R.; Dubourg, V.; et al. Scikit-Learn: Machine Learning in Python. *J. Mach. Learn. Res.* **2011**, *12*, 2825–2830.
74. Géron, A. *Hands-On Machine Learning with Scikit-Learn, Keras, and TensorFlow*, 2nd ed. Available online: https://www.oreilly.com/library/view/hands-on-machine-learning/9781492032632/ (accessed on 13 May 2021).
75. Schmidhuber, J. Deep Learning in Neural Networks: An Overview. *Neural Netw.* **2014**, *61*, 85–117. [CrossRef] [PubMed]
76. Sutskever, I.; Vinyals, O.; Le, Q. Sequence to Sequence Learning with Neural Networks. In Proceedings of the 27th International Conference on Neural Information Processing Systems, Montreal, QC, Canada, 8–13 December 2014; MIT Press: Cambridge, MA, USA, 2014; p. 10.
77. Hassabis, D.; Kumaran, D.; Summerfield, C.; Botvinick, M. Neuroscience-Inspired Artificial Intelligence. *Neuron* **2017**, *95*, 245–258. [CrossRef] [PubMed]

78. Nwadiugwu, M.C. Neural Networks, Artificial Intelligence and the Computational Brain. *arXiv* **2020**, arXiv:2101.08635.
79. Núñez, J.C.; Cabido, R.; Pantrigo, J.J.; Montemayor, A.S.; Vélez, J.F. Convolutional Neural Networks and Long Short-Term Memory for Skeleton-Based Human Activity and Hand Gesture Recognition. *Pattern Recognit.* **2018**, *76*, 80–94. [CrossRef]
80. Vivekanandan, K.; Praveena, N. Hybrid Convolutional Neural Network (CNN) and Long-Short Term Memory (LSTM) Based Deep Learning Model for Detecting Shilling Attack in the Social-Aware Network. *J. Ambient Intell. Hum. Comput.* **2021**, *12*, 1197–1210. [CrossRef]
81. Gulli, A.; Pal, S. *Deep Learning with Keras*; Packt Publishing: Birmingham, UK, 2017; ISBN 978-1-78712-842-2.
82. Weerts, H.J.P.; Mueller, A.C.; Vanschoren, J. Importance of Tuning Hyperparameters of Machine Learning Algorithms. *arXiv* **2020**, arXiv:2007.07588.
83. Yu, T.; Zhu, H. Hyper-Parameter Optimization: A Review of Algorithms and Applications. *arXiv* **2020**, arXiv:2003.05689.

applied sciences

MDPI

Article

Ternion: An Autonomous Model for Fake News Detection

Noman Islam [1], Asadullah Shaikh [2], Asma Qaiser [3], Yousef Asiri [2,*], Sultan Almakdi [2,*], Adel Sulaiman [2,*], Verdah Moazzam [3] and Syeda Aiman Babar [3]

1 Department of Computer Science, Iqra University, Karachi 76400, Pakistan; noman.islam@iuk.edu.pk
2 College of Computer Science and Information Systems, Najran University, Najran 61441, Saudi Arabia; asshaikh@nu.edu.sa
3 Department of Computer Science, NED University of Engineering and Technology, Karachi 76400, Pakistan; pg3200224@cloud.neduet.edu.pk (A.Q.); pg3200165@cloud.neduet.edu.pk (V.M.); pg3200184@cloud.neduet.edu.pk (S.A.B.)
* Correspondence: yasiri@nu.edu.sa (Y.A.); saalmakdi@nu.edu.sa (S.A.); aaalsulaiman@nu.edu.sa (A.S.)

Abstract: In recent years, the consumption of social media content to keep up with global news and to verify its authenticity has become a considerable challenge. Social media enables us to easily access news anywhere, anytime, but it also gives rise to the spread of fake news, thereby delivering false information. This also has a negative impact on society. Therefore, it is necessary to determine whether or not news spreading over social media is real. This will allow for confusion among social media users to be avoided, and it is important in ensuring positive social development. This paper proposes a novel solution by detecting the authenticity of news through natural language processing techniques. Specifically, this paper proposes a novel scheme comprising three steps, namely, stance detection, author credibility verification, and machine learning-based classification, to verify the authenticity of news. In the last stage of the proposed pipeline, several machine learning techniques are applied, such as decision trees, random forest, logistic regression, and support vector machine (SVM) algorithms. For this study, the fake news dataset was taken from Kaggle. The experimental results show an accuracy of 93.15%, precision of 92.65%, recall of 95.71%, and F1-score of 94.15% for the support vector machine algorithm. The SVM is better than the second best classifier, i.e., logistic regression, by 6.82%.

Keywords: fake news detection; natural language processing; machine learning; stance detection; social media

Citation: Islam, N.; Shaikh, A.; Qaiser, A.; Asiri, Y.; Almakdi, S.; Sulaiman, A.; Moazzam, V.; Babar, S.A. Ternion: An Autonomous Model for Fake News Detection. *Appl. Sci.* **2021**, *11*, 9292. https://doi.org/10.3390/app11199292

Received: 22 August 2021
Accepted: 27 September 2021
Published: 6 October 2021

1. Introduction

Fake news detection has always been a problem because of its long-term repercussions and consequences. Its root can be traced back to the 17th century in propaganda, which became misinformation in the cold war [1]. In modern days, this problem has become grave due to the emergence of social media platforms. Specifically, in the past few years, social media channels, such as Facebook, Twitter, and Instagram, have emerged as platforms for quick dissemination and retrieval of information. Figure 1 shows a snapshot of some fake news in recent years. According to various studies [2], almost 50% of the population of developed nations depend on social media for news. The importance of social media cannot be denied, and it has emerged as an effective medium at the time of crises in regard to the role it plays in breaking news, for example [3]. However, one drawback of the convenience provided by social media is the quick dissemination of fake news.

Figure 1. Examples of some fake news [4].

In contrast to conventional mediums such as print media or television, the content of social media can be modified by users, thereby enriching the content with their opinions or biases. This can alter the meaning or context of the news altogether [5]. According to various studies, social media is a fertile ground for quick sharing of information without fact checking [1].

Fake news can be defined as the creation or modification of news content by social media user to deliberately or non-deliberately change its apparent meaning or context, contaminating it with their opinion or biases, where the intent may be to jeopardize or harm a person, organization, or society, monetarily or morally. Examples of fake news are sarcasm, memes, fake advertisements, fake political statements, and rumors [3]. A fakester is a term used for a person responsible for spreading fake news. News can have various degrees based on its credibility, i.e., true, half-true, and false [5]. Fake news can be transmitted in the form of images, video, and text. The life cycle of fake news has been described in [6] as the creation, publication, and propagation of the news.

The impact of fake news spread on social media is immense [7]. It can cause a decline in stock prices, a drop in potential investments, etc. [6]. For instance, the 2016 US election was heavily impacted by fake news [2]. The fake news about the death of President Obama led to the loss of USD 130 billion in the stock market in just a fraction of time. The intent of fake news may be to malign someone for political or personal intent or to mislead people [6]. There are numerous websites used for detecting fake news, such as FactCheck, Snopes, TruthorFiction, and PolitiFact. Moreover, Google has also launched an initiative called Google News Initiative to counter fake news [3]. However, fake news detection is still a cumbersome task. This is because fake news often contains misleading information contaminated with credible facts [2]. The motivation behind fake news can be driven by politics, financial benefit, or ideology [3,5]. In the literature, various approaches based on linguistic features or deep learning techniques, such as the recurrent neural network, convolutional neural network, transformer, bidirectional encoder representations from transformers (BERT), and their combination, have been used for fake news detection [8]. Detection of fake news can be classified as a binary or multi-class classification problem. Alternatively, it can be modeled as a regression problem. A number of datasets are also available for fake news classification, such as Kaggle, ISOT, and LIAR [3].

Despite the extensive studies being carried out, the problem of fake news detection is still very challenging, and it is believed that it requires a comprehensive multi-phased approach. Addressing this problem, this paper proposes a novel approach to validate the authenticity of news. The approach comprises first detecting the stance of the news, then identifying the author's credibility, and finally using machine learning to classify the news as fake or authentic. The objective of the research is to classify news as fake or genuine based on various attributes, such as the text of the news and its author's profile.

The potential implications of the proposed work are multifold. As discussed earlier, fake news related to medical symptoms can have severe consequences if assumed true by its consumer. Similarly, fake news can lead to irreparable damage in rgw health, political, social and economic sectors. By using the proposed approach, this catastrophic effect can be avoided. This study also serves as a baseline and opens up avenues for future research on fake news detection. There is a scarcity of research related to use of a three-pronged approach to fake news classification. Research based on machine learning and deep learning is being extensively carried out to identify a novel solution to the issue of fake news detection. The current paper proposes a three-step solution. We have not found any such study in the past. Finally, based on the proposed work, a commercial tool can be developed that can tag news as fake and also provide appropriate ratings on its credibility.

The remainder of this paper is structured as follows: Section 2 presents related work; Section 3 describes the proposed novel approach to detect fake news; the experimental results are discussed in Section 4; and, finally, Section 5 provides the conclusions and future directions.

2. Related Work

In recent years, several approaches have been identified to establish with a solution to the issue of the detection of fake news. Primarily, they are classified as machine learning approaches, hybrid approaches, topic-agnostic approaches, knowledge-based approaches, and language approaches [1]. The authors of [7] classified the approaches as news content-based learning and social context-based learning. The former is based on the styles of the news being published, while the latter is based on latent information provided to a user by a news article. Users present on social media play an active in the identification of fake news. For example, Facebook ranks the comments on a post based on the number of replies or user engagement for a particular post [6]. An analysis of the existing literature revealed that there is major work in the direction of stance detection, identifying authors' credibility, and using machine learning to classify news as fake or not. Hence, we discuss the work in these three directions below. Interested readers are directed to [9] for a comprehensive survey.

2.1. Stance Detection

Among many natural language processing tasks, stance detection is a very important task. It can be the very first step in fact checking [10,11]. In 2016, an online contest was started known as the fake news challenge [12]. The objective of this challenge was to encourage the improvement of devices that may help human fact checkers to recognize intentional falsehood in reports using artificial intelligence (AI), regular language handling, and artificial knowledge. In this challenge, stance detection is regarded as stage 1 in the identification of fake news. The main aim is to determine the relevancy of a news article headline and its body. Chaudhary [13] et al. discussed numerous deep neural network-based models for stance detection. They found that using a pre-trained global vector for word representation (GloVe) and word embedding along with a long short-term Memory (LSTM)-based bidirectional condition encoding model provided the best performance with 97% accuracy.

Bhatt et al. [14] presented a novel approach combining neural, external, and statistical features. With the help of feature engineering heuristics, handcrafted external features and statistical features from the n-gram bag-of-words model, and the deep recurrent model, the neural embedding was computed. Bourgonje et al. [15] worked on a system that used a lemmatization-based n-gram approach to carry out binary classification of headlines and article sets. They achieved the best accuracy of the system using logistic regression. In [16], the authors proposed a method to detect spam comments on YouTube by using different machine learning algorithms with the n-gram approach, and they proved that this technique is effective in detecting spam comments. García et al. [17] introduced a system for text classification that executes embedded feature elimination via an a priori algorithm. The aim of their study was to speed up the word sequence constructions by minimizing the explored branches' number as much as possible.

In order to classify fake news, Saikh et al. [18] used the technique of stance detection with textual entailment (TE). Moreover, they proposed a system that used a combination of deep learning and statistical machine learning approaches. To detect a stance in fake news, Ghanem et al. [19] combine n-gram, lexical features, and word embedding. They accomplished state-of-the-art results (59.6% Macro F1) on the FNC-1 dataset [20]. In [21], a deep neural network architecture was used to predict the stance of a headline and article body.

2.2. Author Credibility

Research suggests that information related to the authors of articles helps to identify whether the news presented is fake or not. Hence, another area of research is identifying author credibility. Sitaula et al. [2] discussed different attributes that could help to determine author credibility and its role in news. With the attributes explained, they identified 26 features that were obtained in different categories. This paper's results show not only the credibility of a given article but also the credibility of articles published by the same author. According to [22], author credibility plays a very important role in identifying fake reviews online. However, most users do not consider author credibility before sharing news on social media [23].

Research suggests that information related to the authors of articles helps to identify whether the presented news is fake or not. Hence, another area of research is identifying author credibility. Sitaula et al. [2] discussed different attributes that could help to determine author credibility and its role in news. With the attributes explained, they identified 26 features that were obtained in different categories. This paper's results show not only the credibility of a given article but also the credibility of articles published the same author. Another work related to author profiling is mentioned in [24]. A corpus of Twitter data was used for this purpose. According to [22], author credibility plays a very important role in identifying fake reviews online. However, most users do not consider author credibility before sharing news on social media [23]. Therefore, the work on author credibility can be considered to be in the stage of infancy and regarded as an open research challenge in various fields [25].

2.3. Machine Learning-Based Classification

In a considerable amount of research, machine learning algorithms have been used for fake news detection. The credibility of fake news is one of the most important discussions, and many approaches have evolved with time for its detection. To detect fake news in online text, Girgis [26] et al. utilized deep learning algorithms, such as LSTMs and RNN. Models (vanilla and GRU) were implemented on the LIAR dataset. Among all algorithms, GRU showed the best performance, so in order to achieve better accuracy, a hybrid model was developed using the techniques of CNN and GRU on the dataset. For the detection of fake news, Shlok et al. and Gilda [27] applied different machine learning approaches. More machine learning techniques for the detection of fake news can be found in [28–30].

Ajao et al. [31] used a long short-term recurrent neural network and hybrid between convolutional neural network models. They implemented various deep neural networks: (1) LSTM, (2) LSTM along with dropout regularization, and (3) LSTM-CNN. Among all approaches, LSTM stands out and gives 82% accuracy. Sajjad et al. [32] provided a model of decent accuracy to identify fake news using a framed model combined with knowledge engineering and machine learning. In another work, automated discovery of social news is proposed, utilizing three-element extraction procedures, a count vectorizer, term frequency–inverse document frequency, and a hashing vectorizer [4]. An ensemble-based technique for fake news detection is presented in [33]. Ensemble-based approaches combined various weak classifiers to achieve better accuracy for combined classification tasks.

In [34], various machine learning algorithms, such as logistic regression, naive Bayes, and random forest classification, are used.

In [31], a deep learning technique called Fake-BERT was used for the detection of fake news. In [6], a deep learning-based model, EchoFakeD, was proposed with a mix of content and contextual features. The authors proposed an effective tensor factorization scheme. In a number of studies, data augmentation, transfer learning, auto-encoders, and other semi-supervised models have been used for fake news detection [8]. A capsule-based neural network was used in [3] to classify fake news. In [35], the authors used geometric deep learning based techniques for fake news detection. These are an extension of the convolutional neural network that fuses other information, such as user profiles, news propagation, and the actual content. A hybrid deep learning model based on the combination of CNN and RNN was presented in [36]. The proposed model utilizes a combination of embedding, CNN, and RNN layers implemented in Keras and tested on ISO and FA-KES datasets. In [37], blockchain technology was used for the detection of fake news.

In recent years, following the spread of COVID-19, several pieces of fake news have spread in this context. Therefore, numerous studies have focused on the detection of news related to COVID-19. For instance, a novel approach to the detection of fake tweets related to COVID-19 was proposed in [8]. In a similar direction, an analysis of public sentiments based on tweets related to COVID-19 was performed in [38]. In [36], several supervised learning approaches, such as CNN, LSTM, and BERT, were used for the detection of fake news related to COVID-19. Moreover, unsupervised learning techniques, such as model pre-training and distributed word representations, were used.

After an extensive review of the literature, it was found that most of the studies on this topic have focused on stance detection, author credibility, and classification of news. However, existing approaches are limited because of the lack of social or political context awareness underlying the news. Therefore, a multi-stage pipeline is required for the correct classification of the credibility of news. This paper presents a novel approach, combining stance detection, author credibility, and news classification. This approach is motivated by [34], a study in which several machine learning algorithms are used for classification. The objective of this study is to spot fake news on a social medial platform, i.e., Twitter. Similar studies focusing on a specific platform have been conducted [35,39,40].

3. Proposed Approach and Implementation Details

This paper proposed a novel approach to fake news detection. The proposed method comprises the following modules: (1) data collection, (2) pre-processing, (3) feature extraction, and (4) inference engine. The architecture of this fake news detector is depicted in Figure 2.

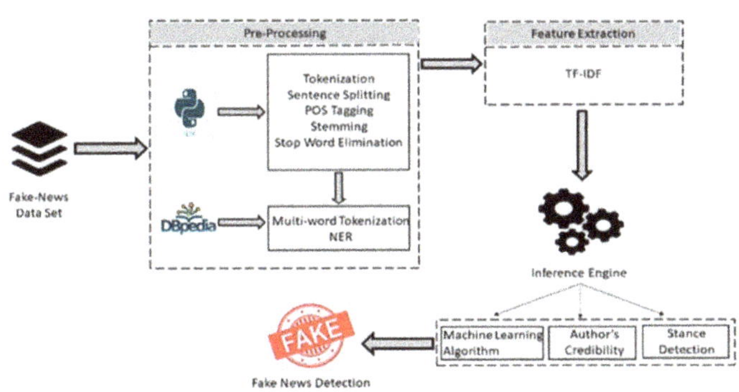

Figure 2. Fake news detection approach.

3.1. Dataset Description

For this paper, a dataset called the fake news dataset [14] is selected from Kaggle. The dataset contains five features, namely "Id", "Title", "Text", "Author", and "Label." The dataset has 20718 entries, of which 10349 entries are deemed fake news and the remaining are real news. A description of the dataset is provided in Table 1. A few records of the dataset are displayed in Figure 3. The extracted data from the dataset were passed through the pre-processing module. By using the Natural Language Tool Kit (NLTK) library [19], the text was divided sentence by sentence in tokens. This was followed by Parts of Speech (PoS) tagging, lemmatization, stop word elimination, and Named Entity Recognition (NER). In this module, the proposed model not only identifies traditional NER (i.e., name, location, and organization), but it also recognizes multiple NER, such as movies, book titles, cartoons, etc. This extension of NER is achieved by utilizing DBpedia.

Table 1. Description of the dataset.

Column	Description
Id	A unique Id assigned to each piece of news
Title	The title of the news
Text	News text
Label	The label of the news

id	title	author	text	label
0	House De	Darrell Lu	House	1
1	FLYNN: Hi	Daniel J. F	Ever get tl	0
2	Why the T	Consortiu	Why the	1
3	15 Civilian	Jessica Pu	Videos	1
4	Iranian wc	Howard P	Print	1
5	Jackie Ma:	Daniel Nu	In these tı	0
6	Life: Life C	nan	Ever	1
7	BenoÃ®t	Alissa J. R	PARIS â€"	0
8	Excerpts F	nan	Donald J.	0
9	A Back-Ch	Megan Tw	A week be	0
10	Obamaâ€	Aaron Kle	Organizin	0
11	BBC Come	Chris Tom	The BBC p	0
12	Russian R	Amando F	The	1
13	US Officia	Jason Ditz	Clinton	1
14	Re: Yes, Tl	AnotherA	Yes,	

Figure 3. A snapshot of the dataset.

A word cloud was made for the headline and body text of fake and real news in the selected dataset, and it is shown in Figure 4. Word cloud is a visualization technique of word frequency. The more regularly terms show up in the content being assessed, the bigger the word in the image created. For machine learning with fake news detection, pre-processed text documents should be represented in vector form. To convert text into features, machine learning provides a variety of options in which classifiers use Bags of Word (BoW) along with the TF-IDF vectorizer. Furthermore, the data were split into train, validation, and test datasets.

(**a**) Fake news word cloud. (**b**) Real news word cloud.

Figure 4. Word cloud of the various news articles.

3.2. Proposed Approach: Inference Engine

This section discusses the proposed multi-stage approach, i.e., (1) stance detection, (2) author credibility verification, and (3) machine learning-based classification.

During stance detection, the very first step in the inference engine, it is determined whether or not the headline and the body of a news article are relevant or not. Listing 1 shows the pseudo-code of stance detection. In order to find relevancy, the cosine similarity technique is implemented, which is used to find similarity between two text documents irrespective of their size. If their headlines and body texts are similar, then one can proceed to the next module, i.e., author credibility; otherwise, the model declares that the examined news is fake news. In NLP, it is a well-informed and popular approach. It allows for detection in favor of the audience, and from the text, it determines whether the audience found the objective to be against, in favor of, or impartial to the target [41]. The objective could be an individual, an association, an administration strategy, a development, an item, and so forth.

Listing 1. Stance detection.

```
def get_vectors(title ,text):
    vocab = [title ,text]
    vectorizer = CountVectorizer(vocab)
    vectorizer.fit(vocab)
    return (vectoriz-er.transform([title]).toarray(),vectorizer.
    transform([text]).toarray())

def stance_detection(row):
    global total, fake
    title, text = get_vectors(row['title'],row['text'])
    total += 1
    if(p.cosine_similarity(title ,text) < 0.25):
        fake += 1

frame.apply(stance_detection ,axis=1)
```

The next step is the verification of author credibility. In this module, the inference engine validates an author's information to judge whether the news is fake or not. Twitter API [42] is used to obtain the author's Twitter profile. It first checks how many followers the author has and then checks how many times this news has been retweeted.

Priya Gupta et al. in [41] described different features of evaluating the believability of client-produced content on Twitter, and a novel continuous framework to survey the trustworthiness of tweets was proposed. The discussed framework was implemented to accomplish this by relegating a score or rating to content on Twitter to show its dependability. The authors of [43] et al. investigated different grouping strategies in order to help versatility, and another solution to the constraints present in previously existing procedures was proposed.

Finally, for fake news detection, four different machine learning algorithms are applied. In this paper, we compare the results of all four algorithms. The selected algorithms are as follows:

- A decision tree is one of the most popular classifiers that helps in prediction and classification, and it is supervised in nature. It splits the dataset by recursively selecting features. The selected features of the dataset can be in nominal or continuous form. This is a well-known classifier for data classification. The most distinct feature is the conversion of the process of complex decisions in order to simplify the process definition, and, as a result, it provides an easy way to understand and interpret the outcome [44].
- Random forest is a regulated AI method that is supervised in nature. On the basis of random element choice, a set of decision trees (base classifiers) is produced, and the dominant party with respect to voting is selected for classification. It generates accurate and diverse decisions that are dynamic algorithms for this classifier [45]. In a random forest, the individual decision trees are an ensemble, and they operate on average to increase the accuracy of the prediction of the model. This model also focuses on the reduction in over-fitting. The sub-samples are drawn with replacement, keeping their size the same as the original input sample size.
- Logistic regression is an AI technique for classification. In this algorithm, the probabilities portraying the potential results of the possible outcomes are demonstrated utilizing a logistic function. It is widely used in circumstances in which humans are not suited to perform the classification and automated functionality is required for this purpose [46].
- The support vector machine (SVM) is known as a supervised learning algorithm that is widely used to predict or classify data. Its classifier is officially characterized by an isolating hyperplane. That is, the labeled dataset for training is required, and the algorithm yields an ideal hyperplane that generates new examples. In two-dimensional space, this hyperplane is a line separating a plane in two sections where each class is located on one of the two sides. SVM carries out generous upgrades and best-performing strategies, and it can be applied to a wide range of learning tasks. Moreover, it is completely programmed, eliminating the requirement for manual parameter tuning [47].

Figure 5 presented below shows the complete workflow of the implemented model.

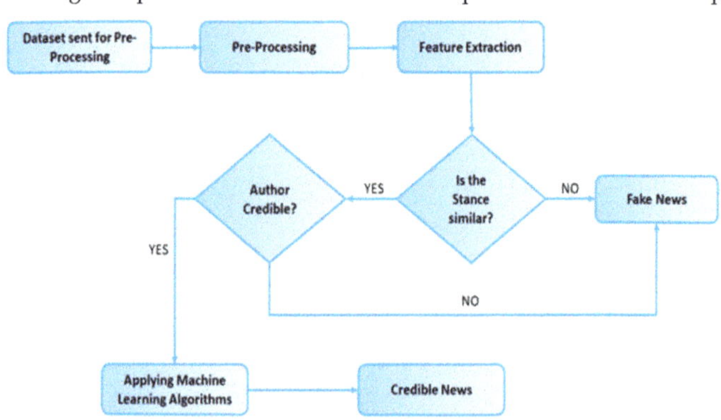

Figure 5. Flow diagram of the architecture.

4. Experimental Results

For experiments, the authors of this paper implemented the proposed approach in Python. To begin the experiment, the selected dataset was passed through the proposed pipeline. Initially, the pre-processing step was performed by using the NLTK library. Stance detection and author credibility were then determined. During the author credibility and stance detection phases, 28.88% of the news was classified as fake, among which 8% was in fact genuine (Figure 6).

```
Total News =  786
Classified as fake correctly       :  0.203
Classified as fake incorrectly     :  0.085
Classified as not fake correctly   :  0.417
Classified as not fake incorrectly :  0.293
```

Figure 6. Result of authors' credibility and stance detection.

In the last step, different machine learning algorithms were applied to the data after the pre-processed text document was converted into vector form using the TF-IDF vectorizer.

Moreover, different machine learning algorithms were applied to the proposed dataset. The first model applied was a decision tree for the detection of fake news. The performance of the decision tree was represented by a confusion matrix. Figure 7 shows the confusion matrix in a heatmap. A confusion matrix shows the true positive, true negative, false positive, and false negative values in the form of a matrix. The definitions of each of these terms are as follows:

- True positive (*TP*): a classifier prediction is true positive if the news is authentic, and the classifier predicts it as authentic.
- False-positive (*FP*): a classifier prediction is false positive if the news is fake, and the classifier predicts it as authentic.
- True negative (*TN*): a classifier prediction is true negative if the news is fake, and the classifier predicts it as fake.
- False-negative (*FN*): a classifier prediction is false negative if the news is authentic, and the classifier predicts it as fake.

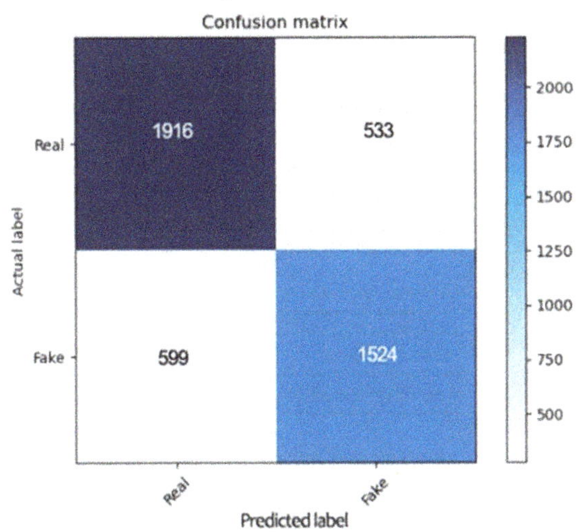

Figure 7. Confusion matrix for decision tree algorithm.

It can be seen that for the decision tree, *TP* is 1916, and *TN* is 1524. Hence, the overall accuracy is as follows:

$$\text{Accuracy} = \frac{TP + TN}{TP + TN + FP + FN} \tag{1}$$

$$\text{Accuracy} = \frac{1916 + 1524}{4572}$$
$$\text{Accuracy} = 75.24\%$$

In many situations, accuracy is not a very good measure. Hence, it is essential to calculate other measures, such as precision, recall, and F1-score. The definitions of these terms are as follows:

- Precision: the ratio of positive examples that were correctly predicted by the classifier to the total number of examples predicted as positive.
- Recall: the ratio of the total number of true positives to the actual number of examples that were positive.
- F1-score: the weighted average score of precision and recall.

The precision of the classifier is defined mathematically as

$$\text{Precision} = \frac{TP}{TP + FP} \tag{2}$$

$$\text{Precision} = \frac{1916}{1916 + 599}$$
$$\text{Precision} = 76.18\%$$

The recall of the classifier is defined mathematically as

$$\text{Recall} = \frac{TP}{TP + FN} \tag{3}$$

$$\text{Recall} = \frac{1916}{1916 + 533}$$
$$\text{Recall} = 78.23\%$$

Finally, F1-score is meant to balance precision and recall. It is defined as

$$\text{F1} = 2 \times \frac{Precision \times Recall}{Precision + Recall} \tag{4}$$

$$\text{F1} = 2 \times \frac{76.18 \times 78.23}{76.18 + 78.23}$$
$$\text{F1} = 77.19\%$$

The confusion matrix for random forest classifier, as illustrated in Figure 8, shows that the accuracy of the classifier is 82.23%, the precision value is 81.95%, the recall is 84.44%, and the F1-score is 83.17%.

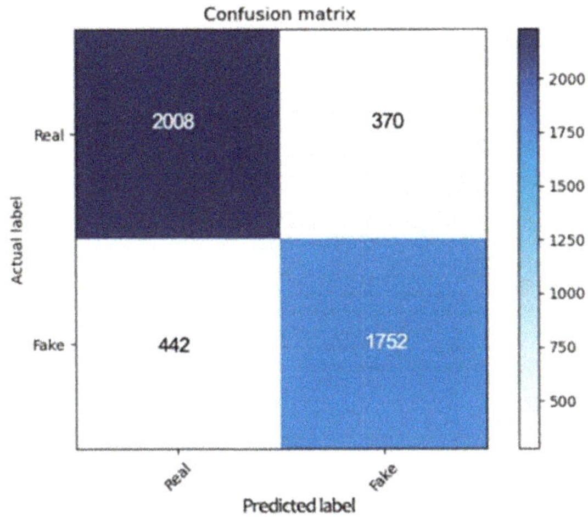

Figure 8. Confusion matrix for random forest algorithm.

The confusion matrix and accuracy of this logistic regression classifier, as illustrated in Figure 9, shows that the accuracy of the classifier is 87.2%, the precision value is 87.90%, the recall is 88.88%, and the F1-score is 88.30%.

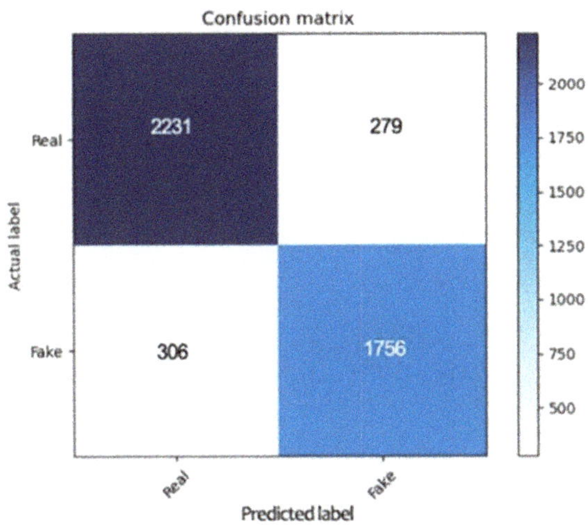

Figure 9. Confusion matrix for logistic regression algorithm.

Lastly, an SVM classifier was applied. The confusion matrix and the accuracy of this classifier are shown in Figure 10, and it can be observed that the accuracy of the classifier is 93.15%, the precision value is 92.65%, the recall value is 95.71%, and the F1-score is 94.15%.

After implementing all of the classifiers, their results were compared, and it was observed that all of the experiments conducted using the support vector machine provide the best accuracy for the proposed fake news detector and perform better than the other classifiers with an accuracy of 93.15%, precision of 92.65%, recall of 95.71%, and F1-score of 94.15%. Table 2 and Figure 11 provide a comparison of various aspects of the classifier. Comparing the SVM with logistic regression, which was the second best classifier, it can be observed that SVM is better than logistic regression in terms of accuracy as follows:

$$\text{Improvement in accuracy} = \frac{93.15 - 87.20}{87.20}$$
$$\text{Improvement in accuracy} = 6.82\%$$

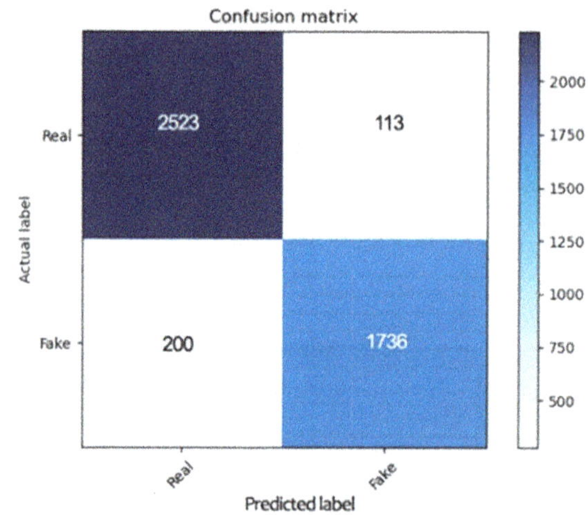

Figure 10. Confusion matrix for SVM.

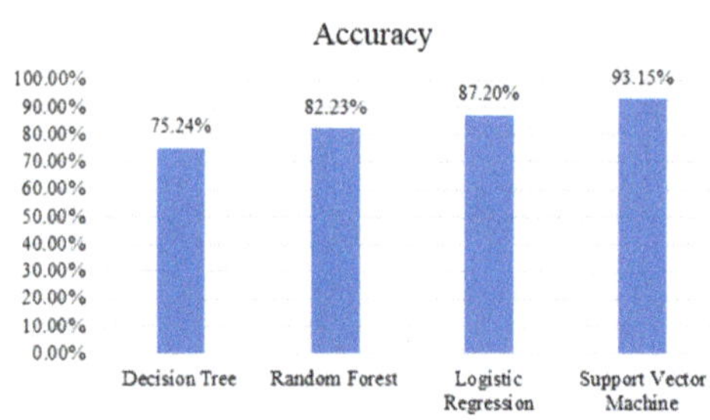

(**a**) Comparison of accuracy of classifiers.

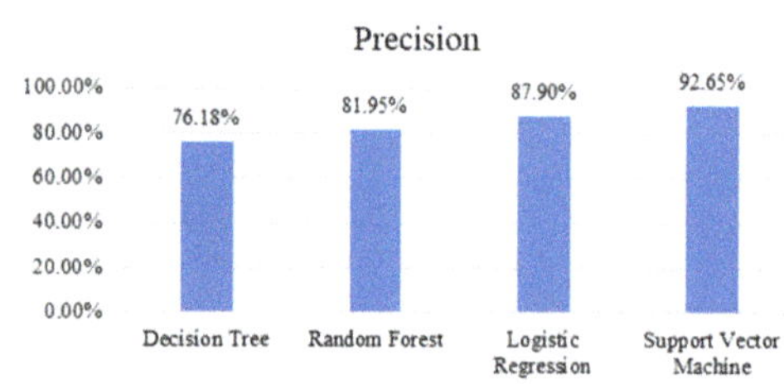

(**b**) Comparison of precision of classifiers.

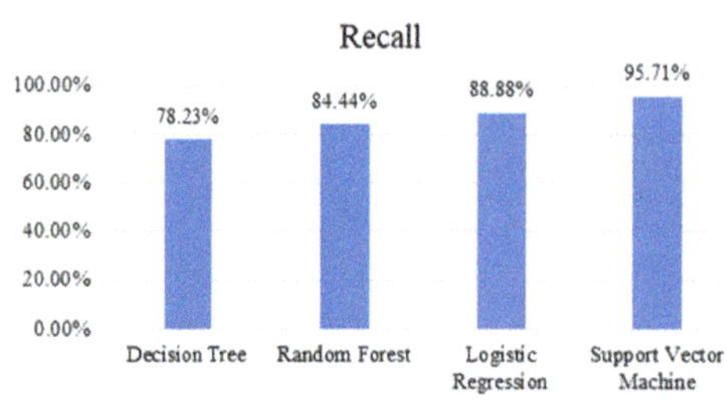

(**c**) Comparison of recall of classifiers.

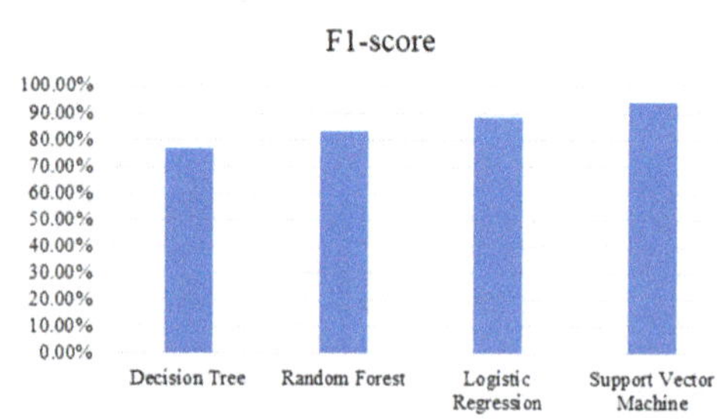

(**d**) Comparison of F1 score of classifiers.

Figure 11. Accuracy, precision, recall, and F1-score of various classifiers.

Table 2. Comparison of classifier performance.

Machine Learning Algorithm	*TP*	*FP*	*FN*	*TN*	Accuracy	Precision	Recall	F1
Decision Tree	1916	599	533	1524	75.24%	76.18%	78.23%	77.19%
Random Forest	2008	442	370	1752	82.23%	81.95%	84.44%	83.17%
Logistic Regression	2231	306	279	1756	87.20%	87.90%	88.88%	88.30%
Support Vector Machine	2523	200	113	1736	93.15%	92.65%	95.71%	94.15%

5. Conclusions and Future Work

The detection of fake news on social media platforms is an essential topic of discussion considering the wide dissemination of news and the number of people consuming information through it. In this paper, a solution is proposed based on natural language processing and machine learning for a fake news dataset produced by Kaggle. The proposed approach is based on stance detection, author credibility, and machine learning algorithms. Stance detection verifies the relevancy between the title and paragraphs of a news article; if there is a match, the next module checks whether the author is authentic in order to determine whether or not the news should be believed. Finally, machine learning algorithms, i.e., logistic regression, support vector machine, decision tree, and random forest, are implemented, and among these, the support vector machine stands out with an accuracy of 93.15%.

In modern day, access to the internet has become ubiquitous. In just one minute on the internet, 18 million text messages are exchanged over WhatsApp, 2.4 million snaps are created on SnapChat, 38 million SMS messages and 187 million emails are sent, and 0.5 million tweets are posted [48]. Unfortunately, most of the population is dependent on the consumption of information from the internet. Hence, fake news detection has become a major concern. Most of the information flow on the internet is unverified and generally assumed true. This can be used to spread misinformation, destabilize a regime, and create riots. It has been predicted that in the next few years, people will consume more false information than true content [21]. Unfortunately, most content analyses cannot address fake news detection because of its challenges. The existing natural language processing techniques are limited because of the absence of the political or social context required to understand the content [35]. Therefore, there is a need for a multi-stage solution that can address this issue in the form of a pipeline. The proposed approach provides a three-pronged solution to verify the authenticity of any news article. After working on the stance and credibility of the author, the solution is then formulated to address a machine learning problem using any of the tested algorithms, such as SVM, random forest, and decision trees. The main advantages of using machine learning are its ability to learn the rules for the detection of fake news by using data and the fact that the end user is not required to explicitly program these rules.

There are several limitations of the proposed approach that can be worked on in the future. The proposed approach does not consider the correlation among news items. The correlation among news articles can assist in determining the credibility of a news article. Moreover, the author credibility check is based on Twitters' information. This can be extended to include other attributes that are generally not available on social media. The proposed approach can also be extended to the use of advanced deep learning algorithms based on convolutional neural networks, LSTM, GRU, or BERT. Currently, the proposed approach is a sequential pipeline, and news passes through each stage one by one. A novel objective function can be developed based on the scores of stance detection, author credibility, and a machine learning classifier to determine if news is fake or not in a joint fashion. The currently available solutions only mark the news as authentic or unauthentic; however, a working solution requires the score or rating on the credibility of news. The detection of fake news is only one aspect of a bigger problem.

Work regarding the fake news evolution process, its mitigation, and later steps of account detection and deletion must also be conducted.

Author Contributions: Conceptualization, N.I., A.S. (Asadullah Shaikh) and A.Q.; methodology, Y.A., S.A. and A.S. (Adel Sulaiman); software, V.M. and S.A.B.; validation, A.S. (Asadullah Shaikh) and V.M.; formal analysis, S.A. and Y.A.; investigation, A.Q., N.I. and A.S. (Asadullah Shaikh); writing—original draft preparation, N.I. and A.S. (Asadullah Shaikh); writing—review and editing, Y.A. and A.S. (Adel Sulaiman); supervision, V.M. and S.A.B.; funding acquisition, A.S. (Asadullah Shaikh). All authors have read and agreed to the published version of the manuscript.

Funding: The authors would like to acknowledge the support of the Deputy for Research and Innovation—Ministry of Education, Kingdom of Saudi Arabia, for this research through a grant (NU/IFC/INT/01/008) under the institutional Funding Committee at Najran University, Kingdom of Saudi Arabia.

Conflicts of Interest: The authors declare that they have no conflicts of interest to report regarding the present study.

References

1. De Beer, D.; Matthee, M. Approaches to identify fake news: A systematic literature review. In *International Conference on Integrated Science, Cambodia*; Springer: Basel, Switzerland, 2020; pp. 13–22.
2. Sitaula, N.; Mohan, C.K.; Grygiel, J.; Zhou, X.; Zafarani, R. Credibility-based fake news detection. In *Disinformation, Misinformation, and Fake News in Social Media*; Springer: Basel, Switzerland, 2020; pp. 163–182.
3. Goldani, M.H.; Momtazi, S.; Safabakhsh, R. Detecting fake news with capsule neural networks. *Appl. Soft Comput.* **2021**, *101*, 106991. [CrossRef]
4. Kaur, S.; Kumar, P.; Kumaraguru, P. Automating fake news detection system using multi-level voting model. *Soft Comput.* **2020**, *24*, 9049–9069. [CrossRef]
5. Bühler, J.; Murawski, M.; Darvish, M.; Bick, M. Developing a Model to Measure Fake News Detection Literacy of Social Media Users. In *Disinformation, Misinformation, and Fake News in Social Media*; Springer: Basel, Switzerland, 2020; pp. 213–227.
6. Kaliyar, R.K.; Goswami, A.; Narang, P. EchoFakeD: Improving fake news detection in social media with an efficient deep neural network. *Neural Comput. Appl.* **2021**, 33, 8597–8613. [CrossRef] [PubMed]
7. Kaliyar, R.K.; Goswami, A.; Narang, P. FakeBERT: Fake news detection in social media with a BERT-based deep learning approach. *Multimed. Tools Appl.* **2021**, *80*, 11765–11788. [CrossRef] [PubMed]
8. Paka, W.S.; Bansal, R.; Kaushik, A.; Sengupta, S.; Chakraborty, T. Cross-SEAN: A cross-stitch semi-supervised neural attention model for COVID-19 fake news detection. *Appl. Soft Comput.* **2021**, *107*, 107393. [CrossRef]
9. Saxena, A.; Saxena, P.; Reddy, H. Fake News Detection Techniques for Social Media. In *Principles of Social Networking*; Springer: Singapore, 2022; pp. 325–354.
10. Mohammad, S.; Kiritchenko, S.; Sobhani, P.; Zhu, X.; Cherry, C. Semeval-2016 task 6: Detecting stance in tweets. In Proceedings of the 10th International Workshop on Semantic Evaluation (SemEval-2016), San Diego, CA, USA, 16–17 June 2016; pp. 31–41.
11. Riedel, B.; Augenstein, I.; Spithourakis, G.P.; Riedel, S. A simple but tough-to-beat baseline for the Fake News Challenge stance detection task. *arXiv* **2017**, arXiv:1707.03264.
12. Pomerleau, D.; Rao, D. Fake News Challenge Stage 1 (fnc-i): Stance Detection. 2017. Available online: www.fakenewschallenge.org (accessed on 10 May 2021).
13. Chaudhry, A.K.; Baker, D.; Thun-Hohenstein, P. Stance detection for the fake news challenge: Identifying textual relationships with deep neural nets. In *CS224n: Natural Language Processing with Deep Learning*; Lecture Notes; Standaford NLP: Stanford, CA, USA, 2017; pp. 1–117. Available online: http://web.stanford.edu/class/cs224n/ (accessed on 2 October 2021).
14. Bhatt, G.; Sharma, A.; Sharma, S.; Nagpal, A.; Raman, B.; Mittal, A. Combining neural, statistical and external features for fake news stance identification. In Proceedings of the WWW '18: Companion Proceedings of the The Web Conference 2018, Geneva, Switzerland, 23–27 April 2018; pp. 1353–1357.
15. Bourgonje, P.; Schneider, J.M.; Rehm, G. From clickbait to fake news detection: An approach based on detecting the stance of headlines to articles. In Proceedings of the 2017 EMNLP workshop: Natural Language Processing Meets Journalism, Copenhagen, Denmark, 2 May 2017; pp. 84–89.
16. Aiyar, S.; Shetty, N.P. N-gram assisted youtube spam comment detection. *Procedia Comput. Sci.* **2018**, *132*, 174–182. [CrossRef]
17. García, M.; Maldonado, S.; Vairetti, C. Efficient n-gram construction for text categorization using feature selection techniques. *Intell. Data Anal.* **2021**, *25*, 509–525. [CrossRef]
18. Saikh, T.; Anand, A.; Ekbal, A.; Bhattacharyya, P. A novel approach towards fake news detection: Deep learning augmented with textual entailment features. In Proceedings of the 24th International Conference on Applications of Natural Language to Information Systems, NLDB 2019, Salford, UK, 26–28 June 2019; pp. 345–358.
19. Ghanem, B.; Rosso, P.; Rangel, F. Stance detection in fake news a combined feature representation. In Proceedings of the First Workshop on Fact Extraction and VERification (FEVER), Brussels, Belgium, 1 November 2018; pp. 66–71.

20. Ferreira, W.; Vlachos, A. Emergent: A novel data-set for stance classification. In Proceedings of the 2016 Conference of the North American Chapter of the Association for Computational Linguistics: Human Language Technologies, San Diego, CA, USA, 12–17 June 2016; pp. 1163–1168.
21. Thota, A.; Tilak, P.; Ahluwalia, S.; Lohia, N. Fake news detection: A deep learning approach. *SMU Data Sci. Rev.* **2018**, *1*, 10.
22. Munzel, A. Assisting consumers in detecting fake reviews: The role of identity information disclosure and consensus. *J. Retail. Consum. Serv.* **2016**, *32*, 96–108. [CrossRef]
23. Xu, W.W.; Sang, Y.; Kim, C. What drives hyper-partisan news sharing: Exploring the role of source, style, and content. *Digit. J.* **2020**, *8*, 486–505.
24. Rangel, F.; Giachanou, A.; Ghanem, B.H.H.; Rosso, P. Overview of the 8th author profiling task at PAN 2020: Profiling fake news spreaders on Twitter. In *CEUR Workshop Proceedings*; Sun SITE Central Europe: Aachen, Germany, 2020; Volume 2696, pp. 1–18.
25. Parikh, S.B.; Atrey, P.K. Media-rich fake news detection: A survey. In Proceedings of the 2018 IEEE Conference on Multimedia Information Processing and Retrieval (MIPR), Miami, FL, USA, 10–12 April 2018; pp. 436–441.
26. Kumar, A.; Upadhyay, M. Rumour Stance Classification using A Hybrid of Capsule Network and Multi-Layer Perceptron. *Turk. J. Comput. Math. Educ. (TURCOMAT)* **2021**, *12*, 4110–4120.
27. Ajao, O.; Bhowmik, D.; Zargari, S. Fake news identification on twitter with hybrid cnn and rnn models. In Proceedings of the 9th International Conference on Social Media and Society, Copenhagen Denmark, 18–20 July 2018; pp. 226–230.
28. Girgis, S.; Amer, E.; Gadallah, M. Deep Learning Algorithms for Detecting Fake News in Online Text. In Proceedings of the 2018 13th International Conference on Computer Engineering and Systems (ICCES), Cairo, Egypt, 18–19 December 2018; pp. 93–97.
29. Gilda, S. Notice of Violation of IEEE Publication Principles: Evaluating machine learning algorithms for fake news detection. In Proceedings of the 2017 IEEE 15th Student Conference on Research and Development (SCOReD), Wilayah Persekutuan Putrajaya, Malaysia, 13–14 December 2017; pp. 110–115.
30. Ahmed, S.; Hinkelmann, K.; Corradini, F. Combining machine learning with knowledge engineering to detect fake news in social networks-a survey. In Proceedings of the AAAI 2019 Spring Symposium, Palo Alto, CA, USA, 25–27 March 2019; Volume 12, p. 8.
31. Library, N. Natural Language Toolkit. 1999. Available online: https://www.nltk.org/ (accessed on 21 August 2021).
32. Kaggle. Fake news Dataset. 2018. Available online: https://www.kaggle.com/c/fake-news/data (accessed on 21 August 2021).
33. Jindal, R.; Dahiya, D.; Sinha, D.; Garg, A. A Study of Machine Learning Techniques for Fake News Detection and Suggestion of an Ensemble Model. In Proceedings of the International Conference on Innovative Computing and Communications, New Delhi, India, 19–20 February 2022; Springer: Berlin/Heidelberg, Germany; pp. 627–637.
34. Shrivastava, S.; Singh, R.; Jain, C.; Kaushal, S. A Research on Fake News Detection Using Machine Learning Algorithm. In *Smart Systems: Innovations in Computing*; Springer: Singapore, 2022; pp. 273–287.
35. Monti, F.; Frasca, F.; Eynard, D.; Mannion, D.; Bronstein, M.M. Fake news detection on social media using geometric deep learning. *arXiv* **2019**, arXiv:1902.06673.
36. Nasir, J.A.; Khan, O.S.; Varlamis, I. Fake news detection: A hybrid CNN-RNN based deep learning approach. *Int. J. Inf. Manag. Data Insights* **2021**, *1*, 100007.
37. Paul, S.; Joy, J.I.; Sarker, S.; Ahmed, S.; Das, A.K. Fake news detection in social media using blockchain. In Proceedings of the 2019 7th International Conference on Smart Computing & Communications (ICSCC), Sarawak, Malaysia, 28–30 June 2019; pp. 1–5.
38. Manguri, K.H.; Ramadhan, R.N.; Amin, P.R.M. Twitter sentiment analysis on worldwide COVID-19 outbreaks. *Kurd. J. Appl. Res.* **2020**, *5*, 54–65. [CrossRef]
39. Helmstetter, S.; Paulheim, H. Weakly supervised learning for fake news detection on Twitter. In Proceedings of the 2018 IEEE/ACM International Conference on Advances in Social Networks Analysis and Mining (ASONAM), Barcelona, Spain, 28–31 August 2018; pp. 274–277.
40. Buntain, C.; Golbeck, J. Automatically identifying fake news in popular twitter threads. In Proceedings of the 2017 IEEE International Conference on Smart Cloud (SmartCloud), New York, NY, USA, 3–5 November 2017; pp. 208–215.
41. Gupta, P.; Pathak, V.; Goyal, N.; Singh, J.; Varshney, V.; Kumar, S. Content credibility check on Twitter. In Proceedings of the International Conference on Application of Computing and Communication Technologies, New Delhi, India, 9–10 March 2018; Springer: Singapore, 2018; pp. 197–212.
42. Twitter, I. Twitter API. 2021. Available online: https://developer.twitter.com (accessed on 21 August 2021).
43. Gupta, P.; Thakral, R.; Aggarwal, M.; Bhatti, S.; Jain, V. A Proposed Framework to Analyze Abusive Tweets on the Social Networks. *Int. J. Mod. Educ. Comput. Sci.* **2018**, *10*, 46–56. [CrossRef]
44. Priyanka.; Kumar, D. Decision tree classifier: A detailed survey. *Int. J. Inf. Decis. Sci.* **2020**, *12*, 246–269. [CrossRef]
45. Kulkarni, V.Y.; Sinha, P.K. Pruning of random forest classifiers: A survey and future directions. In Proceedings of the 2012 International Conference on Data Science & Engineering (ICDSE), Cochin, India, 18–20 July 2012; pp. 64–68.
46. De Menezes, F.S.; Liska, G.R.; Cirillo, M.A.; Vivanco, M.J. Data classification with binary response through the Boosting algorithm and logistic regression. *Expert Syst. Appl.* **2017**, *69*, 62–73. [CrossRef]
47. Joachims, T. Machine Learning: ECML-94. In Proceedings of the European Conference on Machine Learning, Catania, Italy, 6–8 April 1994; Springer Science & Business Media: Singapore, 2005; Volume 784, pp 627–637.
48. Desjardins, J. What Happens in an Internet Minute in 2018? 2018. Available online: https://www.visualcapitalist.com/internet-minute-2018 (accessed on 22 September 2021).

applied sciences

MDPI

Article

Machine Learning Approach for Personality Recognition in Spanish Texts

Yasmín Hernández *, Alicia Martínez *, Hugo Estrada, Javier Ortiz and Carlos Acevedo

Computer Science Department, Tecnológico Nacional de México/Cenidet, Cuernavaca 62490, Mexico; hugo.ee@cenidet.tecnm.mx (H.E.); javier.oh@cenidet.tecnm.mx (J.O.); carlos.acevedo@cenidet.edu.mx (C.A.)
* Correspondence: yasmin.hp@cenidet.tecnm.mx (Y.H.); alicia.mr@cenidet.tecnm.mx (A.M.)

Abstract: Personality is a unique trait that distinguishes an individual. It includes an ensemble of peculiarities on how people think, feel, and behave that affects the interactions and relationships of people. Personality is useful in diverse areas such as marketing, training, education, and human resource management. There are various approaches for personality recognition and different psychological models. Preceding work indicates that linguistic analysis is a promising way to recognize personality. In this work, a proposal for personality recognition relying on the dominance, influence, steadiness, and compliance (DISC) model and statistical methods for language analysis is presented. To build the model, a survey was conducted with 120 participants. The survey consisted in the completion of a personality test and handwritten paragraphs. The study resulted in a dataset that was used to train several machine learning algorithms. It was found that the AdaBoost classifier achieved the best results followed by Random Forest. In both cases a feature selection pre-process with Pearson's Correlation was conducted. AdaBoost classifier obtained the average scores: accuracy = 0.782, precision = 0.795, recall = 0.782, F-measure = 0.786, receiver operating characteristic (ROC) area = 0.939.

Keywords: DISC model; personality recognition; predictive model; text analysis

Citation: Hernández, Y.; Martínez, A.; Estrada, H.; Ortiz, J.; Acevedo, C. Machine Learning Approach for Personality Recognition in Spanish Texts. *Appl. Sci.* **2022**, *12*, 2985. https://doi.org/10.3390/app12062985

Academic Editors: Arturo Montejo-Ráez and Salud María Jiménez-Zafra

Received: 31 January 2022
Accepted: 8 March 2022
Published: 15 March 2022

1. Introduction

Personality has been recognized as a driver of decisions and behavior; it consists of singular characteristics on how individuals think, feel, and behave [1]. Understanding personality provides a way to comprehend how the different traits of an individual merge as a unit, since personality is a mixture of traits and behavior that people have to cope with situations. Personality influences selections and decisions (e.g., movies, music, and books) [2]. Personality guides the interactions among people, relationships, and the conditions around them. Personality has been shown to be related to any form of interaction. In addition, it has been shown to be useful in predicting job satisfaction, success in professional relationships, and even preference for different user interfaces [3].

Previous research on user interfaces and personality has found more receptiveness and confidence in users when the interfaces take personality into account. When personality is predicted from the social media profile of users, applications can use it to personalize presentations and messages [3].

Researchers have recognized that every person has a personality that usually remains consistent over time. Consequently, personality assessment can be used as an important measure. Various psychological models of personality have been proposed, such as the Five-factor model [4], the psychoticism, extraversion, and neuroticism (PEN) model [5], the Myers–Briggs type inventory [4], and the dominance, influence, steadiness, and compliance (DISC) model [6].

Typically, these models propose direct methods such as questionnaires to recognize personality. Conversely, linguistic analysis can be used to detect personality [3,7]. Linguistic analysis can produce useful patterns for establishing relationships between writing characteristics and personality. Researchers in natural language processing have proposed several methods of linguistic analysis to recognize personality, and machine learning has been one of the most investigated approaches.

Machine learning techniques are useful in the recognition of personality since they provide mechanisms to automatize processes that are based on a set of examples. Several proposals for personality recognition based on machine learning can be found in the literature [8,9]. Machine learning algorithms use computational methods to learn directly from data without relying on a predetermined equation as a model. The algorithms adaptively improve their performance as the number of instances available for learning increases [10].

Several efforts in personality prediction from the linguistic analysis approach have been carried out. However, they have focused mostly on the English language and are based on the five-factor model. This model (also called big five model) has been used as a standard for applications that need personality modeling [7].

To contribute to the advancement and understanding of the relationship between personality and language, we have developed a predictive model for personality recognition based on the DISC personality model and a machine learning approach. We performed a personality survey with 120 participants. The participants were asked to complete a demographic form, fill in the DISC test, and handwrite a text on a general topic that they selected.

The model for personality prediction is based on a supervised machine learning approach for multiclass classification. We evaluated six of the most known classifiers: naive Bayes [11], sequential minimal optimization (SMO) [12], k-Nearest neighbors (kNN) [13], AdaBoost [14], J48 [15], and random forest [16]. We conducted preprocess tasks as feature extraction, feature selection and data augmentation to have nine versions of the dataset. We found AdaBoost [14] and random forest [16] had the best performance. Figure 1 presents the overview of our approach.

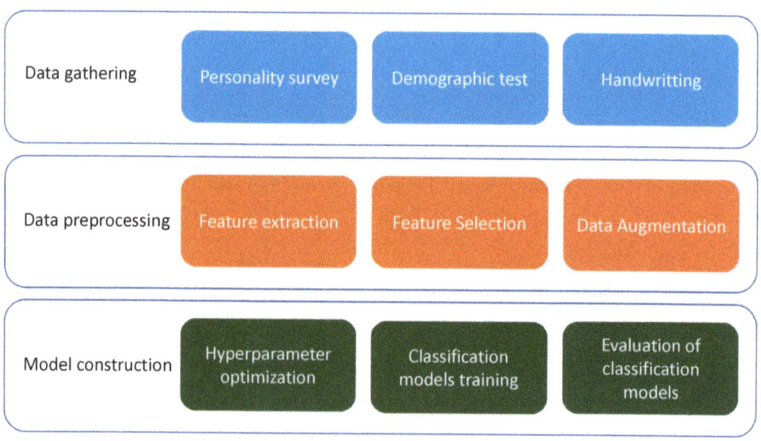

Figure 1. Overview of the construction of the model for personality prediction.

This paper presents the construction of the predictive model for personality recognition. Section 1 presents related work and background. Section 2 describes the protocol for the personality survey. Section 3 presents the machine learning approach for building the predictive model. Section 4 presents the results of this research. Finally, Section 5 discusses the results and outlines future work.

1.1. Related Work

Srinarong and Mongkolnavin [17] developed a model based on machine learning techniques to recognize the personality of the customers of a call center. The model allows the call center to give them an appropriate response. This study is based on the MPI (Maudsley personality inventory) personality model. Audio files of conversational voice were collected from 92 voluntary participants who were instructed to make conversation in the simulated context. Logistic regression, LinearSVC, random forest, and artificial neural networks were used in the modeling process.

Automatic personality recognition based on Twitter in Bahasa Indonesia was proposed by Adi et al. [18]. Tweets were manually annotated by experts in psychology using the big five model. In this study, stacking, gradient boosting, and stochastic gradient descent were evaluated.

A multi-label personality detection model based on neural networks, which combines emotional and semantic features was proposed by Ren et al. [19]. This model relies on bidirectional encoder representation from transformers (BERT) to generate sentence-level embedding for text semantic extraction. A sentiment dictionary is used for text sentiment analysis to consider sentiment information. The performance of the model was evaluated on two public personality datasets for MBTI and big five.

A model for personality prediction from text posts of social network users was developed based on a hierarchical deep neural network by Xue et al. [20]. The model predicts the big five personality by means of traditional regression algorithms and the combination of statistical linguistic features with deep semantic features from the text postings. This approach has achieved the lowest average prediction error of all of the approaches.

A model aiming to assist in recruiting and selecting appropriate personnel by knowing the personality of customers has also been developed by Sher et al. [21]. The XGBoost classifier is used to predict the personality from input text based on the MBTI model. A publicly available benchmark dataset from Kaggle was used in the experiments.

1.2. The DISC Model of Personality

DISC stands for Dominance, Influence, Steadiness, and Compliance. They are the four dimensions of personality proposed by the model that represent the basic behavioral styles. The Dominance and Influence dimensions denote receptiveness and assertiveness. The Steadiness and Compliance dimensions denote control and openness. Personality falls within these four dimensions [6,22].

When a DISC profile shows a high Dominance factor, it is describing someone with an independent attitude and a motivation to succeed on their own terms. Dominant people have the willpower to work under pressure, and they are always ready to take on responsibility [6,22].

When Influence stands out as a major factor, it describes someone with a positive attitude to other people, and the confidence to demonstrate that attitude. People of this kind are comfortable in social situations and interact with others in an open and expressive way [6,22].

Steadiness is related to the natural pace of people and their reactions to change. This factor describes a reticent and careful person. Steady people usually respond to events rather than taking pro-active steps themselves. Steady people are consistent and reliable in their approach. Indeed, they prefer to operate in situations following established patterns and avoid unplanned developments. Therefore, people with high Steadiness tend to be quite resistant to change and will need time to adapt to new situations [6,22].

The Compliance dimension is related to organization, accuracy, and attitudes towards authority. An individual showing high Compliance is concerned with detail and practicality. The key characteristic of this dimension falls in attitudes towards authority. Compliant people are rule oriented. They are also interested in accuracy, structure, and understanding the ways things work [6,22].

The DISC personality test consists of 28 groups of four adjectives. To assess personality, individuals must choose the adjective that identifies them the most and the adjective that identifies them the least. Some examples of the adjective groups of the DISC test are shown in Table 1.

Table 1. Examples of the adjective groups in the dominance, influence, steadiness, and compliance (DISC) personality test.

Group 1	Group 2	Group 3	Group 4
Extroverted	Sociable	Analytical	Daring
Cautious	Impulsive	Bold	Conscientious
Persistent	Determined	Loyal	Talkative
Impatient	Calm	Helpful	Moderate

The DISC model has been used widely in several fields such as education, health, industry, and management. For instance, Milne et al. [23] conducted a study to identify the behavior styles of physiotherapy students and to determine if there is a relationship between students' unique behavior patterns and their clinical placement grades. On the other hand, DISC personality has been considered to be a predictor for the improvement of manageability; Chigova et al. [24] conducted a study to identify impact factors that improve the efficiency of structured interaction in enterprises and organizations.

2. Personality Survey to Gather Data

To obtain the ground-truth data, a personality survey was conducted. The objective of the survey was to gather data to relate writing characteristics and behavior with personality. These relationships are useful for constructing a text classification model. The proposed model for personality prediction is intended to be applied in the selection process of candidates for postgraduate programs. Therefore, the study focused on knowing the personality of undergraduate and graduate students. One hundred and twenty students participated in the survey (49 women and 71 men). The participants ranged in age between 20 and 30 years old.

The survey consisted of three parts: (i) a general information questionnaire; (ii) the DISC personality test; and (iii) handwritten paragraphs. Each participant was contacted individually and was told about the objectives and the procedure of the survey. If they agreed to participate, the three parts of the survey were explained in detail. Additional help was provided if the participants required it, but most of the participants did not need help or explanations during the survey. The participants took between 20 and 30 min to complete the survey. The entire survey was in Spanish.

The first part asked the participants for personal data: age range, gender, schooling, occupation, marital status, preferred social networks, and number of online friends. In the second part, the participants filled in the personality test [5,6]. To complete the DISC personality test, the participants had to do self-inspection and to conclude to what extent the adjectives in the test represented them, as explained in Section 1.2. In the third part of the survey, the participants handwrote some paragraphs on any topic. Suggested topics were provided. These included goals, hobbies, what they did the day before, and so on.

The study showed that Facebook and Twitter are the preferred social networks of the participants, with 105 participants and 15 participants, respectively. The average number of friends of the participants on the social networks was 531 people. Table 2 shows the answers and the results of the personality test for four participants in the survey.

Table 2. Examples of answers of the participants in the survey.

	Gender	Schooling	Civil Status	Occupation	Preferred SN	Friends in Preferred SN [1]	Personality
1	Male	College	College	Student	Twitter	120	Dominance
2	Female	College	College	Student	Facebook	1150	Influence
3	Male	High School	College	Student	Facebook	100	Steadiness
4	Female	College	Married	Student	Facebook	80	Compliance

[1] SN stands for social network.

The results of the personality survey are shown in Table 3. The most frequent personality dimension was Steadiness (62 people), the second most common dimension was Influence (26 people), the next factor was Compliance (18 people), and the least common factor was Dominance (14 people).

Table 3. Results of the personality survey.

Personality	Women	Men	Total
Dominance	8	6	14
Influence	10	16	26
Steadiness	24	38	62
Compliance	7	11	18
	49	71	120

It is noteworthy that the DISC personality model was selected since it is a clean model that only requires a short time for training and assessing answers. The results can be obtained relatively easily, and the model can provide adequate information regardless of whether the people conducting the survey are knowledgeable in psychology [22].

Besides personality and demographic data, a set of 120 handwritten texts by participants was obtained. It was observed that most of the participants chose to write about one of the suggested topics. Just a few decided to write on another topic. It was also observed that the participants used words related to their studies and their desire to be successful and achieve their goals. This could be due to the age and level of studies of the participants. Table 4 presents a sample of a paragraph in Spanish text gathered in the study. The translation of the text in English for purposes of clarity. The complete study and analysis were in the Spanish language. Figure 2 shows the original handwritten text.

To conduct the analysis, the handwriting was transcribed to electronic texts. On average, the texts had 90 words and a lexical diversity of 0.19. To measure lexical diversity, the type–token ratio (TTR) measure was used. This measure is expressed as the number of different words in a document divided by the total number of words in that document [25].

The text processing includes eliminating stop-words since, as is well known, they do not provide relevant information to the analysis because they are common words. There is not a unanimously accepted comprehensive list of stop-words since these words can depend on the context and specific application. However, there is agreement on most words that are considered stop-words. A proposed list of Spanish stop-words was used [26]. This list contains articles, pronouns, adverbs, prepositions, and verbs.

We used AntConc, which is a corpus analysis toolkit for concordance and text analysis which allows the extraction of data such as word frequencies, collocations, concordances, and so on [27]. We eliminated stop-words, computed the number of words with and without stop-words, and the number of different words.

Every word was lemmatized, i.e., it was converted to its root. The FreeLing software suite was used for this process. FreeLing is an open-source software suite for natural language processing. This library provides a wide range of analyzers for several languages. It offers natural language application developers text processing and language annotation facilities [28].

Table 4. Example of a Spanish text gathered in the survey.

Original	Translated
El día de ayer domingo me desperté muy tarde, como a las 10, desperté muy contenta porque como soy foránea únicamente convivo con mis familiares los fines de semana, desperté y encendí la televisión e hice uno de mis pasatiempos favoritos: ver televisión en un canal de animales, me gustan mucho, después llego mi hermana con mi sobrina y junto con ellas seguimos aprendiendo sobre animales, después nos fuimos a almorzar con mi familia completa, después nos pusimos a jugar con mis sobrinos y hermana lotería, después comimos todos juntos y nos pasamos al patio de la casa a ayudar a pintar la casa de una tía, después recordé que hay tarea, encendí la computadora para hacerla, comencé con lo que más me gusta: programación, redes, etc.. Suspendí la computadora para bañarme y después intenté terminar la tarea finalmente se terminó el domingo y mi hermana se fue.	On Sunday, I woke up very late, about 10 o'clock. I woke up very happy because I am from another town, I only live with my family on weekends. I woke up and turned on the television and did one of my favorite hobbies: watch an animal channel. I like it very much. Then my sister arrived with my niece, and I continued learning about animals with them. Then we went to have lunch with my whole family. Then we started playing *lotería*, a table game, with my nieces and nephews, and my sister. Then we all had lunch together and we went to the patio of my aunt to help paint the house. Later, I remembered I had homework. I turned on the computer to do it. I started with what I like the most: programming, networks, etc. I put the computer in energy saving mode to take a bath, and later I tried to finish my homework. Finally, Sunday ended, and my sister left.

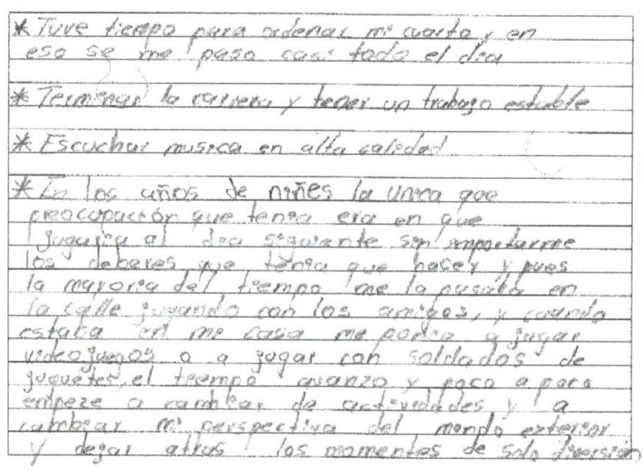

Figure 2. Example of a handwritten text.

With this data, we built an annotated linguistic corpus for Spanish, which was useful for the construction of the predictive model for personality recognition.

3. Supervised Learning Model to Classify Texts

Machine learning is defined as the field of study that gives computers the ability to learn without being explicitly programmed. These algorithms use computational methods to learn from data without relying on a predetermined equation as a model. The algorithms adaptively improve their performance as the number of instances available for learning increases [10].

The model for personality prediction is based on a supervised machine learning approach for multiclass classification. We evaluated six of the most well-known classifiers: naive Bayes [11], sequential minimal optimization [12], k-nearest neighbors [13], AdaBoost [14], J48 [15], and random forest [16].

The construction of the model included a pre-processing data step, since there is often noisy, inconsistent, missing, irrelevant, or imbalanced data. Some of the causes are large databases, multiple and heterogeneous sources of data, and data collected for other objectives other than different to data mining. Techniques for data pre-processing increase the performance of data mining algorithms [10]. Therefore, we applied techniques such as feature extraction, feature selection, and data augmentation. For most of the processes of data mining, we used the Waikato environment for knowledge analysis, WEKA, which is a full implementation of most of the machine learning algorithms [10]. For data augmentation, we used the scikit-learn library in Python programming language.

3.1. Feature Extraction

The text classification problem is challenging since machine learning algorithms prefer well-defined inputs and outputs instead of raw text. Therefore, the text must be converted into an understandable representation. This process is called feature extraction or feature encoding [29]. We used the bag-of-words (BoW) model of text. BoW is a way of extracting features from text for modeling. This model is only concerned with whether known words occur in the document. The intuition is that documents are similar if they have similar content [29]. Every verb and adjective in the text were converted to a nominal feature with two possible values: *Yes* (the word occurs in the text) and *No* (the word does not occur in the text).

3.2. Feature Selection

The dataset is composed of a total of 546 features (540 features representing verbs and adjectives in the text documents, and 6 features representing the demographic data) and a personality label.

Commonly, raw data contains a combination of features, some of which are irrelevant since they do not provide information to the prediction process. The feature selection process takes a subgroup of related features to be included in the training of a learning model. Feature selection techniques are useful because they simplify models and reduce training time. Feature selection aims to establish redundant or irrelevant features which can be eliminated without losing information [10]. We applied two feature selection methods in order to have several versions of the dataset.

We used the correlation feature selection method with a Ranker search. This method evaluates the worth of a feature by measuring the Pearson's correlation between it and the class [30]. This method generated a ranked list of the 546 features.

We also used the Info Gain feature selection method with the Ranker search. This method evaluates the worth of a feature via the information gain with respect to the class. Information gain is computed by the contribution of the feature in decreasing overall entropy [31]. The Info Gain method produced a ranked list of the 546 features.

Additionally, for feature subset selection, we experimented with Wrappers and several classifiers (e.g., AdaBoost and random forest). The Wrappers method evaluates sets of features by means of a learning scheme [32]. However, few features were selected by the Wrappers method; at most, 35 features were selected. Therefore, there was a significant loss of information and the performance of the machine learning decreased.

Cross validation is used to estimate the accuracy of the learning scheme for a set of features. Based on the results of the feature selection process, we built eight datasets from the original dataset. The datasets are detailed below.

3.3. Data Augmentation

From the personality survey, we obtained a dataset with 120 instances where classes are not equally represented (See Table 3). Imbalanced classes could lead to a bias toward the majority class during the model training [33]. To deal with this issue, we resampled the dataset by means of the synthetic minority oversampling technique, SMOTE [33]. SMOTE generates synthetic instances to over-sample the minority class, and it can also under-sample the majority class if necessary. The original dataset was transformed using SMOTE, and the new class distribution is summarized in Table 5. After applying SMOTE, we obtained a dataset with 248 records.

Table 5. Class distribution.

Personality	Original Dataset	After SMOTE Dataset
Dominance	14	62
Influence	26	62
Steadiness	62	62
Compliance	18	62
	120	248

3.4. Datasets

We built eight different datasets base on the results of the feature selection process. In the original dataset there are 546 features, 540 of which represent verbs and adjectives, and six of which represent demographic data. Table 6 describes the nine datasets (including the original dataset). It shows the number of features in each dataset and presents the features representing demographic data.

Table 6. Datasets.

DS	Description	Features	Demographics Features
DS1	Original dataset	546	Gender, Schooling, Civil status, Occupation, Preferred Social Network Friends in Social Network
DS2	The 100 least correlated features with the class were removed, according to the Correlation method	446	Occupation, Preferred Social Network, Friends in Social Network
DS3	The 150 least correlated features with the class were removed, according to the Correlation method	396	Occupation, Friends in Social Network
DS4	The 200 least correlated features with the class were removed, according to Correlation method	346	Occupation, Friends in Social Network
DS5	The 271 least correlated features (about half) with the class were removed, according to Correlation method	275	Occupation, Friends in Social Network
DS6	The 100 least informative features were removed, according to the Info Gain feature selection method	446	Gender, Schooling, Civil status, Occupation, Preferred Social Network, Friends in Social Network
DS7	The 150 least informative features were removed, according to the Info Gain feature selection method	396	Gender, Schooling, Civil status, Occupation, Preferred Social Network, Friends in Social Network
DS8	The 200 least informative features were removed, according to the Info Gain feature selection method	346	Schooling, Civil status, Occupation, Preferred Social Network, Friends in Social Network
DS9	The 265 least informative features (about half) were removed, according to the Info Gain feature selection method	371	Schooling, Civil status, Occupation, Friends in Social Network

To add features to the datasets, we experimented with several characteristics of the text such as TD-IF, lexical diversity, number of words from each word type. However, we do not observe improvement in the learning models. We need to conduct further experiments and undertake processes such as principal components analysis in order to obtain new features that provide relevant information to the model. Consequently, these features were not included in the datasets.

3.5. Hyperparameter Optimization

Some machine learning algorithms have parameters that can be tuned to optimize their behavior. They are called hyperparameters to distinguish them from basic parameters such as the coefficients in linear regression models. An example is the parameter k that determines the number of neighbors considered in a k-nearest neighbor classifier. Usually, best performance on a test set is achieved by adjusting the value of this hyperparameter to suit the characteristics of the data [10].

In the literature, there are some methods to tune hyperparameters such as grid search, random search, and Bayesian optimization, among others [34]. However, there is not a direct way to know how a change in a hyperparameter value will reduce the loss of the model, therefore we must do experimentation.

We conducted an empirical process of hyperparameters based on trial and error. Since our dataset is small, the change of many hyperparameters did not have impact. Mainly our objective with hyperparameters optimization was to have a configuration that allows to have a reliable classification with the nine versions of our small dataset, since some configurations could not evaluate the performance of the learning model because there were few samples. Table 7 presents the hyperparameters configuration for our experiments.

Table 7. Hyperparameter optimization of classification algorithms.

Classifier	Hyperparameters
Naïve Bayes	Use a kernel estimator for numeric attributes = false (use a normal distribution) Number of instances to process with batch prediction = 100
SMO	Kernel = polykernel
kNN	k = 5 Distance function = euclidean distance
AdaBoost	Classifier = random Forest Number of models to create = 10
J48	Pruning = true Minimum number of instances per leaf = 2
Random Forest	Number of features to consider in each split = int (log_2 (#predictors) + 1) Percentage of the raw training dataset = 100 Number of bags = 100

4. Results

After we preprocessed the data and built the datasets, we proceeded to the evaluation of several classifier algorithms to build the predictive model of personality.

In machine learning, classification refers to a predictive modeling problem where a class label is predicted for a given example of input data. A classifier algorithm finds relationships between unknown objects and a set of correctly labeled objects in order to classify the unknown objects [35]. There is an extensive range of classifier algorithms to be used based on the nature of data.

Based on an analysis of recent work on machine learning proposals, the nature of the problem, and the data available, we decided to evaluate six of the most well-known classifiers: naive Bayes [11], sequential minimal optimization (support vector machines) [12], k-nearest neighbors [13], AdaBoost [14], J48 [15], and random forest [16]. A stratified ten times ten-fold cross-validation technique was used in the training and testing of the model, which is the standard when there is limited data [10].

We compared the statistical measures obtained by each one of the classifier algorithms to select the best predictive model. We evaluated the classifier algorithms within the nine datasets for the statistics measures: accuracy, precision, recall, F-measure, and receiver operating characteristic (ROC) area.

Specifically, we focus on F-measure and ROC area. We are interested in F-measure because we want to have a balance between precision and recall. Precision is the fraction of relevant instances among the retrieved instances, while recall is the fraction of relevant instances that have been retrieved over the total amount of relevant instances [36]. The ROC curve is used for the visual comparison of classification models, which shows the tradeoff between the true positive rate and the false positive rate. The area under the ROC curve is a measure of the accuracy of the model. When a model is closer to the diagonal, it is less accurate, and the model with perfect accuracy will have an area of 1.0 [36].

Figure 3 presents the results of the six classifiers within the nine datasets for the five measures. Table 8 depicts the best classifier for each dataset according to F-measure. The best classifier for each dataset according to ROC area is presented in Table 9. Table 10 presents the ten classifiers that have the best performance based on F-measure. Table 11 presents the ten classifiers that have the best performance according to ROC area.

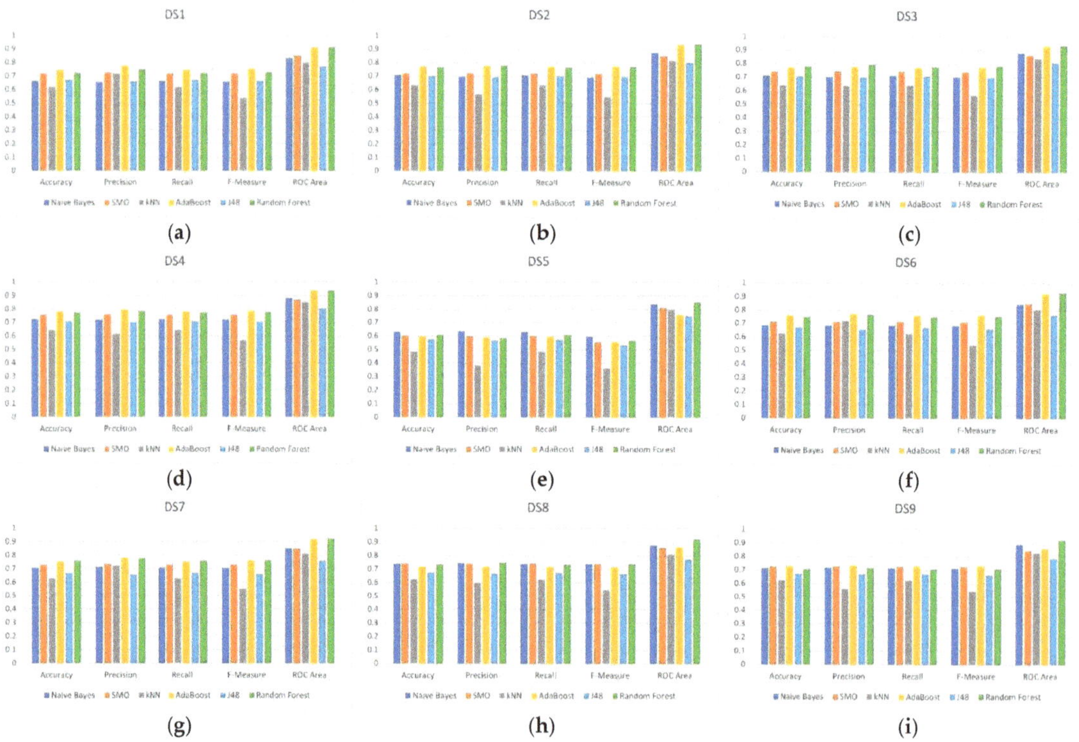

Figure 3. Performance of classifiers in the nine datasets: (**a**) Original dataset; (**b**) without the 100 least correlated features with the class; (**c**) without the 150 least correlated features with the class; (**d**) without the 200 least correlated features with the class; (**e**) without the 271 least correlated features with the class; (**f**) without the 100 least informative features; (**g**) without the 150 least informative features; (**h**) without the 200 least informative features; (**i**) without the 265 least informative features.

Table 8. Best classifier for each dataset according to F-measure.

Dataset	Classifier	Accuracy	Precision	Recall	F-Measure	ROC Area
DS1	AdaBoost	0.745968	0.774	0.746	0.754	0.911
DS2	AdaBoost	0.770161	0.775	0.77	0.772	0.935
DS3	Random Forest	0.778226	0.792	0.778	0.782	0.933
DS4	AdaBoost	0.782258	0.795	0.782	0.786	0.939
DS5	Naïve Bayes	0.629032	0.635	0.629	0.597	0.84
DS6	AdaBoost	0.762097	0.774	0.762	0.766	0.924
DS7	Random Forest	0.758065	0.777	0.758	0.763	0.923
DS8	SMO	0.741935	0.741	0.742	0.74	0.858
DS9	AdaBoost	0.729839	0.734	0.73	0.731	0.858

Table 9. Best classifier for each dataset according to receiver operating characteristic (ROC) area.

Dataset	Classifier	Accuracy	Precision	Recall	F-Measure	ROC Area
DS1	AdaBoost	0.745968	0.774	0.746	0.754	0.911
DS2	Random Forest	0.766129	0.777	0.766	0.769	0.938
DS3	Random Forest	0.778226	0.792	0.778	0.782	0.933
DS4	AdaBoost	0.782258	0.795	0.782	0.786	0.939
DS5	Random Forest	0.608871	0.585	0.609	0.568	0.852
DS6	Random Forest	0.75	0.767	0.75	0.755	0.929
DS7	Random Forest	0.758065	0.777	0.758	0.763	0.923
DS8	Random Forest	0.733871	0.747	0.734	0.738	0.923
DS9	Random Forest	0.705645	0.715	0.706	0.709	0.921

Table 10. Top-ten classifiers according to F-measure.

Dataset	Classifier	Accuracy	Precision	Recall	F-Measure	ROC Area
DS4	AdaBoost	0.782258	0.795	0.782	0.786	0.939
DS3	Random Forest	0.778226	0.792	0.778	0.782	0.933
DS4	Random Forest	0.774194	0.783	0.774	0.777	0.937
DS2	AdaBoost	0.770161	0.775	0.77	0.772	0.935
DS3	AdaBoost	0.770161	0.776	0.77	0.772	0.932
DS2	Random Forest	0.766129	0.777	0.766	0.769	0.938
DS6	AdaBoost	0.762097	0.774	0.762	0.766	0.924
DS7	Random Forest	0.758065	0.777	0.758	0.763	0.923
DS7	AdaBoost	0.754032	0.779	0.754	0.76	0.92

Table 11. Top-ten classifiers according to ROC area.

Dataset	Classifier	Accuracy	Precision	Recall	F-Measure	ROC Area
DS4	AdaBoost	0.782258	0.795	0.782	0.786	0.939
DS2	Random Forest	0.766129	0.777	0.766	0.769	0.938
DS4	Random Forest	0.774194	0.783	0.774	0.777	0.937
DS2	AdaBoost	0.770161	0.775	0.77	0.772	0.935
DS3	Random Forest	0.778226	0.792	0.778	0.782	0.933
DS3	AdaBoost	0.770161	0.776	0.77	0.772	0.932
DS6	Random Forest	0.75	0.767	0.75	0.755	0.929
DS6	AdaBoost	0.762097	0.774	0.762	0.766	0.924
DS7	Random Forest	0.758065	0.777	0.758	0.763	0.923

Tables 8 and 9 show that AdaBoost and random forest are the classifiers with the best performance for most datasets according to F-measure and ROC area. Naive Bayes (DS5) and SMO (DS8) have good performance according to F-measure. The algorithms J48 and kNN have low performance with most datasets.

As can be observed in Tables 8–10, the best classifier is AdaBoost (F-Measure = 0.786 and ROC area = 0.939 for DS4 (276 features selected by Pearson correlation). Table 12 shows the measures for this classifier. The average ROC area of 0.939 indicates that the model separates the four classes very well. Table 12 also shows that measures for Steadiness are low. This phenomenon was observed for every classifier; therefore, this class is the hardest class to predict.

Table 12. Measures for the best classifier.

DS	Classifier	Class	Accuracy	Precision	Recall	F-Measure	ROC Area
DS4	AdaBoost	Steadiness		0.608	0.726	0.662	0.885
		Compliance		0.831	0.790	0.810	0.955
		Influence		0.889	0.774	0.828	0.962
		Dominance		0.852	0.839	0.846	0.954
Avg			0.782258	0.795	0.782	0.786	0.939

DS4 was the dataset that provided the best performance to the classifiers. Tables 10 and 11 shows that the datasets built from correlation feature selection (DS2, DS3 and DS4) provided better performance than info gain feature selection (DS6 y DS7).

Table 13 presents the confusion matrix for AdaBoost with DS4. This confirms the measures in Table 12. There are many true positives and true negatives (diagonal) and a few false positives and false negatives (outside the diagonal).

Table 13. Confusion matrix for AdaBoost classifier with DS4.

Actual	Predicted				
	Steadiness	**Compliance**	**Influence**	**Dominance**	
Steadiness	**45**	7	4	6	62
Compliance	9	**49**	2	2	62
Influence	12	1	**48**	1	62
Dominance	8	2	0	**52**	62
	74	59	54	61	

Error Analysis

We conducted an error analysis of AdaBoost with DS4 (the classifier with the best performance) to identify which personality the model misclassified. We found that the model has trouble in classify the Steadiness personality. Table 14 shows the misclassifications. Most of the errors are related to *Steadiness* personality. The model classified 17 actual Steadiness instances incorrectly and misclassified 29 instances as Steadiness.

Table 14. Classification errors for AdaBoost with DS4.

Actual	Predicted				
	Steadiness	**Compliance**	**Influence**	**Dominance**	
Steadiness	-	7	4	6	17
Compliance	9	-	2	2	13
Influence	12	1	-	1	14
Dominance	8	2	0	-	10
	29	10	6	9	54

Figure 4 shows correct and incorrect classifications for each class and compares the actual personality versus the predicted personality. This shows that the other three personality has more errors with Steadiness personality.

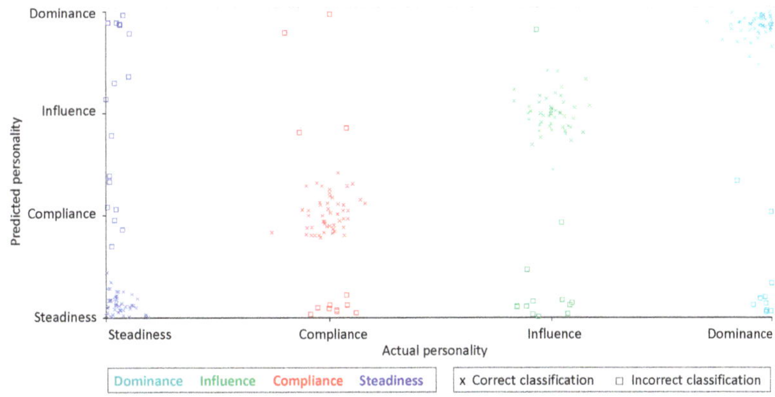

Figure 4. Actual personality versus predicted personality comparation.

Figure 5 compares the prediction margin versus the predicted personality. The prediction margin is defined as the difference between the probability predicted for the actual class and the highest probability predicted for the other classes. We can see that Steadiness personality has a prediction margin very low while the other three personality has many instances with a prediction margin of 1.0.

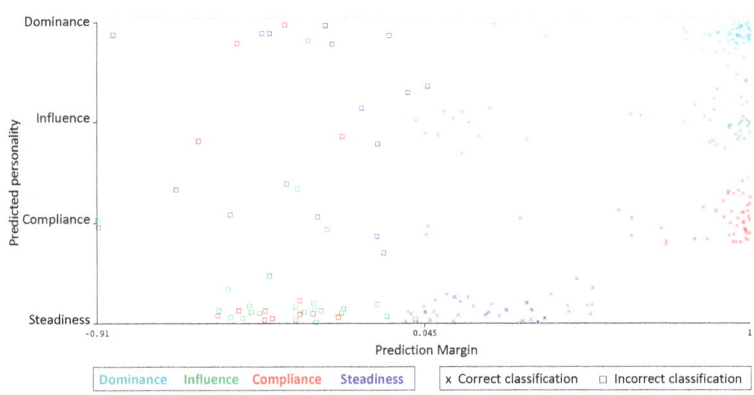

Figure 5. Prediction margin versus predicted personality comparation.

We analyzed some misclassified instances individually. We found that the most common words in Steadiness instances are also common words in other personality instances, therefore when these words are present, the model fails. We also found that *Steadiness* instances has a narrow set of words while the other personalities have a wider range of words, therefore when the instance has just few words and are common word for most of the personalities, the model fails and classify it as Steadiness. Table 15 shows some misclassified instances compared with the actual personality.

Table 15. Examples of misclassifications.

	Predicted Personality	Actual Personality	Words
1	Steadiness	Dominance	To decide, favorite, to do, to play, to be, to smile, to overcome
2	Steadiness	Influence	To have fun, favorite, to play, personal, to prefer, to be, to have
3	Steadiness	Compliance	To create, to write, to listen, to be, to inspire, to get free, older, to publish, to be, to see
4	Dominance	Steadiness	To do, to know, to be
5	Influence	Steadiness	To support, to help, short, to develop, to find, long, medium, personal, main, next, satisfactory, to be, to sustain, to have, to graduate
6	Compliance	Steadiness	To give, to go, to be, to have

5. Discussion

In this paper, a predictive model for personality recognition through text analysis has been proposed. The model was built based on a personality survey. The model relies on a machine learning approach. An annotated linguistic corpus for Spanish was built using the data gathered in the survey. Nine datasets were built using this corpus to train the classification model. Several machine learning algorithms were evaluated. AdaBoost obtained the best performance.

The AdaBoost learning model has a good performance in identifying three of the four classes; as mentioned before, the model has trouble to identify Steadiness. We have reached some conclusions about this weakness of the model. Much research has been conducted on adults who are fully developed, but with adolescents and teenagers, there is still a lot that is unknown; and it is recognized that the personality does not change but it is getting settled as individual grow up. Our population are young adults, they are leaving youth group, therefore they have not developed their personality completely. These results are consistent with the results of another personality test we conducted based on big five model; we found in 58 participants within the same age group (23.2 years old in average) that the 80% are in the middle of the Stability dimension (Neuroticism in big five model), they do not have low Stability neither high Stability [37]. Additionally, we have a population sample with 71 men and 49 women; it is also recognized that younger girls often experience a dip in emotional stability but increase as they near adulthood. For these reasons, we need to conduct a study to know if our benchmark is appropriate for identifying the four classes.

Even though the results are satisfactory, further research is required. At this point, this predictive model is not a replacement for the DISC model for personality analysis. It is important to emphasize that the study was conducted with a very specific group of participants (young people, mostly students) which biases the results. The population sample was also very small.

The DISC model has been extensively used in professional settings, industry, and business organizations. Even DISC is a popular model, this model has not been studied as much as similar models, such as big five and MBTI, and therefore there are less controlled research and relatively little scientific experimentation to support it. Additionally, DISC model is focused on behavior to establish the personality, but there are another deeper thought patterns and characteristics. This makes it less applicable in emotional situations.

In the other hand, data mining is an experimental science, whose results depend on the quality and quantity of the data and the nature of the problem. As a result of the new studies, we will have a bigger and different benchmark, therefore we must set up new experiments to have concluding findings. Additionally, machine learning is a huge field, therefore, there are many techniques that could be useful, and they were not focused on this research.

There are companies which offers predictive analytics for decision makers and technologies to optimize processes through intelligent applications. Such is the case of SOTA solutions (http://sota-solutions.de/wordpress_en/accessed on 6 March 2022), a company that develops big data solutions for producing, the energy, and the services industries. Their products are the results of many years of work on machine learning, statistics, mathematics, and software developing, therefore, they have very good performance. The core of these technologies is the same of our approach, machine learning and data mining. The difference strives in the application domain.

Even though the results are encouraging, there are several points in the research agenda of personality analysis. For example, the DISC model includes 15 patterns that are related to the four dimensions of personality. As future work, we will conduct another survey to obtain more data to recognize personality patterns in addition to the personality dimensions. This will help to provide a more precise prediction. The corpus can also be enriched using other metrics for the texts. For example, it could integrate collocations, use Point Mutual Information, and n-grams in order to obtain the information of associated words. In particular, we want to explore the CollGram technique, which assigns to bigrams in a text two association scores computed on the basis of a large reference corpus to determine the strength of the collocation [38]. This analysis will allow us to deepen into the relationship between writing patterns and personality. CollGram has been used successfully to detect depression in annotated corpus [39]. Our corpus was small; therefore, it would be interesting to compare the performance. However, we are planning to gather more texts in a further study.

The demographic data have not been thoroughly analyzed in the construction of the predictive model and some experimentation is needed to determine its relationship to personality and writing behavior. A future line of research line is to analyze the handwriting.

Additionally, during the results analysis, it was observed that most of the participants chose to write about the suggested topics. Most of the participants used words related to their studies and their desire to be successful. This could be due to the age of the participants. More experimentation is needed with participants of other ages in order to determine if this behavior is more related to the age of the participants or their personality.

In summary, this research provides some insights into the analysis of personality, which will help in the planning of the next steps in the investigation of the relationship between personality and writing characteristics.

Author Contributions: Conceptualization, Y.H. and A.M.; methodology, Y.H. and A.M.; software, C.A.; validation, H.E. and J.O.; formal analysis, Y.H.; investigation, A.M. and C.A.; resources, A.M. and H.E.; data curation, C.A.; writing—original draft preparation, Y.H.; writing—review and editing, Y.H., A.M., H.E. and J.O.; visualization, Y.H.; supervision, A.M.; project administration, A.M.; funding acquisition, A.M. and J.O. All authors have read and agreed to the published version of the manuscript.

Funding: This research was funded by Tecnológico Nacional de México and The APC was funded by PRODEP.

Institutional Review Board Statement: Not applicable.

Informed Consent Statement: Informed consent was obtained from all subjects involved in the study.

Data Availability Statement: The data presented in this study are available on request from the corresponding author.

Conflicts of Interest: The authors declare no conflict of interest.

References

1. Bromme, L.; Rothmund, T.; Azevedo, F. Mapping political trust and involvement in the personality space—A meta-analysis and new evidence. *J. Pers.* **2022**, 1–27. [CrossRef] [PubMed]
2. Stachl, C.; Au, Q.; Schoedel, R.; Gosling, S.D.; Harari, G.M.; Buschek, D.; Völkel, S.T.; Schuwerk, T.; Oldemeier, M.; Ullmann, T.; et al. Predicting personality from patterns of behavior collected with smartphones. *Proc. Natl. Acad. Sci. USA* **2020**, *117*, 17680–17687. [CrossRef] [PubMed]

3. Christian, H.; Suhartono, D.; Chowanda, A.; Zamli, K.Z. Text based personality prediction from multiple social media data sources using pre-trained language model and model averaging. *J. Big Data* **2021**, *8*, 68. [CrossRef]
4. Costa, P.T.; McCrae, R.R. Four ways five factors are basic. *Pers. Individ. Dif.* **1992**, *13*, 653–665. [CrossRef]
5. Eysenck, H.J. *Dimensions of Personality*, 1st ed.; Routledge: New Brunswick, NJ, USA; London, UK, 1997.
6. Marston, W.M. *Emotions of Normal People*; Harcourt Brace & Company: New York, NY, USA, 1928. [CrossRef]
7. Moreno, J.D.; Martínez-Huertas, J.; Olmos, R.; Jorge-Botana, G.; Botella, J. Can personality traits be measured analyzing written language? A meta-analytic study on computational methods. *Pers. Individ. Dif.* **2021**, *177*, 110818. [CrossRef]
8. Amirhosseini, M.H.; Kazemian, H. Machine learning approach to personality type prediction based on the Myers–Briggs type indicator®. *Multimodal Technol. Interact.* **2020**, *4*, 9. [CrossRef]
9. Fu, J.; Zhang, H. Personality trait detection based on ASM localization and deep learning. *Sci. Program.* **2021**, *2021*, 5675917. [CrossRef]
10. Witten, I.H.; Frank, E.; Hall, M.A.; Pal, C.J. *Data Mining: Practical Machine Learning Tools and Techniques*, 4th ed.; Morgan Kaufmann: Cambridge, UK, 2017.
11. John, G.H.; Langley, P. Estimating Continuous Distributions in Bayesian Classifiers. In *Eleventh conference on Uncertainty in Artificial Intelligence, UAI'95*; ACM: New York, NY, USA, 1995; pp. 338–445. [CrossRef]
12. Platt, J.C. Sequential Minimal Optimization: A Fast Algorithm for Training Support Vector Machines. *MSRTR Microsoft Res.* **1998**, *3*, 88–95.
13. Aha, D.W.; Kibler, D.; Albert, M.K. Instance-based learning algorithms. *Mach. Learn.* **1991**, *6*, 37–66. [CrossRef]
14. Freund, Y.; Schapire, R.E. Experiments with a New Boosting Algorithm. In Proceedings of the 13th International Conference on Machine Learning, Bari, Italy, 3–6 July 1996; pp. 148–156.
15. Quinlan, J.R. *C4.5: Programs for Machine Learning*, 1st ed.; Morgan Kaufmann: San Mateo, CA, USA, 1993.
16. Breiman, L. Random Forests. *Mach. Learn.* **2001**, *45*, 5–32. [CrossRef]
17. Srinarong, N.; Mongkolnavin, J. A Development of Personality Recognition Model from Conversation Voice in Call Center Context. In *ACM International Conference Proceeding Series*; Association for Computing Machinery: Bangkok, Thailand, 2021; pp. 1–5. [CrossRef]
18. Adi, G.Y.N.N.; Tandio, M.H.; Ong, V.; Suhartono, D. Optimization for Automatic Personality Recognition on Twitter in Bahasa Indonesia. *Procedia Comput. Sci.* **2018**, *135*, 473–480. [CrossRef]
19. Ren, Z.; Shen, Q.; Diao, X.; Xu, H. A sentiment-aware deep learning approach for personality detection from text. *Inf. Process. Manag.* **2021**, *58*, 102532. [CrossRef]
20. Xue, D.; Wu, L.; Hong, Z.; Guo, S.; Gao, L.; Wu, Z.; Zhong, X.; Sun, J. Deep learning-based personality recognition from text posts of online social networks. *Appl. Intell.* **2018**, *48*, 4232–4246. [CrossRef]
21. Sher Khan, A.; Ahmad, H.; Zubair Asghar, M.; Khan Saddozai, F.; Arif, A.; Ali Khalid, H. Personality Classification from Online Text using Machine Learning Approach. *Int. J. Adv. Comput. Sci. Appl.* **2020**, *11*, 460–476.
22. Agung, A.A.G.; Yuniar, I. Personality assessment website using DISC: A case study in information technology school. In Proceedings of the 2016 International Conference on Information Management and Technology (ICIMTech), Bandung, Indonesia, 16–18 November 2016; pp. 72–77. [CrossRef]
23. Milne, N.; Louwen, C.; Reidlinger, D.; Bishop, J.; Dalton, M.; Crane, L. Physiotherapy students' DiSC behaviour styles can be used to predict the likelihood of success in clinical placements. *BMC Med. Educ.* **2019**, *19*, 1–15. [CrossRef]
24. Chigova, E.A.; Plyushch, I.V.; Leskova, I.V. Organization of structured interaction on the base of psychographic characteristics within the model of personality traits DISC. *IOP Conf. Ser. Mater. Sci. Eng.* **2019**, *483*, 012097. [CrossRef]
25. Jarvis, S. Grounding lexical diversity in human judgments. *Lang. Test.* **2017**, *34*, 537–553. [CrossRef]
26. Bougé, K. Download Stop Words. Available online: https://sites.google.com/site/kevinbouge/stopwords-lists (accessed on 28 January 2022).
27. Anthony, L. Programming for Corpus Linguistics. In *A Practical Handbook of Corpus Linguistics*; Paquot, M., Gries, S.T., Eds.; Springer: Cham, Switzerland, 2020; pp. 181–207. [CrossRef]
28. Padró, L.; Stanilovsky, E. FreeLing 3.0: Towards Wider Multilinguality. In Proceedings of the Eighth International Conference on Language Resources and Evaluation (LREC'12), Istanbul, Turkey, 21–27 May 2012; European Language Resources Association: Paris, France, 2012; pp. 2473–2479.
29. Goldberg, Y. *Neural Network Methods for Natural Language Processing*; Morgan & Claypool: Williston, VT, USA, 2017. [CrossRef]
30. Hall, M.A. Correlation-based Feature Selection for Machine Learning. Ph.D. Thesis, University of Waikato, Hamilton, New Zealand, 1999.
31. Sharma, A.; Dey, S. Performance Investigation of Feature Selection Methods and Sentiment Lexicons for Sentiment Analysis. *IJCA Spec. Issue Adv. Comput. Commun. Technol. HPC Appl.* **2012**, *3*, 15–20.
32. Kohavi, R.; John, G.H. Wrappers for feature subset selection. *Artif. Intell.* **1997**, *97*, 273–324. [CrossRef]
33. Chawla, N.V.; Bowyer, K.W.; Hall, L.O.; Kegelmeyer, W.P. SMOTE: Synthetic Minority Over-sampling Technique. *J. Artif. Intell. Res.* **2002**, *16*, 321–357. [CrossRef]
34. Schratz, P.; Muenchow, J.; Iturritxa, E.; Richter, J.; Brenning, A. Hyperparameter tuning and performance assessment of statistical and machine-learning algorithms using spatial data. *Ecol. Modell.* **2019**, *406*, 109–120. [CrossRef]

35. Sarker, I.H. Machine Learning: Algorithms, Real-World Applications and Research Directions. *SN Comput. Sci.* **2021**, *2*, 160. [CrossRef] [PubMed]
36. Powers, D.M.W. Evaluation: From Precision, Recall And F-Measure to Roc, Informedness, Markedness & Correlation. *J. Mach. Learn. Technol.* **2011**, *2*, 37–63.
37. Hernández, Y.; Arroyo-Figueroa, G.; Sucar, L.E. A model of affect and learning for intelligent tutors. *J. Univers. Comput. Sci.* **2015**, *21*, 912–934. [CrossRef]
38. Bestgen, Y.; Granger, S. Quantifying the development of phraseological competence in L2 English writing: An automated approach. *J. Second Lang. Writ.* **2014**, *26*, 28–41. [CrossRef]
39. Wołk, A.; Chlasta, K.; Holas, P. Hybrid approach to detecting symptoms of depression in social media entries. *arXiv* **2021**, arXiv:2106.10485.

applied sciences

Article

Bert-Based Latent Semantic Analysis (Bert-LSA): A Case Study on Geospatial Data Technology and Application Trend Analysis

Quanying Cheng [1,2], Yunqiang Zhu [1,3,*], Jia Song [1,3], Hongyun Zeng [4], Shu Wang [1], Kai Sun [1] and Jinqu Zhang [5]

1 State Key Laboratory of Resources and Environmental Information System, Institute of Geographic Sciences and Natural Resources Research, Chinese Academy of Sciences, Beijing 100101, China; chengqy.18b@igsnrr.ac.cn (Q.C.); songj@lreis.ac.cn (J.S.); wangshu@igsnrr.ac.cn (S.W.); sunk@lreis.ac.cn (K.S.)
2 College of Resources and Environment, University of Chinese Academy of Sciences, Beijing 100049, China
3 Jiangsu Center for Collaborative Innovation in Geographical Information Resource Development and Application, Nanjing 210023, China
4 School of Earth Sciences, Yunnan University, Kunming 650500, China; hy_zeng@ynu.edu.cn
5 School of Computer Science, South China Normal University, Guangzhou 510000, China; zjq@scnu.edu.cn
* Correspondence: zhuyq@lreis.ac.cn

Abstract: Geospatial data is an indispensable data resource for research and applications in many fields. The technologies and applications related to geospatial data are constantly advancing and updating, so identifying the technologies and applications among them will help foster and fund further innovation. Through topic analysis, new research hotspots can be discovered by understanding the whole development process of a topic. At present, the main methods to determine topics are peer review and bibliometrics, however they just review relevant literature or perform simple frequency analysis. This paper proposes a new topic discovery method, which combines a word embedding method, based on a pre-trained model, Bert, and a spherical k-means clustering algorithm, and applies the similarity between literature and topics to assign literature to different topics. The proposed method was applied to 266 pieces of literature related to geospatial data over the past five years. First, according to the number of publications, the trend analysis of technologies and applications related to geospatial data in several leading countries was conducted. Then, the consistency of the proposed method and the existing method PLSA (Probabilistic Latent Semantic Analysis) was evaluated by using two similar consistency evaluation indicators (i.e., U-Mass and NMPI). The results show that the method proposed in this paper can well reveal text content, determine development trends, and produce more coherent topics, and that the overall performance of Bert-LSA is better than PLSA using *NPMI* and U-Mass. This method is not limited to trend analysis using the data in this paper; it can also be used for the topic analysis of other types of texts.

Keywords: trend analysis; topic modeling; Bert; geospatial data technology and application

Citation: Cheng, Q.; Zhu, Y.; Song, J.; Zeng, H.; Wang, S.; Sun, K.; Zhang, J. Bert-Based Latent Semantic Analysis (Bert-LSA): A Case Study on Geospatial Data Technology and Application Trend Analysis. *Appl. Sci.* **2021**, *11*, 11897. https://doi.org/10.3390/app112411897

Academic Editors: Arturo Montejo-Ráez, Salud María Jiménez-Zafra and Rafael Valencia-Garcia

Received: 7 November 2021
Accepted: 12 December 2021
Published: 14 December 2021

1. Introduction

Geographical data describes a location and its spatial characteristics attributes. With the rapid development of information technology, geospatial data has become an indispensable data resource for research and application in many fields, such as natural resource management, disaster emergency management, climate change and precision agriculture, etc. [1]. The technologies and applications related to geospatial data are also constantly advancing and upgrading, making new ways of thinking possible, so identifying technologies and applications is helpful to foster and fund further innovation. Through topic analysis, we can identify new research hotspots, acquire knowledge transfer processes [2], and quickly analyze the entire development process of research areas, thus benefiting researchers who are interested in a topic. In addition, it can also provide signals for paradigm shifts in discipline development [3]. For individuals, the results of topic analysis provide an overview of the evolution of the research field and are helpful to us in grasping research trends, keeping up-to-date with the latest research trends in the field, and seeking scientific collaborators [4].

The extensive scientific literature provides researchers with a wealth of information, which is also an important data resource for analyzing the development trends. However, the time and cost for understanding and analyzing the complex dynamics of current technical approaches related to geospatial data are increasing [5]. Therefore, researchers try to save time and reduce costs by seeking automated analysis methods, which allow them to quickly find the most important information, in order to make critical decisions without consulting voluminous literature [6]. At present, there are two main methods for identifying topics in texts. One is a qualitative appraisal method used by the academia, which is known as expert overview. The other is a scientometrics-based approach. Expert overview is a comprehensive and effective method for topic identification, but it is highly dependent on expert opinion, which is time- and energy-consuming. In addition, expert overview is becoming an increasingly inefficient means, due to the explosive growth of the scientific literature. In comparison, the bibliometric approach uses related papers for statistical frequency analysis, and simply captures information such as citation statistics to identify topics. An article with a high citation count is considered as a high-value one [7]. The structural and geospatial developments of industrial symbiosis as subfields of industrial ecology have been explored by using bibliometrics [8]. A statistical approach to bibliometric data from U.S. institutions has also been used to identify institutional hotspots on a map where many high-impact papers are published. The bibliometrics-based approach plays a role in identifying the development of trends, but it lacks consideration of the content of the literature texts themselves.

Previous studies are mainly based on traditional methods, which merely review relevant literature or conduct simple frequency analysis without providing insights beyond revealing information about the contents of literature texts [9]. Therefore, it is urgent to conduct comprehensive and in-depth trend analysis of literature texts. Recently, a popular method involves text analysis techniques to identify the main viewpoints and trends of the research [10], for example, by using textual data such as user comments, papers, and patents to analyze keywords or social networks [11–16]. In particular, topic modeling has recently attracted the attention of trend analysis researchers, since the main purpose of trend analysis based on textual data is to detect the upward and downward trends in the frequency of each topic in the target document [17]. Topic modeling originates from early latent semantic analysis (LSA), which aims to discover meaningful semantic structures in the corpus [18], with a focus on keyword extraction. The representative approaches are through the use of TF-IDF, which is based on statistical features [19,20], TextRank, based on word graph models [21,22], and Probabilistic Latent Semantic Analysis (PLSA) and Latent Dirichlet Allocation (LDA), based on topic models [23]. PLSA and LDA are the most widely used probabilistic techniques in topic modeling [24]. PLSA is a latent variable model based on co-occurrence data item-document matrices, also known as the ASPECT model [25]. The superiority of PLSA is demonstrated by its comparison with k-means and LSA [26]. As a variant or extension of PLSA, LDA uses Bayesian methods for parameter estimation to compensate for the incompleteness of PLSA, in terms of topic probability distribution. However, it is difficult to explain LDA without prior knowledge of the underlying topics and hyperparameters. All these approaches mentioned above ignore the most important semantic features of words and the semantic associations between words. Although pre-trained word embeddings are widely used in classification tasks, its application in topic modeling mainly focuses on probability techniques, such as in LDA [27–31], and there is also preliminary work using these embeddings to evaluate the consistency of topic models [32,33].

The probability-based statistical topic modeling methods aforementioned are unable to capture the whole context of a document, as they usually consider only a single graph representation of a word [1]. Alternatively, the n-gram representation that considers multiple words simultaneously can be used, but the efficiency of the model rapidly decreases due to the dimension disaster [34]. Therefore, Bert quantifies words as a vector, which takes into account the context and locates similar words in a similar space to address the limitations of this representation. Although this representation, based on pre-trained models, is widely used and its performance has been validated in recent text analysis, few attempts have yet been made to develop new topic models that are based on Bert. At present, only a few studies have adopted semantic embedding in topic analysis. A recent study uses Bert to generate text semantics as the input for topic classification [35]. While literature abstracts are a core corpus that reveal the distribution of research topics [36,37], these classic scientific journals can extract topic terms to analyze the trends in the research fields [38,39].

In conclusion, both the existing types of studies on geospatial data trend analysis have their limitations. Studies based on screening reviews require a lot of time and energy to screen and summarize all the literature. Methods based on bibliometric analysis are not suitable for discovering potential patterns in fields related to geospatial data. In addition, topic models, which are often used for trend analysis in other fields, are usually based on single-word vector representations that are non-contextual and sparse. In order to overcome these problems, this paper proposes a new topic modeling approach, which applies a new word embedding method in the field of computer linguistics to topic models and can help extract textual topics, namely the Bert-based Latent Semantic Analysis (Bert-LSA) topic modeling approach. It utilizes the Bert contextual word embedding algorithm and spherical k-means clustering to combine context embedding and clustering in a coordinated way, and finally assigns topics to documents. The proposed method is used to conduct a specific trend analysis of the technologies related to geospatial data, which can serve as an advanced and useful alternative method to extract meaningful topics involved in the current trend of geospatial data.

The structure of this paper is as follows. In Section 2, the textual data sources and data pre-processing are introduced. A new topic modeling approach is proposed to compensate for the limitations of existing technologies, which will be discussed in detail in Section 3. Section 4 presents the results of the trend analysis. Section 5 evaluates the proposed method in contrast with existing methods for topic consistency. Section 6 discusses the results and conclusions of this study.

2. Materials

Figure 1 shows the process of data collection and pre-processing used to conduct topic modeling about geospatial data. For the analysis of geospatial data technologies and application trends, abstracts of papers related to geospatial data were collected from two paper databases, namely, ScienceDirect and Scopus. A total of 609 abstracts of papers were collected, which contained terms such as "geospatial data "from 2016 to 2020. In the ScienceDirect database, the query statement was "TITLE-ABS-KEY (geospatial AND data)", and in the Scopus database, the query statement was based on keywords (geospatial data). In the query result, only that were those connected with two words: "geospatial" and "data" were selected. In order to ensure that each abstract contains rich information, only those abstracts with more than 180 characters were selected, and 266 abstracts were finally analyzed. For those collected data, each abstract was used as an input to the Bert model, and obtained the corresponding word vector.

Figure 2 represents the number of papers related to geospatial data published per year and by country. The number of published papers has been consistently increasing from 2016 to 2019, then with a significant decline in 2020 (Figure 2a). The number of papers in the top seven countries, namely, the USA, China, India, Germany, UK, RF (Russian Federation), and Italy, accounts for about 56% of the total number of papers, with the USA having the largest number of published papers (Figure 2b).

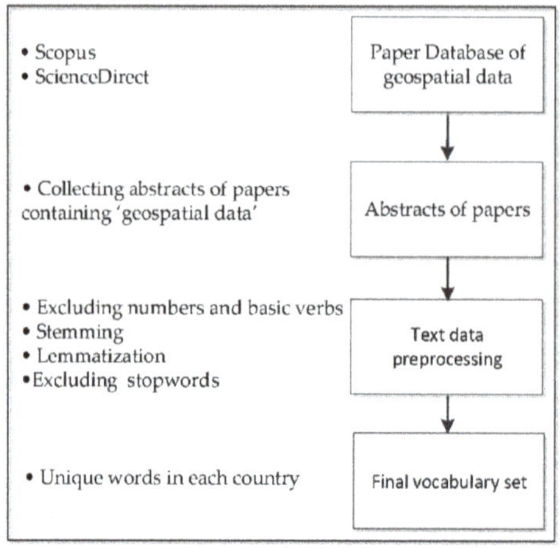

Figure 1. Process of data collection and pre-processing.

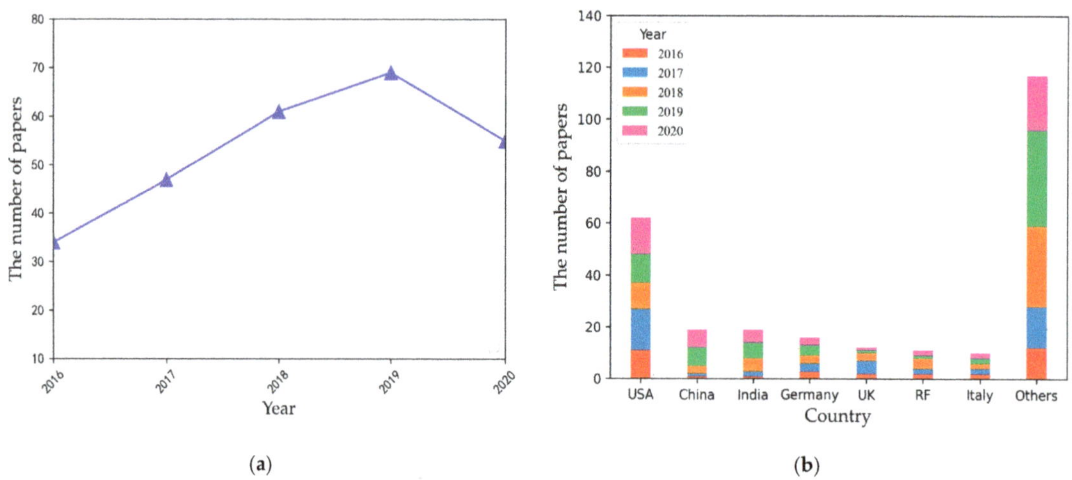

Figure 2. (**a**) Number of papers related to geospatial data published per year; (**b**) Number of papers related to geospatial data published by country.

3. Methodology

3.1. Overall Framework

In this paper, the Bert-LSA topic model is proposed, which combines Bert and spherical k-mean clustering. The model is featured due to its ability to fully take into account the context of documents and to overcome the shortcomings of existing statistical models. Figure 3 depicts the whole process of document topic generation, which is mainly divided into four steps as follows.

- Step 1: All documents are taken as corpus, and the m-dimensional word vector corresponding to the documents is obtained by the Bert model, which is denoted as $v_i \in V^m$, where v_i is the word vector and V^m is the m-dimensional vector space. Note that here the word vectors are obtained after the documents are processed as inputs into the Bert model, rather than being directly obtained from the pre-trained model.

- Step 2: All vectorized words undergo spherical k-means clustering, which first initializes the centroid according to the K value, and then calculates the spherical distance from each word vector v_i to the centroid. According to the distance value, v_i will be assigned to different categories, which will be iterated until convergence. Finally, K clusters are obtained, each of which is called a topic.
- Step 3: The graphical representation of the generation method for each particular document vector d_j, $j = 1, 2, \ldots, D$, is shown in Figure 4, which is obtained by multiplying the m-dimensional vector v_i of all words in the corpus with the term document matrix $Num \times D$, where Num is the number of words in the corpus and D is the number of documents. See Section 3.4 for details.
- Step 4: Figure 5 depicts the process of document topic generation. The cosine distance between each document and the word vector contained in each topic in Step 2 is calculated in turn, and each document is assigned to a different topic by using a topic assignment method. See Section 3.5 for details.

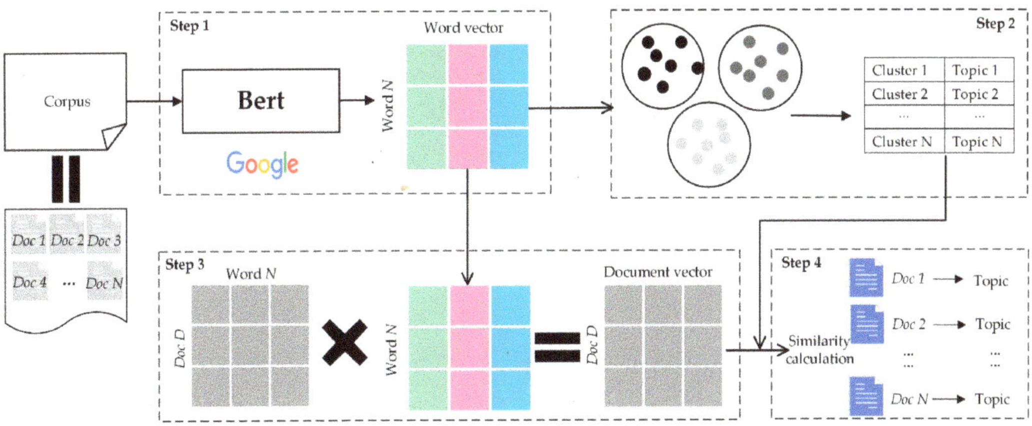

Figure 3. Document topic generation method.

3.2. Word Vector Generation Based on Bert

Bert (Bidirectional Encoder Representations from Transformers) [40] is a pre-trained language model released by Google that has occupied a state-of-the-art position in 11 tasks in the NLP (natural language processing) field. It is based on a multi-layer bidirectional transformer [41], and the framework consists of two steps: pretraining and fine-tuning. In the pretraining stage, it is trained on existing unlabeled text in advance and is released as a general language model. In the fine-tuning stage, it can be fine-tuned using learning data, according to the task to be performed [42,43].

In this paper, the pre-trained Bert model called "Bert-Base, Uncased" [44] is used to generate word vectors and represent the semantics of words. The reason for choosing this model is that the Bert-Base model is smaller than the Bert-Large model, and the language used in the research is only English and does not need to be case-sensitive. By entering sentences of each document, we obtain the word vector corresponding to each word in the sentence, which can accurately represent the semantic meaning of the word in its context. Python executes all the word vector generations mentioned in this paper by using the API released by [45]. It is an open-source Bert service, which allows users to use the Bert model by calling the service without paying attention to the details of Bert implementation. The important parameters max_seq_len and pooling_strategy are set to 512 and NONE respectively.

Bert takes an input of a sequence of no more than 512 tokens and outputs the representation of the sequence that has one or two segments. The first token of the sequence is always [CLS] which contains the special classification embedding, and the other special token [SEP] is used for separating segments. Bert takes the final hidden state, h, of the first token [CLS] as the representation of the whole sequence. WordPiece embedding [46] is used and split word pieces is denoted with ##. So, the statistics of the length of the documents in the datasets are based on the word pieces [47].

3.3. Spherical k-Means Clustering

The spherical k-means method is introduced for clustering sets of sparse text data. This method is based on a vector space model, whose basic principle is to describe the degree to which two vectors point in the same direction by their similarity, rather than their length [48]. For example, in the vector space model V^m, for each word vector $w_i \in V^m, i = 1, 2, \ldots, N$, the inner product (Formula (1)) of two vectors is used to express the semantic similarity, where the column vectors are normalized (Formula (2)) to the unit length of the Euclidean norm, with the aim of assigning equal weights to each of the n points in the data set. Of course, we obtained these vectors after entering the text into the Bert model, rather than directly from the Bert model.

$$\cos(\theta_{x,\ y}) =: x^T y \tag{1}$$

$$\cos(\theta_{x,\ y}) = ||x||_2||y||_2 \tag{2}$$

In Formula (1), it describes the result of the normalization of x and y. In Formula (2), it is the definition of the standard inner product.

Finding clustering centers is also very important. For the clustering vector $v(i) \in 1, 2, \ldots, v$ and w_i, the center of clustering is to find the minimum cosine value between w_i and c_v, $v = 1, 2, \ldots, v$ [49]. To find the number of clusters in the dataset used in the experimentation, we ran the spherical clustering algorithm for a range of multiple values and compared the results obtained for each value.

3.4. Example of Document Vector Generation

The document vector is generated by multiplying the word vector matrix ($\boldsymbol{A}$) and the term document matrix ($\boldsymbol{B}$) in the figure below, i.e., $\boldsymbol{A} \times \boldsymbol{B}$. The word vector matrix ($\boldsymbol{A}$) is composed of m-dimensional word vectors obtained from all the words contained in the document according to the method in Section 3.2. The term document matrix ($\boldsymbol{B}$) is obtained by combining the word frequencies of the words contained in a single document, provided that the order of the words in a single document (i.e., the columns in **B**) needs to be the same as the position of corresponding words in matrix ($\boldsymbol{A}$) (i.e., the rows in **A**), and if the words contained in matrix **A** do not appear in a single document (e.g., ***DOC* 1**), the value of that position is set to 0.

3.5. Method of Document Topic Determination

Section 3.3 describes how we obtained multiple topics after clustering all documents, including Topic 1, Topic 2, Topic 3, and Topic 4 in Figure 5. Section 3.4 describes how we obtained the vector of each document, Doc N, shown in Figure 5. The method of assigning documents to topics is shown in Figure 5. Taking the first document as an example, firstly, the average value of the five words (boldface in Topic 1) with the largest cosine distance between the document and the first topic (Topic 1) was obtained, which were taken as the similarity between the document and Topic 1. By analogy, the similarity values between the document and other topics were then calculated. Finally, the similarity values between the document and all topics were compared, and the document was assigned to its corresponding topic with the maximum similarity value. Other documents were calculated in the same way, and finally, all documents were assigned to different topics.

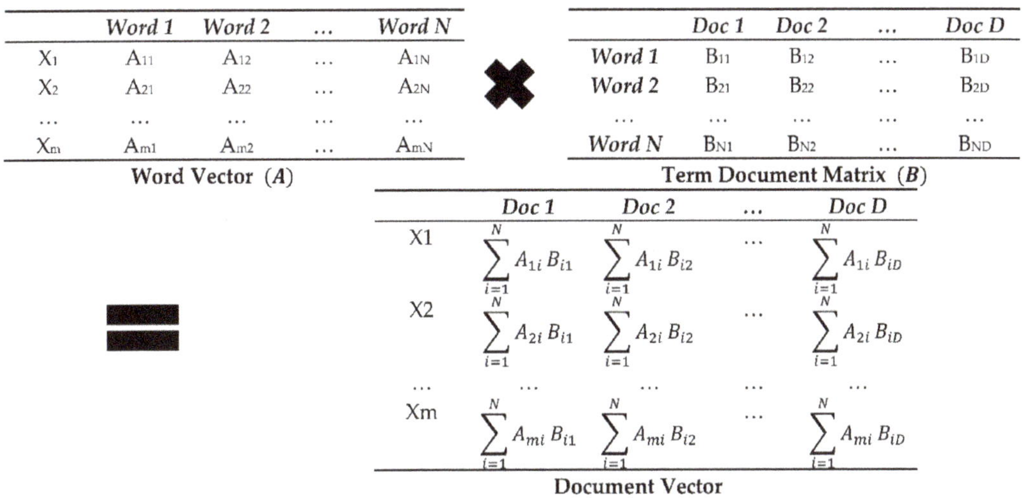

Figure 4. Example of document vector generation.

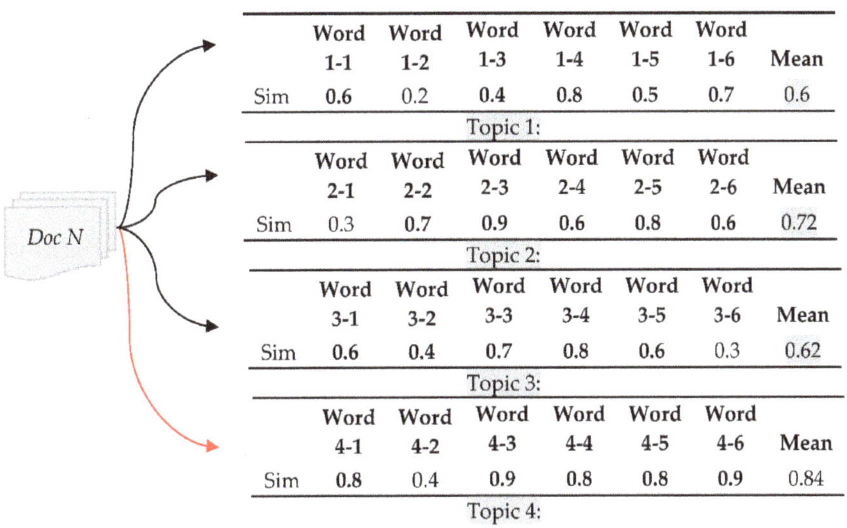

Figure 5. Method of document topic determination.

4. Trend Analysis Based on Bert_LSA

4.1. Topic Selection

In the trend analysis based on Bert_LSA, first, Bert is used to obtain the vector of words contained in the document, and the acquisition method and parameter setting are detailed in Section 3.2. Then, the spherical k-means clustering algorithm is applied for clustering, and the optimal number of clusters k is determined by the elbow method. The core index of the elbow method is Sum of the Squares Errors (SSE), and the formula is as follows:

$$SSE = \sum_{i=1}^{k} \sum_{p \in C_i} |p - m_i|^2 \quad (3)$$

where C_i is the i-th cluster, p is the sample point in C_i, m_i is the centroid of C_i, and SSE is the clustering error of all samples, which represents the clustering effect.

However, when the effect of the elbow method is not obvious, it is combined with the Silhouette Coefficient method to jointly determine the number of clusters. The Silhouette Coefficient is an index to evaluate the degree of density and dispersion of the class. The calculation method is listed as follows, and its value ranges between [−1, 1]. The larger the value is, the more reasonable it is [50].

$$S(i) = \frac{b(i) - a(i)}{\max\{a(i), b(i)\}} \tag{4}$$

where $a(i)$ represents the average value of the dissimilarity of the i vector to other points within the same cluster, $b(i)$ represents the minimum value of the average dissimilarity of the i vector to other clusters.

4.2. The Result of Trend Analysis

According to the topic number selection method in Section 4.1, the cluster numbers of USA, China, India, Germany, UK, Russian Federation, Italy, and Others were finally set to 6, 5, 6, 5, 7,6, 7, and 4, respectively. Table 1 shows the results of topic modeling using Bert_LSA.

Table 1. The topic analysis results based on Bert_LSA model, and the percentage indicates the proportion of the country in all documents.

USA		**China**	
Topic	Ratio (%)	Topic	Ratio (%)
Water/Polarhub/Enviroatlas	21%	Landscape/Livability/Government	29.4%
Building/Air/BIM	19.4%	Extraction/Metadata/Information	23.5%
Fire/Risk/Precipitation	17.7%	Soybean/Crop/Area/Policy	17.6%
GEE/ Framework / Model	17.7%	Geohazards/Landslide/ Anomaly	14.8%
Stream/Land/Temperature	12.9%	Multisource/Search/Metadata	14.7%
Greenery /Heat	11.3%		
India		**Germany**	
Topic	Ratio (%)	Topic	Ratio (%)
Cloud/computing/Hadoop/Share	26.3%	Navigation/Prediction/Street	25%
Flood/Distribution/Coastline	21.1%	Visualization/Database/Datasets	25%
SDI/WPS/Framework	21.1%	Change/Land/Observation	18.8%
Stormwater/ Groundwater/Conserve	10.5%	Stress/Life/Measurement	18.8%
Land/Investor/Vicinity	10.5%	Demand/Heat/Supply	12.4%
School/Platform/location	10.5%		
UK		**Russian Federation**	
Topic	Ratio (%)	Topic	Ratio (%)
Geohazards/Household/Landslide	16.7%	Risk/Environment/Management	27.3%
Point cloud/Framework	16.7%	Network/Generation/Transport	18.2%
Feature/Attribute/Database	16.7%	Customer/Bank/Transaction	18.2%
Mangrove/Fishing/Intensity	16.7%	Monitoring/Change/Climate/Season	18.2%
BIM/Project/Evaluation	16.7%	Client/Cloud/computing/Device	9.1%
Weather/MCSA (Multi-Channel Sequences Analysis)/Condition	8.3%	Image/Anomaly/Validation	9%
Network/Source/Accessibility	8.2%		
Italy		**Others**	
Topic	Ratio (%)	Topic	Ratio (%)
Geo/Disaster/Cluster	30%	Land/Housing/City/Water	41.2%
Challenge/Spiral/OpenGIS	20%	Village/Fire/model/System	23.5%
Crop/Precision/Classification	10%	Datasets/Soil/Accuracy	23.5%
Landslide/Hazard/Flood	10%	SDI/Web/Collection	11.8%
Map/Territory/Accessment	10%		
Location/Behavior/Category	10%		
GNSS/Radar/Remote/sensing	10%		

The results of the trend analysis show that the focus of each country's concern is different. USA, for example, focuses on the environment, buildings, and fires. However, China pays attention to livability, information extraction, and crops, while words like government and policy also appear. India focuses on information technology, such as cloud computing, SDI and WPS, etc., as well as focuses on disaster events such as floods. Italy and others also pay attention to disaster-related content. Germany, UK, and the Russian Federation all focus on content related to climate change. In general, countries pay more attention to disaster events (e.g., fires and floods), and related information technologies, such as cloud computing, Hadoop and GEE, etc., have also received higher attention. This is also a good indication that our proposed method can successfully identify the current technologies and application trends related to the use of geospatial data, which can quickly provide research hotspots for relevant researchers, especially those who are not specialized in GIS. In this way, it considerably saves the time needed to read a large amount of literature, which is of practical significance.

5. Quantitative Evaluation

5.1. Evaluation Method

The methods of evaluating topic models mainly include perplexity and topic consistency. The perplexity has its own merits, as it can evaluate probability-based topic models well, whereas in non-probability-based topic models, these methods do not capture semantic consistency between words [51]. Topic consistency can be used to measure whether words within a topic are coherent, i.e., if a group of terms are consistent with each other, then these terms are coherent. For a specific topic, the semantic similarity between words in the topic determines the degree of coherence of the topic, so topic consistency can be measured by the semantic similarity between words in the topic [52–55]. The greater the consistency value of the topic is, the more coherent the words of each topic will be. To evaluate the non-probabilistic topic model proposed in this paper, the following two consistency measures were used: (1) University of Massachusetts (*U_Mass*) [56], (2) Normalized Pointwise Mutual Information (*NPMI*) [57].

U_Mass is defined as:

$$U\_Mass = \frac{2}{N \times (N-1)} \sum_{i=1}^{N-1} \sum_{j=i+1}^{N} \log \frac{P(w_i, w_j) + \epsilon}{P(w_j)} \tag{5}$$

where $P(w_i, w_j)$ is the joint probability of two words w_i and w_j. A small value for ϵ is chosen to avoid calculating the logarithm of 0.

NPMI is defined as:

$$NPMI = \frac{1}{K} \sum_{K} \frac{2}{T(T-1)} \sum_{1 \le i \le j \le T} \frac{\log_2\left(\frac{p(w_i, w_j)}{p(w_i)p(w_j)}\right)}{-\log_2 p(w_i, w_j)} \tag{6}$$

where K is the number of topics, and each topic consists of the T most relevant word. $p(w_i, w_j)$ is the probability that the word pair (w_i, w_j) co-occurs in a document, and $p(w_i)$ is the probability that the word w_i appears in the document.

5.2. Evaluation Result

The method proposed in this paper was compared with PLSA in terms of its topic consistency, where the PLSA implementation uses open-source code (https://github.com/yedivanseven/PLSA) (accessed on 7 November 2021). Figures 6 and 7 show the average topic consistency calculated by using PLSA and Bert-LSA, respectively, where the abscissa is the number of words N selected in each topic, with values of N ranging from 3 to 13, and the ordinate is the topic consistency value. Here, the value of topic consistency is the average of the corresponding topic consistency values for all countries when different numbers of topics are selected. When evaluated with the U-mass method, the topic consistency of the PLSA model remains almost constant as N increases, and its value is generally low. Similarly, the topic consistency of the Bert-LSA model gradually decreases as N increases, but its value is generally higher than that of PLSA model, which means that the Bert-LSA model performs better than the PLSA model.

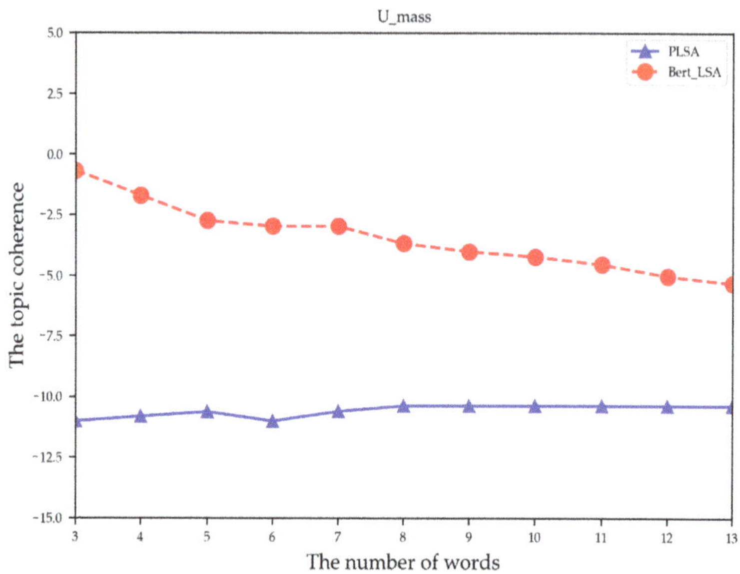

Figure 6. Topic consistency values of PLSA and Bert-LSA models obtained by the U-Mass method.

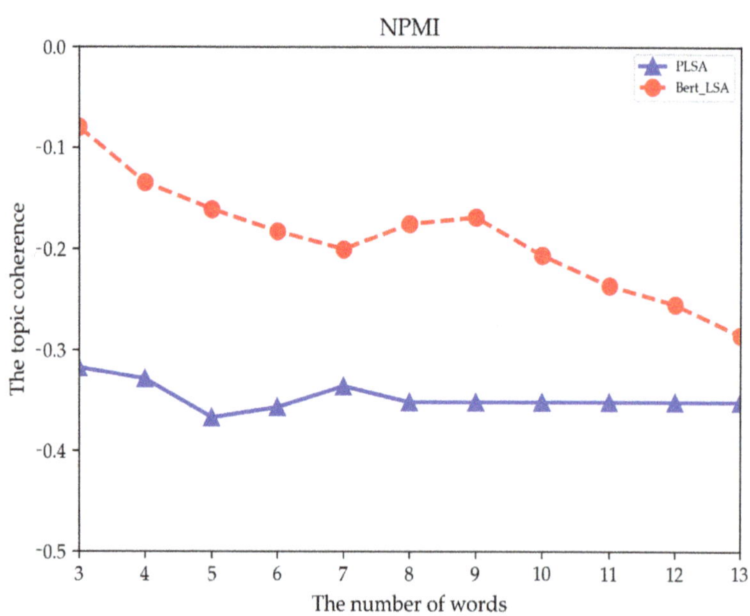

Figure 7. Topic consistency values of PLSA and Bert-LSA models obtained with the *NPMI* method.

When evaluated with the *NPMI* method, the topic consistency of the PLSA model decreases when the value of N is from three to five, increases when the value of N is from five to seven, and then remains basically unchanged thereafter. For the Bert-LSA model, the topic consistency keeps decreasing when the value of N is from three to seven, increases when the value of N is from seven to nine, and then keeps decreasing. On the whole, the Bert-LSA model still outperforms PLSA model.

6. Conclusions and Discussion

In this paper, a new method of topic identification has been proposed. First, a word embedding algorithm was adopted that was based on a pre-trained model, which generates a word representation that can capture the context of a document. After that, we used a spherical k-means clustering algorithm to construct topic clusters. Finally, a topic assignment method was used to assign documents to different topics. The assignment process was in order to calculate the similarity between documents and topics.

The method proposed in this paper was applied to the literature abstracts related to geospatial data. First, it shows the characteristics of geospatial data technology and application development trends in related research in several leading countries. Second, the topic coherence of this method was evaluated by using U-Mass and *NPMI*, and its performance was compared with that of the existing method, PLSA. The results show that the proposed method can produce highly coherent topics. The research in this paper provides new ideas for the trend analysis of technologies and applications related to geospatial data, and helps professionals engaged in research related to geospatial data to identify their future research directions at any time. In addition, this method captures the development trends of related technical fields through text, which can be used as an information tool for anyone who is responsible for strategic decision-making in sectors related to geospatial data, to determine the prospect and market of the fields. This paper also has some shortcomings, for example, it is unable to successfully identify the topic when the number of texts is extremely large. In the future, we will work hard on topic modeling for a large number of texts.

Author Contributions: Conceptualization, methodology, validation, formal analysis, Q.C. and Y.Z.; Software, Q.C.; Supervision, J.S., H.Z., J.Z., K.S. and S.W.; Funding Acquisition, Y.Z.; Writing—original draft preparation, Q.C. and Y.Z.; Writing—review and editing, Q.C. and Y.Z. All authors have read and agreed to the published version of the manuscript.

Funding: This research was funded by National Natural Science Foundation of China (grant numbers: 42050101, 41771430, 41631177) and the Strategic Priority Research Program of the Chinese Academy of Sciences (grant number: XDA23100100).

Institutional Review Board Statement: Not applicable.

Informed Consent Statement: Not applicable.

Data Availability Statement: The data presented in this study are available in Section 2.

Conflicts of Interest: The authors declare no conflict of interest.

References

1. Lu, Y.; Zhai, C.X. Opinion Integration through Semi-Supervised Topic Modeling. In Proceedings of the 17th International Conference on World Wide Web, Beijing, China, 21–25 April 2008; pp. 121–130.
2. Li, F.; Li, M.; Guan, P.; Ma, S.; Cui, L. Mapping publication trends and identifying hot spots of research on Internet health information-seeking behavior: A quantitative and co-word biclustering analysis. *J. Med. Internet Res.* **2015**, *17*, e81. [CrossRef] [PubMed]
3. Ying, D. Community detection: Topological vs. Topical. *J. Informetr.* **2011**, *5*, 498–514.
4. Chen, X.; Wang, S.; Tang, Y.; Hao, T. A bibliometric analysis of event detection in social media. *Online Inf. Rev.* **2019**, *43*, 29–52. [CrossRef]
5. Jacobi, C.; Atteveldt, W.V.; Welbers, K. Quantitative analysis of large amounts of journalistic texts using topic modelling. *Digit. J.* **2016**, *4*, 89–106. [CrossRef]
6. Alami, N.; Meknassi, M.; En-Nahnahi, N.; Adlouni, Y.E.; Ammor, O. Unsupervised Neural Networks for Automatic Arabic Text Summarization Using Document Clustering and Topic modeling. *Expert Syst. Appl.* **2021**, *172*, 114652. [CrossRef]

7. Chertow, M.R.; Kanaoka, K.S.; Park, J. Tracking the diffusion of industrial symbiosis scholarship using bibliometrics: Comparing across Web of Science, Scopus, and Google Scholar. *J. Ind. Ecol.* **2021**, *25*, 913–931. [CrossRef]
8. Bornmann, L.; Angeon, F.D.M. Hot and cold spots in the US research: A spatial analysis of bibliometric data on the institutional level. *J. Inf. Sci.* **2019**, *45*, 84–91. [CrossRef]
9. Kivikunnas, S. Overview of process trend analysis methods and applications. In Proceedings of the Erudit Workshop on Applications in Pulp and Paper Industry, Aachen, Germany, 9 September 1998; pp. 395–408.
10. Song, M.; Kim, S.Y.; Lee, K. Ensemble analysis of topical journal ranking in bioinformatics. *J. Assoc. Inf. Sci. Technol.* **2017**, *68*, 1564–1583. [CrossRef]
11. Hung, J.L. Trends of e-learning research from 2000 to 2008: Use of text mining and bibliometrics. *Br. J. Educ. Technol.* **2012**, *43*, 5–16. [CrossRef]
12. Hung, J.L.; Zhang, K. Examining mobile learning trends 2003–2008: A categorical meta-trend analysis using text mining techniques. *J. Comput. High. Educ.* **2012**, *24*, 1–17. [CrossRef]
13. Kim, H.J.; Jo, N.O.; Shin, K.S. Text Mining-Based Emerging Trend Analysis for the Aviation Industry. *J. Intell. Inf. Syst.* **2015**, *21*, 65–82. [CrossRef]
14. Kim, Y.M.; Delen, D. Medical informatics research trend analysis: A text mining approach. *Health Inform. J.* **2018**, *24*, 432–452. [CrossRef] [PubMed]
15. Terachi, M.; Saga, R.; Tsuji, H. Trends Recognition in Journal Papers by Text Mining. In Proceedings of the 2006 IEEE International Conference on Systems, Man and Cybernetics, Taipei, Taiwan, 8–11 October 2006; IEEE: Taipei, Taiwan, 2006; Volume 6, pp. 4784–4789.
16. Tseng, Y.H.; Lin, C.J.; Lin, Y.I. Text mining techniques for patent analysis. *Inf. Process. Manag.* **2007**, *43*, 1216–1247. [CrossRef]
17. Kang, H.J.; Kim, C.; Kang, K. Analysis of the Trends in Biochemical Research Using Latent Dirichlet Allocation (LDA). *Processes* **2019**, *7*, 379. [CrossRef]
18. Deerwester, S.; Dumais, S.T.; Furnas, G.W.; Landauer, T.K.; Harshman, R. Indexing by latent semantic analysis. *J. Am. Soc. Inf. Sci.* **1990**, *41*, 391–407. [CrossRef]
19. Guo, A.Z.; Tao, Y. Research and improvement of feature words weight based on TFIDF algorithm. In Proceedings of the 2016 IEEE Information Technology, Networking, Electronic and Automation Control Conference, Chongqing, China, 20–22 May 2016; pp. 415–419.
20. Li, J.Z.; Fan, Q.N.; Zhang, K. Keyword Extraction Based on tf/idf for Chinese News Document. *Wuhan Univ. J. Nat. Sci.* **2007**, *12*, 917–921. [CrossRef]
21. Mihalcea, R.; Tarau, P. TextRank: Bringing Order into Text. In Proceedings of the 2004 Conference Empirical Methods in Natural Language Processing, Stroudsburg, PA, USA, 25–26 July 2004; pp. 404–411.
22. Zhang, X.; Wang, Y.; Wu, L. Research on cross language text keyword extraction based on information entropy and TextRank. In Proceedings of the Information Technology, Networking, Electronic and Automation Control Conference, Chengdu, China, 15–17 March 2019; pp. 16–19.
23. Wei, H.X.; Gao, G.L.; Su, X.D. LDA-based word image representation for keyword spotting on historical Mongolian documents. In Proceedings of the International Conference on Neural Information Processing, Kyoto, Japan, 30 September 2016; pp. 432–441.
24. Blei, D.M.; Ng, A.Y.; Jordan, M.I. Latent Dirichlet allocation. *J. Mach. Learn. Res.* **2003**, *3*, 993–1022.
25. Hofmann, T. Probabilistic latent semantic analysis. In Proceedings of the Fifteenth Conference on Uncertainty in Artificial Intelligence, Stockholm, Sweden, 30 July 1999; pp. 289–296.
26. Newman, D.J.; Block, S. Probabilistic topic decomposition of an eighteenth-century American newspaper. *J. Am. Soc. Inf. Sci. Technol.* **2006**, *57*, 753–767. [CrossRef]
27. Xie, P.; Yang, D.; Xing, E. Incorporating word correlation knowledge into topic modeling. In Proceedings of the 2015 Conference of the North American Chapter of the Association for Computational Linguistics: Human Language Technologies, Denver, CO, USA, 31 May–5 June 2015; pp. 725–734.
28. Yang, Y.; Downey, D.; Boyd-Graber, J. Efficient methods for incorporating knowledge into topic models. In Proceedings of the 2015 Conference on Empirical Methods in Natural Language Processing, Lisbon, Portugal, 17–21 September 2015; pp. 308–317.
29. Das, R.; Zaheer, M.; Dyer, C. Gaussian LDA for topic models with word embeddings. In Proceedings of the 53rd Annual Meeting of the Association for Computational Linguistics and the 7th International Joint Conference on Natural Language Processing, Beijing, China, 26–31 July 2015; Volume 1, pp. 795–804.
30. Nguyen, D.Q.; Billingsley, R.; Du, L.; Johnson, M. Improving topic models with latent feature word representations. *Trans. Assoc. Comput. Linguist.* **2015**, *3*, 299–313. [CrossRef]
31. Moody, C.E. Mixing Dirichlet topic models and word embeddings to make lda2vec. *arXiv* **2016**, arXiv:1605.02019.
32. Callaghan, D.; Greene, D.; Carthy, J.; Cunningham, P. An analysis of the coherence of descriptors in topic modeling. *Expert Syst. Appl.* **2015**, *42*, 5645–5657. [CrossRef]
33. Ding, R.; Nallapati, R.; Xiang, B. Coherence-aware neural topic modeling. *Comput. Sci.* **2018**, arXiv:1809.02687.
34. Bengio, Y.; Ducharme, R.; Vincent, P.; Jauvin, C. A neural probabilistic language model. *J. Ma-Chine Learn. Res.* **2003**, *3*, 1137–1155.
35. Zhou, Y.; Li, C.; He, S.; Wang, X.; Qiu, Y. Pre-trained contextualized representation for chinese conversation topic classification. In Proceedings of the 2019 IEEE International Conference on Intelligence and Security Informatics (ISI), Shenzhen, China, 1–3 July 2019; IEEE: Piscataway, NJ, USA; pp. 122–127.

36. Ji, Q.; Pang, X.; Zhao, X. A bibliometric analysis of research on Antarctica during 1993–2012. *Scientometrics* **2014**, *101*, 1925–1939. [CrossRef]
37. Natale, F.; Fiore, G.; Hofherr, J. Mapping the research on aquaculture: A bibliometric analysis of aqua-culture literature. *Scientometrics* **2012**, *90*, 983–999. [CrossRef]
38. Sung, H.Y.; Yeh, H.Y.; Lin, J.K.; Chen, S.H. A visualization tool of patent topic evolution using a growing cell structure neural network. *Scientometrics* **2017**, *111*, 1267–1285. [CrossRef]
39. Qi, Y.; Zhu, N.; Zhai, Y.; Ding, Y. The mutually beneficial relationship of patents and scientific literature: Topic evolution in nanoscience. *Scientometrics* **2018**, *115*, 893–911. [CrossRef]
40. Devlin, J.; Chang, M.W.; Lee, K.; Toutanova, K. BERT: Pre-training of Deep Bidirectional Transformers for Language Understanding. *Comput. Sci.* **2018**, arXiv:1810.04805.
41. Vaswani, A.; Shazeer, N.; Parmar, N.; Uszkoreit, J.; Jones, L.; Gomez, A.N.; Kaiser, L.; Polosukhin, I. Attention is all you need. In Proceedings of the Advances in Neural Information Processing Systems, Long Beach, CA, USA, 4–9 December 2017; pp. 5998–6008.
42. Yoo, S.Y.; Jeong, O.R. Automating the expansion of a knowledge graph. *Expert Syst. Appl.* **2019**, *141*, 112965. [CrossRef]
43. Sun, C.; Qiu, X.; Xu, Y.; Huang, X. How to Fine-Tune BERT for Text Classification? In Proceedings of the China National Conference on Chinese Computational Linguistics, Kunming, China, 13 October 201; pp. 194–206.
44. Available online: https://storage.googleapis.com/bert_models/2018_10_18/uncased_L-12_H-768_A-12.zip (accessed on 7 November 2021).
45. Available online: https://github.com/hanxiao/bert-as-service (accessed on 7 November 2021).
46. Wu, Y.H.; Schuster, M.; Chen, Z.F.; Le, Q.V.; Norouzi, M.; Macherey, W.; Krikun, M.; Cao, Y.; Gao, Q.; Macherey, K.; et al. Google's neural machine translation system: Bridging the gap between human and machine translation. *Comput. Sci.* **2016**, arXiv:1609.08144.
47. Available online: https://spacy.io/ (accessed on 7 November 2021).
48. Dhillon, I.S.; Modha, D.S. Concept decompositions for large sparse text data using clustering. *Mach. Learn.* **2001**, *42*, 143–175. [CrossRef]
49. Buchta, C.; Kober, M.; Feinerer, I.; Hornik, K. Spherical k-means clustering. *J. Stat. Softw.* **2012**, *50*, 1–22.
50. Peter, R.J. Silhouettes: A graphical aid to the interpretation and validation of cluster analysis. *J. Comput. Appl. Math.* **1987**, *20*, 53–65.
51. Chang, J.; Gerrish, S.; Wang, C.; Boyd-Graber, J.L.; Blei, D.M. Reading tea leaves: How humans interpret topic models. In Proceedings of the Advances in Neural Information Processing Systems, Vancouver, BC, Canada, 7–10 December 2009; pp. 288–296.
52. Aletras, N.; Stevenson, M. Evaluating topic coherence using distributional semantics. In Proceedings of the 10th International Conference on Computational Semantics (IWCS 2013)—Long Papers, Potsdam, Germany, 19–22 March 2013; pp. 13–22.
53. Li, C.; Wang, H.; Zhang, Z.; Sun, A.; Ma, Z. Topic modeling for short texts with auxiliary word embed-dings. In Proceedings of the 39th International ACM Sigir Conference on Research and Development in Information Retrieval, Pisa, Italy, 17–21 July 2016; ACM: New York, NY, USA, 2016; pp. 165–174.
54. Mimno, D.M.; Wallach, H.M.; Talley, E.M.; Leenders, M.; McCallum, A. Optimizing semantic coherence in topic models. In Proceedings of the 2011 Conference on Empirical Methods in Natural Language Processing, John McIntyre Conference Centre, Edinburgh, UK, 27–31 July 2011; pp. 262–272.
55. Fu, Q.; Zhuang, Y.; Gu, J.; Zhu, Y.; Guo, X. Agreeing to Disagree: Choosing Among Eight Topic-Modeling Methods. *Big Data Res.* **2021**, *23*, 100173. [CrossRef]
56. Röder, M.; Both, A. Hinneburg, Exploring the space of topic coherence measures. In Proceedings of the 8th ACM International Conference on Web Search and Data Mining, Shanghai, China, 2–6 February 2015; ACM: New York, NY, USA, 2015; pp. 399–408.
57. Verlag, G.N.; Informatik, F. Von der Form zur Bedeutung: Texte automatisch verarbeiten/From Form to Meaning: Processing Texts Automatically. 2009. Available online: http://tubiblio.ulb.tu-darmstadt.de/98069/ (accessed on 7 November 2021).

applied sciences

MDPI

Article

A Corpus-Based Study of Linguistic Deception in Spanish

Ángela Almela

School of Arts, Universidad de Murcia, 30001 Murcia, Spain; angelalm@um.es

Featured Application: Statistical text classification.

Abstract: In the last decade, fields such as psychology and natural language processing have devoted considerable attention to the automatization of the process of deception detection, developing and employing a wide array of automated and computer-assisted methods for this purpose. Similarly, another emerging research area is focusing on computer-assisted deception detection using linguistics, with promising results. Accordingly, in the present article, the reader is firstly provided with an overall review of the state of the art of corpus-based research exploring linguistic cues to deception as well as an overview on several approaches to the study of deception and on previous research into its linguistic detection. In an effort to promote corpus-based research in this context, this study explores linguistic cues to deception in the Spanish written language with the aid of an automatic text classification tool, by means of an ad hoc corpus containing ground truth data. Interestingly, the key findings reveal that, although there is a set of linguistic cues which contributes to the global statistical classification model, there are some discursive differences across the subcorpora, yielding better classification results on the analysis conducted on the subcorpus containing emotionally loaded language.

Keywords: text classification; linguistic corpus; deception; linguistic cues; statistical analysis; discriminant function analysis

Citation: Almela, Á. A Corpus-Based Study of Linguistic Deception in Spanish. *Appl. Sci.* **2021**, *11*, 8817. https://doi.org/10.3390/app11198817

Academic Editors: Arturo Montejo-Ráez and Salud María Jiménez-Zafra

Received: 15 August 2021
Accepted: 17 September 2021
Published: 23 September 2021

1. Introduction

The distinction between truth and deception has garnered considerable attention from domains such as formal logic and psychological research. In the field of human kinetics, non-verbal communication has been claimed to play a key role in the detection of deception. More recently, verbal cues to deception have been also explored, as the investigation of linguistic cues to deception in written language has proved to be of utmost importance not only in the forensic context with statements written by witnesses and people involved in crimes, but also because in the increase seen by computer-mediated communication, where written texts constitute a fundamental element.

In the last decade, the field of natural language processing (NLP) has devoted considerable attention to the automatization of the process of deception detection, developing and employing a wide array of automated and computer-assisted methods for this purpose, (see, for example, Ott et al. [1] and Quijano-Sanchez et al. [2]). Researchers in [3] provide a thorough review of this activity. Similarly, another emerging research area is focusing on computer-assisted deception detection using linguistics, [4,5], with promising results. Thus, some computational approaches supervised by experts in the field are considered an efficient way to supplement and support criminal investigators, being of special interest to linguists, jurists, criminologists, and professionals in the field of communications.

Accordingly, in the present study, an overall review of the state of the art regarding linguistic cues to deception is provided, as well as an overview on several approaches to the study of deception and on previous research into its linguistic detection, describing the main controversies in the area (Section 2). Furthermore, the present author draws a distinction between software packages specifically developed for linguistic deception detection and other verbal assessment tools that are widely used for this and many other purposes (Section 3). Section 4 provides the materials and methods used in the experiment reported, whose results are presented and discussed in Section 5. Lastly, in light of the results obtained, some conclusions are drawn in Section 6 as well as some suggestions for further research.

All in all, this study makes a substantial contribution to the study of computational linguistic tools as an aid to deception detection and deepens the readers' understanding of the linguistic mechanisms underlying deceit. Interestingly, it offers a description of the linguistic cues to deception and promotes a contextualized study of deception, rather than dealing with broader dimensions of analysis.

2. Automated Deception Detection

This section presents the essentials of automated deception detection and advances some prime considerations that, from the present author's viewpoint, should be taken into account when conducting research in this area. For a whole account of theories and controversies in the area of deception detection in general, the reader may resort to [6], which reports past and current research on all aspects of lying and deception, as it is a comprehensive exploration of the state of the art from the combined perspectives of linguistics, philosophy, and psychology.

2.1. Essentials of Linguistic Deception Detection

As stated in [7], context has proved to be an important aspect in research and affects the relation between lying and language. These authors have developed a model called the contextual organization of language and deception (CoLD), which provides a framework including some crucial aspects of context for any deceptive communication. Thus, the nature of the linguistic data in the corpora is worth commenting on. Much has been discussed about the importance of deception in spontaneously produced language. Laboratory-produced lies have been criticized in forensic literature for not being very reliable; for instance, the authors in [8,9] suggest that further research should involve retrospective studies in law enforcement settings to study realistic responses with known outcomes. However, the strength of laboratory-produced data is the possibility for controlling variables and attributes so that the conclusions drawn are experimentally valid. What remains constant during such an experiment are the participants and the topics on which they write, which allows the researcher to avoid confounding intervening variables and to focus on deception in opinions and memories as the only plausible causal factor. Put another way, providing that some variation is observed regarding the dependent variables analyzed, this scientific control will allow the author to assure that the participants' situations are identical until they are asked to lie, and so the potentially new outcome may be attributed to the independent variable. The usefulness of this kind of corpus has indeed been proved in the forensic context, as shown in such studies as [10].

In this respect, it is also worth noting that there are two types of data: low-stakes deception, in which no harm can be done (it is well known that people lie in social situations without intending harm); and high-stakes deception, where real-life damages are possible and likely. This distinction must be considered when drawing conclusions in automated and computer-assisted deception detection research.

Furthermore, a closely related issue in forensic computational linguistics is the importance of working on ground truth data that are forensically feasible. 'Ground truth' data means data for which we know what the correct answers are; thus, for the particular field of deception detection, we need data where we know which texts are true or false. When a method is tested on ground truth data, we can conduct validation testing and accurately report its error rate. In empirical research, validation testing is a technique that determines how well a procedure works, under specific conditions, on a corpus containing texts of known origin [11]. Thus, on a database of ground truth data, the researcher is to apply a replicable analytical method to every text as well as a cross-validation scheme, most typically by building a statistical or a machine learning (ML) model. Last, the error rate is to be computed from the misclassifications in the analysis.

Within the research paradigm of forensic computational linguistics, in the present article, a corpus-based study is presented, attempting to answer the question 'Is this truthful or false?' It is worth noting that automated and computer-assisted methods in other corners, such as author identification, are much more consolidated worldwide and generally admitted in court, such as Chaski's SynAID [12,13], as compared to computer-assisted deception detection, which is not often used for veracity assessment in the legal setting. In other words, in many, if not all, jurisdictions, experts are not allowed to testify that a person is lying, as only the jury or the judge can do it. Thus, deception detection is only an investigative tool, that is to say, its use is restricted to investigation, not trial. However, some expert witnesses, such as the present author, are currently refining specific computational tools, which have proved reliable in research contexts, in order to promote the implementation of empirical investigative methods in real-life forensic settings.

2.2. The Role of Linguistic Variables in the Computational Analysis of Deception

As has been seen, deception detection can play a role in the investigation of different security issues, civil cases, and even some types of crimes, and, according to the Institute for Linguistic Evidence (ILE) (https://linguisticevidence.org/, accessed on 4 July 2021) paradigm, standards for forensic computational linguistic methodology include that forensic linguistics provides an empirical analysis grounded in linguistic theory [11]. Furthermore, the adoption of totally automated deception detection methods and mixed machine–human methods entail some basic stages: choosing an appropriate linguistic level, properly codifying the variables of analysis, engaging in statistical analysis, and conducting validation testing.

These kinds of analyses can make use of variables from different linguistic levels, namely, the phonemic, morphemic, lexical, syntactic, semantic, and pragmatic. As stated in [11], forensic methods dealing with written data have focused on analytical units at the character, word, sentence, and text levels. Specifically, some studies, such as [14], present automated methods for deception detection operating at the character level, whose analytical units include, among others, single characters, punctuation marks, or character-level n-grams (units of adjacent characters). At the word level, analytical units can be word-level n-grams [15], lexical semantics [16], and vocabulary richness [17]. Sentence-level analytical units can include part-of-speech (POS) tags [18], sentence type [19], average sentence length [20], and average number of clauses per utterance [21]. At the textual level, analytical units can include text length [22] and discourse strategies [23], to name but a few. The easiest patterns to detect by machine are character and word level features. On the contrary, at other linguistic levels, automatic pattern detection is harder, especially with forensic data, as they are often messy. For instance, sentence level features can be extracted automatically, but most parsers require human revision of the output to ensure the accuracy of the analysis.

In their meta-analysis of computational deception detection, [24] explored 44 studies and a set of 79 cues, which seemed reasonably consistent across previous literature. Despite some inconsistencies, the authors reported some common conclusions from the poll of studies reviewed: in broad terms, liars experienced greater cognitive load than truth-tellers; using fewer words related to cognitive processes, they used more negative emotion words, detached themselves from the events narrated, and used fewer sensory–perceptual words. Nonetheless, words expressing uncertainty were found indicative neither of deception nor of truth. All in all, the results varied across the studies according to event type, involvement, intensity of interaction, and motivation, among other variables.

3. Description and Explanation of the Most Significant Methodologies

In this section, the main tools for automated deception detection are presented (a schematic overview is provided in Figure 1). The first group is aimed at the automatic extraction of lexical features for different purposes, whereas the second group includes software specifically developed for the computational classification of written statements as true or false.

3.1. Automatic Extraction of Linguistic Features Applied to Detecting Deception

One of the earliest attempts at automated content analysis was the General Inquirer [25,26], and some years later, [27] assessed several linguistic cues, using TEXAN, a computer system that analyzed word frequencies by keypunching the words to map them to different lexical categories, with the main purpose of differentiating truths from lies in the written medium.

In the last 20 years, some more modern content analysis approaches were developed in research contexts on similar grounds, outstandingly the linguistic inquiry and word count, or LIWC [28]. One important difference between LIWC and the General Inquirer is that LIWC focuses on the word as the unit of analysis, while the General Inquirer was based on the sentence, but both systems relate linguistic text to other categories of cognition. Specifically, the categories used in the original version of LIWC were related to standard linguistic processes, psychological processes, relativity, and personal matters; a detailed description of the individual categories can be found in [29]. It has been also adapted and translated into more than 10 languages, including Spanish [30], as will be seen in the exemplary study presented below. In sum, LIWC provides a tool for studying the emotional, cognitive, and structural components contained in language on a word-by-word basis, working out the percentage of words which fall into those categories. Ref. [16] were the first researchers to use this system for deception detection, yielding above-chance accuracy of classifications for different types of lies. Even if LIWC is not entirely unproblematic as an analytical tool in linguistics [31], over the last few years, it has been widely used in such fields as forensic linguistics [15], sentiment analysis [32], and psycholinguistics [33] with considerable success.

Some other automatic corpus classification tools have been developed beyond word frequency analysis, such as CohMetrix [34,35]. It analyzes cohesion relations, taking into account the meaning and context in which words or phrases occur in texts. Ref. [36] was the first piece of research where it was applied to deception detection.

3.2. Software Developed for the Computational Classification of Written Statements as True or False

The software specifically developed for linguistic deception detection is presented in this section. One of the most famous methods for deception detection is scientific content analysis (SCAN). It was developed in 1987 [37], a polygraph examiner, and methods based on it are generally known as statement analysis. Most of the literature published on this type of analysis is merely descriptive (see, for example, Lesce [38] and McClish [39]), although it was automated with reported accuracy results of 71% in [10]. However, as stated in [40], SCAN and other statement analysis systems have been mainly used and taught by practitioners manually, with several studies having examined SCAN with suggestive but inconsistent results [41,42].

Some other computational tools have been specifically developed for deception detection, such as Agent99Analyzer [43], created to extract linguistic cues to deception from texts and videos, iSkim [44], or CueCal [14]. A somewhat different detection deception software is ADAM, or automated deception analysis machine [45], which focuses on editing processes, such as backspace or spacebar while typing messages as well as measuring response latencies. The main methodological drawback of this approach seems to be that it requires a keystroke analyzer to be on the interviewee's machine, which can be seen as an intrusion of privacy.

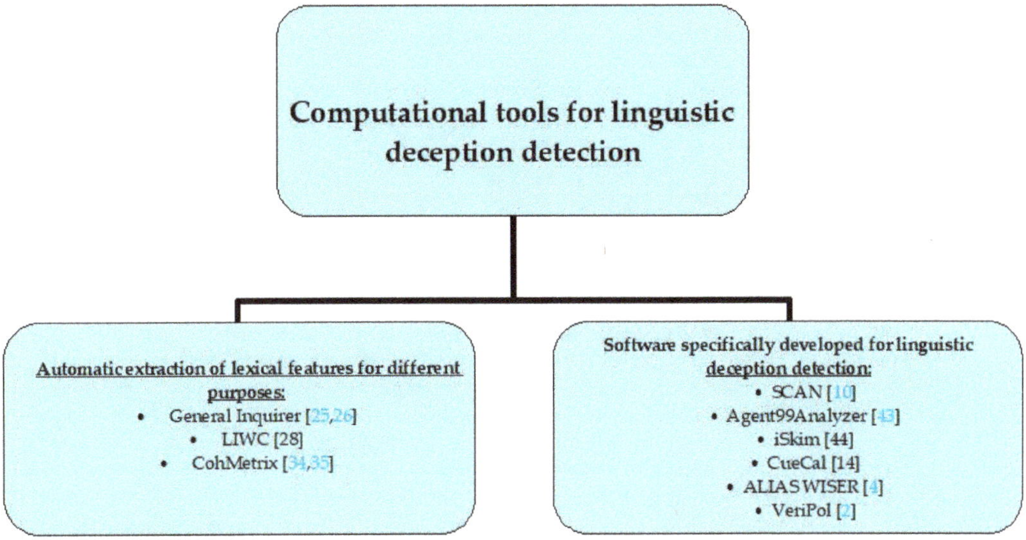

Figure 1. Schematic overview of the main computational tools for linguistic deception detection [2,4,10,14,25,26,28,34,35,43,44].

Remarkably, most previous studies in computerized deception detection have relied exclusively on shallow lexico-syntactic patterns. However, [19] were the first researchers to explore syntactic stylometry. Over four different subcorpora including service reviews and essays on different topics, the authors explore features derived from phrase structure grammar (PSG) parse trees, showing that they consistently improve the detection rate over several baselines that are based only on lexical features. Most relevantly, within the four subcorpora examined, they apply their method to the corpus from TripAdvisor collected for [1], improving the classification results obtained by its collectors by reaching over 91% accuracy.

In this line of linguistic sophistication, a valuable contribution to linguistic deception detection has been made by Witness Statement Evaluation Research (WISER), one of the tools provided by ALIAS Technology (https://aliastechnology.com/, accessed on 4 July 2021), a company which offers forensic linguistics consulting to attorneys, law enforcement, human resources, and security teams. WISER is a project that makes use of automated text analysis and statistical classifiers to determine the best protocol for the computational classification of true and false statements in the forensic-investigative setting. Ref. [4] tested this text analysis tool, based on ALIAS's module Text Analysis Toolkit Toward Linguistic Evidence Research (TATTLER). It combines linguistic analysis at the phonological, syntactic, and lexico-semantic levels and has been applied to deception detection classification on two types of corpora: low-stakes (laboratory) and high-stakes, actual statements in criminal investigations [46]. The low-stakes, laboratory data comprised two narratives of a traumatic experience, one truthful and the other false, from each participant, while the high-stakes data consisted of actual statements from real criminal investigations with non-linguistic evidence of their veracity or falsehood. The WISER method yielded substantially different results, as 71% of the texts in the laboratory corpus were correctly identified, using leave-one-out cross-validation, while the rate reached 93% for high-stakes deception, which can be considered the most successful rate published to date. Furthermore, this brings to light the contrast between lies told in a low-stakes, laboratory setting and those told in a police investigation. All in all, this study shows how TATTLER linguistic variables work better than text analysis tools used for different purposes, such as LIWC or simplistic NLP models, such as bag of words (BoW). The latter is an approach popular among computer scientists working in text classification. The term bag of words was invented by [47] and developed by [48], and in this conception of language, each text is seen as a list of words and their frequencies without regard to any morphosyntax or semantics.

As stated above, context has proved to affect the relation between deception and language (see, for example, Almela et al. [22]). Thus, the development of software designed for specific contextual frameworks is especially valuable in deception detection. An outstanding example of contextualized analysis of deception is VeriPol [2], a model for the detection of false robbery reports in Spanish based only on their text. This tool, developed in collaboration with the Spanish National Police and the Ministry of the Interior, combines NLP and ML methods in a decision support system that provides police officers the probability that a given report is false. The impact of this tool was tested by means of an on-the-field pilot study that took place in 10 Spanish police departments in 2017, specifically on a corpus of 588 false robbery reports and 534 truthful robbery reports, which allowed for a robust validation on ground truth data (see Section 2.1). For the analysis, the authors applied feature selection techniques in their approaches, using model variables, such as POS tags, document statistics (e.g., number of tokens, lemmata, and sentences within a document), and unigram lemmata for the performance of ML and statistical classification techniques [2]. They concluded that, in general, the more details are provided in the report, the more likely it is to be truthful. Empirical results show that it is extremely effective in discriminating between false and true reports with a success rate of more than 91%, improving by more than 15% the accuracy of human expert police officers on the same corpus. The pilot study was so successful that nowadays, it is officially used in all the national police offices in Spain. This fact is indeed significant, as, despite the fact that computer-assisted deception detection is not generally accepted in Spanish courts, it is proved that investigative settings may benefit from its assistance. Indeed, the differences between the situation of forensic linguistics in English- and Spanish-speaking countries are worth noting at this point. As explained in [8], there is an ever-growing respect between British police, criminal psychologists and linguists, probably because of the well-established tradition of these disciplines in English-speaking countries. However, in Spain, these areas do not have such a long tradition, hence the difficulty when it comes to securing comprehensive assistance to conduct realistic lie detection studies in languages other than English.

All in all, computational detection deception in both the WISER and VeriPol studies demonstrate that detection is possible with over 90% accuracy, with high-stakes ground truth data.

4. Materials and Methods

This section will provide the reader with a corpus study of deception in Spanish, an empirical study whose aim is to explore the linguistic cues to deception in written language with the aid of an automatic text classification tool, adopting a forensic computational linguistic approach and testing it on an ad hoc corpus containing ground truth data.

4.1. Contextualizing the Study

Ref. [22] predates the experiment reported here. As stated above, in that study, Almela et al. (2013) conducted a classification experiment, testing the Spanish version of LIWC2001 [30] to classify a corpus similar to that of [15], trained and tested with a support vector machine (SVM) classifier, using the four dimensions of LIWC (standard linguistic dimensions, psychological constructs, general descriptors, and personal concerns) separately and then with the possible combinations of the four dimensions. The authors showed the relatively high performance of the automatic classifier in Spanish written texts through the experiments, conducted on three subcorpora, checking the discriminant power of the variables as to their truth condition, the two first dimensions, linguistic and psychological processes, being the most relevant ones. Specifically, the best performing combinations across all LIWC tests and topics was an F-measure of 84.5%, using the combination of all four categories on the good friend topic. For comparison with the other

LIWC studies that use F-measure, the highest F-measure reported in [1] was 76.9%, using the LIWC features alone on the more lexically constrained hotel reviews, and in [18], it was 79.6%. In [22], the authors state that the higher performance on the good friend topic shows the strong dependence of the task on the topic and attribute the better performance on this topic to the greater emotional involvement that narrators have in describing their best friend.

Building on this previous work, the study presented here is a subsequent experiment conducted on the same corpus, considering some of the authors' suggestions for further research in [22]. Of interest, the novelty of this experiment is twofold:

(1) Regarding the variables for analysis, a fifth dimension is added to the original LIWC set, comprising some stylometric variables which have proved useful in other NLP tasks [49] (described in depth in Section 4.3.2).
(2) Statistical tests are applied to the individual categories instead of the ML algorithms usually employed for automatic deception detection. Specifically, a discriminant function analysis and several logistic regressions is performed so as to assess the discriminant power of the independent variables individually, instead of testing the dimensions as a whole (described in detail in Section 4.4). This rule-based feature extraction is chosen to make the classifier more describable.

4.2. Research Question

The present study addresses the following research question:

How successful are LIWC individual categories and the further stylometric variables analyzed for deception classification on a Spanish ad hoc corpus containing written opinions and emotionally loaded language?

4.3. Methodology

This section outlines the different stages of the present study. It comprises three main issues: an introduction to the nature of the study, an account of the analysis variables, and a full description of the corpus.

4.3.1. Nature of the Study

The present study may be classified as quasi-experimental. Quasi-experiments resemble quantitative and qualitative experiments, but they lack random assignment of groups or proper controls [50]. This feature is sometimes seen as an inherent weakness, especially from the viewpoint of experimental purists in the natural sciences. However, this is a very useful design for measuring social variables since it is not always possible to accomplish a purely random allocation of groups when dealing with human subjects. Thus, the present research takes advantage of the possibilities of this experimental design by comparing two groups of participants under similar circumstances. As explained below, an inter-group comparison is drawn, delving into the similarities and differences of the linguistic profiling of deception in written communication across languages. In addition, an intra-group assessment was undertaken in order to explore differences across topics, using the truthful statements as the control subcorpus against which the untruthful dataset is compared. Due to the quasi-experimental nature of the study, the intention is not to generalize the inferences drawn from the data analysis, but to treat them cautiously.

4.3.2. Variables

Most of the core psychologically meaningful categories contained in LIWC [28] and described above were used. It is worth noting that all the variables selected from LIWC reflect the percentage of total words, with three exceptions: raw word count, words per sentence, and percentage of interrogative sentences.

Interestingly, the LIWC dictionary generally arranges categories hierarchically. Thus, some of the categories are the sum of others. For example, the category 'Total pronouns' comprises '1st person singular', '1st person plural', 'Total 1st person', 'Total 2nd person', and 'Total 3rd person'. The categories '1st person singular' and '1st person plural', in turn, are both subsumed under 'Total 1st person'. Some previous studies, such as [16] and [18], explored categories from different levels in the hierarchy, using the same experiment, which can be considered as a methodological flaw. In ML classification and statistical techniques, this would result in redundancy, which may yield misleading results. As suggested by such authors as [5], in this case, the results might be skewed by counting those variables twice. In order to avoid this, there are two options: either removing the hierarchically superior categories or keeping them and leaving the inferior categories out. In the present study, the first option was selected so as to keep the most specific information. Appendix A shows the LIWC categories removed and their correspondences. The first column contains the highest categories, the second one the subcategories, and the third one the subcategories of the previous subcategories—it is worth noticing that the categories which involve no complexity were not included. Categories in capital letters are the most general ones, which were altogether removed. These categories may comprise either categories in bold, which in turn comprise other lower categories, or just in italics, which are the terminal part of the sequence. Only terminal or most specific categories were kept and counted.

Furthermore, a group of punctuation marks measured by LIWC was also explored in the present study, namely period, comma, colon, semicolon, sentences ending with '?', exclamation, dash, quote, apostrophe, parenthesis, and other punctuation. These variables were not previously explored in [22] because, despite being considered part of Dimension I (Linguistic processes), they were not included as LIWC default predictors.

Last, there are some linguistic features not included in LIWC which were deemed relevant for the present study too, gathered in the fifth dimension of variables. To the best of the author's knowledge, despite having proved useful in areas, such as automated document readability [49], they have not been explored for deception detection yet. They were extracted from the statistics worked out by WordSmith Tools 5.0 (https://www.lexically.net/wordsmith/index.html, accessed on 12 March 2021). The first of these variables is a standardized type/token ratio; it is worth noting that the non-standardized version of this ratio was included in the LIWC standard linguistic dimensions, but it proved to be too size-dependent as an index of lexical richness [51]. Thus, the discriminant power of the original version of the ratio may be greater, due to the disparities among the values for the different texts, so it is not as reliable a measure as the standardized version. On the other hand, word length was considered as well. Despite the fact that a category similar to 'complex words' was already included in LIWC, namely 'Sixltr', all words longer than 6 letters were included. Since the general agreement in corpus linguistics is that complex words should include any word consisting of 8 or more letters [49], their frequency is used for the calculation of one of the independent variables: the ratio of complex words to the number of tokens. Similarly, the ratios of the total amount of 1-, 2-, 3-, 4-, 5-, 6-, and 7-letter words to the number of tokens were worked out. Furthermore, the average word length (in characters) and average text length (in sentences) were considered in this section too. A summary of all the variables is provided in Appendix B, with the variables not previously explored in [22] marked in bold.

4.3.3. Corpus Description

The design of the questionnaire for the compilation of the corpus was focused on three different topics: opinions on homosexual adoption, opinions on bullfighting, and feelings about a good friend. Specifically, the participants received instructions to imagine that they had 10–15 min to express their opinion about the topics. First, they were asked to prepare a text expressing their true opinions on the topics; then, they were asked to prepare a second text expressing the opposite of their opinions, thus lying about their true beliefs. For instance, in the case of the good friend topic, it implied giving positive account on a good friend, and then a false positive account on a bad friend, according to the respondent's personal experience. The guidelines asked for at least 4–5 sentences in as much detail as possible. Regarding the motivation behind the choice of topics, it paralleled that in [15]: the three tasks proposed to participants included two controversial topics (homosexual adoption and bullfighting), sensitive subjects, which caused people to entertain a personal opinion on them. As for the third topic, good friend, it was selected so as to offer a counterpart to the previous topics since it entailed less emotional involvement. Interestingly, the controversial topics dealt with in the present study are likely to generate guilt, preoccupation or remorse, despite not being a high-stakes situation.

The participants (100) were college students, native speakers of European Spanish. Thus, the task was assigned as an exercise for extra credit in a college course and conducted via email over the course of several days. Personal information, such as age and sex, was not taken into account since it was considered irrelevant to the present analysis. It was deemed of utmost importance to avoid overfitting, which may occur when a sample size is too small in relation to the number of variables used, since this could lead to over-optimistic results. It is generally agreed that, for this kind of analysis, it is necessary that the number of cases be twice the number of variables, expressed as $n = 2k$ [52]. In the present study, a set of 76 independent variables was used; thus, in principle a minimum of 152 contributions would be required. In this case, every subcorpus comprises at least 200 contributions—in the case of the subcorpora organized by topics. In line with [15], 600 contributions were collected—100 true and 100 false statements for each topic—with an average of 94 words per statement and a total of 56,882 words, so statistical overfitting should not be a problem in subsequent analyses. A manual check of the quality of the contributions was made, and each one was entered into a separate text file. Appendix C shows a sample of truthful and untruthful language for each of the three topics, and Figure 2 shows the structure of the sample used for the analysis.

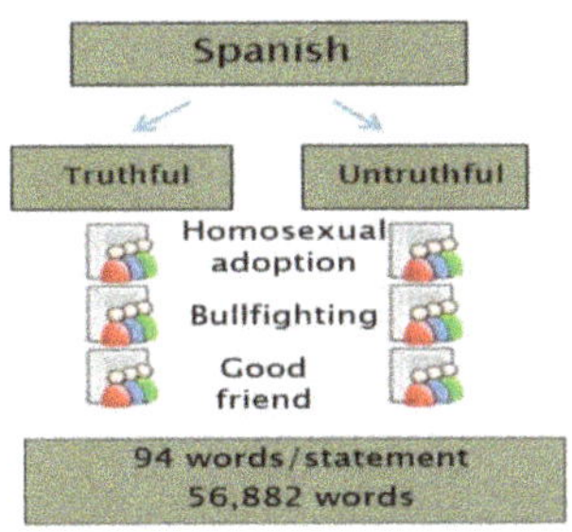

Figure 2. Structure of the dataset.

The dataset was deposited by the present author in a publicly available database, namely https://github.com/angelalm/DeceptionCorpus.

4.4. Data Analysis

As regards the statistical methods applied, discriminant function analysis (DFA) and several binary logistic regressions (LR) were calculated with the software package IBM SPSS (https://www.ibm.com/products/spss-statistics, accessed on 30 March 2021) so as to assess the discriminant power of the variables individually. On the one hand, DFA had been successfully applied in linguistic analysis for the classification of unknown individuals and the probability of their classification into a certain group [53,54]. In principle, DFA is claimed to make more demanding requirements on the data since it assumes that it shares all the usual assumptions of correlation, requiring linear and homoscedastic relationships—homogeneity of variances—and normal distribution of the interval or continuous data. However, DFA is known to be robust, even when these assumptions are violated, as stated in several modern textbooks about multivariate statistics [55]. At any rate, as LR is well known as an alternative to DFA because it makes less stringent requirements of the data, for the three individual subcorpora, a one sample Kolmogorov–Smirnov test provided evidence against the null hypothesis, implying that the samples were not drawn from a normal population. As only a few variables met the requirements of normality and only 100 cases are involved, binary logistic regressions were conducted on the individual subcorpora, where the categorical response has only two possible outcomes (untruthful/truthful). Thus, it can be stated that the analyses reported in the present article explore techniques based on statistical approaches instead of methods based on geometrical properties of the data, such as [4,11–13]. It is worth noting that, for each classifier, a leave-one-out cross-validation was run, all sets having an equal distribution between truthful and untruthful statements. This technique, considered exhaustive cross-validation, is used to evaluate how the results of a statistical analysis would generalize to an independent dataset. As explained in [56], the main difference from non-exhaustive cross validation methods, such as *k*-fold cross-validation, is that the latter does not compute all ways of splitting the original sample. Since the aim of this experiment is the prediction of the truth condition of the texts, a cross-validation was applied in order to estimate the accuracy of the predictive models. It involves partitioning a sample of data into complementary subsets, performing an analysis on the training set and validating the analysis on the testing or validation set [57]. For DFA and logistic regression, cross validation shows how reliable the linear function determined by the original group members is when each member is left out of the group.

5. Results and Discussion

First, the DFA shows a successful discrimination between truthful and untruthful accounts in the general corpus (Wilks' $\lambda = 0.699$, $\chi^2 = 210.7$, $p = 00.000$). Specifically, text length proves to be the best single predictor, as shown in Table 1 and Figure 3. Remarkably, the difference between this predictor and the next one in importance is 20 points. Despite this fact, the F-ratio for the next predictor, 1st person singular, is still rather high. There are some other variables identified as predictors shared with studies for English such as [15], namely 2nd person, friendship, insight, exclusive words, and 3rd person. The remaining predictors are words related to certainty, humans, sexuality, number, anger, semicolon, past, assent, future, and tentative words.

Table 1. F-ratios from DFA.

Predictors	LIWC Abbreviation	Examples	F	Sig.
Word count	WC	-	69.812	0.000
1st person singular	I	*I, my, me*	49.259	0.000
Certainty	Certain	*always, never*	39.199	0.000
Total second person	You	*you, you'll*	33.516	0.000
Friends	Friends	*pal, buddy, coworker*	30.167	0.000
Humans	Humans	*boy, woman, group*	27.682	0.000
Insight	Insight	*think, know, consider*	25.708	0.000
Exclusive	Excl	*but, except, without*	23.601	0.000
Sex and sexuality	Sexual	*lust, penis, suck*	21.871	0.000
Numbers	Number	*one, thirty, million*	20.568	0.000
Anger	Anger	*hate, kill, pissed*	19.397	0.000
Semicolon	SemiC	-	18.329	0.000
Total third person	Other	*she, their, them*	17.495	0.000
Past tense verb	Past	*walked, were, had*	16.643	0.000
Assents	Assent	*yes, OK, mmhmm*	15.909	0.000
Future tense verb	Future	*will, might, shall*	15.239	0.000
Tentative	Tentat	*maybe, perhaps, guess*	14.709	0.000

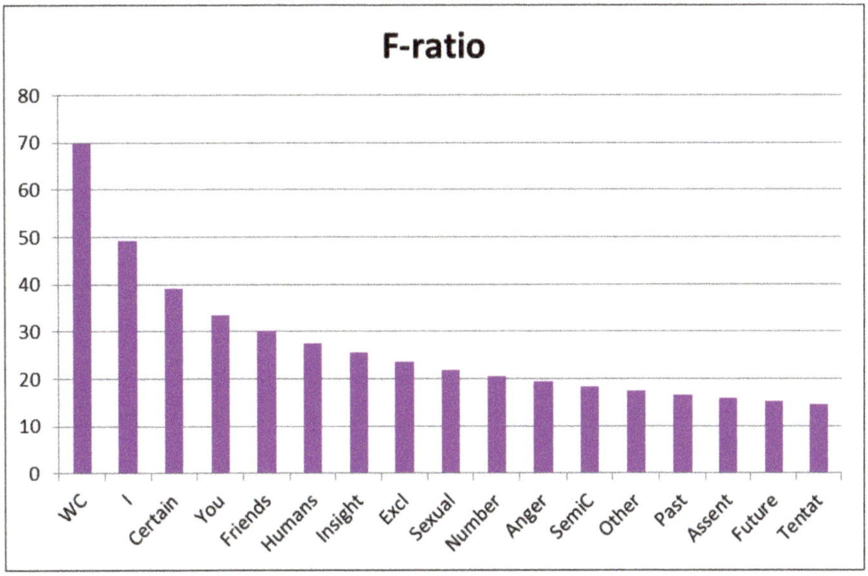

Figure 3. Visual representation of F-ratios from DFA.

As can be seen in Table 2, which gives information about actual group membership vs. predicted group membership, the DFA shows that 76.3% of the original grouped cases were correctly classified, as 77.7% of the truthful statements were correctly classified as truthful (233 out of 300), and 75.0% of the untruthful statements were correctly classified as untruthful (225 out of 300 statements). As regards the leave-one-out classification method, it achieved a success rate of 74%, the percentage of truthful statements correctly classified in the cross-validation being slightly higher than the percentage of untruthful ones (75.7% vs. 72.3%, respectively). Specifically, there is a difference of 10 more statements correctly classified (83 vs. 73 statements).

Table 2. Classification results from DFA (IBM SPSS).

		Deception	Predicted Group Membership		Total
			No	**Yes**	**1**
Original [a]	Count	No	233	67	300
		Yes	75	225	300
	%	**No**	**77.7**	**22.3**	100.0
		Yes	**25.0**	**75.0**	100.0
Cross-validated [b]	Count	No	227	73	300
		Yes	83	217	300
	%	**No**	**75.7**	**24.3**	100.0
		Yes	**27.7**	**72.3**	100.0

[a] 76.3% of original grouped cases correctly classified; [b] 74.0% of cross-validated grouped cases correctly classified.

In order to present a comprehensive picture of the effectiveness of the statistical classification methods employed, a summary of the success rates is provided in Figure 4. The experiment conducted on the good friend subcorpus yielded the best results. In this case, the known bundles of truthful and untruthful texts were differentiated with 84.6% cross-validated accuracy, meaning that 84.6% of the time, we can tell truthful and untruthful texts apart from each other and identify them. Specifically, there is a difference of more than 9 points from the previous subcorpus in terms of success, homosexual adoption (84.6% vs.

75.4%), probably due to the fact that when speakers refer to a good friend, they are more likely to be emotionally involved in the experiment; they are not just giving an opinion on a topic which is alien to them, but relating their personal experience with a dear friend and lying about a person that they really dislike. This personal involvement is probably reflected on the linguistic expression of deception, as suggested by [16].

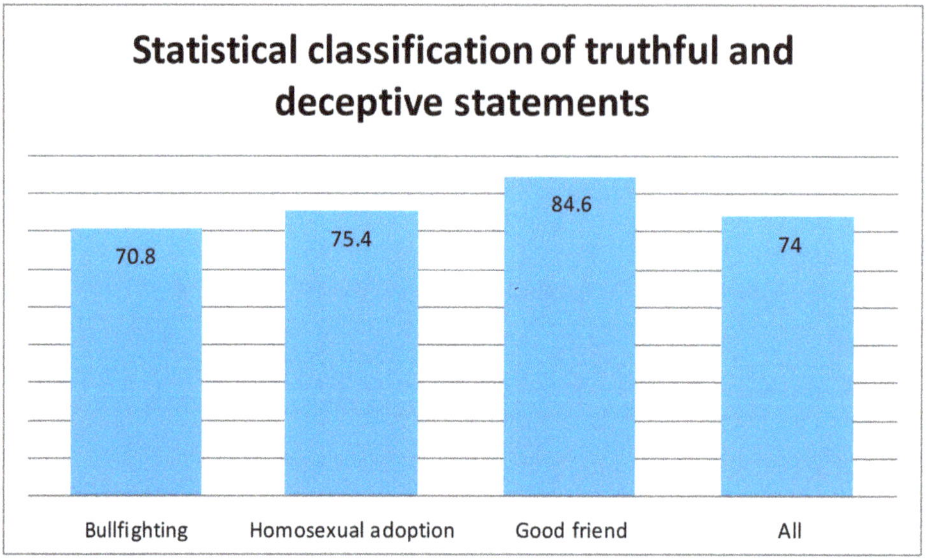

Figure 4. Cross-validated classification of truthful and deceptive statements.

Furthermore, Table 3 shows a collection of the predictors identified for truthful, marked with the initial "T", and untruthful, initial "U", statements across the examined corpora. It is worth noting that the identification of predictors has proved more successful at pinpointing categories indicative of truthful statements, the most widely shared among subcorpora being text length and 1st person singular.

Qualitative Evaluation

Previous research on deception detection has found that, broadly speaking, deceivers provide shorter responses, compared to truth-tellers (see, for example, DePaulo et al. [58]), as creating and managing misinformation is more cognitively demanding than telling the plain truth. This is also the case with participants in synchronous CMC, where time to plan the responses is limited, almost like in oral communication, which is in line with the present results. Regarding 1st person singular, a previous study conducted in Spanish [59] did not find a significant correlation with this feature. Nonetheless, the authors advanced that the communication topic might make a difference since their participants write about trips, which is unlikely to generate guilt, preoccupation or remorse. On the contrary, the controversial topics dealt with in the present study are more likely to arouse these feelings, despite not being a high-stakes situation.

On the other hand, the strongest predictors for untruthfulness are 2nd and 3rd person. The latter is clearly in line with previous research [14,60]. This cue entails detachment from the self when providing false or imprecise information, indicating the leading role of non-immediacy in deception. Accordingly, there is also a significant 2nd person orientation in untruthful statements, as in [15]. Interestingly enough, it has proved a predictor of deception in the subcorpora of good friend and in the whole corpus, confirming the preference of deceivers for non-immediacy.

Table 3. Predictors identified for truthful and untruthful statements across the subcorpora.

	Bullfighting	Homosexual Adoption	Friend	All
WC	T	T	T	T
1st p. sing.	T	T	T	T
2nd p.			U	U
3rd p.			U	U
Semicolon				T
Number			T	T
Anxiety				T
Insight				T
Sadness			T	
Friends			T	T
Humans		U		U
Posfeel		T		
Certainty			U	U
Achievement			U	
Inhibition			T	
Assent				U
Tentative				T
Future				T
Past				T
Inclusive		U		
Exclusive	T			T
Sexuality				T
Motion		U		

As for the rest of predictors, the results seem to be in line with previous research on the English corpora, with liars experiencing a greater cognitive load than truth-tellers, using fewer words related to cognitive processes and more negative emotion words, as well as fewer sensory–perceptual words [24].

Finally, a novel feature proved significant for the model in Spanish: the semicolon. As mentioned above, it was not previously explored in [22], as neither this one nor the other punctuation marks were included as LIWC default predictors. Although the average sentence length does not appear in any of the discriminant models, both variables are integrally related. As explained above, participants produced a larger number of words when telling the truth, especially the Spanish ones, hence the discriminant power of the semicolon in this language. Significantly, this is one of the novel findings in this study.

Overall, statistical classification methodologies with individual categories have performed better than the ML techniques with whole dimensions reported in [22]. Furthermore, the distribution of the classification results parallels that from the experiment with whole categories.

6. Conclusions and Suggestions for Further Research

All in all, the computational detection of verbal deception has come a long way in a short time, with accuracy scores ranging from 60% on laboratory data [60] to 93% accuracy on high-stakes corpora, as reported in [4]. Remarkably, research on high-stakes, real-life type of data has proved far more successful than results on low-stakes, laboratory data, although some relatively successful experiments using this kind of corpus were reported in this work, which represents a step forward. Specifically, as regards the percentage of untruthful statements correctly classified in the cross-validation, the classifier yielded 74% accuracy for the whole corpus (DFA), 70.8% for the bullfighting subcorpus (LR), and 75.4% for the homosexual adoption subcorpus (LR). As regards the experiment conducted on the good friend subcorpus, untruthful texts were differentiated with 84.6% cross-validated accuracy (LR). As was stated, the main factor leading to success in these cases seems to be the delimitation of the topic and the communicative context, due to the strong dependence of the task on the topic and on the author's degree of emotional involvement. Thus, the highest degree of accuracy on the last dataset may be attributed to the fact that when referring to a good friend, the participants are more likely to be emotionally involved in the experiment; they are not just voicing an opinion on a topic which is alien to them, but relating their personal experience with a dear friend and lying about a person that they really dislike. This personal involvement is probably reflected on the linguistic expression of deception, as suggested in some previous studies [16,22].

Thus, even if the classification results from the experiments reported in the present article are not as high as those obtained on high-stakes datasets, the relative strength compared to earlier work on low-stakes corpora is worth noting. Furthermore, although the results may seem not good enough to use forensically, basing on the literature review conducted, it can be assumed that a classification method that proves acceptably successful on low-stakes deception will work even better on high-stakes data.

New methods for automated deception detection are continually being developed, especially in the computational paradigm, and in order for the area to move in the right direction, the availability of data tagged for ground truth seems crucial [40,61]. In this sense, collaboration with law enforcement may be of utmost importance. Significantly, within the ILE paradigm, the present author is currently involved in a project for the refining of WISER, given its successful classification performance, as well as its adaptation to Spanish from English.

As a further proposal for future research, a deeper comparison and analysis of other existing methods for deception detection on the same dataset could strengthen the contributions of the newly introduced predictors, as in the outstanding case of semicolon.

All things considered, the use of corpus tools developed out of linguistic theory is of the utmost importance as is the adoption of reliable scientific methods. Researchers should keep on testing methods on real life data, deploying their knowledge of linguistics—theory, corpus linguistics, and computational linguistics—to improve both low-stakes and high-stakes deception detection.

Funding: This research received no external funding.

Institutional Review Board Statement: The study was conducted according to the guidelines of the Declaration of Helsinki, and approved by the Ethics Committee of Universidad de Murcia.

Informed Consent Statement: Informed consent was obtained from all subjects involved in the study.

Data Availability Statement: According to suggested Data Availability Statements in section "MDPI Research Data Policies" at https://www.mdpi.com/ethics, the dataset has been deposited by the present author in a publicly available database, namely https://github.com/angelalm/DeceptionCorpus.

Acknowledgments: I would like to express my gratitude to the anonymous referees for their careful review and insightful comments. Furthermore, I am also grateful to Carole E. Chaski, PhD for critically reading a previous version of this manuscript and stimulating discussions during the preparation of this article.

Conflicts of Interest: The authors declare no conflict of interest.

Appendix A

Table A1. Selection of redundant LIWC categories for the experiment.

I. Linguistic dimensions		
TOTAL PRONOUNS	**Total 1st person**	*1st person singular*
		1st person plural
	Total 2nd person	-
	Total 3rd person	-
II. Psychological processes		
AFFECTIVE OR EMOTIONAL PROCESSES	**Positive emotions**	*Positive feelings*
		Optimism and energy
	Negative emotions	*Anxiety or fear*
		Anger
		Sadness or depression
COGNITIVE PROCESSES	*Causation*	-
	Insight	-
	Discrepancy	-
	Inhibition	-
	Tentative	-
	Certainty	-
SENSORY AND PERCEPTUAL PROCESSES	*Seeing*	-
	Hearing	-
	Feeling	-
SOCIAL PROCESSES	*Communication*	-
	Other references to people	*1st person plural*
		Total 2nd person
		Total 3rd person
	Friends	-
	Family	-
	Humans	-
III. Relativity		
TIME	*Past tense verb*	-
	Present tense verb	-
	Future tense verb	-
SPACE	*Up*	-
	Down	-
	Inclusive	-
	Exclusive	-

Table A1. *Cont.*

IV. Personal concerns		
OCCUPATION	*School*	-
	Job or work	-
	Achievement	-
LEISURE ACTIVITY	*Home*	-
	Sports	-
	Television and movies	-
	Music	-
	Money and financial issues	-
METAPHYSICAL ISSUES	*Religion*	-
	Death and dying	-
PHYSICAL STATES AND FUNCTIONS	*Body states, symptoms*	-
	Sex and sexuality	-
	Eating, drinking, dieting	-
	Sleeping, dreaming	-
	Grooming	-
	Swearing	-

Appendix B

Table A2. Variables in the experiment.

Variables	Class
Word count	LIWC
Words per sentence	LIWC
Words longer than 6 letters	LIWC
Period	**LIWC**
Comma	**LIWC**
Colon	**LIWC**
Semicolon	**LIWC**
Sentences ending with '?'	**LIWC**
Exclamation	**LIWC**
Dash	**LIWC**
Quote	**LIWC**
Apostrophe	**LIWC**
Parenthesis	**LIWC**
Other punctuation	**LIWC**
1st person singular	LIWC
1st person plural	LIWC
2nd person	LIWC
3rd person	LIWC
Negations	LIWC
Assents	LIWC
Articles	LIWC
Prepositions	LIWC
Numbers	LIWC
Positive feelings	LIWC
Optimism and energy	LIWC
Anxiety or fear	LIWC
Anger	LIWC
Sadness or depression	LIWC

Table A2. *Cont.*

Variables	Class
Causation	LIWC
Insight	LIWC
Discrepancy	LIWC
Inhibition	LIWC
Tentative	LIWC
Certainty	LIWC
Seeing	LIWC
Hearing	LIWC
Feeling	LIWC
Communication	LIWC
Friends	LIWC
Family	LIWC
Humans	LIWC
Past tense verb	LIWC
Present tense verb	LIWC
Future tense verb	LIWC
Up	LIWC
Down	LIWC
Inclusive	LIWC
Exclusive	LIWC
Motion	LIWC
School	LIWC
Job or work	LIWC
Achievement	LIWC
Home	LIWC
Sports	LIWC
Television and movies	LIWC
Music	LIWC
Money and financial issues	LIWC
Religion	LIWC
Death and dying	LIWC
Body states, symptoms	LIWC
Sex and sexuality	LIWC
Eating, drinking, dieting	LIWC
Sleeping, dreaming	LIWC
Grooming	LIWC
Swearing	LIWC
Standardized type/token ratio	**Styl.**
Mean word length	**Styl.**
Sentences/WC	**Styl.**
1-letter words/WC	**Styl.**
2-letter words/WC	**Styl.**
3-letter words/WC	**Styl.**
4-letter words/WC	**Styl.**
5-letter words/WC	**Styl.**
6-letter words/WC	**Styl.**
7-letter words/WC	**Styl.**
Complex words/WC	**Styl.**

Appendix C

Table A3. Random sample 1 of truthful and untruthful statements in Spanish.

TRUTH	LIE
HOMOSEXUAL ADOPTION	
Para mí no está clara la repercusión que tendría sobre los niños el hecho de que las parejas homosexuales adopten. Sería necesario un estudio previo de las posibles consecuencias o secuelas psicológicas, o de la ausencia de ellas, en el mejor de los casos.	*La familia es y ha sido siempre la formada por un hombre y una mujer. No debemos cambiar esto, pues es un claro síntoma de la degeneración de la sociedad. Hemos de defender las tradiciones que llevan funcionando bien durante miles de años.*
Translation into English:	**Translation into English:**
It is not clear to me what the repercussions would be for children if homosexual couples were to adopt. A prior study of the possible psychological consequences or sequelae, or the absence of them at best, would be necessary.	The family is and has always been the one formed by a man and a woman. We must not change this, as it is a clear symptom of the degeneration of society. We must defend the traditions that have been working well for thousands of years.
BULLFIGHTING	
Es una salvajada. Regodearse en el sufrimiento de un animal, disfrutar viendo cómo realiza sus últimos movimientos, agotado y herido. ¿Cómo puede ser un arte esto? Sin duda hay muchas personas que están familiarizadas con las corridas de toros.Es para ellos una situación normal.	*Los espectáculos relacionados con los toros son una tradición antiquísima y un arte. Es más, los toros de lidia se pasan la vida al aire libre y son bien mimados por sus criadores, disfrutando así de una vida muchísimo mejor que la que se les ofrece a los animales de granja.*
Translation into English:	**Translation into English:**
It is a savagery. To wallow in the suffering of an animal, to enjoy watching it make its last movements, exhausted and wounded. How can this be art? Undoubtedly, there are many people who are familiar with bullfighting. For them, it is a normal situation.	Bullfighting shows are an ancient tradition and an art. Moreover, fighting bulls spend their lives outdoors and are well pampered by their breeders, enjoying a much better life than that offered to farm animals.
GOOD FRIEND	
Cuando conocí a José María pensé que era uno más, que incluso no nos podríamos llevar bien. Qué equivocación más grande, ¡y qué afortunada! Es hoy uno de mis mejores amigos, que me encontré de casualidad en una de mis muchas andanzas por el mundo.	*Sergio es un chaval inteligente, que sabe lo que quiere. Es realmente una buena persona, con la que puedes contar para todo. Su principal cualidad es su simpatía y amabilidad con todos, no importa que no te conozca de nada, siempre te da una oportunidad.*
Translation into English:	**Translation into English:**
When I first met José María I thought he was just another guy, and that we might not even get along. What a big mistake, and how fortunate! Today he is one of my best friends, whom I met by chance in one of my many wanderings around the world.	Sergio is an intelligent guy, who knows what he wants. He is a really good person, you can count on him for everything. His main quality is his sympathy and kindness with everyone, it doesn't matter if he doesn't know you at all, he always gives you a chance.

Table A4. Random sample 2 of truthful and untruthful statements in Spanish.

TRUTH	LIE
HOMOSEXUAL ADOPTION	
Yo pienso que es un tema muy delicado y tal vez ahora mismo los hijos de parejas homosexuales podrían ser discriminados en el colegio, tendrá que cambiar la sociedad poco a poco pero aun así pienso que es importante tener un referente masculino y otro femenino en la educación de un niño.	*Me gustaría decir estoy cansado de las discriminaciones que sufren las parejas homosexuales en la sociedad hoy en día. Son parejas como cualquier otra y sienten lo mismo que las demás. Por lo tanto pienso que sería correcto que pudieran adoptar ya que querrían a su hijo de la misma manera que las parejas heterosexuales. El respeto a los demás y la tolerancia es uno de los valores centrales de la educación en una familia.*
Translation into English:	**Translation into English:**
I think it is a very delicate issue and maybe right now the homosexual couples' children could be discriminated at school; society will have to change little by little, but I still think it is important to have a male and female reference in the education of a child.	I would like to say that I am tired of the discrimination that homosexual couples suffer in today's society. They are couples like any other and feel the same as others. Therefore, I think it would be right for them to be able to adopt since they would love their child in the same way as heterosexual couples. Respect for others and tolerance is one of the core educational values in a family.
BULLFIGHTING	
El animal agoniza en una sopa de sangre, siente miedo, dolor, angustia, desesperación. No tiene posibilidades reales de defenderse, no tiene noción de lo que sucede a su alrededor, no tiene capacidad de razonar y por ende, de imaginarse cuándo cesarán todas esas desagradables sensaciones. El toro no lucha por su vida. Es sometido a una serie de torturas sistemáticas que lo humillan, lo denigran y lo hacen padecer infinito dolor.	*Los toros y las corridas como acto o evento social me parece algo que está hace muchísimos años y da de comer a muchísimas familias, a pesar que dicen que es cruento, piensen que si se quitaran las corridas mucha gente quedaría en paro y lo más señalado es que nos comeríamos los toros igualmente, así que no es interesante el acabar con la famosa fiesta taurina y algo más, ¿toda la carne que comemos todos que pasa? ¿Es sintética?*
Translation into English:	**Translation into English:**
The animal dies in a soup of blood, feels fear, pain, anguish, despair. It has no real possibility of defending itself, it has no notion of what is happening around it, it has no capacity to reason and, therefore, to imagine when all these unpleasant sensations will cease. The bull does not fight for its life. It is subjected to a series of systematic tortures that humiliate it, denigrate it and make it suffer from infinite pain.	Bullfighting as a social act or event seems to me something that has been around for many years and feeds many families, even though they say it is cruel; think that if bullfighting were banned, many people would be unemployed, and the most important issue is that we would eat bulls anyway, so it is not interesting to ban the famous bullfighting tradition, and something else: What happens with the meat that we all eat? Is it synthetic?
GOOD FRIEND	
Mi mejor amigo es la persona con la que paso prácticamente todo mi tiempo libre. Es la persona con la que siempre puedo contar, sea cual sea el problema que tenga. Siempre solemos tener los mismos gustos y aficiones. Nos conocemos desde el colegio y a pesar de los años siempre hemos mantenido una amistad, aunque durante los dos últimos años está siendo mi prioridad.Espero que no se acabe nunca.	*Mi amigo X es una de esas personas con las que siempre te lo pasas bien, tiene una gran capacidad para hacerte sentir bien y que eres especial. Es una persona muy sociable y abierta con todo el mundo.Aunque si hay una cualidad que lo distingue es su fidelidad y confianza.*
Translation into English:	**Translation into English:**
My best friend is the person I spend practically all my free time with. He is the person I can always count on, no matter what problem I have. We always tend to have the same tastes and hobbies. We have known each other since school and, despite the years, we have always maintained a friendship, although for the last two years he has been my priority. I hope it never ends.	My friend X is one of those people with whom you always have a good time, he has a great ability to make you feel good and feel that you are special. He is a very sociable and open person with everyone. Although if there is one quality that distinguishes him it is his loyalty and trust.

References

1. Ott, M.; Choi, Y.; Cardie, C.; Hancock, J.T. Finding deceptive opinion spam by any stretch of the imagination. In Proceedings of the 49th Annual Meeting of the Association for Computational Linguistics: Human Language Technologies, Portland, OR, USA, 19–24 June 2011; pp. 309–319.
2. Quijano-Sánchez, L.; Liberatore, F.; Camacho-Collados, J.; Camacho-Collados, M. Applying automatic text-based detection of deceptive language to police reports: Extracting behavioral patterns from a multi-step classification model to understand how we lie to the police. *Knowl. Based Syst.* **2018**, *149*, 155–168. [CrossRef]
3. Vogler, N.; Pearl, L. Using linguistically defined specific details to detect deception across domains. *Nat. Lang. Eng.* **2019**, *26*, 349–373. [CrossRef]
4. Chaski, C.E.; Almela, A.; Holness, G.; Barksdale, L. WISER: Automatically Classifying Written Statements as True or False. Oral communication presented. In Proceedings of the American Academy of Forensic Sciences 67th Annual Scientific Meeting, Orlando, FL, USA, 16–21 February 2015; pp. 576–577.
5. Picornell, I. Cues to Deception in a Textual Narrative Context: Lying in Written Witness Statements. Ph.D. Dissertation, Aston University, Birmingham, UK, 2012.
6. Meibauer, J. (Ed.) *The Oxford Handbook of Lying*; Oxford University Press: Oxford, UK, 2018.
7. Markowitz, D.M.; Hancock, J.T. Deception and Language: The Contextual organization of Language and Deception (CoLD) framework. In *The Palgrave Handbook of Deceptive Communication*; Docan-Morgan, T., Ed.; Palgrave Macmillan: New York, NY, USA, 2019; pp. 193–212.
8. Bull, R.; Cook, C.; Hatcher, R.; Woodhams, J.; Bilby, C.; Grant, T. *Criminal Psychology: A Beginner's Guide*; Oneworld Publications: Oxford, UK, 2006.
9. Sporer, S.L.; Manzanero, A.L.; Masip, J. Optimizing CBCA and RM research: Recommendations for analyzing and reporting data on content cues to deception. *Psychol. Crime Law* **2020**, *27*, 1–39. [CrossRef]
10. Fitzpatrick, E.; Bachenko, J. Detecting Deception across Linguistically Diverse Text Types. In Proceedings of the Linguistic Society of America Annual Meeting, Boston, MA, USA, 3–6 January 2013.
11. Chaski, C.E. Author Identification in the Forensic Setting. In *The Oxford Handbook of Language and Law*; Solan, L.M., Tiersma, P.M., Eds.; Oxford University Press: Oxford, UK, 2012.
12. Chaski, C.E. Empirical Evaluations of Language-based Author Identification Techniques. *Forensic Linguist.* **2001**, *8*, 1–66. [CrossRef]
13. Chaski, C.E. Best Practices and Admissibility of Forensic Author Identification. *J. Law Policy* **2013**, *21*, 233.
14. Zhou, L.; Burgoon, J.K.; Nunamaker, J.F.; Twitchell, D. Automating linguistics-based cues for detecting deception in text-based asynchronous computer-mediated communications. *Group Decis. Negot.* **2004**, *13*, 81–106. [CrossRef]
15. Mihalcea, R.; Strapparava, C. The lie detector: Explorations in the automatic recognition of deceptive language. In Proceedings of the ACL-IJCNLP 2009 Conference Short Papers, Association for Computational Linguistics, Singapore, 4 August 2009; pp. 309–312.
16. Newman, M.; Pennebaker, J.; Berry, D.; Richards, J. Lying words: Predicting deception from linguistic styles. *Personal. Soc. Psychol. Bull.* **2003**, *29*, 665–675. [CrossRef]
17. Almela, Á.; Alcaraz-Mármol, G.; Cantos, P.Y. Analysing deception in a psychopath's speech: A quantitative approach. *DELTA Doc. Estud. Lingüíst. Teór. Apl.* **2015**, *31*, 559–572. [CrossRef]
18. Fornaciari, T.; Poesio, M. Automatic deception detection in Italian court cases. *Artif. Intell. Law* **2013**, *21*, 303–340. [CrossRef]
19. Feng, S.; Banerjee, R.; Choi, Y. Syntactic stylometry for deception detection. In Proceedings of the 50th Annual Meeting of the Association for Computational Linguistics, Association for Computational Linguistics, Jeju Island, Korea, 8–14 July 2012; pp. 171–175.
20. Pérez-Rosas, V.; Mihalcea, R. Experiments in open domain deception detection. In Proceedings of the 2015 Conference on Empirical Methods in Natural Language Processing, Association for Computational Linguistics, Lisbon, Portugal, 17–21 September 2015; pp. 1120–1125.
21. Yancheva, M.; Rudzicz, F. Automatic detection of deception in child-produced speech using syntactic complexity features. In Proceedings of the 51st Annual Meeting of the Association for Computational Linguistics, Association for Computational Linguistics, Stroudsburg, PA, USA, 4–9 August 2013; pp. 944–953.
22. Almela, A.; Valencia-García, R.; Cantos, P. Seeing through Deception: A Computational Approach to Deceit Detection in Spanish Written Communication. In Proceedings of the Workshop on Computational Approaches to Deception Detection, Association for Computational Linguistics, Avignon, France, 23 April 2012; pp. 15–22.
23. Rubin, V.L.; Vashchilko, T. Identification of truth and deception in text: Application of vector space model to rhetorical structure theory. In Proceedings of the Workshop on Computational Approaches to Deception Detection, Association for Computational Linguistics, Avignon, France, 23 April 2012; pp. 97–106.
24. Hauch, V.; Blandón-Gitlin, I.; Masip, J.; Sporer, S.L. Are Computers Effective Lie Detectors? A Meta-Analysis of Linguistic Cues to Deception. *Personal. Soc. Psychol. Rev.* **2015**, *19*, 307–342. [CrossRef]
25. Stone, P.J.; Bales, R.F.; Namenwirth, J.Z.; Ogilvie, D.M. The general inquirer: A computer system for content analysis and retrieval based on the sentence as a unit of information. *Behav. Sci.* **1962**, *7*, 484–494. [CrossRef]

26. Stone, P.J.; Dunphy, D.; Smith, M.S.; Ogilvie, D.M. *The General Inquirer: A Computer Approach to Content Analysis*; MIT Press: Cambridge, MA, USA, 1966.
27. Knapp, M.L.; Hart, R.P.; Dennis, H.S. An exploration of deception as a communication construct. *Hum. Commun. Res.* **1974**, *1*, 15–29. [CrossRef]
28. Pennebaker, J.W.; Francis, M.E.; Booth, R.J. *Linguistic Inquiry and Word Count*; Erlbaum Publishers: Mahwah, NJ, USA, 2001.
29. Pennebaker, J.W.; Graybeal, A. Patterns of natural language use: Disclosure, personality, and social integration. *Curr. Dir. Psychol. Sci.* **2001**, *10*, 90–93. [CrossRef]
30. Ramírez-Esparza, N.; Pennebaker, J.W.; García, F.A.; Suriá, R. La psicología del uso de las palabras: Un programa de computadora que analiza textos en español. *Rev. Mex. Psicol.* **2007**, *24*, 85–99.
31. Hunt, D.; Brookes, G. *Corpus, Discourse and Mental Health*; Bloomsbury Publishing: London, UK, 2020.
32. Salas-Zárate, M.P.; López-López, E.; Valencia-García, R.; Aussenac-Gilles, N.; Almela, A.; Alor-Hernández, G. A study on LIWC categories for opinion mining in Spanish reviews. *J. Inf. Sci.* **2014**, *40*, 749–760. [CrossRef]
33. Almela, A.; Alcaraz-Mármol, G.; Garcia, A.; Pallejá-López, C. Developing and Analyzing a Spanish Corpus for Forensic Purposes. *LESLI Linguist. Evid. Secur. Law Intell.* **2019**, *3*, 1–13. [CrossRef]
34. Graesser, A.C.; McNamara, D.S.; Louwerse, M.M.; Cai, Z. Coh-Metrix: Analysis of text on cohesion and language. *Behav. Res. Methods Instrum. Comput.* **2004**, *36*, 193–202. [CrossRef]
35. McNamara, D.S.; Graesser, A.C.; McCarthy, P.M.; Cai, Z. *Automated Evaluation of Text and Discourse with Coh-Metrix*; Cambridge University Press: Cambridge, MA, USA, 2014.
36. Bedwell, J.S.; Gallagher, S.; Whitten, S.N.; Fiore, S.M. Linguistic correlates of self in deceptive oral autobiographical narratives. *Conscious. Cogn.* **2011**, *20*, 547–555. [CrossRef]
37. Sapir, A. *Scientific Content Analysis (SCAN)*; Laboratory of Scientific Investigation: Phoenix, AZ, USA, 1987.
38. Lesce, T. SCAN: Deception Detection by Scientific Content Analysis. *Law Order* **1990**, *38*, 8. Available online: http://www.lsiscan.com/id37.htm (accessed on 25 November 2020).
39. McClish, M. *I Know You Are Lying. Detecting Deception through Statement Analysis*; The Marpa Group, Inc.: Winterville, GA, USA, 2001.
40. Fitzpatrick, E.; Bachenko, J.; Fornaciari, T. *Automatic Detection of Verbal Deception*; Morgan and Claypool Publishers: Williston, VT, USA, 2015. [CrossRef]
41. Adams, S.H.; Jarvis, J.P. Indicators of veracity and deception: An analysis of written statements made to police. *Speech Lang. Law* **2006**, *13*, 1–22. [CrossRef]
42. Kang, S.M.; Lee, H. Detecting deception by analyzing written statements in Korean. *Linguist. Evid. Secur. Law Intell.* **2014**, *2*, 1–10. [CrossRef]
43. Fuller, C.M.; Biros, D.P.; Burgoon, J.K.; Adkins, M.; Twitchell, D.P. An analysis of text-based deception detection tools. In Proceedings of the 12th Americas Conference on Information Systems, Acapulco, Mexico, 4–6 August 2006; pp. 3465–3472.
44. Zhou, L.; Booker, Q.E.; Zhang, D. ROD: Towards rapid ontology development for underdeveloped domains. In Proceedings of the 35th Hawaii International Conference on System Sciences, Honolulu, HI, USA, 10 January 2002; pp. 957–965. [CrossRef]
45. Derrick, D.; Meservy, T.; Burgoon, J.; Nunamaker, J. An experimental agent for detecting deceit in chat-based communication. In *Proceedings of the Rapid Screening Technologies, Deception Detection and Credibility Assessment Symposium*; Jensen, M., Meservy, T., Burgoon, J., Nunamaker, J., Eds.; Grand Wailea: Maui, HI, USA, 2012; pp. 1–21. [CrossRef]
46. Chaski, C.E.; Barksdale, L.; Reddington, M.M. Collecting Forensic Linguistic Data: Police and Investigative Sources of Data for Deception Detection Research. In Proceedings of the Linguistic Society of America Annual Meeting, Minneapolis, MN, USA, 2–5 January 2014.
47. Harris, Z. Distributional Structure. *Word* **1954**, *10*, 146–162. [CrossRef]
48. Salton, G.; McGill, M. *Introduction to Modern Information Retrieval*; McGraw-Hill: New York, NY, USA, 1983.
49. Cantos, P.; Almela, A. Readability indices for the assessment of textbooks: A feasibility study in the context of EFL. *Vigo Int. J. Appl. Linguist.* **2019**, *16*, 31–52. [CrossRef]
50. Shadish, W.R.; Cook, T.D.; Campbell, D.T. *Experimental and Quasi-Experimental Designs for Generalized Causal Inference*; Houghton Mifflin Company: Boston, MA, USA, 2002.
51. Chipere, N.; Malvern, D.; Richards, B.J. Using a corpus of children's writing to test a solution to the sample size problem affecting Type-Token Ratios. In *Corpora and Language Learners*; Aston, G., Bernardini, S., Stewart, D., Eds.; John Benjamins: Amsterdam, The Netherlands, 2004; pp. 139–147.
52. Kline, P. *A Handbook of Test Construction*; Methuen: New York, NY, USA, 1986.
53. Berber-Sardinha, T.; Veirano, M. (Eds.) *Multidimensional Analysis*; Bloomsbury Publishing: London, UK, 2019.
54. Cantos, P. *Statistical Methods in Language and Linguistic Research*; Equinox: London, UK, 2013.
55. Tabachnick, B.G.; Fidell, L.S. *Using Multivariate Statistics, New International Edition*, 6th ed.; Pearson Education Limited: Harlow, UK, 2013.
56. Molinaro, A.M.; Simon, R.; Pfeiffer, R.M. Prediction error estimation: A comparison of resampling methods. *Bioinformatics* **2005**, *21*, 3301–3307. [CrossRef]
57. Kohavi, R. A study of cross-validation and bootstrap for accuracy estimation and model selection. In *Proceedings of the Fourteenth International Joint Conference on Artificial Intelligence*; Morgan Kaufmann: San Mateo, CA, USA, 1995; pp. 1137–1143.

58. DePaulo, B.M.; Lindsay, J.J.; Malone, B.E.; Muhlenbruck, L.; Charlton, K.; Cooper, H. Cues to deception. *Psychol. Bull.* **2003**, *129*, 74–118. [CrossRef]
59. Masip, J.; Bethencourt, M.; Lucas, G.; Sánchez-San Segundo, M.; Herrero, C. Deception detection from written accounts. *Scand. J. Psychol.* **2012**, *53*, 103–111. [CrossRef]
60. Burgoon, J.K.; Blair, J.P.; Qin, T.; Nunamaker, J.F. Detecting deception through linguistic analysis. *Intell. Secur. Inform.* **2003**, *2665*, 91–101. [CrossRef]
61. Vivancos-Vicente, P.J.; García-Díaz, J.A.; Almela, A.; Molina, F.; Castejón-Garrido, J.A.; Valencia-García, R. Transcripción, indexación y análisis automático de declaraciones judiciales a partir de representaciones fonéticas y técnicas de lingüística forense. *Proces. Leng. Nat.* **2020**, *65*, 109–112.

applied sciences

Article

Incorporating Concreteness in Multi-Modal Language Models with Curriculum Learning

Erhan Sezerer and Selma Tekir *

Department of Computer Engineering, Izmir Institute of Technology, 35430 Izmir, Turkey; erhansezerer@iyte.edu.tr
* Correspondence: selmatekir@iyte.edu.tr

Abstract: Over the last few years, there has been an increase in the studies that consider experiential (visual) information by building multi-modal language models and representations. It is shown by several studies that language acquisition in humans starts with learning concrete concepts through images and then continues with learning abstract ideas through the text. In this work, the curriculum learning method is used to teach the model concrete/abstract concepts through images and their corresponding captions to accomplish multi-modal language modeling/representation. We use the BERT and Resnet-152 models on each modality and combine them using attentive pooling to perform pre-training on the newly constructed dataset, which is collected from the Wikimedia Commons based on concrete/abstract words. To show the performance of the proposed model, downstream tasks and ablation studies are performed. The contribution of this work is two-fold: A new dataset is constructed from Wikimedia Commons based on concrete/abstract words, and a new multi-modal pre-training approach based on curriculum learning is proposed. The results show that the proposed multi-modal pre-training approach contributes to the success of the model.

Keywords: multi-modal dataset; Wikimedia Commons; multi-modal language model; concreteness; curriculum learning

Citation: Sezerer, E.; Tekir, S. Incorporating Concreteness in Multi-Modal Language Models with Curriculum Learning. *Appl. Sci.* **2021**, *11*, 8241. https://doi.org/10.3390/app11178241

Academic Editors: Emanuele Carpanzano, Arturo Montejo-Ráez and Salud María Jiménez-Zafra

Received: 1 August 2021
Accepted: 27 August 2021
Published: 6 September 2021

1. Introduction

After the success of contextual representations, language model pre-training and fine-tuning the model for downstream tasks have been common practices in natural language processing (NLP) . The wide-spread adoption of BERT [1] led to several pre-trained language models that are described as BERT variants [2–5]. Putting BERT at the core, these models provide extensions with different viewpoints, cross-lingual, multi-task, multi-modal, and world knowledge, to name a few. Among these models, Albert [3] targets efficiency by using weight sharing and decreasing memory consumption, RoBERTa [2] increases the amount of training data and times and removes the next sentence prediction objective, XLNet [4] uses permutation instead of masking to capture the bidirectional context and combines BERT with autoregressive language modeling, and ERNIE [5] aims to exploit world knowledge by masking named entities and phrases rather than random words, and, in its updated version [6], the pre-training task is organized as a multi-task objective to capture different relations, such as lexical, syntactic, and semantic.

The earlier approaches to bridge vision and language relied on architectures with a visual feature extractor, a text encoder, a multi-modal fusion component, and a classification layer to perform the given multi-modal task, e.g., visual question answering. The robust pre-trained language models have caused a shift from a task-specific perspective to a task-agnostic one, multi-modal language model pre-training.

Multi-modality, especially with vision and language, has been implemented in some BERT variants [7–9], as well. VisualBERT [7] and VideoBERT [8] use similar transformer-based architectures. The former processes image captions together with image regions to discover implicit alignments between language and vision. On the other hand, the latter works with spoken words paired with a series of images to learn a similar alignment. Distinctively, ViLBERT [9] has a two-stream transformer model, which processes vision and language separately but learns their relationships through co-attentions between them.

The primary motivation for combining vision and language in these models has been visual grounding to learn visual features under the guidance of textual descriptions. Apart from it, we can leverage visual and language features to mimic human language acquisition.

There have been studies that indicate we can mainly attribute language acquisition in children to experiential information in early ages [10–12]. It is mentioned in those works that the language acquisition in children starts with experiential information, where we mostly learn about concrete concepts in languages and continue with the textual information in later ages where we mostly know about abstract concepts. Thus, many researchers tried to build language models with multi-modal information (Refs. [9,13,14], and many more), leveraging both textual and visual inputs.

This work aims to create a multi-modal language model that uses both textual and visual features, similar to what humans do. First, we feed the image model concrete examples. Then, we train the textual model with all of the samples concrete and abstract combined, in a curriculum learning fashion [15,16]. We rely on University of Western Australia The Medical Research Council (UWA MRC) Psycholinguistic Dataset [17] for the lists of the abstract/concrete words. The contribution of this work is two-fold: A new dataset is constructed from Wikimedia Commons based on concrete/abstract terms, and a new multi-modal pre-training approach that is based on curriculum learning [15,16] is proposed.

The results show that the proposed multi-modal pre-training method contributes to the success of the model in downstream tasks, e.g., visual question answering. In addition, it can be seen from the ablation study that this increase in performance is consistent among all fusion techniques used in this work. We obtained the best results when the multi-modal pre-training scheme is used with attentive pooling as the fusion mechanism. In addition to the tests mentioned above, we performed several tests for measuring the informativeness of the newly constructed dataset.

The rest of the manuscript is structured as follows: In Section 2, we give background information on the task of language modeling/representation. Model details and the new dataset are explained in Section 3. We share the experimental results in Section 4, along with the descriptions of the datasets used. In addition, finally, in Section 5, final remarks are made with possible future directions.

2. Related Work

The idea of building word representations from frequency statistics comes from the Distributional Hypothesis [18,19]. The distributional hypothesis states that one can determine the meaning of a word through the words that co-occur with it in the same context. Famously, Harris (1954 [19]) states that the "words that occur in the same context tend to have similar meanings".

Although the count-based methods can leverage the distributional model to learn the representations of words, they suffer from several drawbacks: lack of word order, unable to retrieve representations from partial information (generalization power), and the curse of dimensionality (they create millions, if not trillions, of different possible n-grams which are very unlikely to be observed in the training data, which leads to a very sparse matrix with a lot of uninformative zero entries).

Neural network solutions emerged to solve these issues. In such a first attempt, Hinton et al., in 1986 [20], utilized the idea of distributed representations for concepts. They proposed to use patterns of hidden layer activations (which are only allowed to be 0 or 1) as the representation of meanings instead of representing words with discrete entities, such as the number of occurrences, together. They argued that the most critical evidence of distributed representations is their degree of similarity to the weaknesses and strengths of the human mind.

Elman (1990) [21] was the first to implement the distributional model proposed by Reference [20] in a language model. He presents a specific recurrent neural network structure with memory, called the Elman network, to predict bits in temporal sequences. Memory is provided to the network through context units that are fully connected with hidden units.

Although these models build the basis of neural word representations, Bengio et al., in 2003 [22], popularized the distributional representation idea by realizing it through a language model and lead to numerous other studies that are built on it. Their model architecture uses a feed-forward network with a single hidden layer and optional direct connections from the input layer to the softmax layer. The weights of the hidden layer are then taken as the representations of words.

Once it is shown that neural language models are efficiently computable by Bengio et al., as in 2003 [22], newer language models, along with better word embeddings, are developed successively. In such an effort, Mikolov et al., in 2013 [23], proposed word2vec to learn high-quality word vectors. The authors removed the non-linearity in the hidden layer in the proposed model architecture of Bengio et al., in 2003 [22], to gain an advantage in computational complexity. Due to this change, the system can be trained using billions of words efficiently. Thus, it is considered as the initiator of early word embeddings [24].

Despite the success of these earlier word embeddings, there were still many limitations in terms of the accuracy of representations (lack of polysemy, unable to account for morphology, antonymy/synonymy problem). Many methods have been proposed for solving the deficiencies of embedding methods. Each of them is specialized on a single problem, such as sense representations [25,26], morpheme representations [27,28], etc., while none of them could combine different aspects into a single model, a single solution. It is the idea of contextual representations to provide a solution that covers each element successfully. The main idea behind contextual representations is that words should not have a single representation to be used in every context. Instead, one should calculate a representation separately for different contexts. Contextual representation methods calculate the embedding of a word from the surrounding words each time the word is seen. This characteristic leads to an implicit solution to many problems, such as sense representations, since multi-sense words can now have different representations according to their contexts. Furthermore, character-level processing has been proposed to incorporate the sub-word information into embeddings. Therefore, contextual representation models described below can incorporate different aspects together into a single model.

In such a first attempt to create contextual representations, Melamud et al., in 2016 [29], developed a neural network architecture based on bidirectional-LSTMs to learn context embeddings with the target word embeddings jointly. CoVe [30] uses Glove [24] as the initial word embeddings and feeds them into a machine translation architecture to learn contextual representations. The authors argue that pre-training the contextual representations on machine learning tasks, where there are vast amounts of data, can lead to better contextual representations to transfer learning to other downstream tasks. Using language modeling and learning word representations as a pre-training objective then fine-tuning the architecture to downstream tasks is first proposed by References [31,32]. ELMO [33] improves on the character-aware neural language model by Reference [34]. The architecture takes characters as input to a CNN network from where it is fed to a 2-layer bidirectional-LSTM network to predict a target word. They show that this architecture can learn various aspects of semantic, syntactic, and sub-word information. Instead of using words as input, Flair [35] uses a character-level language model to learn contextual word representations. Unlike ELMO, where character-level inputs are later converted into word features, authors propose using characters only in this work. BERT [1] uses a bidirectional transformer [36] architecture to learn contextual word representations. XLNet [4] is an autoregressive method that combines the advantages of two language modeling methods: Autoregressive models (i.e., transformer-XL [37]) and autoencoder models (i.e., BERT). ALBERT [3] aims at lowering the memory consumption and training times of BERT [1]. To accomplish this, they perform two changes on the original BERT model: They factorize the embeddings into two matrices to use smaller dimensions, and they apply weight sharing to decrease the number of parameters.

The success of uni-modal language models drives the researchers into studies that examine the use of visual information for training language models. They base this decision on the advances in cognitive science where it is shown that language acquisition in children mostly relies on experiential data [10–12]. While some of those studies focused on producing better representations, [12,38–42], most of these models produce multi-modal embeddings as a side-product of a multi-modal task. These tasks include image retrieval with text and caption [43,44], image-text alignment [45,46], image segmentation using a target text [47], visual question answering [13,14,48], visual common-sense reasoning [49], and image captioning [42]. Some other studies also contributed to the field of multi-modal language modeling by encompassing many of these models similar to contextual embeddings [9] or by enhancing the existing models [50]. As the field is relatively new, most of these works focus on the fusion of modalities more than the individual models.

Curriculum learning [15,16] used in this study is a progressive training method that puts the samples in a meaningful order instead of random shuffling. Training is done in learning steps where, in each step, the difficulty of the examples is increased. Curriculum learning provides two benefits: faster convergences of neural methods and finding a better local minimum. Many aspects of multi-modal language models are well studied, and curriculum learning methods are applied to other NLP subjects. However, to the best of our knowledge, there has not been a study that explored curriculum learning approaches in multi-modal language modeling.

3. Method

In this section, we introduce the details of the proposed model and dataset. First, a newly created dataset from Wikimedia Commons is described in Section 3.1. In the following Sections 3.2 and 3.3, the proposed model, along with the training method, is explained.

3.1. Wikimedia Commons Dataset

Wikimedia Commons (https://commons.wikimedia.org/wiki/Main_Page, accessed on through 1 January 2020 to 13 April 2020) is a repository of free-to-use images that is a part of Wikimedia Foundation. Wikimedia Commons files are used across all Wikimedia projects in all languages, including Wikipedia, Wiktionary, Wikibooks, Wikivoyage, Wikispecies, Wikisource, Wikinews, or downloaded offsite use. It comprises approximately 65 million images that take about 250 TB of space. The images also contain captions, descriptions, and timestamps.

To retrieve the images, one must send queries to the Wikimedia Commons website. To this end, we have used two different sets of query words to construct datasets. For retrieving the entire dataset, the dictionary of the BERT model [1] is used. As for getting the subset that we primarily used in this work, UWA MRC psycholinguistic dataset words are used.

UWA MRC Psycholinguistic Dataset [17] contains 98538 words and their properties, such as type, meaningfulnes, concreteness, part-of-speech, familiarity, and many more. Concreteness scores which are used in this research are derived from merging the two datasets provided by References [51,52].

In this dataset, 4293 out of 98538 words have a concreteness rating, rated by human annotators. Human annotators are asked to rate the concreteness of words between (including) 1 and 7, where the higher the score, the more concrete the word is. The mean of all users' scores is the final concreteness rating of the word, which is scaled between 100 and 700. Overall, the most abstract term in the dataset is "as" with a rating of 158, and the most concrete word is "milk" with a score of 670. The mean rating of all terms is 438, and the standard deviation is 120.

To successfully integrate this dataset into our task, some processing is required. Although the UWA MRC Psycholinguistic dataset successfully identifies the concreteness of words, it considers the words in isolation, unlike this work, where contextual embeddings and language models regard words in their context. Therefore, all the stop-words are removed (stop-words from the NLTK library are used) from the dataset, considering that they can appear in various contexts with different levels of concreteness and therefore can lead to misleading results. It is observed from the dataset that the lowest-rated words are usually stop-words, such as "as", "therefore", and "and". Thus, a lot of abstract words are removed in the lower bound. The most abstract word in the dataset after the removal is "apt" with a rating of 183. The final version of the dataset contains 1674 abstract and 2434 concrete words.

For each word, a query is sent to the Wikimedia Commons website with 1000 as a maximum threshold for the number of results. As a result, we have images, their corresponding captions, descriptions, and concreteness labels. Figure 1 shows the number of images returned for each query word in UWA MRC psycholinguistic dataset. As seen from the graph, most of the query words returned less than 100 results despite a large threshold. Only around a hundred words have more than 500 images associated with them. The number of samples collected is shown in Table 1. More than 43 million images are collected using the dictionary of BERT, while approximately 3.2 million images are collected using the words in UWA MRC psycholinguistic dataset. We can also observe that not all images have a description and/or caption associated with them. Some images contain only captions, some images contain descriptions but no caption, and, finally, some images do not contain any textual information at all. In total, 630,000 images contain captions, and approximately 2 million images contain descriptions. Overall, there is an overlap between both sets which means that some images contain both captions and descriptions.

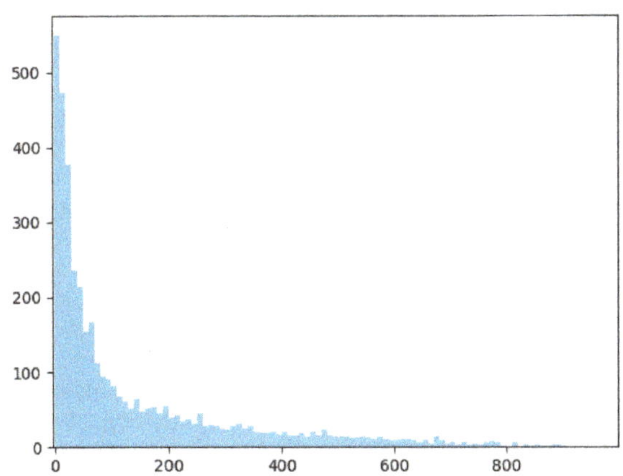

Figure 1. Histogram of the samples retrieved for words. Horizontal axis shows the number of images retrieved, while the vertical axis shows the amount of words which have that many images associated with them.

Table 1. Wikimedia Commons dataset statistics.

Dataset	# of Images	# of Captions	# of Descriptions
Complete Dataset	43,726,268	1,022,829	17,767,000
Subset (queried w/UWA MRC words)	3,206,765	629,561	1,961,567

The retrieved images have many formats, such as .jpeg, .jpg, .jpe .png, .apng, .gif, .tif, .tiff, .xcf, .webp, and many image modes, such as RGB (3 × 8-bit pixels, true color), CMYK (4 × 8-bit pixels, color separation), I (32-bit signed integer pixels), I;16 (16-bit unsigned integer pixels). Although many of these formats and modes are supported, we eliminated some of them. Images with the extension .xcf and .webp are filtered because mainstream image processing libraries do not support them. In addition to this, images with mode I (and other modes of I, such as I;16, I;16L, I16B, and so on) are eliminated because they are single-channel image modes, and the neural network models that process these images run with multi-channel inputs. Nearly 26,000 images are eliminated after this filtering. In the final version of the dataset, there are approximately 603,000 images with captions, where 177,000 belongs to abstract concepts, while 425,000 belongs to concrete concepts.

Many images in Wikimedia Commons have a very high resolution (resolutions, such as 3000×5000, 6000×6000, are very common), therefore requiring huge storage space. In addition to the filters applied above, a resize operation is performed to cope with this storage problem. All images are converted to a resolution of 224×224 since all the image models (GoogleNet [53], VGG [54], Resnet [55]) run with those.

Figure 2 shows some example images and their corresponding captions and descriptions from the collected Wikimedia Commons dataset. The selected images have captions and descriptions, except for the bottom-left image where a description does not exist.

One thing to be observed from these images is, indeed, the images and the texts convey different information on the relationship of concepts. For example, there is no textual information in the top-left image, neither in the caption nor in the description, about the buildings that can be seen in the image. However, streets are primarily located near buildings (almost 70% of all images from Wikimedia Commons contains buildings when you search for the keyword "street"), which is captured by the image. Therefore the system can learn a relationship of concrete concepts, such as "street" and "building", from the pictures without relying on the text. Similarly, the image contains no clue about its location, but it is understandable from both the caption and the description that it is in Mogadishu, Somalia. In the same vein, in the bottom-left image; there is no mention of a sea/lake in the text, but the lighthouse and the sea/lake can be seen together (which occur with almost no exception in real life) in the image, which will help the model to learn their relationships better. So, a language model trained with both images and text can help to improve the performances of language models.

Although the collected dataset contains captions and descriptions, captions are used to train the multi-modal language model. The reason is two-fold. We observed that descriptions in Wikimedia Commons are unclean. They include many additional texts, such as copyright notices, information about the photographer, or information about how the photograph is taken (such an example can be seen in the last sentence of the top-right image of Figure 2). On the other hand, captions are already cleaned and contain information only about the picture itself. Because of the requirement of tedious cleaning, we relied on captions.

The second but most important reason is the image-text alignment issues. Captions are written to describe the images briefly without giving any other information or making any further comment classified as common-sense knowledge or real-world knowledge. Contrarily, descriptions contain much information that cannot be seen in or referred from the images. Although these additional pieces of knowledge can be essential and valuable in other tasks, they break the image-text alignment and lead to learning noisy contexts in language modeling. If we take the top-right image in Figure 2 as an example, we can see how this can affect the language models. The description of the top-right image provides many semantically similar words to the context of the image, which is sheep lounging in a field, such as "breeding", "slaughtered", and "vegetation". However, it also provides a lot of different or unrelated words, such as "castle", "ruin", "municipality", which has very little to do with the image itself. Consequently, this leads to learning from an accidental relationship, for example, between the context of "sheep" and the context of "municipality". On account of this fact, captions are used in all language modeling tasks in this work to provide a better image-text alignment in training samples.

Caption	A man carries a huge hammerhead through the streets of Mogadishu	Sheep lounging in the shade of a tree with matriarch standing guard
Description	Mogadishu, Somalia. 10/10/2015. A man carries a huge hammerhead shark through the streets of Mogadishu. A recent escalation of plunders of Somali waters by foreign fishing vessels could mean the return of hijackings, locals warn. The country's waters have been exploited by illegal fisheries and the economic infrastructure that once provided jobs has been ravaged. Somalia has been at war for the last 25 years, but 2017 is a turning point. This country in the Horn of Africa is holding its first free elections since 1969; a whole culture is being overturned. Those who created it have shot and killed, but finally, they are on the losing side.	A flock of sheep (*Ovis aries*) lounging in the shade of a tree with the matriarch of the flock standing outside the shade. The flock was kept in the enclosed area of Röe Castle ruin to keep the vegetation in check. The standing matriarch is tagged in both ears meaning that she is selected for breeding and will not be slaughtered after her first year. The rest of the flock have tags in only one ear and will be slaughtered withing twelve months after their birth. Röe Castle ruin, Röe, Lysekil Municipality, Sweden. The image is stacked manually from two photos (handheld) for focus and light.

Caption	Aniva lighthouse on a rocky promontory in Sakhalin, Russia, with a flock of gulls circling in the surrounding mists	A Javan Slow Loris (Nycticebus javanicus) clings to a branch.
Description	-	The Javan slow loris (*Nycticebus javanicus*) is a strepsirrhine primate and a species of slow loris native to the western and central portions of the island of Java, in Indonesia. Although originally described as a separate species, it was considered a subspecies of the Sunda slow loris (*N. coucang*) for many years, until reassessments of its morphology and genetics in the 2000s resulted in its promotion to full species status. It is most closely related to the Sunda slow loris and the Bengal slow loris (*N. bengalensis*). The species has two forms, based on hair length and, to a lesser extent, coloration.

Figure 2. Example images and their corresponding captions and descriptions from the Wikimedia Commons Dataset.

There have been several other multi-modal datasets proposed in the literature that consist of image-text pairs, such as Flickr [56], MS COCO [57], Wikipedia, British Library, and ESP Game[58]. Table 2 shows the collected dataset in comparison with these multi-modal datasets. The Flickr dataset and MS COCO dataset contain image-caption pairs, while the Wikipedia dataset provides the images in Wikipedia with their corresponding articles. The British Library book dataset, on the other hand, contains historical books and the pictures depicted in them. Finally, the ESP game dataset consists of 5 words for each image labeled by human annotators. Although both Wikipedia and BL datasets provide much longer texts, they lack the image-text alignment of caption datasets. Therefore, caption datasets, such as MS COCO, Flickr, or the proposed dataset in this work, are more suited to the task of multi-modal language modeling. Compared with these image captioning datasets, the size of the collected dataset is much greater. As deep neural representations have massive data requirements, it is preferable to have such a large amount of data. Recently, the WIT [59] dataset was also proposed, with a large number of image-text pairs that can be used for multi-lingual, multi-modal pre-training. It contains 11.4 million unique images with captions and descriptive text from Wikipedia articles for various languages. Among them, 3.98 million images have textual information in English, where 568,000 of them have captions. In addition to captions, the collection also includes contextual data, such as page titles, page descriptions, section titles, etc., with their descriptions. However, the most significant benefit of the proposed dataset is the concreteness labels provided for each image-text pair which might be very useful for various tasks, especially for the multi-modal language modeling. The other datasets mentioned in this section, including WIT, do not contain that information.

Table 2. Comparison of Wikimedia Commons to other multi-modal datasets.

Dataset	# of Images	Textual Source	Ave. Word Length	Additional Info.
Flickr [56]	32,000	Captions	9	-
COCO [57]	123,000	Captions	10.5	-
Wikipedia	549,000	Articles	1397.8	-
BL	405,000	Books	2269.6	-
ESP[58]	100,000	Object Annotations	5	-
WIT[59]	11.4 million	Captions/Articles	-	-
	3.98 million	Captions/Article (En)	-	-
	568,000	Captions (En)	-	-
Wikimedia Commons (ours)	3.2 million	-	-	Concreteness Ratings
	629,000	Captions	10.2	Concreteness Ratings
	1.96 million	Descriptions	57.4	Concreteness Ratings

3.2. Model

The overall architecture of the proposed model can be seen in Figure 3. The model is comprised of three main parts: text processing part, image processing part, and a fusion mechanism where the outputs of text and image models are combined. Each piece is explained below in its respective subsection.

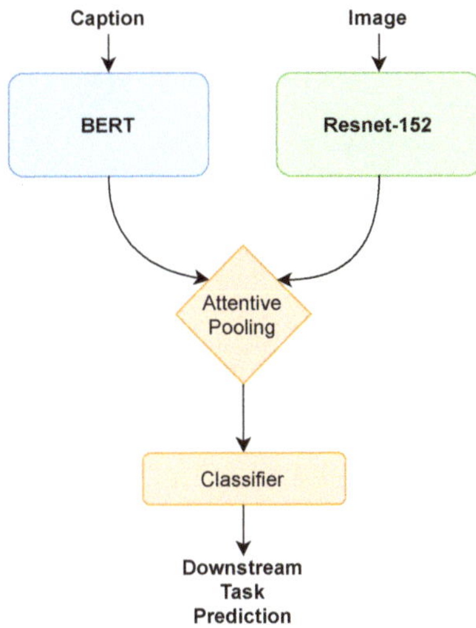

Figure 3. Proposed black-box model architecture.

3.2.1. Text Model

In this work, BERT is primarily used for processing text input, while we also utilized DistilBERT in some of the tests.

BERT [1] is a neural network model that uses a bidirectional transformer architecture [36], a self-attention mechanism to learn contextual word embeddings. It has multiple layers of transformers (12 in BERT-base, 24 in BERT-large) where each layer has 12 attention heads that span the entire sentence from both right-to-left and left-to-right, learning "where to look" by producing probabilistic weights for each word.

Different from the earlier language modeling approaches, BERT does not use next word prediction as an objective. Instead, it uses two training objectives: Masked Language Modeling (MLM) and Next Sentence Prediction (NSP). For the MLM objective, randomly selected words are occluded from the model and labeled as masks. The model tries to predict the masked word as the training objective. Attention heads do not span these masked words since it would create a bias for the prediction. Using MLM enables the model to learn contextual dependencies among words very successfully. The embedding of a word is computed depending on the surrounding terms instead of using the same vector in the embedding space for every instance of that word. For the NSP objective, the model tries to predict whether the two sentences provided to the model belong to the same context or not. It helps BERT to consider multiple sentences as context and to represent inter-sentence relations.

In addition to the token (word) embeddings, BERT also uses segment (sentence) embeddings and position embeddings (words' position in segments) as input. While sentence embedding determines which sentence the word is in, positional embedding acknowledges the word order. Therefore, a word's embedding is fed to the model as the average of its token embedding, sentence embedding, and positional embedding. This input structure has many benefits: Positional embeddings raise the model's awareness of word order, while segment embeddings help the NSP objective. In addition, giving multiple sentences as input helps BERT be integrated into most downstream tasks requiring inter-sentence connections, such as Question Answering and Natural Language Inference (NLI), easily, without requiring any other architecture.

To integrate BERT to downstream tasks, an additional fully connected layer is used on top of transformer layers to predict the given text's class instead of the target (masked) word. Usually, the Wikipedia dataset is used to pre-train the model on MLM and NSP objectives. The resulting parameters are fine-tuned on the downstream task with the addition of the aforementioned fully connected layer.

In this study, we performed some tests using the DistilBERT language model. DistilBERT [60] is based on the original BERT model. It is a more efficient version of BERT in expense for a minor deficiency in classification performance. It retains 97% of BERT's performance while using 40% fewer parameters. To accomplish this, they use knowledge distillation, where a small model is trained to reproduce the behavior of a larger model (DistilBERT and BERT, respectively, in this case). Knowledge distillation aims to make the student model (DistilBERT) predict the same values as the teacher model (BERT) using fewer parameters. This way, one can transfer the knowledge learned by the teacher model to more efficient student models. Parameter reduction from BERT to DistilBERT comes from the removal of some of the transformer layers in BERT. The authors of DistilBERT show that some of the parameters of BERT are not used in the prediction, therefore, do not contribute to learning downstream tasks. Consequently, they suggest removing some layers and use the knowledge distillation technique to create a more efficient language model.

3.2.2. Image Model

We used Resnet [55] as the image model due to its success in many image processing tasks. It is a very deep neural network model that relies on convolutional neural network architecture. At the time it is published, it was the state-of-the-art model in the ImageNet [61] object classification challenge.

Resnet has several different variations in network depth: 34-layered model Resnet34, 50-layered model Resnet50, 101-layered model Resnet101, and, finally, the largest model with 152-layers Resnet152. Each layer consists of several 1×1 and 3×3 convolutions. Each model starts and ends with an average pooling operation before the first layer and after the last layer.

Stacking so many layers in deep neural networks naively does not immediately lead to better results; instead, it causes performance degradation problems. An increase in the depth of a model causes an increase in training errors, and accuracy is saturated. To deal with this issue and build substantially deeper networks, authors needed a workaround. Therefore, shortcut connections called residual connections are used. These shortcut connections are used after every two layers in the architecture, propagating the inputs to the outputs of those two layers. They are parameter-free, which means that they do not perform any operation on the inputs, such as pooling, convolution, or multiplication; therefore, they do not contain any learnable parameters. It is shown that these shortcut connections can overcome the performance degradation problem in very deep neural network architectures, making models, such as Resnet, very successful at stacking many layers and capturing more features than the prior models.

In this work, Resnet152 is used because it outperforms the smaller Resnet models, and the Wikimedia Commons dataset was large enough to tune such a large model.

3.2.3. Text-Image Combination Method

Combining multiple modalities can be problematic and risks breaking the learned semantic relationship of words by individual models. Thus, many studies in this field focus on the fusion of modalities.

We used attentive pooling networks [62] to combine the text and vision parts of the model. It is a two-way attention mechanism that is aware of both modalities and jointly learns to attend over them through matrix multiplications and pooling operations.

Attentive pooling takes the hidden states of each word in BERT as textual input and takes the last layer of Resnet in the form of a matrix as visual input. These inputs are multiplied with the matrix U , which is composed of parameters to learn and passed through $tanh$ activation. The result is a single matrix of visual features on the rows and textual features on the columns. This representation scheme allows features from different modalities to be jointly represented in a single matrix where max-pooling operation is performed over each row and column to find out the most important feature dependent upon the other modality. Two vectors, I_{output} and T_{output}, are the outcomes of the attentive pooling mechanism. For fine-tuning this model on downstream tasks, these two outputs are concatenated and passed through an additional fully connected layer to reduce the dimension to the number of classes.

3.3. Multi-Modal Language Model Training

The idea of pre-training neural language models is borrowed from the advances in image processing models [32]. It is shown in both vision and text models that pre-training a model on a preliminary image/text understanding task improves the performance vastly.

For image processing, the pre-training task is usually the object classification task on the ImageNET dataset [61]. ImageNET dataset has 1.2 million images that are hand-labeled into 1000 categories. Respective models are trained to predict the objects in each image by adding a fully connected layer on top to reduce the feature vectors' size to 1000. The aim here is to teach the model basic image understanding: Identifying objects and entities in images. It is shown by many vision models that they are even able to differentiate images of 120 different dog breeds in the imageNET dataset, such as "Australian terrier" and "Airedale terrier". They manage to do this by using the shapes and colors of entities in the pictures.

The process is similar for language models, with the only difference in pre-training objectives. Earlier models (before BERT) used next word prediction in huge unlabeled text, such as Wikipedia and Common Crawl text. The aim was to predict the next word given the previous set of words. Starting from BERT and onward, the pre-training objective changed from the next word prediction to masked language modeling. This method allowed the text models to successfully grasp language understanding by training them on massive datasets containing billions of words. They learned the meaning and semantic/syntactic relations of words (due to distributional hypothesis), which are fundamental to any downstream task.

Once the pre-training objective is completed and the image/text model gained basic image/language understanding, respectively, the last fully connected layer is removed from the model and replaced with an appropriate classification layer according to the task at hand. The model is, then, fine-tuned for the downstream task. For image models, downstream tasks can be object detection, semantic segmentation, etc., while, on the textual models, they are composed of sentiment analysis, sentence classification, natural language inference, and so on.

In this work, we adopt a novel multi-modal pre-training objective. The idea is inspired from the advances in cognitive psychology. It is shown that language acquisition in children starts with experiential information and continues with textual information [11,12]. As Kiela et al., in 2015 [63], stated, perceptual information is more relevant for, e.g., elephants than it is for happiness. In other words, we first learn the language through images and learn concrete concepts, and then we start learning abstract concepts from textual sources.

Advancements in computational linguistics also reinforce this idea by showing that concrete examples in language are easier to learn, while abstract ones are more challenging. Hessel et al., in 2018 [64], showed that the more concrete the downstream task gets, the easier it becomes for language models. Bruni et al., in 2014 [38], showed that the semantic/syntactic similarities of concrete examples on the MEN dataset are easier to learn, while the abstract words can get ambiguous. They prove this by showing that the concrete examples have a 0.78 Spearman correlation rank, while the abstract examples have 0.52 (contributing to an overall 0.76).

To adopt this learning scheme to this project, the Wikimedia Commons Dataset (see Section 3.1) is divided into two categories: Abstract samples and concrete samples. We determined concrete/abstract examples based on the concreteness levels of words from the UWA MRC Psycholinguistic Database. First, we fed the image model concrete examples. Then, we trained the textual model with all of the samples concrete and abstract combined, in a curriculum learning fashion [15,16]. Therefore, the learning model mimics humans through this pre-training process.

4. Experiments

The first step of experimentation was to measure the informativeness of the collected dataset. To meet this objective, we selected concreteness classification and tested the performance of captions in this task. Moreover, to show the expressiveness of captions relative to regular texts, we did the same classification with the regular Wikipedia articles. We worked with the June 2020 version of wikidumps, which consists of 6, 957, 578 documents in total.

To prepare the dataset for comparison, we search for articles in the Wikipedia dataset using UWA MRC Psycholinguistic dataset words. Specifically, each article titled with the corresponding words is retrieved. We concatenated the captions that corresponded to the same word and removed the terms that do not have a Wikipedia article to match captions with the Wikipedia articles further. After this, there are 4108 samples remaining in the dataset, which is partitioned into the train (70%), dev (10%), and test (20%) sets randomly.

Table 3 shows the results of DistilBERT and BERT along with the random baselines on these datasets. The results show that, although the Wikimedia captions give us worse than the Wikipedia articles, results are not far off, making the Wikimedia captions almost as informative as the Wikipedia text itself.

Table 3. Results comparing the informativeness of the proposed dataset.

Model	Wikimedia Captions				Wikipedia Articles			
	Accuracy	Precision	Recall	F1	Accuracy	Precision	Recall	F1
Random	0.5171	0.5171	0.5171	0.5171	0.5255	0.5255	0.5255	0.5255
DistilBERT	80.91 (−1.47 + 2.28)	80.89 (−1.47 + 2.31)	80.91 (−1.47 + 2.28)	80.83 (−1.41 + 2.36)	86.54 (−1.97 + 0.53)	86.69 (−1.08 + 0.83)	86.54 (−1.97 + 0.53)	86.58 (−1.99 + 0.50)
BERT	82.37 (−1.88 + 1.19)	82.35 (−1.96 + 1.10)	82.37 (−1.88 + 1.19)	82.31 (−1.97 + 1.12)	85.60 (−1.91 + 1.35)	85.69 (−1.89 + 1.24)	85.60 (−1.91 + 1.35)	85.45 (−1.07 + 1.49)

Table 4 shows the experimental results of the multi-modal pre-training task on the test set. As stated before, we performed this pre-training in a curriculum learning fashion. Our image model is further pre-trained with concrete samples of the training set, and then the text model is trained on all the samples on the training set, concrete, and abstract combined. The results show the performance of each model on the test set of the pre-training dataset. While the image model obtained 0.8147 F1 on the concrete samples, the text model obtained 0.8707 and 0.6518 F1 on the concrete and abstract samples. Although we did not pre-train the image model on abstract samples, we also show its results to give an idea.

Table 4. Experimental results of the multi-modal pre-training task.

Model	Accuracy	Precision	Recall	F1	F1-abs	F1-Conc
Bert	0.8116	0.8057	0.8116	0.8069	0.6518	0.8708
Resnet	0.7001	0.6472	0.7001	0.6383	0.2144	0.8147

We can draw several conclusions from the results. Firstly, the results comply with References [38,64]: Identifying concrete concepts is much easier than identifying abstract concepts. Both the Resnet and BERT models perform above 0.8 in terms of F1 scores for the concrete class. On the other hand, the F1 score of Resnet on the abstract class turns out to be significantly lower, with a value of 21.5. These results show that both image and text models struggle more with abstract concepts than concrete ones.

Secondly, the results of Resnet agree with the scientific work (i.e., References [11,12]) on human language acquisition. Thus, they also comply with the curriculum learning objectives in this work: Experiential information is used early in language acquisition on concrete concepts, while leaving its place to textual information for learning abstract ones.

It can be argued that, no matter how abstract an idea is, one needs to find a concrete example to show that in an image. For example, the image/caption pairs returned for the search word "dream" frequently contain pictures of places. Although the term itself can safely be considered abstract, one needs to find a particular and concrete idea/object to represent it as an image. Therefore, we can conclude that images almost always contain concrete concepts. To determine abstractness, one should use a diverse set of images belonging to a particular concept instead of individual images (the variance in images for the word "tomato" is very low, with the first 25 results are all images of single or a couple of red tomatoes, while the variance in images for the word "dream" is very high, ranging from the picture of places, famous people to screenshots of literary work).

To validate the effectiveness of the proposed multi-modal pre-training scheme, we tested the model's performance on a downstream NLP task. As a multi-modal task, Visual Question Answering fits nicely with our objective. Visual Question Answering dataset is a multi-modal dataset that was proposed by Antol et al., in 2015 [65]. It includes approximately 200,000 images from the COCO dataset [57]. Each image in this dataset has multiple questions associated with it in various forms, such as yes/no questions and open-ended questions. Yes/No questions are binary questions, such as "Is the umbrella upside down?", while the open-ended questions, such as "Who is wearing glasses?", require more diverse answers. Close to 40% of all questions are yes/no questions, and the rest is open-ended. Open-ended questions have a variety of types, including but not limited to "What is ... ?", "How many ... ?", and "Who is ... ?".

Although the dataset requires a lot of inference between modalities, Agrawal et al., in 2018 [13], stated that the dataset includes bias towards some question/answer pairs. In their work, they showed that questions related to colors ("What is the color of ... ?" or "is ... white?") almost always lead to the answers of white/no for open-ended and yes/no questions, respectively. Similarly, Goyal et al., in 2017 [66], suggested that answering the questions that are starting with the phrase "Do you see a ...?" with yes blindly leads to an accuracy of 87% among those questions. Therefore, using language priors alone, a model can correctly predict a significant amount of questions. The authors develop the second version of the dataset to overcome this problem, which has additional samples to balance the biased question/answer pairs. This update increased the dataset size to 443 thousand, 214 thousand, and 453 thousand pairs (question, image) for train, dev, and test sets, respectively. The results reported in this manuscript refer to this new dataset as v2, while they refer to the former as v1.

Table 5 shows the model's performance on VQA. The best result is obtained when both multi-modal pre-training and attentive pooling mechanisms are used, although the performance is consistent across all configurations. In terms of accuracy, there is a 1.01% difference between the best performing model (with multi-modal pre-training and attentive pooling) and the worst (with fully connected layer and without multi-modal pre-training). Performance difference becomes more significant in F1: a 3.37% increase can be observed between the best and worst-performing models (model with multi-modal pre-training and attentive pooling, and model without multi-modal pre-training with a fully connected layer, respectively, similar to the previous case).

Table 5. Model performance on VQA dataset v2. (FC = Fully-connected, AP = Attentive pooling).

Model	Multi-Modal Pre-Training	Combination Method	Accuracy	F1	Precision	Recall
Bert + Resnet	✗	FC	53.12	50.71	54.07	53.12
Bert + Resnet	✓	FC	53.17	52.79	53.34	53.17
Bert + Resnet	✗	AP	53.56	52.91	53.69	53.56
Bert + Resnet	✓	AP	54.13	54.08	54.07	54.13

One can better analyze performance differences with ablation studies. Table 6 reports the relative improvements of each component. Each column represents the percentage increase in relative performance when the feature/component in the row is replaced or enhanced by the feature/component in the column. The results show that multi-modal pre-training increases the model's performance regardless of the underlying fusion mechanism (Fully-connected or attentive pooling). It leads to a 4.1% increase when used with fully connected layers and leads to a 2.21% increase when used with attentive pooling networks. Similarly, the attentive pooling mechanism improves the performance of the model in both cases: When the fully-connected layer is replaced with attentive pooling, it amounts to an increase of 4.34% without multi-modal pre-training and an increase of 2.44% with multi-modal pre-training. Additionally, from the first row, we can conclude that replacing FC with an attentive pooling mechanism is slightly more beneficial than using FC together with multi-modal pre-training. Overall, as the results suggest, using both attentive pooling and multi-modal pre-training proved to be useful and led to an increase in performance up to 6.65% compared to the baseline model.

Table 6. Results of the ablation study. Relative performance improvements (%) of each component in terms of F1. MMPT = Multi-modal pre-training, FC = Fully-connected, AP = Attentive pooling.

	FC	**MMPT + FC**	**AP**	**MMPT + AP**
FC	0	4.10	4.34	6.65
MMPT + FC	-	0	0.23	2.44
AP	-	-	0	2.21
MMPT + AP	-	-	-	0

Table 7 shows the performance of the multi-modal models described in Section 2 on the VQA task. We share the results on version 1 and version 2, though it would only be fair to compare the models that run on the same version. The models that run on both versions (stacked attention network (SAN) and GVQA) suggest that a performance difference between 3–7% can be expected between the versions, most likely due to the effect of language priors. Human baselines, obtained on the 3000 samples in the training set of the v1 dataset, are also provided in the top part.

Table 7. Experimental results on VQA task. Top part shows human baselines.

Model	**Dataset Version**	**Accuracy**
Question	v1	40.81
Question + Caption	v1	57.47
Question + Image	v1	83.30
SAN [67]	v1	58.9
GVQA [13]	v1	51.12
SAN [67]	v2	52.2
GVQA [13]	v2	48.24
Anderson et al., 2018 [14]	v2	70.34
DFAF [48]	v2	70.34
VilBERT [9]	v2	70.92
ours	v2	54.13

Although human baselines are on v1 and our performance is on the v2 version of the dataset, our 54.13% accuracy indicates that the model can perform similarly to humans when given only questions and corresponding captions without images. Compared to the other models, ours performed better than the earlier models but cannot reach the success obtained by the state-of-the-art model (VilBERT), which has 70.92% accuracy. VilBERT processes paired visiolinguistic data in the architecture of BERT to exploit visual grounding in a task-agnostic way.

It should be noted that there are subtle but vital differences between our model and the VilBERT model. The main focus of VilBERT is to process text and image streams in parallel under the transformer architecture to encode their relationship in a pre-trained model to have optimized performance in downstream tasks. On the other hand, the main focus of this work is to optimize the model for the fusion of modalities and curriculum learning. Although our work is much similar to earlier multi-modal works in this regard, our model is a language pre-training model, not a task-specific architecture. The main difference in our work is to add curriculum learning methodology on top of the pre-trained models.

Other than the main focus described above, several reasons might lead to the performance discrepancy between the proposed model and the state-of-the-art models, such as VilBERT. First, the number of learnable parameters in VilBERT is much greater than the proposed model (~600 million versus ~170 million). Second, VilBERT uses the Faster-RCNN [68] model to match each word in the text with the corresponding image patch, while our model uses the Resnet-152 model on the entire image. One could argue that the better alignment provided by the faster-RCNN method might lead to better learning since the model also learns which part in the image a particular word corresponds to. Providing such an alignment could also benefit the proposed model for catching up with the performance of the state-of-the-art models.

5. Conclusions

This study aims to contribute to one of the oldest and most predominant subjects in computer science: language modeling. Since the distributional hypothesis in the early 1950s, many models with many different architectures and methodologies have been introduced in this field. Until recently, models focused on a single modality where a language learner is trained with plain text. Lately, however, the focus is shifted from single modality to multi-modal language models. An increase in the success of neural models, cheaper and more powerful hardware sources, and advances in cognitive science were the major driving forces behind this change.

Similar to this latest trend, this work aims to create a language model/representation technique inspired by the advances in cognitive science, which states that language acquisition in humans starts with the experiential information for concrete concepts and continues with distributional information for abstract concepts. To this end, we combined the BERT and Resnet models with the attentive pooling mechanism to construct a multi-modal language model and embeddings. The image model is trained with the concrete samples from Wikimedia samples first, and then the text model is trained with concrete and abstract examples combined in a curriculum learning fashion. Additionally, we constructed a new dataset composed of image caption pairs from Wikimedia Commons based on concrete/abstract metadata.

The contribution of this work is two-fold: First, a new dataset, created from Wikimedia Commons, is introduced, which has approximately 3.2 million images, with 630,000 captions, 1.96 million descriptions, and concreteness labels. Second, a new training scheme for multi-modal pre-training is introduced. We inspired this novel learning scheme from the curriculum learning approaches in artificial intelligence. The results show that, although the model could not outperform state-of-the-art results, the multi-modal pre-training objective can significantly increase the models' performance. Our results also confirm the findings in the literature by showing that it is harder to detect and classify abstract samples.

Author Contributions: Conceptualization, E.S. and S.T.; methodology, E.S. and S.T.; software, E.S.; validation, E.S. and S.T.; formal analysis, E.S.; investigation, E.S.; resources, E.S. and S.T.; data curation, E.S.; writing—original draft preparation, E.S. and S.T.; writing—review and editing, E.S. and S.T.; visualization, E.S.; supervision, S.T.; project administration, S.T. Both authors have read and agreed to the published version of the manuscript.

Funding: This research received no external funding.

Institutional Review Board Statement: Not applicable.

Informed Consent Statement: Not applicable.

Data Availability Statement: UWA MRC dataset is available from here: https://websites.psychology.uwa.edu.au/school/MRCDatabase/uwa_mrc.htm (accessed on 31 March 2020), Wikipedia dumps can be downloaded from here: https://dumps.wikimedia.org/backup-index.html (accessed on 16 June 2020). The data collected in this study are available on request from the corresponding author.

Acknowledgments: The Titan V used for the experiments in this work is donated by the NVIDIA Corporation.

Conflicts of Interest: The authors declare no conflict of interest.

References

1. Devlin, J.; Chang, M.; Lee, K.; Toutanova, K. BERT: Pre-training of Deep Bidirectional Transformers for Language Understanding. In Proceedings of the 2019 Conference of the North American Chapter of the Association for Computational Linguistics: Human Language Technologies (NAACL-HLT 2019), Minneapolis, MN, USA, 2–7 June 2019; Volume 1, pp. 4171–4186.
2. Liu, Y.; Ott, M.; Goyal, N.; Du, J.; Joshi, M.; Chen, D.; Levy, O.; Lewis, M.; Zettlemoyer, L.; Stoyanov, V. RoBERTa: A Robustly Optimized BERT Pretraining Approach. *arXiv* **2019**, arXiv:1907.11692.
3. Lan, Z.; Chen, M.; Goodman, S.; Gimpel, K.; Sharma, P.; Soricut, R. ALBERT: A Lite BERT for Self-supervised Learning of Language Representations. In Proceedings of the International Conference on Learning Representations, Addis Ababa, Ethiopia, 26–30 April 2020.
4. Yang, Z.; Dai, Z.; Yang, Y.; Carbonell, J.; Salakhutdinov, R.R.; Le, Q.V. XLNet: Generalized Autoregressive Pretraining for Language Understanding. In *Advances in Neural Information Processing Systems 32*; Wallach, H., Larochelle, H., Beygelzimer, A., d'Alché-Buc, F., Fox, E., Garnett, R., Eds.; Curran Associates, Inc.: Red Hook, NY, USA, 2019; pp. 5753–5763.
5. Sun, Y.; Wang, S.; Li, Y.; Feng, S.; Chen, X.; Zhang, H.; Tian, X.; Zhu, D.; Tian, H.; Wu, H. ERNIE: Enhanced Representation through Knowledge Integration. *arXiv* **2019**, arXiv:1904.09223.
6. Sun, Y.; Wang, S.; Li, Y.; Feng, S.; Tian, H.; Wu, H.; Wang, H. ERNIE 2.0: A Continual Pre-training Framework for Language Understanding. *arXiv* **2020**, arXiv:1907.12412.
7. Li, L.H.; Yatskar, M.; Yin, D.; Hsieh, C.; Chang, K. VisualBERT: A Simple and Performant Baseline for Vision and Language. *arXiv* **2019**, arXiv:1908.03557.
8. Sun, C.; Myers, A.; Vondrick, C.; Murphy, K.; Schmid, C. VideoBERT: A Joint Model for Video and Language Representation Learning. In Proceedings of the 2019 IEEE/CVF International Conference on Computer Vision (ICCV), Seoul, Korea, 27 October–2 November 2019; pp. 7463–7472. [CrossRef]
9. Lu, J.; Batra, D.; Parikh, D.; Lee, S. ViLBERT: Pretraining Task-Agnostic Visiolinguistic Representations for Vision-and-Language Tasks. In *Advances in Neural Information Processing Systems*; Wallach, H., Larochelle, H., Beygelzimer, A., d'Alché-Buc, F., Fox, E., Garnett, R., Eds.; Curran Associates, Inc.: Red Hook, NY, USA, 2019; Volume 32.
10. Griffiths, T.L.; Tenenbaum, J.B.; Steyvers, M. Topics in semantic representation. *Psychol. Rev.* **2007**, *114*, 2007.
11. Vigliocco, G.; Meteyard, L.; Andrews, M.; Kousta, S. Toward a theory of semantic representation. *Lang. Cogn.* **2009**, *1*, 219–247.
12. Andrews, M.; Vigliocco, G.; Vinson, D. Integrating experiential and distributional data to learn semantic representations. *Psychol. Rev.* **2009**, *116*, 463–498.
13. Agrawal, A.; Batra, D.; Parikh, D.; Kembhavi, A. Don't Just Assume; Look and Answer: Overcoming Priors for Visual Question Answering. In Proceedings of the IEEE Conference on Computer Vision and Pattern Recognition (CVPR), Salt Lake City, UT, USA, 18–22 June 2018.
14. Anderson, P.; He, X.; Buehler, C.; Teney, D.; Johnson, M.; Gould, S.; Zhang, L. Bottom-up and top-down attention for image captioning and visual question answering. In Proceedings of the IEEE Conference on Computer Vision and Pattern Recognition, Salt Lake City, UT, USA, 18–22 June 2018; pp. 6077–6086.
15. Elman, J.L. Learning and development in neural networks: The importance of starting small. *Cognition* **1993**, *48*, 71–99. [CrossRef]
16. Bengio, Y.; Louradour, J.; Collobert, R.; Weston, J. Curriculum Learning. In Proceedings of the 26th Annual International Conference on Machine Learning, Association for Computing Machinery (ICML'09), New York, NY, USA, 19–24 June 2009; pp. 41–48. [CrossRef]
17. Coltheart, M. The MRC Psycholinguistic Database. *Q. J. Exp. Psychol. Sect. A* **1981**, *33*, 497–505.[CrossRef]
18. Wittgenstein, L. *Philosophical Investigations*; Basil Blackwell: Oxford, UK, 1953.
19. Harris, Z.S. Distributional Structure. *Word* **1954**, *10*, 146–162. [CrossRef]
20. Hinton, G.E.; McClelland, J.L.; Rumelhart, D.E. *Parallel Distributed Processing: Explorations in the Microstructure of Cognition*; Chapter Distributed Representations; MIT Press: Cambridge, MA, USA, 1986; Volume 1, pp. 77–109.
21. Elman, J.L. Finding structure in time. *Cogn. Sci.* **1990**, *14*, 179–211.
22. Bengio, Y.; Ducharme, R.; Vincent, P.; Janvin, C. A Neural Probabilistic Language Model. *J. Mach. Learn. Res.* **2003**, *3*, 1137–1155.
23. Mikolov, T.; Chen, K.; Corrado, G.; Dean, J. Efficient Estimation of Word Representations in Vector Space. *arXiv* **2013**, arXiv:1301.3781.
24. Pennington, J.; Socher, R.; Manning, C.D. Glove: Global Vectors for Word Representation. In Proceedings of the 2014 conference on empirical methods in natural language processing (EMNLP), Doha, Qatar, 25–29 October 2014; pp. 1532–1543.

25. Reisinger, J.; Mooney, R.J. Multi-prototype Vector-space Models of Word Meaning. In *Human Language Technologies, Proceedings of the 2010 Annual Conference of the North American Chapter of the Association for Computational Linguistics, HLT'10, Los Angeles, CA, USA, 1–6 June 2010*; Association for Computational Linguistics: Stroudsburg, PA, USA, 2010; pp. 109–117.
26. Huang, E.H.; Socher, R.; Manning, C.D.; Ng, A.Y. Improving Word Representations via Global Context and Multiple Word Prototypes. In Proceedings of the 50th Annual Meeting of the Association for Computational Linguistics (ACL'12), Jeju, Korea, 8–14 July 2012; Volume 1, pp. 873–882.
27. Luong, T.; Socher, R.; Manning, C. Better Word Representations with Recursive Neural Networks for Morphology. In Proceedings of the Seventeenth Conference on Computational Natural Language Learning, Sofia, Bulgaria, 8–9 August 2013; pp. 104–113.
28. Rothe, S.; Schütze, H. AutoExtend: Extending Word Embeddings to Embeddings for Synsets and Lexemes. In Proceedings of the 53rd Annual Meeting of the Association for Computational Linguistics and the 7th International Joint Conference on Natural Language Processing, Beijing, China, 26–30 July 2015; Volume 1, pp. 1793–1803. [CrossRef]
29. Melamud, O.; Goldberger, J.; Dagan, I. context2vec: Learning Generic Context Embedding with Bidirectional LSTM. In Proceedings of the 20th SIGNLL Conference on Computational Natural Language Learning (CoNLL 2016), Berlin, Germany, 11–12 August 2016; pp. 51–61.
30. McCann, B.; Bradbury, J.; Xiong, C.; Socher, R. Learned in Translation: Contextualized Word Vectors. In Proceedings of the Conference on Neural Information Processing Systems, Long Beach, CA, USA, 4–9 December 2017; pp. 6297–6308.
31. Dai, A.M.; Le, Q.V. Semi-Supervised Sequence Learning. In Proceedings of the 28th International Conference on Neural Information Processing Systems, Montreal, QC, Canada, 8–13 December 2015; Volume 2, pp. 3079–3087.
32. Howard, J.; Ruder, S. Universal Language Model Fine-tuning for Text Classification. In Proceedings of the 56th Annual Meeting of the Association for Computational Linguistics, Melbourne, Australia, 15–20 July 2018; Volume 1, pp. 328–339. [CrossRef]
33. Peters, M.E.; Neumann, M.; Iyyer, M.; Gardner, M.; Clark, C.; Lee, K.; Zettlemoyer, L. Deep Contextualized Word Representations. In Proceedings of the 2018 Conference of the North American Chapter of the Association for Computational Linguistics: Human Language Technologies (NAACL-HLT 2018), New Orleans, LA, USA, 1–6 June 2018; Volume 1, pp. 2227–2237.
34. Kim, Y.; Jernite, Y.; Sontag, D.A.; Rush, A.M. Character-Aware Neural Language Models. In Proceedings of the Thirtieth AAAI Conference on Artificial Intelligence, Phoenix, AZ, USA, 12–17 February 2016; pp. 2741–2749.
35. Akbik, A.; Blythe, D.; Vollgraf, R. Contextual String Embeddings for Sequence Labeling. In Proceedings of the COLING 2018, 27th International Conference on Computational Linguistics, Santa Fe, NM, USA, 21–25 August 2018; pp. 1638–1649.
36. Vaswani, A.; Shazeer, N.; Parmar, N.; Uszkoreit, J.; Jones, L.; Gomez, A.N.; Kaiser, L.u.; Polosukhin, I. Attention is All you Need. In *Advances in Neural Information Processing Systems 30*; Guyon, I., Luxburg, U.V., Bengio, S., Wallach, H., Fergus, R., Vishwanathan, S., Garnett, R., Eds.; Curran Associates, Inc.: Red Hook, NY, USA, 2017; pp. 5998–6008.
37. Dai, Z.; Yang, Z.; Yang, Y.; Carbonell, J.; Le, Q.; Salakhutdinov, R. Transformer-XL: Attentive Language Models beyond a Fixed-Length Context. In Proceedings of the 57th Annual Meeting of the Association for Computational Linguistics, Florence, Italy, 28 July–2 August 2019; pp. 2978–2988. [CrossRef]
38. Bruni, E.; Tran, N.K.; Baroni, M. Multimodal Distributional Semantics. *J. Artif. Int. Res.* **2014**, *49*, 1–47.
39. Kiros, R.; Salakhutdinov, R.; Zemel, R. Multimodal Neural Language Models. In Proceedings of the 31st International Conference on International Conference on Machine Learning (ICML'14), Beijing, China, 21–26 June 2014; Volume 32, p. II-595–II-603.
40. Liu, Y.; Guo, Y.; Bakker, E.M.; Lew, M.S. Learning a Recurrent Residual Fusion Network for Multimodal Matching. In Proceedings of the 2017 IEEE International Conference on Computer Vision (ICCV), Venice, Italy, 22–29 October 2017; pp. 4127–4136. [CrossRef]
41. Hill, F.; Korhonen, A. Learning Abstract Concept Embeddings from Multi-Modal Data: Since You Probably Can't See What I Mean. In Proceedings of the 2014 Conference on Empirical Methods in Natural Language Processing (EMNLP), Doha, Qatar, 25–29 October 2014; pp. 255–265. [CrossRef]
42. Kiros, R.; Salakhutdinov, R.; Zemel, R. Unifying Visual-Semantic Embeddings with Multimodal Neural Language Models. *arXiv* **2014**, arXiv:1411.2539.
43. Karpathy, A.; Joulin, A.; Li, F.-F. Deep Fragment Embeddings for Bidirectional Image Sentence Mapping. In Proceedings of the 27th International Conference on Neural Information Processing Systems (NIPS'14), Montreal, QC, Canada, 8–13 December 2014; Volume 2, pp. 1889–1897.
44. Wang, L.; Li, Y.; Lazebnik, S. Learning Deep Structure-Preserving Image-Text Embeddings. In Proceedings of the 2016 IEEE Conference on Computer Vision and Pattern Recognition (CVPR), Las Vegas, NV, USA, 27–30 June 2016; pp. 5005–5013. [CrossRef]
45. Lee, K.H.; Chen, X.; Hua, G.; Hu, H.; He, X. Stacked cross attention for image-text matching. In Proceedings of the European Conference on Computer Vision (ECCV), Munich, Germany, 8–14 September 2018; pp. 201–216.
46. Socher, R.; Li, F.-F. Connecting modalities: Semi-supervised segmentation and annotation of images using unaligned text corpora. In Proceedings of the 2010 IEEE Computer Society Conference on Computer Vision and Pattern Recognition, San Francisco, CA, USA, 13–18 June 2010; pp. 966–973. [CrossRef]
47. Yu, L.; Lin, Z.; Shen, X.; Yang, J.; Lu, X.; Bansal, M.; Berg, T.L. MAttNet: Modular Attention Network for Referring Expression Comprehension. In Proceedings of the IEEE Conference on Computer Vision and Pattern Recognition (CVPR), Salt Lake City, UT, USA, 18–22 June 2018.
48. Gao, P.; Jiang, Z.; You, H.; Lu, P.; Hoi, S.C.H.; Wang, X.; Li, H. Dynamic Fusion With Intra- and Inter-Modality Attention Flow for Visual Question Answering. In Proceedings of the 2019 IEEE/CVF Conference on Computer Vision and Pattern Recognition (CVPR), Long Beach, CA, USA, 15–20 June 2019; pp. 6632–6641.[CrossRef]

49. Zellers, R.; Bisk, Y.; Farhadi, A.; Choi, Y. From Recognition to Cognition: Visual Commonsense Reasoning. In Proceedings of the IEEE Conference on Computer Vision and Pattern Recognition (CVPR), Long Beach, CA, USA, 16–20 June 2019.
50. Shi, H.; Mao, J.; Xiao, T.; Jiang, Y.; Sun, J. Learning Visually-Grounded Semantics from Contrastive Adversarial Samples. In Proceedings of the 27th International Conference on Computational Linguistics, Santa Fe, NM, USA, 21–25 August 2018; pp. 3715–3727.
51. Paivio, A.; Yuille, J.C.; Madigan, S.A. Concreteness, imagery, and meaningfulness values for 925 nouns. *J. Exp. Psychol.* **1968**, *76*, 1.
52. Gilhooly, K.J.; Logie, R.H. Age-of-acquisition, imagery, concreteness, familiarity, and ambiguity measures for 1944 words. *Behav. Res. Methods Instrum.* **1980**, *12*, 395–427.
53. Szegedy, C.; Liu, W.; Jia, Y.; Sermanet, P.; Reed, S.; Anguelov, D.; Erhan, D.; Vanhoucke, V.; Rabinovich, A. Going Deeper with Convolutions. In Proceedings of the Computer Vision and Pattern Recognition (CVPR), Boston, MA, USA, 7–12 June 2015.
54. Simonyan, K.; Zisserman, A. Very Deep Convolutional Networks for Large-Scale Image Recognition. *arXiv* **2015**, arXiv:1409.1556.
55. He, K.; Zhang, X.; Ren, S.; Sun, J. Deep Residual Learning for Image Recognition. In Proceedings of the 2016 IEEE Conference on Computer Vision and Pattern Recognition (CVPR), Las Vegas, NV, USA, 27–30 June 2016; pp. 770–778.[CrossRef]
56. Young, P.; Lai, A.; Hodosh, M.; Hockenmaier, J. From image descriptions to visual denotations: New similarity metrics for semantic inference over event descriptions. *Trans. Assoc. Comput. Linguist.* **2014**, *2*, 67–78. [CrossRef]
57. Lin, T.Y.; Maire, M.; Belongie, S.; Hays, J.; Perona, P.; Ramanan, D.; Dollár, P.; Zitnick, C.L. Microsoft COCO: Common Objects in Context. In Proceedings of the European Conference on Computer Vision, Zurich, Switzerland, 6–12 September 2014; pp. 740–755.
58. von Ahn, L.; Dabbish, L. Labeling Images with a Computer Game. In Proceedings of the SIGCHI Conference on Human Factors in Computing Systems (CHI'04), Vienna, Austria, 25 April 2004; pp. 319–326. [CrossRef]
59. Srinivasan, K.; Raman, K.; Chen, J.; Bendersky, M.; Najork, M. WIT: Wikipedia-Based Image Text Dataset for Multimodal Multilingual Machine Learning. In Proceedings of the 44th International ACM SIGIR Conference on Research and Development in Information Retrieval (SIGIR'21), Montreal, QC, Canada, 11–15 July 2021; pp. 2443–2449. [CrossRef]
60. Sanh, V.; Debut, L.; Chaumond, J.; Wolf, T. DistilBERT, a distilled version of BERT: smaller, faster, cheaper and lighter. *arXiv* **2019**, arXiv:1910.01108.
61. Deng, J.; Dong, W.; Socher, R.; Li, L.J.; Li, K.; Li, F.-F. Imagenet: A large-scale hierarchical image database. In Proceedings of the 2009 IEEE Conference on Computer Vision and Pattern Recognition, Miami, FL, USA, 20–25 June 2009; pp. 248–255.
62. Santos, C.d.; Tan, M.; Xiang, B.; Zhou, B. Attentive pooling networks. *arXiv* **2016**, arXiv:1602.03609.
63. Kiela, D.; Rimell, L.; Vulić, I.; Clark, S. Exploiting Image Generality for Lexical Entailment Detection. In Proceedings of the 53rd Annual Meeting of the Association for Computational Linguistics and the 7th International Joint Conference on Natural Language Processing, Beijing, China, 26–30 July 2015; Volume 2, pp. 119–124. [CrossRef]
64. Hessel, J.; Mimno, D.; Lee, L. Quantifying the Visual Concreteness of Words and Topics in Multimodal Datasets. In Proceedings of the 2018 Conference of the North American Chapter of the Association for Computational Linguistics: Human Language Technologies, New Orleans, LA, USA, 1–6 June 2018; Volume 1, pp. 2194–2205. [CrossRef]
65. Antol, S.; Agrawal, A.; Lu, J.; Mitchell, M.; Batra, D.; Zitnick, C.L.; Parikh, D. VQA: Visual Question Answering. In Proceedings of the International Conference on Computer Vision (ICCV), Santiago, Chile, 7–13 December 2015.
66. Goyal, Y.; Khot, T.; Summers-Stay, D.; Batra, D.; Parikh, D. Making the V in VQA Matter: Elevating the Role of Image Understanding in Visual Question Answering. In Proceedings of the 2017 IEEE Conference on Computer Vision and Pattern Recognition (CVPR), Honolulu, HI, USA, 21–26 July 2017; pp. 6325–6334. [CrossRef]
67. Yang, Z.; He, X.; Gao, J.; Deng, L.; Smola, A. Stacked Attention Networks for Image Question Answering. In Proceedings of the 2016 IEEE Conference on Computer Vision and Pattern Recognition (CVPR), Las Vegas, NV, USA, 27–30 June 2016; pp. 21–29. [CrossRef]
68. Ren, S.; He, K.; Girshick, R.; Sun, J. Faster R-CNN: Towards Real-Time Object Detection with Region Proposal Networks. In Proceedings of the 28th International Conference on Neural Information Processing Systems (NIPS'15), Montreal, QC, Canada, 8–13 December 2015; pp. 91–99.

applied sciences

MDPI

Article

Improving Entity Linking by Introducing Knowledge Graph Structure Information

Qijia Li [1,2,3], Feng Li [1,2,4,*], Shuchao Li [1,2], Xiaoyu Li [1,2], Kang Liu [1,2], Qing Liu [1,2] and Pengcheng Dong [1,2]

1 Aerospace Information Research Institute, Chinese Academy of Sciences, Beijing 100190, China; liqijia19@mails.ucas.ac.cn (Q.L.); lisc@aircas.ac.cn (S.L.); lixy01@aircas.ac.cn (X.L.); lkwnsh615@163.com (K.L.); liuqing1@aircas.ac.cn (Q.L.); dongpc@aircas.ac.cn (P.D.)
2 Key Laboratory of Network Information System Technology (NIST), Aerospace Information Research Institute, Chinese Academy of Sciences, Beijing 100190, China
3 School of Electronic, Electrical and Communication Engineering, University of Chinese Academy of Sciences, Beijing 100190, China
4 QILU Research Institute, Aerospace Information Research Institute, Chinese Academy of Sciences, Jinan 250000, China
* Correspondence: lifeng@mail.ie.ac.cn

Abstract: Entity linking involves mapping ambiguous mentions in documents to the correct entities in a given knowledge base. Most of the current methods are a combination of local and global models. The local model uses the local context information around the entity mention to independently resolve the ambiguity of each entity mention. The global model encourages thematic consistency across the target entities of all mentions in the document. However, the known global models calculate the correlation between entities from a semantic perspective, ignoring the correlation information between entities in nature. In this paper, we introduce knowledge graphs to enrich the correlation information between entities and propose an entity linking model that introduces the structural information of the knowledge graph (KGEL). The model can fully consider the relations between entities. To prove the importance of the knowledge graph structure, extensive experiments are conducted on multiple public datasets. Results illustrate that our model outperforms the baseline and achieves superior performance.

Keywords: entity linking; knowledge graph; entity embedding; global model

Citation: Li, Q.; Li, F.; Li, S.; Li, X.; Liu, K.; Liu, Q.; Dong, P. Improving Entity Linking by Introducing Knowledge Graph Structure Information. *Appl. Sci.* **2022**, *12*, 2702. https://doi.org/10.3390/app12052702

Academic Editors: Arturo Montejo-Ráez and Salud María Jiménez-Zafra

Received: 27 January 2022
Accepted: 3 March 2022
Published: 5 March 2022

1. Introduction

The named entity linking (NEL) task refers to correctly linking entity mentions in text to entities in a structured knowledge base (such as Wikipedia, Freebase [1], or YAGO [2]), which can solve the ambiguity of mentions in natural language processing. In Figure 1, for example, a mention of "Michael Jordan" may correspond to entity entries in the knowledge base (KB) such as "Michael Jordan", "Michael I. Jordan", "Michael Jordan (footballer)", "Michael B. Jordan", etc. The entity linking (EL) involves linking the mention "Michael Jordan" to the correct entity "Michael I. Jordan" in the KB. Entity linking is also the basis of many other natural language processing tasks, such as knowledge base question and answer [3], information retrieval [4], and content analysis [5].

Given a document, the named entity mentions are recognized in advance by a named entity recognition (NER) method. Generally speaking, a typical entity linking system consists of two steps: (1) candidate entity generation, in which a model retrieves a set of candidate entities, which contains the entities that the mention may refer to; and (2) candidate entity ranking, in which a model ranks the entities in the candidate set and selects the entity that the mention is most likely to link to. Recently, some methods such as techniques based on a named dictionary and techniques based on surface form expansion have achieved high candidate recalls, and thus most work focuses on methods for downstream candidate entity ranking, as described in this paper.

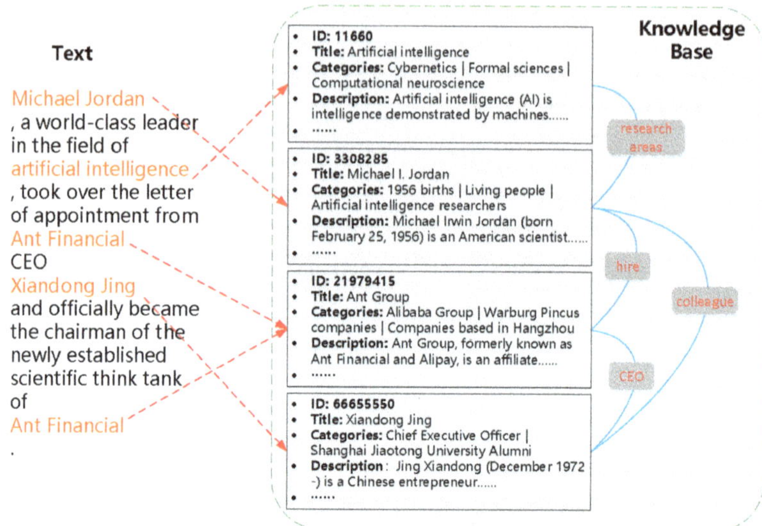

Figure 1. An example of NEL whose goal is to link each mention to an entity in the KB (e.g., "Michael Jordan" is linked to Michael I. Jordan; "Artificial intelligence" is linked to Artificial intelligence). Note that there are various relations between entities in the KB.

In early work, prior distribution and local contexts played important roles in disambiguating different candidate entities. However, in many cases, local features alone cannot provide sufficient information for disambiguation. Therefore, many global models have emerged to solve the task of entity linking. For example, Ganea and Hofmann [6] combine local and global information. First, the word-entity co-occurrence counts are used to train the entity embeddings, then the local scores between contexts of mentions and the entity embeddings are calculated in the local model, and the scores between candidate entities of all mentions in the document are calculated in the global model. On the basis of [6], Le and Titov [7] model the latent relations between mentions. Based on [7], Hou et al. [8] inject fine-grained semantic information into entity embeddings. In addition, Yang et al. [9] propose the dynamic context augmentation method, which uses the entity embedding in [6].

However, the above methods still have some shortcomings. They essentially calculate the similarity between entity embeddings when obtaining global scores, which only consider the semantic proximity between entities. While there are real relations between some entity mentions in a document, these relations are contained in some knowledge graphs, and comprise the so-called knowledge graph structural information. As shown in Figure 1, there is an association relation of "colleague" between entity "Michael I. Jordan" and entity "Xiandong Jing" in the knowledge base. In addition, although there are also some works [10–13] that involve knowledge graphs, this is because their target knowledge base is a knowledge graph, and our method is different from them essentially. For example, Cetoli et al. [12] use bi-directional long short-term memory (Bi-LSTM) to encode graph triplets. Mulang et al. [13] develop a context-aware attentive neural network approach on Wikidata. Instead, on the basis of Wikipedia, we introduce the structural information of other knowledge graphs to complement the semantic information of Wikipedia, which is somewhat similar to the fusion of information from different knowledge bases.

To address the limitations of existing methods, we propose an entity linking model that introduces knowledge graph structural information (KGEL). First, under the premise that the target knowledge base is Wikipedia, we obtain the entities and triples in the knowledge graph Wikidata corresponding to the candidate entities. Then, the knowledge graph embedding method is used to train entity embeddings and relation embeddings.

Finally, according to the different characteristics of local and global models, we use the previously trained entity embeddings and relation embeddings only for the global model of entity linking; that is, the global scores are computed from the perspective of the graph structure and fused with the Ment–Norm [7] model. Existing methods have been able to achieve more than 90% F1 on the standard AIDA-CoNLL dataset; for example, Ment–Norm achieves 93.07% F1. Our KGEL method achieves an improvement of 0.4% F1 on the basis of Ment–Norm, and the average result of KGEL on the five out-of-domain datasets is also 0.2% higher than Ment–Norm, which indicates that our model also has better generalization. Our method can also further improve the performance when using a more superior baseline.

The main contributions of our paper can be summarized as follows. (1) We propose to introduce knowledge graph structure information into the entity linking model, so as to complement the semantic information. (2) We obtain the Wikipedia–Wikidata mappings of entities and the required triples, and then obtain the entity and relation embeddings containing the graph structure through the knowledge graph embedding method. This provides a new idea for information fusion between different knowledge bases (graphs). (3) Extensive experiments on multiple datasets show the excellent performance of our method and demonstrate the effectiveness of the knowledge graph structure for entity linking.

2. Background and Related Work

2.1. Problem Definition

Given a knowledge base containing a set of entities $E_s = \{e_1, \ldots, e_t\}$ and a set of entity mentions $M = \{m_1, \ldots, m_n\}$ in corpus $\mathcal{D}$, the goal of entity linking is to map each entity mention $m_i \in M$ in the text to its corresponding entity $e_i^* \in E_s$. Because a KB may contain a large number of entities, in order to reduce complexity, we usually use a heuristic to choose potential candidates, thus obtaining candidate set $C_i = (e_{i1}, \ldots, e_{il_i})$, which is the candidate entity generation we mentioned earlier. Then, we select gold entities on the candidate set in the candidate entity ranking stage.

2.2. Entity Linking

As it is an important task in natural language processing, there is a lot of work in the field of entity linking. Most of the early work comprises methods based on manually designed features and rule-based methods, which are not enough to capture the potential dependence and interaction in the data. With the rapid development of deep learning, a large number of deep-learning-based methods have appeared in the field of entity linking, and they have achieved better results than previous methods. Topics related to the work of this article are as follows.

Local model. The local model uses the local text context information around the entity mention to independently resolve the ambiguity of each entity mention. He et al. [14] were early adopters of deep learning for entity linking. They learned distributed representations of entities to measure similarity, avoiding manually designed features, so that words and entities could be in the joint semantic space, and then candidate entities could be sorted based on vector similarity. Subsequently, Sun et al. [15] used neural networks to encode mentions, contexts of mentions, and entities. Among them, contexts of mentions are encoded by convolutional neural networks (CNN), which are combined with representations of the mention titles to obtain the final mention representations. The entity representations are obtained from the entity titles and entity categories. Finally, the similarities between the mention representations and the entity representations are calculated to obtain local scores. Based on [15], Francis-Landau et al. [16] used CNN and stacked denoising autoencoders to encode different granular information of mentions and entities to enhance the representation. In addition, Gupta et al. [17] cascaded the output of two long short-term memory (LSTM) [18] networks. The two LSTM networks independently encode the left and right context of the entity mention, including the entity mention itself. Kolitsas et al. [19] expressed entity mention as a combination of LSTM hidden states contained in the span of entity mention. Eshel et al. [20] used a variant of LSTM-GRU [21]. Ganea and Hofmann [6] introduced an attention mechanism in the local model. They assumed that a context word was important if it was strongly related to at least one candidate entity, and the context words were hard pruned. The local model in this paper is based on Ganea and Hofmann [6].

Global model. The global model links all the mentions in a document at the same time and considers that the target entities of all the mentions are consistent on the subject. The previous global methods usually executed RandomWalk [22] or PageRank [23] algorithms on the graph containing candidate entities. Another solution is to maximize the conditional random field [24], but the problem is NP-hard. Ganea and Hofmann [6] used loopy belief propagation (LBP) [25] to iteratively propagate entity scores to reduce complexity. Based on [6], Le and Titov [7] modeled the latent relations between mentions and added them to the global model in the form of features, achieving better results. Some recent studies have defined the global entity linking problem as a sequential decision task, where the linking of the new entity is based on the already linked entity. Fang et al. [26] used LSTM to maintain long-term memory for previous decisions; Yang et al. [9] proposed a dynamic context integration method that uses previous decisions as dynamic context to improve subsequent decisions; Yamada et al. [27] calculated the confidence scores based on the previous decisions. In addition, graph neural networks (GNNs) can also be used for the global model of entity linking. Wu et al. [28] proposed a dynamic graph convolutional network model, in which the graph structure is dynamically calculated and changed during training, and fusion of knowledge through dynamically linked nodes can effectively obtain the theme consistency in the document. Fang et al. [29] proposed a sequential graph attention network to synthesize the advantages of the graph model and the sequence model, which dynamically encodes the preceding and following entity mentions, and assigns different weights to these entity mentions. The global model of this article refers to the work of [7].

Entity embedding. Entity embedding is a key component in entity linking to avoid manual features and enhance model effects. There is also a lot of work for entity embedding. Yamada et al. [30] proposed to map words and entities to the same continuous vector space. They used two models to extend the skip-gram model. The KB graph model uses the link structure in the KB to learn the relevance of entities. The anchor context model aims to use KB anchor text and context words to align vectors so that similar words and entities are close in the vector space. Yamada et al. [31] further proposed to jointly learn distributed representations of text and entities. Given a piece of text in the knowledge base, a model is trained to predict entities related to the text; that is, using a large amount of text extracted from Wikipedia and their entity annotations to train the model. Ganea and Hofmann [6] used pre-trained word embeddings and word-entity co-occurrence counts to obtain entity embeddings so that words and entities were represented in the same low-dimensional vector space. Ling et al. [32] proposed a fill-in-the-blank task to learn context-independent entity representations from the text context. Hou et al. [8] proposed incorporating fine-grained semantic information into entity embedding to reduce uniqueness and promote the learning of contextual commonality. Yamada et al. [27] used the pre-trained model BERT [33] to generate the representation of words and entities, and the results were greatly improved compared to the previous method. This paper also uses the entity embeddings of [6].

2.3. Knowledge Graph Embedding

The knowledge graph is a multi-relational graph composed of entities (nodes) and relations (edges), and each edge is in the form of a triple (head entity, relation, tail entity). The existing knowledge graphs include Freebase [1], DBpedia [34], Wikidata, etc. Knowledge graph embedding [35] involves embedding the entities and relations in the knowledge graph into a continuous vector space. In general, knowledge graph embedding methods can be divided into two groups: translational distance models and semantic matching models [36–38]. The former use distance-based scoring functions, and the latter similarity-based ones. Among translational distance models, TransE [39] is the most representative. The main idea is to give a triple (h, r, t), the goal is $\mathbf{h} + \mathbf{r} \approx \mathbf{t}$, where h, r, t are the head entity, relation, and tail entity, respectively, and $\mathbf{h}, \mathbf{r}, \mathbf{t}$ are, respectively, vector representations. To solve the limitations of the TransE model in dealing with 1-to-N, N-to-1, and N-to-N complex relations, TransH [40] introduces relation-specific hyperplanes that allow an entity to have different representations under different relations. In order to further improve the representation ability, TransR [41] introduces relation-specific spaces, rather than hyperplanes. TransD [42] simplifies TransR by further decomposing the projection matrix into a product of two vectors. TransM [43] assigns specific relation weight to each triple (h, r, t).

There are also recent knowledge graph embedding methods with better performance. Zhang et al. [44] proposed the hierarchy-aware knowledge graph embedding model (HAKE), which maps entities into a polar coordinate system. PairRE [45] has paired vectors for each relation representation, which can adaptively adjust the margin in a loss function to fit for complex relations. Additionally, PairRE can encode three relation patterns: symmetry/antisymmetry, inverse, and composition. DualE [46] introduces dual quaternions into knowledge graph embedding, where a dual quaternion is similar to a "complex quaternion" with its real and imaginary part all being quaternar. DualE universally models relations as the combination of a series of translation and rotation operations. EIGAT [47] allows correct incorporation of global information into the graph attention network (GAT) family of models by using scaled entity importance, which is computed by an attention-based global random walk algorithm. In order to focus on the importance of the knowledge graph structure for the entity linking task, the knowledge graph embedding method used in this article is the most basic TransE model.

3. Learning Entity Embeddings KGEmbs

3.1. Wikipedia–Wikidata Mappings

Since the target knowledge base of the dataset we use is Wikipedia, and we want to introduce the structural information of other knowledge graphs, for the Wikipedia entities used, we need to obtain their corresponding Wikidata entities, i.e., obtain the Wikipedia–Wikidata mappings. In the entity's Wikipedia page, there is a corresponding Wikidata hyperlink, as shown in Figure 2. Therefore, we can obtain the Wikidata ID of the Wikipedia entity through the crawler. Examples of the Wikipedia–Wikidata mappings are shown on the left side of Table 1.

Table 1. Examples of Wikipedia–Wikidata mappings and triples.

Wikipedia–Wikidata Mappings		**Triples**		
en.wikipedia.org/wiki/Universe	Q1	Q1	P2670	Q523
en.wikipedia.org/wiki/Star	Q523	Q1	P2184	Q136407
en.wikipedia.org/wiki/Big_Bang	Q323	Q1	P793	Q323
en.wikipedia.org/wiki/Happiness	Q8	Q8	P31	Q331769
en.wikipedia.org/wiki/Mood_(psychology)	Q331769	Q8	P31	Q9415
...		...		
en.wikipedia.org/wiki/Toledo,_Minas_Gerais	Q22065023	Q22065023	P131	Q39109
en.wikipedia.org/wiki/Minas_Gerais	Q39109	Q22065023	P17	Q155

3.2. Triple Knowledge

We can obtain the triple knowledge of Wikidata from OpenKE: http://139.129.163.161/index/toolkits (accessed on 1 March 2022), including 20,982,733 entities, 594 relations, and 68,904,773 triples. According to the work of [7], we obtain 274,474 entities in the candidate entity generation stage to filter relations and triples, and finally obtain 486 relations and 807,587 triples. The triple format is shown on the right side of Table 1. For example, (Q1, P2670, Q523) is a triple, where Q1 is the head entity and its corresponding entity is "universe", Q523 is the tail entity and its corresponding entity is "star", and P2670 is the relation between entities Q1 and Q523; that is, "instance has part(s) of the class". Therefore, the triple can be represented as (universe, instance has part(s) of the class, star).

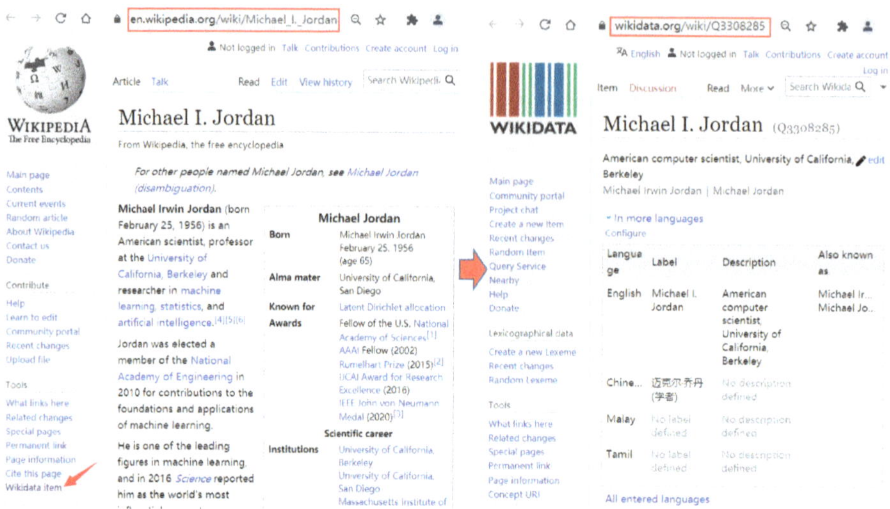

Figure 2. Example for the Wikipedia–Wikidata mapping. We can obtain the corresponding Wikidata ID through the entity's Wikipedia page.

3.3. *Entity and Relation Embeddings*

In order to demonstrate more intuitively the effectiveness of the knowledge graph structure for entity linking, and also considering the speed differences of each model, we use the TransE model to train entity and relation embeddings on triples, where $h, t \in E$ (the set of entities) and $r \in R$ (the set of relations). The main idea is that the functional relation obtained from the edges labeled by r corresponds to the translation of the embedding; that is, we hope that $h + r \approx t$ when (h, r, t) holds, while $h + r$ should be far away from t otherwise.

In order to learn entity and relation embeddings, we minimize the following loss:

$$\mathcal{L} = \sum_{(h,r,t)\in S} \sum_{(h',r,t')\in S'_{(h,r,t)}} \left[\gamma_1 + d(\mathbf{h}+\mathbf{r},\mathbf{t}) - d\left(\mathbf{h}'+\mathbf{r},\mathbf{t}'\right)\right]_+ \tag{1}$$

where $[x]_+$ denotes the positive part of x, $\gamma > 0$ is a margin hyperparameter, and $d(\mathbf{h}+\mathbf{r},\mathbf{t})$ is an indicator to measure similarity. Here we use the L_1-norm, and

$$S'_{(h,r,t)} = \left\{\left\{h', r, t\right\} \middle| h' \in E\right\} \cup \left\{\left(h, r, t'\right) \middle| t' \in E\right\} \tag{2}$$

The optimization is performed by stochastic gradient descent, and an additional constraint is that the L_2-norm of the embeddings of the entities is 1.

4. Model

The entity linking model in this paper integrates local and global features and is a conditional random field model in form. Figure 3 provides an overview of our model. Specifically, a scoring function g is defined to evaluate the mappings from entity mentions $m_1, \ldots, m_n$ to the entities $e_1, \ldots, e_n$ in a document D:

$$g(e_1, \ldots, e_n) = \sum_{i=1}^{n} \Psi(e_i) + \sum_{i \neq j} \Phi(e_i, e_j, D) \tag{3}$$

where n represents the number of entity mentions in the document. The first part of Equation (3) is the local score, which is the matching score between the local context of the entity mention and the candidate entity, and the second part is the global score, which is the score between entities in the document. The local model and the global model are described below.

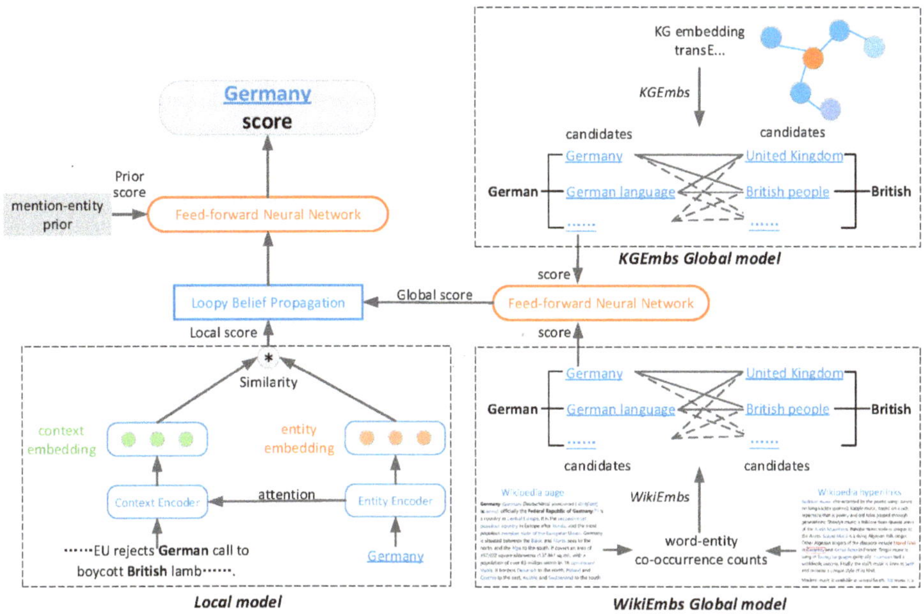

Figure 3. The architecture of the proposed KGEL model. It contains three parts: *Local model*, *WikiEmbs Global model*, and *KGEmbs Global model*. Specifically, in the *Local model*, the similarity calculated by context embedding and entity embedding is used as the local score. In the Global model, the scores between the candidate entities of all mentions in the document are taken as the global score. Among them, in the *WikiEmbs Global model*, entity embedding is obtained through word-entity co-occurrence counts, which consider the semantic information. In the *KGEmbs Global model*, entity embedding is obtained through triples, considering the structural information of the knowledge graph.

4.1. Local Model

According to Ganea and Hofmann [6], this paper takes the local model as an attention model based on entity embedding. For an entity mention m, if a word in the context is strongly related to at least one candidate entity, the word is considered important.

In the candidate generation stage, we can obtain the candidate entity set $C_i = (e_{i1}, \ldots, e_{il_i})$. Then we calculate the score of each candidate entity $e \in C_i$ according to the P-word window local context $c = \{w_1, ..., w_p\}$ around m. First, we calculate the unnormalized support score of each word in the context; that is, the weight of each word

$$u(w) = \max_{e \in C_i} \mathbf{e}^T \mathbf{A} \mathbf{w} \tag{4}$$

where $\mathbf{A}$ is a parameterized diagonal matrix, $\mathbf{w}$ is the word embedding (we use the pre-trained word2vec word embedding), and $\mathbf{e}$ is the candidate entity embedding, which is trained based on the co-occurrence counts of the word-entity in Wikipedia [6]. If the word is strongly related to at least one candidate entity, its weight score is relatively high. In addition, it is observed that some words with insufficient information will introduce noise to the local model, so the hard pruning method is used to select $Q \leq P$ words with the highest weight scores:

$$\bar{c} = \{w \in c | u\{w\} \in topQ\{\mathbf{u}\}\} \tag{5}$$

Therefore, the final attention weight is:

$$\beta(w) = \begin{cases} \frac{\exp[u[w]]}{\sum\limits_{v \in \bar{c}} \exp[u[v]]} & if\ w \in \bar{c} \\ 0 & otherwise \end{cases} \tag{6}$$

Finally, we can obtain the local scores of the candidate entities:

$$\Psi(e) = \sum_{w \in \bar{c}} \beta(w) \mathbf{e}^T \mathbf{B} \mathbf{w} \tag{7}$$

where **B** is another diagonal matrix that can be trained.

4.2. Global Model

Ganea and Hofmann [6] mainly considered the consistency between entities. However, Le and Titov [7] proposed that there is not only consistency between entities, but there are also some latent relations that can support the constraints on entities. Assuming that there are K latent relations, each relation k corresponds to a pair (m_i, m_j), so the second term of Equation (3) can be written as:

$$\Phi(e_i, e_j, D) = \sum_{k=1}^{K} \alpha_{ijk} \Phi_k(e_i, e_j, D) \tag{8}$$

That is, the paired score (m_i, m_j) is the weighted sum of the corresponding scores of each relation, and α_{ijk} is the weight corresponding to the relation k. Here, each relation k is a diagonal matrix $\mathbf{R}_k \in \mathbb{R}^{d \times d}$, and

$$\Phi_k(e_i, e_j, D) = \mathbf{e}_i^T \mathbf{R}_k \mathbf{e}_j \tag{9}$$

The weight α_{ijk} is the normalized score:

$$\alpha_{ijk} = \frac{1}{Z_{ijk}} \exp\left\{ \frac{f^T(m_i, c_i) \mathbf{D}_k f(m_j, c_j)}{\sqrt{d}} \right\} \tag{10}$$

where Z_{ijk} is the normalization factor, $\mathbf{D}_k \in \mathbb{R}^{d \times d}$ is a diagonal matrix, and $f(m_i, c_i)$ is a single-layer neural network, which is used to obtain the local context representation of the mention m_i. For c_i, we first obtain the average c_l of the word embeddings of the context words on the left of the mention m_i, then obtain the average c_r of the word embeddings of the context words on the right, and finally take the concatenation of c_l and c_r. In addition, Le and Titov [7] proposed two normalization methods of Z_{ijk}: normalization over relations and normalization over mentions. We adopt the method of normalization over mentions, then

$$Z_{ijk} = \sum_{\substack{j'=1 \\ j' \neq i}}^{n} \exp\left\{ \frac{f^T(m_i, c_i) \mathbf{D}_k f(m_{j'}, c_{j'})}{\sqrt{d}} \right\} \tag{11}$$

Now $\sum_{j=1, j \neq i}^{n} \alpha_{ijk} = 1$, which means that for each relation k and mention m_i, we want to find another mention that has a relation k with the mention m_i. The entity embeddings $\mathbf{e}_i, \mathbf{e}_j$ here are obtained by training using word-entity co-occurrence counts in Wikipedia, so the global model is called the WikiEmbs model, and there is $\Phi_{wiki}(e_i, e_j, D) = \Phi(e_i, e_j, D)$. The WikiEmbs model essentially only uses the semantic information of the entities; that is, the more semantically related entities have a greater probability of appearing in the same document. However, the structural information in the knowledge graph is ignored, so we propose the KGEmbs model, which explicitly uses the knowledge graph structure information in the global model. Our motivation is that the knowledge graph structure should be maintained when the entity mentions in a document are mapped to the knowledge base. Assuming that there are R_n relations (Section 3.2), the second term in Equation (3) can be written as:

$$\Phi_{KG}(e_i, e_j, D) = \max_{r \in R_n} f_{KG}(e_i, e_j, r) \tag{12}$$

where $f_{KG}(e_i, e_j, r)$ is the scoring function of the knowledge graph embedding method; that is, for all relations R, the score of (e_i, e_j) must be calculated, and then the maximum value is taken. The TransE [39] model is used here, and because the head entity and tail entity in (e_i, e_j) cannot be distinguished, there is:

$$f_{KG}(e_i, e_j, r) = \max(\gamma_1 - d(e_i + r, e_j), \gamma_1 - d(e_j + r, e_i)) \tag{13}$$

where γ_1 is consistent with γ_1 in Equation (1), and

$$d(h + r, t) = \|\mathbf{h} + \mathbf{r} - \mathbf{t}\|_1 \tag{14}$$

Among them, the smaller $d(h + r, t)$, the greater the probability that the entities h and t have the relation r. In addition, $\mathbf{h}, \mathbf{t}$ are the entity embeddings obtained by the TransE model, and $\mathbf{r}$ is the relation embedding. Finally, we combine the two global scores obtained above:

$$\Phi(e_i, e_j, D) = f_{global}(\Phi_{wiki}(e_i, e_j, D), \Phi_{KG}(e_i, e_j, D)) \tag{15}$$

where f_{global} is a two-layer neural network.

4.3. Model Training

The solution of Equation (3) is NP-hard. Following Le and Titov [7], we also adopt max-product loopy belief propagation (LBP) to estimate the max-marginal probability:

$$\hat{g}_i(e|D) \approx \max_{\substack{e_1,\ldots,e_{i-1} \\ e_{i+1},\ldots,e_n}} g(e_1, \ldots, e_n) \tag{16}$$

Then we obtain the final score of mention m_i

$$\rho_i(e) = f_{final}(\hat{g}_i(e|D), \hat{p}(e|m_i)) \tag{17}$$

The one with the highest score is the candidate entity to be linked to, f_{final} is another two-layer neural network, and $\hat{p}(e|m)$ is the mention-entity prior. We optimize the parameters in the model by minimizing the ranking loss as follows:

$$L(\theta) = \sum_{D \in \mathcal{D}} \sum_{m_i \in D} \sum_{e \in C_i} h(m_i, e) \tag{18}$$

$$h(m_i, e) = \max(0, \gamma_2 - \rho_i(e_i^*) + \rho_i(e)) \tag{19}$$

where θ denotes the model parameters, $\mathcal{D}$ is the training corpus, D is a document, and e_i^* is the gold entity.

5. Experiments

5.1. Datasets

To prove the effectiveness of our method, we conducted experiments on six popular open-source datasets, including an in-domain dataset and five out-domain datasets. For the in-domain dataset, we used the AIDA-CoNLL dataset [48], which contains AIDA-train, AIDA-A, and AIDA-B, which were used for training, verification, and testing, respectively. For out-domain datasets, we used MSNBC (MSB), AQUAINT (AQ), and ACE2004 (ACE), which are cleaned and updated by Guo and Barbosa [22]; and WNED-WIKI (WW) and WNED-CWEB (CWEB), which are automatically extracted from ClueWeb and Wikipedia corpora by Guo and Barbosa [22]. Among them, the latter two datasets are larger in scale and noisier, making linking of entities more difficult. Statistics of these datasets are summarized in Table 2. The target knowledge base is Wikipedia. Based on previous work [6,7], we do not consider mentions that have no corresponding entities in the KB.

Table 2. Statistics of experiment datasets. Gold recall is the probability that the candidate sets of mentions contain the ground truth entities.

Dataset	Number Mentions	Number Docs	Mentions per Doc	Gold Recall
AIDA-train	18,448	946	19.5	-
AIDA-A	4791	216	22.1	97.3
AIDA-B	4485	231	19.4	98.3
MSNBC	656	20	32.8	98.5
AQUAINT	727	50	14.5	94.2
ACE2004	257	36	7.1	90.6
CWEB	11,154	320	34.8	91.1
WIKI	6821	320	21.3	92.4

5.2. Candidate Entity Generation

To ensure fairness and comparable results, we use the candidate generation method of Le and Titov [7]. First, we select the top 30 candidate entities for each mention m_i based on the prior $\hat{p}(e|m_i)$, and then select 7 from them. Among them, the top 4 entities are selected based on $\hat{p}(e|m_i)$, and the top 3 entities are selected based on the score $\mathbf{e}^T(\sum_{w\in d_i}\mathbf{w})$, where $\mathbf{e},\mathbf{w}\in\mathbb{R}^d$ are entity and word embeddings, respectively, and d_i is the 50-word local context surrounding m_i. The quality of the candidate set obtained by the above method is shown in Table 2.

5.3. Hyper-Parameter Setting

Our models are implemented in the Pytorch framework. For the *Local model*, according to Ganea and Hofmann [6], we use the following hyper-parameters: $P = 100, Q = 25$ (Equation (5)). We set the dimensions of word embedding and entity embedding to 300, where word embedding and entity embedding are from [6]. For the *WikiEmbs Global model*, when calculating f (Equation (10)), we use the word embedding in Le and Titov [7] and the entity embedding in [6], both of which have a dimension of 300. In addition, according to [7], the number of LBP loops is set to 10, the dropout rate for f is set to 0.3, the window size c_i of the local context used when calculating pairwise score functions is 6, and the number of relations in *Ment-norm* is 3. For the *KGEmbs Global model*, we use the TransE model to train entity embeddings and relation embeddings, where learning rate $\lambda = 0.0001$, margin $\gamma_1 = 24$ (Equation (1)), batch size is 1024, hidden size is 300, and the dimensions of entity embedding and relation embedding are 300. When training the model, we set $\gamma_2 = 0.01$ (Equation (19)). When the F1 score of the model on the validation set reaches 91%, we adjust the learning rate from 1×10^{-4} to 1×10^{-5}, and we stop learning if the F1 on the validation set does not improve after 20 epochs.

5.4. Main Results

The following methods are selected as baselines.

1. AIDA [48] combines the previous methods into a comprehensive framework that contains three measures: the prior probability of an entity being mentioned, the similarity between the context of mention and the candidate entity, and the consistency among candidate entities for all mentions. It constructs a weighted graph whose nodes are mentions and candidate entities and calculates a dense subgraph to obtain an approximately optimal mention-entity mapping.
2. GLOW is a global entity disambiguation system proposed by [49], which formulates the entity disambiguation task as an optimization problem with local and global variants.

3. RI [50] combines statistical methods to perform richer relational analysis on the text. It proposes a modular formulation that includes the entity-relation inference problem. It also proves that the recognition of relations in the text is not only helpful for candidate entities, but also the subsequent ranking stage.
4. PBoH [51] uses a graphical model to perform global entity disambiguation. It simultaneously disambiguates mentions in a document by using the co-occurrence probability between entities in the document and the local context information of the mentions. It uses LBP to perform approximate inference.
5. Deep-ED [6] introduces an attention mechanism into the local model, and the context words of mentions are hard pruned. Its global model is a fully-connected pairwise conditional random field. Because the problem is NP-hard, it uses LBP to iteratively propagate entity scores to reduce complexity.
6. Ment-Norm [7] models the latent relations between mentions and adds them to the global model in the form of features. There are two options for normalization, where it is normalization over mentions.
7. DCA-SL [9] regards entity linking as a sequence decision task and uses the previous decision as dynamic contexts to improve the later decisions. It explores supervised learning strategies for learning the DCA model.
8. DCA-RL [9] involves the use of reinforcement-learning strategies to learn the DCA model.

Table 3 shows micro F1 scores on AIDA-B and five out-domain test sets. Compared with Deep-ED [6], our method achieves a substantial improvement on both the in-domain dataset AIDA-B and the average result on five out-domain datasets. Moreover, KGEL's F1 score is still 0.4% higher than Ment–Norm on the AIDA-B dataset, and for the average result on the five out-domain datasets, KGEL also has an improvement of 0.2% F1 on Ment–Norm. It should be noted that although the DCA-SL model has good results on the datasets AIDA-B and MSNBC, it has poor results on the dataset CWEB, so its average result on the out-domain datasets is not good. The same is true for DCA-RL. This indicates that our method has better generalization. Therefore, overall, our method achieves very competitive results on the AIDA-B dataset. Moreover, KGEL achieves higher F1 scores than previous methods on the ACE2004 dataset as well as on the average of out-domain datasets. This fully demonstrates the effectiveness of our method, i.e., the importance of knowledge graph structure for entity linking.

Table 3. F1 scores on AIDA-B and five out-domain test sets. The last column is the average of F1 scores on the five out-domain datasets. The best results are in bold.

Model	AIDA-B	MSNBC	AQUAINT	ACE2004	CWEB	WIKI	Avg
AIDA	-	79	56	80	58.6	63	67.32
GLOW	-	75	83	82	56.2	67.2	72.68
RI	-	90	**90**	86	67.5	73.4	81.38
PBoH	87.6	91	89.2	88.7	-	-	-
Deep-ED	92.22	93.7	88.5	88.5	**77.9**	77.5	85.22
Ment–Norm	93.07	93.9	88.3	89.9	77.5	78	85.5
DCA-SL	**94.64**	**94.57**	87.38	89.44	73.47	78.16	84.6
DCA-RL	93.73	93.80	88.25	90.14	75.59	**78.84**	85.32
KGEL(ours)	93.47	94.26	88.11	**90.54**	77.21	78.40	**85.7**

5.5. Ablation Study

In order to study the role of each module of the model, an ablation study was also performed in this research, and the experimental results are shown in Table 4. We utilize the following variants:

1. *KGEL* is our proposed method, which includes three modules: Local model, WikiEmbs Global model, and KGEmbs Global model.

2. *-KGEmbs* represents the results on each dataset after removing the KGEmbs global model.
3. *-WikiEmbs* represents the experimental results after removing the WikiEmbs global model.
4. *-local-WikiEmbs* is the result of removing the Local model and WikiEmbs Global model at the same time.

Table 4. F1 scores of the ablation experiments.

Model	AIDA-B	MSNBC	AQUAINT	ACE2004	CWEB	WIKI	Avg
KGEL	93.47	94.26	88.11	90.54	77.21	78.40	85.7
- KGEmbs	93.07	93.9	88.3	89.9	77.5	78.0	85.5
- WikiEmbs	87.16	92.12	81.54	87.73	72.84	68.96	80.64
- local *- WikiEmbs*	84.86	91.05	79.16	86.92	70	64.46	78.32

As can be seen in Table 4, when the KGEmbs Global model is removed, the results on four datasets and the average result on the out-domain datasets drop dramatically. This proves the validity of the KGEmbs Global model, i.e., the necessity of introducing knowledge graph structural information. Similarly, we can find that the results on each dataset drop more significantly when the WikiEmbs Global model is removed, indicating that using only the structural information in the knowledge graph is insufficient because there is a certain sparsity in the knowledge graph, i.e., not every pair of entities has a clear relationship with each other, so the structural information of the knowledge graph has a certain guiding effect on the linking of entities, but cannot be used independently. After removing the Local model based on *-WikiEmbs*, we find that the results on each dataset have further decreased, which illustrates the necessity of the local model. Thus, the entire ablation experiment shows that all modules of the model are valid.

5.6. Other Ways of Using KG Structure

In addition to using knowledge graph embedding methods such as TransE on triples, we also try to use triples directly. We consider two entities to be related if there is a relation between them, i.e., two entities that can form a triple are related. Therefore, for entity e_1, we obtain the entity set E_r related to it from the triples. For example, in Table 1, the related entity set of entity Q1 is {Q523, Q136407, Q323}. To incorporate information about its related entities in the representation of entity e_1, we perform the following operations:

$$\mathbf{e}_r = \frac{1}{a}\sum_{i=1}^{a} \mathbf{e}_i \tag{20}$$

$$\mathbf{e} = \alpha \mathbf{e}_1 + (1-\alpha)\mathbf{e}_r \tag{21}$$

where $\mathbf{e}_i \in E_r$ is the entity associated with entity e_1, a is the size of the entity set E_r, $\mathbf{e}_r$ is the average embedding of entities associated with entity e_1, $\mathbf{e}_1$ is the original embedding of entity e_1, $\mathbf{e}$ is the embedding of entity e_1 after fusing information, and α is a hyperparameter. This operation is equivalent to using 1-hop information of the knowledge graph.

In order to determine the optimal value of α, we performed a lot of experiments for different α; that is, directly replacing the original entity embedding with the entity embedding after fusion, and the model structure is consistent with Le and Titov [7]. The experimental results are shown in Figure 4.

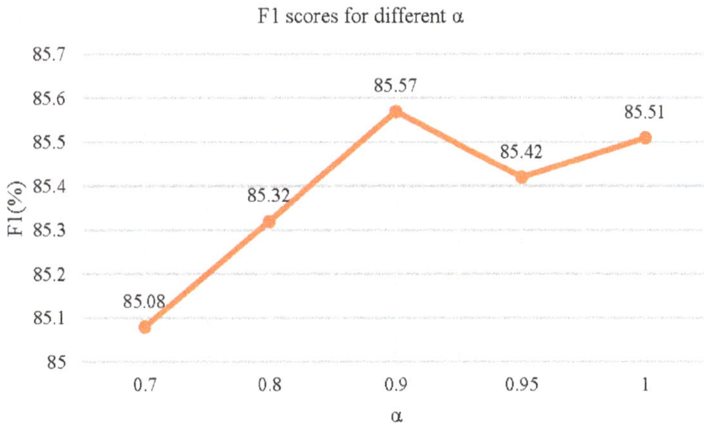

Figure 4. F1 scores for different α, where F1 is the average result on five out-domain datasets.

From the figure, it is clear that the best results are obtained when $\alpha = 0.9$. In addition, we also tried some other variants:

1. Ment-Norm is the model of Le and Titov [7] and also our basic model.
2. *KGEL* is our main model; that is, the entity and relation embeddings obtained by the knowledge graph embedding method are used in the global model of entity linking.
3. *Related-Fixed* refers to the method of using related entities mentioned in this section, in which the parameter α is fixed at 0.9.
4. *Related-Vari* means that the parameter α is variable; that is, it changes during training.
5. Based on *Related-Vari*, *Related-Vari-diff* makes the α in the global model and the local model different.
6. *Related-nn* indicates the use of a neural network to fuse $\mathbf{e}_1$ and $\mathbf{e}_r$.

From the Table 5, it can be seen that the parameter α fixed to 0.9 is the optimal result when using related entities. The result of *Related-Fixed* is slightly better than that of *Ment–Norm*, indicating that the knowledge graph structure is beneficial for the effect of entity linking. However, the result of *Related-Fixed* is worse than that of *KGEL*, which shows that how the knowledge graph structure is used is also very important. Obviously, it is better for us to use the entity embedding obtained by the knowledge graph embedding for the characteristics of the global model considering the correlations between entities.

Table 5. F1 scores of different variants on out-domain datasets.

Model	MSNBC	AQUAINT	ACE2004	CWEB	WIKI	Avg
Ment–Norm	93.9	88.3	89.9	77.5	78.0	85.5
KGEL	94.26	88.11	90.54	77.21	78.40	85.7
Related-Fixed	94.26	88.39	89.74	77.41	78.06	85.57
Related-Vari	93.65	88.25	88.13	77.07	77.82	84.98
Related-Vari-diff	93.8	87.41	87.73	77.01	77.39	84.67
Related-nn	92.58	86.43	88.13	74.87	71.67	82.74

5.7. Better Baseline

To further prove the importance of the knowledge graph structure to the entity linking, we used the *KGEmbs* module for a better baseline. *FGS2EE* [8] is an improvement of *Ment–Norm* [7], which introduces fine-grained semantic information into the original entity embedding to improve the model performance. *KGEL-FGS2EE* adds the *KGEmbs* module on the basis of *FGS2EE*. The experimental results are shown in Figure 5. We can find that for the average F1 score, *KGEL-FGS2EE* can further improve the performance based on

FGS2EE. This shows that the *KGEmbs* module we proposed is effective. Similarly, the *KGEmbs* module can also be used in other methods. In other words, it should be useful to introduce knowledge graph structure based on other methods.

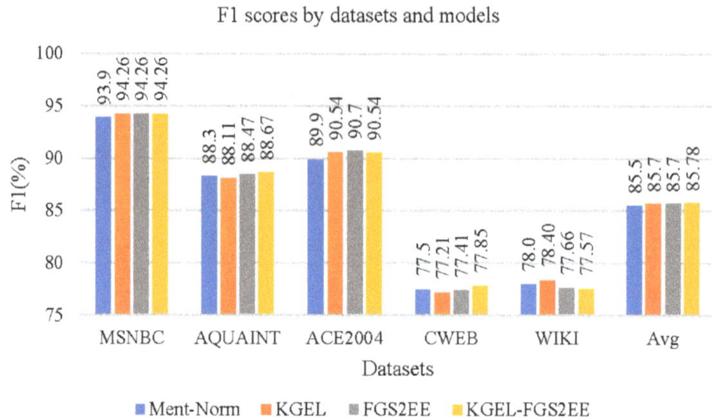

Figure 5. F1 scores of different baselines on out-domain datasets.

5.8. Case Study

Table 6 shows the mentions and their real entities, as well as the results predicted by the model. Examples of incorrect model predictions are shown in red, e.g., "Scotland" is predicted to be "Scotland_national_cricket_team". This shows that in some cases, only semantic information cannot complete the link to the entity. We note that a document contains a knowledge graph structure. As shown in Figure 6, there is a certain connection between the entities "Scotland" and "England". When calculating the global score, the score between "Scotland" and "England" will be higher than the scores between other entities, indicating that mentions "English" and "Scotland" are more likely to refer to entities "England" and "Scotland", respectively. Therefore, we can guide the prediction of mention "Scotland" based on this connection. Similarly, we can use the knowledge graph structure between "Edgbaston" and "Birmingham" to guide the prediction of "Edgbaston". In summary, the introduction of the knowledge graph structure solves the problem of incorrect prediction of some mentions.

Table 6. The examples predicted by the baseline model. The bold font in the first column denotes the mention, the second column is the entity predicted by the model, and the last column is the real entity corresponding to the mention.

Mention	Pred	Gold
...Arrive in **London** May 14...	London	London
...matches against **English** county sides...	England	England
...Counties and **Scotland** Tour itinerary...	Scotland_national_cricket_team	Scotland
...match (at **Edgbaston**,Birmingham)...	Edgbaston_Cricket_Ground	Edgbaston
...Edgbaston,**Birmingham**) June...	Birmingham	Birmingham
...international (at **The Oval**, London)...	The_Oval	The_Oval
...Sussex or **Surrey** (three days)...	Surrey_County_Cricket_Club	Surrey_County_Cricket_Club

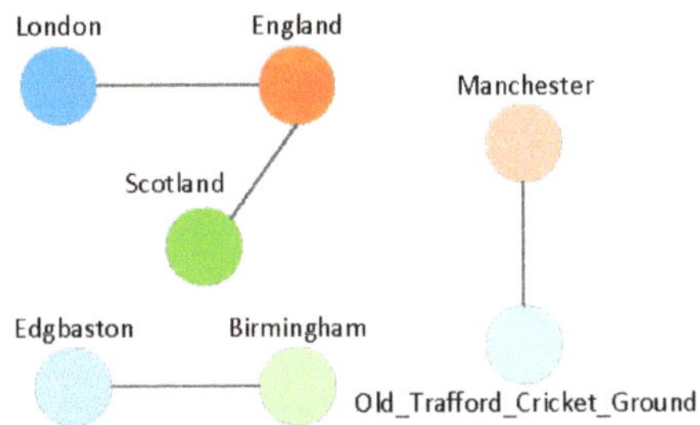

Figure 6. The knowledge graph structure contained in the example.

5.9. Execution Times of the Models

To investigate the complexity of the method, we conducted experiments on the training and inference time of the model. Among them, the model was trained on the AIDA-train dataset and inference was performed on AIDA-B and five out-of-domain datasets. The results are shown in Table 7, where the second column indicates the time spent for one epoch during model training, and the third column indicates the total time spent by the model for inference on several datasets. As can be seen from the table, under the same experimental conditions, our proposed model KGEL is close to the model Ment–Norm [7] in both training and inference time, because we calculated the scores between entities in the KGEmbs Global model offline. In addition, the epochs required for KGEL and Ment–Norm to converge are similar, so the introduced knowledge graph structure does not have much impact on the execution times.

Table 7. The execution times of the models.

Model	Train Time/Epoch	Inference Time
Ment-Norm	23 s	9 s
KGEL	25 s	10 s

6. Conclusions

In this work, we proposed a simple but effective method, KGEL, to introduce knowledge graph structure information into entity linking. In addition to considering the relevance of entities at the semantic level, the relations between entities were also considered from the perspective of structure. We first obtained the triples and then trained them using the knowledge graph embedding method to obtain the entity embeddings and relation embeddings that contained the graph structure. Finally, the entity embeddings and relation embeddings obtained above were used in the calculation of the global score. Extensive experiments on multiple datasets prove the effectiveness of our method; that is, the knowledge graph structure is useful for entity linking tasks. In addition, KGEmbs can be used as a module to enhance the effects of other baseline models.

In future work, we will solve the sparsity problem of the knowledge graph. Not every entity has a corresponding triple, nor is there a relation between every pair of entities. In addition, we will try to use better methods to utilize the knowledge graph structure, such as other knowledge graph embedding methods. As introduced in Section 2.3, some recent knowledge graph embedding methods such as HAKE [44], PairRE [45], DualE [46], and EIGAT [47] can better encode entities and relations in knowledge graphs, and theoretically they should further improve the performance of entity linking.

Author Contributions: Conceptualization, Q.L. (Qijia Li) and F.L.; methodology, Q.L. (Qijia Li) and S.L.; validation, Q.L. (Qijia Li), X.L. and K.L.; formal analysis, Q.L. (Qijia Li) and S.L.; investigation, Q.L. (Qijia Li); resources, Q.L. (Qing Liu); data curation, P.D.; writing—original draft preparation,

Q.L. (Qijia Li); writing—review and editing, F.L. and S.L.; visualization, X.L.; funding acquisition, F.L. All authors have read and agreed to the published version of the manuscript.

Funding: This research was funded by the Strategic Priority Research Program of the Chinese Academy of Sciences (grant number Y835120378).

Institutional Review Board Statement: Not applicable.

Informed Consent Statement: Not applicable.

Data Availability Statement: https://github.com/lephong/mulrel-nel (accessed on 1 March 2022).

Conflicts of Interest: The authors declare that they do not have any conflict of interest. This research does not involve any human or animal participation. All authors have checked and agreed with the submission.

References

1. Bollacker, K.; Evans, C.; Paritosh, P.; Sturge, T.; Taylor, J. Freebase: A collaboratively created graph database for structuring human knowledge. In Proceedings of the 2008 ACM SIGMOD International Conference on Management of Data, Vancouver, BC, Canada, 9–12 June 2008; pp. 1247–1250.
2. Fabian, M.; Gjergji, K.; Gerhard, W. Yago: A core of semantic knowledge unifying wordnet and wikipedia. In Proceedings of the 16th International World Wide Web Conference, WWW, Banff, AB, Canada, 8–12 May 2007; pp. 697–706.
3. Yih, S.W.t.; Chang, M.W.; He, X.; Gao, J. Semantic parsing via staged query graph generation: Question answering with knowledge base. In Proceedings of the Joint Conference of the 53rd Annual Meeting of the ACL and the 7th International Joint Conference on Natural Language Processing of the AFNLP, Beijing, China, 26–31 July 2015.
4. Hoffmann, R.; Zhang, C.; Ling, X.; Zettlemoyer, L.; Weld, D.S. Knowledge-based weak supervision for information extraction of overlapping relations. In Proceedings of the 49th Annual Meeting of the Association for Computational Linguistics: Human Language Technologies, Portland, OR, USA, 19–24 June 2011; pp. 541–550.
5. Michelson, M.; Macskassy, S.A. Discovering users' topics of interest on twitter: A first look. In Proceedings of the Fourth Workshop on Analytics for Noisy Unstructured Text Data, Toronto, ON, Canada, 26 October 2010; pp. 73–80.
6. Ganea, O.E.; Hofmann, T. Deep joint entity disambiguation with local neural attention. *arXiv* **2017**, arXiv:1704.04920.
7. Le, P.; Titov, I. Improving entity linking by modeling latent relations between mentions. *arXiv* **2018**, arXiv:1804.10637.
8. Hou, F.; Wang, R.; He, J.; Zhou, Y. Improving entity linking through semantic reinforced entity embeddings. *arXiv* **2021**, arXiv:2106.08495.
9. Yang, X.; Gu, X.; Lin, S.; Tang, S.; Zhuang, Y.; Wu, F.; Chen, Z.; Hu, G.; Ren, X. Learning dynamic context augmentation for global entity linking. *arXiv* **2019**, arXiv:1909.02117.
10. Fang, W.; Zhang, J.; Wang, D.; Chen, Z.; Li, M. Entity disambiguation by knowledge and text jointly embedding. In Proceedings of the 20th SIGNLL Conference on Computational Natural Language Learning, Berlin, Germany, 11–12 August 2016; pp. 260–269.
11. Luo, A.; Gao, S.; Xu, Y. Deep semantic match model for entity linking using knowledge graph and text. *Procedia Comput. Sci.* **2018**, *129*, 110–114. [CrossRef]
12. Cetoli, A.; Akbari, M.; Bragaglia, S.; O'Harney, A.D.; Sloan, M. Named entity disambiguation using deep learning on graphs. *arXiv* **2018**, arXiv:1810.09164.
13. Mulang, I.O.; Singh, K.; Vyas, A.; Shekarpour, S.; Sakor, A.; Vidal, M.E.; Auer, S.; Lehmann, J. Context-aware entity linking with attentive neural networks on wikidata knowledge graph. *arXiv* **2019**, arXiv:1912.06214.
14. He, Z.; Liu, S.; Mu, L.; Ming, Z.; Wang, H. Learning Entity Representation for Entity Disambiguation. In Proceedings of the 51st Annual Meeting of the Association for Computational Linguistics (Volume 2: Short Papers), Sofia, Bulgaria, 4–9 August 2013.
15. Sun, Y.; Lin, L.; Tang, D.; Yang, N.; Ji, Z.; Wang, X. Modeling mention, context and entity with neural networks for entity disambiguation. In Proceedings of the Twenty-Fourth International Joint Conference on Artificial Intelligence, Buenos Aires, Argentina, 25–31 July 2015.
16. Francis-Landau, M.; Durrett, G.; Klein, D. Capturing semantic similarity for entity linking with convolutional neural networks. *arXiv* **2016**, arXiv:1604.00734.
17. Gupta, N.; Singh, S.; Roth, D. Entity linking via joint encoding of types, descriptions, and context. In Proceedings of the 2017 Conference on Empirical Methods in Natural Language Processing, Copenhagen, Denmark, 7–11 September 2017; pp. 2681–2690.
18. Hochreiter, S.; Schmidhuber, J. Long short-term memory. *Neural Comput.* **1997**, *9*, 1735–1780. [CrossRef]
19. Kolitsas, N.; Ganea, O.E.; Hofmann, T. End-to-end neural entity linking. *arXiv* **2018**, arXiv:1808.07699.
20. Eshel, Y.; Cohen, N.; Radinsky, K.; Markovitch, S.; Yamada, I.; Levy, O. Named entity disambiguation for noisy text. *arXiv* **2017**, arXiv:1706.09147.
21. Chung, J.; Gulcehre, C.; Cho, K.; Bengio, Y. Empirical evaluation of gated recurrent neural networks on sequence modeling. *arXiv* **2014**, arXiv:1412.3555.
22. Guo, Z.; Barbosa, D. Robust named entity disambiguation with random walks. *Semant. Web* **2018**, *9*, 459–479. [CrossRef]

23. Pershina, M.; He, Y.; Grishman, R. Personalized page rank for named entity disambiguation. In Proceedings of the 2015 Conference of the North American Chapter of the Association for Computational Linguistics: Human Language Technologies, 2015; pp. 238–243.
24. Lafferty, J.; McCallum, A.; Pereira, F.C. Conditional Random Fields: Probabilistic Models for Segmenting and Labeling Sequence Data. 2001. Available online: https://repository.upenn.edu/cis_papers/159/?ref=https://githubhelp.com (accessed on 1 March 2022).
25. Murphy, K.; Weiss, Y.; Jordan, M.I. Loopy belief propagation for approximate inference: An empirical study. *arXiv* **2013**, arXiv:1301.6725.
26. Fang, Z.; Cao, Y.; Li, Q.; Zhang, D.; Zhang, Z.; Liu, Y. Joint entity linking with deep reinforcement learning. In Proceedings of the World Wide Web Conference, San Francisco, CA, USA, 13–17 May 2019; pp. 438–447.
27. Yamada, I.; Washio, K.; Shindo, H.; Matsumoto, Y. Global entity disambiguation with pretrained contextualized embeddings of words and entities. *arXiv* **2019**, arXiv:1909.00426.
28. Wu, J.; Zhang, R.; Mao, Y.; Guo, H.; Soflaei, M.; Huai, J. Dynamic graph convolutional networks for entity linking. In Proceedings of the Web Conference 2020, Ljubljana, Slovenia, 19–23 April 2020; pp. 1149–1159.
29. Fang, Z.; Cao, Y.; Li, R.; Zhang, Z.; Liu, Y.; Wang, S. High quality candidate generation and sequential graph attention network for entity linking. In Proceedings of the Web Conference 2020, Ljubljana, Slovenia, 19–23 April 2020; pp. 640–650.
30. Yamada, I.; Shindo, H.; Takeda, H.; Takefuji, Y. Joint learning of the embedding of words and entities for named entity disambiguation. *arXiv* **2016**, arXiv:1601.01343.
31. Yamada, I.; Shindo, H.; Takeda, H.; Takefuji, Y. Learning distributed representations of texts and entities from knowledge base. *Trans. Assoc. Comput. Linguist.* **2017**, *5*, 397–411. [CrossRef]
32. Ling, J.; FitzGerald, N.; Shan, Z.; Soares, L.B.; Févry, T.; Weiss, D.; Kwiatkowski, T. Learning cross-context entity representations from text. *arXiv* **2020**, arXiv:2001.03765.
33. Devlin, J.; Chang, M.W.; Lee, K.; Toutanova, K. Bert: Pre-training of deep bidirectional transformers for language understanding. *arXiv* **2018**, arXiv:1810.04805.
34. Lehmann, J.; Isele, R.; Jakob, M.; Jentzsch, A.; Kontokostas, D.; Mendes, P.N.; Hellmann, S.; Morsey, M.; Van Kleef, P.; Auer, S.; et al. Dbpedia–a large-scale, multilingual knowledge base extracted from wikipedia. *Semant. Web* **2015**, *6*, 167–195. [CrossRef]
35. Wang, Q.; Mao, Z.; Wang, B.; Guo, L. Knowledge graph embedding: A survey of approaches and applications. *IEEE Trans. Knowl. Data Eng.* **2017**, *29*, 2724–2743. [CrossRef]
36. Nickel, M.; Tresp, V.; Kriegel, H.P. A Three-Way Model for Collective Learning on Multi-Relational Data. 2011. Available online: https://openreview.net/forum?id=H14QEiZ_WS (accessed on 1 March 2022).
37. Trouillon, T.; Welbl, J.; Riedel, S.; Gaussier, É.; Bouchard, G. Complex embeddings for simple link prediction. In Proceedings of the International Conference on Machine Learning, PMLR, New York, NY, USA, 19–24 June 2016; pp. 2071–2080.
38. Bordes, A.; Glorot, X.; Weston, J.; Bengio, Y. A semantic matching energy function for learning with multi-relational data. *Mach. Learn.* **2014**, *94*, 233–259. [CrossRef]
39. Bordes, A.; Usunier, N.; Garcia-Duran, A.; Weston, J.; Yakhnenko, O. Translating embeddings for modeling multi-relational data. *Adv. Neural Inf. Process. Syst.* **2013**, *26*, 9.
40. Wang, Z.; Zhang, J.; Feng, J.; Chen, Z. Knowledge graph embedding by translating on hyperplanes. In Proceedings of the AAAI Conference on Artificial Intelligence, Quebec City, QC, Canada, 27–31 July 2014; Volume 28.
41. Lin, Y.; Liu, Z.; Sun, M.; Liu, Y.; Zhu, X. Learning entity and relation embeddings for knowledge graph completion. In Proceedings of the Twenty-Ninth AAAI Conference on Artificial Intelligence, Austin, TX, USA, 25–30 January 2015.
42. Ji, G.; He, S.; Xu, L.; Liu, K.; Zhao, J. Knowledge graph embedding via dynamic mapping matrix. In Proceedings of the 53rd Annual Meeting of the Association for Computational Linguistics and the 7th International Joint Conference on Natural Language Processing (Volume 1: Long Papers), Beijing, China, 26–31 July 2015; pp. 687–696.
43. Fan, M.; Zhou, Q.; Chang, E.; Zheng, F. Transition-based knowledge graph embedding with relational mapping properties. In Proceedings of the 28th Pacific Asia Conference on Language, Information and Computing, Phuket, Thailand, 12–14 December 2014; pp. 328–337.
44. Zhang, Z.; Cai, J.; Zhang, Y.; Wang, J. Learning hierarchy-aware knowledge graph embeddings for link prediction. In Proceedings of the AAAI Conference on Artificial Intelligence, New York, NY, USA, 7–12 February 2020; Volume 34, pp. 3065–3072.
45. Chao, L.; He, J.; Wang, T.; Chu, W. PairRE: Knowledge graph embeddings via paired relation vectors. *arXiv* **2020**, arXiv:2011.03798.
46. Cao, Z.; Xu, Q.; Yang, Z.; Cao, X.; Huang, Q. Dual quaternion knowledge graph embeddings. In Proceedings of the AAAI Conference on Artificial Intelligence, Vancouver, BC, Canada, 2–9 February 2021; Volume 35, pp. 6894–6902.
47. Zhao, Y.; Zhou, H.; Xie, R.; Zhuang, F.; Li, Q.; Liu, J. Incorporating Global Information in Local Attention for Knowledge Representation Learning. In Proceedings of the Findings of the Association for Computational Linguistics: ACL-IJCNLP 2021, Online Event, 1–6 August 2021; pp. 1341–1351.
48. Hoffart, J.; Yosef, M.A.; Bordino, I.; Fürstenau, H.; Pinkal, M.; Spaniol, M.; Taneva, B.; Thater, S.; Weikum, G. Robust disambiguation of named entities in text. In Proceedings of the 2011 Conference on Empirical Methods in Natural Language Processing, Scotland, UK, 27–31 July 2011; pp. 782–792.
49. Ratinov, L.; Roth, D.; Downey, D.; Anderson, M. Local and global algorithms for disambiguation to wikipedia. In Proceedings of the 49th Annual Meeting of the Association for Computational Linguistics: Human Language Technologies, Portland, OR, USA, 19–24 June 2011; pp. 1375–1384.

50. Cheng, X.; Roth, D. Relational inference for wikification. In Proceedings of the 2013 Conference on Empirical Methods in Natural Language Processing, Seattle, DC, USA, 18–21 October 2013; pp. 1787–1796.
51. Ganea, O.E.; Ganea, M.; Lucchi, A.; Eickhoff, C.; Hofmann, T. Probabilistic bag-of-hyperlinks model for entity linking. In Proceedings of the 25th International Conference on World Wide Web, Montreal, QC, Canada, 11–15 April 2016; pp. 927–938.

applied sciences

MDPI

Article

BERT-Based Transfer-Learning Approach for Nested Named-Entity Recognition Using Joint Labeling

Ankit Agrawal [1], Sarsij Tripathi [2], Manu Vardhan [1], Vikas Sihag [3], Gaurav Choudhary [4] and Nicola Dragoni [4,*]

1 Department of Computer Science & Engineering, National Institute of Technology Raipur, Raipur 492010, Chhattisgarh, India; aagrawal.phd2017.cse@nitrr.ac.in (A.A.); mvardhan.cs@nitrr.ac.in (M.V.)
2 Department of Computer Science & Engineering, Motilal Nehru National Institute of Technology Allahabad, Prayagraj 211004, Uttar Pradesh, India; sarsij@mnnit.ac.in
3 Department of Cyber Security, Sardar Patel University of Police, Security and Criminal Justice, Jodhpur 342037, Rajasthan, India; vikas.sihag@policeuniversity.ac.in
4 DTU Compute, Department of Applied Mathematics and Computer Science, Technical University of Denmark (DTU), 2800 Kongens Lyngby, Denmark; gauravchoudhary7777@gmail.com
* Correspondence: ndra@dtu.dk

Abstract: Named-entity recognition (NER) is one of the primary components in various natural language processing tasks such as relation extraction, information retrieval, question answering, etc. The majority of the research work deals with flat entities. However, it was observed that the entities were often embedded within other entities. Most of the current state-of-the-art models deal with the problem of embedded/nested entity recognition with very complex neural network architectures. In this research work, we proposed to solve the problem of nested named-entity recognition using the transfer-learning approach. For this purpose, different variants of fine-tuned, pretrained, BERT-based language models were used for the problem using the joint-labeling modeling technique. Two nested named-entity-recognition datasets, i.e., GENIA and GermEval 2014, were used for the experiment, with four and two levels of annotation, respectively. Also, the experiments were performed on the JNLPBA dataset, which has flat annotation. The performance of the above models was measured using F1-score metrics, commonly used as the standard metrics to evaluate the performance of named-entity-recognition models. In addition, the performance of the proposed approach was compared with the conditional random field and the Bi-LSTM-CRF model. It was found that the fine-tuned, pretrained, BERT-based models outperformed the other models significantly without requiring any external resources or feature extraction. The results of the proposed models were compared with various other existing approaches. The best-performing BERT-based model achieved F1-scores of 74.38, 85.29, and 80.68 for the GENIA, GermEval 2014, and JNLPBA datasets, respectively. It was found that the transfer learning (i.e., pretrained BERT models after fine-tuning) based approach for the nested named-entity-recognition task could perform well and is a more generalized approach in comparison to many of the existing approaches.

Keywords: named-entity recognition; transfer learning; BERT model; conditional random field; pre-trained model; fine-tuning

Citation: Agrawal, A.; Tripathi, S.; Vardhan, M.; Sihag, V.; Choudhary, G.; Dragoni, N. BERT-Based Transfer-Learning Approach for Nested Named-Entity Recognition Using Joint Labeling. *Appl. Sci.* **2022**, *12*, 976. https://doi.org/10.3390/app12030976

Academic Editors: Arturo Montejo-Ráez and Salud María Jiménez-Zafra

Received: 15 November 2021
Accepted: 14 January 2022
Published: 18 January 2022

Publisher's Note: MDPI stays neutral with regard to jurisdictional claims in published maps and institutional affiliations.

1. Introduction

There is much focus on identifying and classifying important words present in text into their respective semantic classes, such as DNA, RNA, cell, or protein [1]. These important words are known as named entities (NEs), and the task is known as named-entity recognition (NER). The task of named-entity recognition is important because it further helps in different natural language processing (NLP) tasks such as question answering [2], machine translation [3], relation extraction [4], and many more [5,6]. It is often the case that one entity resides within or overlaps with another entity. The text data of different domains commonly contain overlapping entities. However, most of the research work focuses on flat entities only, i.e., they cannot identify the overlapping or nested entities present in the text [7]. In flat named-entity recognition (or named-entity recognition), each token within the text corpus can be determined as anyone entity type only. In the overlapping or nested-entity recognition problem, each token can be classified as more than one entity type. Due to this, there is a potential loss of information in the flat entity recognition task, which also negatively impacts the subsequent natural language processing tasks. The solution is to try and identify overlapping entities. An example of overlapping entities within a sentence is illustrated in Figure 1.

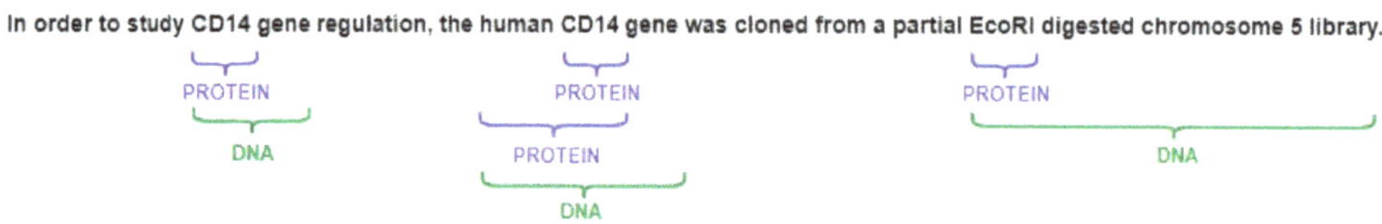

Figure 1. Example of overlapping entities from GENIA dataset.

In the above figure, the word "CD14" is recognized as a PROTEIN type. However, the phrase "CD14 gene" is identified as a DNA type. Similarly, two other overlapping entities can be seen among the PROTEIN and DNA entities in the above example. The annotation was made at multiple levels for each word of the sentence to correctly capture the overlapping entities. For this research work, the experiments were performed using two different nested-entity datasets (GENIA and GermEval 2014) and a flat named-entity-recognition dataset (JNLPBA).

The Bidirectional Encoder Representation from Transformers (BERT) language model recently came into the picture and achieved state-of-the-art results over 11 different natural language processing problems, including named-entity recognition. In the past, many researchers have proposed to solve the problem of nested named-entity recognition using very complex neural network architectures. In recent times, transfer learning has achieved great success and is very widely used to solve different problems in the fields of computer vision and natural language processing. Transfer learning is a well-known approach in which a deep learning model is trained on a large unlabeled dataset (pretrained model) and is further trained on the downstream task dataset (labeled dataset) to fine-tune the pretrained parameters of the pretrained model. The main idea in this research work was to evaluate the effectiveness of the BERT-based transfer-learning approach for the problem of nested entity recognition. The proposed transfer-learning approach (fine-tuning a pretrained BERT-based model) is easy to implement and is simpler than the other models based on complex neural network architecture, which were mostly used earlier to solve the nested named-entity-recognition problem. There are several variants of the pretrained BERT models that are different based on their pretraining on domain-specific texts, or using different vocabulary (word case also matters), etc. Since the pretrained BERT model can be fine-tuned to solve the flat named-entity-recognition problem, we converted the problem of nested entity recognition to the flat named-entity-recognition problem by using the joint labeling technique. The conventional flat named-entity-recognition models could be used without any modification once the labels of different levels were joined together into a single level.

The contributions of this research work are as follows:

- We proposed to solve the nested named-entity-recognition problem using the transfer-learning approach, i.e., by fine-tuning pretrained, BERT-based language models.
- The proposed transfer-learning approach (fine-tuning a pretrained BERT language model) could outperform many of the existing heavily engineered, complex-architecture-based approaches for the nested named-entity-recognition problem.

- The nested datasets were jointly labeled so that conventional named-entity-recognition models could also be used, which treated the nested named entity problem as the flat named-entity-recognition task.
- The experiment was carried with two other well-known machine-learning models (conditional random field and the Bi-LSTM-CRF) for the performance comparison. In addition, the performance of the best-performing proposed model was compared with the existing research work.
- This research work compared the performance of different variants of the pretrained BERT models for the nested named-entity-recognition problem based on domain, size of models (base or large), and cased and uncased versions.
- The results were analyzed and discussed in detail while clarifying the factors that were important in the variants of the pretrained BERT models for different categories, which further led to providing good results for the nested named-entity-recognition problem.

The sections of this paper are arranged as follows: Section 2 covers related works in which similar existing research works have been discussed. Section 3 presents the proposed transfer learning based approach, followed by existing machine-learning models in Section 4, datasets in Section 5, and the evaluation tool used in this research work in Section 6. Section 7 discusses the experimental results and compares the performance by comparing the result of the other models and the existing approaches. Section 8 concludes the paper.

2. Related Works

As discussed above, the annotation was performed at multiple levels to capture the nested information in the named-entity-recognition dataset. Apart from the machine learning model, different modeling techniques must be used to solve the problem of nested named-entity recognition in most cases. This modeling technique includes three different approaches: layering (inside-out and outside-in), cascading, and joint labeling [8,9]. In most of the research works, the layering approach was used. In this research work, the joint labeling modeling technique was used, as it allowed the use of the conventional named-entity-recognition models for identifying the nested entities, and joined the different levels of the nested dataset such that they could be treated as flat entities. Examples of jointly labeled sentences from the nested datasets can be found in Section 5.

Recently, Plank et al., 2021 [10] experimented with a comparison between the cross-language (i.e., German) and in-language for Danish nested entity recognition with different variants of the BERT model. They also presented a new multidomain named entity dataset and experimented with the domain shift problem. They found that BERT-based language models could not perform well for the out-of-domain setup. In another work by Mulyar et al., 2021 [11], a new variant of the BERT model was presented that could perform eight different tasks of clinical information extraction at the same time. It was found that the BERT fine-tuning baseline model performed well in comparison to the proposed multitask model, as a single-task-specific model could better exploit the dataset and its properties. Similarly, Bang et al., 2021 [12] proposed an approach to detect "fake news" related to COVID-19 using different versions of fine-tuned, pretrained, BERT-based language models with the robust loss function.

In the past, the nested named-entity-recognition problem has been solved using one of the following: the neural-network-based approach, the non-neural-network-based approach, and the graph-based approach [13,14].

A new model based on a layered neural network model was presented by Wang et al., 2020 [15] in which pyramid-shaped layers were present, and each layer length was reduced by one when moving from bottom to top. The word embeddings were passed, and each layer l represented the l-gram of the input text. The above model produced good results for different nested named-entity-recognition datasets. Another outside-to-inside approach was proposed by Shibuya et al., 2020 [16]; it also was a neural-network-based approach in which a new objective function and a decoding method that worked iteratively was presented. The model performed similarly to the above neural-network-based model for the nested named-entity-recognition dataset. Another work by Wang et al., 2020 [17] proposed an approach based on a head-tail detector to detect the boundary tokens explicitly. In addition, they have proposed a token-interaction tagger to determine the internal connection among the tokens present within the boundary. There are a number of other neural-network-based approaches that have obtained good results using complex architecture to solve the problem of nested named-entity recognition, such as those presented in [18–20], and many more.

The approaches based on a non-neural network include the constituency-parser-based approach and the graph-based approach. Initially, the constituency-parser-based approach was used by Finkel et al., 2009 [21], in which they represented the sentences using a constituency tree and proposed a CRF-based constituency parser. Recently, a similar approach was proposed by Fu et al., 2020 [22] in which the nested named-entity-recognition problem was solved using a partially observed Tree-CRF model by proposing a new MASKED INSIDE algorithm for computation of probability of partial trees.

Different graph-based approaches have also been used widely for the problem of nested named-entity recognition. They began with the hypergraph-based representation proposed by Lu et al., 2015 [23] to detect correct head, type, and boundary information using a single framework. A similar approach was presented by Wang et al., 2019 [24] and Muis et al., 2018 [25], in which a new segmental hypergraph and mention separator and a multigraph were used for modeling and representation of nested entities, respectively. There are also hybrid models in which graph-based approaches were combined with neural networks to identify the overlapping entities, as in Luo et al., 2020 [26].

Overall, different types of approaches have been used in the past to solve the problem of classification of nested named entities that can provide a good result. However, all the above approaches are either complex in nature or have a complex architecture. In addition, there is a need to explore the transfer-learning approach (i.e., by using the fine-tuned, pretrained language model) for solving the nested named-entity-recognition problem using whichever one has more generalization capabilities compared to any of the existing approaches. Moreover, there are very few existing research works that used a joint labeling modeling technique for nested entity recognition. Hence, in this research work, we proposed to solve this problem using transfer learning (by fine-tuning different variants of pretrained BERT language models) using joint labeling of the nested tags for the nested entity recognition. We also implemented the conditional random field model and the Bi-LSTM-CRF model for comparison of the performance of both models using different NER datasets.

3. Transfer-Learning Approach

In this research work, the nested named-entity-recognition problem was solved using the transfer-learning approach. In this approach, a pretrained language model is used that is already trained on a large unlabeled text dataset. The pretrained model is further trained on a small task-specific text dataset to fine-tune the pretrained parameters. The main motive of using the above-mentioned transfer-learning approach is that it enhances the generalization capability of the model for the low-resource, task-specific text dataset while leveraging the high-resource dataset. The language model is pretrained on the high resource dataset, which is unlabeled (i.e., a plain-text dataset) and is available in abundance. The pretraining task requires a significant computational resource, as a large model is trained on a large plain-text dataset for a considerable amount of time (usually days). However, the fine-tuning of the downstream task is very easy and can be done quickly.

Moreover, prominent NLP researchers have released different pretrained BERT-based language models for public use. In this work, the experiments were performed with different variants of the pretrained BERT language models that fell broadly in the three different categories: Google AI, SciBERT, and the BioBERT pretrained BERT language models. The pretrained models belonging to these categories differed based on the domain of the datasets on which they were pretrained. In addition, there were multiple variants of the pretrained BERT language model in each category that differed based on the case, vocabulary size, language, etc. The experiments were also performed using the conditional random field (CRF) and Bi-LSTM-CRF models so that their results could also be compared with the results of the different variants of the pretrained BERT-based language models. The details of the models and the parameters used were as follows.

3.1. Pretrained BERT Models Used in the Transfer-Learning-Based Approach

Bidirectional Encoder Representations from Transformers (BERT) is a new unsupervised contextualized language representation model that is highly popular for natural language processing tasks. It has been shown that the requirement of heavily engineered task-specific architectures has been reduced significantly by using pretrained representations [27]. This was the first fine-tuning based model to achieve state-of-the-art results on 11 different natural language processing tasks. For natural language processing tasks, the pretraining of the language models has already proved to be effective [27,28]. The pretrained language representation can be applied to downstream natural language processing tasks in two ways: (a) using a fine-tuning-based approach; and (b) using a feature-based approach. The fine-tuning-based approach is minimally dependent on task-specific parameters, i.e., training is performed over downstream tasks while fine-tuning the pretrained parameters. In the feature-based approach, task-specific architectures, including pretrained parameters, are used as additional features. However, during pretraining, the same objective function is used by both approaches, in which language representations are learned using unidirectional models [27]. The BERT model uses the "masked language model" (MLM) and "next sentence prediction" (NSP) pretraining objectives, which mixes the left and right context, allowing the pretraining of a bidirectional deep transformer while removing unidirectional constraints.

In this paper, we used the fine-tuning-based approach, which used already-pretrained models. The pretrained models were trained for different pretraining tasks on unlabeled data from scratch. The details of the pretrained models used in this research work are discussed in further subsections. While performing fine-tuning for downstream tasks, the BERT model began with the parameters of the pretrained models. These parameters were fine-tuned as per the downstream task, which here was nested named-entity recognition. The pretraining and the fine-tuning scheme discussed above can be seen in Figure 2, which was inspired by [27,29].

For this research work, we used the scikit-learn wrapper provided by [30] for fine-tuning the BERT-based models belonging to different categories. For fine-tuning of each of the pretrained BERT models over the named-entity-recognition datasets, the number of epochs was set to 3 (as overfitting was observed for epochs more than 3), the maximum sequence length was set according to the max token length of the wordpiece tokenizer in the training set (plus two to the max token length for the '[CLS]' and '[SEP]' delimiter tokens that BERT uses, so that no data were truncated), the gradient accumulation step was set to 2, the batch size was 8, the validation fraction was set to 0.05, ignore_label was set to other tags according to the dataset, and num_mlp_layers was set to 0 so that linear classifier was used for classification along with the cross entropy loss function for single-label classification. Note that for most of the above and remaining other hyperparameters, the default values were used (the same as in the original BERT paper). For each model, the experiment was conducted three times, with learning rates of 3×10^{-5}, 4×10^{-5}, and 5×10^{-5}. The average result of each run and the standard deviation are reported in the Results section. The scikit-learn wrapper for fine-tuning BERT and the default settings for named-entity-recognition problems were used for the rest of the parameters [30]. The BERT base and large models have 12 and 24 layers (or transformer blocks), 768 and 1024 hidden sizes, and 12 and 16 self-attention heads, respectively. The Adam optimization algorithm and gelu activation function was used in the original BERT model [27]. No manual feature extraction is required in BERT-based models.

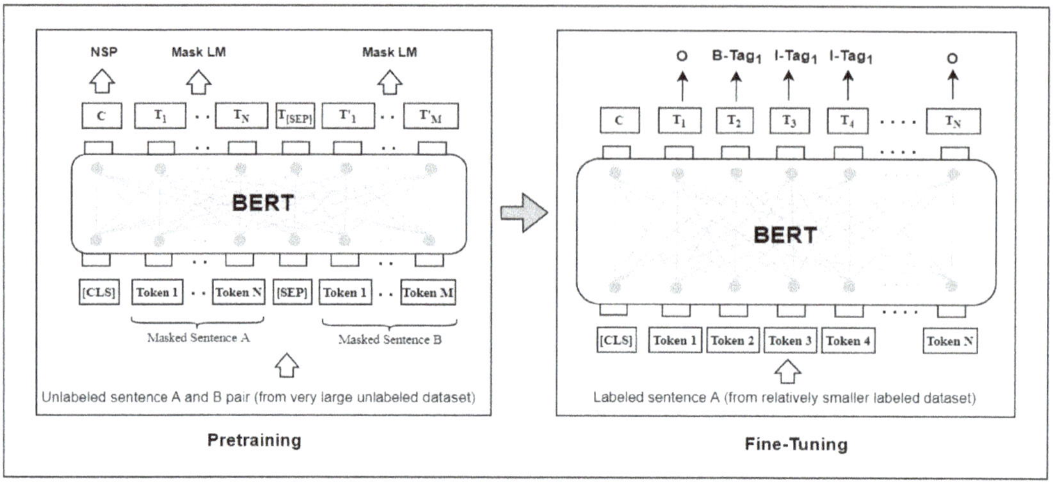

Figure 2. The pretraining and fine-tuning scheme for the BERT model. The same pretrained BERT model can be used for various natural language processing tasks.

The pretrained BERT-based models used for the experiment can be broadly classified into three categories, which are outlined below.

3.1.1. Google AI's Pretrained BERT Models

The basic details of the BERT models belonging to this category were already discussed. They have a multilayer, bidirectional, transformer encoder architecture. Here, the input is an arbitrary span of contiguous text passed as a sequence of tokens. WordPiece embedding, which has a vocabulary of 30,000 tokens, is used [27,31]. This model was trained over the corpus of a general domain, i.e., on BooksCorpus (800 million words) and English Wikipedia (2.5 billion words). The details of parameters used for pre-training the $BERT_{BASE}$ and $BERT_{LARGE}$ models are given in [27,30,32]. In addition, a multilingual BERT was used for named entity recognition, as one of the datasets was in the German language [30,32,33]. We experimented with both cased and uncased versions of the above models.

3.1.2. SciBERT Pretrained BERT Models

SciBERT follows the architecture of the BERT model, but was pretrained using scientific text. The designers used the vocabulary provided by BERT as BASEVOCAB. In addition, they constructed their own WordPiece vocabulary named SCIVOCAB using the scientific text corpus with the same vocabulary size. Similar to the above, they also produced both the cased and the uncased version of models. The SciBERT model was trained over scientific text corpus from the Semantic Scholar, which has 1.14 million full-text papers and a total of 3.17 billion tokens [34].

3.1.3. BioBERT Pretrained BERT Models

BioBERT is another pretrained, domain-specific language model, and was pretrained on large-scale biomedical text corpora for the purpose of biomedical text mining. It also has an architecture similar to that of the BERT model. It has been shown that BioBERT significantly outperformed in the three different biomedical text mining tasks, which included: biomedical named-entity recognition, biomedical question answering and biomedical relation extraction. In this paper, we experimented with five different versions of the pretrained BioBERT models. These models used the $BERT_{BASE}$ pretrained model and were further pretrained over combinations of PubMed and PMC corpora for the different numbers of steps. Further details on the models that were used in the experiment can be found in [30,35].

4. Existing Machine-Learning Models

4.1. Conditional Random Field Model

The conditional random field model is commonly used for sequence labeling, as it allows both the flow of probabilistic information across the sequence and discriminative training. Given some observation sequences, the conditional random field represents the probability of hidden state sequences. The non-independent and overlapping features in the observation sequence can be modeled using a conditional random field (CRF). Other theoretical details of the conditional random field model have been skipped, but can be found in [36,37]. Figure 3 shows the basic workflow diagram for the named-entity recognition using the conditional random field model used in this research work. For implementing the conditional random field model, python's sklearn-crfsuite library was used [38]. The training and testing dataset were initially available in the CoNLL 2002 format, which was further preprocessed and stored as a list of lists of tuples. The outermost list contained all the list of sentences; the individual sentences were also stored in the list data structure of python, and the words along with their respective features (if any) and correct labels were stored in the tuple data structure of python before passing the dataset for manual feature extraction. The feature extraction is discussed in detail in a further section. The complete extracted features were stored as a list of lists of dictionaries. Similar to above, the outermost list contained complete features of all the sentences; inner lists contained features of individual sentences, and the dictionaries inside the inner list contained the features of a particular word of a sentence in order. All the parameters used for the conditional random field (CRF) model for this research work were set according to [38].

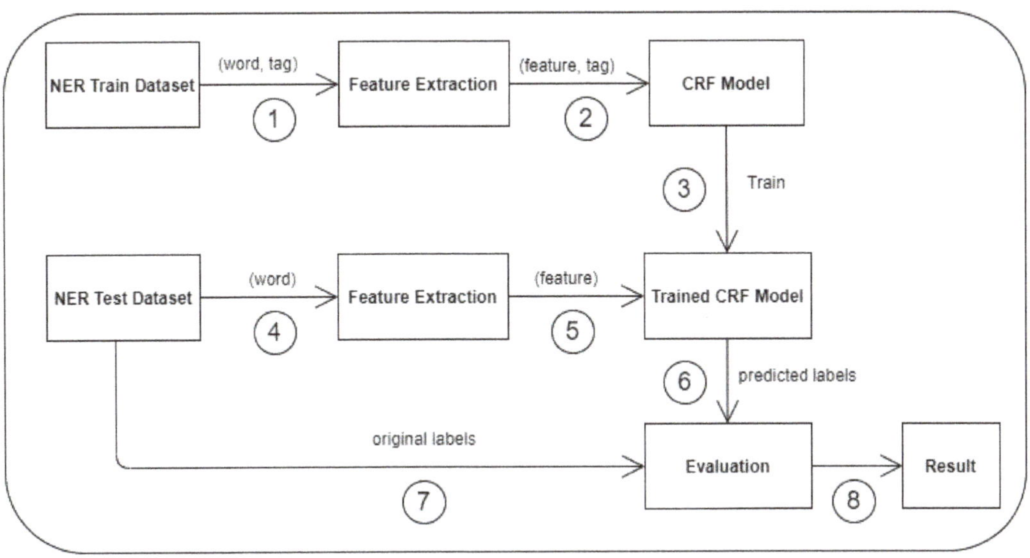

Figure 3. The workflow used for NER using conditional random field (CRF) model.

Table 1. Sample features extracted from a sentence of GermEval 2014 train dataset.

Sentence	Barauszahlungen	Sind	Grundsätzlich	Nicht	Möglich	.
Label	O + O	O + O	O + O	O + O	O + O	O + O
Features	{'word.lower()': 'barauszahlungen', 'word.len()': 15, 'word.hasHyphen()': False, 'word[−4:]': 'ngen', 'word[−3:]': 'gen', 'word[−2:]': 'en', 'word[:2]': 'Ba', 'word[:3]': 'Bar', 'word[:4]': 'Bara', 'word.type': 'Alpha', 'word.case()': 'Title', 'word.pattern()': 'ULLLLLLLLLLLLLL', 'stem': 'barauszahl', 'BOS': True, '+1:word.lower()': 'sind', '+1:word.len()': 4, '+1:word.hasHyphen()': False, '+1:word.type': 'Alpha', '+1:word.case()': 'Lower', '+1:word.pattern()': 'LLLL', '+2:word.lower()': 'grundsätzlich', '+2:word.len()': 13, '+2:word.hasHyphen()': False, '+2:word.type': 'Alpha', '+2:word.case()': 'Lower', '+2:word.pattern()': 'LLLLLLLLLLLLL'}	{'word.lower()': 'sind', 'word.len()': 4, 'word.hasHyphen()': False, 'word[−4:]': 'sind', 'word[−3:]': 'ind', 'word[−2:]': 'nd', 'word[:2]': 'si', 'word[:3]': 'sin', 'word[:4]': 'sind', 'word.type': 'Alpha', 'word.case()': 'Lower', 'word.pattern()': 'LLLL', 'stem': 'sind', '−1:word.lower()': 'barauszahlungen', '−1:word.len()': 15, '−1:word.hasHyphen()': False, '−1:word.type': 'Alpha', '−1:word.case()': 'Title', '−1:word.pattern()': 'ULLLLLLLLLLLLLL', '+1:word.lower()': 'grundsätzlich', '+1:word.len()': 13, '+1:word.hasHyphen()': False, '+1:word.type': 'Alpha', '+1:word.case()': 'Lower', '+1:word.pattern()': 'LLLLLLLLLLLLL', '+2:word.lower()': 'nicht', '+2:word.len()': 5, '+2:word.hasHyphen()': False, '+2:word.type': 'Alpha', '+2:word.case()': 'Lower', '+2:word.pattern()': 'LLLLL'}	{'word.lower()': 'grundsätzlich', 'word.len()': 13, 'word.hasHyphen()': False, 'word[−4:]': 'lich', 'word[−3:]': 'ich', 'word[−2:]': 'ch', 'word[:2]': 'gr', 'word[:3]': 'gru', 'word[:4]': 'grun', 'word.type': 'Alpha', 'word.case()': 'Lower', 'word.pattern()': 'LLLLLLLLLLLLL', 'stem': 'grundsatz', '−1:word.lower()': 'sind', '−1:word.len()': 4, '−1:word.hasHyphen()': False, '−1:word.type': 'Alpha', '−1:word.case()': 'Lower', '−1:word.pattern()': 'LLLL', '−2:word.lower()': 'barauszahlungen', '−2:word.len()': 15, '−2:word.hasHyphen()': False, '−2:word.type': 'Alpha', '−2:word.case()': 'Title', '−2:word.pattern()': 'ULLLLLLLLLLLLLL', '+1:word.lower()': 'nicht', '+1:word.len()': 5, '+1:word.hasHyphen()': False, '+1:word.type': 'Alpha', '+1:word.case()': 'Lower', '+1:word.pattern()': 'LLLLL', '+2:word.lower()': 'möglich', '+2:word.len()': 7, '+2:word.hasHyphen()': False, '+2:word.type': 'Alpha', '+2:word.case()': 'Lower', '+2:word.pattern()': 'LLLLLLL'}	{'word.lower()': 'nicht', 'word.len()': 5, 'word.hasHyphen()': False, 'word[−4:]': 'icht', 'word[−3:]': 'cht', 'word[−2:]': 'ht', 'word[:2]': 'ni', 'word[:3]': 'nic', 'word[:4]': 'nich', 'word.type': 'Alpha', 'word.case()': 'Lower', 'word.pattern()': 'LLLLL', 'stem': 'nicht', '−1:word.lower()': 'grundsätzlich', '−1:word.len()': 13, '−1:word.hasHyphen()': False, '−1:word.type': 'Alpha', '−1:word.case()': 'Lower', '−1:word.pattern()': 'LLLLLLLLLLLLL', '−2:word.lower()': 'sind', '−2:word.len()': 4, '−2:word.hasHyphen()': False, '−2:word.type': 'Alpha', '−2:word.case()': 'Lower', '−2:word.pattern()': 'LLLL', '+1:word.lower()': 'möglich', '+1:word.len()': 7, '+1:word.hasHyphen()': False, '+1:word.type': 'Alpha', '+1:word.case()': 'Lower', '+1:word.pattern()': 'LLLLLLL', '+2:word.lower()': '.', '+2:word.len()': 1, '+2:word.hasHyphen()': False, '+2:word.type': 'None', '+2:word.case()': 'None', '+2:word.pattern()': '.'}	{'word.lower()': 'möglich', 'word.len()': 7, 'word.hasHyphen()': False, 'word[−4:]': 'lich', 'word[−3:]': 'ich', 'word[−2:]': 'ch', 'word[:2]': 'mö', 'word[:3]': 'mög', 'word[:4]': 'mögl', 'word.type': 'Alpha', 'word.case()': 'Lower', 'word.pattern()': 'LLLLLLL', 'stem': 'moglich', '−1:word.lower()': 'nicht', '−1:word.len()': 5, '−1:word.hasHyphen()': False, '−1:word.type': 'Alpha', '−1:word.case()': 'Lower', '−1:word.pattern()': 'LLLLL', '−2:word.lower()': 'grundsätzlich', '−2:word.len()': 13, '−2:word.hasHyphen()': False, '−2:word.type': 'Alpha', '−2:word.case()': 'Lower', '−2:word.pattern()': 'LLLLLLLLLLLLL', '+1:word.lower()': '.', '+1:word.len()': 1, '+1:word.hasHyphen()': False, '+1:word.type': 'None', '+1:word.case()': 'None', '+1:word.pattern()': '.'}	{'word.lower()': '.', 'word.len()': 1, 'word.hasHyphen()': False, 'word[−4:]': '.', 'word[−3:]': '.', 'word[−2:]': '.', 'word[:2]': '.', 'word[:3]': '.', 'word[:4]': '.', 'word.type': 'None', 'word.case()': 'None', 'word.pattern()': '.', 'stem': '.', 'EOS': True, '−1:word.lower()': 'möglich', '−1:word.len()': 7, '−1:word.hasHyphen()': False, '−1:word.type': 'Alpha', '−1:word.case()': 'Lower', '−1:word.pattern()': 'LLLLLLL', '−2:word.lower()': 'nicht', '−2:word.len()': 5, '−2:word.hasHyphen()': False, '−2:word.type': 'Alpha', '−2:word.case()': 'Lower', '−2:word.pattern()': 'LLLLL'}

Feature Extraction for the CRF Model

Feature extraction is an essential step for machine-learning models. The features describe the dataset effectively, and can be passed to the machine learning model for training and testing, respectively. The CRF model is trained over the extracted features from the training dataset, and later, extracted features from the test dataset are passed to the trained CRF model to predict the correct tags according to input test dataset features. In this subsection, the features used for the CRF model for different datasets are described in detail after their introduction. The base form (of any word) is the root of the verb without any suffixes (such as -ed, -s, and -ing). Similarly, stemming is the process of reducing any word to its stem but not necessarily to its dictionary root. Part-of-speech (POS) tagging is used to identify how the words are used in a sentence. There are different parts of speech tags, such as noun, verb, pronoun, adverb, etc. Chunking helps in the identification of phrases present in unstructured text. It has labels such as noun phrase (NP), verb phrase (VP), etc. For the GENIA and JNLPBA datasets, we used the GENIA tagger [39,40] to provide the base form, POS tagging, and chunking. Here, chunking also followed the Begin–Inside–Outside (BIO) format. The base form, POS tags, and chunking tags were appended to the original GENIA and JNLPBA datasets, to be used as a feature. Since the GENIA tagger would not have provided good results on the German dataset, we used nltk's snowball German stemmer and appended its outputs to the original GermEval 2014 dataset. The other features for all the three datasets included: begin of sentence (BOS) and end of sentence (EOS) markers for the beginning and end of sentences as Boolean type; word in lowercase; length of word; suffix and prefix of word; type of word (i.e., whether the word was a type of digit, alphanumeric, alphabetic, or none of above); whether the word has a hyphen (as Boolean type); pattern present in word (i.e., pattern obtained after replacing the following: uppercase characters present in the word with "U", lowercase characters with "L", full-stop and comma characters with a full-stop character, digits with "D", symbols ("_", "+", "*", "/", "=", "|") with "#", symbols (":", ";", "!", "?") with ";", and braces (">", ")", "}", "]") with ")" and braces ("<", "(", "{", "[") with "("; case information of the word (i.e., whether the word was in title case, uppercase, lowercase, or none of these); and finally, context information of the word having features mentioned as above, with a window size of 2. An example of a sample feature extracted from a sentence of the training set of the GermEval 2014 dataset is presented in Table 1. The English meaning (according to Google Translate) of the sample sentence presented in Table 1 is "Cash payments are generally not possible".

4.2. Bidirectional LSTM-CRF Model

The Bi-LSTM-CRF model is a combination of the bidirectional LSTM and CRF layer. Here, the model had access to the sentence-level label information, as well as the past and future input features. The GLOVE-based pretrained word vectors, which were trained over 840 billion tokens with 300 dimensions, have been used for word embedding [41]. The FastText German word embeddings [42] were only used for the GermEval 2014 dataset, as the results using above word embeddings were not good for the German dataset. Firstly, the word embedding for each word was obtained using the vocabulary and the pretrained word vectors. Secondly, the contextual word representation was obtained by passing the token representation to the Bi-LSTM layer. Finally, the decoding of the contextual word representation was done for the prediction. The existing code from [43,44] was used for implementation.

5. Datasets

Three different datasets were used in this research work for experiments; namely, the GENIA, GermEval 2014 (German dataset), and JNLPBA datasets. All the above three datasets were divided into training and testing sets only to keep uniformity. However, the last 10% of sentences were taken from the training set of each dataset to be used as the validation set for the Bi-LSTM-CRF model only. The first two datasets (i.e., GENIA and GermEval 2014) have nested entities, and the JNLPBA dataset has flat entities. The GENIA and the JNLPBA datasets are from the biomedical domain. All the datasets were having named entities labeled in Begin–Inside–Outside (BIO2) format in which the first word of any entity starts with 'B-' indicating the beginning of the label and other remaining words of that entity begins with 'I-' indicating inside of the label. Also, the word (or the token) that are labeled with 'O' are not named entities. For the nested dataset, different levels of annotations were jointly labeled. A sample sentence is shown as an example for each of the nested datasets. The only disadvantage of joint labeling was that there was a significant increase in the number of classes in which each word could be identified. However, the advantage was that all the conventional models used for the flat named-entity recognition could be used for nested named-entity recognition. Further details of all three datasets are discussed below.

5.1. GermEval 2014 Dataset

This dataset is a nested dataset for German named-entity recognition and was presented by [45] for the GermEval 2014 Named-Entity Recognition Shared Task [46]. This dataset consists of around 31,000 manually labeled sentences from German online news and German Wikipedia. There are 12 categories of labels in the dataset, out of which 4 main categories are: ORGanization, LOCation, PERson, and OTHer. They also used two fine-grained labels for each of the above four main categories, i.e., -part and -deriv for partial and derived named entities [45]. For example, "EU" belongs to an Organization category; but "EU-Verwaltung" (English meaning: EU administration) is identified as Organization_part. There are many other examples in which phrases partly contains names, such as "deutschlandweit" (English meaning: Germany-wide). Similarly, the derivations are separately identified. For example, "österreichischen" (English meaning: Austrian) is identified as Location_deriv in the dataset [45,46]. This dataset has two levels of nested labeling. A sample sentence is presented in Table 2.

Table 2. Sample sentence from GermEval 2014 dataset along with nested level annotation (L1 and L2) and joint labeling. The English translation of the sentence (according to Google Translate) was: "From 4 p.m., the pursuers Aston Villa and Tottenham Hotspur will be challenged".

Sentence	Label L1	Label L2	Joint Label
Ab	O	O	O + O
16	O	O	O + O
Uhr	O	O	O + O
sind	O	O	O + O
dann	O	O	O + O
die	O	O	O + O
Verfolger	O	O	O + O
Aston	B-ORG	B-LOC	B-ORG + B-LOC
Villa	I-ORG	O	I-ORG + O
und	O	O	O + O
Tottenham	B-ORG	B-LOC	B-ORG + B-LOC
Hotspur	I-ORG	O	I-ORG + O
gefordert	O	O	O + O
.	O	O	O + O

5.2. GENIA Dataset

The GENIA dataset is a semantically annotated dataset that contains 2000 abstracts from the MEDLINE database [47]. It has four levels of nesting and five types of entities after simplification (DNA, Protein, cell-line, RNA, and cell-type). We followed [21,23] and kept about 90% of data in the training set, and about 10% of the data were present in the testing set. A sample sentence showing labels with four nested levels is presented in Table 3.

Table 3. Sample sentence from GENIA dataset along with nested level annotation (L1, L2, L3, and L4) and joint labeling.

Sentence	Label L1	Label L2	Label L3	Label L4	Joint Label
In	O	O	O	O	O + O + O + O
order	O	O	O	O	O + O + O + O
to	O	O	O	O	O + O + O + O
study	O	O	O	O	O + O + O + O
CD14	B-protein	B-DNA	O	O	B-protein + B-DNA + O+O
gene	O	I-DNA	O	O	O + I-DNA + O+O
regulation	O	O	O	O	O + O + O + O
,	O	O	O	O	O + O + O + O
the	O	O	O	O	O + O + O + O
human	O	B-protein	B-DNA	O	O + B-protein + B-DNA + O
CD14	B-protein	I-protein	I-DNA	O	B-protein + I-protein + I-DNA + O
gene	O	O	I-DNA	O	O + O + I-DNA + O
was	O	O	O	O	O + O + O + O
cloned	O	O	O	O	O + O + O + O
from	O	O	O	O	O + O + O + O
a	O	O	O	O	O + O + O + O
partial	O	O	O	O	O + O + O + O
EcoRI	B-protein	B-DNA	O	O	B-protein + B-DNA + O+O
digested	O	I-DNA	O	O	O + I-DNA + O+O
chromosome	O	I-DNA	O	O	O + I-DNA + O+O
5	O	I-DNA	O	O	O + I-DNA + O+O
library	O	I-DNA	O	O	O + I-DNA + O+O
	O	O	O	O	O + O + O + O

5.3. *JNLPBA Dataset*

The GENIA project organized the BioNLP Shared Task 2004 [48], in which the JNLPBA dataset was introduced. Like the GENIA dataset, it also has five types of entities (DNA, Protein, cell-line, RNA, and cell-type). In the training set of the JNLPBA dataset, there are about 2000 MEDLINE abstracts, and in the testing dataset, there are 404 MEDLINE abstracts. Further details of all the above datasets are presented in Table 4.

Table 4. Named-entity recognition datasets used in this research work.

Dataset	Training Dataset			Testing Dataset			No. of Entity Types (Except Others)
	No. of Abstracts	No. Sent	No. of Tokens	No. of Abstracts	No. Sent	No. of Tokens	
GENIA	1800 (approx.)	16,692	503,857	200 (approx.)	1854	57,024	5
GermEval 2014	-	26,202	494,506	-	5100	96,499	12
JNLPBA	2000	20,546	494,551	404	4260	101,443	5

6. Evaluation Tool and Metrics

In this research work, the F1-score was reported, as it is a standard metric used to evaluate the performance in the problem of named-entity recognition. The F1-score is the harmonic mean of the two other metrics, precision and recall. The F1-score strikes a balance between the precision and the recall. For this research work, we used the third-party tool used in [49] and many others, provided during a CoNLL 2000 shared task for evaluating the F1-score [50].

7. Results and Discussion

This section discusses the performances of the CRF model, Bi-LSTM-CRF model, and different pre-trained BERT models belonging to different categories. The above performances were evaluated using the GENIA dataset (nested biomedical NER dataset), GermEval 2014 dataset (the German language nested NER dataset), and JNLPBA dataset (flat biomedical NER dataset). Since the labels of different levels in the nested datasets were joined together into a single label level, all three different datasets were treated as flat named entity datasets. In addition, a comparison of the best-performing pretrained BERT models for each dataset was made with the existing approaches.

7.1. Discussion of Results for BERT-Based Models

The results for the pretrained BERT models belonging to different categories are discussed in this subsection for the above three different named-entity-recognition datasets. The average F1-score and the standard deviation of three runs are presented for each of the models in Table 5 below.

For the GENIA dataset, the overall best F1-score of 74.38 was obtained by the biobert-base-cased pretrained BERT model (C.1). In category A of the pretrained BERT models (i.e., Google AI's pretrained BERT models), the large and cased version performed better in comparison to the base and the uncased versions of the pretrained models. The best-performing model was the large-cased model. The worst performance in this category was obtained by both the cased and uncased multilingual BERT models. In category B of the pretrained BERT models (i.e., the SciBERT pretrained BERT models), the uncased model with scivocab (B.1) performed best, with an F1-score of 74.07, followed by the remaining models in this category. It is important to note that here, the uncased model performed better than the cased model, and all the models in this category performed better than the models in category A. In category C of the pretrained BERT models (i.e., the BioBERT pretrained BERT models), the base cased model (C.1) performed the best, with an overall F1-score of 74.38 for the GENIA dataset; its performance was followed by models C.2, C.4, and C.5. Since the BioBERT model obtained the overall best results, it was clear that this result was obtained due to domain-based pretraining. In addition, most of the models in this category performed better than the models in other two categories.

For the GermEval 2014 dataset (German language NER dataset), the overall best F1-score of 85.29 was obtained by Google's multilingual base cased model (A.6). In category A of the pretrained BERT models (i.e., Google AI's pretrained BERT models), the performances of both the multilingual BERT models were far better than any of the other models. Their performance was followed by the large BERT models (cased and uncased), and then similarly by the base models. In addition, the results were dependent on both the domain of the pretraining dataset and the model size in this case, as the large model performed slightly better than the base models. In category B of the pre-trained BERT models (i.e., the SciBERT pretrained BERT models), the models with the BASEVOCAB vocabulary performed better than the SCIVOCAB-vocabulary-based models. In addition, the performances of the cased models were significantly better than that of the uncased models. The best results in the category were obtained by the basevocab cased model (B.4); i.e., an F1-score of 79.05. In category C of the pretrained BERT models (i.e., the BioBERT pretrained BERT models), the best-performing model in this category (C.5) had a F1-score of 77.02, which was far behind the overall best F1-score of the model (A.6). Here, it was observed that the results for the cased model were slightly better than the uncased for all the models. It is important to note that all the models in category A performed better than the models in the other two categories.

For the JNLPBA dataset, the overall best F1-score of 80.68 was obtained by the scivocab-uncased pretrained BERT model (B.1). In category A of the pre-trained BERT models (i.e., Google AI's pretrained BERT models), the base cased model (A.1) had the best F1-score. Its performance was followed by both the large cased and uncased versions of the pretrained model. The difference in results was not that significant among other models. In category B and category C of the pretrained BERT models (i.e., the SciBERT and BioBERT pretrained BERT models, respectively), almost all the models had an F1-score greater than 80. However, as discussed above, the overall best F1-score of 80.68 was obtained by the B.1 model, followed by the C.4 model, which attained an F1-score of 80.48 among the pre-trained BERT models in the C category. It was observed for the JNLPBA dataset, in most of the cases, the uncased version of the models performed slightly better than the cased version of the model. In category B, the SCIVOCAB-vocabulary-based models also performed better in comparison to the BASEVOCAB-based pretrained models. In addition, most of the time, the models in categories B and C performed better in comparison to the models in category A.

7.2. Discussion of Results for the CRF Model

The results obtained by the CRF model are also presented in Table 5. This model is still the most popular for the named-entity-recognition problem and can be used for both nested and non-nested datasets. The named-entity recognition for all three datasets could be treated as a flat named-entity-recognition problem. The CRF model used the feature described above obtained an F1-score of 65.15, 68.93, and 74.23 for the GENIA, GermEval 2014, and JNLPBA datasets, respectively. The results obtained for all three datasets had a very significant difference from any of the pretrained BERT models. This model obtained the worst results, even after the manual feature extraction for each of the datasets.

7.3. Discussion of Results for the Bi-LSTM-CRF Model

The results obtained by the Bi-LSTM-CRF model are recorded in Table 5. This model is also very widely used for sequence-tagging tasks such as named-entity recognition. As mentioned before, the GLOVE (and the FastText) word embeddings were used for obtaining the word embeddings. The Bi-LSTM-CRF model obtained an F1-score of 70.19, 76.14, and 77.56 for the GENIA, GermEval 2014 (using German FastText word vectors), and JNLPBA datasets, respectively. An F1-score of 70.21 was obtained using the GLOVE word vectors (for the English language) for the GermEval 2014 dataset, which was very poor, and hence was not included in the results table. The results obtained for all three datasets were much better than for the CRF model, but they were still significantly worse than for the fine-tuning-based pretrained BERT models for all the datasets.

Table 5. Result for the CRF, Bi-LSTM-CRF, and fine-tuned pretrained BERT models for different categories of three different NER datasets. For all the datasets, the best results of the pretrained BERT model in each category are shown in bold. The results of the overall best-performing model for each of the NER datasets are in bold and underlined.

S. No.		Model Details		Dataset Details		
				Nested Dataset		Non-Nested Dataset
		Model Category	Model Name	F1 Score (GENIA Test Dataset)	F1 Score (GermEval 2014 Test Dataset)	F1 Score (JNLPBA Test Dataset)
A.	1.	Google AI's pretrained BERT models	bert-base-uncased	72.91 $\pm$ 0.09	79.08 $\pm$ 0.24	**80.13 $\pm$ 0.04**
	2.		bert-large-uncased	73.11 $\pm$ 0.11	80.76 $\pm$ 0.24	79.98 $\pm$ 0.14
	3.		bert-base-cased	73.19 $\pm$ 0.08	79.76 $\pm$ 0.25	79.51 $\pm$ 0.24
	4.		bert-large-cased	**73.38 $\pm$ 0.09**	81.37 $\pm$ 0.24	80.11 $\pm$ 0.16
	5.		bert-base-multilingual-uncased	72.49 $\pm$ 0.24	84.72 $\pm$ 0.17	79.42 $\pm$ 0.11
	6.		bert-base-multilingual-cased	72.44 $\pm$ 0.24	**85.29 $\pm$ 0.23**	79.65 $\pm$ 0.12
B.	1.	SciBERT pretrained BERT models	scibert-scivocab-uncased	**74.07 $\pm$ 0.18**	76.35 $\pm$ 0.14	**80.68 $\pm$ 0.22**
	2.		scibert-scivocab-cased	73.56 $\pm$ 0.13	77.38 $\pm$ 0.13	80.60 $\pm$ 0.07
	3.		scibert-basevocab-uncased	73.34 $\pm$ 0.05	78.66 $\pm$ 0.11	80.33 $\pm$ 0.18
	4.		scibert-basevocab-cased	73.57 $\pm$ 0.21	**79.05 $\pm$ 0.23**	79.82 $\pm$ 0.18
C.	1.	BioBERT pretrained BERT models	biobert-base-cased	**74.38 $\pm$ 0.14**	75.67 $\pm$ 0.26	80.43 $\pm$ 0.14
	2.		biobert-v1.1-pubmed-base-cased	74.29 $\pm$ 0.07	75.76 $\pm$ 0.32	80.42 $\pm$ 0.14
	3.		biobert-v1.0-pubmed-base-cased	73.63 $\pm$ 0.07	76.32 $\pm$ 0.36	80.25 $\pm$ 0.15
	4.		biobert-v1.0-pubmed-pmc-base-cased	73.79 $\pm$ 0.21	76.62 $\pm$ 0.11	**80.48 $\pm$ 0.04**
	5.		biobert-v1.0-pmc-base-cased	73.84 $\pm$ 0.18	**77.02 $\pm$ 0.25**	80.44 $\pm$ 0.24
D.		CRF model		65.15 $\pm$ 0.21	68.93 $\pm$ 0.23	74.23 $\pm$ 0.38
E.		Bi-LSTM-CRF		70.19 $\pm$ 0.56	76.14 $\pm$ 0.31	77.56 $\pm$ 0.36

7.4. Comparison of the Results with Other Existing Approaches

The best results of the pretrained BERT models for each of the three datasets were compared with the results of existing approaches. The comparison results for the GENIA dataset with the existing approaches are presented in Table 6.

Similarly, a comparison was made in terms of performance for the GermEval 2014 dataset with the existing approaches. The comparison results for the GermEval 2014 dataset are presented in Table 7.

A similar comparison was made for the JNLPBA dataset with the existing approaches. The comparison results for this dataset are presented in Table 8.

A few important points observed from the above discussion of results are as follows:

- On comparing the CRF, Bi-LSTM-CRF, and BERT-based language models, it was found that almost all the BERT-based models performed better than both the other models. The performance of the Bi-LSTM-CRF models was better than that of the CRF model, but not the fine-tuning-based, pretrained BERT-based models.
- There was a huge impact of the language on the BERT-based model, which was clear from the results for the GermEval 2014 (German) nested NER dataset. Even the Bi-LSTM-CRF model performed poorly if the English GLOVE word vectors were used for the word embedding (due to which the German FastText word vectors were used only for the GermEval 2014 dataset).
- The transfer-learning-based approach without any modifications or any external resources performed well on the GENIA, GermEval 2014, and JNLPBA datasets compared to many of the existing approaches. In Tables 6–8, comparisons were made with existing research work. There were a number of other research works in this area that achieved better results than the presented transfer-learning approach. Note that we are still far from the state-of-the-art results for the above three datasets. Our approach did not possess any kind of complexity in architecture or implementation. The same was not true for the other existing research works. In this study, we wanted to compare the performances of the pretrained, BERT-based transfer-learning approach without using any external resources such as embeddings, unsupervised training on the new dataset, etc. The study was conducted for a performance comparison between the pretrained BERT models based on domain, model size (base or large), and cased and uncased versions.
- Domain-based pretrained models could perform significantly better than the other BERT models pretrained on different domains. For example, the BioBERT-based models performed better on the GENIA dataset, Google's multilingual BERT-based model performed better on the GermEval 2014 dataset, and the SciBERT-based model performed better on the JNLPBA dataset (followed by the BioBERT).
- The model size of the pretrained BERT model can also put some impact on the results (in most cases). However, the result difference may not be very significant between the base and large models in all the cases.
- In most cases (for the GENIA and GermEval 2014 datasets), the performance of the cased version of the model was better than that of the uncased version of the model. However, the uncased versions of the BERT language model performed better on the JNLPBA dataset.
- Some of the common postprocessing methods from the existing research works have been carried to improve the prediction of best-performing models. However, the performance declined, rather than improving. So, postprocessing is not recommended for the named-entity-recognition problem.

Table 6. Comparison of results with existing approaches for GENIA dataset.

Source	Used Approach	F1-Score
[21]	Parser-based	70.33
[23]	Mention-hypergraph-based	68.70
[25]	Multigraph-based	70.80
[51]	Neural-network-based (LSTM, hypergraph features)	73.80
[52]	Neural-network-based (LSTM-CRF, seq2seq, contextual embeddings)	73.90
[13]	Neural-network-based (boundary aware Bi-LSTM)	73.90
This Paper	Transfer-learning-based (best BERT model)	74.38
[16]	Neural-network-based (Bi-LSTM-CRF, contextual embeddings)	77.36
[14]	Neural-network-based (seq2seq, contextual embeddings)	78.31

Table 7. Comparison of results with existing approaches for GermEval 2014 dataset.

Source	Used Approach	F1-Score
[13]	Neural-network-based (boundary aware Bi-LSTM)	71.7
[53]	Neural-network-based (feed forward, Bi-LSTM, Win-bi-LSTM)	76.12
[54]	Neural-network-based (Bi-LSTM-CRF)	75.3
[17]	Neural-network-based (head–tail pair, token interaction tagger)	72.6
This Paper	Transfer-learning-based (best BERT model)	85.29
[55]	Neural-network-based (PolDeepNer2)	87.69
[56]	Transfer-learning-based (unsupervised pretraining, pretrained BERT)	88.6

Table 8. Comparison of results with existing approaches for JNLPBA dataset.

Source	Used Approach	F1-Score
[57]	Neural-network-based (Bi-LSTM, embeddings)	78.4
[17]	Neural-network-based (head–tail pair, token interaction tagger)	74.9
[58]	Neural-network-based (Bi-LSTM, embeddings)	75.87
This Paper	Transfer-learning-based (best BERT model)	80.48
[59]	Neural-network-based (BLSTM-CNN-Char and Spark NLP)	81.29
[60]	Transformer-based	82.0

It is important to note that the existing approaches for the nested named-entity-recognition problem are complex. At the same time, the presented transfer-learning-based approach is much simpler than any of the other existing approaches and can be easily used for similar problems. In addition, it is important to note that the presented transfer-learning-based approach had no requirement for manual feature extraction or the word vectors, while these were needed for the conditional random field and Bi-LSTM-CRF models.

8. Conclusions

In this research work, the transfer-learning approach was used to solve the nested named-entity-recognition problem. The presented transfer-learning approach fine-tuned the pretrained BERT language models for the NER task. The experiments were conducted with different variants of the pretrained BERT-based language models belonging to three popular categories based on the domain. The performance comparison has been done with the existing approaches for each of the datasets. In addition, the experiments were conducted using the conditional random field (CRF) and the Bi-LSTM-CRF models for performance comparison. Manual feature extraction and word embeddings were required for the CRF and Bi-LSTM-CRF models. However, there were no such requirements for the presented transfer-learning approach. The performance was evaluated using two biomedical datasets and a German language NER dataset, out of which one biomedical dataset (i.e., the GENIA dataset) and the German language dataset (i.e., GermEval 2014 dataset) contained nested annotations. The different levels of annotation were joined together for the nested datasets so that the nested named-entity-recognition problem could be treated as a flat named-entity-recognition problem. It was found that the performance of the presented transfer-learning approach was much better than that of the other two models and many of the existing approaches. The presented transfer-learning approach achieved better results than many of the existing research works for the nested and non-nested NER datasets. This research work presented a performance comparison between the pretrained BERT models based on domain, size of models, and cased and uncased versions. It was found that the performance of the presented BERT-based language model depended on the domain and the language of the downstream task. In addition, the presented transfer-learning-based approach had more generalization capability and was much simpler than any of the existing approaches. The presented transfer-learning approach can be used for any of the similar downstream natural language processing tasks. In the future, we will conduct a similar study of several different natural language processing tasks other than named-entity recognition to further test the performance and generalization capabilities of the presented transfer-learning approach.

Author Contributions: Conceptualization, A.A., S.T. and M.V.; methodology, A.A.; software, A.A.; validation, A.A.; investigation, A.A.; resources, A.A., S.T. and M.V.; data curation, A.A.; writing—original draft preparation, A.A.; writing—review and editing, S.T., M.V., V.S., G.C. and N.D.; visualization, A.A.; supervision, S.T. and M.V.; project administration, V.S., G.C. and N.D.; funding acquisition, G.C. and N.D. All authors have read and agreed to the published version of the manuscript.

Funding: This work was partially supported by the Danish Industry Foundation through "CIDI-Cybersecure IoT in Danish Industry" under Project 2018-0197.

Institutional Review Board Statement: Not applicable.

Informed Consent Statement: Not applicable.

Data Availability Statement: The data that are used in this research work are openly available at the following links: 1. GermEval 2014 dataset: https://sites.google.com/site/germeval2014ner/data (accessed on: 4 March 2021); 2. GENIA dataset: http://www.geniaproject.org/genia-corpus (accessed on: 5 March 2021); 3. JNLPBA dataset: http://www.geniaproject.org/shared-tasks/bionlp-jnlpba-shared-task-2004 (accessed on: 5 March 2021). All the above datasets can also be found at the GitHub repositories of [13,57].

Conflicts of Interest: The authors declare no conflict of interest.

References

1. Li, C.; Wang, G.; Cao, J.; Cai, Y. A Multi-Agent Communication Based Model for Nested Named Entity Recognition. *IEEE/ACM Trans. Audio Speech Lang. Process.* **2021**, *29*, 2123–2136. [CrossRef]
2. Alzubi, J.A.; Jain, R.; Singh, A.; Parwekar, P.; Gupta, M. COBERT: COVID-19 Question Answering System Using BERT. *Arab. J. Sci. Eng.* **2021**, 1–11. [CrossRef]

3. Chauhan, S.; Saxena, S.; Daniel, P. Fully unsupervised word translation from cross-lingual word embeddings especially for healthcare professionals. *Int. J. Syst. Assur. Eng. Manag.* **2021**, 1–10. [CrossRef]
4. Kumar, N.; Kumar, M.; Singh, M. Automated ontology generation from a plain text using statistical and NLP techniques. *Int. J. Syst. Assur. Eng. Manag.* **2016**, *7*, 282–293. [CrossRef]
5. Kumar, R.B.; Suresh, P.; Raja, P.; Sivaperumal, S. Artificial intelligence powered diagnosis model for anaesthesia drug injection. *Int. J. Syst. Assur. Eng. Manag.* **2021**, 1–9. [CrossRef]
6. Parthasarathy, J.; Kalivaradhan, R.B. An effective content boosted collaborative filtering for movie recommendation systems using density based clustering with artificial flora optimization algorithm. *Int. J. Syst. Assur. Eng. Manag.* **2021**, 1–10. [CrossRef]
7. Dai, X. Recognizing Complex Entity Mentions: A Review and Future Directions. In Proceedings of the ACL 2018, Student Research Workshop, Melbourne, Australia, 15–20 July 2018; pp. 37–44. Available online: https://aclanthology.org/P18-3006.pdf (accessed on 15 March 2021).
8. Alex, B.; Haddow, B.; Grover, C. Recognising Nested Named Entities in Biomedical Text. In Proceedings of the Biological, Translational, and Clinical Language Processing, Prague, Czech Republic, 29 June 2007; pp. 65–72. Available online: https://aclanthology.org/W07-1009.pdf (accessed on 15 March 2021).
9. Chen, Y.; Zheng, Q.; Chen, P. A Boundary Assembling Method for Chinese Entity-Mention Recognition. *IEEE Intell. Syst.* **2015**, *30*, 50–58. [CrossRef]
10. Plank, B.; Jensen, K.N.; Van Der Goot, R. DaN+: Danish Nested Named Entities and Lexical Normalization. In Proceedings of the 28th International Conference on Computational Linguistics, Bacelona, Spain, 8–13 December 2020; pp. 6649–6662.
11. Mulyar, A.; Uzuner, O.; McInnes, B. MT-clinical BERT: Scaling clinical information extraction with multitask learning. *J. Am. Med. Inform. Assoc.* **2021**, *28*, 2108–2115. [CrossRef] [PubMed]
12. Bang, Y.; Ishii, E.; Cahyawijaya, S.; Ji, Z.; Fung, P. Model Generalization on COVID-19 Fake News Detection. *arXiv* **2021**, arXiv:2101.03841.
13. Zheng, C.; Cai, Y.; Xu, J.; Leung, H.-F.; Xu, G. A Boundary-aware Neural Model for Nested Named Entity Recognition. In Proceedings of the 2019 Conference on Empirical Methods in Natural Language Processing and the 9th International Joint Conference on Natural Language Processing (EMNLP-IJCNLP), Hong Kong, China, 3–7 November 2019; pp. 357–366. Available online: https://aclanthology.org/D19-1034.pdf (accessed on 11 March 2021).
14. Straková, J.; Straka, M.; Hajic, J. Neural Architectures for Nested NER through Linearization. In Proceedings of the 57th Annual Meeting of the Association for Computational Linguistics, Florence, Italy, 28 July 2019; pp. 5326–5331. Available online: http://aclanthology.lst.uni-saarland.de/P19-1527.pdf (accessed on 13 March 2021).
15. Wang, J.; Shou, L.; Chen, K.; Chen, G. Pyramid: A Layered Model for Nested Named Entity Recognition. In Proceedings of the 58th Annual Meeting of the Association for Computational Linguistics, Online, 5–10 July 2020; pp. 5918–5928. Available online: https://aclanthology.org/2020.acl-main.525.pdf (accessed on 11 March 2021).
16. Shibuya, T.; Hovy, E. Nested Named Entity Recognition via Second-best Sequence Learning and Decoding. *Trans. Assoc. Comput. Linguistics* **2020**, *8*, 605–620. [CrossRef]
17. Wang, Y.; Li, Y.; Tong, H.; Zhu, Z. HIT: Nested Named Entity Recognition via Head-Tail Pair and Token Interaction. In Proceedings of the 2020 Conference on Empirical Methods in Natural Language Processing (EMNLP), Online, 16–20 November 2020; pp. 6027–6036. Available online: https://aclanthology.org/2020.emnlp-main.486.pdf (accessed on 12 March 2021).
18. Chen, Y.; Wu, L.; Deng, L.; Qing, Y.; Huang, R.; Zheng, Q.; Chen, P. A Boundary Regression Model for Nested Named Entity Recognition. *arXiv* **2020**, arXiv:2011.14330.
19. Dadas, S.; Protasiewicz, J. A Bidirectional Iterative Algorithm for Nested Named Entity Recognition. *IEEE Access* **2020**, *8*, 135091–135102. [CrossRef]
20. Tan, C.; Qiu, W.; Chen, M.; Wang, R.; Huang, F. Boundary Enhanced Neural Span Classification for Nested Named Entity Recognition. *Proc. AAAI Conf. Artif. Intell.* **2020**, *34*, 9016–9023. [CrossRef]
21. Finkel, J.R.; Manning, C.D. Nested Named Entity Recognition. In Proceedings of the 2009 Conference on Empirical Methods in Natural Language Processing, Singapore, 6–7 August 2009; pp. 141–150. Available online: https://aclanthology.org/D09-1015.pdf (accessed on 7 March 2021).
22. Fu, Y.; Tan, C.; Chen, M.; Huang, S.; Huang, F. Nested Named Entity Recognition with Partially-Observed TreeCRFs. In Proceedings of the AAAI Conference on Artificial Intelligence, Online, 2–9 February 2021.
23. Lu, W.; Roth, D. Joint Mention Extraction and Classification with Mention Hypergraphs. In Proceedings of the 2015 Conference on Empirical Methods in Natural Language Processing, Lisbon, Portugal, 17–21 September 2015; pp. 857–867. Available online: https://aclanthology.org/D15-1102.pdf (accessed on 14 March 2021).
24. Wang, B.; Lu, W. Neural Segmental Hypergraphs for Overlapping Mention Recognition. In Proceedings of the 2018 Conference on Empirical Methods in Natural Language Processing, Brussels, Belgium, 31 October–4 November 2018.
25. Muis, A.O.; Lu, W. Labeling Gaps Between Words: Recognizing Overlapping Mentions with Mention Separators. In Proceedings of the 2017 Conference on Empirical Methods in Natural Language Processing, Copenhagen, Denmark, 9–11 September 2018.
26. Luo, Y.; Zhao, H. Bipartite Flat-Graph Network for Nested Named Entity Recognition. In Proceedings of the 58th Annual Meeting of the Association for Computational Linguistics, Online, 5–10 July 2020; pp. 6408–6418.
27. Devlin, J.; Chang, M.-W.; Lee, K.; Toutanova, K. BERT: Pre-training of Deep Bidirectional Transformers for Language Understanding. In Proceedings of the 2019 Conference of the North American Chapter of the Association for Computational

Linguistics: Human Language Technologies, Minneapolis, MN, USA; 2019; Volume 1, pp. 4171–4186. Available online: https://aclanthology.org/N19-1423.pdf (accessed on 19 March 2021).

28. Howard, J.; Ruder, S. Fine-tuned Language Models for Text Classification. *arXiv* **2018**, arXiv:1801.06146.
29. Kang, M.; Lee, K.; Lee, Y. Filtered BERT: Similarity Filter-Based Augmentation with Bidirectional Transfer Learning for Protected Health Information Prediction in Clinical Documents. *Appl. Sci.* **2021**, *11*, 3668. [CrossRef]
30. Nainan, C. Scikit-Learn Wrapper to Finetune BERT. Available online: https://github.com/charles9n/bert-sklearn (accessed on 5 January 2021).
31. Wu, Y.; Schuster, M.; Chen, Z.; Le, Q.V.; Norouzi, M.; Macherey, W.; Krikun, M.; Cao, Y.; Gao, Q.; Macherey, K.; et al. Google's Neural Machine Translation System: Bridging the Gap between Human and Machine Translation. *arXiv* **2016**, arXiv:1609.08144.
32. Rush, A.; Wolf, T.; Debut, L.; Sanh, V.; Chaunmond, J.; Delangue, C.; Moi, A.; Cistac, P.; Rault, T.; Louf, R.; et al. Transformers: State-of-the-Art Natural Language Processing. In Proceedings of the 2020 Conference on Empirical Methods in Natural Language Processing: System Demonstrations, Online, October 2020; pp. 38–45.
33. Team, T.H. Multi-Lingual Models. Available online: https://huggingface.co/transformers/multilingual.html (accessed on 25 August 2021).
34. Beltagy, I.; Cohan, A.; Lo, K. SciBERT: Pretrained Contextualized Embeddings for Scientific Text. In Proceedings of the 2019 Conference on Empirical Methods in Natural Language Processing and the 9th International Joint Conference on Natural Language Processing (EMNLP-IJCNLP), Hong Kong, China, 3–7 November 2019.
35. Lee, J.; Yoon, W.; Kim, S.; Kim, D.; Kim, S.; So, C.H.; Kang, J. BioBERT: A pre-trained biomedical language representation model for biomedical text mining. *Bioinformatics* **2019**, *36*, 1234–1240. [CrossRef] [PubMed]
36. Wallach, H.M. Conditional Random Fields: An Introduction. 2004. Available online: http://www.inference.org.uk/hmw26/papers/crf_intro.pdf (accessed on 7 March 2021).
37. Zhu, X. CS838-1 Advanced NLP: Conditional Random Fields. 2007. Available online: http://pages.cs.wisc.edu/~{}jerryzhu/cs838/CRF.pdf (accessed on 7 March 2021).
38. Korobov, M. Sklearn-Crfsuite Docs. Available online: https://sklearn-crfsuite.readthedocs.io/en/latest/tutorial.html (accessed on 11 August 2021).
39. Tsuruoka, Y.; Tateishi, Y.; Kim, J.-D.; Ohta, T.; McNaught, J.; Ananiadou, S.; Tsujii, J. Developing a Robust Part-of-Speech Tagger for Biomedical Text. In Proceedings of the Advances in Informatics 10th Panhellenic Conference on Informatics, PCI 2005, Volos, Greece, 11–13 November 2005; Bozanis, P., Houstis, E.N., Eds.; Springer: Berlin/Heidelberg, Germany, 2005; pp. 382–392.
40. Tsuruoka, Y.; Tsujii, J. Bidirectional Inference with the Easiest-First Strategy for Tagging Sequence Data. In Proceedings of the Human Language Technology Conference and Conference on Empirical Methods in Natural Language Processing, Vancouver, BC, Canada, 6–8 October 2005; Association for Computational Linguistics: Vancouver, BC, Canada, 2005; pp. 467–474.
41. Pennington, J.; Socher, R.; Manning, C.D. GloVe: Global Vectors for Word Representation. Available online: https://nlp.stanford.edu/projects/glove/ (accessed on 5 January 2021).
42. Inc, F. Word Vectors for 157 Languages. Available online: https://fasttext.cc/docs/en/crawl-vectors.html (accessed on 7 August 2021).
43. Huang, Z.; Xu, W.; Yu, K. Bidirectional LSTM-CRF models for sequence tagging. *arXiv* **2015**, arXiv:1508.01991.
44. Genthial, G. Intro to Tf.Estimator and Tf.Data. Available online: https://guillaumegenthial.github.io/introduction-tensorflow-estimator.html (accessed on 6 August 2021).
45. Benikova, D.; Biemann, C.; Reznicek, M. NoSta-D Named Entity Annotation for German: Guidelines and Dataset. In Proceedings of the Ninth International Conference on Language Resources and Evaluation (LREC'14), Reykjavik, Iceland, 26–31 May 2014; pp. 2524–2531. Available online: http://www.lrec-conf.org/proceedings/lrec2014/pdf/276_Paper.pdf (accessed on 4 March 2021).
46. Benikova, D.; Biemann, C.; Kisselew, M.; Pado, S. GermEval 2014 Named Entity Recognition Shared Task: Companion Paper. 2014. Available online: https://www.inf.uni-hamburg.de/en/inst/ab/lt/publications/2014-benikovaetal-germeval2014.pdf (accessed on 10 March 2021).
47. Kim, J.-D.; Ohta, T.; Tateisi, Y.; Tsujii, J. GENIA corpus—a semantically annotated corpus for bio-textmining. *Bioinformatics* **2003**, *19*, i180–i182. [CrossRef] [PubMed]
48. Project, G. BioNLP/JNLPBA Shared Task. 2004. Available online: http://www.geniaproject.org/shared-tasks/bionlp-jnlpba-shared-task-2004 (accessed on 11 March 2021).
49. Nguyen, T.-S.; Nguyen, L.-M. Nested Named Entity Recognition Using Multilayer Recurrent Neural Networks BT-Computational Linguistics. In Proceedings of the NAACL-HLT 2018, New Orleans, LA, USA, 1–6 June 2018; Hasida, K., Pa, W.P., Eds.; Springer: Singapore, 2018; pp. 233–246.
50. Tjong Kim Sang, E.F.; Buchholz, S. Introduction to the CoNLL-2000 Shared Task: Chunking. In Proceedings of the Fourth Conference on Computational Natural Language Learning and the Second Learning Language in Logic Workshop, Lisbon, Portugal, 13–14 September 2000; pp. 127–132. Available online: https://www.clips.uantwerpen.be/conll2000/pdf/12732tjo.pdf (accessed on 8 March 2021).
51. Katiyar, A.; Cardie, C. Nested Named Entity Recognition Revisited. In Proceedings of the 2018 Conference of the North American Chapter of the Association for Computational Linguistics: Human Language Technologies, New Orleans, LA, USA; 2018; Volume 1 (Long Papers), pp. 861–871. Available online: https://aclanthology.org/N18-1079.pdf (accessed on 16 March 2021).
52. Wang, B.; Lu, W.; Wang, Y.; Jin, H. A Neural Transition-based Model for Nested Mention Recognition. *arXiv* **2018**, arXiv:1810.01808.

53. Shao, Y.; Hardmeier, C.; Nivre, J. Multilingual Named Entity Recognition using Hybrid Neural Networks. In Proceedings of the Sixth Swedish Language Technology Conference (SLTC); 2016. Available online: https://uu.diva-portal.org/smash/get/diva2:1055627/FULLTEXT01.pdf (accessed on 13 March 2021).
54. Pikuliak, M.; Simko, M.; Bielikova, M. Towards Combining Multitask and Multilingual Learning. In Proceedings of the SOFSEM 2019: Theory and Practice of Computer Science, Nový Smokovec, Slovakia, 27–30 January 2019; Catania, B., Královič, R., Nawrocki, J., Pighizzini, G., Eds.; Springer International Publishing: Cham, Switzerland, 2019; pp. 435–446.
55. Marcińczuk, M.; Radom, J. A Single-run Recognition of Nested Named Entities with Transformers. *Procedia Comput. Sci.* **2021**, *192*, 291–297. [CrossRef]
56. Labusch, K.; Neudecker, C.; Zellhöfer, D. BERT for Named Entity Recognition in Contemporary and Historic German. In Proceedings of the KONVENS, Erlangen, Germany, 9–11 October 2019; pp. 8–11.
57. Sohrab, M.G.; Miwa, M. Deep exhaustive model for nested named entity recognition. In Proceedings of the 2018 Conference on Empirical Methods in Natural Language Processing, Brussels, Belgium, 31 October–4 November 2018; pp. 2843–2849.
58. Gridach, M. Character-level neural network for biomedical named entity recognition. *J. Biomed. Inform.* **2017**, *70*, 85–91. [CrossRef] [PubMed]
59. Kocaman, V.; Talby, D. Biomedical Named Entity Recognition at Scale. *Intell. Comput. Theor. Appl.* **2021**, 635–646. [CrossRef]
60. Yuan, Z.; Liu, Y.; Tan, C.; Huang, S.; Huang, F. Improving Biomedical Pretrained Language Models with Knowledge. In Proceedings of the 20th Workshop on Biomedical Language Processing, Online, 11 June 2021.

applied sciences

MDPI

Article

Fine-Grained Named Entity Recognition Using a Multi-Stacked Feature Fusion and Dual-Stacked Output in Korean

Hongjin Kim [1] and Harksoo Kim [2,*]

[1] Artificial Intelligence, Konkuk University, 120 Neungdong-ro, Gwangjin-gu, Seoul 05029, Korea; jin3430@konkuk.ac.kr

[2] Computer Science and Engineering, Konkuk University, 120 Neungdong-ro, Gwangjin-gu, Seoul 05029, Korea

* Correspondence: nlpdrkim@konkuk.ac.kr; Tel.: +82-2-450-3499

Abstract: Named entity recognition (NER) is a natural language processing task to identify spans that mention named entities and to annotate them with predefined named entity classes. Although many NER models based on machine learning have been proposed, their performance in terms of processing fine-grained NER tasks was less than acceptable. This is because the training data of a fine-grained NER task is much more unbalanced than those of a coarse-grained NER task. To overcome the problem presented by unbalanced data, we propose a fine-grained NER model that compensates for the sparseness of fine-grained NEs by using the contextual information of coarse-grained NEs. From another viewpoint, many NER models have used different levels of features, such as part-of-speech tags and gazetteer look-up results, in a nonhierarchical manner. Unfortunately, these models experience the feature interference problem. Our solution to this problem is to adopt a multi-stacked feature fusion scheme, which accepts different levels of features as its input. The proposed model is based on multi-stacked long short-term memories (LSTMs) with a multi-stacked feature fusion layer for acquiring multilevel embeddings and a dual-stacked output layer for predicting fine-grained NEs based on the categorical information of coarse-grained NEs. Our experiments indicate that the proposed model is capable of state-of-the-art performance. The results show that the proposed model can effectively alleviate the unbalanced data problem that frequently occurs in a fine-grained NER task. In addition, the multi-stacked feature fusion layer contributes to the improvement of NER performance, confirming that the proposed model can alleviate the feature interference problem. Based on this experimental result, we conclude that the proposed model is well-designed to effectively perform NER tasks.

Keywords: fine-grained named entity recognition; *k*-stacked feature fusion; dual-stacked output; unbalanced data problem

Citation: Kim, H.; Kim, H. Fine-Grained Named Entity Recognition Using a Multi-Stacked Feature Fusion and Dual-Stacked Output in Korean. *Appl. Sci.* **2021**, *11*, 10795. https://doi.org/10.3390/app112210795

Academic Editor: Arturo Montejo-Ráez

Received: 21 September 2021
Accepted: 11 November 2021
Published: 15 November 2021

1. Introduction

Named entity recognition (NER), a well-known task in natural language processing (NLP), identifies word sequences in texts and classifies them into predefined categories. NER was initially studied as a subtask of information extraction, when coarse-grained NER systems that extract the names of people, locations, and organizations from texts were widely used. Growing interest in NLP tasks, such as relation extraction, answering questions, and knowledge base construction, has increased the demand for fine-grained NER systems. Although early NER systems performed well in coarse-grained NER, they often required well-designed features in the form of language-dependent human knowledge. To address this issue, many NER systems have adopted deep learning methods to yield state-of-the-art (SOTA) performance. Although these models based on deep learning delivered good performance, it was restricted to tasks involving coarse-grained classification. In fine-grained NER for English language tasks, certain systems based on deep learning performed satisfactorily and were between 80% and 85% accurate. However, in languages with a large number of characters that do not use capitalization or word boundary by spacing (e.g., a spacing unit is not a word in Korean), the performance of these systems is unsatisfactory, that is, between 65% and 75% [1]. The main reason for the lower performance is that the training data for fine-grained NER are more unbalanced than those for coarse-grained NER. It is easy to find fine-grained NEs that seldom occur in training data. To alleviate this sparse data problem, we propose an NER model that compensates for the sparseness of fine-grained NEs with the contextual information of coarse-grained NEs that semantically include the fine-grained NEs. Table 1 presents examples of coarse-grained NE categories and their fine-grained NE categories.

Table 1. Example of two-level NE categories.

Coarse-Grained NE		Fine-Grained NE	
Class	**Example**	**Class**	**Example**
Location	USA, Washington, D.C., Memorial park	Country Province City	Korea Gangwon-do Seoul
Date	Thanksgiving day, 24 April 2020	Year Duration	2020 2019–2021

These examples show that the classes of coarse-grained NEs are supersets that are tightly associated with the classes of fine-grained NEs. If the fine-grained NE "Seoul" in Table 1 does not occur in the training data, the contextual information of the coarse-grained NE "Washington, D.C." could be helpful in that it would enable the NE class "City" of "Seoul" to be inferred because they are both capital cities.

Many NER systems actively use various linguistic and domain-specific features to improve their performance. For example, part-of-speech (POS) tags play an important role in detecting NE boundaries, and domain-specific gazetteers play a decisive role in determining NE categories. Previous NER models based on deep neural networks embedded various types of linguistic and domain-specific knowledge into vector spaces. Then, they used the embedded vectors as nonhierarchical features of input layers, although the embedded vectors imply different levels of knowledge (e.g., POS tags imply a grammatical level of linguistic knowledge, and entities in a gazetteer imply a semantic level of domain knowledge). In addition, some previous works have shown that different layers of deep RNNs encode different types of information [2]. In other words, the embedded vectors are simply concatenated to the word embeddings being used as input, and the concatenated vectors are simply input into the NER models. Therefore, in the case of NER models with deep architecture, such as a multi-stacked recurrent neural network (RNN), different levels of features are mixed and interfere with each other. To alleviate the feature interference problem, we propose a NER model in which a multi-stacked RNN layer hierarchically uses different levels of features.

The remainder of this paper is organized as follows. In Section 2, we review the previous NER models. In Section 3, we describe our model to alleviate the sparse data problem in fine-grained NER. In Section 4, we explain our experimental setup and report some of our experimental results. In Section 5, we provide the conclusion of our study.

2. Previous Studies

NER tasks were previously resolved by considering them as sequence labeling problems. In this regard, most previous NER systems adopted machine learning (ML) models, such as decision trees [3], maximum entropy [4], and conditional random fields [5]. To improve the NER performance, these ML-based systems focused on feature engineering methods, such as word n-grams, part-of-speech n-grams, lexical clues, and knowledge look-up (to determine whether an input word exists in an external knowledge base) [6]. With the recent success of deep learning (DL), many DL-based NER systems have been proposed to reduce the labor required for feature engineering [7,8]. These DL-based systems performed reasonably by using various distributed representations (e.g., word embeddings and character embeddings) instead of expensive knowledge features. Although many researchers have studied NER, it is not easy to find studies on fine-grained NER with hundreds of NE classes, especially for non-English languages. The authors in [9] presented a fine-grained entity recognizer that solved a multi-label, multi-class classification problem by adapting a Perceptron model. The Ref. [1] conducted an empirical study to develop a fine-grained NER model that is robust across various settings, such as the number of NE classes and the size of the training dataset, by using a bidirectional long short-term memory model with a conditional random field layer (BI-LSTM-CRF) [10]. They reported that a fine-grained NER model that is effective for English is not necessarily effective for Japanese. This showed that a fine-grained NER task has language-dependent characteristics. To overcome a lack of training data, [11] proposed a method using a language model and an expensive knowledge base in a fine-grained NER task. The Ref. [12] proposed a novel adversarial multitask learning framework in which POS tagging is performed together with NER in Chinese. The Ref. [13] proposed a sequence-to-sequence model to consider the entire meaning of an input sentence. They used BI-LSTM as the encoder to equally process the past and future information of an input sentence. Then, they added a self-attention mechanism to address the long-term dependency problem in a long sequence. The Ref. [14] showed that the embeddings of decomposed NE labels can be effectively used to improve the performance of instances with low-frequency NE labels.The Ref. [15] proposed a model that included the initial encoding layer, the enhanced encoding layer, and the decoding layer, combining the advantages of pre-training model encoding, dual bidirectional long short-term memory (BiLSTM) networks, and a residual connection mechanism. The Ref. [16] proposed a Chinese fine-grained NER method based on a language model and model transfer considering active learning (MTAL) to research a few labeled data. The Ref. [17] proposed a Cognitive Impairment model that can filter, study, analyze, and interpret written communications from social media platforms. The Ref. [18] proposed a label attention network (LAN) that captured possible long-term label dependency by utilizing an attention mechanism and label embedding. We adopted this LAN and proposed a dual-stacked LAN, with the lower for coarse-grained NER and the upper for fine-grained NER.

3. Fine-Grained NER Model

To detect word boundaries (i.e., morpheme boundaries) in Korean, many NER models perform morphological analysis in advance. Then, they generally use morphemes and POS tags of an input sentence as inputs. Under this kind of pipeline architecture, errors of morphological analysis directly lead to diminished performance in NER models. To overcome this limitation, we use character n-grams as inputs. Given n characters, $C_{1,n}$, in a sentence S, let $E^{c}_{1,n}$ and $E^{f}_{1,n}$ denote sequences of coarse-grained NE tags and fine-grained NE tags in S, respectively. Table 2 presents NE tags that are defined according to the well-known begin-inner-outer (BIO) character-level tagging scheme.

Table 2. Character-unit NE tags.

NE Tag	Description
B-(PER | LOC | ORG | ...)	Beginning character of an NE with the category following "B-"
I-(PER | LOC | ORG | ...)	Inner character of an NE with the category following "I-"
O	Character out of any NE boundary

The fine-grained NER model named FG-NER can then be formally expressed in the following equation.

$$FG-NER(S) \stackrel{\text{def}}{=} argmaxP(E^{c}_{1,n}, E^{f}_{1,n}|C_{1,n}) \tag{1}$$

According to the chain rule, (1) can be rewritten as the following equation.

$$FG-NER(S) \stackrel{\text{def}}{=} argmaxP(E^c_{1,n}|C_{1,n})P(E^f_{1,n}|C_{1,n}) \quad (2)$$

As shown in (2), coarse-grained NEs depend on input characters, and fine-grained NEs depend on input characters and the given coarse-grained NEs. To simplify (2), we adopt the following two assumptions: a first-order Markov assumption that a current tag is dependent on the previous tag, and a conditional independent assumption that a current tag is dependent only on its current observational information. Based on these assumptions, we rewrite (2) as the following equation. Note that the reason why we used two assumptions is to simplify Equation (2).

$$FG-NER(S) \stackrel{\text{def}}{=} argmax \prod_{i=1}^{n} \left\{ \frac{P(E^c_i|C_i)P(E^c_i|E^c_{i-1})}{P(E^f_i|C_i, E^c_i)P(E^f_i|E^f_{i-1})} \right\} \quad (3)$$

To obtain the sequence labels, $E^f_{1,n}$, that maximize (3) by using coarse-grained NEs as additional contextual information, we adopt a stacked BI-LSTM-LAN [17]. Figure 1 shows the architecture of the proposed fine-grained NER (FG-NER) model. This model comprises a k-stacked feature fusion layer (shown on the left) and a dual-stacked output layer (shown on the right). The feature fusion layer shown in Figure 1 accepts different levels of input embeddings that are fed into each layer of a three-stack BI-LSTM to yield a sequence of forward hidden and backward hidden states, respectively. Subsequently, these two states of each layer are concatenated to reflect bidirectional contextual information, as shown in the following equation.

$$\begin{aligned}
\overrightarrow{h}^k_i &= LSTM(Emb^k_i, \overrightarrow{h}^k_{i-1}) \\
\overleftarrow{h}^k_i &= LSTM(Emb^k_i, \overleftarrow{h}^k_{i-1}) \\
\overleftrightarrow{h}^k_i &= [\overrightarrow{h}^k_i, \overleftarrow{h}^k_i] \\
\overleftrightarrow{H}^k &= \{\overleftrightarrow{h}^k_1, \overleftrightarrow{h}^k_2, \overleftrightarrow{h}^k_3, \ldots, \overleftrightarrow{h}^k_n\}
\end{aligned} \quad (4)$$

where Emb^k_i is the i-th input embedding in the k-th stacked LSTM. Then, $\overleftrightarrow{h}^k_i = [\overrightarrow{h}^k_i; \overleftarrow{h}^k_i]$ is the concatenation of the forward hidden state $\overrightarrow{h}^k_i$ and the backward hidden state $\overleftarrow{h}^k_i$ of the i-th input in the k-th stacked LSTM. In the first stacked feature fusion layer (i.e., the lowest layer) of Figure 1, C_0, C_i, and C_{n+1} are a special beginning symbol of a sentence, the i-th one of n input characters, and a special ending symbol of a sentence, respectively. Then, $C_{0,n+1}$ is a randomly-initialized character embedding of each character. A concatenation of three successive character embeddings (i.e., a character tri-gram embedding; $[C_{i-1}; C_i; C_{i+1}]$ is used as an input embedding Emb^1_i for the first stacked feature fusion layer. In the second stacked feature fusion (i.e., the middle layer) of Figure 1, POS_{C_i} is a character-unit POS tag of the i-th input character according to a BIO-tagging scheme similar to Table 1. Then, $Emb(POS_{C_i})$ is a randomly-initialized POS embedding of the i-th character. To enrich input characters with grammatical information, a POS tri-gram embedding, $[Emb(POS_{C_{i-1}}); Emb(POS_{C_i}); Emb(POS_{C_{i+1}})]$, is concatenated with the tri-gram character embedding, Emb^1_i, by using residual connections. The concatenated embedding is used as the input embedding Emb^2_i for the second stacked feature fusion layer. In the last stacked feature fusion layer (i.e., the uppermost layer) of Figure 1, $DIC_{C_{i-1};C_i;C_{i+1}}$ is a dictionary look-up feature on whether a character tri-gram, $[C_{i-1}; C_i; C_{i+1}]$, exists in a dictionary including character trigrams of predefined NE lists (i.e., NE lists in a training data). Then, $Emb(DIC_{C_{i-1};C_i;C_{i+1}})$ is a randomly-initialized dictionary embedding of the i-th character. To enrich input characters with domain knowledge, the dictionary embedding is concatenated with the character trigram embedding, Emb^1_i, by using residual connections. The concatenated embedding is used as the input embedding Emb^3_i for the last stacked feature fusion layer. The hidden states of each stacked layer are concatenated as $\overleftrightarrow{H} = [\overleftrightarrow{H}^1; \overleftrightarrow{H}^2; \overleftrightarrow{H}^3]$. The lower output layer shown in Figure 1, calculates the degrees of association between H and the coarse-grained NE tag embeddings $Emb(NE^c) = \{Emb(NE^c_1), Emb(NE^c_2), \ldots, Emb(NE^c_m)\}$ based on a multi-head attention mechanism [17], as shown in the following equation.

$$\begin{aligned} head_j &= attention(QW_j^Q, KW_j^K, VW_j^K) = \alpha_j * VW_j^V, \\ Where Q &= \overleftrightarrow{H}, K = V = Emb(NE^c), \\ \alpha_j &= softmax(\frac{QW_j^Q * (KW_j^k)^T}{\sqrt{d_h}}), \\ A(c_i^c) &= head_1 \oplus head_2 \oplus \ldots \oplus head_k \end{aligned} , \quad (5)$$

where $W_j^Q \in R^{d_h \times \frac{d_h}{k}}$, $W_j^K \in R^{d_h \times \frac{d_h}{k}}$, and $W_j^V \in R^{d_h \times \frac{d_h}{k}}$ are the weighting parameters of the j-th parameter among k heads to be learned during training. Then, $Emb(NE^c)$ represent the embedding vectors of m coarse-grained NE tags that are randomly initialized and fine-tuned during training. The attention score α_j is calculated using a scaled-dot product, where d_h is a dimension of H (same as the dimension of Coarse-grained NE embedding). The attention score vector $A(C_i^c)$ represents the degrees of associations between the contextualized input embedding $\overleftrightarrow{h}_i$ of the i-th input character and each coarse-grained NE tag. In other words, the vector can be considered as a potential distribution of coarse-grained tags associated with an input character. In the prediction phase, the lower output layer returns coarse-grained NE tags, as shown in the following equation.

$$\hat{E}_i^c = argmax(\hat{A}_i^1, \hat{A}_i^2, ..., \hat{A}_i^m) \quad (6)$$

where $\hat{A}_i^j$ denotes the j-th one among m attention scores in the trained attention vector $\hat{A}_i$. In the upper output layer in Figure 1, each coarse-grained attention score vector $A(C_i^c)$ is concatenated with the hidden states of each stacked layer H to enrich fine-grained NEs with the contextual information of coarse-grained NEs. Except that the concatenated vector is used as a query vector of the multi-head attention mechanism, the upper output layer follows the same procedure as the lower output layer, as shown in the following equation.

$$\begin{aligned} head_j &= attention(QW_j^Q, KW_j^K, VW_j^V) = \alpha_j * VW_j^V, \\ Where Q &= [LSTM(A(C^c); \overleftrightarrow{H})], K = V = Emb(NE^f), \\ \alpha_j &= softmax(\frac{QW_j^Q * (KW_j^k)^T}{\sqrt{d_h}}), \\ A(c_i^f) &= head_1 \oplus head_2 \oplus \ldots \oplus head_k \end{aligned} , \quad (7)$$

where $A(C^c)$, $Emb(NE^f)$, and $A(C_i^f)$ are the coarse-grained attention score vectors of n input characters, fine-grained NE tag embeddings, and a fine-grained attention score vector of the i-th input character, respectively. In the prediction phase, the upper output layer follows the same process as the lower output layer, as shown in the following equation.

$$\hat{E}_i^f = argmax(\hat{A}_i^1, \hat{A}_i^2, ..., \hat{A}_i^l), \quad (8)$$

where $\hat{A}_i^j$ denotes the j-th attention score of l fine-grained NE categories.

In general, coarse-grained training data are less unbalanced than fine-grained training data because coarse-grained NEs constitute a superset of fine-grained NEs. This led us to use a two-phase training scheme to optimize the weighting parameters of the proposed model. We first train the lower output layer to minimize the cross-entropy between the correct coarse-grained NE tags, E_i^c, and the outputs of the lower output layer, $\hat{E}_i^c$, as shown in the following equation.

$$H_{\hat{E}^c}(E^c) = -\sum_i \hat{E}_i^c log(E_i^c). \tag{9}$$

In this phase, the weighting parameters in the lower output layer are considered to be pre-trained because they were trained by using less unbalanced training data. Then, we train the upper output layer to minimize the cross-entropy between the correct fine-grained NE tags, E_i^f, and the outputs of the upper output layer, $\hat{E}_i^f$, as shown in the following equation.

$$H_{\hat{E}^f}(E^f) = -\sum_i \hat{E}_i^f log(E_i^f). \tag{10}$$

In this second phase, we expect the weighting parameters in the lower output layer to be fine-tuned to specific values associated with the fine-grained NE tags.

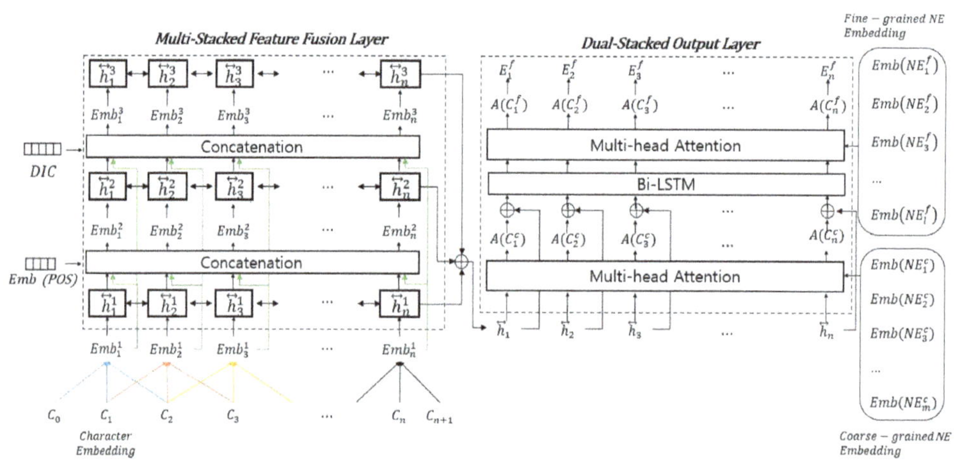

Figure 1. Overall architecture of FG-NER.

4. Evaluation

4.1. Datasets and and Experimental Settings

In our experiments, we used a gold-labeled corpus annotated with 14 coarse-grained NE tags and 147 fine-grained NE tags. This corpus was constructed by ETRI (Electronics and Telecommunications Research Institute, https://www.etri.re.kr/eng/main/main.etri, accessed on 15 September 2021). This corpus has been tagged with the coarse-grained named entity and fine-grained named entity in the sentences in the encyclopedia. In addition, it is the only training data for fine-grained NER in Korean. Table 3 presents the distribution of NE tags found in the gold-labeled corpus.

We converted the gold-labeled corpus into an NE dataset in which each character was annotated with the NE tags in Table 2. Then, we divided the NE dataset into training, validation, and test datasets, respectively, to obtain a ratio of 8:1:1. Finally, we evaluated the proposed model using the following evaluation measures: precision, recall rate, and F1-score.

$$Precision = \frac{\#\ of\ correct\ NE's}{\#\ of\ NE's\ returned\ by\ a\ system}. \tag{11}$$

$$Recall = \frac{\#\ of\ correct\ NE's\ returned\ by\ asystem}{\#\ of\ correct\ NE's\ in\ a\ test\ data}. \tag{12}$$

$$F1 - score = \frac{2 \times Precision \times Recall}{Precision + Recall}. \tag{13}$$

To calculate the precision, recall rate, and F1-score, the proposed model automatically generates NE sequences by concatenating the characters with B tags and successive I tags.

Table 3. Distribution of NE tags which mainly occurred in the corpus.

Coarse-Grained NE Tag	Description	Percent
QT	Quantity	15.9%
DT	Date	14.0%
OG	Organization	12.6%
TR	Theroy	8.7%
CV	Civilization	8.2%
Fine-Grained NE Tag	**Description**	**Percent**
TR-Technology	The technology of Theory	7.6%
DT-Year	The Year of Date	5.3%
DT-Month	The month of Date	4.8%
PS-Name	The name of Person	4.7%
OG-Business	The business of Organization	3.9%

4.2. Implementation

We implemented the proposed model using the Pytorch [19]. Training and prediction occurred on a per-sentence level. Table 4 lists the parameter settings we used to train the model.

Table 4. Model parameters.

Parameter	Value
The dimension of character embedding	50
The dimension of POS embedding	16
The dimension of hidden node in the feature fusion layer	128
The dimension of hidden node in the output layer	256
The dimension of Coarse-grained NE embedding	768
The dimension of Fine-grained NE embedding	512
Batch size	64
Learning rate	0.001
Epoch	100

4.3. Experimental Results

First, we evaluated the effectiveness of the k-stacked feature fusion layer and the dual-stacked output layer; the results are summarized in Table 5.

Table 5. Performance comparison depending on changes in the architecture.

Model	Precision	Recall	F1-Score
1-In+1-Out	0.783	0.681	0.728
3-In+1-Out	0.831	0.730	0.777
3-In(H)+1-Out	0.844	0.752	0.795
1-In+2-Out	0.808	0.701	0.750
3-In+2-Out	0.849	0.760	0.801
3-In(H)+2-Out	0.865	0.769	0.814

In Table 5, "k-In", "k-In(H)", "1-Out", and "2-Out" denote a k-stacked feature fusion layer in which a flat concatenation of all input embeddings (i.e., a character trigram embedding, a POS trigram embedding, and a dictionary embedding) is fed into the first layer, the proposed feature fusion layer in which different levels of input embeddings are hierarchically fed into the k-stacked LSTMs, a single output layer (i.e., only the upper output layer), and a dual-stacked output layer, respectively. The results in Table 5 show that the models with a dual-stacked output layer always outperformed the models with a single output layer. This reveals that the proposed dual-stacked output layer contributes to alleviate the problem of unbalanced training data. In addition, "3-In(H)+2-Out" delivered the best performance. This reveals that the hierarchical feature embeddings in a stacked feature fusion layer are able to effectively hand over different levels of linguistic features to an output layer.

The second experiment was conducted to compare the performance of the proposed model with those of the previous fine-grained NER models; the results are summarized in Table 6. In this table, "KoELECTRA-NER" is an NER model in which the character-based ELECTRA model [20] in Korean is fine-tuned to a sequence-labeling task. ELECTRA is a pre-trained language model with SOTA performance in many downstream NLP tasks, such as span prediction, sequence labeling, and text classification. We carried out a Korean NER task by pre-training KoELECTRA by using 96M sentences with 2.6B tokens. Then, we fine-tuned KoELECTRA by using the training NE dataset. "BI-LSTM-LAN" [21] is an NER model that simultaneously performs morphological analysis and coarse-grained NER in Korean. This model achieved SOTA performance by outperforming Korean NER models that did not use large pre-trained language models, such as BERT [22], ALBERT [23], and ELECTRA.

Table 6. Performance comparison with the previous models.

Task	Model	Precision	Recall	F1-Score
Fine-grained NER	KoELECTRA-NER	0.855	0.757	0.802
	3-In(H)+2-Out	0.865	0.769	0.814
Coarse-grained NER	Bi-LSTM-LAN	0.855	0.813	0.833
	KoELECTRA-NER	0.879	0.838	0.857
	3-In(H)-1-Out	0.861	0.831	0.845

The results presented in Table 6 show that "3-In(H)+2-Out" outperformed "KoELECTRA-NER" in the fine-grained NER task. We attribute the improved performance to the well-formed neural network architecture with the stacked feature fusion layer and its ability to effectively reflect contextual information and features. On the coarse-grained NER task, the performance of "3-In(H)+2-Out" was slightly less accurate than that of "KoELECTRA-NER". However, "3-In(H)+2-Out" was 25 times lighter than "KoELECTRA-NER".

4.4. Discussion

In Table 5, "3-In+1-Out" and "3-In+2-Out" were significantly outperformed by "1-In+1-Out" and "1-In+2-Out", respectively. This suggests that the tri-gram character vector that concatenates three uni-gram character vectors has much more enriched information than the uni-gram character vector. In addition, "3-In(H)+1-Out" and "3-In(H)+2-Out" were outperformed by "3-In+1-Out" and "3-In+2-Out", respectively. This suggests that our hierarchical feature embeddings in a stacked feature fusion layer are more practical to capture POS and dictionary look-up features. Finally, "3-In(H)+2-Out" was outperformed by "3-In(H)+1-Out". This suggests that using coarse-grained NE information contributes to the recognization of fine-grained NE.

5. Conclusions

We proposed a fine-grained NER model that compensates for the sparseness of fine-grained NEs by using the contextual information of coarse-grained NEs. In addition, the model uses a hierarchical approach to alleviate the interference of features at different levels with each other. The proposed model consists of a multi-stacked feature fusion layer and a dual-stacked output layer. The feature fusion layer generates multiple levels of sentence representations and word representations by using multi-stacked BI-LSTMs. Based on the multilevel representations, the output layer returns fine-grained NE tags by using dual-stacked BI-LSTMs in which the lower layer is trained for coarse-grained NER. In the experiments, the proposed model delivered SOTA performance. Based on the experimental results, we concluded that the proposed model can effectively alleviate the problem caused by unbalanced data in fine-grained NER tasks. In addition, we concluded that the feature fusion architecture of the proposed model can contribute to the alleviation of the feature interference problem.

Author Contributions: Conceptualization, H.K. (Harksoo Kim); methodology, H.K. (Harksoo Kim); software, H.K. (Hongjin Kim); validation, H.K. (Hongjin Kim); formal analysis, H.K. (Harksoo Kim); investigation, H.K. (Harksoo Kim); resources, H.K. (Hongjin Kim); data curation, H.K. (Hongjin Kim); writing—original draft preparation, H.K. (Hongjin Kim); writing—review and editing, H.K. (Harksoo Kim); visualization, H.K. (Harksoo Kim); supervision, H.K. (Harksoo Kim); project administration, H.K. (Harksoo Kim); funding acquisition, H.K. (Harksoo Kim). All authors have read and agreed to the published version of the manuscript.

Funding: This work was supported by Institute of Information & communications Technology Planning & Evaluation (IITP) grant funded by the Korea government (MSIT) (No.2020-0-00368, A Neural-Symbolic Model for Knowledge Acquisition and Inference Techniques). Also, This work was supported by the National Research Foundation of Korea(NRF) grant funded by the Korea government(MSIT) (No. 2020R1F1A1069737).

Institutional Review Board Statement: Not applicable.

Informed Consent Statement: Not applicable.

Acknowledgments: We thank the menbers of the NLP laboratory at Konkuk University for their technical support.

Conflicts of Interest: The authors declare no conflict of interest.

References

1. Mai, K.; Pham, T.H.; Nguyen, M.T.; Nguyen, T.D.; Bollegala, D.; Sasano, R.; Sekine, S. An empirical study on fine-grained named entity recognition. In Proceedings of the 27th International Conference on Computational Linguistics, Santa Fe, NM, USA, 20–26 August 2018; pp. 711–722.
2. Peters, E.M.; Neumann, M.; Iyyer, M.; Gradner, M. Deep contexualized word representation. In Proceedings of the NAACL-HLT 2018, New Orleans, LA, USA, 1–6 June 2018; pp. 2227–2237.
3. Sekine, S.; Grishman, R.; Shinnou, H. A decision tree method for finding and classifying names in Japanese texts. In Proceedings of the 6th Workshop on Vary Large Corpora, Montreal, QC, Canada, 15–16 August 1998; pp. 171–178.
4. Borthwick, A.; Sterling, J.; Agichtein, E.; Grishman, R. NYU: Description of the MENE named entity system as used in MUC-7. In Proceedings of the Seventh Message Understanding Conference, Fairfaxm VA, USA, 29 April–1 May 1998.
5. Cohen, W.W.; Sarawagi, S. Exploiting dictionaries in named entity extraction: combining semi-markov extraction processes and data integration methods. In Proceedings of the KDD 2004, Seattle, WA, USA, 22–25 August 2004; pp. 89–98.
6. Nadeau, D.; Sekine, S. A survey of named entity recognition and classification. *J. Linguist. Investig.* **2007**, *30*, 3–26. [CrossRef]
7. Lample, G.; Ballesteros, M.; Subramanian, S.; Kawakami, K.; Dyer, C. Neural architectures for named entity recognition. In Proceedings of the NAACL-HLT 2016, San Diego, CA, USA, 12–17 June 2016; pp. 260–270.
8. Chiu, J.P.; Nichols, E. Named entity recognition with bidirectional LSTM-CNNs. *Trans. Assoc. Comput. Linguist.* **2016**, *40*, 357–370. [CrossRef]
9. Ling, X.; Weld, D.S. Fined-Grained Entity Recognition. In Proceedings of the 26th AAAI Conference on Artificial Intelligence, Toronto, ON, Canada, 22–26 July 2012; pp. 94–100.
10. Ma, X.; Hovy, E. End-to-end sequence labeling via bi-directional lstmcnns-crf. In Proceedings of the Association Computational Linguistics, Berlin, Germany, 18–24 May 2016; pp. 1064–1074.

11. Dogan, C.; Dutra, A.; Gara, A.; Gemma, A.; Shi, L.; Sigamani, M.; Walters, E. Fine-grained named entity recognition using elmo and wikidata. *arXiv* **2019**, arXiv:1904.10503.
12. Man, X.;Yang, P. Fine-grained Chinese Named Entity Recognition in Entertainment News Using Adversarial Multi-task Learning. In Proceedings of the 2019 IEEE 5th International Conference on Computer and Communications (ICCC), Chengdu, China, 6–9 December 2019; pp. 1671–1675.
13. Zhu, H.; He, C.; Fang, Y.; Xiao, W. Fine Grained Named Entity Recognition via Seq2seq Framework. *IEEE Access* **2020**, *8*, 53953–53961. [CrossRef]
14. Kato, T.; Abe, K.; Ouchi, H.; Miyawaki, S.; Suzuki, J.; Inui, K. Embeddings of Label Components for Sequence Labeling: A Case Study of Fine-grained Named Entity Recognition. In Proceedings of the Association Computational Linguistics, Virtual, 5–10 July 2020; pp. 222–229.
15. Liu, J.; Xia, C.; Yan, H.; Xu, W. Innovative Deep Neural Network Modeling for Fine-Grained Chinese Entity Recognition. *Electronics* **2020**, *9*, 1001. [CrossRef]
16. Yao, L.; Huang, H.; Wang, K.-W.; Chen, S.-H.; Xiong, Q. Fine-Grained Mechanical Chinese Named Entity Recognition Based on ALBERT-AttBiLSTM-CRF and Transfer Learning. *Symmetry* **2020**, *12*, 1986. [CrossRef]
17. Thakur, N.; Han, C.Y. A Multimodal Approach for Early Detection of Cognitive Impairment from Tweets. In *Human Interaction, Emerging Technologies and Future Systems V. IHIET 2021. Lecture Notes in Networks and Systems*; Ahram, T., Taiar, R., Eds.; Springer: Cham, Switzerland, 2021; Volume 319._2. [CrossRef]
18. Cui, L.; Zhang, Y. Hierarchically-refined label attention network for sequence labeling. In Proceedings of the EMNLP, Hong Kong, China, 3–7 November 2019; pp. 4113–4126.
19. Paszke, A.; Gross, S.; Massa, F.; Lerer, A.; Bradbury, J.; Chanan, G.; Killeen, T.; Lin, Z.; Gimelshein, N.; Antiga, L.; et al. Pytorch: An imperative style, high-performance deep learning library. *Adv. Neural Inf. Process. Syst.* **2019**, *32*, 8026–8037.
20. Clark, K.; Luong, M.-T.; Le, Q.V.; Manning, C.D. ELECTRA: Pre-training text encoders as discriminators rather than generators. In Proceedings of the ICLR, Virtual, 26 April–1 May 2020; pp. 1–14.
21. Kim, H.; Kim, H. Integrated Model for Morphological Analysis and Named Entity Recognition Based on Label Attention Networks in Korean. *Appl. Sci.* **2020**, *10*, 3740. [CrossRef]
22. Devlin, J.; Chang, M.-W.; Lee, K.; Toutanova, K. BERT: Pre-training of deep bidirectional transformers for language understanding. In Proceedings of the NAACL-HLT 2019, Minneapolis, MN, USA, 2–7 June 2019; pp. 4171–4186.
23. Lan, Z.; Chen, M.; Goodman, S.; Gimpel, K.; Sharma, P.; Soricut, R. ALBERT: A lite BERT for self-supervised learning of language representations. In Proceedings of the ICLR, Virtual, 26 April–1 May 2020.

applied sciences

MDPI

Review

A Survey on Recent Named Entity Recognition and Relationship Extraction Techniques on Clinical Texts

Priyankar Bose [1,*], Sriram Srinivasan [2,3], William C. Sleeman IV [1,2,3], Jatinder Palta [2,3], Rishabh Kapoor [2,3] and Preetam Ghosh [1,2]

1 Department of Computer Science, Virginia Commonwealth University, Richmond, VA 23284, USA; william.sleemaniv@vcuhealth.org (W.C.S.IV); pghosh@vcu.edu (P.G.)
2 Department of Radiation Oncology, Virginia Commonwealth University, Richmond, VA 23284, USA; sriram.srinivasan@vcuhealth.org (S.S.); jatinder.palta@vcuhealth.org (J.P.); rishabh.kapoor@vcuhealth.org (R.K.)
3 National Radiation Oncology Program, Department of Veteran Affairs, Richmond, VA 23249, USA
* Correspondence: bosep@vcu.edu

Abstract: Significant growth in Electronic Health Records (EHR) over the last decade has provided an abundance of clinical text that is mostly unstructured and untapped. This huge amount of clinical text data has motivated the development of new information extraction and text mining techniques. Named Entity Recognition (NER) and Relationship Extraction (RE) are key components of information extraction tasks in the clinical domain. In this paper, we highlight the present status of clinical NER and RE techniques in detail by discussing the existing proposed NLP models for the two tasks and their performances and discuss the current challenges. Our comprehensive survey on clinical NER and RE encompass current challenges, state-of-the-art practices, and future directions in information extraction from clinical text. This is the first attempt to discuss both of these interrelated topics together in the clinical context. We identified many research articles published based on different approaches and looked at applications of these tasks. We also discuss the evaluation metrics that are used in the literature to measure the effectiveness of the two these NLP methods and future research directions.

Keywords: electronic health records; clinical text; natural language processing; named entity recognition; relationship extraction; machine learning

Citation: Bose, P.; Srinivasan, S.; Sleeman, W.C., IV; Palta, J.; Kapoor, R.; Ghosh, P. A Survey on Recent Named Entity Recognition and Relationship Extraction Techniques on Clinical Texts. *Appl. Sci.* **2021**, *11*, 8319. https://doi.org/10.3390/app11188319

Academic Editors: Arturo Montejo-Ráez and Salud María Jiménez-Zafra

Received: 7 August 2021
Accepted: 2 September 2021
Published: 8 September 2021

1. Introduction

The amount of text generated every day is increasing drastically in different domains such as health care, news articles, scientific literature, and social media. Since 2010, the International Data Corporation (IDC) has predicted that the amount of data can potentially grow 50-fold to 40 billion terabytes by 2020 [1]. Textual data is very common in most domains, but automated comprehension is difficult due to its unstructured nature and has led to the design of several text mining (TM) techniques in the last decade.

TM refers to the extraction of interesting and nontrivial patterns or knowledge from text [2]. Common text mining tasks include text preprocessing, text classification, question-answering, clustering, and statistical techniques.

TM has become extremely popular and useful in the biomedical and healthcare domains. In healthcare, about 80% of the total medical data is unstructured and untapped after its creation [3]. This unstructured data from hospitals, healthcare clinics, or biomedical labs can come in many forms such as text, images, and signals. Out of the various text mining tasks and techniques, our goal in this paper is to review the current state-of-the-art in Clinical Named Entity Recognition (NER) and Relationship Extraction (RE)-based techniques. Clinical NER is a natural language processing (NLP) method used for extracting important medical concepts and events i.e., clinical NEs from the data [4]. Relationship

Extraction (RE) is used for detecting and classifying the annotated semantic relationships between the recognized entities. Significant research on NER and RE has been carried out in the past both on clinical narratives and other types of text. For example, in the sentence, "**Her white count** remained **elevated** despite discontinuing **her G-CSF**", the words in bold are the various entities in the sentence. After the entities are recognized, the relationship between two or more entities is extracted. In this case, "**her white count**" and "**elevated**" are found to be related to each other in a manner dissimilar to the nature of the relationship between "**elevated**" and "**her G-CSF**". In the sentence "**Atorvastatin** is found to have therapeutic effects in **breast cancer** although no clinical trials are performed at present", the NE of interest includes the name of the drug (atorvastatin) and the disease name (breast cancer), whereas the drug–disease relation (atorvastatin–breast cancer) is the relationship of interest. Figure 1 shows a pictorial representation of the association between NER and RE.

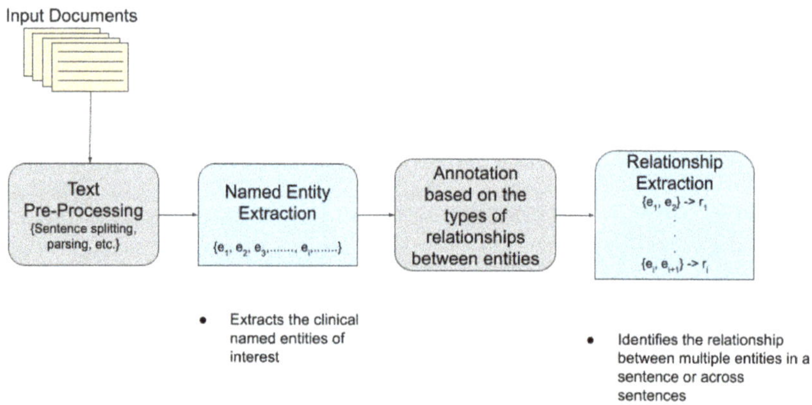

Figure 1. Association between Named Entity Recognition and Relationship Extraction.

2. Background

Over the years, many toolkits and applications have been introduced to address different NLP tasks in the clinical domain, including NER and RE. The WEKA Data Mining Software [5] first came into existence in the late nineties. It was updated several times over the years to include NLP systems for language identification, tokenization, sentence boundary detection, and named entity recognition. Later on, the clinical NLP toolkit, CLAMP (Clinical Language Annotation, Modeling, and Processing) [6] was introduced in 2018 and provides a GUI-based state-of-the-art NLP system. CLAMP achieved good performance on NER and concept encoding and is also publicly available for research use. Comprehend Medical, a NER- and RE-related Web Service (2019) [7], is a very recent effort that introduces an NLP service launched under Amazon Web Services (AWS). Likewise, other research works have also addressed these topics, which motivates this review. A high-level overview of machine learning, neural networks, and evaluation metrics is presented below before we review clinical NER- and RE-related tasks.

2.1. Machine Learning

Machine learning (ML) is a type of data-driven Artificial Intelligence (AI) that provides the ability to learn about a system without explicit programming. ML algorithms are applied in many scientific domains and the most common applications include recommendation systems, data mining, and pattern recognition. ML is classified into one of the four subdomains:

- Supervised Learning: With these algorithms, the training data are given ground-truth labels, which can be used for learning the underlying patterns in the dataset.

Classification and regression algorithms are most commonly used, including Naive Bayes [8], Support Vector Machines (SVM) [9], and Decision Trees [10].

- Unsupervised Learning: In this case, the training dataset is not given labels and, thus, many of the solutions attempt to find patterns without any prior guidance. Commonly used algorithms in this category are association rules and clustering methods, such as K-Means [11] or DBSCAN [12].
- Semi-Supervised Learning: Here, only some of the training data is labeled, putting these solutions in a space somewhere between fully supervised and unsupervised learning. Text classification [13] is one of the most common applications for semi-supervised learning.
- Reinforcement Learning: Using a reward system, a reinforcement learning agent optimizes future returns based on prior results. This iterative, continuous learning process mirrors how humans learn from their experiences when interacting with an environment. Deep Adversarial Networks [14] and Q-Learning [15] are well known reinforcement learning algorithms.

2.2. Neural Networks

The traditional machine learning algorithms often perform well with structured data but can struggle with unstructured or semi-structured data, i.e., human information processing mechanisms such as vision and speech [16]. Neural networks, specifically deep learning algorithms, have shown promising results with NLP and image analysis tasks. In neural networks, the input is processed through different layers of the network, where each layer transforms the features of the dataset following some mathematical function. The concept of neural networks follows the mechanism that the human brain uses to solve a problem. Once the data is processed through different layers within a neural network, the output layer performs the classification. In general, this approach does not require as much human intervention as the nested layers using different hierarchies try to find the hidden patterns on their own.

2.3. Common Evaluation Metrics

The F1-score is a popular evaluation metric for the two NLP functions reviewed in this paper. Comparisons can be classified as exact or relaxed match [17]. Relaxed match only considers the correct type and ignores the boundaries as long as there is an overlap with ground truth boundaries. In the case of an exact match, it is expected that the entity identified correctly should also detect boundary and type correctly at the same time [17]. The following keys are used to calculate the F-score, precision, and recall.

- True Positive (TP): A perfect match between the entity obtained by NER system and the ground truth.
- False Positive (FP): Entity detected by the NER system but not present in the ground truth.
- False Negative (FN): Entity not detected by the NER system but present in the ground truth.
- True Negative (TN): No match between the entity obtained by NER system and the ground truth.

Precision provides the number of correct results detected correctly whereas recall provides the total entities correctly detected; they are calculated as follows:

$$\text{Precision} = \frac{\text{TP}}{\text{TP} + \text{FP}}$$

$$\text{Recall} = \frac{\text{TP}}{\text{TP} + \text{FN}}$$

$$\text{F1-Score} = 2 * \frac{\text{Precision} * \text{Recall}}{\text{Precision} + \text{Recall}}$$

$$\text{Accuracy} = \frac{\text{TP} + \text{TN}}{\text{TP} + \text{FP} + \text{FN} + \text{TN}}$$

2.4. Named Entity Recognition

Named Entity Recognition (NER) is the task of identifying named entities such as specific location, treatment plan, medicines/drug, and critical health information from the clinical text. NER was first introduced in 1995 [18] where the three categories (Entity, Name, and Number) were defined. The original design idea for NER was to parse the text, to identify proper nouns from the text, and to categorize them.

NER is an extremely popular machine learning method and is also considered a base technique for many of the NLP tasks. Prior to 2011, all work on NERs was domain-specific and was designed to performing specific tasks based on ontologies. Collobert et al. [19] introduced a neural network-based NER, which for the first time, made it domain independent. This approach is now quite common, and there are many variations proposed over the last decade that leverage Recurrent Neural Networks (RNN) and word embeddings among others.

2.5. Relationship Extraction

A relationship can be extracted between any combination of named entities. An RE task is basically a classification of the semantic relationship between entities from textual data. RE between entities in any text is a vital task that facilitates its automated natural language understanding. The abundance and heterogeneity of unstructured data in any domain are hard to be fathomed by humans alone. Hence, the conversion of unstructured text into structured data by annotating its semantics needs to be automated. RE tasks are thus very useful in automating the process of identification of different relations from clinical data. Some important applications of clinical RE include gene–disease, drug–effect, disease–mutation, and disease–symptom relationships. In general, the pair-wise association between entities is considered, but in many cases, more than two entities are also involved. The process of checking whether a relationship exists between entities is a classification problem that can also be extended to multi-class classification or multi-label classification. In [20], a relation is defined as a tuple $t = (e_1, e_2, e_3, \ldots, e_n)$, where e_i are the entities with a predefined relationship r within the document D. Similarly, all of the different relationships in a document can be defined.

Similar to NER, RE has been applied to many domains, including the healthcare domain. One of the oldest works on RE was published in 1999, which extracted informative patterns and relations from the World Wide Web [21]. In the following year, relationship extraction from the large plain text was conducted, where a system named Snowball introduced novel strategies for pattern generation [22]. Kernel-based methods such as dependency tree kernel-based technique [23], shortest path dependency kernel-based technique [24], and subsequence kernel-based techniques [25] were proposed. The integration of probabilistic models and data mining were also proven to be good techniques for extracting relations and patterns from text [26]. Although there are innumerable RE methods in place, the models and algorithms are very domain- and data-specific. The absence of generalized algorithms to perform RE makes it challenging to define and perform a new RE task; the state-of-the-art models vary between different datasets and from one domain to another. In general, RE is most commonly viewed as a supervised learning technique performing classification [27]. In such cases, a machine learning (ML) algorithm, either traditional ML or deep learning-based methods, is used. RE can also be achieved by using unsupervised learning and rule-based methods. In the following sections, we discuss the various RE tasks and techniques applied to the clinical and biomedical domains.

2.6. Motivation

The significant growth in Electronic Health Records (EHR) over the last decade has resulted in a rich availability of clinical text, which is unfortunately stored in an unstructured format. For example, in the radiation oncology domain, when analyzed using ML techniques, a lot of valuable information such as physician clinical assessments, which includes pre-existing conditions, clinical and social history, and clinical disease status embedded in free text and entered in clinical notes, can help physicians provide better treatment. Hence, there is a need to explore robust techniques to extract such information from the clinical text. NER and RE are the key components in information extraction. In this paper, our goal is to highlight the present status of NER and RE by evaluating the models and their performance and by discussing the challenges and factors affecting the NER and RE models that need to be considered while designing a clinical decision support system.

3. Methodology

We used Google Scholar to search for articles related to NER and RE and specifically papers used in the context of clinical text. We also checked for publications where the above mentioned techniques are used in the radiation oncology domain. We discovered that there is very limited work on NER and RE in the radiation oncology domain; however, we did notice that there are a plethora of publications in using NER and RE in the clinical text in general.

Figure 2 provides a high-level overview of the steps carried out to select research articles for the survey. For clinical NER, search terms such as 'Clinical Named Entity Recognition', 'NER in Radiation Oncology', 'Deep learning Clinical NER', and 'Machine Learning based clinical NER' were used. From the resulting articles, we categorized them based on the language used for NER i.e., English, Chinese, and Italian, among others. Next, we classified the articles based on the type of approach used for NER; we found that a majority of them used ML-based approaches, and only a few articles within the machine learning class used deep learning-based methods. Overall, for clinical NER, we selected around 23 papers, out of which 19 articles used machine learning-based approaches and 3 articles used rule-based methods while 1 article used a dictionary-based approach. Since 2018, most of the clinical NER models used only ML models, we discuss such methods in greater detail. Figure 3a shows a representation of various clinical NER models that were identified; we came across ∼8 papers that use ML approaches to develop NER models for clinical text. Figure 3b represents the distribution of ML-based clinical NER models.

For clinical RE, we used the search term, 'Clinical Relationship Extraction' and obtained a number of research papers on clinical information extraction. After going through them, we found out that most people consider this to be a classification problem using machine learning models. Hence to filter out more of these papers, we again used the search term, 'Supervised Clinical Relationship Extraction'. Next, we used our judgement to use the search term, 'Unsupervised Clinical Relationship Extraction' to see if the community focuses on clinical RE without data-annotation. The last search term for clinical RE is 'Rule based Clinical Relationship Extraction' as we found out from the first search that rule-based methods are also used to some extent besides ML-based methods. From the top results of this search process, we manually identified the relevant papers based on their closeness to clinical RE and considering the diversity of the presented methods. We also kept the search results mostly limited to papers after 2016; however, this filter could not effectively find clinical RE-based articles using rule-based and unsupervised learning-based approaches. Not much work was conducted on clinical RE using unsupervised learning-based approaches because, in the clinical domain, most datasets are annotated and the supervised approaches are able to outperform these approaches in most cases, which are discussed later; we could only find two papers in this area. Rule-based methods have been used for clinical RE to some extent, but most of the noteworthy work was conducted before 2015–2016. After that, the application of supervised learning-based approaches for clinical RE started escalating distinctly and the focus on other approaches diminished. Hence, we manually identified two papers using rule-based approaches after 2016; both were published in 2021. We also manually chose three earlier papers using rule-based methods as they were popular in the past. We manually chose the 16 top, relevant, diverse papers using supervised learning-based approaches after 2016. We also considered another noteworthy supervised learning-based method for clinical RE before 2016. Out of these 17 articles, 15 papers used traditional ML and deep learning-based approaches and 4 papers used language models, with 2 papers using both language models and ML. Overall, we were able to choose 23 papers for clinical RE that used rule-based, deep learning-based, or language model-based methods. Figure 4 shows the distribution of the clinical RE research articles considered here on the basis of the methods therein using a bar chart and a Venn diagram.

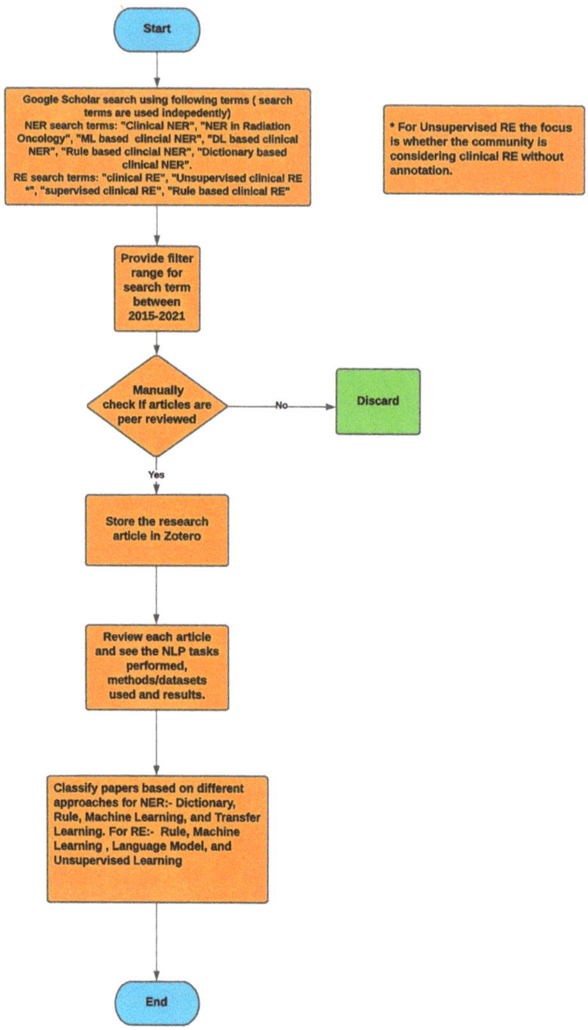

Figure 2. Methodology flowchart used here for both NER and RE to select articles.

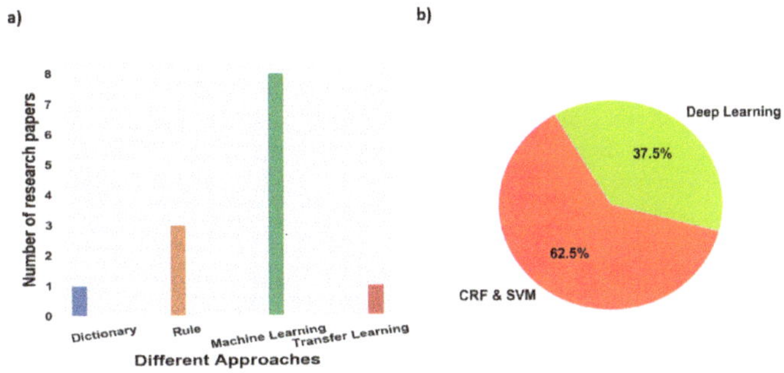

Figure 3. (**a**) Representation of the various clinical NER models based on different approaches for this survey paper and (**b**) percentage of NLP models identified based on different machine learning approaches.

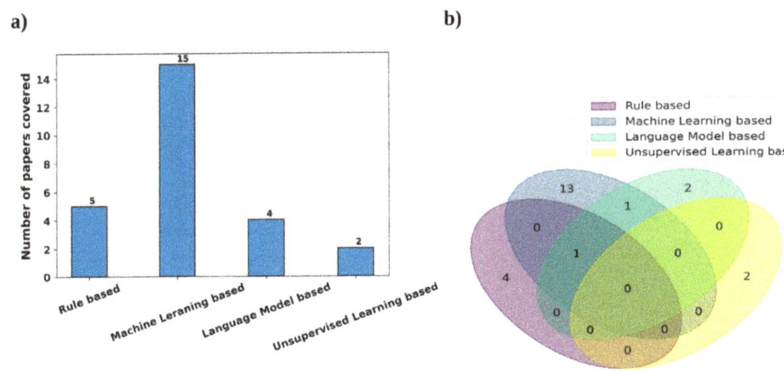

Figure 4. Representation of the clinical RE research papers used, based on the variety of the methods by using (**a**) a bar chart and (**b**) a Venn diagram.

We used software tools such as Zotero to collect all of the papers and to perform the literature survey. The next step was to categorize all of the articles and to prepare an outline for this survey. We evaluated the architectures used, how the results were reported, and the data used in the experiments. In total, we came across 51 articles (28 for clinical NER, and 23 for RE), and 46 of them were used for this survey paper; only peer reviewed articles were considered. It is worth mentioning that a couple of survey papers [17,28] also provide an in-depth view of each topic separately; however, we did not find any such survey that discusses these two related topics together specifically with respect to the clinical domain such as radiation oncology. To the best of our knowledge, this paper surveys clinical NER and RE for the first time and discusses various approaches along with their outcomes and limitations. The paper is organized as follows: Section 5 discusses the tasks associated with clinical NER, followed by a brief overview of various approaches and their results. Similarly for RE, we review the various approaches and their performance in Section 6. Finally, in Section 7, we provide our inference about the latest trends, state-of-the-art techniques, and what we believe the community (both for clinical NER and RE) needs to focus on in the future.

4. NLP Competitions and Datasets for Clinical Text

In this section, we review the different NLP competitions and datasets that are more geared towards clinical text.

4.1. Competitions

Competitions and datasets are considered assets in NLP tasks. Although most of these challenges are for data from the general domains, clinical domain-related challenges have come up in the past. Clinical-NER based competitions were mostly focused on the de-identification of Personal Health Information (PHI). In 2014, there was a i2b2 UTHealth challenge that had longitudinal data [29], and the goal of the competition was to perform de-identification on clinical narratives, with a second track focused on determining risk factors for heart disease over time. Stubbs et al. [30] provides a comprehensive review of a workshop that includes how data were released and how the submissions were evaluated. The 2016-CEGS N-GRID shared tasks that the workshop used in gathering psychiatric data [31] for addressing text de-identification, symptom severity detection, and the proposal of new research questions. Stubbs et al. [31] explained how the data were generated; discussed the challenges with psychiatric data as it contains higher occurrence of PHI; and the outcomes, which showcase the best performing systems and how the submitted models were evaluated. There was also another competition on clinical NER for de-identification on Japanese text (2012 NTCIR-10) [32]. Coffman et al. [33] organized a competition, which was also a final deliverable for the Applied NLP course taught at UC Berkeley. The objective of the competition was to develop an algorithm that predicts/assigns an ICD-9 (International Classification of Diseases, 9th revision) code to clinical free text [33]. MADE1.0 [34] is a competition for detecting Adverse Drug Events (ADEs) from EHR. The goal of the NLP task is to detect medication names and other attributes such as frequency and duration. Around 11 teams participated in at least one of the three tasks. There was a total of 41 submissions, among which Wunnava et al. [35] ranked first for the NER task, with a micro-averaged F1-score of 0.892. SemEval-2014 [36] Task 7 was another competition on analyzing clinical text; it had two subtasks, namely, identification and normalization of disease and disorders in a clinical text from the ShARe [37] corpus. Around 21 teams participated in the identification task, and the best F1-score reported was 81.3, while for the normalization task, 18 teams participated, reporting a best accuracy of ~74.1.

National NLP Clinical Challenges, also called n2c2, is a very popular competition for different clinical NLP tasks. Between 2004 and 2014, the competition was called Informatics for Integrating Biology and the Bedside (i2b2) but was then changed to n2c2 in 2018. They introduced the following clinical RE tasks over the years, with datasets generated by the NIH-funded National Centers for Biomedical Computing (NCBC).

- 2010 i2b2/VA Challenge on Concepts, Assertions, and Relations in Clinical Text [38]: In this competition, 16 teams participated in the relationship extraction task that showed that rule-based methods can be augmented with machine learning-based methods. SVM-based supervised learning algorithm performed the best with an F1-score of 0.737 [39].
- 2011 Evaluating the state-of-the-art in co-reference resolution for electronic medical records [40]: In this competition, 20 teams participated and rule-based and machine learning-based approaches performed best, with an augmentation of the external knowledge sources (coreference clues) from the document structure. The best results on the co-reference resolution on the ODIE corpus with the ground truth concept mentions and the ODIE clinical records were provided by Glinos et al. [41], with an F1-score of 0.827. The best results on both the i2b2 and the i2b2/UPMC data were provided by Xu et al. [42], with F1-scores of 0.915 and 0.913, respectively.
- Evaluating temporal relations in clinical text, 2012 i2b2 Challenge [43]: 18 teams participated in this challenge, where for the temporal relations task, the participants first determined the event pairs and temporal relations exhibiting temporal expressions and then identified the temporal relation between them. This competition also showed that hybrid approaches based on machine learning and heuristics performed the best for the relationship classification. Rule-based pair selection with CRF and SVM by Vanderbilt University provided the best results here (F1-score: 0.69).

- 2018 n2c2 shared task on adverse drug events and medication extraction in electronic health records [44]: a total of 21 teams participated in the relationship classification task on adverse drug events (ADEs) and medication. Team UTHealth/Dalian (UTH) [45] designed a BiLSTM–CRF-based joint relation extraction system that performed the best (F1-score: 0.9630).

4.2. Datasets

The datasets are important in understanding the different entities and relations extracted in the clinical domain. This subsection gives an overview of the different datasets used for clinical NER and RE tasks for a better understanding of the challenges in the domain.

We came across a few publicly available datasets for clinical NER; however, these datasets are restricted to specific NLP tasks in clinical domain. Below is a list of datasets that were used in NER challenges or used as training for NER models, which are discussed in Section 5.3 for training, testing, and validation:

- Mayo Clinic EMR: It has around 273 clinical notes, which includes 61 consult, 4 educational visits and general medical examinations, and a couple of exam notes. A few models, such as Savova et al. [46], generated a clinical corpus from Mayo Clinic EMR [47].
- MADE1.0 Data set: This dataset consists of 1092 medical notes from 21 randomly selected cancer patients' EHR notes at the University of Massachusetts Memorial Hospital.
- FoodBase Corpus: It consists of 1000 recipes annotated with food concepts. The recipes were collected from a popular recipe sharing social network. This is the first annotated corpora with food entities and was used by Popovski et al. [48] to compare food-based NER methods and to extract food entities from dietary records for individuals that were written in an unstructured text format.
- Swedish and Spanish Clinical Corpora [49]: This dataset consists of annotated corpora clinical texts extracted from EHRs; the Spanish dataset consists of annotated entities for disease and drugs, while the Swedish dataset has entities annotated for body parts, disorder, and findings. This dataset is mostly used for training and validation for NER on Swedish and Spanish clinical text.
- i2b2 2010 dataset [38]: This dataset includes discharge data summaries from Partners Healthcare, Beth Israel Deaconess Medical center, and University of Pittsburgh (also contributed progress reports). It consists of 394 training, 477 test, and 877 unannotated reports. All of the information are de-identified and released for challenge. These datasets are used for training and validation in many of the NER models used for clinical text.
- MIMIC-III Clinical Database [50]: This is a large and freely available dataset consisting of de-identified clinical data of more than 40,000 patients who stayed at the Beth Israel Deaconess Medical Center between 2001 and 2012. This dataset also consists of free-text notes, besides also providing a demo dataset with information for 100 patients.
- Shared Annotated Resources (shARe) Corpus [37]: This dataset consists of a corpus annotated with disease/disorder in clinical text.
- CanTeMiST [51]: It comprises 6933 clinical documents that does not contain any PHI. The dataset is annotated for the synonyms of tumor morphology and was used for clinical NER on a Spanish text by Vunkili et al. [51].

Specific relations annotated in the datasets from the various clinical RE challenges mentioned in Section 4.1 are as follows:

1. 2010 i2b2/VA Challenge on Concepts, Assertions, and Relations in Clinical Text [38]: A wide variety of relations were identified as follows:
    - Medical problem–treatment relations:
        - *TrIP* indicates that treatment improves medical problems, such as *hypertension* being controlled by *hydrochlorothiazide*.

  - *TrWP* indicates that treatment worsens medical conditions, such as the *tumor* growing despite the available *chemotherapeutic regimen*.
  - *TrCP* indicates that treatment causes medical problems, such as *Bactrium* possibly being a cause of *abnormalities*.
  - *TrAP* indicates that treatment is administered for medical problems, e.g., periodic *Lasix* treatment preventing *congestive heart failure*.
  - *TrNAP* indicates that treatment is not administered because of medical problems e.g., *Relafen* being contraindicated because of *ulcers*.
  - *Others* that do not fit into medical problem–treatment relations.
- Medical problem–test relations:
  - *TeRP* indicates that the test reveals medical problems, such as an *MRI* revealing a *C5-6 disc herniation*.
  - *TeCP* indicates that the test was conducted to investigate a medical problem, such as a *VQ scan* being performed to investigate a *pulmonary embolus*.
  - *Others* that do not fit into medical–test relations.
- Medical problem–medical problem relations:
  - *PIP* indicates any kind of medical problem such as a C5–6 *disc herniation* with *cord compression*.
  - *Other* relations with respect to medical problems that do not fit into the PIP relationship.

2. 2011 Evaluating the state-of-the-art in coreference resolution for electronic medical records [40]: The data for this challenge was similar to the 2010 i2b2/VA challenge as the dataset contained two separate corpora, i.e., the i2b2/VA corpus and the Ontology Development and Information Extraction (ODIE) corpus, which contained de-identified clinical reports, pathology reports, etc.
3. Evaluating temporal relations in clinical text, 2012 i2b2 Challenge [43]: The temporal relations or links in the dataset indicate how two events or two time expressions or an event and a time expression is related to each other. The possible links annotated in the dataset were BEFORE, AFTER, SIMULTANEOUS, OVERLAP, BEGUN_BY, ENDED_BY, DURING, and BEFORE_OVERLAP.
   Ex: OVERLAP -> She denies any *fever* or *chills*.
   Ex: ENDED_BY -> His *nasogastric tube* was discontinued on *05-26-98*.
4. 2018 n2c2 shared a task on adverse drug events and medication extraction in electronic health records [44]: The different relations identified between two entities in this case are either of the following types: Strength–Drug, Form–Drug, Dosage–Drug, Frequency–Drug, Route–Drug, Duration–Drug, Reason–Drug, and ADE–Drug.

5. Discussion on Clinical Named Entity Recognition

The goal of using NER on clinical text is to extract entities or subjects of interest from the clinical text. The clinical text, in general, has many medical terms such as the disease name, location, and medical procedures, and hence, the named entities can help in finding useful patterns. The nature of the clinical text, in general, is dictated by notes from physicians based on their interaction with the patients. In most cases, it is in free text format, which can be split into multiple paragraphs, and is mostly narrative in nature. For example, the clinical text written by physicians in the consultation notes from the radiation oncology domain has the following information:

- Physical Exam: This section can have both structured and unstructured information such as toxicity and review of systems, where we try to store information such as dizziness, cough, and rectal bleeding.
- Past Medical History: This has all of the allergy information, medications, prior military service, prior surgery information, and prior diseases for patients and is mostly stored as unstructured free text.

- Oncologic History: This includes all of the prior oncologic information in unstructured format and varies based on the types of cancer.
- Diagnostic Test: Various tests may be performed on patients and vary based on cancer types. They are mostly in structured format; however, some tests may be specific to patients that can be documented and stored in unstructured free text format such as Bone Scan and CT Pelvis.

Clinical NER is very common these days due to the massive growth in EHRs and is considered the first step in processing clinical text. The output of clinical NER is further used for other tasks such as decision-making in precision treatment. Due to the unstructured nature of the clinical text, there are challenges in designing effective clinical NER systems, as discussed below. We observed that many clinical NER models are developed for different languages such as Chinese and Italian; Figure 5 shows the number of clinical NER models that we came across for different languages. Due to the strict privacy rules in the EU (European Union) and HIPAA compliance in the US, it is difficult to disseminate medical information. We found very few articles that use clinical NER models for de-identification where medical document is parsed and any Protected/Personal Health Information (PHI) is removed; for example, recently Catelli et al. [52] developed a clinical NER model for Italian COVID-19 clinical text.

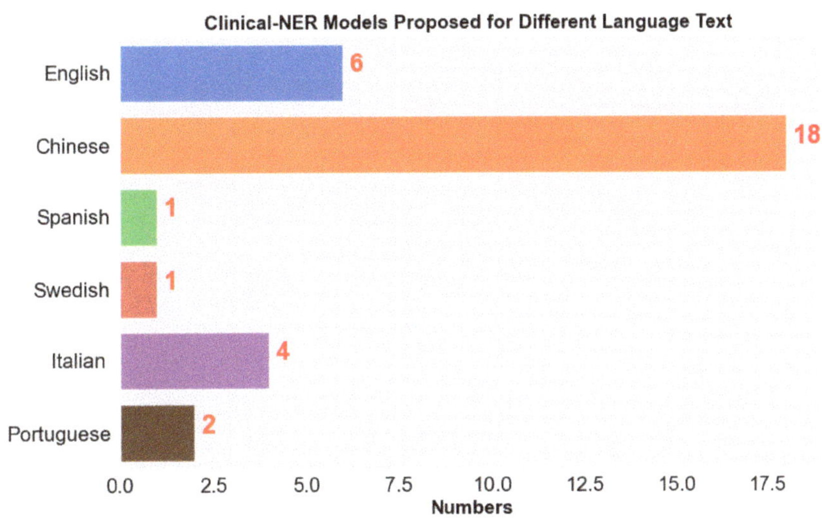

Figure 5. Clinical NER models available for text in different languages.

5.1. Challenges in Clinical NER

- Nested Entities and Ambiguity: Most clinical terms are often confusing as there is no common ontology. Physicians often use abbreviations or acronyms, which makes it very difficult to standardize clinical text. In the radiation oncology domain, a common challenge is that physicians dictate their clinical assessment based on the style they were trained in and it varies significantly for different types of cancers, which makes it very difficult to develop a standard NER model for processing radiation oncology notes that cater to all of the different types.
- Meaning of Context: The clinical terms used can have different meanings, which vary based on the context. Although this problem mostly applies to non-clinical notes, for clinical NER, this becomes more challenging as the model should understand the complete clinical context along with the entity. A common issue is negative medical findings, where text is written in such a manner that it reports findings in a negative context; however, the NER considers that as a positive.

To address the nested entities and ambiguity, there are efforts to standardize the nomenclature of clinical entities [53,54]. However, this is still in an initial phase, and to be successful, it needs to be widely adopted.

5.2. Clinical NER Methods

One of the important challenges in designing a clinical NER is how to extract meaningful information without much human effort. Prior to NER, the NLP techniques used required a lot of human effort to process the text. There are various NER models proposed over the last decade to extract information from the clinical text that can be broadly classified into four types of approaches:

- Dictionary-Based Approach: In this approach, a predefined set of named entities are defined that are later used as a lookup while parsing the clinical text for entities. For example, Savova et al. [46] used a dictionary-based approach to detect NERs from clinical text using their NLP toolkit.
- Rule-Based Approach: Here, the rules/entities are predefined by domain experts. Most of the rules are handcrafted and are used to detect entities in a specific text. The limitation of this approach is generalizability or extensibility, as most of them are applicable to the domain they were defined in. This approach certainly requires a lot of effort where experts spend time defining the entities, and then, it is used as a lookup while parsing the clinical notes.
- Machine Learning-Based Approach: The purpose of this approach is to completely automate the NER process. Commonly used ML algorithms such as Random Forest (RE), Support Vector Machines (SVMs), and Neural Networks (NN) are used to learn the pattern (entities and boundaries) using the training set. Once the training is over, the model can classify the clinical text into predefined classes. This approach is garnering much attention due to recent advancements in ML and the easy availability of computational resources. The majority of the articles collected for this survey used this approach.
- Conditional Random Field (CRF)-Based Approach: The CRF approaches fall under the ML category and mostly solve a label sequencing problem, where for a given input sequence $X = x_1x_2x_3$, CRF tries to find the best label sequence $Y^* = y_1y_2y_3$. At first, the entities are annotated with tags; in general, the BIO (Beginning, Inside, and Outside of Entity) schema is used for annotation, where each word is assigned to a label. The input for CRF models is mostly designed by humans and represented as a bag-of-words style vector. Wu et al. [4] introduced seven tags and three CRF baselines using different features. All of the commonly used CRF-based implementations in clinical NER can be found in the CRF++ package. In Tables A1 and A2, we observe that there are many models using CRFs for NER with good accuracy.
- Deep Learning-Based Methods: This is similar to the CRF label sequencing problem using the BIO schema, where the input is a raw sequence of words. An added layer performs the word embedding by converting words into densely valued vectors. In the training phase, it learns the dependencies and features to determine entities. Deep learning methods are very popular for clinical NER as they achieve state-of-the-art results and can also detect hidden features automatically. The first neural network architecture for NER was proposed by Collobert et al. [19], with a convolution layer, several standard layers, and a non-linear layer. This architecture achieved state-of-the-art performance in clinical NER. Details on the CNN model for clinical NER can be found in [17]. New studies have recently shown that RNNs (Recurrent Neural Networks) perform much better than CNNs and are capable of capturing long-term dependencies for sequence data. Lample et al. [55] introduced Long Short-Term Memory (LSTM), a popular implementation of RNN architecture, for this problem. Wu et al. [4] evaluated the performance of CNNs, RNNs, and CRFs with different features and concluded that the RNN implementation outperformed the other two.

- Hybrid Approaches: here, any of the above approaches are combined and then used to determine entities.

5.3. Clinical NER Models

- Savova et al. [46] proposed a dictionary look up algorithm, where each named entity is mapped to a terminology. The dictionary was constructed using the terms from UMLS, SNOMED CT, and RxNORM. This implementation also involves a parser in which the output is used further to search for noun phrases. The limitation of this implementation is that it fails to resolve ambiguities while working with results from multiple terms in the same text. They datasets for NER are derived from Mayo clinic EMR. For exact and overlapping matches F1-score reported were 0.715 and 0.824 respectively.
- Skeppstedt et al. [56] used CRF model and a rule-based approach to detect NER on Swedish health records and identified four entities: Drug, Finding, Disorder, and Body structure. They also compared it on English clinical text. They reported precision and recall for all of their findings: 0.88 and 0.82 for body structure, 0.80 and 0.82 for disorders, 0.72 and 0.65 for findings, and 0.95 and 0.83 for pharmaceutical drugs.
- Chen et al. [57] developed a rule-based NER system that was designed to detect patients for clinical trial. They used the n2c2-1 challenge dataset for training and achieved an F1-score of 0.90.
- Eftimov et al. [48] developed a rule-based approach to detect extraction of food, nutrient, and dietary recommendations from text. They discussed four methods FoodIE, NCBO, NCBO (OntoFood), and NCBO(FoodON). Based on their comparison, they identified that FoodIE performs well. Their model was trained on the FoodBase Corpus and was able to identify entities from dietary recommendation.
- Xu et al. [58] developed a joint model based on which CRF performs word segmentation and NER. Generally, both systems are developed independently, but the joint model used to detect Chinese discharge summaries performed well. There was no score reported in this publication; they only reported that the joint model performance is better when they compared it with the two individual tasks.
- Magge et al. [59] developed an NLP pipeline, which processed clinical notes and performed NER using bi-directional LSTM coupled with CRF in the output layer. They used 1092 notes from 21 cancer patients, from which 800 notes were used for NER training. They reported NER precision, recall, F1-score for the entities individually and reported a macro-averaged F1-score of 0.81.
- Nayel et al. [60] proposed a novel ensemble approach using the strength of one approach to overcome the weakness of other approaches. In their proposed two-stage approach, the first step is to identify base classifiers using SVM, while in the second phase, they combined the outputs of base classifiers based on voting. They used the i2b2 dataset and reported an F1-score of 0.77.
- Wu et al. [4] performed a comparison study between two well-known deep learning architectures, CNN and RNN, with three other implementations: CRFs and two state-of-the-art NER systems from the i2b2 2010 competition to extract components from clinical text. The comparison created a new state-of-the art performance for the RNN model and achieved an F1-score of 85.94%.
- Wang et al. [61] proposed a model to study symptoms from Chinese clinical text. They performed an extensive set of experiments and compared CRF with HMM and MEMM for detecting symptoms. They also used label sequencing and the CRF approach outperformed the other methods.
- Yadav et al. [17] provided a comprehensive survey of deep neural architectures for NER and compared it with other approaches including supervised and semi-supervised learning algorithms. Their experiments showed good performance when they include neural networks, and they claim that integrating neural networks with earlier work on NER can help obtain better results.

- Vunikili et al. [51] used Bidirectional Encoder Representations from Transformers (BERT) [62] and Spanish BERT (BETO) [63] for transfer learning. This model is used to extract tumor information from clinical reports written in Spanish. They reported an F1-score of 73.4%.
  Jiang et al. [64] developed ML-based approaches to extract entities such as discharge summaries, medical problems, tests, and treatment from the clinical text. They used a dataset comprising 349 annotated notes for training and evaluated their model on 477 annotated notes to extract entities. They reported an F1-score of 0.83 for concept extraction.
- Yang et al. [65] proposed a deep learning model to extract family history, and they compared LSTM, BERT, and ensemble models using a majority voting.

All of the NER models discussed above are summarized and presented in Tables A1 and A2.

5.4. Clinical NER Evaluation Metrics

The outputs from clinical NER systems are usually compared with human annotations. In general, a comparison can be either exact or relaxed matches [17]. A relaxed match only considers the correct type and ignores the boundaries as long as there is an overlap with ground truth boundaries. We observed from our cohort of selected articles on clinical NER that all of them reported exact matches, which is the F1-score and variations such as macro F1-score. In the case of an exact match, it is expected that the entity identified correctly should also detect boundary and type correctly at the same time [17]. We also observed that a few NER models report performance in macro- and micro-average. In macro-averaging, the F1-scores of all entities are calculated independently and then averaged. In micro-averaging, the sums of the false positive, false negative, and true positive across all entities are taken. Other commonly used metrics in ML such as sensitivity, specificity, ROC (Receiver Operator Characteristic), and AUC (Area Under the Curve) were not used in the clinical NER articles reviewed here. There are however many studies such as [66] that point to the limitations of using F1-score as an evaluation metric in NLP; one of the major issues is that the F1-score metric is biased towards the majority class. The class imbalance problem has been recently garnering attention for both binary and multi-class classification. Accuracy and precision scores are relevant if we focus on majority classes; for a minority, those metrics evaluations do not have any significant incluence. Branco et al. [67] provided a comprehensive list of metrics both for binary class and multi-class classification such as classes average accuracy, and Matthews Correlation Coefficient. Along with the list of metrics provided, they claim that the metrics available are not suitable for all cases. We also found very few papers that tested for statistical significance between experimental methods.

6. Discussion on Clinical Relationship Extraction

RE is a specialized task of collecting meaningful structured information from unstructured text. In clinical and biomedical domains, RE has been applied to drug–gene relationships [68], disease–gene relationships [69], semantic classes for radiology report text identification [70,71], relation extraction for biological pathway construction [72], relationship between lexical contexts and category of medical concepts [73], and disease–mutation relationship from biomedical literature [74]. Temporal relationship extraction from clinical texts is another important RE task [75]. In all these of different tasks, the NLP-based methods that are used to extract the relations between different entities are very much specific to the particular dataset, i.e.,the particular combination of feature representation and learning algorithm is very distinct from any other case. Due to this fact, the methods that are used to extract relations such as an ML problem are not very generalizable. However, RE from clinical texts has also been performed by using a domain invariant convolutional neural network (CNN) [76]. Most RE tasks are based on finding the relationship between entities inside the same sentence but there are some instances of RE tasks across sentences as well [77–79]. Since, in most cases, RE is treated as a classification problem, both multi-label classification [80] and multi-class classification [76,81] were proposed to extract clinical relations. An RE task consists of syntactic processing modules, which deals with the process of text representation and feature generation such as tokenization, word embeddings, etc., and semantic processing modules, which deals with meaningful information collection such as relationship identification and classification, in this case. In clinical texts, a variety of feature generation techniques are used to extract relations from various data, which can range from contextualized word embeddings, part-of-speech (POS) tagging, etc. The next or the final step is to select a learning algorithm such as supervised, unsupervised, or even rule-based methods on the features in order to identify the relations. The various feature representation and learning methods used in the clinical and biomedical text are discussed next. A pictorial representation of the different learning methods used to learn the different relations from clinical texts is shown in Figure 6.

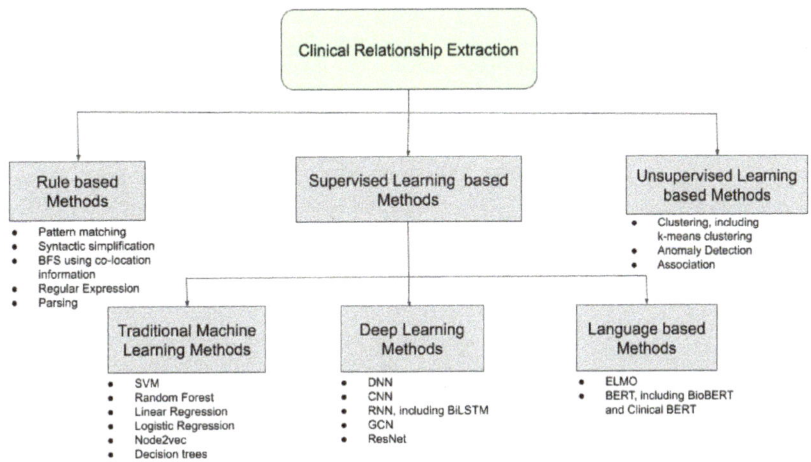

Figure 6. Different learning methods used for clinical RE.

6.1. Feature Generation

Feature generation is an important step for RE, where features are extracted from the unstructured text and then represented only with numbers. This step is particularly very important for the supervised and the unsupervised learning methods because these methods require inputs in the form of numbers only. The performance of these ML models depends not only on the actual algorithm but also on how the input features were represented. The first step before representing the features is preprocessing and tokenizing the text. In many deep learning-based approaches, the whole instance is considered the input, which is basically a featureless representation. The various features that can be considered for RE tasks are the word, the words distance from both the entities, chunk tag of the word, POS tag of the word, type of the word, n-grams, etc. Sahu et al. (2016) [76] introduced a domain-invariant RE technique using CNN, where the inputs were represented with the word, its distance from the first entity, its distance from the second entity, a Part-of-Speech (POS) tag, chunk, and the type of the word. Singhal et al. (2016) [79] used Nearness to Target Disease Score, Target Disease Frequency Score, Other Disease Frequency Score, Same Sentence Disease-Mutation Co-occurrence Score, Within Text Sentiment Score, and Text Sentiment Subjectivity Score as input features to the decision trees to extract the disease–mutation relationship. Hasan et al. (2020) [82] used word embedding, POS embedding, IOB embedding, relative distance, concept embedding, and dependency tree to represent the input features. Alimova et al. (2020) [83] compared the performance of BERT with that of random forest based on a multitude of features such as word distance; character distance; sentence distance; punctuation distance; position distance; bag of words; bag of entities;

entity types; entities embedding; concept embedding; sentence embedding; the similarity between entities; and some knowledge features such as UMLS, MeSH, etc. Mahendran et al. (2021) [70] also divided the sentences into five segments based on the location of the context to represent the input features of the segment-CNN model. Textual input features used for various ML algorithms are mostly a combination of the features mentioned above.

6.2. Rule-Based Methods

Though rule-based methods are not the most popular method nowadays to extract relations from clinical texts, they are still being used and have been used in the past in good numbers. These methods require defining some rules in the beginning based on the nature of the input dataset. These methods of extracting information by using well-defined rules and patterns are often not very computationally efficient such as the machine learning models with respect to their performance, and hence, these methods are not very popular these days. Segura-Bedmar et al. (2011) [84] developed a linguistic hybrid rule-based method to extract drug–disease interactions via the combined use of shallow parsing, syntactic simplification, and pattern matching. A pharmacist defined the domain-specific lexical patterns of the drug–disease interactions that were matched with the generated sentences. This method did not perform well with an average precision and a very low recall. Xu et al. (2011) [85] combined rule-based methods with ML to engineer features for structured RE from clinical discharge summaries as provided by the i2b2 2010 challenge. The RE task received a micro-averaged F1-score of 0.7326. Li et al. (2015) [86] matched the drug names to their attributes in a prescription list, and then the matching was confirmed by means of the co-location information and RxNorm dictionary. It helped in identifying the medication discrepancies with very high performances. Veena et al. (2021) [87] used NLP-based regular expressions to extract the words from the text document of different medical data using scraping and POS tagging. Then, the relations between different medical terms were extracted using a path similarity analysis. Mahendran et al. (2021) [70] used the co-location information between the drug and the non-drug entity types by using a breadth-first-search (BFS) algorithm to find the adverse drug effects. The left-only rule-based approach (macro-average F1-score: 0.83) eclipsed the performance of other rule-based models. Overall, the rule-based approaches for clinical RE can perform well, depending on how the rules are defined. Some clinical RE tasks using rule-based methods are tabulated in Table A3.

6.3. Supervised Learning Methods

As mentioned before, supervised learning algorithms have been extensively considered for RE. This method uses a classifier to determine the presence or absence of a relationship between two entities. Computers cannot understand the unstructured text, and hence, this kind of learning method requires features about the text as an input. As a result of this, there is an absolute necessity to annotate the clinical texts by domain experts. Annotating or labeling examples is a time-consuming procedure as it takes a lot of effort to manually annotate the data. This is an important limitation of these methods although they have high accuracy. These methods used in clinical RE however suffer from the difficulty of adding new relations. Supervised learning algorithms can also be extended to include distantly supervised RE or weakly supervised learning or semi-supervised learning.

6.3.1. Traditional Machine Learning and Deep Learning-Based Methods

Supervised learning is defined as an ML task to learn a function that maps the input to the output of each input–output data point [88]. This requires the annotated data to be divided into training and testing samples. The model learns the function based on the values of the inputs and the outputs of the training examples. Analyzing the inputs and the outputs, the model comes up with an inferred function. Then, the efficiency of the inferred function is analyzed by testing the function on the testing set. Supervised ML algorithms can be classified into two categories: (i) traditional supervised learning algorithms and (ii) Artificial Neural Network (ANN) based algorithms. The traditional methods are heavily dependant on the well-defined features, and hence, their performance relies on the efficacy of the feature extraction process. Moreover, these shallow algorithms are found to be overshadowed by ANNs where data is large and of high dimension. Still, the shallow traditional ML algorithms perform better in the case of low-dimensional data or data with a limited number of training samples. ANNs can be very deep, depending on the number of hidden layers between the input and the output, leading to deep learning-based methods. The differences between the traditional shallow methods and ANNs are surveyed by Janiesch et al. (2021) [89]. Examples of traditional algorithms include but are not limited to Support Vector Machines (SVM), Linear Regression, Logistic Regression, Naive Bayes, Linear Discriminant Analysis, Decision Trees, K-Nearest Neighbor (KNN), Node2vec, etc., whereas Dense Neural Networks (DNN), Convolutional Neural Networks (CNN), Recurrent Neural Networks (RNN), Graph Neural Networks (GNN), autoencoders, etc. are some of the Deep Learning algorithms. These algorithms have been extensively used in clinical domain for a variety of tasks [90–94].

Swampillai et al. (2011) [78] first used an SVM-based approach on adapted features to extract relations between entities spread across different sentences. Their work showed that the structured features used for intra-sentential RE can be adopted for inter-sentential RE as they both performed comparably. Later on, inter-sentential RE tasks were defined on clinical notes too. In the 2010 i2b2/VA challenge on concepts, assertions, and relations in clinical text [38], SVM-based supervised learning algorithm performed the best with an F1-score of 0.737 [39]. The domain-invariant CNN on multiple features for clinical RE used by Sahu et al. (2016) [76] showed a decent performance with various filter length combinations; filter combination of {4, 6} performed the best (precision: 0.7634, recall: 0.6735, and F1-score: 0.7116). Singhal et al. (2016) [74] used a C4.5 decision tree because of its superior performance on the features extracted from various biomedical literature for disease-mutation RE. It demonstrated improved performance when compared with the previous state-of-the-art models with F1-scores of 0.880 and 0.845 for prostate and breast cancer mutations, respectively. The performance of a sparse deep autoencoder-based model introduced by Lv et al. (2016) [95] outperformed the performance of a deep autoencoder on most of the clinical relation types. Lin et al. (2017) [80] presented a multi-label structured SVM for Disorder Recognition in the 2013 Conference and Labs of the Evaluation Forum (CLEF) textual dataset. This model achieved an F1-score of 0.7343, i.e., 0.1428 higher than their baseline BIOHD1234 scheme. Mondal et al. (2017) [73] compared the performance of a rule-based approach with a feature-oriented SVM-based supervised learning approach for clinical RE, where the supervised learning model reported higher F1-scores. Magge et al. (2018) [59] used a bidirectional LSTM-CRF for the clinical NER and a random forest-based binary classifier for the clinical RE. The various features used for RE as an input to the random forest classifier such as entity types, number of words in the entity, and an average of the entity word embeddings resulted in a micro-averaged F1-score of 0.88 (precision: 0.82; recall: 0.94). Kim et al. (2018) [72] used node2vec to learn the features from texts in networks in order to extract relations for biological pathways, which outshone the previous methods to detect relationships in the type 2 diabetes pathway. Munkhdalai et al. (2018) [96] compared the performance of an SVM model with a deep learning-based LSTM model to extract relations towards drug surveillance. SVM showed better performance (89.1% F1-score) on the test data compared with that of LSTM. Li et al. (2019) [97] introduced a novel approach for RE in clinical texts by using neural networks to model the shortest dependency path between the target entities along with the sentence sequence. This approach used on the 2010 i2b2 relation extraction dataset improved the performance to an F1-score of 74.34%. The multi-class SVM model on this dataset, introduced by Minard et al. (2019) [81] achieved an F1-score of 0.70, which is lower than the previous models. Christopoulou et al. (2020) [79] proposed an ensemble deep learning method to extract the adverse drug events and medications relations, which achieved a micro-averaged F1-score of 0.9472 and 0.8765 for RE and end-to-end RE, respectively. Hasan et al. (2020) [82] compared the performance of different deep learning methods such as CNN, GCN, GCN-CDT, ResNet, and BiLSTM on various combinations of features, as mentioned in the previous subsection for clinical RE. BiLSTM achieved the highest 9 class F1-score of 0.8808 in that dataset. Both CNN models used by Mahendran et al. (2021) [70], segment-CNN and the sentence-CNN, failed to surpass the performance of the rule-based model proposed for this dataset. Research has shown that the traditional ML methods have outperformed deep learning methods in many clinical RE tasks where the dataset has limited data instances, whereas in some cases where more data is present, deep learning methods given better performance. Additionally, the level of performance depends on the complexity of the data. Currently, it is not possible to generalize whether traditional ML methods or deep learning methods perform the best for clinical RE as the performance is very data-dependent. Some clinical RE tasks using traditional machine learning and deep learning-based methods are tabulated in Table A4.

6.3.2. Language Model-Based Methods

Language model-based approaches have shown improved performance in many NLP tasks as these language models use contextual information into account to represent the features. Then, a classifier is added on top of the language model output to perform the classification of relationships in the end. It is also a supervised learning model as the inputs are well defined for each instance. The language models popularly used in NLP tasks are ULMFit, ELMO, BERT, etc. Out of them, BERT [62], introduced by Google in 2019, has become extremely popular for various NLP tasks including RE. Its breakthrough has resulted in improved performance in many NLP tasks because of its strong ability to pretrain deep bidirectional representations of any unlabelled text by conditioning on its context on both sides in all the 12 transformer layers. For biomedical clinical texts, two BERT-based models were later introduced such as BioBERT [98], trained on biomedical PubMed corpus, and Clinical BERT [99], trained on a biomedical corpus, clinical notes, and only discharge summaries. These models have the same model architecture as that of BERT, but they were trained on a medical corpus.

BERT and the biological and clinical versions of BERT gained high popularity for RE tasks on clinical texts. Since these are language models, there is no need to generate and represent the features. The entire text, i.e., the complete sentence or the complete paragraph of each instance, is taken as input to the model. Lin et al. (2019) [77] established state-of-the-art results in temporal RE in clinical domain using pretrained domain-specific as well as fine-tuned BERT: 0.684F for in-domain texts and 0.565F for cross-domain texts. Alimova et al. (2020) [83] used BERT-based models, including BioBERT and Clinical BERT. The BERT models used there performed really well for some of the classes, but for other classes, the Random Forest Classifier using different input features performed better. Wei et al. (2020) [100] established that the Fine-Tuned BERT eclipsed the performance of other models for RE on clinical narratives. Overall, the language models have shown superior performances than other models on clinical RE tasks. BERT (cased and uncased), BioBERT and Clinical BERT were the language models used by Mahendran et al. (2021) [70]. All of the BERT models, with an impressive macro-averaged F1-score of 0.93, outshone the performance of all of the other rule-based or deep learning methods on this dataset. Therefore, in most cases, language models such as BERT have outshone other ML and deep learning methods for clinical RE due to their capability to learn from the context. Some clinical RE tasks using traditional language model-based methods are tabulated in Table A5.

6.4. *Unsupervised Learning Methods*

Unsupervised Learning is defined as an ML technique where users are not required to supervise the model, but it allows the model to run and learn by itself to excavate interesting patterns that were earlier undetected. These methods do not require annotated texts as they are capable of working on unlabelled data on their own. The level of processing needed for these kind of tasks is very high, but due to their simplicity, these algorithms are more suitable for simpler tasks and, hence, unsupervised learning algorithms can be unpredictable for RE. The different types of unsupervised learning techniques are Clustering, Anomaly Detection, and Association. In clinical RE, unsupervised learning algorithms have been used to identify the different types of relations in the text that needs to be later reviewed and annotated by domain specialists to evaluate the performance of the model. In real life, the text contains a lot of noise and unsupervised learning is not always effective in identifying the different relations with a high level of accuracy. However, this method is less time expensive and is preferred in some cases.

Unsupervised learning has been the least popular in RE on clinical texts because of the limitations of the unsupervised algorithms to identify relation patterns from complex textual data. Without proper clinical annotations by the clinicians, this learning task is far more ambiguous, which might result in the decreased accuracy of these models. Out of the very few works, Quan et al. (2014) [101] were the pioneers in proposing an unsupervised text mining method for RE on clinical data. The unsupervised clustering-based method that is a combination of dependency and phase structure parsing for RE performed moderately with respect to the previous models but their proposed semi-supervised model surpassed its performance to become the second-best model on this dataset. Alicante et al. (2016) [102] used unsupervised methods for entity and relation extraction from Italian clinical records. The performance of the unsupervised clustering algorithm in the space of entity pairs, being represented by an ad hoc feature vector, is found to be promising in labeling the clinical records by using the most significant features. Since the dataset was not annotated here, similarity measures such as Manhattan, Binary, and Cosine similarities are used to measure the goodness of the clustering models. Not many other unsupervised methods have been proposed for RE on clinical notes. Some clinical RE tasks using unsupervised learning-based methods are tabulated in Table A6.

7. Trends and Future Research Directions

Our main observation from this review is that the clinical-NER community is more focused on deep learning as it has shown promising results. The other approaches such as dictionary or rule-based methods have lost popularity in the last few years. We believe that the upcoming research on clinical NER will develop models using hybrid approaches where the ML-based and rule/dictionary-based approaches can be combined. One of the major challenges while evaluating different clinical NER models was how to measure their effectiveness. The F1-score measure has its own limitations, as mentioned earlier; simply comparing the F1-score does not give much insight into the models. We have seen recently that there are few attempts to address the limitations of F1-score and suggest alternative metrics such as [103]. However, currently, we did not see any attempts to standardize an evaluation metric for clinical NER. For the class imbalance problem discussed in this survey paper, we believe that the community should consider using metrics that address the multi-class imbalance problem. We did see multiple metrics available; however, the selection of correct metric is based on the user interest towards majority or minority classes. Alternatively, we recommend using multiple metrics to obtain a better idea of the balanced performance. We have seen many recent works published on performing clinical NER on text from different languages apart from English and Chinese text such as [52] in Italian text. There are attempts to use transfer learning from the text in different languages to improve the performance such as [52]; although this is still in an initial phase, we believe that, in the next few years, more work will follow this approach. As mentioned in the Clinical NER section, one of the major issues in clinical NER is that most of the models developed are only limited to specific clinics or centers, and specific domains. In order to address this and to make clinical NER models widely available for usage, the clinical terms should be standardized and widely adopted. We found a few attempts on the standardization of clinical terms such as [53]; however, there is not much work currently available that attempts to perform clinical NER on standardized clinical terms and is available for adoption. We believe that the community will move towards a standardization of clinical terms and that future models developed will aim to use those terms. We also noticed that the clinical NER tasks performed vary based on different domains; our survey found that none of them have used transfer learning approaches to train their models from different domains. We believe that, with the success of transfer learning in [52], the community will be looking to develop their deep learning models using transfer learning from different clinical NER tasks.

Most of the clinical NER tasks that we came across aimed to identify the entities from clinical text and then to use them for other NLP tasks. Given the sensitive nature of the clinical text, it is becoming difficult to publish models that are developed for clinical NER. The community is trying to overcome this by developing clinical NER models that identify sensitive terms/entities from clinical text, remove them, and make them available for publishing. Recently, other ML communities are using GANs (Generative Adversarial Networks) [104], which automatically discover patterns in the data and can develop synthetic data that looks similar to the actual data. This approach has many benefits such as handling privacy as no real data is compromised or used in a training phase, and it is capable of handling under sampling and oversampling for multiple classes. We believe that, in the future, clinical NER models will use GANs to develop more robust and scalable models. Likewise, this approach can be one of the potential approaches for clinical RE.

NER reconciliation is a process of collecting data from multiple sources, gathering and mapping them to a real-world object. In clinical NER, this problem can be more severe, as in the radiation oncology domain, different physicians can assign different names to the same structure. Most of the datasets discussed in this paper are annotated and follow the standard naming convention, but this process is not scalable if multiple data sources are used for integration. We performed an extensive search to find any literature on clinical NER reconciliation. To date, we did not find any attempts to perform clinical NER reconciliation. However, we found a few attempts for NER reconciliation in other domains such as Isaac et al. [105] and Van Holland et al. [106]; these approaches are geared towards vocabulary reconciliation. We believe that clinical NER reconciliation is an open research problem. As mentioned earlier, there are ongoing attempts to standardize the clinical terms, and if such a standardization is widely adopted by physicians, it can make the integration process a lot simpler.

After surveying the clinical RE papers, it was found that, lately, the community is most interested in investigating traditional ML-based approaches, deep learning-based approaches, and language models to perform clinical RE. Very little research using rule-based approaches are coming up but unsupervised learning-based methods for clinical RE have become somewhat dormant because of the uncertainty in the results generated by these methods. Rule-based methods were used in many research works before 2016. With the introduction of newer techniques and newer research over the years, the performance of the clinical RE tasks kept on improving. Later on, traditional ML-based methods and deep learning methods along with different feature representation techniques were adopted for this purpose. It was observed that the traditional methods outperformed the deep learning methods in many cases. In some cases, deep learning methods performed poorer than rule-based methods. This may be due to the limited data used in most of these works. Deep learning methods generally perform better than traditional methods in case of a large amount of data, but clinical data is often limited. This is a practical limitation of using deep learning methods for clinical RE. In this era of supervised learning on clinical texts, it was found that the language models such as BERT and its variations vastly perform the best in extracting relations from clinical texts. This shows that the language models are somewhat capable of understanding the intricacies of the language better. However, experimentation with newer and advanced supervised algorithms for relationship classification in the clinical domain should continue in the future as the performance of the algorithms often vary with the data.

In all of the articles we found on clinical RE, F1-score is the metric used for evaluating the performance of the methods. Although other statistical metrics can be used for this purpose, these works chose to only use the F1-score perhaps because of its popularity. When the dataset is not annotated and unsupervised learning-based algorithms have to be used [102], only then other statistical measures are used to quantify the goodness of those measures; for example, Manhattan, Binary, and Cosine similarities were used for comparing the performance of the various clustering models such as Model-Based, K-Means, and Hierarchical Clustering. However, these measures are only used for assessing the goodness of unsupervised learning-based clustering algorithms to provide high-level model performance estimates as they do not serve as a direct evaluation metric for NER/RE tasks. It was observed that most of the clinical RE tasks from a computational point of view are multi-class classification tasks. However, multi-label classification tasks are not used in large numbers for clinical RE because most datasets are annotated into multiple classes but not into multiple labels most of the time.

8. Conclusions

In this paper, we present the first review of the various interrelated NER and RE methods in the context of clinical text. Our literature survey highlights the increasing popularity of various traditional machine learning-based approaches and deep learning models over the past few years, which has somewhat led to a sharp decline in the usage of rule-based methods for both NER and RE or dictionary-based methods for NER only. Hence, hybrid approaches by combining machine learning-based and rule/dictionary-based approaches have the potential to be one of the dominant approaches for these tasks in the future. On top of that, various other machine learning approaches, deep learning approaches, and language model-based approaches for clinical NER and RE will most probably continue to come up in good numbers in the next few years. GANs, which can automatically discover patterns in the data, can potentially also be a good architecture for clinical NER and RE.

In the case of both NER and RE, the F1-score is the most frequentlly used evaluation metric. For unsupervised clinical RE, some work used different similarity measures such as Manhattan, Binary, and Cosine similarities to measure the goodness of the various unsupervised clustering approaches. A few clinical NER papers have mentioned the usage of t-tests on the models to find out their statistical significance. Other popular metrics used in ML-like sensitivity, specificity, ROC, and AUC can also be used in the future to evaluate the performance of the different approaches used for both NER and RE.

We also believe that the community will move towards a standardization of clinical terms and that the future models developed will aim to use these terms. Standardization will help us integrate data from multiple sources and will also help in NER reconciliation. Clinical NER tasks vary based on different domains; we observed that none of them use transfer learning approaches to train their models from different domains. Developing deep learning models using transfer learning from different clinical NER tasks can be a promising future research direction. In the case of clinical RE, relationships are mostly extracted between entities present in a sentence and the types of relationships are mostly multiclass but not multilabel in most cases. Therefore, from the computational angle, it may be worthwhile to carry out more research on RE across sentences besides also multilabel RE but these tasks require data preparation and annotation in some specific formats.

Author Contributions: The contributions of the authors are listed as follows: conceptualization, P.B., S.S. and P.G.; methodology, P.B. and S.S.; investigation, P.B. and S.S.; resources, W.C.S.IV, R.K., J.P. and P.G.; writing—original draft preparation, P.B. and S.S.; writing—review and editing, P.G., W.C.S.IV and R.K.; supervision, R.K., J.P. and P.G.; project administration, R.K., J.P. and P.G.; funding acquisition, R.K. and J.P. All authors have read and agreed to the published version of the manuscript.

Funding: This research has been funded by the US Veterans Health Administration-National Radiation Oncology Program (VHA-NROP). The results, discussions, and conclusions reported in this paper are completely those of the authors and are independent from the funding sources.

Institutional Review Board Statement: Not applicable.

Informed Consent Statement: Not applicable.

Data Availability Statement: Not applicable.

Conflicts of Interest: The authors declare no conflicts of interest.

Abbreviations

The following abbreviations are used in this manuscript:

IDC	International Data Corporation
TM	Text Mining
NLP	Natural Language Processing
NE	Named Entity
NER	Named Entity Recognition
RE	Relationship Extraction
ML	Machine Learning
AI	Artificial Intelligence
EHR	Electronic Health Record
WEKA	Waikato Environment for Knowledge Analysis
CLAMP	Clinical Language Annotation, Modeling, and Processing
AWS	Amazon Web Services
EU	European Union
HIPAA	Health Insurance Portability and Accountability Act
n2c2	National NLP Clinical Challenges
i2b2	Informatics for Integrating Biology and the Bedside
NIH	National Institutes of Health
NCBC	National Centers for Biomedical Computing
EMR	Electronic Medical Record
SVM	Support Vector Machine
RF	Random Forest
NN	Neural Network
CRF	Conditional Random Field
ME	Maximum Entropy
BERT	Bidirectional Encoder Representations from Transformers
BETO	SPanish BERT
BIO	Beginning, Inside, Outside of Entity
POS	Parts of Speech
BFS	Breadth First Search
CNN	Convolutional Neural Network
RNN	Recurrent Neural Network
LSTM	Long Short Term Memory
GCN	Graph Convolutional Network
CDT	Concept Dependency Tree
GNN	Graph Neural Network
TP	True Positive
TN	True Negative
FP	False Positive
FN	False Negative
ROC	Receiver Operator Characteristic
AUC	Area Under the Curve
ODIE	Ontology Development and Information Extraction
ADE	Adverse Drug Events
PHI	Personal Health Information

Appendix A

Table A1. Summary of previous works in clinical NER.

Publication	Task	Methods	Performance
Savova et al. [46]	Extraction of entities from EMR using NLP tools	Dictionary look-up algorithm	Conducted multiple performance evaluation on different NLP tasks; for NER, the F1-scores reported were 0.71 (exact match) and 0.82 (overlapping matches).
Skeppstedt et al. [56]	Detecting disorders, findings, and body structures from Swedish clinical text	Rule-based and CRF approach	Precision and recall for detecting body structure are 0.88 and 0.82, respectively, while for disorder, they were reported as 0.72 and 0.65; for finding, they are 0.72 and 0.65; and for drug, they are 0.95 and 0.83
Chen et al. [57]	Detecting patients who are qualified for clinical trial	Rule-Based approach using knowledge input defined by lexical, syntactic, or meta-level tasks	F1-score reported was 0.90
Eftimov et al. [48]	Extraction of food entity, nutrient entity, and quantity/unit from dietary recommendations	Rule-based approach	TP for food, nutrient, and quantity was reported as 538, 557, and 86. FN for food, nutrient, and quantity was reported as 25, 17, 11. FP for food, nutrient, and quantity was reported as 5, 2, and none.
Xu et al. [58]	Combined Segmentation and NER on Chinese text	CRF using three features	96% F1-score was recorded as the best performance; the authors also provided a comparison between individual, incremental, and joint models.
Magge et al. [59]	Identification of specific entities from clinical notes such as drug, dose, and route; a total of nine terms were used for identification	Machine ;earning-based approach: bidirectional LSTM-CRF	F1-score average for all nine terms is 0.81; they used the standard gold annotated dataset available at the University of Massachusetts comprising about 1092 medical notes. Around 800 notes were used for training, 76 was for validation, and the rest was used for testing.

Table A2. Summary of previous work for clinical NER.

Publication	Task	Methods	Performance
Nayel et al. [60]	Detection of annotated data from clinical text	Designed an ensemble approach which combined the results of base classifiers and used SVM for learning base classifiers	The proposed ensemble learning model reported an F1-score of 77%.
Wu et al. [4]	Concept extraction from clinical text by using and comparing CNN and RNN	Deep learning-based approach	RNN model performed better when compared with CNN and achieved an F1-score of 86%.

Table A2. *Cont.*

Publication	Task	Methods	Performance
Wang et al. [61]	Studying symptoms and parthenogenesis in Chinese EHR	ML-based approach used CRF, SVM, and Maximum Entropy (ME)	Among all three methods applied, CRF outperformed the others.
Yadav et al. [17]	Advancement and improvement in NER from deep learning models	ML-based approach but focus was more on using deep learning	Better performance reported using deep learning compared with other supervised and semi-supervised learning algorithms.
Vunikili et al. [51]	NER on Spanish Clinical Text to extract tumor morphology	Transfer learning using BERT and BETO	73% F1-score was reported without any features.
Jiang et al. [64]	Extraction of clinical entities from 349 clinical annotated notes with different features	ML-based approach (SVM and CRF)	CRF outperformed SVM and their hybrid system achieved an F1-score of 0.84 for concept extraction and 0.93 for assertion classification.
Yang et al. [65]	Extraction of family history from clinical narratives	Deep learning-based models such as LSTM, BERT, and ensemble models using majority voting strategy	Micro-averaged F1-score of 0.7944 for concept extraction.

Table A3. Summary of the rule-based approaches for clinical RE.

Publication	Task	Methods	Performance
Segura-Bedmar et al. (2011) [84]	Drug–disease interaction extraction from clinical texts	Linguistic hybrid rule-based method using shallow parsing, syntactic simplification, and pattern matching	Did not perform well with an average precision and a very low recall
Xu et al. (2011) [85]	Clinical RE on 2010 i2b2 dataset	Combination of Rule-based and ML methods	Model performed decently with a micro-average F1-score of 0.7326
Li et al. (2015) [86]	Automated extraction of medication discrepancy	Matching of drug names with their attributes from a prescription list and confirming it by means of co-location information	Performed well in identifying the medical discrepancies
Veena et al. (2021) [87]	RE between different clinical words	Path similarity analysis on the terms extracted by scraping and POS tagging	Successfully converted the data into a classified form
Mahendran et al. (2021) [70]	Adverse drug event extraction on 2018 n2c2 dataset	BFS based on the co-location information between the drug and the non-drug entity types	Left-only rule-based approach (macro-average F1-score: 0.83) performed the best amongst other rule-based models

Table A4. Summary of the machine learning-based approaches for clinical RE.

Publication	Task	Methods	Performance
Roberts et al. (2011) [39]	2010 i2b2/VA challenge on concepts, assertions, and relations in clinical text [38]	SVM-based supervised learning algorithm	Best performance with an F1-score of 0.737
Sahu et al. (2016) [76]	Clinical RE on 2010 i2b2 dataset	Domain invariant CNN on multiple features	Decent performance: filter combination of [4, 6] performed the best (F1-score: 0.7116) amongst CNNs
Singhal et al. (2016) [79]	Disease-mutation RE on biomedical texts	C4.5 decision trees on various features	State-of-the art performance thus far; F1-score of 0.880 and 0.845 on prostate and lung disease mutations
Lv et al. (2016) [95]	Clinical RE on 2010 i2b2 dataset	Deep autoencoder-based model and sparse deep autoencoder-based model	Sparse deep autoencoder-based model performed better with an F1-score above 80%
Lin et al. (2017) [80]	Disorder Recognition in the 2013 CLEF task-1 dataset	multi-label structured SVM	Improved Performance: F1-score: 0.7343, i.e., 0.1428 more than the baseline BIOHD1234 scheme.
Mondal et al. (2017) [73]	Clinical RE based on the categories of medical concepts	Feature-oriented SVM-based supervised learning	Better performance (F1-score: 0.86) than the rule-based approach (F1-score: 0.79)
Kim et al. (2018) [72]	Clinical RE for biological pathway	Node2vec to learn the features from texts in networks	Best performance for type 2 diabetes pathway
Munkhdalai et al. (2018) [96]	Clinical RE towards drug surveillance	SVM model and a deep learning-based LSTM model	SVM performed better (89.1% F1-score) than all of the LSTM models
Li et al. (2019) [97]	Clinical RE on 2010 i2b2 dataset	NNs to model the shortest dependency path between entities and sentences	Resulted in an improved performance with an F1-score of 74.34%
Minard et al. (2019) [81]	Clinical RE on 2010 i2b2 dataset	Multi-class SVM	Poor performance (F1-score: 0.70) compared with the previous models
Christopoulou et al. (2020) [79]	Extraction of the adverse drug events and medications relations	An ensemble deep learning method	Achieved a micro-averaged F1-score of 0.9472 and 0.8765 for RE and end-to-end RE, respectively
Hasan et al. (2020) [82]	Clinical RE on 2010 i2b2 dataset	Deep learning methods such as CNN, GCN, GCN-CDT, ResNet, and BiLSTM	BiLSTM performed the best with a nine-class F1-score of 0.8808 and a six-class F1-score of 0.8894
Mahendran et al. (2021) [70]	Adverse drug event extraction on 2018 n2c2 dataset	Sentence-CNN and segment-CNN	The CNN models did not perform better (micro-average F1-score: 0.78 and macro-average F1-score: 0.77) than the other models mentioned

Table A5. Summary of the language model-based approaches for clinical RE.

Publication	Task	Methods	Performance
Lin et al. (2019) [77]	Temporal RE in clinical domain	Pretrained domain-specific as well as fine-tuned BERT	State-of-the art performance; 0.684 F1-score for in-domain texts and 0.565 F1-score for cross-domain texts
Alimova et al. (2020) [83]	Drug–disease RE from biomedical and clinical texts	BERT, BioBERT and Clinical BERT and Random Forest	The BERT models performed much better on the MADE corpus
Wei et al. (2020) [100]	RE on two clinical corpus: 2018 n2c2 dataset and 2010 i2b2 dataset	Fine-tuned and feature-combined BERT along with some deep learning methods	MIMIC fine-tuned BERT performed the best: F1-score of 0.9409 and 0.7679 on the n2c2 and the i2b2 datasets, respectively
Mahendran et al. (2021) [70]	Adverse drug event extraction on 2018 n2c2 dataset	BERT (cased and uncased), BioBERT, and Clinical BERT along with other methods	All of the BERT models performed the best, with a micro-averaged F1-score of 0.94 and a macro-averaged F1-score of 0.93

Table A6. Summary of the unsupervised learning approaches for clinical RE.

Publication	Task	Methods	Performance
Quan et al. (2014) [101]	Protein–protein interactions and gene–suicide association extraction	Clustering based on dependency and phased structure parsing	Performed moderately but the proposed semi-supervised model surpassed its performance
Alicante et al. (2016) [102]	Domain-relevant entities and RE from Italian clinical records	Model Based, K-Means, and Hierarchical Clustering for pattern discovery	Promising performance to introduce a semi-automatic relation labelling

References

1. Gantz, J.; Reinsel, D. The digital universe in 2020: Big data, bigger digital shadows, and biggest growth in the far east. *IDC IView IDC Anal. Future* **2012**, *2007*, 1–16.
2. Tan, A.H. Text mining: The state of the art and the challenges. In Proceedings of the pakdd 1999 Workshop on Knowledge Disocovery from Advanced Databases, Bejing, China, 26–28 April 1999; Volume 8, pp. 65–70.
3. Kong, H.J. Managing unstructured big data in healthcare system. *Healthc. Inform. Res.* **2019**, *25*, 1–2. [CrossRef] [PubMed]
4. Wu, Y.; Jiang, M.; Xu, J.; Zhi, D.; Xu, H. Clinical Named Entity Recognition Using Deep Learning Models. *AMIA Annu. Symp. Proc.* **2017**, *2017*, 1812–1819.
5. Hall, M.; Frank, E.; Holmes, G.; Pfahringer, B.; Reutemann, P.; Witten, I.H. The WEKA Data Mining Software: An Update. *SIGKDD Explor. Newsl.* **2009**, *11*, 10–18. [CrossRef]
6. Soysal, E.; Wang, J.; Jiang, M.; Wu, Y.; Pakhomov, S.; Liu, H.; Xu, H. CLAMP—A toolkit for efficiently building customized clinical natural language processing pipelines. *J. Am. Med. Inform. Assoc.* **2017**, *25*, 331–336. [CrossRef] [PubMed]
7. Bhatia, P.; Celikkaya, B.; Khalilia, M.; Senthivel, S. Comprehend Medical: A Named Entity Recognition and Relationship Extraction Web Service. In Proceedings of the 2019 18th IEEE International Conference On Machine Learning and Applications (ICMLA), Boca Raton, FL, USA, 16–19 December 2019; pp. 1844–1851. [CrossRef]
8. Rish, I. An empirical study of the naive Bayes classifier. In Proceedings of the IJCAI 2001 Workshop on Empirical Methods in Artificial Intelligence, Seattle, WA, USA, 4–6 August 2001.
9. Vishwanathan, S.; Murty, M.N. SSVM: A simple SVM algorithm. In Proceedings of the 2002 International Joint Conference on Neural Networks, IJCNN'02 (Cat. No. 02CH37290), Honolulu, HI, USA, 12–17 May 2002; Volume 3, pp. 2393–2398.
10. Quinlan, J.R. Induction of decision trees. *Mach. Learn.* **1986**, *1*, 81–106. [CrossRef]
11. Hartigan, J.A.; Wong, M.A. Algorithm AS 136: A k-means clustering algorithm. *J. R. Stat. Society. Ser. C (Appl. Stat.)* **1979**, *28*, 100–108. [CrossRef]

12. Schubert, E.; Sander, J.; Ester, M.; Kriegel, H.P.; Xu, X. DBSCAN revisited, revisited: Why and how you should (still) use DBSCAN. *ACM Trans. Database Syst. (TODS)* **2017**, *42*, 1–21. [CrossRef]
13. Aggarwal, C.C.; Zhai, C. A survey of text classification algorithms. In *Mining Text Data*; Springer: Boston, MA, USA, 2012; pp. 163–222.
14. Derr, T.; Karimi, H.; Liu, X.; Xu, J.; Tang, J. Deep Adversarial Network Alignment. *arXiv* **2019**, arXiv:cs.SI/1902.10307.
15. Watkins, C.J.; Dayan, P. Q-learning. *Mach. Learn.* **1992**, *8*, 279–292. [CrossRef]
16. Abiodun, O.I.; Jantan, A.; Omolara, A.E.; Dada, K.V.; Mohamed, N.A.; Arshad, H. State-of-the-art in artificial neural network applications: A survey. *Heliyon* **2018**, *4*, e00938. [CrossRef]
17. Yadav, V.; Bethard, S. A survey on recent advances in named entity recognition from deep learning models. *arXiv* **2019**, arXiv:1910.11470.
18. Grishman, R.; Sundheim, B. Message understanding conference-6: A brief history. In Proceedings of the 1995 International Conference on Computational Linguistics (COLING), Copenhagen, Denmark, 5–9 August 1995.
19. Collobert, R.; Weston, J.; Bottou, L.; Karlen, M.; Kavukcuoglu, K.; Kuksa, P. Natural language processing (almost) from scratch. *J. Mach. Learn. Res.* **2011**, *12*, 2493–2537.
20. Bach, N.; Badaskar, S. A review of relation extraction. *Lit. Rev. Lang. Stat. II* **2007**, *2*, 1–15.
21. Brin, S. Extracting Patterns and Relations from the World Wide Web. In *The World Wide Web and Databases, Proceedings of the International Workshop WebDB'98, Valencia, Spain, 27–28 March 1998*; Atzeni, P., Mendelzon, A., Mecca, G., Eds.; Springer: Berlin/Heidelberg, Germany, 1999; pp. 172–183.
22. Agichtein, E.; Gravano, L. Snowball: Extracting Relations from Large Plain-Text Collections. In Proceedings of the Fifth ACM Conference on Digital Libraries (DL'00), San Antonio, TX, USA, 2–7 June 2000; Association for Computing Machinery: New York, NY, USA, 2000; pp. 85–94. [CrossRef]
23. Culotta, A.; Sorensen, J. Dependency Tree Kernels for Relation Extraction. In Proceedings of the 42nd Annual Meeting of the Association for Computational Linguistics (ACL-04), Barcelona, Spain, 21–26 July 2004; pp. 423–429. [CrossRef]
24. Bunescu, R.C.; Mooney, R.J. A Shortest Path Dependency Kernel for Relation Extraction. In Proceedings of the HLT/EMNLP, Vancouver, BC, Canada, 6–8 October 2005; pp. 724–731.
25. Bunescu, R.C.; Mooney, R.J. Subsequence Kernels for Relation Extraction. In Proceedings of the 18th International Conference on Neural Information Processing Systems (NIPS'05), Vancouver, BC, Canada, 5–8 December 2005; MIT Press: Cambridge, MA, USA, 2005; pp. 171–178.
26. Culotta, A.; McCallum, A.; Betz, J. Integrating Probabilistic Extraction Models and Data Mining to Discover Relations and Patterns in Text. In Proceedings of the HLT-NAACL, New York, NY, USA, 4–9 June 2006.
27. Allahyari, M.; Pouriyeh, S.; Assefi, M.; Safaei, S.; Trippe, E.D.; Gutierrez, J.B.; Kochut, K. A Brief Survey of Text Mining: Classification, Clustering and Extraction Techniques. *arXiv* **2017**, arXiv:cs.CL/1707.02919.
28. Hedderich, M.A.; Lange, L.; Adel, H.; Strötgen, J.; Klakow, D. A survey on recent approaches for natural language processing in low-resource scenarios. *arXiv* **2020**, arXiv:2010.12309.
29. Stubbs, A.; Uzuner, Ö. Annotating longitudinal clinical narratives for de-identification: The 2014 i2b2/UTHealth corpus. *J. Biomed. Inform.* **2015**, *58*, S20–S29. [CrossRef]
30. Stubbs, A.; Kotfila, C.; Xu, H.; Uzuner, Ö. Identifying risk factors for heart disease over time: Overview of 2014 i2b2/UTHealth shared task Track 2. *J. Biomed. Inform.* **2015**, *58*, S67–S77. [CrossRef]
31. Stubbs, A.; Filannino, M.; Uzuner, Ö. De-identification of psychiatric intake records: Overview of 2016 CEGS N-GRID Shared Tasks Track 1. *J. Biomed. Inform.* **2017**, *75*, S4–S18. [CrossRef]
32. Goto, I.; Chow, K.P.; Lu, B.; Sumita, E.; Tsou, B.K. Overview of the Patent Machine Translation Task at the NTCIR-10 Workshop. In Proceedings of the NTCIR, Tokyo, Japan, 18–21 June 2013.
33. Coffman, A.; Wharton, N. Clinical Natural Language Processing: Auto-Assigning ICD-9 Codes. Overview of the Computational Medicine Center's. 2007. Available online: https://courses.ischool.berkeley.edu/i256/f09/Final%20Projects%20write-ups/coffman_wharton_project_final.pdf (accessed on 2 September 2012).
34. Jagannatha, A.; Liu, F.; Liu, W.; Yu, H. Overview of the first natural language processing challenge for extracting medication, indication, and adverse drug events from electronic health record notes (MADE 1.0). *Drug Saf.* **2019**, *42*, 99–111. [CrossRef]
35. Liu, F.; Jagannatha, A.; Yu, H. Towards Drug Safety Surveillance and Pharmacovigilance: Current Progress in Detecting Medication and Adverse Drug Events from Electronic Health Records. *Drug Saf.* **2019**, *42*, 95–97. [CrossRef]
36. Pradhan, S.; Chapman, W.; Man, S.; Savova, G. Semeval-2014 task 7: Analysis of clinical text. In Proceedings of the 8th International Workshop on Semantic Evaluation (SemEval 2014), Dublin, Ireland, 23–24 August 2014.
37. Pradhan, S.; Elhadad, N.; South, B.R.; Martinez, D.; Christensen, L.; Vogel, A.; Suominen, H.; Chapman, W.W.; Savova, G. Evaluating the state of the art in disorder recognition and normalization of the clinical narrative. *J. Am. Med. Inform. Assoc.* **2015**, *22*, 143–154. [CrossRef]
38. Uzuner, Ö.; South, B.R.; Shen, S.; DuVall, S.L. 2010 i2b2/VA challenge on concepts, assertions, and relations in clinical text. *J. Am. Med. Inform. Assoc.* **2011**, *18*, 552–556. [CrossRef]
39. Roberts, K.; Rink, B.; Harabagiu, S. Extraction of medical concepts, assertions, and relations from discharge summaries for the fourth i2b2/VA shared task. In Proceedings of the 2010 i2b2/VA Workshop on Challenges in Natural Language Processing for Clinical Data, Washington, DC, USA, 12–13 November 2010.

40. Uzuner, O.; Bodnari, A.; Shen, S.; Forbush, T.; Pestian, J.; South, B.R. Evaluating the state of the art in coreference resolution for electronic medical records. *J. Am. Med. Inform. Assoc.* **2012**, *19*, 786–791. [CrossRef]
41. Glinos, D. A search based method for clinical text coreference resolution. In Proceedings of the 2011 i2b2/VA/Cincinnati Workshop on Challenges in Natural Language Processing for Clinical Data, Washington, DC, USA, 21–22 October 2011.
42. Xu, Y.; Liu, J.; Wu, J. EHUATUO: A mention-pair coreference system by exploiting document intrinsic latent structures and world knowledge in discharge summaries: 2011 i2b2 challenge. In Proceedings of the 2011 i2b2/VA/Cincinnati Workshop on Challenges in Natural Language Processing for Clinical Data, Washington, DC, USA, 21–22 October 2011.
43. Sun, W.; Rumshisky, A.; Uzuner, O. Evaluating temporal relations in clinical text: 2012 i2b2 Challenge. *J. Am. Med. Inform. Assoc.* **2013**, *20*, 806–813. [CrossRef]
44. Henry, S.; Buchan, K.; Filannino, M.; Stubbs, A.; Uzuner, O. 2018 n2c2 shared task on adverse drug events and medication extraction in electronic health records. *J. Am. Med. Inform. Assoc.* **2019**, *27*, 3–12. [CrossRef]
45. Xu, J.; Lee, H.J.; Ji, Z.; Wang, J.; Wei, Q.; Xu, H. *UTH_CCB System for Adverse Drug Reaction Extraction from Drug Labels at TAC-ADR 2017*; TAC: Gaithersburg, MD, USA, 2017.
46. Savova, G.K.; Masanz, J.J.; Ogren, P.V.; Zheng, J.; Sohn, S.; Kipper-Schuler, K.C.; Chute, C.G. Mayo clinical Text Analysis and Knowledge Extraction System (cTAKES): Architecture, component evaluation and applications. *J. Am. Med. Inform. Assoc.* **2010**, *17*, 507–513. [CrossRef] [PubMed]
47. Olson, J.E.; Ryu, E.; Johnson, K.J.; Koenig, B.A.; Maschke, K.J.; Morrisette, J.A.; Liebow, M.; Takahashi, P.Y.; Fredericksen, Z.S.; Sharma, R.G.; et al. The Mayo Clinic Biobank: A building block for individualized medicine. *Mayo Clin. Proc.* **2013**, *88*, 952–962. [CrossRef]
48. Popovski, G.; Seljak, B.K.; Eftimov, T. A survey of named-entity recognition methods for food information extraction. *IEEE Access* **2020**, *8*, 31586–31594. [CrossRef]
49. Weegar, R.; Pérez, A.; Casillas, A.; Oronoz, M. Recent advances in Swedish and Spanish medical entity recognition in clinical texts using deep neural approaches. *BMC Med. Inform. Decis. Mak.* **2019**, *19*, 274. [CrossRef]
50. Johnson, A.E.; Pollard, T.J.; Shen, L.; Li-Wei, H.L.; Feng, M.; Ghassemi, M.; Moody, B.; Szolovits, P.; Celi, L.A.; Mark, R.G. MIMIC-III, a freely accessible critical care database. *Sci. Data* **2016**, *3*, 1–9. [CrossRef]
51. Vunikili, R.; SH, N.; Marica, G.; Farri, O. Clinical NER using Spanish BERT Embeddings. In Proceedings of the Iberian Languages Evaluation Forum (IberLEF 2020), Malaga, Spain, 23 September 2020.
52. Catelli, R.; Gargiulo, F.; Casola, V.; De Pietro, G.; Fujita, H.; Esposito, M. Crosslingual named entity recognition for clinical de-identification applied to a COVID-19 Italian data set. *Appl. Soft Comput.* **2020**, *97*, 106779. [CrossRef]
53. Nalluri, J.; Kapoor, R.; Sleeman, W.; Soni, P.; Ghosh, P.; Khajamoinuddin, S.; Hagan, M.; Palta, J. Health Information and Gateway Exchange (HINGE): Big Data Curation Tool for Radiation Oncology. *Int. J. Radiat. Oncol. Biol. Phys.* **2019**, *105*, E132. [CrossRef]
54. Kapoor, R.; Sleeman, W.C., IV; Nalluri, J.J.; Turner, P.; Bose, P.; Cherevko, A.; Srinivasan, S.; Syed, K.; Ghosh, P.; Hagan, M.; et al. Automated data abstraction for quality surveillance and outcome assessment in radiation oncology. *J. Appl. Clin. Med. Phys.* **2021**, *22*, 177–187. [CrossRef] [PubMed]
55. Lample, G.; Ballesteros, M.; Subramanian, S.; Kawakami, K.; Dyer, C. Neural architectures for named entity recognition. *arXiv* **2016**, arXiv:1603.01360.
56. Skeppstedt, M.; Kvist, M.; Dalianis, H. Rule-based Entity Recognition and Coverage of SNOMED CT in Swedish Clinical Text. In Proceedings of the LREC, Istanbul, Turkey, 23–25 May 2012; pp. 1250–1257.
57. Chen, L.; Gu, Y.; Ji, X.; Lou, C.; Sun, Z.; Li, H.; Gao, Y.; Huang, Y. Clinical trial cohort selection based on multi-level rule-based natural language processing system. *J. Am. Med. Inform. Assoc.* **2019**, *26*, 1218–1226. Available online: https://academic.oup.com/jamia/article-pdf/26/11/1218/36089031/ocz109.pdf (accessed on 13 July 2019). [CrossRef]
58. Xu, Y.; Wang, Y.; Liu, T.; Liu, J.; Fan, Y.; Qian, Y.; Tsujii, J.; Chang, E.I. Joint segmentation and named entity recognition using dual decomposition in Chinese discharge summaries. *J. Am. Med. Inform. Assoc.* **2014**, *21*, e84–e92. [CrossRef] [PubMed]
59. Magge, A.; Scotch, M.; Gonzalez-Hernandez, G. Clinical NER and relation extraction using bi-char-LSTMs and random forest classifiers. In Proceedings of the International Workshop on Medication and Adverse Drug Event Detection, Virtual, 4 May 2018; pp. 25–30.
60. Nayel, H.; Shashirekha, H. Improving NER for clinical texts by ensemble approach using segment representations. In Proceedings of the 14th International Conference on Natural Language Processing (ICON-2017), Kolkata, India, 18–21 December 2017; pp. 197–204.
61. Wang, Y.; Yu, Z.; Chen, L.; Chen, Y.; Liu, Y.; Hu, X.; Jiang, Y. Supervised methods for symptom name recognition in free-text clinical records of traditional Chinese medicine: An empirical study. *J. Biomed. Inform.* **2014**, *47*, 91–104. [CrossRef] [PubMed]
62. Devlin, J.; Chang, M.W.; Lee, K.; Toutanova, K. Bert: Pre-training of deep bidirectional transformers for language understanding. *arXiv* **2018**, arXiv:1810.04805.
63. Canete, J.; Chaperon, G.; Fuentes, R.; Pérez, J. Spanish pre-trained bert model and evaluation data. In Proceedings of the PML4DC, ICLR 2020, Addis Aboba, Ethiopia, 26 April 2020.
64. Jiang, M.; Chen, Y.; Liu, M.; Rosenbloom, S.T.; Mani, S.; Denny, J.C.; Xu, H. A study of machine-learning-based approaches to extract clinical entities and their assertions from discharge summaries. *J. Am. Med. Inform. Assoc.* **2011**, *18*, 601–606. [CrossRef]
65. Yang, X.; Zhang, H.; He, X.; Bian, J.; Wu, Y. Extracting Family History of Patients From Clinical Narratives: Exploring an End-to-End Solution With Deep Learning Models. *JMIR Med. Inform.* **2020**, *8*, e22982. [CrossRef]

66. Hsu, T.C.; Feldt, L.S. The effect of limitations on the number of criterion score values on the significance level of the F-test. *Am. Educ. Res. J.* **1969**, *6*, 515–527.
67. Branco, P.; Torgo, L.; Ribeiro, R.P. Relevance-based evaluation metrics for multi-class imbalanced domains. In Proceedings of the Pacific-Asia Conference on Knowledge Discovery and Data Mining, Jeju, Korea, 23–26 May 2017; Springer: Cham, Switzerland, 2017; pp. 698–710.
68. Law, V.; Knox, C.; Djoumbou, Y.; Jewison, T.; Guo, A.C.; Liu, Y.; Maciejewski, A.; Arndt, D.; Wilson, M.; Neveu, V.; et al. DrugBank 4.0: Shedding new light on drug metabolism. *Nucleic Acids Res.* **2013**, *42*, D1091–D1097. Available online: https://academic.oup.com/nar/article-pdf/42/D1/D1091/3559045/gkt1068.pdf (accessed on 13 July 2019). [CrossRef] [PubMed]
69. Hebbring, S.J. The challenges, advantages and future of phenome-wide association studies. *Immunology* **2014**, *141*, 157–165. Available online: https://onlinelibrary.wiley.com/doi/pdf/10.1111/imm.12195 (accessed on 13 July 2019). [CrossRef] [PubMed]
70. Mahendran, D.; McInnes, B.T. Extracting Adverse Drug Events from Clinical Notes. *arXiv* **2021**, arXiv:cs.CL/2104.10791.
71. Sarker, A.; Gonzalez, G. Portable automatic text classification for adverse drug reaction detection via multi-corpus training. *J. Biomed. Inform.* **2015**, *53*, 196–207. [CrossRef]
72. Kim, M. Relation extraction for biological pathway construction using node2vec. *BMC Bioinform.* **2018**, *19*, 206. [CrossRef] [PubMed]
73. Mondal, A.; Das, D.; Bandyopadhyay, S. Relationship Extraction based on Category of Medical Concepts from Lexical Contexts. In Proceedings of the 14th International Conference on Natural Language Processing (ICON-2017), Kolkata, India, 18–21 December 2017; NLP Association of India: Kolkata, India, 2017; pp. 212–219.
74. Singhal, A.; Simmons, M.; Lu, Z. Text mining for precision medicine: Automating disease-mutation relationship extraction from biomedical literature. *J. Am. Med. Inform. Assoc.* **2016**, *23*, 766–772. [CrossRef]
75. Lim, C.G.; Choi, H.J. Temporal Relationship Extraction for Natural Language Texts by Using Deep Bidirectional Language Model. In Proceedings of the 2020 IEEE International Conference on Big Data and Smart Computing (BigComp), Busan, Korea, 19–22 February 2020; pp. 555–557. [CrossRef]
76. Sahu, S.K.; Anand, A.; Oruganty, K.; Gattu, M. Relation extraction from clinical texts using domain invariant convolutional neural network. *arXiv* **2016**, arXiv:cs.CL/1606.09370.
77. Lin, C.; Miller, T.; Dligach, D.; Bethard, S.; Savova, G. A BERT-based universal model for both within-and cross-sentence clinical temporal relation extraction. In Proceedings of the 2nd Clinical Natural Language Processing Workshop, Minneapolis, MN, USA, 7 June 2019; pp. 65–71.
78. Swampillai, K.; Stevenson, M. Extracting relations within and across sentences. In Proceedings of the International Conference Recent Advances in Natural Language Processing 2011, Hissar, Bulgaria, 12–14 September 2011; pp. 25–32.
79. Christopoulou, F.; Tran, T.T.; Sahu, S.K.; Miwa, M.; Ananiadou, S. Adverse drug events and medication relation extraction in electronic health records with ensemble deep learning methods. *J. Am. Med. Inform. Assoc.* **2019**, *27*, 39–46. [CrossRef] [PubMed]
80. Lin, W.; Ji, D.; Lu, Y. Disorder recognition in clinical texts using multi-label structured SVM. *BMC Bioinform.* **2017**, *18*, 75. [CrossRef] [PubMed]
81. Minard, A.L.; Ligozat, A.L.; Grau, B. Multi-class SVM for relation extraction from clinical reports. In Proceedings of the Recent Advances in Natural Language Processing, Varna, Bulgaria, 2–4 September 2011.
82. Hasan, F.; Roy, A.; Pan, S. Integrating Text Embedding with Traditional NLP Features for Clinical Relation Extraction. In Proceedings of the 2020 IEEE 32nd International Conference on Tools with Artificial Intelligence (ICTAI), Baltimore, MD, USA, 9–11 November 2020; pp. 418–425. [CrossRef]
83. Alimova, I.; Tutubalina, E. Multiple features for clinical relation extraction: A machine learning approach. *J. Biomed. Inform.* **2020**, *103*, 103382. [CrossRef]
84. Segura-Bedmar, I.; Martínez, P.; de Pablo-Sánchez, C. A linguistic rule-based approach to extract drug-drug interactions from pharmacological documents. *BMC Bioinform.* **2011**, *12*, S1. [CrossRef]
85. Xu, Y.; Hong, K.; Tsujii, J.; Chang, E.I.C. Feature engineering combined with machine learning and rule-based methods for structured information extraction from narrative clinical discharge summaries. *J. Am. Med. Inform. Assoc.* **2012**, *19*, 824–832. [CrossRef]
86. Li, Q.; Spooner, S.A.; Kaiser, M.; Lingren, N.; Robbins, J.; Lingren, T.; Tang, H.; Solti, I.; Ni, Y. An end-to-end hybrid algorithm for automated medication discrepancy detection. *BMC Med. Inform. Decis. Mak.* **2015**, *15*, 37. [CrossRef]
87. Veena, G.; Hemanth, R.; Hareesh, J. Relation Extraction in Clinical Text using NLP Based Regular Expressions. In Proceedings of the 2019 2nd International Conference on Intelligent Computing, Instrumentation and Control Technologies (ICICICT), Kannur, India, 5–6 July 2019; Volume 1, pp. 1278–1282. [CrossRef]
88. Russell, S.; Norvig, P. *Artificial Intelligence: A Modern Approach*; Prentice Hall: Hoboken, NJ, USA, 2002.
89. Janiesch, C.; Zschech, P.; Heinrich, K. Machine learning and deep learning. *Electron. Mark.* **2021**. [CrossRef]
90. Bose, P.; Sleeman, W.C.; Syed, K.; Hagan, M.; Palta, J.; Kapoor, R.; Ghosh, P. Deep Neural Network Models to Automate Incident Triage in the Radiation Oncology Incident Learning System. In Proceedings of the 12th ACM International Conference on Bioinformatics, Computational Biology and Health Informatics (BCB'21), Gainesville, FL, USA, 1–4 August 2021; Association for Computing Machinery: New York, NY, USA, 2021. [CrossRef]
91. Watson, D.S.; Krutzinna, J.; Bruce, I.N.; Griffiths, C.E.; McInnes, I.B.; Barnes, M.R.; Floridi, L. Clinical applications of machine learning algorithms: Beyond the black box. *BMJ* **2019**, *364*, l886. [CrossRef]

92. Weng, S.F.; Reps, J.; Kai, J.; Garibaldi, J.M.; Qureshi, N. Can machine-learning improve cardiovascular risk prediction using routine clinical data? *PLoS ONE* **2017**, *12*, e0174944. [CrossRef]
93. Sleeman, W.; Bose, P.; Ghosh, P.; Palta, J.; Kapoor, R. Using CNNs to Extract Standard Structure Names While Learning Radiomic Features. In *Medical Physics*; Wiley: Hoboken, NJ, USA, 2021; Volume 48.
94. Bose, P.; Sleeman, W.; Srinivasan, S.; Palta, J.; Kapoor, R.; Ghosh, P. Integrated Structure Name Mapping with CNN. In *Medical Physics*; Wiley: Hoboken, NJ, USA, 2021; Volume 48.
95. Lv, X.; Guan, Y.; Yang, J.; Wu, J. Clinical relation extraction with deep learning. *Int. J. Hybrid Inf. Technol.* **2016**, *9*, 237–248. [CrossRef]
96. Munkhdalai, T.; Liu, F.; Yu, H. Clinical Relation Extraction Toward Drug Safety Surveillance Using Electronic Health Record Narratives: Classical Learning Versus Deep Learning. *JMIR Public Health Surveill.* **2018**, *4*, e29. [CrossRef]
97. Li, Z.; Yang, Z.; Shen, C.; Xu, J.; Zhang, Y.; Xu, H. Integrating shortest dependency path and sentence sequence into a deep learning framework for relation extraction in clinical text. *BMC Med. Inf. Decis. Mak.* **2019**, *19*, 22. [CrossRef]
98. Lee, J.; Yoon, W.; Kim, S.; Kim, D.; Kim, S.; So, C.H.; Kang, J. BioBERT: A pre-trained biomedical language representation model for biomedical text mining. *Bioinformatics* **2019**, *36*, 1234–1240. [CrossRef]
99. Alsentzer, E.; Murphy, J.R.; Boag, W.; Weng, W.H.; Jin, D.; Naumann, T.; McDermott, M.B.A. Publicly Available Clinical BERT Embeddings. *arXiv* **2019**, arXiv:cs.CL/1904.03323.
100. Wei, Q.; Ji, Z.; Si, Y.; Du, J.; Wang, J.; Tiryaki, F.; Wu, S.; Tao, C.; Roberts, K.; Xu, H. Relation Extraction from Clinical Narratives Using Pre-trained Language Models. *AMIA Annu. Symp. Proc.* **2020**, *2019*, 1236–1245.
101. Quan, C.; Wang, M.; Ren, F. An Unsupervised Text Mining Method for Relation Extraction from Biomedical Literature. *PLoS ONE* **2014**, *9*, e102039. [CrossRef] [PubMed]
102. Alicante, A.; Corazza, A.; Isgrò, F.; Silvestri, S. Unsupervised entity and relation extraction from clinical records in Italian. *Comput. Biol. Med.* **2016**, *72*, 263–275. [CrossRef] [PubMed]
103. Hand, D.J.; Christen, P.; Kirielle, N. F*: An interpretable transformation of the F-measure. *Mach. Learn.* **2021**, *110*, 451–456. [CrossRef]
104. Aggarwal, A.; Mittal, M.; Battineni, G. Generative adversarial network: An overview of theory and applications. *Int. J. Inf. Manag. Data Insights* **2021**, *1*, 100004.
105. Isaac, A.; Schlobach, S.; Matthezing, H.; Zinn, C. Integrated access to cultural heritage resources through representation and alignment of controlled vocabularies. *Libr. Rev.* **2008**, *57*, 187–199. [CrossRef]
106. Van Hooland, S.; Verborgh, R.; De Wilde, M.; Hercher, J.; Mannens, E.; Van de Walle, R. Evaluating the success of vocabulary reconciliation for cultural heritage collections. *J. Am. Soc. Inf. Sci. Technol.* **2013**, *64*, 464–479. [CrossRef]

applied sciences

MDPI

Article

Question Difficulty Estimation Based on Attention Model for Question Answering

Hyun-Je Song [1,†], Su-Hwan Yoon [2,†] and Seong-Bae Park [2,*]

[1] Department of Information Technology, Jeonbuk National University, Jeonju 54896, Korea; hyunje.song@jbnu.ac.kr
[2] Department of Computer Science and Engineering, Kyung Hee University, Youngin 17104, Korea; yunsh3432@khu.ac.kr
* Correspondence: sbpark71@khu.ac.kr
† These authors contributed equally to this work.

Abstract: This paper addresses a question difficulty estimation of which goal is to estimate the difficulty level of a given question in question-answering (QA) tasks. Since a question in the tasks is composed of a questionary sentence and a set of information components such as a description and candidate answers, it is important to model the relationship among the information components to estimate the difficulty level of the question. However, existing approaches to this task modeled a simple relationship such as a relationship between a questionary sentence and a description, but such simple relationships are insufficient to predict the difficulty level accurately. Therefore, this paper proposes an attention-based model to consider the complicated relationship among the information components. The proposed model first represents bi-directional relationships between a questionary sentence and each information component using a dual multi-head co-attention, since the questionary sentence is a key factor in the QA questions and it affects and is affected by information components. Then, the proposed model considers inter-information relationship over the bi-directional representations through a self-attention model. The inter-information relationship helps predict the difficulty of the questions accurately which require reasoning over multiple kinds of information components. The experimental results from three well-known and real-world QA data sets prove that the proposed model outperforms the previous state-of-the-art and pre-trained language model baselines. It is also shown that the proposed model is robust against the increase of the number of information components.

Keywords: attention model; dual multi-head attention; inter-information relationship; question answering; question difficult estimation

Citation: Song, H.-J.; Yoon, S.H.; Park, S.-B. Question Difficulty Estimation Based on Attention Model for Question Answering. *Appl. Sci.* **2021**, *11*, 12023. https://doi.org/10.3390/app112412023

Academic Editors: Arturo Montejo-Ráez and Salud María Jiménez-Zafra

Received: 7 November 2021
Accepted: 14 December 2021
Published: 17 December 2021

1. Introduction

Question-Answering (QA) is an important natural language processing task in which a model understands questions and answers them based on its understanding of the questions. Several QA tasks such as ARC [1], SQuAD [2], and HotpotQA [3] were recently proposed, and many QA models based on a pre-trained language model have been developed to solve these QA tasks [4–7]. In these QA tasks, the questions are in general prepared without consideration of difficulty. Therefore, the QA models attacking the tasks do not recognize the difficulty of each question even though the difficulty is important information to answer the questions [8]. As a result, a difficulty level is tagged in new QA tasks such as DramaQA [9] in conjunction with Piaget's theory [10].

All QA tasks do not contain information about question difficulty, but the difficulty exists latently in their questions. The questions in a QA task can be regarded as easy if they are correctly answered by many answering models, and they can be considered as difficult if few models give a correct answer for them. When investigating (This investigation was done on 10 November 2020) the questions in the QuAC task with top three single models from the leaderboard of the task, we found out that only 10% of the questions are answered correctly by all three models while about 50% are not answered correctly by any of the models. Besides, some QA tasks have intrinsic question difficulty. For instance, the RACE data set was collected by two subgroups of middle school examinations and high school examinations, respectively. Thus, the questions from middle school examinations are easier than those from high school examinations.

This paper deals with a question difficulty estimation of which goal is to estimate the difficulty level of a given question. Predicting the difficulty level of the question helps create adversarial QA datasets [11] or identify the way in which QA models challenges. Most previous studies on this task extracted some difficulty features from questions and then predicted the difficulty level of the questions with the features using machine learning algorithms [12–16]. These features were designed to model the relationship between a questionary sentence and associated information components such as a passage or candidate answers. However, some recent QA studies have shown that inter-information relationship is vital since many difficult questions can be answered through reasoning over multiple kinds of information components [17]. For such an example, Figure 1 shows a question in the RACE task. To answer this question, an answering model has to identify the relationship between the passage and a candidate answer (marked in cyan) as well as the relationship between the questionary sentence and the passage (marked in green). As in the question answering, these relationships are important factors also in estimating the question difficulty. Especially, the inter-information relationship should be considered explicitly because they are directly related to the question difficulty, but no previous studies made many efforts to consider the relationship.

questionary sentence		Why did Mami experience culture shock in Japan?
Information components	passage	A Japanese student called Mami told me about her own experiences in British. She spent 10 months in the UK last year, studying English at a language school. She really enjoyed her first two weeks in the UK. But soon she started to miss things of her own country. (...) To comfort herself Mami began to spend many hours on the Internet chatting with her friends back home. She spent a couple of weeks in the countryside in Kent. She went to a social club for British people who were interested in Japan and started to make some friends there. In addition, she took a short course in calligraphy to get an opportunity of mixing with local people. A few months later, Mami's impression of the UK had greatly changed. She found that most of the British were friendly, witty and fun. However, once Mami was back in Japan, she experienced "culture shock" again. She said, "I missed the friends I had made in England. My way of thinking had changed. Sometimes I was annoyed by the views of people in my country—for example, about the value of money and time. I thought people around me lived in such a small world." Mami noticed some changes in her behaviour: "I kept the habit of always carrying an umbrella with me, even on a fine day—my friends thought I was crazy!"
	candidate answers	1. She didn't like Japanese culture any more. 2. The Japanese behaviour had changed a lot. 3. The world in Japan was too small for her. 4. She had got used to British culture and life.

Figure 1. An example question in the RACE data set that is difficult to answer without inter-information inference. The inter-information clues are marked in green and cyan, and the underline in the candidate answers implies a correct answer. (best view in color).

This paper proposes an attention-based model that estimates the difficulty of a question. The proposed attention model is designed to consider the inter-information relationship as well as the relationships between a questionary sentence and each information component. To be specific, the proposed model represents each type of the relationships consecutively and adopts the attention mechanism to capture both types of relationships. That is, the relationships between a questionary sentence and each information component are first identified by the dual multi-head attention designed to capture a bi-directional relationship with two multi-head attentions. Since a single directional relationship is not sufficient for QA tasks [18], the proposed model captures bi-directional relationships between a questionary sentence and information components through the dual multi-head attention. Note that the bi-directional relationships do not reflect an inter-relationship among various information components fully. Thus, the proposed model represents the inter-information relationship by applying a multi-head attention again to the outputs of the dual multi-head attentions. That is, it first expresses the bi-directional relationships between a questionary sentence and each information component, and then accumulates the inter-information relationship onto the concatenation of the bi-directional relationship representations using the transformer encoder. Finally, it determines the difficulty of the question from the accumulated representation since the representation contains all information about the question components and their relationships.

The proposed model is verified with three QA data sets of RACE, QuAC, and DramaQA. Note that not all datasets are attached with the difficulty levels. DramaQA is manually tagged with four difficulty levels but RACE and QuAC are not tagged. For RACE dataset, we regard the middle school examinations as easy questions and high school examinations as hard questions. For QuAC, the difficulty levels are tagged using the results of multiple QA models [19]. The experimental results show the effectiveness of the proposed model in two folds. One is that the proposed model outperforms current state-of-the-art and pre-trained language models, and the other is that the performance of question difficulty estimation is improved by considering inter-information relationship. In particular, the proposed model achieves 68.37 of F1-score in QuAC. This is 8.5 higher than the F1-scores of the state-of-the-art pre-trained language models. It is also shown that the performance of the proposed model improves monotonically as the number of information components increases. The major performance improvement of the proposed model is made from difficult questions, since the proposed model is robust against the increase of the number of information components.

The major contributions of this paper can be summarized as follows:

- We formally define the question difficult estimation as estimating the difficulty level of a given question in question-answering tasks. The question difficult estimation for any question answering tasks can be formulated using the proposed definition.
- We design an attention-based model for question difficulty estimation. The proposed attention-based model captures the relationship among the information components as well as the inter-relationships between a questionary sentence and each information component.
- We examine the performance of the proposed model with intensive experiments on three real-world QA data sets. The intensive experiments validate the effectiveness of the proposed model.
- We empirically show that the performance of question answering is improved by adding the difficulty level.

The rest of this paper is organized as follows. Section 2 reviews related studies on question difficult estimation, and Section 3 presents the proposed model, the attention-based question difficulty estimator. The experimental results and discussions are given in Section 4. Finally, Section 5 draws some conclusions.

2. Related Work

Question answering is a task of answering a question where the question consists of a questionary sentence written in the natural language and a set of information components. Depending on the domain of the main information component, QA tasks are categorized into text-based [2,3], table-based [20,21], image-based [22,23], video-based [8,24], and so on. All QA tasks require an understanding of a question to answer it regardless of QA types. One key factor for the question understanding is the difficulty of the question [8], so that there have been many efforts to measure the difficulty of questions [14,16].

The efforts for the question difficulty estimation can be clustered into two types. The first type defines hand-crafted features from given QA materials. For instance, a question and its associated passage are usually given in the reading comprehension, where the passage provides background information of the question. Thus, the question difficulty is estimated with the information residing in the question and the passage. Desai and Moldovan defined, as such information, six features that are question length, cosine similarity between a question and a passage, the nature of a question and its answer, the number of clauses and prepositional phrases in a question, and existence of discourse connectives in a question [12]. On the other hand, Ha et al. defined the features for multiple-choice examinations [13]. Since they focused on medical examinations, they do not include only lexical, syntactic, and semantic features from a question and candidate answers, but also some cognitively-motivated features from a medical database. The main problem of these studies is that it is extremely difficult to design the features without profound knowledge about the reading materials.

The other type is to adopt a machine learning method to predict question difficulty without manual features. Since every QA task has its own idiosyncratic circumstances, the previous studies attacked question difficulty estimation by focusing on a specific task. Huang et al. estimated question difficulty for standard English tests in which each problem consists of a question, a reading passage, and candidate answers [14]. They proposed a CNN-like architecture to represent all sentences in the question, the passage, and the candidate answers as vectors, and adopted an attention mechanism to reflect the relevancy of the sentences in the passage and candidate answers to the question. Qiu et al. estimated question difficulty for multiple-choice problems at medical examinations [16]. Unlike English tests, the problems of medical examinations do not have a passage, but a set of documents related to a question. Thus, they measured two kinds of difficulties: the difficulty of searching the documents for potential answers of a question and the confusion difficulty among candidate answers. Then, the final difficulty of a question is determined by their weighted sum. Xue et al. expressed a question and candidate answers as embedding vectors by a pre-trained language model, ELMo, and then predicted the difficulty of the question using a simple linear regression of which input is the embedding vectors [25].

Note that many QA tasks provide some information components of a question as well as the question itself. Thus, the studies about representing inter-information have been performed [26], and they are grouped into two types according to the approach to expressing inter-information. One is to adopt a graph of which nodes are the entities appearing at information components and edges are a relation between the entities. Cao et al. expressed the relations among supporting documents in a multi-hop QA as a graph [27]. The nodes of this graph are the named entities in the documents and the edges are the co-reference or same-matching relation between entities. Then, they represented the graph as a vector reflecting the relations using the graph convolutional network. Song et al. also expressed the named entities as the nodes of a graph [28], but they added the window relation for the edges where two entities are regarded to have a window relation if they both appear within a word window. After that, they represented the graph as a vector for solving a multi-hop QA with the graph recurrent network.

The other approach to expressing inter-information is to obtain attention among information components. In the multi-passage reading comprehension, the candidate answers as well as the multiple passages can be regarded as information components. Thus, Wang et al. represented the candidate answers as vectors and expressed the relationship among all candidate answers as an attention matrix by applying an attention mechanism to the vectors [29]. On the other hand, Zhuang and Wang represented the relationships between a questionary sentence and its associated passages as vectors using Bi-DAF [17]. Then, they expressed the relationship among the passage vectors with the proposed dynamic self-attention. In the open-domain QA, Dehghani et al. used the universal transformer to represent the inter-information among the documents related to a question [30]. In the multi-evidence QA, Zhong et al. expressed the inter-information among a questionary sentence, candidate answers, and associated documents [31]. In this work, they adopted the co-attention to express the relationship among the information components since the co-attention allows the representation of bidirectional relationships.

3. Attention-Based Question Difficulty Estimation

This paper defines a question difficulty estimation as determining the optimal difficulty level $y^* \in \mathcal{Y}$ of a given question, where $\mathcal{Y}$ is a set of difficulty levels. It assumes that a question consists of a questionary sentence q and a set of information components $A = \{a_1, \ldots, a_n\}$. An information component can be a passage associated with q, candidate answers in multiple-choice QAs, or a video-clip description in video QAs. Then, the difficulty estimation becomes a classification problem in which a classifier $f(\cdot;\theta)$ parameterized by θ determines y^* given q and A. According to Figure 1, q is "Why did Mami experience culture shock in Japan?" and the passage "A Japanese student ..." and five candidate answers become the elements of information component set, A. Then, the classifier f determines the question difficulty given q and A.

The proposed model of which architecture is given in Figure 2 implements $f(\cdot;\theta)$ with two kinds of attention modules. It takes q and A as its input and encodes them using a pre-trained language model. Then, it represents the bi-directional relationships between q and every $a_i \in A$ with the dual multi-head attention and the relationship among a_i's with the transformer encoder. Indeed, the representation of the relationships are accomplished in two steps, since the relationship among a_i's can be expressed after the relationships between q and every $a_i \in A$ are all represented. After that, it predicts the difficult level of q using the relationships.

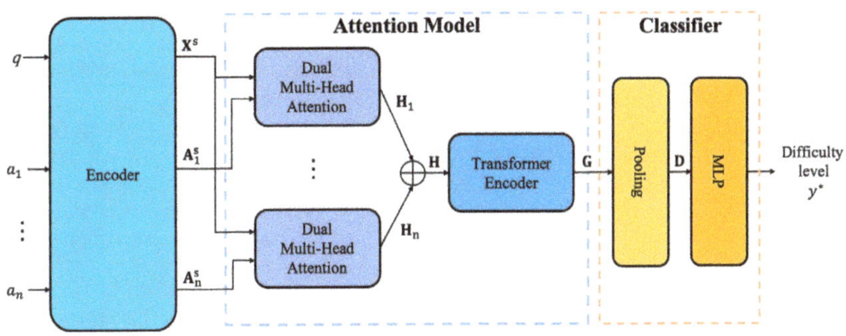

Figure 2. The overall architecture of the proposed model for question difficulty estimation.

3.1. Encoding Question Components

The proposed model first encodes the questionary sentence q and a set of information components $A = \langle a_1, \ldots, a_n \rangle$ into vector representations. As the first step of vector representation, q and all a_i's are expressed in the standard format for BERT [32] using special tokens of [CLS] and [SEP] (This paper assumes that all components in a question are represented in a text form. The question difficulty estimation for the QAs that require analysis of a video or audio stream is out of the scope of this paper). For instance, when q is "Why did Mami experience culture shock in Japan?", it is expressed as "[CLS] why did ma ##mi experience culture shock in japan ? [SEP]". Then, the formatted q and a_i's are encoded into vector representations using the BERT-Base. That is,

$$\begin{aligned} \mathbf{X}^p, \mathbf{X}^s &= BERT(q), \\ \mathbf{A}_i^p, \mathbf{A}_i^s &= BERT(a_i), \end{aligned} \tag{1}$$

where $\mathbf{X}^p$ and $\mathbf{A}_i^p$ are the pooled representations corresponding to the [CLS] token of q and a_i respectively, while $\mathbf{X}^s$ and $\mathbf{A}_i^s$ represent the sequence representations of the whole tokens in q and a_i. This paper uses only $\mathbf{X}^s$ and $\mathbf{A}_i^s$ in the following steps because the individual tokens deliver more information than the special token in solving QA tasks.

3.2. Representing Relationships Using Attention Model

The attention model is responsible for capturing the relationships between q and A, and the model consists of two attention modules: a dual multi-head co-attention and a transformer encoder based on the multi-head attention. The proposed model first represents the relationships between q and every $a_i \in A$ directly since the questionary sentence q is a key factor in the question-and-answering. Thus, all information components should be represented in accordance with the questionary sentence. However, these representations do not express the relationship among a_i's sufficiently. Although the inter-information among a_i's is reflected indirectly and slightly through the relationships between q and a_i's, a direct inter-information relationship plays an important role in estimating the question difficulty and thus the second attention module is designed to consider the inter-information relationship directly.

In order to identify the bi-directional relationship between q and a_i $(1 \leq i \leq n)$, the proposed model adopts the dual multi-head co-attention (DUMA) [18]. DUMA is composed of two multi-head attentions where each multi-head attention captures a single directional attention representation. Thus, it captures both representations from q to a_i and from a_i to q. Then, it fuses these two representations to obtain a final unified representation. That is, the relationship between q and a_i, denoted as $\mathbf{H}_i$, is obtained by applying DUMA to the representations of $\mathbf{X}^s$ and $\mathbf{A}_i^s$ in Equation (1).

$$\begin{aligned} \mathbf{H}_i &= \mathrm{DUMA}(\mathbf{X}^s, \mathbf{A}_i^s) \quad (2)\\ &= Fuse(\mathrm{MHA}(\mathbf{X}^s, \mathbf{A}_i^s, \mathbf{A}_i^s), \mathrm{MHA}(\mathbf{A}_i^s, \mathbf{X}^s, \mathbf{X}^s)), \quad (3) \end{aligned}$$

where $\mathrm{MHA}(\cdot,\cdot,\cdot)$ denotes a multi-head attention and $Fuse(\cdot,\cdot)$ is a function for fusing two representations dynamically.

The multi-head attention $\mathrm{MHA}(\cdot,\cdot,\cdot)$ is an attention mechanism to obtain a representation by paying attention jointly to the information from different representations at different positions [33], where the attention is obtained by applying the scaled dot-product attention several times in parallel and then concatenating the results of the attention. Formally, the multi-head attention maps a sequence of query $\mathbf{Q}$ and a set of key-value pairs of $\mathbf{K}$ and $\mathbf{V}$ to a representation by

$$\begin{aligned} \mathrm{MHA}(\mathbf{Q},\mathbf{K},\mathbf{V}) &= \mathrm{Concat}(head_1,\ldots,head_h)W^O \\ head_i &= \mathrm{Attention}(\mathbf{Q}W_i^{\mathbf{Q}}, \mathbf{K}W_i^{\mathbf{K}}, \mathbf{V}W_i^{\mathbf{V}}), \end{aligned}$$

where $W_i^{\mathbf{Q}}$, $W_i^{\mathbf{K}}$, $W_i^{\mathbf{V}}$, and W^O are all learnable parameters. Here, $\mathrm{Attention}(\mathbf{Q},\mathbf{K},\mathbf{V})$ represents the scaled dot-product attention. It is a weighted sum of the values of which weight is determined by the dot product of the query with all the keys. Thus, it is defined as

$$\mathrm{Attention}(\mathbf{Q},\mathbf{K},\mathbf{V}) = \mathrm{softmax}\left(\frac{\mathbf{Q}\mathbf{K}^\top}{\sqrt{d_k}}\right)\mathbf{V},$$

where d_k is a key dimensionality that works for a scaling factor.

Among several candidates of $Fuse(\cdot,\cdot)$ function in Equation (3), the performance of using the concatenation is higher than that of using the element-wise summation according to our experiments below (see Section 4.2). This result complies with the results of the previous study by Zhu et al. [18], and thus the concatenation is used as a fuse function in this paper.

After obtaining n $\mathbf{H}_i$'s by applying Equation (3) to $\mathbf{X}^s$ and every $\mathbf{A}_i^s$, the proposed model applies a transformer encoder based on the multi-head attention [33] to them in order to capture inter-information relationship directly. For this, all $\mathbf{H}_i$'s are concatenated as $\mathbf{H} = [\mathbf{H}_1; \ldots; \mathbf{H}_n]$, and then the transformer encoder is applied to $\mathbf{H}$ to produce the direct representation $\mathbf{G}$ of inter-information relationship. That is,

$$\mathbf{G} = TransEncoder(\mathbf{H}), \tag{4}$$

where *TransEncoder* denotes the transformer encoder. The transformer encoder is a stack of transformer blocks. The l-th transformer block is composed of two layers of a multi-head attention (MHA) and a feed-forward network (FFN). That is, the two layers of $\mathbf{g}^l$ and $\mathbf{h}^l$ are

$$\begin{aligned} \mathbf{g}^l &= \text{LayerNorm}(\text{MHA}(\mathbf{h}^{l-1}, \mathbf{h}^{l-1}, \mathbf{h}^{l-1}) + \mathbf{h}^{l-1}), \\ \mathbf{h}^l &= \text{LayerNorm}(\text{FFN}(\mathbf{g}^l) + \mathbf{g}^l), \end{aligned}$$

where LayerNorm$(\cdot)$ is a layer normalization [34], and $\mathbf{h}^l$ and $\mathbf{h}^{l-1}$ are the outputs of the l-th and $(l-1)$-th transformer block, respectively. The output of the 0-th transformer block is set as $\mathbf{H}$. That is, $\mathbf{h}^0 = \mathbf{H}$.

Note that *TransEncoder* forces every $\mathbf{H}_i$ to consider all other $\mathbf{H}_j$'s ($i \neq j$), since it is based on the self-attention of which query is $\mathbf{H}_i$, and both key and value are other $\mathbf{H}_j$'s. As a result, $\mathbf{G}$ gets able to reflect the inter-information relationship. Therefore, $\mathbf{G}$ becomes the representation that does not reflect only the relationships between the questionary sentence q and information components $a_i \in A$, but also the inter-relationship among all pairs of information components.

3.3. Difficulty Prediction and Implementation

After all relationships between q and A are represented as $\mathbf{G} \in \mathbb{R}^{|hidden| \times |n|}$ where *hidden* is the hidden dimension of *TransEncoder* in Equation (4), the difficulty of a question is determined by a MLP classifier of which input is $\mathbf{G}$. The classifier first summarizes $\mathbf{G}$ into a single dense representation $\mathbf{D}$. There are several operators for this summarization such as max-pooling, average-pooling, and attention. This paper adopts max-pooling for summarizing $\mathbf{G}$ because it is known to be effective in obtaining representative features [35] and shows higher performance than others in our preliminary experiments. After obtaining the final representation $\mathbf{D}$, the MLP predicts the final difficulty level y^* of q. The proposed model is trained to minimize the standard cross-entropy loss.

The proposed model can be applied to most well-known question answering tasks. In the machine reading comprehension tasks such as SQuAD, a question is composed of a questionary sentence, an associated passage, and an answer span. The tasks meet our problem formulation in that the questionary sentence is q, the associated passage is a_1, and the answer span is a_2. Thus, the proposed model can be applied to this type of tasks without any change. In the multiple-choice QAs such as RACE, a question is composed of a questionary sentence, an associated passage, and multiple answer candidates. The difference between the multiple-choice QAs and the machine reading comprehension is that the multiple-choice QAs have multiple answer candidates instead of a single answer. To encode the multiple candidate answers, the proposed model concatenates all candidate answers into one sentence. That is, it regards the multiple candidate answers as one information component. The rest is the same as the machine reading comprehension.

4. Experiments

4.1. Experimental Setting

Three QA tasks are used for the verification of the proposed model: RACE [36], QuAC [37], and DramaQA [9]. RACE is a data set for the multiple choice QA where a question is composed of a questionary sentence, an associated passage, and a set of candidate answers. This data set was collected from English examinations designed for 12∼15-year-old middle school students and those for 15∼18-year-old high school students in China. Thus, there are two subgroups in this data set with a difficulty gap: RACE-M and RACE-H. RACE-M includes middle school examinations and RACE-H contains high school ones. QuAC is a data set for the machine reading comprehension like SQuAD, and is designed to model information-seeking dialogues. Given a section (in a text form) from a Wikipedia article, two annotators are involved to construct the data set as teacher-student interactions. That is, one annotator (student) asks a sequence of questions to learn about the article, and the other annotator (teacher) answers them by providing excerpts from the article. Since it follows an interactional form, the questions are context-dependent and open-ended so that it is more challenging than SQuAD. On the other hand, the DramaQA data set is constructed for a video QA task to measure the level of machine intelligence for video understanding. It is based on the South Korean television show *'Another Miss Oh.'* Each query in this data set consists of a sequence of video frames, a description of the video frames to deliver background information of the frames, character utterances, and a pair of a questionary sentence and candidate answers. Since this paper assumes that all components in a question are texts, the video frames are excluded from the information components. That is, a question in DramaQA is composed of a questionary sentence, candidate answers, a description of the video frames, and the utterances of the characters. Table 1 summarizes the simple statistics of these data sets.

Table 1. A simple statistics on the data sets used in the experiments.

Data Set	No. of Questions	No. of Information Components
RACE	97,687	2 (passage, candidate answers)
QuAC	7354	2 (passage, candidate answers)
DramaQA	16,191	3 (description, candidate answers, utterance)

DramaQA is manually tagged with four difficulty levels, but RACE and QuAC are not tagged with a difficulty level. Recall that the RACE data set consists of RACE-M and RACE-H. Since RACE-M is about middle school examinations, it is naturally regarded as easy (level 1) questions. RACE-H is then considered as difficult (level 2) questions. For QuAC, we followed the protocol by Gao et al. [19] to label the difficulty level of questions, where the protocol is to assess the difficulty of a question with multiple QA models. This paper employs top three single models (RoBERTa, BERT, and XLNet) from the leaderboard of the QuAC task. A question is labeled as level 1 if at least one model answers it correctly, and is labeled as level 2 if all models give a wrong answer for it. Figure 3 depicts the distributions of difficulty levels in these tasks. In QuAC and DramaQA, the level-1 questions account for about half of the whole questions. On the other hand, the ratio of the level-1 questions is just approximately 20% in RACE.

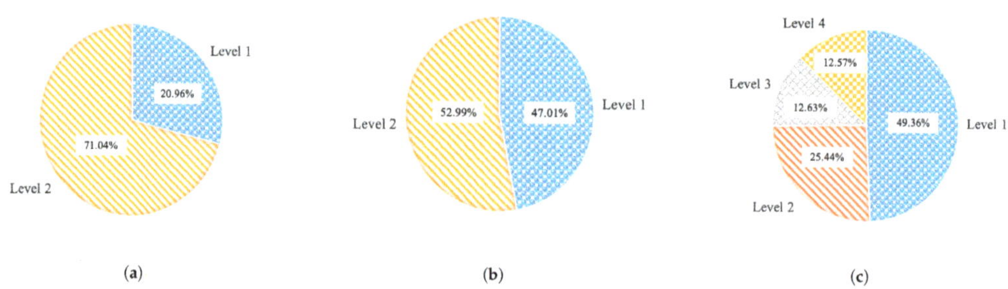

Figure 3. Distributions of difficulty levels in each data set. (**a**) RACE. (**b**) QuAC. (**c**) DramaQA.

For the evaluation of the proposed model, the official data split is used for RACE and DramaQA. In QuAC, the data set is split with the ratio of 80:10:10, where 80% are used for training, 10% are for validation, and the remaining 10% are for test. All hyper-parameters are searched using a grid search and the best hyper-parameters are selected over the validation set. The hyper-parameters used in the experiments are given at Table 2.

BERT-Base model is used for the encoder in Equation (1). The *Fuse* function in Equation (3) is set as the concatenation function. Adam optimizer [38] with default settings is used to train all models, and early stopping over the validation set is executed where 100 is the maximum number of epochs.

Table 2. Parameter values used in the experiments.

	Parameters	RACE	QuAC	DramaQA
Encoder	Model		BERT-Base	
DUMA	Hidden dim.	1536	1536	1536
	No. head	8	6	6
	dropout	0.2	0.2	0.2
	Fuse		*concat*	
TransEncoder	Hidden dim.	3072	3072	4608
	No. head	4	4	4
	No. layers	6	6	6
	dropout	0.2	0.2	0.2

The proposed model is mainly compared with TACNN [14] which is widely used as a main baseline for question difficulty estimation. TACNN uses CNN [39] to obtain the representations of a questionary sentence and a set of information components. Then, it constructs the relationships between the questionary sentence and each information component using a simple attention model, but does not consider the inter-information among the components. Some pre-trained language models are also adopted as baselines of the proposed model, since the language models achieve top performances in many QA tasks. The baseline language models adopted are BERT [32], RoBERTa [40], and XLNet [41]. They concatenate a questionary sentence and all information components with a special token [SEP] and then convert the concatenated sequence to the standard input format of each language model. After that, the formatted sequence is encoded to embedding vectors by each language model. Finally, the embedding vector for the [CLS] token is used to predict the difficulty level in BERT and RoBERTa, while the embedding vector for the last token is used in XLNet. All the models are evaluated with F1-score and accuracy.

4.2. Experimental Results

We first investigate the reliability of labeling the difficulty level on QuAC dataset. The reliability is measured by the agreement between the labels tagged by multiple QA models and the human-annotated labels. To do this, we first randomly sampled 50 data samples. Then, two annotators labeled the difficulty level manually for each sample. The Kappa coefficient [42] between the annotators is 0.52, which falls under the category of 'Moderate'. This implies that the annotators have an agreement to a degree. To obtain the final level of a question from human annotations, we performed an additional procedure as done in the automatic labeling protocol. That is, a question is labeled as level 2 if at least one annotator labels it as level 2 and is labeled as level 1 if both annotators label it as level 1. We have achieved 76% agreement which implies that the labeling of the difficulty level is reliable.

We also investigate the adequateness of the implementation options for $Fuse(\cdot,\cdot)$ in Equation (3) and the direction of the relationships between a questionary sentence and information components. Both have two options. That is, $Fuse(\cdot,\cdot)$ can be implemented by the concatenation or the element-wise summation, and the direction of the relationships can be single or dual. Table 3 summarizes the F1-scores according to the options. The F1-score of the concatenation is generally higher than that of the element-wise summation. Even if the F1-score of the concatenation is 0.26 lower in QuAC, it is much higher in both RACE and DramaQA. Thus, the concatenation is used for $Fuse(\cdot,\cdot)$ in all the experiments below for the sake of consistency.

Table 3. The F1-scores according to the implementation options for $Fuse(\cdot,\cdot)$ and the direction of relationships between a questionary sentence and information components.

Implementation Option		RACE	QuAC	DramaQA
$Fuse(\cdot,\cdot)$	*concat*	**89.56**	70.13	**89.15**
	summation	88.40	**70.39**	88.15
Relationship direction	single (MHA)	89.26	69.74	88.45
	dual (DUMA)	**89.56**	**70.13**	**89.15**

The effectiveness of the bi-directional relationships between a questionary sentence and information components is investigated by replacing DUMA in Equation (2) with a single directional multi-head attention (MHA). That is, $\mathbf{H}_i$, the relationship between a questionary sentence q and each information component a_i, is computed by

$$\mathbf{H}_i = \text{MHA}(\mathbf{X}^s, \mathbf{A}_i^s, \mathbf{A}_i^s).$$

As shown in Table 3, the F1-score of DUMA is higher than that of MHA for all data sets, where the largest difference is 0.7 in DramaQA. This result implies that the bi-directional relationships between a questionary sentence and information components are helpful in improving the performance of question difficulty estimation.

Table 4 compares the performances of the proposed model and its baselines. The first thing to note is that TACNN shows the worst performance in RACE and DramaQA. This is because TACNN does not utilize any pre-trained contextual representation even if the contextual representation is one of the key factors to improve the performance of the natural language tasks. On the other hand, it achieves slightly higher F1-score and accuracy than other pre-trained language models in QuAC. This is due to the fact that TACNN considers the relationships between a questionary sentence and each information component using an attention model explicitly, while the language models do not.

Table 4. Performances of question difficulty estimation.

Data Set	RACE		QuAC		DramaQA	
	Acc. (%)	F1-Score	Acc. (%)	F1-Score	Acc. (%)	F1-Score
BERT	87.75	87.55	58.31	58.25	87.95	87.89
RoBERTa	**89.82**	**89.84**	58.72	58.34	88.81	88.73
XLNet	89.02	88.22	58.51	58.48	89.07	89.05
TACNN	87.27	87.12	60.71	59.87	84.46	84.72
Proposed model	89.81	**89.84**	**68.23**	**68.37**	**89.53**	**89.59**

Among the three pre-trained language models, BERT shows the worst performances for all data sets. RoBERTa and XLNet report similar performances on average. Especially, RoBERTa achieves the best performance in RACE with 89.82% of accuracy and 89.84 of F1-score, respectively. XLNet is the best baseline with 89.07% of accuracy and 89.05 of F1-score in DramaQA. However, the proposed model outperforms all the baselines in QuAC and DramaQA, and achieves a similar performance to RoBERTa in RACE. The F1-score of the proposed model is up to 8.5 higher than those of baselines in QuAC and up to 0.5 higher in DramaQA. These results prove that the proposed model is effective in estimating the difficulty of questions.

4.3. Ablation Study

We investigate the effectiveness of DUMA and *TransEncoder* in the proposed model. Table 5 shows the result of an ablation study over the validation set. The F1-scores of the proposed model over the validation sets of each task are 89.56 for RACE, 70.13 for

QuAC, and 89.15 for DramaQA. The '–' symbol in front of a module indicates exclusion of the module. Thus, '– DUMA' implies that DUMA is excluded from the proposed model. Without DUMA, the F1-score drops up to 4.91 from that of the proposed model, which implies that the bi-directional relationships between a questionary sentence and information components represented by DUMA helps improve the performance of the proposed model.

Table 5. Ablation study of the proposed over validation data.

Model Variations	RACE	QuAC	DramaQA
Proposed model	**89.56**	**70.13**	**89.15**
– DUMA	88.54	65.22	86.77
– TransEncoder	88.94	68.81	88.58
– DUMA and TransEncoder	87.81	63.53	85.38

A similar phenomenon is observed with *TransEncoder* in Equation (4). '– TransEncoder' implies that the concatenation $\mathbf{H}$ of bi-directional relationships $\mathbf{H}_i$'s is directly used as an input of the pooling layer of the final classifier. Its F1-score also drops up to 1.32 from that of the proposed model, which proves the consideration of inter-information relationship is helpful in boosting the performance of the proposed model. In order to take a close look at this result, the F1-scores of each difficulty level are further investigated. Figure 4 depicts the F1-scores for every difficulty level of the questions in DramaQA. When comparing F1-scores of the proposed model with those without *TransEncoder*, the improvement in difficult (level 3 and level 4) questions is larger than that in easy (level 1 and level 2) questions. Especially at level 4, the proposed model achieves 97.40 of F1-score, but the model without *TransEncoder* shows just 94.50. Finally, the model without both DUMA and *TransEncoder* demonstrates the worst performance for all data sets. From these results, we can conclude that the adoption of DUMA for bi-directional relationships and *TransEncoder* for inter-information relationship are effective to predicting the level of question difficulty.

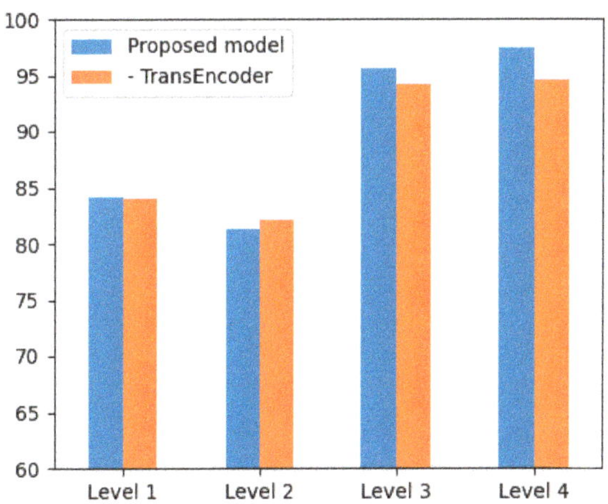

Figure 4. F1-scores for the difficulty levels of the questions in DramaQA.

4.4. Performance Change according to No. of Information Components

There are different numbers of information components depending on the QA tasks and the proposed model is designed to consider a various number of information components. Thus, one consequential question about the proposed model is how the performance of the proposed model changes as the number of information components increases. Figure 5 depicts the performance changes according to the number of information components. The X-axis of this figure denotes the information components used and the Y-axis represents F1-score. In the QA tasks of our experiments, a description is the most common and important information component. Thus, it is used as a base information component for all models. The candidate answers and utterances are added consecutively in DramaQA, while only the candidate answers are added in RACE and an answer span is added in QuAC. The performances of all models in RACE increase monotonically as new components of candidate answers are added. This result seems natural because a questionary sentence (QS), a description, and candidate answers all provide somewhat information for predicting the level of question difficulty. On the other hand, the performances do not improve large in QuAC even though a new information component of an answer span is added. This is because the answer span is extracted from a description so that the information of the answer span might be already reflected by the description.

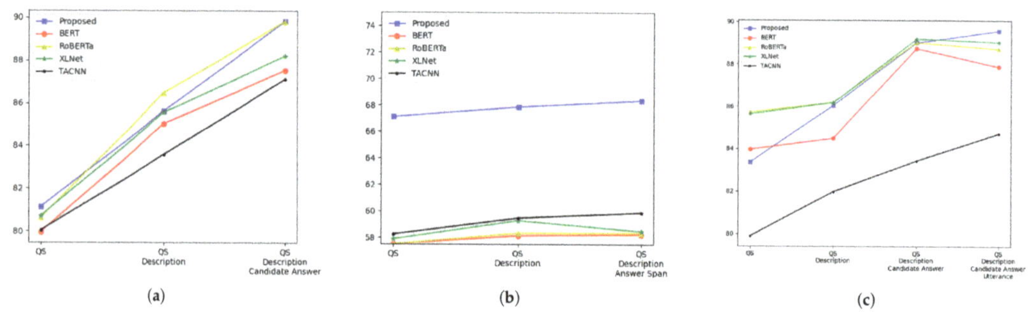

Figure 5. Performance change of the proposed model according to the number of information components. The X-axis denotes the information components used where QS stands for a questionary sentence. (**a**) RACE. (**b**) QuAC. (**c**) DramaQA.

An interesting fact is found in DramaQA. As in RACE, when a description, candidate answers, and utterances are added in order, the performances of the proposed model and TACNN increase monotonically but those of the pre-trained language models do not. Especially when utterances are newly added, the performances of the language models rather decrease. This is because the language models regard all information components as a single sequence, not as individual sequences. Although the sequence differentiates each information component with a special token [SEP], some individuality among the information components might be lost. Due to this loss of individuality, their performances decrease though the utterances are considered. On the other hand, the proposed model and TACNN treat every information component separately. Furthermore, the proposed model is superior to TACNN because it utilizes the pre-trained contextual representations and considers additional inter-information relationship. These results imply that the proposed model predicts the difficulty of questions well even when the number of information components increases.

4.5. Performance of Question Answering with Difficulty Level

In this section, we solve the question answering with a predicted difficulty level to verify that the performance of question answering is improved with the difficulty level. We choose the multi-level context matching model [9] as a question answering model, since is currently the state-of-the-art model for the DramaQA QA task. The multi-level context matching model is designed to understand the multimodal story of a drama. This QA model consists of two streams for a vision and a textual modality. Each stream of modality is combined with embeddings from a questionary sentence and information components using a context matching module and then predicts a score for each answer. Since it does not adopt any difficulty level, we modify it to use the proposed difficulty level by regarding the difficulty level as an additional modality of the question answering (There will be several methods to utilize the difficulty level in the QA model. However, this experiment focuses on showing that the difficulty level helps the QA model to get better performance than the model without the level). That is, the modified QA model consists of three streams including the difficulty level information.

Table 6 shows the question answering on the drama QA dataset is improved by adding the difficulty level. The '+ Difficulty level' indicates the inclusion of the difficulty level to the multi-level context matching model. The QA model with the difficulty level achieves better performance than the QA model without the level. With the difficulty level, the accuracy rises to 73.83% which is higher up to 2.69% than that of the QA model. These results imply that the question difficulty estimation helps the performance of question answering tasks improved. Especially, the improvement in difficult (level 3 and level 4) questions is larger than that in easy (level 1 and level 2) questions. This is because the proposed method has achieved better performances on difficult questions than on easy questions (refer to Section 4.3 and Figure 4). From these results, we verify the usefulness of the question difficulty estimation.

Table 6. Accuracy of question answering with the question difficulty estimation.

QA Model	Diff. 1	Diff. 2	Diff. 3	Diff. 4	Overall	Diff. Avg.
Multi-level context matching model [9]	75.96	74.65	57.36	56.63	71.14	66.15
+ Difficulty level	76.12	74.82	59.12	57.33	73.83	66.85

4.6. Performance Change according to Data Ratio

The proposed model is based on the transformer encoder designed to consider the relationships among all components in a question. It is known that a number of training examples are required to train the transformer encoder. Thus, one possible question about the proposed model is whether the training data in QA data sets are sufficient enough to train it. Since the proposed model adopts a pre-trained language model, BERT-Base, and fine-tunes it, it does not require too many training examples actually. This is proved empirically by showing the performance change according to the ratio of data used to train the proposed model.

Figure 6 depicts the performance changes, where the X-axis is the ratio of data used to train the proposed model and the Y-axis represents F1-score. In all QA data sets, the more the training data are used, the better the predictions of the proposed model are. In QuAC and RACE, the performances of the proposed model converge after 90% data are consumed. This implies that the proposed model is trained well for the data sets. However, a different phenomenon is observed in DramaQA data set with which the performance increases continually. This continual increase is believed to be affected by a larger number of information components in DramaQA. The number of information components in DramaQA is three, while it is two in other data sets. In addition, the F1-score is around 90 when 100% of data are used to train the proposed model. Thus, even if more data are provided, the improvement by them would not be great.

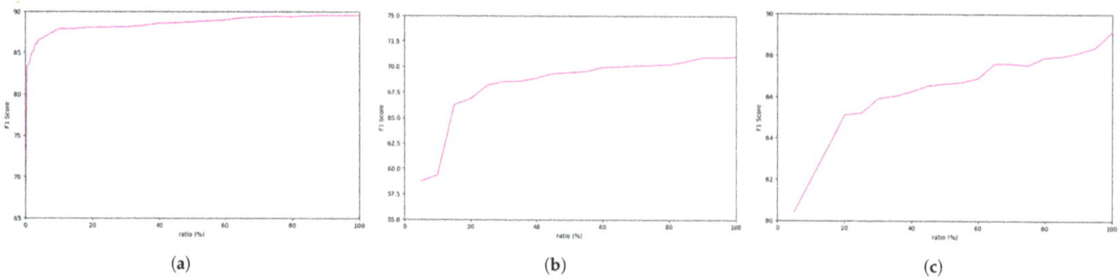

Figure 6. Performance change model according to the ratio of data sets. The X-axis denotes the ratio of data used to train the proposed model and the Y-axis is F1-score. (**a**) RACE. (**b**) QuAC. (**c**) DramaQA.

5. Conclusions

In this paper, we have proposed an attention model for question difficulty estimation. The proposed attention model first represents bi-directional relationships between a questionary sentence and information components, and then accumulates the inter-information relationship over the concatenated bi-directional relationships. As a result, the proposed method can model complicated relationships among the questionary sentence and information components.

The contributions of this paper are three folds. The first is that the proposed model achieves the state-of-the-art performance in this task. It outperforms the existing model and pre-trained language models. The second is that the proposed model predicts the difficulty of high-level questions accurately. It is required to reason over multiple kinds of information components to predict the difficulty of high-level questions. Since the proposed model is designed to consider the complicated relationships among information components, the reasoning is taken place properly in the proposed model. The last is that the proposed method works efficiently and can be applied to any text-based QA tasks. The proposed method is based on the simple attention model and does not require any other pre-training models except the BERT. Furthermore, it is free from the number of information components.

Through intensive experiments with three well-known QA data sets, it has been shown empirically that the proposed model achieves higher performances than all the previous study and pre-trained language models. Moreover, it is also shown that the proposed attention is essential for accurate prediction of the difficulty level for more difficult questions. Through these experiments, we have proven that the proposed model is plausible for predicting the question difficulty and helps to improve the performances of the question answering.

Author Contributions: Conceptualization, H.-J.S. and S.-H.Y.; methodology, H.-J.S. and S.-H.Y.; visualization, S.-H.Y.; validation, H.-J.S., S.-H.Y. and S.-B.P.; funding acquisition, S.-B.P.; writing—original draft preparation, H.-J.S. and S.-H.Y.; writing—review and editing, H.-J.S. and S.-B.P. All authors have read and agreed to the published version of the manuscript.

Funding: This work was supported by Institute of Information & Communications Technology Planning & Evaluation (IITP) grant funded by the Korea government (MSIT) (No.2017-0-01772, Development of QA systems for Video Story Understanding to pass the Video Turing Test) and National Research Foundation of Korea (NRF) grant funded by the Korea government (MSIT) (No. 2020R1A4A1018607).

Institutional Review Board Statement: Not applicable because this study is involved with neither humans nor animals.

Informed Consent Statement: Not applicable.

Data Availability Statement: Publicly available datasets were analyzed in this study. The data can be found here: RACE: https://www.cs.cmu.edu/~glai1/data/race/ (accessed on 1 November

2021), QuAC: https://quac.ai (accessed on 1 November 2021), DramaQA: https://dramaqa.snu.ac.kr (accessed on 1 November 2021).

Conflicts of Interest: The authors declare no conflict of interest.

References

1. Clark, P.; Cowhey, I.; Etzioni, O.; Khot, T.; Sabharwal, A.; Schoenick, C.; Tafjord, O. Think you have Solved Question Answering? Try ARC, the AI2 Reasoning Challenge. *arXiv* **2018**, arXiv:1803.05457.
2. Rajpurkar, P.; Jia, R.; Liang, P. Know What You Don't Know: Unanswerable Questions for SQuAD. In Proceedings of the 56th Annual Meeting of the Association for Computational Linguistics, Melbourne, Australia, 15–20 July 2018; pp. 784–789. [CrossRef]
3. Yang, Z.; Qi, P.; Zhang, S.; Bengio, Y.; Cohen, W.W.; Salakhutdinov, R.; Manning, C.D. HOTPOTQA: A Dataset for Diverse, Explainable Multi-hop Question Answering. In Proceedings of the 2018 Conference on Empirical Methods in Natural Language Processing, Brussels, Belgium, 31 October–4 November 2018; pp. 2369–2380. [CrossRef]
4. Cao, Q.; Trivedi, H.; Balasubramanian, A.; Balasubramanian, N. DeFormer: Decomposing Pre-trained Transformers for Faster Question Answering. In Proceedings of the 58th Annual Meeting of the Association for Computational Linguistics, Online, 5–10 July 2020; pp. 4487–4497. [CrossRef]
5. Saxena, A.; Tripathi, A.; Talukdar, P. Improving Multi-hop Question Answering over Knowledge Graphs using Knowledge Base Embeddings. In Proceedings of the 58th Annual Meeting of the Association for Computational Linguistics, Online, 5–10 July 2020; pp. 4498–4507. [CrossRef]
6. Zhu, M.; Ahuja, A.; Juan, D.C.; Wei, W.; Reddy, C.K. Question Answering with Long Multiple-Span Answers. In Proceedings of the 2020 Conference on Empirical Methods in Natural Language Processing: Findings, Online, 16–20 November 2020; pp. 3840–3849. [CrossRef]
7. He, Y.; Zhu, Z.; Zhang, Y.; Chen, Q.; Caverlee, J. Infusing Disease Knowledge into BERT for Health Question Answering, Medical Inference and Disease Name Recognition. In Proceedings of the 2020 Conference on Empirical Methods in Natural Language Processing, Online, 16–20 November 2020; pp. 4604–4614. [CrossRef]
8. Heo, Y.J.; On, K.W.; Choi, S.; Lim, J.; Kim, J.; Ryu, J.K.; Bae, B.C.; Zhang, B.T. Constructing Hierarchical Q&A Datasets for Video Story Understanding. *arXiv* **2019**, arXiv:1904.00623.
9. Choi, S.; On, K.W.; Heo, Y.J.; Seo, A.; Jang, Y.; Lee, M.; Zhang, B.T. DramaQA: Character-Centered Video Story Understanding with Hierarchical QA. In Proceedings of the 35th AAAI Conference on Artificial Intelligence, Virtually, 2–9 February 2021; pp. 1166–1174.
10. Collis, K.F. *A Study of Concrete and Formal Operations in School Mathematics: A Piagetian Viewpoint*; Australian Council for Educational Research: Camberwell, Australia, 1975.
11. Bartolo, M.; Roberts, A.; Welbl, J.; Riedel, S.; Stenetorp, P. Beat the AI: Investigating Adversarial Human Annotation for Reading Comprehension. *Trans. Assoc. Comput. Linguist.* **2020**, *8*, 662–678. [CrossRef]
12. Desai, T.; Moldovan, D.I. Towards Predicting Difficulty of Reading Comprehension Questions. In Proceedings of the 32th International Flairs Conference, Melbourne, FL, USA, 21–23 May 2018; pp. 8–13.
13. Ha, L.A.; Yaneva, V.; Baldwin, P.; Mee, J. Predicting the Difficulty of Multiple Choice Questions in a High-stakes Medical Exam. In Proceedings of the Fourteenth Workshop on Innovative Use of NLP for Building Educational Applications, Florence, Italy, 2 August 2019; pp. 11–20. [CrossRef]
14. Huang, Z.; Liu, Q.; Chen, E.; Zhao, H. Question Difficulty Prediction for READING Problems in Standard Tests. In Proceedings of the 31th AAAI Conference on Artificial Intelligence, San Francisco, CA, USA, 4–9 February 2017; pp. 1352–1359.
15. Liu, J.; Wang, Q.; Lin, C.Y.; Hon, H.W. Question Difficulty Estimation in Community Question Answering Services. In Proceedings of the 2013 Conference on Empirical Methods in Natural Language Processing, Seattle, WA, USA, 18–21 October 2013; pp. 85–90.
16. Qiu, Z.; Wu, X.; Fan, W. Question Difficulty Prediction for Multiple Choice Problems in Medical Exams. In Proceedings of the 28th ACM International Conference on Information and Knowledge Management, Beijing, China, 3–7 November 2019; pp. 139–148. [CrossRef]
17. Zhuang, Y.; Wang, H. Token-level Dynamic Self-Attention Network for Multi-Passage Reading Comprehension. In Proceedings of the 57th Annual Meeting of the Association for Computational Linguistics, Florence, Italy, 28 July–2 August 2019; pp. 2252–2262. [CrossRef]
18. Zhu, P.; Zhao, H.; Li, X. DUMA: Reading Comprehension with Transposition Thinking. *arXiv* **2020**, arXiv:2001.09415.
19. Gao, Y.; Bing, L.; Chen, W.; Lyu, M.R.; King, I. Difficulty Controllable Generation of Reading Comprehension Questions. In Proceedings of the 28th International Joint Conference on Artificial Intelligence, Macao, China, 10–16 August 2019; pp. 4968–4974.
20. Pasupat, P.; Liang, P. Compositional Semantic Parsing on Semi-Structured Tables. In Proceedings of the 53rd Annual Meeting of the Association for Computational Linguistics and the 7th International Joint Conference on Natural Language Processing, Beijing, China, 26–31 July 2015; pp. 1470–1480. [CrossRef]
21. Herzig, J.; Nowak, P.K.; Müller, T.; Piccinno, F.; Eisenschlos, J. TaPas: Weakly Supervised Table Parsing via Pre-training. In Proceedings of the 58th Annual Meeting of the Association for Computational Linguistics, Online, 5–10 July 2020; pp. 4320–4333. [CrossRef]
22. Antol, S.; Agrawal, A.; Lu, J.; Mitchell, M.; Batra, D.; Lawrence Zitnick, C.; Parikh, D. Vqa: Visual question answering. In Proceedings of the IEEE International Conference on Computer Vision, Santiago, Chile, 7–13 December 2015; pp. 2425–2433.

23. Ionescu, B.; Müller, H.; Villegas, M.; de Herrera, A.G.S.; Eickhoff, C.; Andrearczyk, V.; Cid, Y.D.; Liauchuk, V.; Kovalev, V.; Hasan, S.A.; et al. Overview of ImageCLEF 2018: Challenges, datasets and evaluation. In Proceedings of the International Conference of the Cross-Language Evaluation Forum for European Languages, Avignon, France, 10–14 September 2018; pp. 309–334.
24. Ye, Y.; Zhang, S.; Li, Y.; Qian, X.; Tang, S.; Pu, S.; Xiao, J. Video question answering via grounded cross-attention network learning. *Inf. Process. Manag.* **2020**, *57*, 102265. doi: 10.1016/j.ipm.2020.102265. [CrossRef]
25. Xue, K.; Yaneva, V.; Christopher Runyon, P.B. Predicting the Difficulty and Response Time of Multiple Choice Questions Using Transfer Learning. In Proceedings of the 15th Workshop on Innovative Use of NLP for Building Educational Applications, Seattle, WA, USA, 10 July 2020; pp. 193–197. [CrossRef]
26. Zheng, J.; Cai, F.; Chen, H.; de Rijke, M. Pre-train, Interact, Fine-tune: A novel interaction representation for text classification. *Inf. Process. Manag.* **2020**, *57*, 102215. doi: 10.1016/j.ipm.2020.102215. [CrossRef]
27. Cao, N.D.; Aziz, W.; Titov, I. Question answering by reasoning across documents with graph convolutional networks. In Proceedings of the 2019 Conference of the North American Chapter of the Association for Computational Linguistics: Human Language Technologies, Minneapolis, MN, USA, 2–7 June 2019; pp. 2306–2317.
28. Song, L.; Wang, Z.; Yu, M.; Zhang, Y.; Florian, R.; Gildea, D. Exploring Graph-structured Passage Representation for Multi-hop Reading Comprehension with Graph Neural Networks. *arXiv* **2018**, arXiv:1809.02040.
29. Wang, Y.; Liu, K.; Liu, J.; He, W.; Lyu, Y.; Wu, H.; Li, S.; Wang, H. Multi-passage machine reading comprehension with cross-passage answer verification. In Proceedings of the 56th Annual Meeting of the Association for Computational Linguistics, Melbourne, Australia, 15–20 July 2018; pp. 1918–1927. [CrossRef]
30. Dehghani, M.; Azarbonyad, H.; Kamps, J.; de Rijke, M. Learning to transform, combine, and reason in open-domain question answering. In Proceedings of the Twelfth ACM International Conference on Web Search and Data Mining, Melbourne, Australia, 11–15 February 2019; pp. 681–689. [CrossRef]
31. Zhong, V.; Xiong, C.; Keskar, N.S.; Socher, R. Coarse-grain fine-grain coattention network for multi-evidence question answering. In Proceedings of the 7th International Conference on Learning Representations, New Orleans, LA, USA, 6–9 May 2019.
32. Devlin, J.; Chang, M.W.; Lee, K.; Toutanova, K. BERT: Pre-training of Deep Bidirectional Transformers for Language Understanding. In Proceedings of the 2019 Conference of the North American Chapter of the Association for Computational Linguistics: Human Language Technologies, Minneapolis, MN, USA, 2–7 June 2019; pp. 4171–4186. [CrossRef]
33. Vaswani, A.; Shazeer, N.; Parmar, N.; Uszkoreit, J.; Jones, L.; Gomez, A.N.; Kaiser, L.U.; Polosukhin, I. Attention is All you Need. *Adv. Neural Inf. Process. Syst.* **2017**, *30*, 5998–6008.
34. Ba, J.L.; Kiros, J.R.; Hinton, G.E. Layer Normalization. *arXiv* **2016**, arXiv:1607.06450.
35. Scherer, D.; Müller, A.; Behnke, S. Evaluation of Pooling Operations in Convolutional Architectures for Object Recognition. In Proceedings of the 20th International Conference on Artificial Neural Networks, Thessaloniki, Greece, 15–18 September 2010; pp. 92–101.
36. Lai, G.; Xie, Q.; Liu, H.; Yang, Y.; Hovy, E. RACE: Large-scale Reading Comprehension Dataset From Examinations. In Proceedings of the 2017 Conference on Empirical Methods in Natural Language Processing, Copenhagen, Denmark, 7–11 September 2017; pp. 785–794. [CrossRef]
37. Choi, E.; He, H.; Iyyer, M.; Yatskar, M.; Yih, W.T.; Choi, Y.; Liang, P.; Zettlemoyer, L. QuAC: Question Answering in Context. In Proceedings of the 2018 Conference on Empirical Methods in Natural Language Processing, Brussels, Belgium, 31 October–4 November 2018; pp. 2174–2184. [CrossRef]
38. Kingma, D.P.; Ba, J. Adam: A Method for Stochastic Optimization. In Proceedings of the 3rd International Conference on Learning Representations, San Diego, CA, USA, 7–9 May 2015.
39. Collobert, R.; Weston, J.; Bottou, L.; Karlen, M.; Kavukcuoglu, K.; Kuksa, P. Natural Language Processing (Almost) from Scratch. *J. Mach. Learn. Res.* **2011**, *12*, 2493–2537.
40. Liu, Y.; Ott, M.; Goyal, N.; Du, J.; Joshi, M.; Chen, D.; Levy, O.; Lewis, M.; Zettlemoyer, L.; Stoyanov, V. RoBERTa: A Robustly Optimized BERT Pretraining Approach. *arXiv* **2019**, arXiv:1907.11692.
41. Yang, Z.; Dai, Z.; Yang, Y.; Carbonell, J.; Salakhutdinov, R.R.; Le, Q.V. XLNet: Generalized Autoregressive Pretraining for Language Understanding. *Adv. Neural Inf. Process. Syst.* **2019**, *32* , 5753–5763.
42. Carletta, J. Assessing Agreement on Classification Tasks: The Kappa Statistic. *Comput. Linguist.* **1996**, *22*, 249–254.

applied sciences

MDPI

Article

Enhance Text-to-Text Transfer Transformer with Generated Questions for Thai Question Answering

Puri Phakmongkol and Peerapon Vateekul *

Department of Computer Engineering, Faculty of Engineering, Chulalongkorn University, Bangkok 10300, Thailand; puri.pmk@gmail.com
* Correspondence: peerapon.v@chula.ac.th

Abstract: Question Answering (QA) is a natural language processing task that enables the machine to understand a given context and answer a given question. There are several QA research trials containing high resources of the English language. However, Thai is one of the languages that have low availability of labeled corpora in QA studies. According to previous studies, while the English QA models could achieve more than 90% of F1 scores, Thai QA models could obtain only 70% in our baseline. In this study, we aim to improve the performance of Thai QA models by generating more question-answer pairs with Multilingual Text-to-Text Transfer Transformer (mT5) along with data preprocessing methods for Thai. With this method, the question-answer pairs can synthesize more than 100 thousand pairs from provided Thai Wikipedia articles. Utilizing our synthesized data, many fine-tuning strategies were investigated to achieve the highest model performance. Furthermore, we have presented that the syllable-level F1 is a more suitable evaluation measure than Exact Match (EM) and the word-level F1 for Thai QA corpora. The experiment was conducted on two Thai QA corpora: Thai Wiki QA and iApp Wiki QA. The results show that our augmented model is the winner on both datasets compared to other modern transformer models: Roberta and mT5.

Keywords: natural language processing; question answering; machine reading comprehension

Citation: Phakmongkol, P.; Vateekul, P. Enhance Text-to-Text Transfer Transformer with Generated Questions for Thai Question Answering. *Appl. Sci.* **2021**, *11*, 10267. https://doi.org/10.3390/app112110267

Academic Editor: Arturo Montejo-Ráez

Received: 20 September 2021
Accepted: 27 October 2021
Published: 1 November 2021

1. Introduction

One of the Natural Language Processing (NLP) tasks that allow machines to understand the information in text format and answer given questions is Question Answering (QA). Many researchers aim to develop QA systems in many languages because QA systems have many benefits and can be used as a part of many intelligent systems such as chat bots, or answer highlighters in search engines. One of the most popular languages developed in QA tasks is English. There are many techniques and machine learning models as well as many language resources that contribute to QA system development in the English language. For example, the Text-to-Text Transfer Transformer model [1], a Transformer-based model [2] that was trained with the huge English dataset called Colossal Clean Crawled Corpus (C4) [3], achieved state-of-the-art results in SQuAD 1.1 [4] with an F1 score of 96.22%. These contributions can support English QA models to reach higher performance than other languages.

There are several research works about Thai QA, for example, using heuristic functions to extract the answer, developed by Hatsanai Decha et al. [5], and using the Bi-Directional Attention Flow (BiDAF) model [6] developed by Theerit Lapchaicharoenkit et al. [7]. However, one of the most important limitations of Thai QA is a lack of availability of training data. There are currently only two datasets of Thai QA: Thai Wiki QA [8] and iApp Wiki QA. Each sample of both datasets consists of a context, a question, and a ground truth answer. Both datasets are span extraction type such that the answer to the question is the span of text in the corresponding context. Thai Wiki QA contains 15,000 samples while iApp Wiki QA contains 7242 samples. Each dataset has a small number of samples compared to an English span extraction dataset such as SQuAD 1.1, which contains more than 100 thousand samples. With this limitation, directly using the same techniques or models of the English language such as deep learning models with Thai corpora might not be able to utilize the capability of models to raise the performance.

In this paper, we aim to improve the Thai QA model performance by presenting an enhanced QA framework tailored for the Thai language, with low training resources. First, the limitation of data is overcome by generating synthesized data using Raul Puri et al.'s method [9]. We further investigated and improved their technique in many aspects: the synthesized data selection (all vs. filtered data) and the fine-tuning strategies (merge and sequence). Second, we employed recent transformer models, where the pretrained weights supported the Thai language. There are two chosen models in our comparison: WanchanBERTa [10] and Multilingual Text-to-Text Transfer Transformer (mT5) [11]. Third, we presented preprocessing methods for the Thai language to reduce the misspelling words as well as to improve the quality of data. Final, the metrics that are widely used in QA tasks for evaluating model performance, such as Exact Match (EM) and F1 score, are not sufficient due to inabilities of the word tokenizer and the ambiguity of the Thai language. To obtain nearer-correct scores, we proposed a Syllable-level F1 that calculates the F1 score with syllable-tokens of prediction and the ground truth instead of word-tokens. In this work, we evaluated the models with syllable-level F1 along with word-level F1. The details of each module in our framework are explained in Chapter 3. The experiment was conducted on two Thai QA corpora: Thai Wiki QA and iApp Wiki QA. The results showed that the synthesized data along with a sequence fine-tuning strategy outperformed the original Transformer based models.

In summary, our contributions are as follows:

- We present a data preprocessing method for the Thai Language.
- We demonstrate fine-tuning of two Transformer based models, WangchanBERTa and mT5, for the QA task, with synthesized data and real human-labeled corpus, and achieve higher EM and F1 scores than those when using only the real human-labeled data.
- We compare the quality of the generated question-answer pairs used in the QA models as well as training strategies.
- We propose new metrics: Syllable-level F1 to evaluate the models along with the original Word-level F1.

We organize the rest of this paper as follows. Related works are introduced in Section 2, followed by the presentation of our proposed framework in Section 3. We then explain our experiment settings in Section 4. The result and discussion are presented in Sections 5 and 6, and finally the conclusion of our work in Section 7.

2. Literature Review

In this section, we introduce related research to our work. This section is divided into five parts as follows: recent research on QA, research on Thai QA, data augmentation methods, the Text-to-Text Transfer Transformer, and the WangchanBERTa model.

2.1. Recent Research in Question Answering

Most recent research works on NLP focus on developing language models to use with many tasks including the QA task. Most language models use the Transformer model as a part of their processing because the Transformer model has proved that it can reach higher performance than older-style NLP models, such as BiDAF [6], that use Long Short-Term Memory (LSTM) [12]. BERT [13] is the first Transformer based language model that uses only the encoder part of the Transformer. There are two sizes of BERT models: $BERT_{Base}$ with 12 layers of Encoder and $BERT_{Large}$ with 24 layers of Encoder. In the experiment, BERT could achieve state-of-the-art performance in QA tasks. $BERT_{Base}$ could reach 80.8 and 88.5 EM and F1 scores, respectively, $BERT_{Large}$ could also reach 84.1 and 90.9 EM and F1 scores, respectively, when tested with SQuAD 1.1, while BiDAF could achieve only 68.0 and 77.3 EM and F1 scores, respectively.

There were several trials to develop a BERT model to better predict span, which is more appropriate with QA tasks. SpanBERT [14] is one of these. SpanBERT changed some functions of the pretraining process by using span masking instead of token masking, and adding a Span Boundary Objective to train the model to predict the masked span with adjoining words. SpanBERT used $BERT_{Large}$ architecture and applied a described method. SpanBERT was tested with the SQuAD 1.1 dataset to evaluate its performance, and achieved 88.8 and 94.6 EM and F1 scores, respectively.

However, neither of those developments was chosen to use in our work because WangchanBERTa was considered a better model than BERT, and SpanBERT must be pre-trained with Thai documents before using, which is not convenient to use.

2.2. Researches in Thai Question Answering

Hatsanai Decha et al. [5] developed a QA system in Thai with a keyword extraction method by finding keywords from questions and using extracted keywords to find candidate answers from a set of contexts, then finding the best answer with a heuristic function called word order consistency, which functions in a manner that measures similarity between contexts and questions. This work does not use deep learning model, it is thus not directly related to our work.

Theerit Lapchaicharoenkit et al. [7] modified the BiDAF model to support two types of questions, span extraction type and yes-no question type, by adding a question type classifier to the model. The model also used contextualized word embedding from the BERT model that was pretrained with only Thai documents. The model was tested in a competition called the National Software Contest organized in Thailand in 2018–2019. This competition dataset consisted of 15,000 samples of span extraction tasks and 2000 samples of yes-no question tasks.

Nevertheless, we did not use both above-mentioned works in our research because the first method was not related to our work, and Transformer based models have proved that they could achieve better performance than BiDAF in QA tasks.

2.3. Data Augmentation Methods

There are several research works in data augmentation for improving the performance in QA tasks. Bhuwan Dhingra et al. [15] presented a cloze-style question generation method by extracting questions and answers using the document structure of English articles that mostly provides the summary of articles in the introduction. They used the BiDAF model as a QA model. This method was able to raise the EM and F1 evaluation scores by 0.32% and 0.11%, respectively.

Raul Puri et al. [9] introduced a Question Generation pipeline with three Transformer based models inside. There are three steps of the pipeline including (1) answer generation, (2) question generation, and (3) question filtration. Answer generation is performed by a BERT model trained to select the candidate answer from a given context. Question generation is performed by a GPT-2 model [16] trained to create a proper question to a given context and answer. The last step, Question filtration, is performed by a BERT model trained with question answering objectives with human-labeled data. The researchers used this model to predict an answer from the generated question and context. If the answer from this model was equivalent to the answer from the answer generation step, they considered the generated question-answer pair to be an admissible sample. With this pipeline, they were able to generate more than 19 million question-answer pairs from Wikipedia articles, and used them to train the BERT model. The result achieved more than their baseline EM and F1 scores by 1.7% and 1.2%, respectively.

To conclude, the first method cannot be used with Thai articles because the Thai article structure is more ambiguous than English. It cannot simply extract the answers and questions by using heuristic rules. Given this limitation, using a deep learning model to extract answers and questions is a more appropriate method to synthesize the data.

2.4. The Text-to-Text Transfer Transformer Model

The Text-to-Text Transfer Transformer (T5) [1] is one of the Transformer based models that uses the same architecture of Transformer as shown in Figure 1. The objective of T5 models is to support every NLP task by treating every text processing problem as a "text-to-text" task, by taking the given text as input and producing new text as output. With this method, many tasks could be used with this model, for example, Question Answering, document summarization, or sentiment classification. There is research work that uses a set of documents containing 101 languages, including Thai, to pretrain T5 models called Multilingual Text-to-Text Transfer Transformer (mT5) [11].

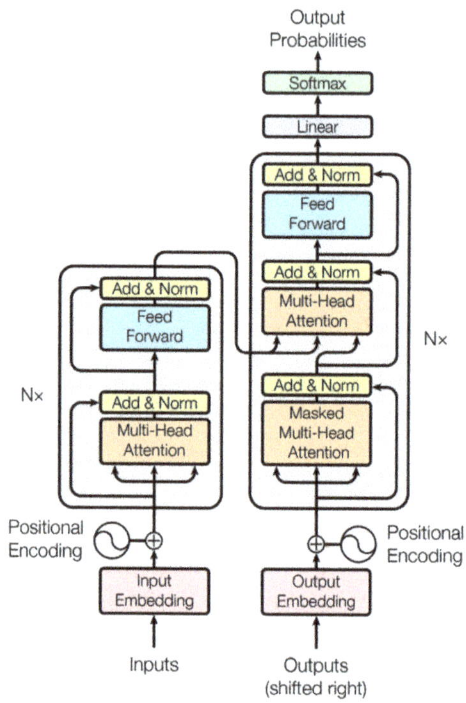

Figure 1. Model architecture of Transformer [2] that was used in the mT5 model.

Due to the model's ability to be used with various tasks, this model was used in our research in both the question generation and question answering parts.

2.5. The WangchanBERTa Model

WangchanBERTa [10] is a pretrained language model based on the Roberta [17] configuration. The architecture of Roberta is the same as that of BERT in terms of using only the Encoder part of the Transformer model as shown in Figure 2. WangchanBERTa was pretrained on a large set of Thai documents including social media texts, news, and public articles. In addition, the appropriate methods were applied to the texts before training. The result showed that this model beat other Thai supported Transformer based models, such as Multilingual BERT, on many downstream tasks. We used this model in question answering part to compare with the mT5 model.

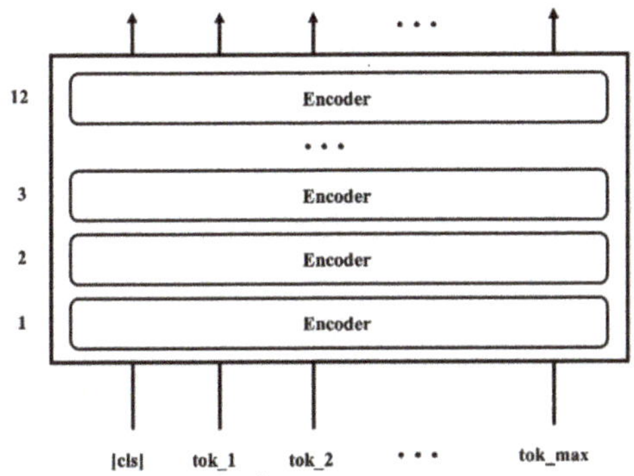

Figure 2. Model architecture of BERT that uses the Encoder part of the Transformer model. (Note: This architecture is also used in the Roberta model.)

3. Proposed Method

This section explains the components of the proposed QA framework. For example, preprocessing methods for Thai texts, question-answer pairs generation, training strategies of QA models and model evaluation. The components of the framework are illustrated in Figure 3.

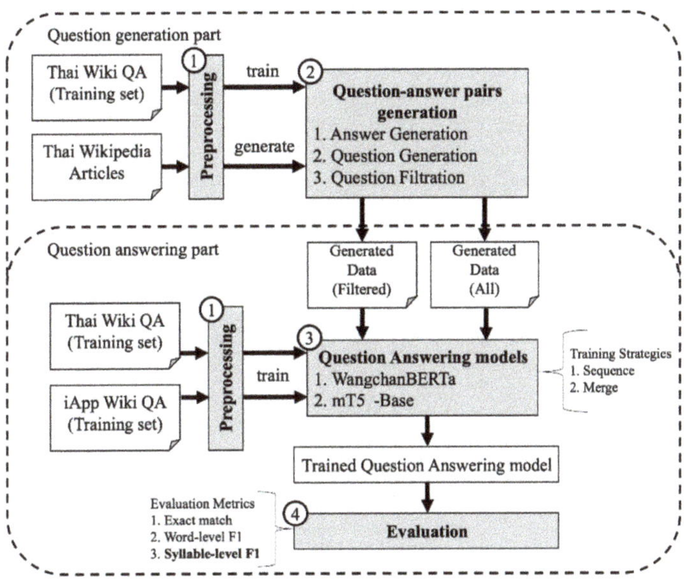

Figure 3. Illustration of the proposed overall QA framework.

3.1. Preprocessing Methods for Thai Texts

All Thai texts must be preprocessed with appropriate methods before being used in training and testing with the models. The first step is applying lowercase characters to the text in case there are English characters in the text. The second step is normalizing the text into the correct and standard form by removing duplicate characters, and changing the order of word typing to the correct one. With this step, we could reduce misspelled words in the datasets, which enables the model to work more accurately. We used the implementation of PyThaiNLP's normalization function [18] for normalizing texts as described.

3.2. *Question-Answer Pairs Generation*

The method for generating question-answer pairs is based on Raul Puri et al.'s method [9], which consists of three steps: (1) Answer Generation, (2) Question Generation, and (3) Question Filtration. This method is able to generate a set of triplets which include Context c, Question q and Answer a by using a given set of Articles A, pursuant to Probability $p(q, a|c)$.

The difference of implementation between Raul Puri et al.'s work and our work is the selection of the base models in the Question Generation pipeline. In our work, we used the same type of models corresponding to the original work, WangchanBERTa for BERT and mT5 for GPT-2, but our models support the Thai language. The summaries of the different models are shown in Table 1. However, we used the mT5 model instead of WangchanBERTa in the Answer Generation step because we found that the mT5 model could generate more appropriate answers than WangchanBERTa.

Table 1. Difference of implementation between Raul Puri et al.'s work and ours in the Question Generation pipeline.

Implementation	Answer Generation	Question Generation	Question Filtration
Raul Puri et al.	BERT	GPT-2	BERT
Our	mT5-Large	mT5-Large	WangchanBERTa

3.2.1. Step 1: Answer Generation

Due to the difficulty and ambiguity of the Thai Language, extracting answer candidates from heuristic rules is not sufficient to select the high-quality answers because natural language processing tools for Thai do not perform correctly in every word or sentence; sometimes word features from the given text are extracted incorrectly.

To overcome this limitation, the Answer Generation Model—$p(a|c)$ was used to select an appropriate word to be an Answer $\hat{a}$ of a sample. Unlike Raul Puri et al.'s implementation, we fine-tuned the mT5-Large model by using Context c as an input of the model to learn the answer distribution of the dataset as shown in Figure 4. The answer was selected by the highest probability score.

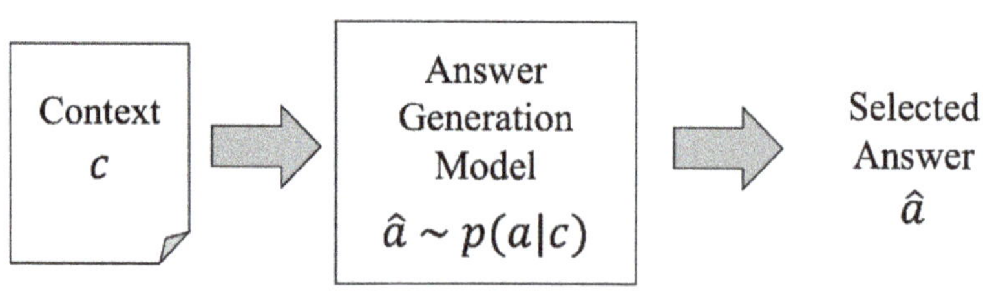

Figure 4. The input and the output of the Answer Generation Model.

3.2.2. Step 2: Question Generation

In this step, the Question Generation Model—$p(q|\hat{a}, c)$ was trained to a generated question in accordance with a given context and answer. We fine-tuned the mT5-Large model by using Context c and selected Answer $\hat{a}$ from Answer Generation Model as inputs as shown in Figure 5.

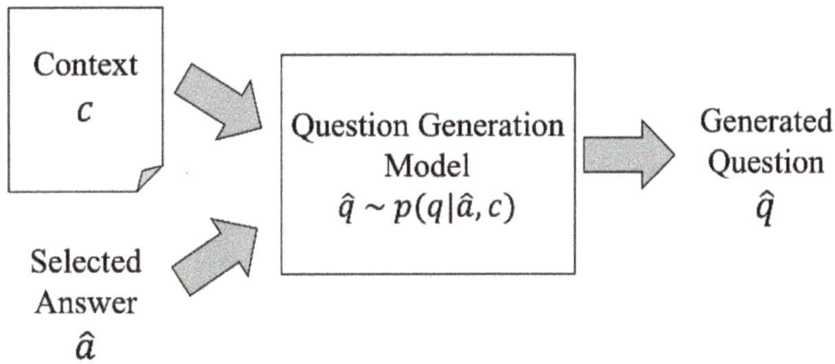

Figure 5. The inputs and the output of the Question Generation Model.

3.2.3. Step 3: Question Filtration

After obtaining a Generated Question $\hat{q}$ from Question Generation Model and an Answer $\hat{a}$ from Answer Generation Model, we already obtained a triplet of generated data $(c, \hat{q}, \hat{a})$. Before using a sample from the generated data, we must verify if this triplet is admissible. To achieve this, we trained a Question Filtration model in the question answering task with labeled training data. After that, we applied the generated Question $\hat{q}$ and Context c to the Question Filtration model for predicting the Answer $\tilde{a}$ as shown in Figure 6. We then compared the Answer $\tilde{a}$ from the model with Answer $\hat{a}$ from the triplet. If these two answers are equivalent, then this triplet is considered an admissible and high-quality sample. Thus, the process of generating a question-answer pair is illustrated in Figure 7.

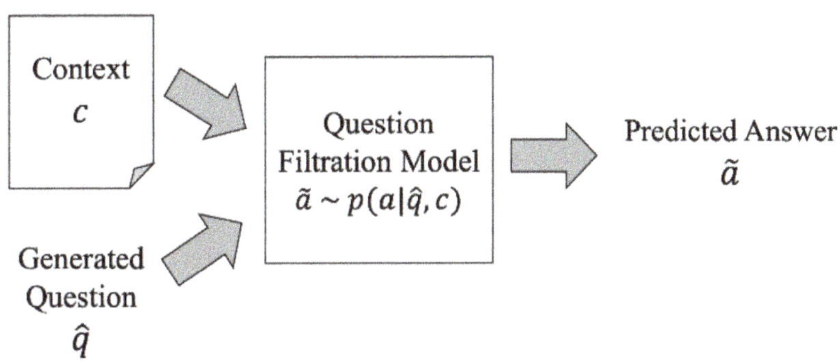

Figure 6. The inputs and the output of the Question Filtration model.

In this part, we selected to use WangchanBERTa as a base model because this model is similar to Raul Puri et al.'s work that used BERT as a base model. Moreover, the WangchanBERTa model is more proper to use in Thai because it was pretrained with Thai documents. Before using it, we fine-tuned the question answering task to this model with Thai QA datasets as we describe in Section 4.1.

In conclusion, we compared two types of generated data: (1) filtered generated data, and (2) all generated data in the experiment. The 'filtered generated data' is the set of samples $(c, \hat{q}, \hat{a})$ that passes the Question Filtration step while the 'all generated data' is the set of triplets $(c, \hat{q}, \hat{a})$ after passing Question Generation step, whether it passes the Question Filtration step or not.

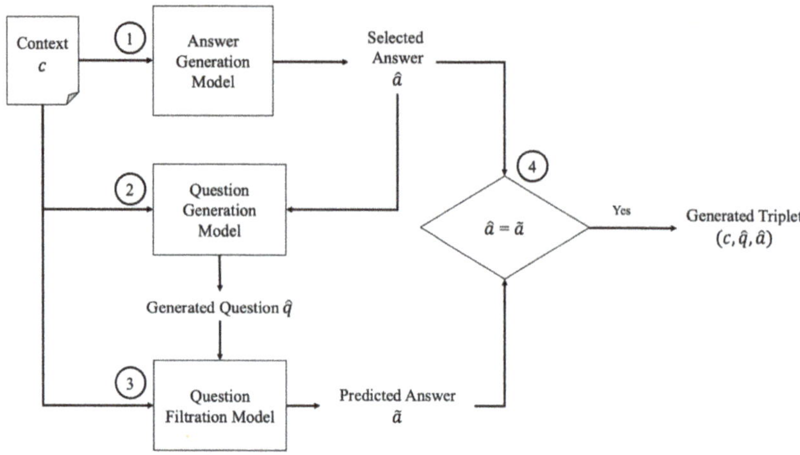

Figure 7. Illustration of the Question Generation pipeline.

3.3. Question Answering Models Training

In QA model training, we selected two Transformer based models as a baseline QA model: WangchanBERTa and mT5. In addition, we compared two training strategies for fine-tuning QA models with generated data and real human-labeled data.

The first training strategy is the Sequence Strategy, which involves sequentially fine-tuning the generated data, followed by the real human-labeled training data. The Sequence Strategy process is illustrated in Figure 8.

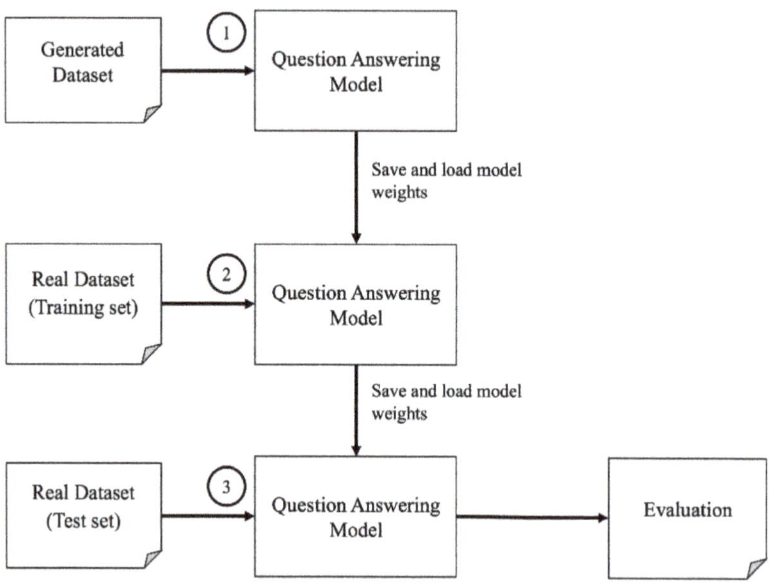

Figure 8. Illustration of the training flow of the Sequence Strategy.

The other training strategy is Merge Strategy, which merges the generated data and the real training data, and fine-tunes at the same time, as illustrated in Figure 9.

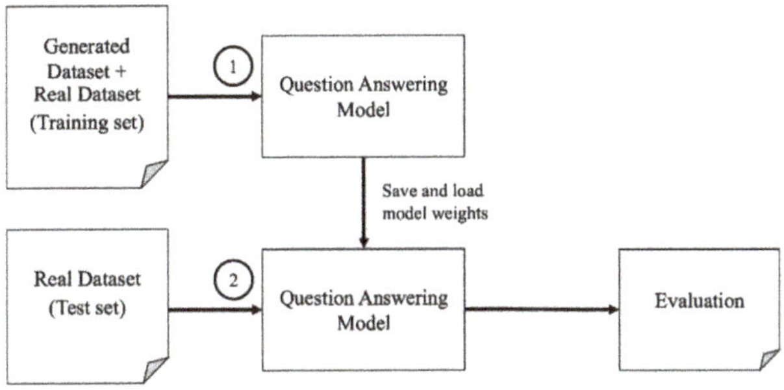

Figure 9. Illustration of the training flow of the Merge Strategy.

3.4. Model Evaluation

We used the F1 score and Exact Match (EM), which are widely used in span extraction Question Answering tasks [19] to evaluate the performance of models. The Exact Match measures how much the model is able to retrieve the exact ground truth span correctly in the whole dataset. The F1 score is a harmonic mean of precision and recall of prediction compared to the ground truth. Originally, to measure the precision and the recall, we count the number of words found in both the prediction and the ground truth.

To calculate the F1 score, the equations below were used. *TP* refers to 'True Positive', that counts the tokens appearing in both the prediction and the ground truth. *FP* refers to 'False Positive' that counts the tokens that appear only in the prediction. *FN* refers to 'False Negative', which means the number of the tokens that appear only in the ground truth. The F1 score of the dataset is an average of the F1 score of every sample.

$$Precision = \frac{TP}{TP + FP} \tag{1}$$

$$Recall = \frac{TP}{TP + FN} \tag{2}$$

$$\mathrm{F1} = \frac{2 \times Precision \times Recall}{Precision + Recall} \tag{3}$$

$$\mathrm{F1} = \frac{\sum_{i=1}^{N} F1_i}{N} \tag{4}$$

Due to the imperfections of the Thai word tokenizer, measuring at the word-level might not be sufficient. In English, there are space separators between words that make English easier to be tokenized into words. On the other hand, the Thai language is more ambiguous as there is no space between words. Thus, the Word-level F1 depends on the quality of the tokenizer used. To overcome this, we also calculated the F1 score at the syllable-level.

The Syllable-level F1 score, the F1 score that calculates based on syllable tokens, is a more appropriate metric than the Word-level F1 score for a language ambiguous to segment because of the following reasons. First, due to the quality of word tokenizers, using different word tokenizers may result in different F1 scores and cause the score to be unable to be compared with other works. Secondly, because of the imperfection of word tokenizers, there are still mistakes when segmenting some similar words. Lastly, due to the ambiguity of the Thai language, some Thai words can be tokenized in many ways, especially the proper nouns. In contrast, using syllables to calculate scores is less ambiguous because there is only a way to segment a word into syllables that maintains a unit of pronunciation.

In this experiment, we used the 'newmm' tokenizer [18] that is currently one of the fastest and the most reliable word tokenizers for the Thai language. However, as shown inthe example in Table 2, the 'newmm' tokenizer could not tokenize the word into a proper form. To address this problem, we evaluated the predictions with Syllable-level F1 along with Word-level F1. Using syllable tokens to calculate the F1 score could obtain a more accurate score to the linguistic word segmentation than word tokens because the syllable tokenizer can extract overlapping words into pieces of syllables while the word tokenizer cannot.

Table 2. Example of tokenization used in this work.

	Original Word	Human-Tokenized	Use Word Tokenizer (newmm)	Syllable Tokenizer
Ground truth	*มลายู* (Malay language in short term)	*มลายู*	*มลายู*	*ม / ลา / ยู*
Prediction	*ภาษามลายู* (Malay language)	*ภาษา / มลายู*	*ภาษามลายู*	*ภา / ษา / ม / ลา / ยู*
F1 Score		**66.6**	**0.0**	**77.49**

Similar to the English language, Thai words can have one or more syllables. Tokenizing a word into syllables means dividing the word by a unit of pronunciation that has one vowel sound. For example, in Table 2, the word "*มลายู*" (Malay language in the short term) can be pronounced as /mala:ju:/ which has three syllables as "*ม / ลา / ยู*"; each piece can be pronounced as /ma/, /la:/ and /ju:/ respectively. Another example is the word "*ภาษา*" (language), which can be pronounced as /pa:sa:/. This word has two syllables as "*ภา / ษา*"; each piece can be pronounced as /pa:/ and /sa:/ sequentially.

4. Experiment Setup

In this section, we describe the datasets used in the experiments, tools and parameter setup as follows.

4.1. Datasets

There are two Thai QA corpora used in our experiments: Thai Wiki QA and iApp Wiki QA. The dataset statistics of both datasets are shown in Table 3.

Table 3. Datasets splitting for experiments.

Dataset	No. of Training Set	No. of Validation Set	No. of Test Set
Thai Wiki QA	9045	1005	4950
iApp Wiki QA	5761	742	439

Thai Wiki QA [8] is a SQuAD-like dataset in the Thai language. It was used as a QA competition dataset in Thailand National Software Contest (NSC), during 2018–2019. This dataset consists of 15,000 question-answer pairs with contexts from Thai Wikipedia and annotated by 15 native Thai speakers with many kinds of expertise and education levels. The publisher of Thai Wiki QA also published 125,302 Thai Wikipedia articles to support this dataset as an open domain QA task. In this study, we also used the published articles for generating more question answering samples.

iApp Wiki QA (https://github.com/iapp-technology/iapp-wiki-qa-dataset (accessed on 10 September 2021)) is a SQuAD-like dataset published by iApp Technology Company Limited. This dataset includes 7242 question-answer pairs made with Thai Wikipedia articles. However, the publisher of this dataset does not provide information about the data annotation method.

4.2. Tools and Parameter Setup

For all implementations of models including Question Generation and Question Answering parts, we used HuggingFace's Transformers [20] for model developments and training, including model architecture, model configuration and model weights. HuggingFace also provided model training tools. All models in our research used default training arguments provided by HuggingFace, except the learning rate, batch size, weight decay, and number of epochs for training. We changed the value of the learning rate to 10^{-6}, the weight decay to 0.01, and the number of epochs to 25. We also changed the value of batch size to 12 if the trained model was WangchanBERTa, and to 4 if the trained model was mT5-Base and mT5-Large.

We selected the number of batch size configurations based on technical reasons. Our system, DGX A100 with NVIDIA A100 GPU, could use only a batch size of 4 for training mT5-Base and mT5-Large models due to its enormous trainable parameters; 580 M for mT5-Base and 1.2 B for mT5-Large, while the WangchanBERTa model has only 110 M of parameters. Thus, we selected a batch size of 12 for training the WangchanBERTa model to decrease disparities and make them comparable.

The other apparatus used in this research was PyThaiNLP, a Thai natural language processing toolkit. We used this tool for applying text preprocessing before applying the text to the models. In addition, this tool provides the Thai syllable tokenizer and text tokenizers used in this research, such as the 'newmm' tokenizer, which is a fast and reliable Thai word tokenizer.

5. Results

In this section, we report the results of the experiments in several aspects. First, we explain the overall results by comparing every combination of QA models, training strategies, and generated data. The overall results correspond to Tables 4 and 5, which are the main results, and Table 6, which shows the dataset statistics of the augmented data. Secondly, we explain the performance related to training strategies. Thirdly, we describe the comparison of the generated data used. Fourthly, we present the comparison of base QA models. Lastly, we explain the results of the Syllable-level F1 score and provide some samples from the test set calculated Syllable-level F1 score.

Table 4. The experiment result of our method compared to baseline models of Thai Wiki QA. (EM, W-F1, and S-F1 refer to Exact Match, word-level F1 and syllable-level F1 respectively. Boldface refers to the winner.).

Thai Wiki QA	Baseline			+ Filtered Generated Pairs (FLT)			+ All Generated Pairs (ALL)		
	EM	W-F1	S-F1	EM	W-F1	S-F1	EM	W-F1	S-F1
WangchanBERTa (WBT)									
Sequence Strategy (SEQ)	43.92	70.73	74.71	45.90	73.35	77.55	46.48	74.40	78.60
Merge Strategy (MRG)				44.63	71.11	75.15	43.35	71.09	75.26
mT5-Base (mT5)									
Sequence Strategy (SEQ)	64.14	78.24	80.35	**70.42**	**83.64**	**85.29**	69.03	83.02	84.74
Merge Strategy (MRG)				69.01	82.88	84.68	63.66	79.30	81.17

Table 5. The experiment result of our method compared to baseline models of iApp Wiki QA. (EM, W-F1, and S-F1 refer to Exact Match, word-level F1 and syllable-level F1 respectively. Boldface refers to the winner.).

iApp Wiki QA	Baseline			+ Filtered Generated Pairs (FLT)			+ All Generated Pairs (ALL)		
	EM	W-F1	S-F1	EM	W-F1	S-F1	EM	W-F1	S-F1
WangchanBERTa (WBT)									
Sequence Strategy (SEQ)	30.58	69.68	71.81	32.88	71.81	74.42	33.15	72.98	75.14
Merge Strategy (MRG)				29.77	69.59	71.97	31.66	70.37	72.82
mT5-Base (mT5)									
Sequence Strategy (SEQ)	29.36	63.46	63.71	56.02	81.31	82.58	**58.05**	**81.97**	**83.15**
Merge Strategy (MRG)				57.10	81.42	82.57	53.45	79.53	80.81

Table 6. Dataset statistics after combining with generated data.

Datasets	No. Training Set	No. Training Set + Filtered Generated Pairs	No. Training Set + All Generated Pairs
Thai Wiki QA	9045	62,610 (+592.2%)	119,813 (+1224.6%)
iApp Wiki QA	5761	59,326 (+929.8%)	116,529 (+1922.7%)

5.1. Overall Results

The experiment was conducted based on two Thai QA datasets: Thai Wiki QA and iApp Wiki QA. We compared the results in three aspects: quality of the synthesized question-answer pairs, training strategies, and baseline models used in this experiment—WangchanBERTa and mT5-Base. Furthermore, we also evaluated the results in exact match (EM), word-level F1 (W-F1) and Syllable-level F1 (S-F1). The results are summarized in Table 4 for Thai Wiki QA and Table 5 for iApp Wiki QA. The training set statistics, including the generated dataset, are illustrated in Table 6.

In the result description, we created the combination name for readily referring to the tested model. The combination name consists of three parts: the QA models, training strategies, and augmented data used. The QA models have two possible types: WangchanBERTa (WBT) and mT5-Base (mT5). Training strategies have two possible strategies: Sequence (SEQ) and Merge (MRG). Lastly, the generated data used in the experiments have two types: Filtered (FLT) and ALL. All three parts connect together with the dash symbol (-). For example, mT5-SEQ-FLT refers to using the mT5 model fine-tuned with filtered generated data and the Sequence strategy.

We compared the results with the baseline models which are the question answering models that were trained with real human-labeled data only. In most cases, using generated data could improve the performance of every metric. In the Thai Wiki QA dataset, using the mT5-SEQ-FLT provided the best performance combination that beat the result of the baseline of mT5-Base by 6.28%, 5.14%, and 4.94% for EM, word-level F1, and syllable-level F1, respectively. In the iApp Wiki QA dataset, the best performance combination used the mT5-SEQ-ALL, which beat the baseline result of mT5-Base by 28.69%, 18.51%, and 19.44% EM, word-level F1, and syllable-level F1, respectively. Using all generated data with iApp Wiki QA could slightly improve performance compared to using the filtered generated data because the human-labeled training set of iApp Wiki QA has a smaller number of samples, compared to those of Thai Wiki QA, and it needs more data to fine-tune. However, the results between using the filtered generated pairs and all the generated pairs are not much different. We can conclude that using mT5-SEQ-FLT is the best combination that outperforms both datasets.

5.2. Comparison of the Training Strategies

The difference between the two training strategies is the fine-tuning steps. Sequence Strategy is sequentially fine-tuned on the generated data before the real data while Merge Strategy fine-tunes the combination of the generated data and the real data at the same time. As a result, in most combinations, using Sequence Strategy explicitly outperforms Merge Strategy, as shown in Tables 4 and 5 in the Sequence Strategy rows.

5.3. Comparison of Using Different Qualities of the Generated Question-Answer Pairs

In our trial, fine-tuning with the filtered generated pairs versus doing so with all generated pairs has no difference between the numbers of the outperforming cases. However, if we consider only the Sequence Strategy cases, the number of the outperforming cases of fine-tuning with the filtered generated pairs is greater than that of fine-tuning with all generated pairs. We can therefore conclude that using the filtered generated pairs is preferrable compared to using all generated pairs.

However, according to Table 5, which shows the result of the mT5-Base model with the Sequence Strategy, using all generated data (mT5-SEQ-ALL) is slightly better than using the filtered generated data (mT5-SEQ-FLT), but the number of training samples is around twice as much, which consumes longer training time. In this case, we can summarize that using the filtered generated data is better than using all generated data in terms of the training duration.

5.4. Comparison of the Baseline Models

In the baseline experiments, the tables show that the results of outperforming models are different depending on the datasets. However, when we apply our method, as listed in Tables 4 and 5 in the columns "+ Filtered Generated Pairs" and "+ All Generated Pairs," compared to the column "Baseline," the mT5-Base model outperforms WanchanBERTa in all cases.

5.5. Comparison of the Syllable-Level F1

From Tables 4 and 5, the syllable-level F1 scores of all combinations are greater than the word-level F1 score. The mT5-SEQ-FLT in Thai Wiki QA has a syllable-level F1 score greater than the word-level F1 score by 1.65%, and the mT5-SEQ-ALL in iApp Wiki QA has a syllable-level F1 score greater than the word-level F1 score by 1.18%. The reason is that there are some predicted answers and/or ground truth answers that the word tokenizer used in the F1 score calculation does not perform correctly due to inability of tokenizer itself. This results in some overlapping words not counted to calculate the F1 score. However, the syllable tokenizer can tokenize those overlapping words into syllables and is able to calculate a nearer-correct F1 score. We provide some examples of the predicted answer and ground truth answer in Table 7, showing that the syllable F1 score gives a more accurate result than the word-level F1 score.

Table 7. Examples of the predicted answer and the ground truth answer that show the syllable-level F1; all results get 0 of the word-level F1.

Predicted Answer	Ground Truth Answer	Syllable-Level F1
มหาวิทยาลัยรามคำแหง (Ramkhamhaeng University)	*รามคำแหง* (Ramkhamhaeng)	66.67
คณะนิติศาสตร์ (Faculty of Law)	*นิติศาสตร์* (Law)	85.71
คมนาคม (Transportation)	*กระทรวงคมนาคม* (Ministry of Transportation)	75.00
ไม้ล้มลุกขนาดเล็ก (Small biennial plant)	*ล้มลุก* (Biennial)	57.14

6. Discussion

In this section we further analyze the improvement of the models in detail. First, we explain the analysis of model improvements by using the distribution of the word-level F1 score of both datasets. Secondly, we report the model performance by question types. Questions are classified by keyword extraction from Table 8. The results of each question type are illustrated in Tables 9 and 10. Lastly, we present the comparison of word tokenizers in terms of word-level F1 score calculation in Tables 11 and 12.

Table 8. Example of question word categories in Thai.

Who	What	Where	Year	Date	Number
ใคร (Who)	*อะไร* (What)	*ที่ไหน* (Where)	*ปีใด* (What year)	*วันที่เท่าใด* (Which date)	*เท่าไร* (How many/much)
คนใด (Which one)	*...ใด* (What/Which ...)	*ที่ใด* (Where)	*ปีไหน* (What year)	*วันใด* (Which date)	*เท่าใด* (How many/much)
คนไหน (Which one)	*...ไหน* (What/Which ...)	*ประเทศใด* (Which country)	*พ.ศ ใด* (What B.E. year)	*เมื่อใด* (When)	*กี่..* (How many/much of)
		จังหวัดใด (Which province)	*ค.ศ ใด* (What C.E. year)		

Table 9. Model performance evaluated on the Thai Wiki QA dataset classified by question types.

Question Type	mT5 Baseline			mT5-SEQ-FLT			% of Improvement		
	EM	W-F1	S-F1	EM	W-F1	S-F1	EM	W-F1	S-F1
Overall Performance	64.14	78.24	80.35	70.42	83.64	85.29	9.79	6.90	6.15
Who	69.16	82.03	81.40	73.05	85.87	85.30	5.62	4.68	4.79
What	62.90	77.45	80.48	68.82	82.89	85.24	9.41	7.02	5.91
Where	67.54	83.23	83.02	76.61	88.82	88.93	13.42	6.72	7.12
Year	76.04	83.06	85.03	83.83	88.75	89.53	10.24	6.85	5.29
Date	58.88	79.53	79.21	72.08	90.13	89.43	22.42	13.33	12.90
Number	61.99	74.14	76.67	68.19	78.94	81.32	10.00	6.47	6.07

Table 10. Model performance evaluated on the iApp Wiki QA dataset classified by question types.

Question Type	mT5 Baseline			mT5-SEQ-ALL			% of Improvement		
	EM	W-F1	S-F1	EM	W-F1	S-F1	EM	W-F1	S-F1
Overall Performance	29.36	63.46	63.71	58.05	81.97	83.15	97.72	29.17	30.51
Who	48.57	67.03	66.30	65.71	86.10	87.77	35.29	28.45	32.38
What	32.18	65.00	66.56	56.02	81.03	82.64	74.08	24.66	24.16
Where	33.33	68.22	71.85	55.56	79.72	85.17	66.69	16.85	18.54
Year	62.50	84.17	79.17	62.50	82.08	81.25	0.00	−2.48	2.63
Date	9.68	63.91	58.22	62.90	87.97	86.43	549.79	37.64	48.45
Number	13.83	55.49	53.70	60.64	80.90	80.76	338.47	45.79	50.39

Table 11. The result of Word-level F1 with different types of tokenizers of the Thai Wiki QA dataset.

Model	'Newmm' (Word-Level F1)	AttaCut (Word-Level F1)	Syllable Tokenizer (Syllable-Level F1)
WangchanBERTa Baseline	70.73	64.46	74.71
mT5 Baseline	78.24	73.58	80.35
mT5-SEQ-FLT	83.64	78.98	85.29

Table 12. The result of Word-level F1 with different types of tokenizers of the iApp Wiki QA dataset.

Model	'Newmm' (Word-Level F1)	AttaCut (Word-Level F1)	Syllable Tokenizer (Syllable-Level F1)
WangchanBERTa Baseline	69.68	65.42	71.81
mT5 Baseline	63.46	59.75	63.71
mT5-SEQ-ALL	81.97	78.87	83.15

6.1. Analysis of Model Improvement

To investigate what the improvement to the result is, we use the charts of the F1 score distribution to visualize how the score increases. We report only the result of the mT5-Base models because the results from Tables 4 and 5 show that the mT5-Base model outperforms the WangchanBERTa model. Based on the Thai Wiki QA dataset illustrated in Figure 10, the result of the baseline of the mT5-Base model (a) shows that there are more than 500 erroneous samples of F1 = 0, which means that those predicted answers are not overlapping with the ground truth answers. After using the combination mT5-SEQ-FLT (b), the result shows that the number of the perfect answers of F1 = 100 increases while the number of samples of F1 = 0 decreases. This is a proof that this method can increase the QA model performance.

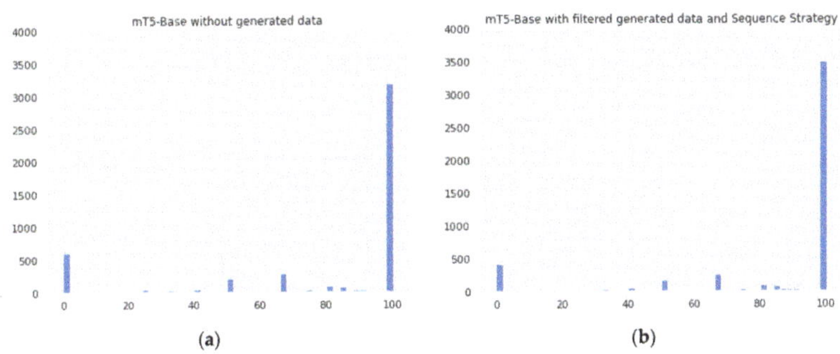

Figure 10. F1 distribution of the Thai Wiki QA dataset: (**a**) the result from the mT5-Base baseline model, and (**b**) the result from the combination mT5-SEQ-FLT of the Thai Wiki QA dataset.

The F1 distribution of the iApp Wiki QA dataset is illustrated in Figure 11. It represents a similar result to that of Thai Wiki QA; after applying the generated data and the Sequence strategy with mT5-Base model (mT5-SEQ-ALL), the number of perfect answers of F1 = 100 increases while the number of F1 = 0 cases decreases. This is also a proof that this method can increase the QA model performance even though the dataset is changed.

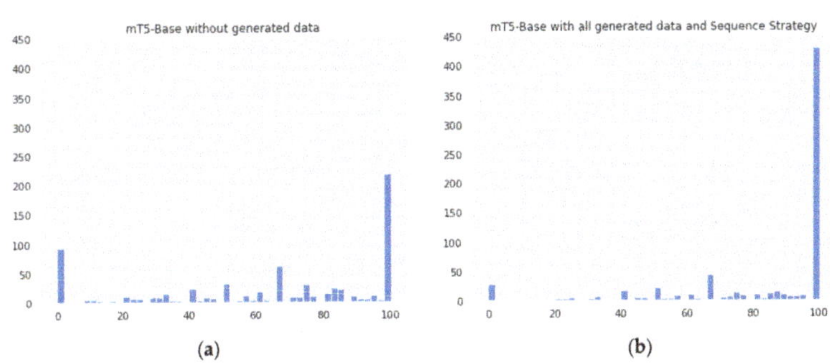

Figure 11. F1 distribution of the iApp Wiki QA dataset: (**a**) the result from the mT5-Base baseline model, and (**b**) the result from the combination mT5-SEQ-ALL of the iApp Wiki QA dataset.

6.2. Model Performance Analysis by Question Types

From the questions in the datasets, they can be classified into six groups based on the types of answers as follows.

- Who: a group of questions that requires an answer as a person name
- What: a group of questions that requires an answer as a thing or a name of things
- Where: a group of questions that requires an answer as a name of places, for instance, countries, provinces, or states
- Year: a group of questions that requires an answer as a year, either Common Era (C.E.) or Buddhist Era (B.E.)
- Date: a group of questions that requires an answer as a date
- Number: a group of questions that requires an answer as a number

We classified questions in a test set by keyword detection. The keywords that were used to categorize the questions are listed in Table 8. We next evaluated the model performance of each question type. The results from both datasets are listed in Tables 9 and 10. The results show that the best combination of both datasets can raise the performance above the baseline in most question types, except the question type 'Year' of iApp Wiki QA, which has the Word-level F1 score slightly lower than the baseline. After investigating, we found that there were only eight samples in the group 'Year' of iApp Wiki QA, which made this group of samples sensitive to the change of F1 score. However, the Syllable-level F1 score of this group improved after applying our method. This indicates that the best combination of iApp-Wiki-QA (mT5-SEQ-ALL) could predict more accurate answers, compared to the baseline model. This is further evidence that our method can increase the QA model performance.

6.3. Word Tokenizer Choices

In this work, we used 'newmm' for calculating the Word-level F1 score. However, there are several choices of Thai word tokenizers that can be used. We conducted experiments to compare two Thai word tokenizers; we selected AttaCut [21], a deep learning-based word tokenizer for Thai, to compare with 'newmm'. The results are shown in Tables 11 and 12.

From the results, the Word-level F1 scores from 'newmm' are higher than the Word-level F1 scores from AttaCut in all cases. This means that the AttaCut tokenizer has more tokenization mistakes on ambiguous words than 'newmm', which caused the drop of F1 scores. This proves that changing tokenizers may result in different Word-level F1 scores. In contrast, using Syllable-level F1 can address this problem by tokenizing a word into syllables, which is less ambiguous.

7. Conclusions

In this paper, we propose to employ transformer-based models for Thai QA, which aims to improve the performance of the Thai question answering model. The limitation of the low resource Thai QA corpora can be overcome by using a data generation composed of three steps: answer generation, question generation, and question filtration. To utilize the generated question-answer pairs, different fine-tuning strategies were investigated. Apart from the model improvement, all challenges in Thai were addressed in data preprocessing. We also propose a new evaluation metric at the syllable-level, which is more suitable for the Thai language because there is no ambiguity in syllable tokenization. We conducted experiments on two Thai question answering datasets, the Thai Wiki QA and the iApp Wiki QA. The results showed that the generated data can explicitly enhance the model performance from 78.24 to 83.64 in Thai Wiki QA and 63.46 to 81.97 in iApp Wiki QA, in terms of the Word-level F1 score.

However, a limitation of our work is that our data generation technique is appropriate with a span-extraction question answering task; the answer of the given question is part of the given context only.

Author Contributions: Conceptualization, P.P.; methodology, P.P.; software, P.P.; validation, P.V.; data curation, P.P.; writing—original draft preparation, P.P.; writing—review and editing, P.V. All authors have read and agreed to the published version of the manuscript.

Funding: This research received no external funding.

Conflicts of Interest: The authors declare no conflict of interest.

References

1. Raffel, C.; Shazeer, N.; Roberts, A.; Lee, K.; Narang, S.; Matena, M.; Zhou, Y.; Li, W.; Liu, P.J. Exploring the limits of transfer learning with a unified text-to-text transformer. *arXiv* **2019**, arXiv:1910.10683.
2. Vaswani, A.; Shazeer, N.; Parmar, N.; Uszkoreit, J.; Jones, L.; Gomez, A.N.; Kaiser, Ł.; Polosukhin, I. Attention is all you need. In Proceedings of the Advances in Neural Information Processing Systems, Long Beach, CA, USA, 4–9 December 2017; pp. 5998–6008.
3. Dodge, J.; Sap, M.; Marasovic, A.; Agnew, W.; Ilharco, G.; Groeneveld, D.; Gardner, M. Documenting the english colossal clean crawled corpus. *arXiv* **2021**, arXiv:2104.08758.
4. Rajpurkar, P.; Zhang, J.; Lopyrev, K.; Liang, P. Squad: 100,000+ questions for machine comprehension of text. *arXiv* **2016**, arXiv:1606.05250.
5. Decha, H.; Patanukhom, K. Development of thai question answering system. In Proceedings of the 3rd International Conference on Communication and Information Processing, Tokyo, Japan, 24–26 November 2017; pp. 124–128.
6. Seo, M.; Kembhavi, A.; Farhadi, A.; Hajishirzi, H. Bidirectional attention flow for machine comprehension. *arXiv* **2016**, arXiv:1611.01603.
7. Lapchaicharoenkit, T.; Vateekul, P. Machine Reading Comprehension on Multiclass Questions Using Bidirectional Attention Flow Models with Contextual Embeddings and Transfer Learning in Thai Corpus. In Proceedings of the 8th International Conference on Computer and Communications Management, Singapore, 17–19 July 2020; pp. 3–8.
8. Trakultaweekoon, K.; Thaiprayoon, S.; Palingoon, P.; Rugchatjaroen, A. The first wikipedia questions and factoid answers corpus in the thai language. In Proceedings of the 2019 14th International Joint Symposium on Artificial Intelligence and Natural Language Processing (iSAI-NLP), Chiang Mai, Thailand, 7–9 November 2019; pp. 1–4.
9. Puri, R.; Spring, R.; Patwary, M.; Shoeybi, M.; Catanzaro, B. Training question answering models from synthetic data. *arXiv* **2020**, arXiv:2002.09599.
10. Lowphansirikul, L.; Polpanumas, C.; Jantrakulchai, N.; Nutanong, S. WangchanBERTa: Pretraining transformer-based Thai Language Models. *arXiv* **2021**, arXiv:2101.09635.
11. Xue, L.; Constant, N.; Roberts, A.; Kale, M.; Al-Rfou, R.; Siddhant, A.; Barua, A.; Raffel, C. mt5: A massively multilingual pre-trained text-to-text transformer. *arXiv* **2020**, arXiv:2010.11934.
12. Hochreiter, S.; Schmidhuber, J. Long short-term memory. *Neural Comput.* **1997**, *9*, 1735–1780. [CrossRef] [PubMed]
13. Devlin, J.; Chang, M.-W.; Lee, K.; Toutanova, K. Bert: Pre-training of deep bidirectional transformers for language understanding. *arXiv* **2018**, arXiv:1810.04805.
14. Joshi, M.; Chen, D.; Liu, Y.; Weld, D.S.; Zettlemoyer, L.; Levy, O. Spanbert: Improving pre-training by representing and predicting spans. *Trans. Assoc. Comput. Linguist.* **2020**, *8*, 64–77. [CrossRef]
15. Dhingra, B.; Pruthi, D.; Rajagopal, D. Simple and effective semi-supervised question answering. *arXiv* **2018**, arXiv:1804.00720.
16. Radford, A.; Wu, J.; Child, R.; Luan, D.; Amodei, D.; Sutskever, I. Language models are unsupervised multitask learners. *OpenAI Blog* **2019**, *1*, 9.
17. Liu, Y.; Ott, M.; Goyal, N.; Du, J.; Joshi, M.; Chen, D.; Levy, O.; Lewis, M.; Zettlemoyer, L.; Stoyanov, V. Roberta: A robustly optimized bert pretraining approach. *arXiv* **2019**, arXiv:1907.11692.
18. Phatthiyaphaibun, W.; Suriyawongkul, A.; Chormai, P.; Lowphansirikul, L.; Siwatammarat, P.; Tanruangporn, P.P.; Charoen-chainetr, P.; Udomcharoenchaikit, C.; Janthong, A.; Chaovavanichet, K.; et al. PyThaiNLP/pythainlp: PyThaiNLP v2.3.2 Release! *Zenodo* **2021**. [CrossRef]
19. Zeng, C.; Li, S.; Li, Q.; Hu, J.; Hu, J. A Survey on Machine Reading Comprehension—Tasks, Evaluation Metrics and Benchmark Datasets. *Appl. Sci.* **2020**, *10*, 7640. [CrossRef]
20. Wolf, T.; Chaumond, J.; Debut, L.; Sanh, V.; Delangue, C.; Moi, A.; Cistac, P.; Funtowicz, M.; Davison, J.; Shleifer, S. Transformers: State-of-the-art natural language processing. In Proceedings of the 2020 Conference on Empirical Methods in Natural Language Processing: System Demonstrations, Online, 16–20 November 2020; pp. 38–45.
21. Chormai, P.; Prasertsom, P.; Rutherford, A. AttaCut: A Fast and Accurate Neural Thai Word Segmenter. *arXiv* **2019**, arXiv:1911.07056.

applied sciences

MDPI

Article

FedQAS: Privacy-Aware Machine Reading Comprehension with Federated Learning

Addi Ait-Mlouk *, Sadi A. Alawadi, Salman Toor and Andreas Hellander

Department of Information Technology, Division of Scientific Computing, Uppsala University, 75236 Uppsala, Sweden; sadi.alawadi@it.uu.se (S.A.A.); salman.toor@it.uu.se (S.T.); andreas.hellander@it.uu.se (A.H.)
* Correspondence: addi.ait-mlouk@it.uu.se

Abstract: Machine reading comprehension (MRC) of text data is a challenging task in Natural Language Processing (NLP), with a lot of ongoing research fueled by the release of the Stanford Question Answering Dataset (SQuAD) and Conversational Question Answering (CoQA). It is considered to be an effort to teach computers how to "understand" a text, and then to be able to answer questions about it using deep learning. However, until now, large-scale training on private text data and knowledge sharing has been missing for this NLP task. Hence, we present FedQAS, a privacy-preserving machine reading system capable of leveraging large-scale private data without the need to pool those datasets in a central location. The proposed approach combines transformer models and federated learning technologies. The system is developed using the FEDn framework and deployed as a proof-of-concept alliance initiative. FedQAS is flexible, language-agnostic, and allows intuitive participation and execution of local model training. In addition, we present the architecture and implementation of the system, as well as provide a reference evaluation based on the SQuAD dataset, to showcase how it overcomes data privacy issues and enables knowledge sharing between alliance members in a Federated learning setting.

Keywords: machine reading comprehension; natural language processing; question answering; data privacy; federated learning; transformer

Citation: Ait-Mlouk, A.; Alawadi, S.; Toor, S.; Hellander, A. FedQAS: Privacy-Aware Machine Reading Comprehension with Federated Learning. *Appl. Sci.* **2022**, *12*, 3130. https://doi.org/10.3390/app12063130

Academic Editors: Arturo Montejo-Ráez and Salud María Jiménez-Zafra

Received: 25 January 2022
Accepted: 14 March 2022
Published: 18 March 2022

1. Introduction

Machine reading comprehension (MRC) is a sub-field of natural language understanding (NLU) that aims to teach machines to read and understand human languages (text). A user can ask the machine to answer questions based on a given paragraph or text document. Generally, MRC requires modeling complex interactions between the context and the query in a specific domain. It could be used in many NLP applications such as dialogue systems and search engines as shown in Figure 1—a Google search engine with MRC techniques can directly return the correct answers to questions rather than a list of content and web pages. These kinds of techniques have been based on hand-crafted rules that need substantial human effort and resources. However, with the rise of artificial intelligence, there has been an explosion of various MRC benchmark datasets and models that contribute to a better understanding of the task and show their ability to exceed human performance. Despite this rapid progress on MRC datasets and models, most of the existing work has focused on algorithms for improving model performance.

At present, several MRC models have already surpassed human performance on many of the MRC datasets [1], but there is still a limit in terms of data availability due to privacy concerns, collaborative training, resource consumption, and communication overhead due to data transfer. Hence, there is a need for extending existing MRC models in a way that keeps data private on its generated location and allows several participants to train a machine learning model by sharing only model parameters. This will let cross-silo (companies) or cross-device (phones, IoT devices) participate in the training process and let the model learn from large distributed datasets by sharing knowledge between different local models. To address these gaps, we proposed a privacy-aware approach based on federated learning to learn new global models in a geographically distributed manner using our FEDn framework [2], build more challenging MRC models by integrating with private data generation and labeling for local accurate training as well as an incremental learning approach to strengthening the model performances during collaborative training without compromising the data.

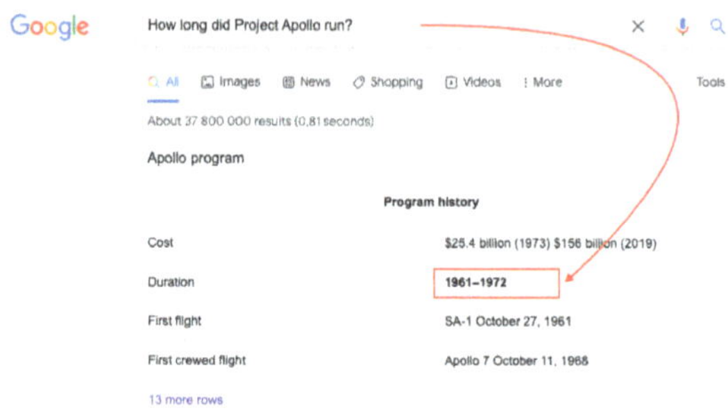

Figure 1. An example of Google search engine with machine reading comprehension techniques.

The remainder of this paper is organized as follows. Section 2 surveys related work. Section 3 details the proposed approach and architecture of FedQAS, with an emphasis on its privacy and scalability properties. In Section 4, we demonstrate the frameworks potential in an evaluation based on the SQuAD dataset. Finally, Section 4 concludes the work and outlines future work.

2. Related Work

Machine reading comprehension was proposed for the first time in 1977 by Lehnert, who built a question answering program called the QUALM [3]. In 1999, Hirschman et al. [4] built a reading comprehension system using a corpus of 60 development and 60 test stories of 3rd to 6th grade material. Because of the lack of benchmark datasets in that period, most MRC systems were rule-based or statistical models [5,6]. In recent years, many benchmark datasets have been released and focused on MRC by answering questions, see Table 1. Since these datasets were made available, there has been considerable progress on MRC tasks. The Stanford Question Answering Dataset (SQuAD) is one of the most well-known reading comprehension datasets, consisting of 100,000 questions posed by crowd-workers on Wikipedia articles, where the answer to every question is a segment of text, or span, from the corresponding reading context. Important progress based on SQuAD concerns include the attention method [7] and Bi-Directional Attention Flow (BiDAF) [8], which considerably improved the question answering performance. These two methods compute Context to Question attention and Question to Context attention using a similarity matrix computed directly from context and question. Authors in [9] describe a novel hierarchical attention network for reading comprehension style question answering, which aims to answer questions for a given narrative paragraph. In their work, attention and fusion are conducted horizontally and vertically across layers at different levels of granularity between question and paragraph. In recent work in language modeling, authors in [10] incorporate explicit contextual semantics from pre-trained semantic role labeling and introduce an improved language representation model, Semantics-aware BERT (SemBERT), which is capable of explicitly absorbing contextual semantics over a BERT backbone. Moreover, Zhuosheng et al. [11] propose using syntax to guide the text modeling by incorporating explicit syntactic constraints into the attention mechanism for better linguistically motivated word representations. Recently, there has been an explosion of various MRC benchmark datasets that leads to a variety of models such as

BiDAF [12] and other models based on BERT [13], RoBERTa [14], XLNet [15], ELMo [16] and transformer [17]. Other relevant works have been proposed, including [18–20].

Table 1. List of some existing MRC datasets.

Dataset	Answer Type	Domain-Specific
MCTest [8]	Multiple choice	Children's stories
CNN/Daily Mail [21]	Spans	News
Children's book [22]	Spans	Children's stories
MS MARCO [23]	Free-form text	Web Search
NewsQA [24]	Spans	News
SearchQA [25]	Spans	Jeopardy
TriviaQA [26]	Spans	Trivia
SQuAD [27]	Spans	Wikipedia
SQuAD 2.0 [28]	Spans, Unanswerable	Wikipedia
CoQA [29]	Free-form text,	News, Reddit Wikipedia

All these proposed approaches required a very large amount of data for training, which is not always available in some cases, in particular when the text data is sensitive, private (medical text, business, social media), and very big. In this context, we here propose the use of federated learning as a method for distributed and collaborative machine learning. Organizations maintain and govern their data locally and participate in learning a new global, federated model by sending only their model updates (model weights) to a server for aggregation into the global model. Hence, all participants (clients) can benefit from a newly trained model without exposing their data publicly.

Our contributions in this paper can be summarized as follows: (1) We propose federated learning models for MRC using a transformer architecture, (2) we design and develop the FedQAS system for collaborative training, (3) we preserve data privacy (4) we improve the local training with incremental learning scheme and private data generation, and (5) our analyses of the models respect data privacy regulations and outperforms the baseline model on SQuAD after a couple of rounds.

3. Proposed Approach

The overall architecture of our proposed FedQAS system is shown in Figure 2. The main modules are private data pipeline, federated learning settings, question answering, and incremental learning. The private data pipeline module allows local users (clients) to process and prepare their data locally to be used by federated learning methods. The federated learning module enables multiple private clients to form an alliance to collaboratively train machine learning/deep learning models and send parameters to the server for global model generation (aggregation of local models). Afterward, the system allows the client to add new data locally and train the model incrementally through a defined number of rounds to improve the performance using incremental learning techniques. Finally, participating clients can use the global model for question answering system. The system is implemented using the FEDn federated learning framework [2] and a web interface using Flask (https://flask.palletsprojects.com/, accessed on 24 January 2022). FEDn provides a highly scalable federated learning run-time, and Flask is used to develop interactive and user-friendly interfaces for the different processes in the workflow. The list of available datasets related to question answering used for the demo is placed in the local data sources. Moreover, the developed system is scalable, flexible, and can be expanded with new clients/data sets on-demand (without the need to re-train the federated model).

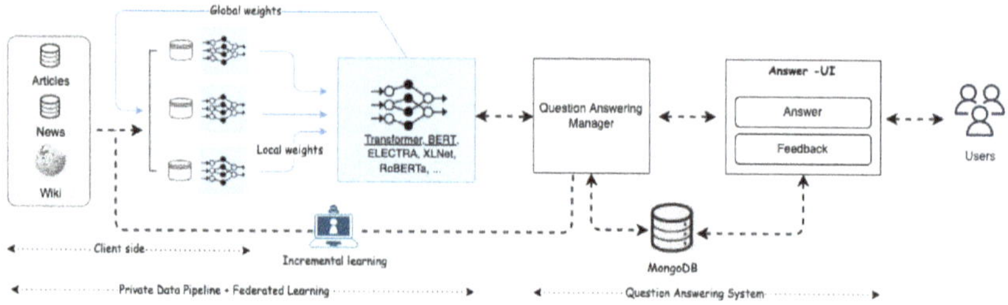

Figure 2. Overview of the FedQAS architecture. The FedQAS approach is organized in three main logical layers, the first one is for collaborative privacy-preserving training, the second one is a federated question answering manager, and the third is for incremental learning and private data generation.

3.1. Data Processing (Client Side)

Stanford Question Answering Dataset (SQuAD) [30] is a machine reading comprehension dataset, consisting of questions posed by crowd-workers on a set of Wikipedia articles, where the answer to every question is a segment of text, or span, from the corresponding reading passage, alternatively the question might be unanswerable. SQuAD2.0 combines the 100.000 questions in SQuAD1.1 with over 50,000 unanswerable questions written adversarially by crowd-workers to look similar to answerable ones. To do well on SQuAD2.0, systems must not only answer questions when possible, but also determine when no answer is supported by the paragraph and abstain from answering (https://rajpurkar.github.io/SQuAD-explorer/, accessed on 24 January 2022), see Figure 3 for an example of passage, questions, and answers. Consider the question "How many countries does Shell operate in?" posed in the passage. To answer the question, one might first locate the relevant part of the passage "It has operations in over 90 countries", then reason that "under" refers to a cause (not location), and thus determine the correct answer: "over 90".

Shell was vertically integrated and is active in every area of the oil and gas industry, including exploration and production, refining, distribution and marketing, petrochemicals, power generation and trading. It has minor renewable energy activities in the form of biofuels and **wind**. It has operations in **over 90** countries, produces around 3.1 million barrels of oil equivalent per day and has **44,000 service stations** worldwide. Shell Oil Company, its subsidiary in the United States, is one of its largest businesses.

Question 1: Aside from biofuels what other renewable energy activities is Shell involved with? wind
Question 2: How many countries does Shell operate in? over 90
Question 3: How many services stations does Shell have? 44,000 service stations

Figure 3. A paragraph from Wikipedia and three associated questions together with their answers, taken from the SQuAD dataset.

3.2. Private Data Pipeline Module

To train a model with a high level of accuracy, machine reading comprehension models require large datasets to 'learn' from, however, data might be sensitive and private. To preserve data privacy, different anonymization techniques have been used. The most relevant are k-anonymity [31], l-diversity [32], and t-closeness [33]. In k-anonymity, specific columns (e.g., name, religion, sex) are removed or altered (e.g., replacing a specific age with an age span). L-diversity and t-closeness are extensions of k-anonymity, which are used to protect attribute disclosure, these anonymization techniques are applied before data is shared for training. However, with the rise of AI, this form of anonymizing personal data is not enough to protect privacy because the data can often be reverse-engineered using machine learning to re-identify individuals [34]. In question answering systems, there might be sensitive documents, personal data that needs to be processed for MRC task, without exposing data. To handle this issue, we propose a question answering system based on federated learning methodology to protect data leakage and ensure secure collaborative training. The proposed system follows a federated learning paradigm in which participating clients are required to train their local models and then share the gradient (model parameters) for an eventual aggregation strategy in a central server. This approach ensures input data privacy, enables collaborative training, low-cost training by distributing the workload across clients instead of training a large model individually, sharing the local learning model within the alliance (training clients) without compromising private data, and improving local learning by using incremental learning and local data generation pipeline.

3.3. Federated Machine Learning Module

Federated learning is an emerging technology enabling multiple parties to jointly train machine learning models on private data. These parties could be mobile and IoT devices (cross-device FL), or organizations (cross-silo). Data remain locally at each party, only the parameter updates are communicated with a server and other parties. In our system, we use FL to develop FedQAS based on transformer architecture for question answering. FedQAS trains a global model (Algorithm 1) on large amounts of data from multiple geographically distributed parties. Each party trains a local transformer model on its data (Algorithm 2) and sends parameters W_t to the central server for aggregation (FedAVG [35]) instead of the whole model. In the aggregation part, the aggregator (running in the *combiner* in FEDn [2]) combines parameters and generates a single global model $M(W_t)$ for each round using federated incremental averaging [35].

Algorithm 1: Incremental FedAVG algorithm. **k**: Number of clients, **r**: Number of rounds, W_i: Local model weights and **M**: Global model weights.

Input: W_t
Output: $M(W_t)$

Server executes:
initialized W_0
Function `IncrementealFedAVG`(k, W_{t-1}, W_t):
    foreach $t \leftarrow 1$ **to** r **do**
        $S_t \leftarrow$ (sample a random set of clients)
        foreach *client* $k \in S_t$ ***in parallel*** **do**
            $W^k_{t+1} \leftarrow ClientUpdate(k, W_t, N_l)$
            $W_{t+1} \leftarrow \sum_{k=1}^{k} \frac{n_k}{n} W^k_{t+1}$
        end
        $W_t \leftarrow (W_{t-1} + (W_t - W_{t-1})/t)$
    end
    return $M(W_t)$

Algorithm 2: Local client update, **k**: Number of clients, $\boldsymbol{D^k}$: Client k local dataset, **e**: Number of local epochs, and η is the learning rate.

Output: W_t
// Run on client k
Function `ClientUpdate`(k, W_t):
    $\beta \leftarrow$ (split D^k into mini batches)
    for *local epoch* $e_i \in 1, \ldots e$ **do**
        for *batch* $b \in \beta$ **do**
            $W_t \leftarrow W_t - \eta \nabla l(W_t, b)$
        end
    end
    return W_t

3.4. Transformer Model

In this paper, we develop a FedQAS using a Transformer architecture [17] to handle questions. The transformer is a model architecture eschewing recurrence and instead relying on an attention mechanism to draw global dependencies between input and output. The transformer follows the encoder and decoder architecture using stacked self-attention. The encoder maps an input sequence of symbol representations $(x_1, \ldots, x_n)$ to a sequence of continuous representations $z = (z_1, \ldots, z_n)$. The decoder then generates an output sequence $(y_1, \ldots, y_m)$ of symbols. Both encoder and decoder are composed of layers and sub-layers that can be stacked on top of each other multiple times. The first is a multi-head self-attention mechanism and the second is a simple, position-wise fully connected feed-forward network. In this paper, by applying the self-attention mechanism, we aim at capturing the long dependencies in the input sentence, the inputs and outputs are first embedded into an n-dimensional space.

3.5. Incremental Learning Module

Incremental learning is a machine learning case in which input data is continuously used to extend the existing model's knowledge, i.e., to further train the model. It attempts to improve a model's performance while adding the fewest samples possible. In the proposed system, adding data locally by clients is an important task to improve the local model performance first, then propagating these improvements into the global model after new training rounds in a privacy-preserving manner. We have engineered an intuitive process for each local client to contribute to the adding of new samples on top of their local data (Figure 4). The first step is to add a new data point that will remain on the local site, this allows the user to add their private data, questions, and correct answers. The incremental learning module will process and transform the private data locally and generate training points to be used in the local training. This process enables collaborative data generation between organizations in a private way in order to strengthen data protection and avoid unnecessary sharing within the alliance. In addition, a database layer is used to store user queries and global model predictions as feedback to enhance and improve the performance for further usage.

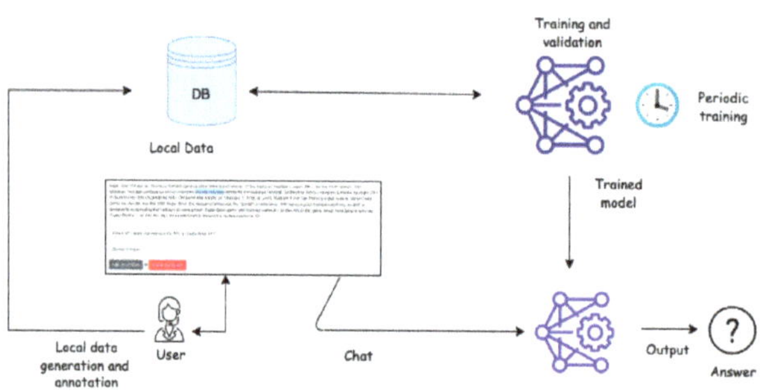

Figure 4. Federated incremental learning process. Clients can add data continuously to extend the existing model's knowledge locally while training a global model, the generated data can be stored locally on the client-side.

4. Experiment and Results

SQuAD is a reading comprehension dataset made up of questions posed by crowd workers on a collection of high-quality Wikipedia articles. It covers a wide range of topics from music celebrities to abstract notions. When comparing SQuAD with other datasets, SQuAD is one of the most popular question answering datasets (it's been cited over 4096 times) because it's well-created and improves on many aspects that other datasets fail to address. Other reading comprehension datasets such as MCTest [8] and Deep Read [4] are too small to support intensive and complex models. Hence, we conduct our experiment on the SQuAD 1.0 dataset, which contains 100.000+ question-answer pairs. To ensure collaborative training, we randomly select and split data over 5 clients with 20% for validation dataset for all clients. We used FedAVG [35] for the aggregation of model parameters, see Table 2.

Table 2. Federated training configuration.

Rounds	Total Number of Clients	Update Size	Total Number of Parameters
5	5	400 MB	109.483.776

For the fine-tuning in our task, we used the BERT base as an encoder to build our model and the implementations are based on the public TensorFlow implementation from Keras (https://github.com/tensorflow/tensorflow, accessed on 24 January 2022) we set the initial learning rate to 5×10^{-5}. The batch size is set to 8. The maximum number of epochs is set to 1. Texts are tokenized using Wordpieces [36] with a maximum length of 384. Table 3 presents the hyperparameters used in our experiments. We used three input layers, two dense layers, two flatten layers, and Adam optimizer. We run the model for optimizing the cross-entropy loss between the output probabilities and the output answers.

Table 3. Experimental model parameters.

Hyper-Parameter	Range	Value
Epochs	[1–3]	1
Batch size	[8–128]	8
Learning rate	[0.001–0.004]	5×10^{-5}
Optim. method	Adam, SGD, RMSProp	Adam
MAX_SEQ_LENGTH	[1–1000]	384

4.1. Framework Evaluation

For the evaluation, we used exact match (EM) and F1 score, the main metrics commonly used for question answering systems. These metrics are computed on individual (question, answer) pairs. In case of multiple correct answers for a given question, the maximum score over all possible correct answers is computed. In the EM metric, for each pair (question, answer), if the characters of the model's prediction exactly match the characters of (one of) the True Answer(s), EM = 1, otherwise EM = 0.

The Accuracy represents the percentage of the questions that an MRC system accurately answers. Each question corresponds to one correct answer. For the span prediction task, the accuracy is the same as Exact Match and can be computed by the Formula (1) as follows:

$$Accuracy = EM = \frac{Number\ of\ correct\ answers}{Number\ of\ questions} \quad (1)$$

The precision represents the percentage of token overlap between the tokens in the correct answer and the tokens in the predicted answer, while the recall is the percentage of tokens in a correct answer that have been correctly predicted in a question. The True Positive (TP) denotes the same tokens between the predicted answer and the correct answer, the False Positive (FP) denotes the tokens which are not in the correct answer but the predicted answer, while the False Negative (FN) presents the tokens that are not in the predicted answer but the correct answer. Precision and Recall can be computed by the Formulas (2) and (3) as follows:

$$Precision = \frac{N(TP)}{N(TP) + N(FP)} \quad (2)$$

$$Recall = \frac{N(TP)}{N(TP) + N(FN)} \tag{3}$$

The F1 score is a measure of a test's accuracy. It is the weighted average between precision and recall. The formula for this score is given in (4). In our case, it's computed over the individual words in the prediction against those in the True Answer. The number of shared words between the prediction and the truth is the basis of the F1 score.

$$F1\ score = 2 \times \frac{(Precision \times Recall)}{(Precision + Recall)} \tag{4}$$

To demonstrate the benefit of FedQAS, we partitioned the SQuAD dataset into 5 equal chunks, so that each client has "20%" of the total dataset. We then compare the federated scenario to centralized model training. Table 4 lists the available metrics for different training rounds of the global model. Our implemented model baselines show similar EM and F1 scores with the global model during the first rounds and slightly outperform the baseline with respect to data privacy and knowledge sharing across participants. Overall, the result shows that question-answering in federated learning settings performs well compared to centralized settings. This is due to the used hyper-parameters in the federated learning setting, see Figure 5 for the convergence of accuracy (EM) and Figure 6 for the convergence of F1.

Table 4. Comparisons with equivalent parameters on the validation set of SQuAD1.0.

Model	F1 Score	Accuracy (EM)
Baseline	0.31	0.75
FedQAS	0.33	0.81

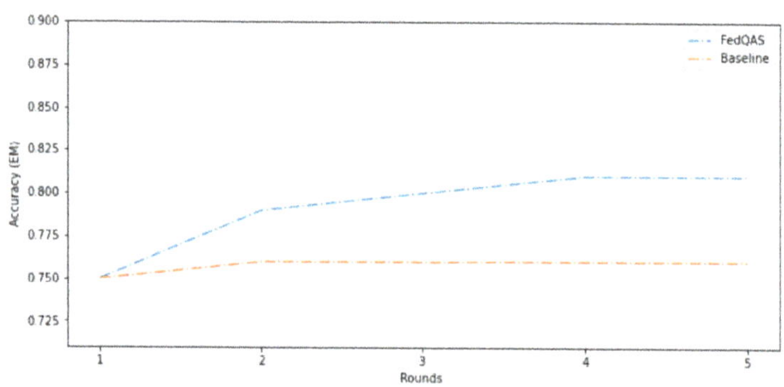

Figure 5. Convergence of accuracy (Exact Match) on the SQUAD dataset with 1 combiner, 5 clients and 5 rounds.

In terms of resources, the result proves the fact that model architecture affects client training time and combiner round time. Hence, training a large model (400 MB) in a centralized way requires more resources than the federated setting. For demonstration, we consider a FEDn network consisting of a single, high-powered combiner (8 VCPU, 32 GB RAM) with connected clients (8 VCPU, 32 GB RAM) instances in SSC (SNIC Science Cloud [37]) and measure the average round time over five global rounds. Figure 7 shows round time for global model training, since the model size affects both the training time at clients and the cost for data transfer and model aggregation, we show the mean training time for reference.

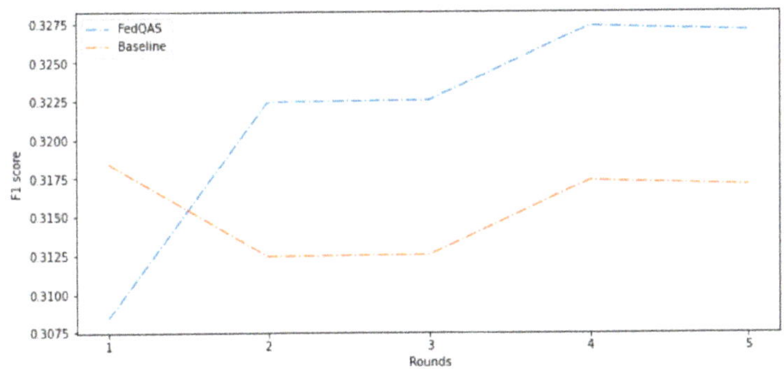

Figure 6. Convergence of F1 score on SQUAD dataset with 1 combiner, 5 clients and 5 rounds.

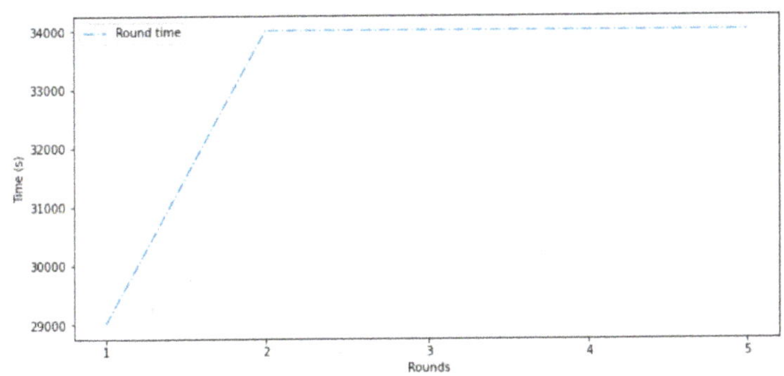

Figure 7. Round times for global model training (FEDn network).

To gain an intuitive observation of the predictions, we give a prediction example on SQuAD1.0 from both the baseline and federated model in Table 5, which shows that FedQAS works better at answering the question on a given passage. Hence, the proposed approach has contributed overall to a better understanding of QA, preserving data privacy, and contributed to low-cost training as well as a collaborative question answering system task using large models.

Table 5. Comparison of answer prediction on test data.

Title: Project Apollo
Passage: The Apollo program, also known as Project Apollo, was the third United States human spaceflight program carried out by the National Aeronautics and Space Administration (NASA), which accomplished landing the first humans on the Moon from 1969 to 1972. First conceived during Dwight D. Eisenhower's administration as a three-man spacecraft to follow the one-man Project Mercury which put the first Americans in space, Apollo was later dedicated to President John F. Kennedy's national goal of landing a man on the Moon and returning him safely to the Earth by the end of the 1960s, which he proposed in a 25 May 1961, address to Congress. Project Mercury was followed by the two-man Project Gemini. The first manned flight of Apollo was in 1968. Apollo ran from 1961 to 1972, and was supported by the two man Gemini program which ran concurrently with it from 1962 to 1966...
Question 1: How long did Project Apollo run?
Gold answer (human): 1961 to 1972 **Google search engine answer:** see Figure 1 **Baseline model answer:** 1961 to 1972 **FedQAS answer:** 1961 to 1972

Table 5. *Cont.*

Question 2: What program was created to carry out these projects and missions?
Gold answer (human): Apollo program **Baseline model answer:** National Aeronautics and Space Administration **FedQAS answer:** Apollo program
Question 3: What year did the first manned Apollo flight occur?
Gold answer (human): 1968 **Baseline model answer:** 1968 **FedQAS answer:** 1968 **Question 4:** What President is credited with the original notion of putting Americans in space?
Gold answer (human): John F. Kennedy **Baseline model answer:** John F. Kennedy **FedQAS answer:** John F. Kennedy
Question 5: Who did the U.S. collaborate with on an Earth orbit mission in 1975?
Gold answer (human): Soviet Union **Baseline model answer:** Soviet Union **FedQAS answer:** Soviet Union
Question 6: How long did Project Apollo run?
Gold answer (human): 1962 to 1966 **Baseline model answer:** 1961 to 1972, and was supported by the two man Gemini program which ran 1966 **FedQAS answer:** 1962 to 1966
Question 7: What program helped develop space travel techniques that Project Apollo used?
Gold answer (human): Gemini **Baseline model answer:** Gemini **FedQAS answer:** Gemini
Question 8: What space station supported three manned missions in 1973–1974?
Gold answer (human): Skylab **Baseline model answer:** Skylab **FedQAS answer:** Skylab

4.2. Implementation and Demo Environment

Designing and developing question answering in a privacy-aware manner is not a trivial task. It requires design strategies to comply with data governance and privacy regulations. Several third-party frameworks have been proposed for federated learning; providing open-source building blocks that help to collaborate in training machine learning models. The present QAS application framework needs to provide scalability, large models training, and production-grade features such as robustness to failure. Based on these requirements, we chose to design and develop our proposed FedQAS system on top of private data using the FEDn framework [2]. FEDn is an open-source, modular, and model agnostic framework for federated machine learning. We developed interactive and user-friendly interfaces using the Flask framework (https://flask.palletsprojects.com, accessed on 24 January 2022), which make it easy for a third party to contribute to data annotation and then participate in training global models directly from their location site. The proposed FedQAS has the following features:

- Privacy-preserving: sharing only model parameters with a central server (cloud) and keeping data private on the client side,
- Incremental learning: improving the global models by attaching more clients and adding new data points,
- Robust: robust enough to deal with natural language tasks (e.g., question answering, chatbot, etc.) and large models in a geographically distributed manner,

- Multilingual: language agnostic, can be trained on any language,
- Standalone: multiple platforms (i.e., guarantee for low disk and memory footprint). It can be run production-grade on a standard laptop having two cores and 2GB of RAM,
- Accuracy and F1 score: achieve competitive performance compared with centralized training and the used baseline model (see experiment and evaluation section).

The proposed FedQAS is composed of three main components: FEDn for collaborative training, MongoDB (https://www.mongodb.com, accessed on 24 January 2022) as a NoSQL [38] database and question answering UI for prediction and local incremental learning. The system is interactive, scalable, suitable for secure collaborative training and data privacy-preserving, and can be used both in the cloud, on edge nodes and in a standalone mode. It is accessible from different platforms to engage a wide range of users, and it is also optimized for both desktop and mobile. The source code is publicly available on Github via this link https://github.com/aitmlouk/FEDn-client-FedQAS-tf.git, accessed on 24 January 2022.

FedQAS is, to the best of our knowledge, the only approach for question answering that supports data privacy and knowledge sharing through federated learning. Its main value is to provide an environment to quickly ensure data privacy and low-cost training by collaborative training. Nearly every deep learning application can benefit from data privacy and knowledge sharing across a different client in a federated learning setting. The transformer model used in FedQAS can be improved by tuning parameters and using transfer learning for new pre-trained models (GPT-2 [39], GPT-3 [40], etc.).

5. Conclusions

In this paper, we have proposed FedQAS, a high-quality question answering (FedQAS) approach, to address the data-sharing issue in machine learning. Validation experiments for FedQAS was implemented based on 5 rounds of training with a transformer neural network. The system consists of several components including the private data pipeline, collaborative training and private incremental learning. Experiments on the SQuAD dataset using the transformer architecture demonstrate that our FedQAS significantly outperforms the baseline model performances, protecting data privacy and sharing knowledge within an alliance. The proposed FedQAS allows collaborators (collaborative training participants) to have overall control of their sensitive and private data while collaboratively training question answering models. The integration of federated learning within machine reading comprehension provides a sustainable solution by preserving data privacy and ensuring low-cost training. We conclude that the application of FL to NLP tasks such as question answering can contribute to solving the problem that arises when using machine learning in the context of data protection and privacy. The system actively supports end-users in joining training and improving the performance through incremental learning on a various range of local clients.

In future work, we aim to extend the FedQAS to cover more datasets in a geographically distributed manner and test the model with other aggregation algorithms (e.g., FedOPT, FedProx, etc.) for federated learning. We also plan to fine-tune the pre-trained models such as BERT large, GPT-2 for MRC and particularly investigate their effectiveness in federated learning settings especially when it comes to large private documents (text).

Author Contributions: Formal analysis, A.A.-M.; investigation, A.A.-M.; methodology, A.A.-M.; project administration, A.H.; writing—review and editing, S.A.A., S.T. and A.H. All authors have read and agreed to the published version of the manuscript.

Funding: Funding has been provided by the eSSENCE strategic collaboration on eScience (Ait-Mlouk, Alawadi, Toor, and Hellander) and the Swedish Innovation Agency Vinnova grant no. 2019-02819 (awarded to Scaleout Systems AB).

Institutional Review Board Statement: Not applicable.

Informed Consent Statement: Not applicable.

Data Availability Statement: The data used to support the findings of this work can be found on this link https://github.com/aitmlouk/FEDn-client-FedQAS-tf/tree/main/data, accessed on 24 January 2022.

Acknowledgments: The computations were enabled by resources provided by the Swedish National Infrastructure for Computing (SNIC) at C3SE partially funded by the Swedish Research Council through grant agreement no. 2018-05973.

Conflicts of Interest: The authors declare no conflict of interest regarding the design of the study; the collection, analyses, or interpretation of data; the writing of the manuscript, or the decision to publish the results.

References

1. Rajpurkar, P.; Zhang, J.; Lopyrev, K.; Liang, P. SQuAD: 100,000+ Questions for Machine Comprehension of Text. *arXiv* **2016**, arXiv:1606.05250.
2. Ekmefjord, M.; Ait-Mlouk, A.; Alawadi, S.; Åkesson, M.; Stoyanova, D.; Spjuth, O.; Toor, S.; Hellander, A. Scalable federated machine learning with FEDn. *arXiv* **2021**, arXiv:2103.00148.
3. Lehnert, W. The Process of Question Answering. Ph.D. Thesis, Yale University, New Haven, CT, USA, 1977.
4. Hirschman, L.; Light, M.; Breck, E.; Burger, J.D. Deep Read: A Reading Comprehension System. In Proceedings of the 37th Annual Meeting of the Association for Computational Linguistics on Computational Linguistics, ACL '99, College Park, MD, USA, 20–26 June 1999; Association for Computational Linguistics: Stroudsburg, PA, USA, 1999; pp. 325–332. [CrossRef]
5. Riloff, E.; Thelen, M. A Rule-Based Question Answering System for Reading Comprehension Tests. In Proceedings of the 2000 ANLP/NAACL Workshop on Reading Comprehension Tests as Evaluation for Computer-Based Language Understanding Sytems—Volume 6, ANLP/NAACL-ReadingComp '00, Seattle, WA, USA, 4 May 2000; Association for Computational Linguistics: Stroudsburg, PA, USA, 2000; pp. 13–19. [CrossRef]
6. Charniak, E.; Altun, Y.; de Salvo Braz, R.; Garrett, B.; Kosmala, M.; Moscovich, T.; Pang, L.; Pyo, C.; Sun, Y.; Wy, W.; et al. Reading Comprehension Programs in a Statistical-Language-Processing Class. In Proceedings of the ANLP-NAACL 2000 Workshop: Reading Comprehension Tests as Evaluation for Computer-Based Language Understanding Systems, Seattle, WA, USA, 4 May 2000; Association for Computational Linguistics: Stroudsburg, PA, USA, 2000.
7. Wang, Z.; Hamza, W.; Florian, R. Bilateral Multi-Perspective Matching for Natural Language Sentences. In Proceedings of the Twenty-Sixth International Joint Conference on Artificial Intelligence, IJCAI-17, Melbourne, Australia, 19–25 August 2017; pp. 4144–4150. [CrossRef]
8. Richardson, M.; Burges, C.J.; Renshaw, E. MCTest: A Challenge Dataset for the Open-Domain Machine Comprehension of Text. In Proceedings of the 2013 Conference on Empirical Methods in Natural Language Processing, Seattle, WA, USA, 18–21 October 2013; Association for Computational Linguistics: Stroudsburg, PA, USA, 2013; pp. 193–203.
9. Wang, W.; Yan, M.; Wu, C. Multi-Granularity Hierarchical Attention Fusion Networks for Reading Comprehension and Question Answering. In Proceedings of the 56th Annual Meeting of the Association for Computational Linguistics, Melbourne, Australia, 15–20 July 2018; Association for Computational Linguistics: Stroudsburg, PA, USA, 2018; Volume 1: Long Papers, pp. 1705–1714. [CrossRef]
10. Zhang, Z.; Wu, Y.; Zhao, H.; Li, Z.; Zhang, S.; Zhou, X.; Zhou, X. Semantics-aware BERT for Language Understanding. *arXiv* **2020**, arXiv:1909.02209.
11. Zhang, Z.; Wu, Y.; Zhou, J.; Duan, S.; Zhao, H.; Wang, R. SG-Net: Syntax-Guided Machine Reading Comprehension. *arXiv* **2019**, arXiv:1908.05147.
12. Seo, M.; Kembhavi, A.; Farhadi, A.; Hajishirzi, H. Bidirectional Attention Flow for Machine Comprehension. *arXiv* **2018**, arXiv:1611.01603.
13. Devlin, J.; Chang, M.W.; Lee, K.; Toutanova, K. BERT: Pre-training of Deep Bidirectional Transformers for Language Understanding. In Proceedings of the 2019 Conference of the North American Chapter of the Association for Computational Linguistics: Human Language Technologies, Minneapolis, MN, USA, 2–7 June 2019; Association for Computational Linguistics: Stroudsburg, PA, USA, 2019; Volume 1 (Long and Short Papers), pp. 4171–4186. [CrossRef]
14. Zhuang, L.; Wayne, L.; Ya, S.; Jun, Z. A Robustly Optimized BERT Pre-training Approach with Post-training. In Proceedings of the 20th Chinese National Conference on Computational Linguistics, Hohhot, China, 13–15 August 2021; Chinese Information Processing Society of China: Beijing, China, 2021; pp. 1218–1227.
15. Yang, Z.; Dai, Z.; Yang, Y.; Carbonell, J.; Salakhutdinov, R.; Le, Q.V. XLNet: Generalized Autoregressive Pretraining for Language Understanding. *arXiv* **2020**, arXiv:1906.08237.
16. Peters, M.E.; Neumann, M.; Iyyer, M.; Gardner, M.; Clark, C.; Lee, K.; Zettlemoyer, L. Deep Contextualized Word Representations. In Proceedings of the 2018 Conference of the North American Chapter of the Association for Computational Linguistics: Human Language Technologies, New Orleans, LA, USA, 1–6 June 2018; Volume 1 (Long Papers); Association for Computational Linguistics: Stroudsburg, PA, USA, 2018; pp. 2227–2237. [CrossRef]
17. Vaswani, A.; Shazeer, N.; Parmar, N.; Uszkoreit, J.; Jones, L.; Gomez, A.N.; Kaiser, L.; Polosukhin, I. Attention Is All You Need. *arXiv* **2017**, arXiv:1706.03762.

18. Yamada, I.; Asai, A.; Shindo, H.; Takeda, H.; Matsumoto, Y. LUKE: Deep Contextualized Entity Representations with Entity-aware Self-attention. *arXiv* **2020**, arXiv:2010.01057.
19. Lan, Z.; Chen, M.; Goodman, S.; Gimpel, K.; Sharma, P.; Soricut, R. ALBERT: A Lite BERT for Self-supervised Learning of Language Representations. *arXiv* **2020**, arXiv:1909.11942.
20. Zhang, Z.; Yang, J.; Zhao, H. Retrospective Reader for Machine Reading Comprehension. *arXiv* **2020**, arXiv:2001.09694.
21. Hermann, K.M.; Kočiský, T.; Grefenstette, E.; Espeholt, L.; Kay, W.; Suleyman, M.; Blunsom, P. Teaching Machines to Read and Comprehend. *arXiv* **2015**, arXiv:1506.03340.
22. Hill, F.; Bordes, A.; Chopra, S.; Weston, J. The Goldilocks Principle: Reading Children's Books with Explicit Memory Representations. *arXiv* **2016**, arXiv:1511.02301.
23. Bajaj, P.; Campos, D.; Craswell, N.; Deng, L.; Gao, J.; Liu, X.; Majumder, R.; McNamara, A.; Mitra, B.; Nguyen, T.; et al. MS MARCO: A Human Generated MAchine Reading COmprehension Dataset. *arXiv* **2018**, arXiv:1611.09268.
24. Trischler, A.; Wang, T.; Yuan, X.; Harris, J.; Sordoni, A.; Bachman, P.; Suleman, K. NewsQA: A Machine Comprehension Dataset. In Proceedings of the 2nd Workshop on Representation Learning for NLP, Vancouver, BC, Canada, 3 August 2017; Association for Computational Linguistics: Stroudsburg, PA, USA, 2017; pp. 191–200. [CrossRef]
25. Dunn, M.; Sagun, L.; Higgins, M.; Guney, V.U.; Cirik, V.; Cho, K. SearchQA: A New Q&A Dataset Augmented with Context from a Search Engine. *arXiv* **2017**, arXiv:1704.05179.
26. Joshi, M.; Choi, E.; Weld, D.S.; Zettlemoyer, L. TriviaQA: A Large Scale Distantly Supervised Challenge Dataset for Reading Comprehension. *arXiv* **2017**, arXiv:1705.03551.
27. Rajpurkar, P.; Zhang, J.; Lopyrev, K.; Liang, P. SQuAD: 100,000+ Questions for Machine Comprehension of Text. In Proceedings of the 2016 Conference on Empirical Methods in Natural Language Processing, Austin, TX, USA, 1–5 November 2016; Association for Computational Linguistics: Stroudsburg, PA, USA, 2016; pp. 2383–2392. [CrossRef]
28. Rajpurkar, P.; Jia, R.; Liang, P. Know What You Don't Know: Unanswerable Questions for SQuAD. *arXiv* **2018**, arXiv:1806.03822.
29. Reddy, S.; Chen, D.; Manning, C.D. CoQA: A Conversational Question Answering Challenge. *Trans. Assoc. Comput. Linguist.* **2019**, *7*, 249–266. [CrossRef]
30. Rajpurkar, P.; Jia, R.; Liang, P. Know What You Don't Know: Unanswerable Questions for SQuAD. In Proceedings of the 56th Annual Meeting of the Association for Computational Linguistics, Melbourne, Australia, 15–20 July 2018; Volume 2: Short Papers; Association for Computational Linguistics: Stroudsburg, PA, USA, 2018; pp. 784–789. [CrossRef]
31. Sweeney, L. K-Anonymity: A Model for Protecting Privacy. *Int. J. Uncertain. Fuzziness Knowl.-Based Syst.* **2002**, *10*, 557–570. [CrossRef]
32. Machanavajjhala, A.; Kifer, D.; Gehrke, J.; Venkitasubramaniam, M. L-Diversity: Privacy beyond k-Anonymity. *ACM Trans. Knowl. Discov. Data* **2007**, *1*, 1–52. [CrossRef]
33. Li, N.; Li, T.; Venkatasubramanian, S. t-Closeness: Privacy Beyond k-Anonymity and l-Diversity. In Proceedings of the 2007 IEEE 23rd International Conference on Data Engineering, Istanbul, Turkey, 15–20 April 2007; pp. 106–115. [CrossRef]
34. Rocher L., H.J.; de Montjoye YA. Estimating the success of re-identifications in incomplete datasets using generative models. *Nat. Commun.* **2019**, *10*, 3069. [CrossRef]
35. McMahan, H.B.; Moore, E.; Ramage, D.; Hampson, S.; y Arcas, B.A. Communication-Efficient Learning of Deep Networks from Decentralized Data. *arXiv* **2017**, arXiv:1602.05629.
36. Wu, Y.; Schuster, M.; Chen, Z.; Le, Q.V.; Norouzi, M.; Macherey, W.; Krikun, M.; Cao, Y.; Gao, Q.; Macherey, K.; et al. Google's Neural Machine Translation System: Bridging the Gap between Human and Machine Translation. *arXiv* **2016**, arXiv:1609.08144.
37. Toor, S.; Lindberg, M.; Falman, I.; Vallin, A.; Mohill, O.; Freyhult, P.; Nilsson, L.; Agback, M.; Viklund, L.; Zazzik, H.; et al. SNIC Science Cloud (SSC): A National-Scale Cloud Infrastructure for Swedish Academia. In Proceedings of the 2017 IEEE 13th International Conference on e-Science (e-Science), Auckland, New Zealand, 24–27 October 2017; pp. 219–227. [CrossRef]
38. Pokorny, J. *NoSQL Databases: A Step to Database Scalability in Web Environment*; iiWAS '11; Association for Computing Machinery: New York, NY, USA, 2011; pp. 278–283. [CrossRef]
39. Radford, A.; Wu, J.; Child, R.; Luan, D.; Amodei, D.; Sutskever, I. Language Models are Unsupervised Multitask Learners. *OpenAI Blog* **2019**, *1*, 9.
40. Brown, T.B.; Mann, B.; Ryder, N.; Subbiah, M.; Kaplan, J.; Dhariwal, P.; Neelakantan, A.; Shyam, P.; Sastry, G.; Askell, A.; et al. Language Models are Few-Shot Learners. *arXiv* **2020**, arXiv:2005.14165.

applied sciences

MDPI

Article

Named Entity Correction in Neural Machine Translation Using the Attention Alignment Map

Jangwon Lee [1,2], Jungi Lee [3], Minho Lee [3] and Gil-Jin Jang [2,4,*]

1 SK Holdings C&C, Gyeonggi-do, Suwon City 13558, Korea; saraitne11@naver.com
2 School of Electronic and Electrical Engineering, Kyungpook National University, Daegu 41566, Korea
3 Department of Artificial Intelligence, Kyungpook National University, Daegu 41566, Korea; darbams77@naver.com (J.L.); mholee@gmail.com (M.L.)
4 School of Electronics Engineering, Kyungpook National University, Daegu 41566, Korea
* Correspondence: gjang@knu.ac.kr; Tel.: +82-53-950-5517

Featured Application: machine translation; information retrieval; text-to-speech.

Abstract: Neural machine translation (NMT) methods based on various artificial neural network models have shown remarkable performance in diverse tasks and have become mainstream for machine translation currently. Despite the recent successes of NMT applications, a predefined vocabulary is still required, meaning that it cannot cope with out-of-vocabulary (OOV) or rarely occurring words. In this paper, we propose a postprocessing method for correcting machine translation outputs using a named entity recognition (NER) model to overcome the problem of OOV words in NMT tasks. We use attention alignment mapping (AAM) between the named entities of input and output sentences, and mistranslated named entities are corrected using word look-up tables. The proposed method corrects named entities only, so it does not require retraining of existing NMT models. We carried out translation experiments on a Chinese-to-Korean translation task for Korean historical documents, and the evaluation results demonstrated that the proposed method improved the bilingual evaluation understudy (BLEU) score by 3.70 from the baseline.

Keywords: neural networks; recurrent neural networks; natural language processing; neural machine translation; named entity recognition

Citation: Lee, J.; Lee, J.; Lee, M.; Jang, G.-J. Named Entity Correction in Neural Machine Translation Using the Attention Alignment Map. *Appl. Sci.* **2021**, *11*, 7026. https://doi.org/10.3390/app11157026

Academic Editor: Arturo Montejo-Ráez

Received: 1 July 2021
Accepted: 26 July 2021
Published: 29 July 2021

1. Introduction

Neural machine translation (NMT) models based on artificial neural networks have shown successful results in comparison to traditional machine translation methods [1–5]. Traditional methods usually consist of sequential steps, such as morphological, syntactic, and semantic analyses. On the contrary, NMT aims to construct a single neural network and jointly train the entire system. Therefore, NMT requires less prior knowledge than traditional methods if a sufficient amount of training data is provided. Early NMT models, called sequence-to-sequence [6–9], are based on encoder–decoder architectures implemented with recurrent neural networks (RNNs) [10], such as long short-term memory (LSTM) [11] and the gated recurrent unit (GRU) [12]. The attention mechanism is usually used in RNN-based machine translation systems with variable lengths. The network generates an output vector, as well as its importance, called attention, to allow the decoder focus on the important part of the output [13–15]. Recently, a new NMT model called the transformer [16] has been proposed based on an attention mechanism with feedforward networks and without RNNs. Using the transformer, the learning time is reduced greatly with the help of non-RNN-type networks.

One of the problems in machine translation is the lack of training data. This problem was reported by Seljan [17] and Dunđer [18,19] for the problem of the automatic translation of poetry with a low-resource language pair. It was reported that the fluency and adequacy of the translation results were skewed to higher scores. Especially for old literature translation where the machine translation is of great importance, obtaining reliable training data is much more difficult. The types of errors in the machine translation were extensively analyzed by Brkić [20]. They were wrong word mapping, omitted or surplus words, morphological and lexical errors, and syntactic errors such as word order and punctuation errors. There have been several methods to successfully solve these problems using transfer learning [21], contrastive learning [22], and open vocabularies [23].

Another major problem in NMT are out-of-vocabulary (OOV) words [24,25]. This is often called the rare word problem as well [26,27]. The words in the training dataset are converted into indices to the word dictionary or a predefined set of vectors, and a sequence of the converted numbers or vectors is used as an input to NMT systems. When a new word that is not in the dictionary is observed, the behavior of the trained network is unpredictable because there are no training sentences with the OOV words. It is almost impossible to include all of them in the dictionary because of the complexity limit for efficient translation. One of the solutions to this problem is subword tokenization using byte pair encoding (BPE) [27]. In this work, the unknown words are broken into reasonable subunits. Another solution is the unsupervised learning of the OOVs [28]. However, most of the OOV words are for named entities: human names, city names, and newly coined academic terms, and subword tokenization [27] and unsupervised learning [28] are not able to handle the named entities because they do not contain any meaningful information in them. As a solution, conventional systems use special labels for such OOV words (often as "UNK") and include them in the training data [24–26], so that the NMT model would distinguish them from ordinary words. Table 1 shows examples of translation outputs with an "UNK" symbol. The first named entity in the first example, "李周鎭," is mistranslated into "이진," although the expected output is "이주진." The second named entity, "元景淳," is not translated, but replaced with an "UNK" symbol because the true translation "원경순" is an OOV or rarely occurring word for the trained NMT model. Moreover, there are many similar cases in the subsequent named entities.

There have been several attempts to build open-vocabulary NMT models to deal with OOV words. Ling et al. [29] used a sub-LSTM layer that takes a sequence of characters to produce a word embedding vector. In the decoding process, another LSTM cell also generates words character-by-character. Luong and Manning [25] proposed a hybrid word–character model. This model adopts a sub-LSTM layer to use the information at the character level when it finds *unknown* words both in the encoding and decoding steps. Although character-based models show a translation quality comparable to word-based models and achieve open-vocabulary NMT, they require a huge amount of training time when compared with word-based models. This is because, if words are split into characters, their sequence lengths are increased to the number of characters, so the model complexity grows significantly. There are other approaches to use character-based models such as using convolutional neural networks [30,31]. However, it is hard to directly apply fully character-based models to Korean, because a Korean character is made by combining consonants and a vowel. Luong et al. [26] augmented a parallel corpus to allow NMT models to learn the alignments of "UNK" symbols between the input and output sentences. However, this method is difficult to apply to language pairs with extremely different structures, such as English–Chinese, English–Korean, and Chinese-to-Korean. Luong [26] and Jean [24] effectively addressed "UNK" symbols in translated sentence. However, mistranslated words, which often appear for rare input words, still were not considered.

In this work, we propose a postprocessing method that corrects mistranslated named entities in the target language using a named entity recognition (NER) model and an attention alignment mapping (AAM) between an input and an output sentence by using the attention mechanism (to the best of our knowledge, first proposed by Bahdanau et al. [13]). The proposed method can be directly applied to pretrained NMT models that use an attention mechanism by appending the postprocessing step to its output, without retraining the existing NMT models or modifying the parallel corpus. Our experiments on the Chinese-to-Korean translation task of historical documents, the *Annals of the Joseon Dynasty* (http://sillok.history.go.kr/main/main.do, last access date: 1 July 2021) demonstrate that the proposed method is effective. In a numerical evaluation, the proposed method shows that the bilingual evaluation understudy (BLEU) score [32] was improved up to 3.70 compared to the baseline when the proposed method was not applied. Our work is available in a Git repository https://bitbucket.org/saraitne76/chn_nmt/src/master/, last access date: 1 July 2021).

Table 1. Examples of Chinese-to-Korean translation results with OOV words. Input and Truth: raw sentence pairs from the Chinese-to-Korean parallel corpus. English translation: translation of the Korean sentence to an English expression. NMT output: Korean translation results of a typical NMT model, with OOV words represented by the "UNK" symbol. The underlined words are named entities. Among those words, red-colored ones are human names; blue-colored ones are place names; green-colored ones are book names.

Input	以李周鎭爲平安監司, 元景淳爲副校理, 尹敬周爲正言。
Truth	이주진을 평안 감사로, 원경순을 부교리로, 윤경주를 정언으로 삼았다.
English Translation	Lee Joo Jin is assigned as the Pyeongan inspector, Won Kyung Soon as the vice dictator, Yun Gyeong Joo as the dictator.
NMT output	*이진을 평안 감사로, UNK을 부교리로, 윤주를 정언으로 삼았다.*
Input	分遣暗行御史李允明, 金夢臣, 李宇謙等, 廉察諸道。
Truth	암행어사 이윤명 · 김몽신 · 이우겸 등을 나누어 파견하여 여러 도를 검찰하게 하였다.
English Translation	The secret royal inspectors Lee Yun Myeong, Kim Mong Shin, and Lee Woo Gyeom were dispatched to investigate various provinces.
NMT output	암행어사 UNK · UNK · UNK 등을 나누어 보내어 두루 제도를 살피게 하였다.
Input	江原道 楊口縣民家九十九戶, 一時燒燼。道臣以聞, 上命行恤典。
Truth	강원도 양구현의 민가 99호가 한꺼번에 불타 없어졌는데, 도신이 계문하니, 임금이 휼전을 시행하라고 명하였다.
English Translation	99 civil houses in Yanggu Gangwon province were burnt down all at once, Do Shin requested the king to distribute food tickets to civilians.
NMT output	강원도 UNK민가 99호가 한꺼번에 불에타버렸다. 도주가 아뢰니, 상이 휼전을 행하라고 명하였다.
Input	上詣永禧殿展謁, 仍詣儲慶宮, 毓祥宮, 延祜宮, 宣禧宮展拜。
Truth	임금이 영희전에 나아가 전알하고, 이어서 저경궁 · 육상궁 · 연호궁 · 선희궁에 나아가 전배하였다.
English Translation	The king went to Yeonghuijeon and perform a rites, and then to Jeogyeonggung, Sokseonggung, Yeonhogung, and Seonhuigung and performed rites.
NMT output	임금이 영모전에 나아가서 전알하고, 이어서 경복궁 · UNK · UNK · 경희전에 나아가 참배하였다.
Input	行召對, 講≪名臣奏議≫。
Truth	소대를 행하고 ≪명신주의≫를 강론하였다.
English Translation	Conducted a So Dae and lectured on ≪Myungshinism≫.
NMT output	소대를 행하고 ≪UNK≫를 강하였다.

The remainder of the paper is organized as follows: Section 2 provides a review of the conventional machine translation and NER methods that are related to the proposed method. The named entity matching using the attention alignment map that forms the core of the current study is introduced in Section 3, along with the implementation details of the transformer and the proposed NER algorithm. Section 4 describes a series of experiments that were carried out to evaluate the performance of the proposed NER method. In Section 5, the output of the NER results is further analyzed, and Section 6 concludes the paper.

2. Machine Translation

2.1. Neural Machine Translation

NMT maps a source sentence to a target sentence with neural networks. In a probabilistic representation, the NMT model is required to map a given source sentence $\mathbf{X} = [\mathbf{x}_1\ \mathbf{x}_2\ \cdots\ \mathbf{x}_n] \in \mathbb{B}^{v_s \times n}$ to a target sentence $\mathbf{Y} = [\mathbf{y}_1\ \mathbf{y}_2 \cdots\ \mathbf{y}_m] \in \mathbb{B}^{v_t \times m}$, where $\mathbb{B} = \{0, 1\}$, a binary domain space, v_s and v_t are the source (input) and target (output) vocabulary size, and n and m represent the sequence lengths of the input and output sentences, respectively. A vocabulary is usually defined by a set of tokens, which is a minimum processing unit for natural language processing (NLP) models. From a linguistic point of view, words or characters are the most popular mapping units for the tokens, depending on the grammar of the source and target languages. Each element of the encoding vector is assigned a positive integer index that uniquely identifies a single token in the corresponding vocabulary, so we can construct source vectors $\mathbf{x}_k \in \mathbb{B}^{v_s}$ by the following one-hot representation:

$$x_{k,i} = \begin{cases} 1 & \text{if } i = \text{index of } k\text{th token} \\ 0 & \text{otherwise} \end{cases}, \tag{1}$$

where $x_{k,i}$ is the ith element of $\mathbf{x}_k$. We can also construct a one-hot representation for the target vector $\mathbf{y}_k$ in a similar manner as well. The one-hot representation is extremely sparse, and the dimensions of input and target vector, n and m, may become too large to handle for the large vocabulary sizes. The embedding method, a general approach in natural language processing, is introduced to produce dense vector representations for the one-hot encoding vectors [33–36]. For given dimensions of the source and target, d_s and d_t with $d_s \ll v_s$ and $d_t \ll v_t$, linear embeddings from a higher dimensional binary space to a lower dimensions real domain are defined as follows:

$$\begin{aligned} \mathbf{E}^s &\in \mathbb{R}^{d_s \times v_s}, & \mathbf{E}^t &\in \mathbb{R}^{d_t \times v_t}, \\ \tilde{\mathbf{x}}_i &= \mathbf{E}_s \mathbf{x}_i \in \mathbb{R}^{d_s}, & \tilde{\mathbf{y}}_j &= \mathbf{E}_t \mathbf{y}_i \in \mathbb{R}^{d_t}, \end{aligned} \tag{2}$$

where $\mathbb{R}$ is the real number space and $\mathbf{E}^s$ and $\mathbf{E}^t$ are the source and target embedding matrices, respectively. Applying the linear embedding in (2), dense representation for the source and the target sentences are obtained by multiplying $\mathbf{E}_s$ and $\mathbf{E}_t$ to $\mathbf{X}$ and $\mathbf{Y}$,

$$\tilde{\mathbf{X}} = [\tilde{\mathbf{x}}_1\ \tilde{\mathbf{x}}_2\ \cdots \tilde{\mathbf{x}}_n] = \mathbf{E}_s \mathbf{X} \in \mathbb{R}^{d_s \times n} \tag{3}$$

$$\tilde{\mathbf{Y}} = [\tilde{\mathbf{y}}_1\ \tilde{\mathbf{y}}_2\ \cdots \tilde{\mathbf{y}}_n] = \mathbf{E}_t \mathbf{Y} \in \mathbb{R}^{d_t \times n}. \tag{4}$$

This linear transformation is one of the Word2Vec methods [33]. In our paper, we use this embedding for all the input and target one-hot vectors.

The target of machine translation is finding a mapping that maximizes the conditional probability $p(\mathbf{Y}|\mathbf{X})$. The direct approximation of $p(\mathbf{Y}|\mathbf{X})$ is intractable due to the high dimensionality, so most of recent NMT models are based on an encoder–decoder architecture [34]. The encoder reads an input sentence $\tilde{\mathbf{X}}$ in a dense representation and encodes it into an intermediate, contextual representation $\mathbf{C}$.

$$\mathbf{C} = \text{Encoder}(\tilde{\mathbf{X}}), \tag{5}$$

where "Encoder" is a neural network model for deriving contextual representation. After the encoding process, the decoder starts generating a translated sentence. At the first decoding step, it takes encoded contextual representation $\mathbf{C}$ and the "START" symbol, which means the start of the decoding process, and generates the first translated token. Second, the token generated previously is fed back into the decoder. It produces the next token based on the tokens generated previously and contextual representation **C**. These decoding processes are conducted recursively until an "EOS" symbol is generated, which denotes the end of the sentence. The decoding process can be formulated by the following Markovian equation,

$$p(\mathbf{y}_j) \quad = \quad g(\{\mathbf{y}_1, \mathbf{y}_2, \ldots, \mathbf{y}_{j-1}\}, \mathbf{C}), \tag{6}$$

where j is the symbol index to be generated, $\mathbf{y}_i$ is ith symbol, and $g(\cdot)$ is decoding step function, which generates a conditional probability if $\mathbf{y}_j$ given the previous outputs, $\{\mathbf{y}_1, \ldots, \mathbf{y}_{j-1}\}$, and the encoder output of the input sequence. Figures 1 and 2 illustrate the framework of the "sequence-to-sequence with attention mechanism" model (seq2seq) [13] and the framework of the "the transformer" model [16], which are NMT models used in the proposed methods.

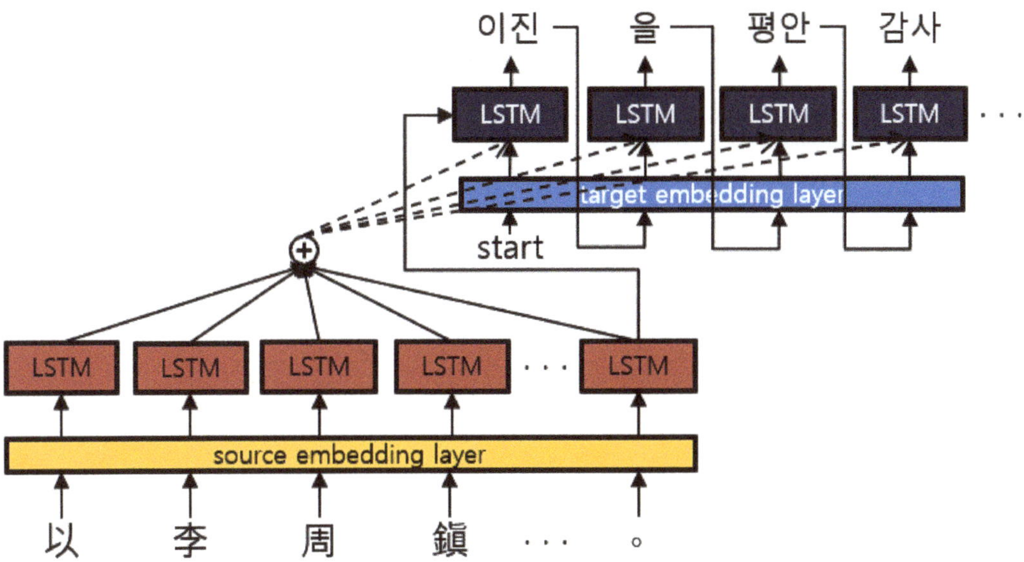

Figure 1. Framework of the seq2seq model.The red LSTM cell is the encoder and the blue is the decoder. The encoder encodes a source sentence into the context vector. The decoder is initialized with the context vector and generates a translated token, receiving a previous token and an attention vector. The translation results in this figure are given just to show that the input and output are Chinese text and Korean text, respectively. The Korean and the Chinese text do not have one-to-one correspondence.

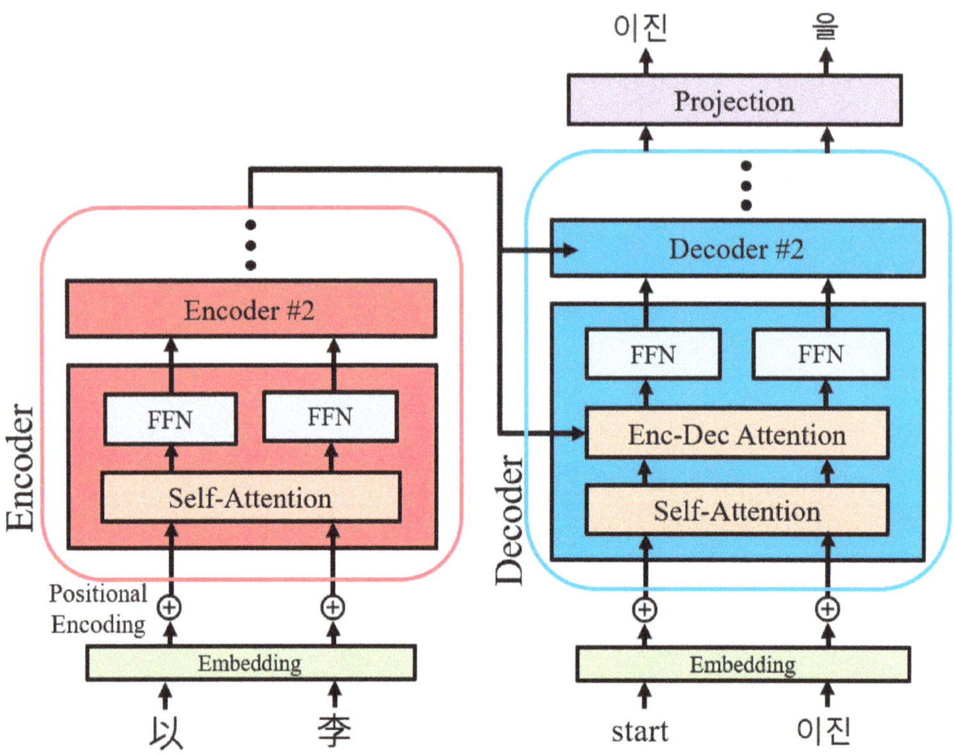

Figure 2. Framework of the transformer model. The red box is the encoder, and the blue one is the decoder. The outputs of the encoder are fed into the encoder–decoder attention layers. All residual connections and normalization layers are omitted.

2.2. Conventional Named Entity Recognition

There have been many studies for named entity recognition (NER) based on recurrent neural networks (RNNs) [37,38]. Similar to neural machine translation, the input is a sequence of tokens, $\tilde{\mathbf{X}} = [\tilde{\mathbf{x}}_1 \, \tilde{\mathbf{x}}_2 \, \cdots \, \tilde{\mathbf{x}}_n] \in \mathbb{R}^{v_s \times n}$, and the output is a sequence of binary labels indicating which tokens are named entities, so the length of the output is the same as that of the input sequence: $\mathbf{t} = [t_1 \, t_2 \, \cdots t_n] \in \mathbb{B}^n$. The example target encoding is shown in Table 1. Each word in the "Truth" and "NMT output" is underlined if it is a named entity. In those cases, the target labels are assigned one. The objective of named entity recognition is finding a sequence that maximizes the posterior probability of $\mathbf{t}$ given the input,

$$\mathbf{t}^* = \arg\max_{\mathbf{t}} p(\mathbf{t}|\mathbf{X}), \tag{7}$$

where $\mathbf{t}^*$ is an optimal NER result. Recently, a novel model for NER based only on attention mechanisms and feedforward networks achieved state-of-the-art performance on the CoNLL-2003 NER task [39].

3. Proposed Method

3.1. AAM in Sequence-to-Sequence Models

In this subsection, we describe the seq2seq model [13] used in our method and explain how to obtain an AAM from it. The seq2seq model consists of an LSTM-based [11] encoder–decoder and an attention mechanism. The encoder encodes a sequence of input tokens $\tilde{\mathbf{x}} = (\tilde{x}_1, \tilde{x}_2, \ldots, \tilde{x}_n)$, represented as dense vectors, into a context vector c, which is a fixed-length vector. We used a bidirectional LSTM (BiLSTM) [40–43] as the encoder to capture bidirectional context information of the input sentences.

$$\overrightarrow{h_i} = f(\tilde{x}_i, \overrightarrow{h_{i-1}}), \tag{8}$$
$$\overleftarrow{h_i} = f(\tilde{x}_i, \overleftarrow{h_{i+1}}), \tag{9}$$

where f are stacked unidirectional LSTM cells and $\overrightarrow{h_i} \in \mathbb{R}^d$ and $\overleftarrow{h_i} \in \mathbb{R}^d$ are the hidden states of the top forward LSTM cell and the top backward cell, respectively. Moreover, i indicates the encoding steps, and d is the number of hidden units of the top LSTM cell in the encoder.

$$c = [\overrightarrow{h_n}; \overleftarrow{h_1}] \tag{10}$$

Hidden states at the last encoding step for both directions $\overrightarrow{h_n}$ and $\overleftarrow{h_1}$ are concatenated to obtain $c \in \mathbb{R}^{2d}$. Stacked unidirectional LSTM cells are used for the decoder. Once the encoder produces context vector c, the bottom LSTM cell of the decoder is initialized with c.

$$s_{j=1} = c, \tag{11}$$

where $s_j \in \mathbb{R}^{2d}$ are the hidden states of the bottom LSTM cell in the decoder, and subscript j denotes the index of the decoding step. Next, the decoder starts the process of decoding:

$$p(y_j) = g(y_{j-1}, s_j, o_j). \tag{12}$$

Each decoding step computes the probability of the next token using three components. The first is the previously generated token y_{j-1}; the second is the current hidden state s_j; the third is an attention output vector o_j. An attention output allows the decoder to retrieve hidden states of encoder $\{h_1, \cdots, h_n\}$, where $h_i = [\overrightarrow{h_i}; \overleftarrow{h_i}]$.

$$o_j = \sum_{i=1}^{n} a_{ij} h_i, \tag{13}$$
$$a_{ij} = \frac{exp(e_{ij})}{\sum_{k=1}^{n} exp(e_{kj})}, \tag{14}$$
$$e_{ij} = v_a^T \tanh(W^a s_{j-1} + U^a h_i + b^a). \tag{15}$$

Attention scores e_{ij} indicate how related s_{j-1} is to h_i. Here, $W^a \in \mathbb{R}^{d_a \times 2d}$, $U^a \in \mathbb{R}^{d_a \times 2d}$, $v_a \in \mathbb{R}^{d_a}$, and $b^a \in \mathbb{R}^{d_a}$ are trainable parameters, where d_a is the hidden dimension of the attention mechanism. Attention weights are computed by the softmax function across the attention scores. The attention output o_j is a weighted sum of hidden states from the encoder $\{h_1, \cdots, h_n\}$. It tells the decoder where to focus on the input sentence when the decoder generates the next token.

In the seq2seq model, an attention alignment map (AAM) $\mathbf{A} \in \mathbb{R}^{n \times m}$, where n and m represent the sequence lengths of the input and output sentences, can be easily computed by stacking up the results of (14) while the model generates the translation. Figure 1 illustrates the framework of the seq2seq model used in this paper.

3.2. Transformer

In this subsection, we describe the transformer model [16] used in our method and explain how to obtain an AAM from it. The transformer also consists of an encoder and a decoder. Unlike the seq2seq model, it introduces a position encoding [16,44] to add positional information into the model, because it does not have any recurrent units that would model positional information automatically.

$$x'_i = \tilde{x}_i + PE(i)\,, \tag{16}$$

$$y'_j = \tilde{y}_j + PE(j)\,. \tag{17}$$

Here, $PE(k) \in \mathbb{R}^{d_{model}}$ produces a position encoding vector that corresponds to position k, d_{model} is the dimensionality of the model, and i and j are the positional indices of the input and output sentence, respectively. The positional encoding vectors are added to a sequence of input tokens $\tilde{\mathbf{x}} = (\tilde{x}_1, \tilde{x}_2, \ldots, \tilde{x}_n)$ represented as dense vectors as in (2). The sum of embedding vectors and positional encoding $\mathbf{x}' = (x'_1, x'_2, \ldots, x'_n) \in \mathbb{R}^{d_{model} \times n}$ are fed into the bottom encoding layer.

The encoder is a stack of encoding layers, where each encoding layer is composed of a self-attention layer and a feedforward layer. The self-attention layer of the bottom encoding layer takes $\mathbf{x}'$, and the others receive the outputs of the encoding layer right below them. Self-attention layers allow the model to refer to other tokens in the input sequence.

$$A_h^{enc} = \text{softmax}\left(\frac{Q_h K_h^T}{\sqrt{d_k}}\right) \tag{18}$$

$$\text{Attention}(Q_h, K_h, V_h) = A_h^{enc} V_h \tag{19}$$

$$\text{head}_h = \text{Attention}(Q_h, K_h, V_h) \tag{20}$$

$$\text{MultiHead}(Q, K, V) = \text{Concat}(\text{head}_1, \cdots, \text{head}_{h_n}) W^O\,. \tag{21}$$

The "multihead scaled dot-product attention" is computed by the above equations and was proposed by Vaswani et al. [16]. Here, Q_h, K_h, and V_h are linear transformations of its input. $Q_h = \mathbf{z_e^T} \cdot W_h^Q$, $K_h = \mathbf{z_e^T} \cdot W_h^K$, and $V_h = \mathbf{z_e^T} \cdot W_h^V$, where $\mathbf{z_e} \in \mathbb{R}^{d_{model} \times n}$ is the input of each attention layer in the encoder. A_h^{enc} denotes an AAM of the encoder self-attention layer on the hth head. Moreover, $W_h^Q \in \mathbb{R}^{d_{model} \times d_k}$, $W_h^K \in \mathbb{R}^{d_{model} \times d_k}$, $W_h^V \in \mathbb{R}^{d_{model} \times d_v}$, and $W^O \in \mathbb{R}^{h_n d_v \times d_{model}}$ are trainable parameters, and $d_k = d_v = d_{model}/h_n$, where h_n is the number of heads.

$$\text{FFN}(x) = \max(0, xW_1 + b_1)W_2 + b_2\,. \tag{22}$$

The outputs of the self-attention layers pass through the feedforward network. Each position is processed independently and identically. Here, $W_1 \in \mathbb{R}^{d_{model} \times d_{ff}}$, $W_2 \in \mathbb{R}^{d_{ff} \times d_{model}}$, $b_1 \in \mathbb{R}^{d_{ff}}$, and $b_2 \in \mathbb{R}^{d_{model}}$ are learnable parameters, where d_{ff} is the dimensionality of the inner linear transformation.

The final output of the encoder is considered as contextual representation $\mathbf{c} = (c_1, c_2, \ldots, c_n) \in \mathbb{R}^{d_{model} \times n}$ as in (5). It is fed into the encoder–decoder attention layers of the decoder. The decoder has a stack of decoding layers, where each decoding layer consists of a self-attention layer, encoder–decoder attention, and a feedforward network. By analogy to the encoder, the bottom decoding layer takes the sum of embedding vectors and positional encoding $\mathbf{y}' = (y'_1, y'_2, \ldots, y'_m) \in \mathbb{R}^{d_{model} \times m}$, as in (17), and the others receive the outputs of the decoding layer right below them. Self-attention layers in the decoder are similar to those in the encoder. However, the model can only retrieve the earlier positions at the current step. Hence, the model cannot attend to tokens not yet generated in the prediction phase. An encoder–decoder attention layer receives contextual representation

c and the output of self-attention layers located below in the decoder $\mathbf{z_d}^T \in \mathbb{R}^{d_{model} \times m}$ as in Figure 2.

$$Q_h = \mathbf{z_d}^T W_h^Q \tag{23}$$

$$K_h = \mathbf{c}^T W_h^K \tag{24}$$

$$V_h = \mathbf{c}^T W_h^V \tag{25}$$

$$A_h^{enc-dec} = \text{softmax}\left(\frac{Q_h K_h^T}{\sqrt{d_k}}\right) \tag{26}$$

$$\text{Attention}(Q_h, K_h, V_h) = A_h^{enc-dec} V_h \,. \tag{27}$$

This layer helps the decoder concentrate on the proper context in an input sequence when the decoder generates the next token. For every sublayer, residual connection [45] and layer normalization [46] are applied. Although we did not annotate layer indices for the trainable parameters, each layer does not share them. There are h_n encoder–decoder AAMs $A_h^{enc-dec}$ for each decoding layer. To obtain $\mathbf{A} \in \mathbb{R}^{n \times m}$, we reduced the mean across layers l and attention heads h.

$$\mathbf{A} = \underset{l}{\text{mean}}(\underset{h}{\text{mean}}(A_{hl}^{enc-dec})) \tag{28}$$

Figure 2 illustrates the framework of the transformer model in our study. The input Chinese characters, "以李", if directly translated, can be mapped to Korean "이을". However, due to the embedding in the decoder with the context information, the output of the transformer becomes "이진을".

3.3. NER Model

The detection of named entities of the input Chinese sentence is required to improve the quality of the translation. The NER model used in our study was based on stacked BiLSTM [40–42] and the conditional random field (CRF) [47,48]. We considered each Chinese character as a token and assigned a tag to each token. The tagging scheme was the IOB format [47]. As shown in Figure 3, to each of the input characters, it was given a label that was composed of one or two tags according to the membership of the input character to the named entities. The first tag is one of I, O, or B, for the inside, outside, or beginning of named entity words, respectively. The I-*tag* denotes the inside part of the named entity, but not the first character. The B-*tag* is the beginning character of the named entity. The O-*tag* means that a corresponding character is not inside a named entity. In our implementation, there were 4 types of named entities: *Person*, *Location*, *Book*, and *Era*. This type of information corresponds to B-*tag* and I-*tag*. Therefore, the NER model is asked to assign one of the nine tags to each token.

$$t_i \in \{\text{BP, BL, BB, BE, IP, IL, IB, IE, O}\}, \tag{29}$$

where BP, BL, BB, and BE are B-*tags* for *Person*, *Location*, *Book* and *Era*, respectively, and IP, IL, IB, and IE are I-*tags* for the same 4 named entity types. Table 2 shows an example of the input and output of the NER model. The NER model receives n Chinese tokens (characters) and predicts n named entity tags. A named entity "楊口縣" can be extracted by taking characters from the Chinese input from the index of B-Location to the index of the last I-Location. To separate the consecutive named entities, we used B-*tag* and I-*tag* together. If we only classify whether a character is within a named entity or not, it is impossible to separate "江原道楊口縣" into "江原道" and "楊口縣."

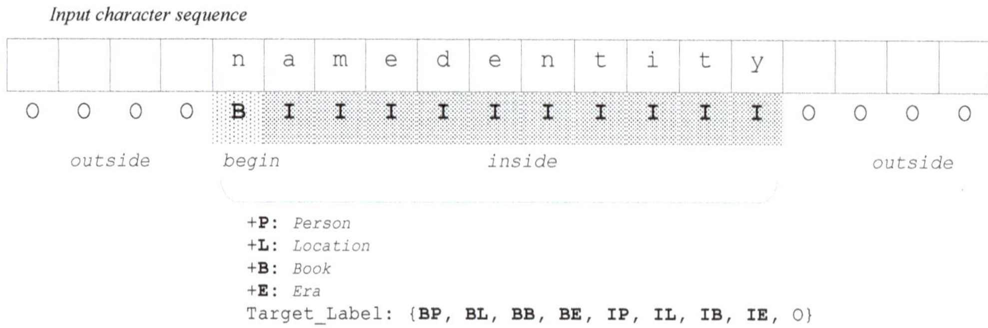

Figure 3. Target labeling of the NER model. All the individual input characters are assigned a target label of one or two tags. Characters not belonging to named entities are labeled by a single tag ''O'', meaning "outside" of the named entities. The first character of a named entity word is assigned the ''B''-tag, whose meaning is "beginning" of the named entity. All the other characters of the named entity word are assigned the ''I''-tag ("inside"). To each of the first tags of the the named entities, ''B'' and ''I'', an extra tag from {P, L, B, E} is concatenated according to the types of the named entities, {Place, Location, Book, Era}, respectively.

Table 2. Example of an input and an output of the NER model. Input: Chinese sentences that are fed into the NER model. The underlined words are named entities. Among those, human names are red; place names are blue. Output: named entity tags for each Chinese character that should be predicted by the model. BL and IL are the B-*tag* and I-*tag* for the *Location* named entity occurrence, and BP and IP are for the *Person* named entities, while O is for *Outside*.

Input	江原道 楊口縣民家九十九戶, 一時燒燼。
Output	BL IL IL BL IL IL O O O O O O O O O O O O
Input	道臣以聞, 上命行恤典。
Output	BP IP O O O O O O O O O

As in the NMT model, a Chinese sentence $\mathbf{x} = (x_1, x_2, \ldots, x_n) \in \mathbb{R}^{v_s \times n}$ represented as one-hot encoding vectors is converted into dense vector representations $\tilde{\mathbf{x}} = (\tilde{x}_1, \tilde{x}_2, \ldots, \tilde{x}_n) \in \mathbb{R}^{d_s \times n}$ by using the embedding method [33–36] as in (2). Next, $\mathbf{x}$ are fed into the BiLSTM sequentially, and the BiLSTM captures bidirectional contextual information from input sequence $\mathbf{x}$, as in (8) and (9).

$$\begin{aligned} h_i &= [\overrightarrow{h_i}; \overleftarrow{h_i}] \\ p(t_i|x_i) &= \mathrm{CRF}(W^N h_i + b^N). \end{aligned} \tag{30}$$

Hidden state $h_i \in \mathbb{R}^{2d}$, which is the output of BiLSTM, is a concatenation of both directional LSTM hidden states $\overrightarrow{h_i} \in \mathbb{R}^d$ and $\overleftarrow{h_i} \in \mathbb{R}^d$. A linear transformation layer and a CRF [47,48] layer are applied to $\mathbf{h} = (h_1, h_2, \ldots, h_n)$, and the CRF layer predicts named entity tags for each input token x_i, where i is the time step of tokens and d is the number of hidden units of a top LSTM cell. Here, $W^N \in \mathbb{R}^{2d \times d_t}$ and $b^N \in \mathbb{R}^{d_t}$ are trainable parameters, where $d_t = 9$ is the number of tag classes.

Finally, we can extract a list of named entities from the combination between the input sentence and the predicted tags. Figure 4 illustrates the NER framework used in our study.

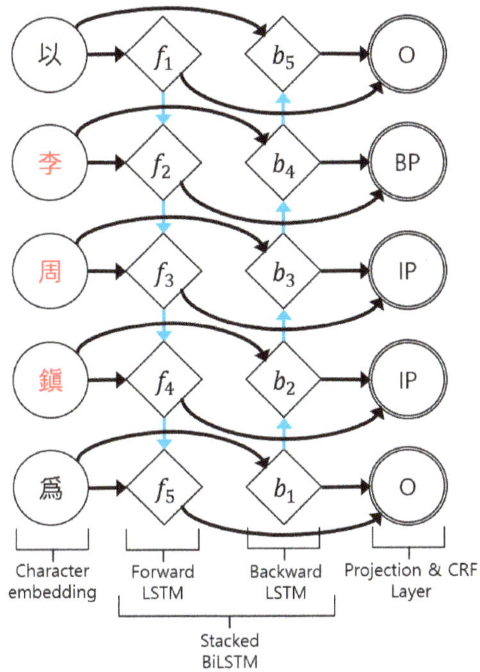

Figure 4. Framework of the NER model. The model predicts tags for each Chinese character. f_i and b_i represent the forward and backward LSTM cells, and i indicates the time steps. BP and IP are the B-*tag* and I-*tag* for the *Person* named entity occurrence, and O is for *Outside*.

3.4. Named Entity Correction with AAM

In Table 1, we can see that mistranslated words in the output of the NMT model correspond to named entities in the input sentences. The reason for this is that these named entities are OOV words or rarely occur in the training corpus. The NMT system cannot model these named entities well. In this section, we describe the proposed method that corrects mistranslated words in the output sentences through an example.

First, the NMT model translates a given Chinese sentence to a Korean sentence. In Table 3, it cannot accurately predict named entities that are names of persons.

Table 3. Neural machine translation. English translation is to explain the meaning of the text.

Input	以李周鎭爲平安監司, 元景淳爲副校理, 尹敬周爲正言。
Output	이진을 평안 감사로, UNK을 부교리로, 윤주를 정언으로 삼았다.
English	Lee Joo Jin is assigned as the Pyeongan inspector, Won Kyung Soon as the vice dictator, Yun Gyeong Joo as the dictator.

Second, the NER model finds named entities in the given Chinese sentence. In Table 4, red-colored words denote the named entities found by the NER model.

$$\mathbf{A} = \begin{pmatrix} a_{11} & a_{12} & \cdots & a_{1m} \\ a_{21} & a_{22} & \cdots & a_{2m} \\ \vdots & \vdots & \ddots & \vdots \\ a_{n1} & a_{n2} & \cdots & a_{nm} \end{pmatrix} \tag{31}$$

Third, we computed AAM $\mathbf{A} \in \mathbb{R}^{n \times m}$ from the NMT model, using (14) and (28). Here, n and m are the sequence length of the input and output sentence, respectively. Figure 5 shows examples of the attention alignment map. Each element a_{ij} of $\mathbf{A}$ is the amount of related information between input token x_i and output token y_j.

Table 4. Named entity recognition. The detected named entities of human names are underlined colored red.

Input	以李周鎭爲平安監司, 元景淳爲副校理, 尹敬周爲正言。
Output	以李周鎭爲平安監司, 元景淳爲副校理, 尹敬周爲正言。

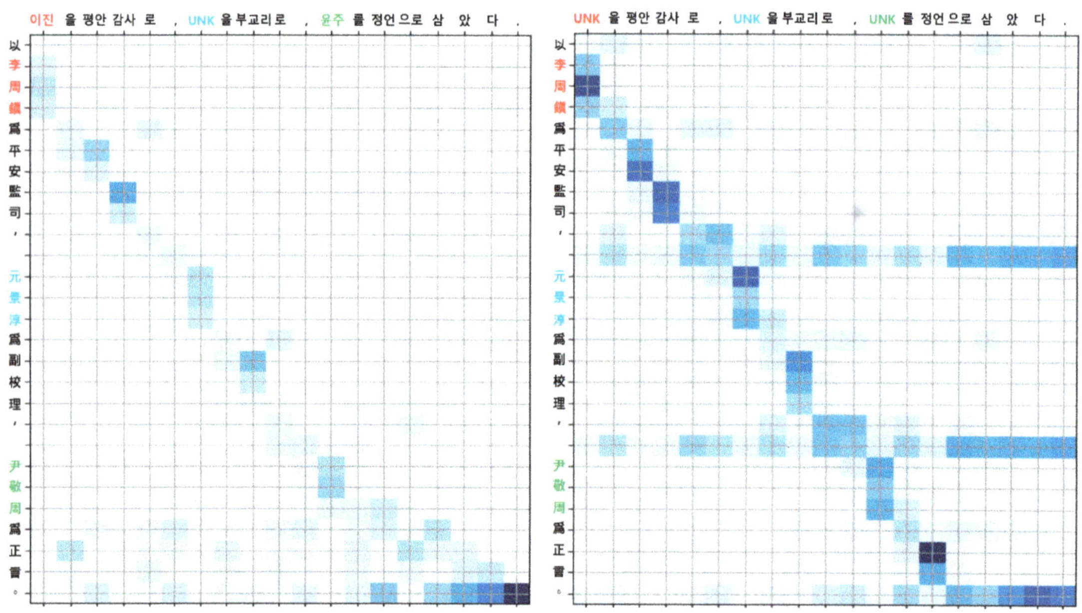

(**a**) Attention alignment map from the seq2seq model.

(**b**) Attention alignment map from the transformer model.

Figure 5. Attention alignment maps. Labels of columns and rows correspond to the tokens in the input sentences (Chinese) and the output sentences (Korean), respectively. The postprocessing has not yet been applied to the output. Colored tokens on the input side are the named entities predicted by the NER model. Colored tokens on the output side are aligned with equally colored named entities on the input side by AAM.

Fourth, we took the row vectors of AAM corresponding to indices of the Chinese named entities. Figure 6 illustrates a part of the AAM. In this example, the indices of the

Chinese named entities "李周鎭" are 2, 3, 4, so we took row vectors $\mathbf{a}_2, \mathbf{a}_3$ and $\mathbf{a}_4$, where $\mathbf{a}_2 = (a_{21}, a_{22}, \ldots, a_{2m})$.

$$\hat{j} = \arg\max_{j} \sum_{i \in \{2,3,4\}} \mathbf{a}_i \quad (32)$$

Fifth, summation across the columns of $\mathbf{a}_2, \mathbf{a}_3$ and $\mathbf{a}_4$ was implemented to obtain the vector form. The index of the Korean token aligned with the Chinese named entity was found by the arg max function, where $\hat{j}$ is the index of Korean token "이진" aligned with Chinese named entity "李周鎭." The NER matching results are shown in Table 5.

Repeating the above process, we can align all Chinese named entities found by the NER model with the Korean tokens in the sentence translated by the NMT model.

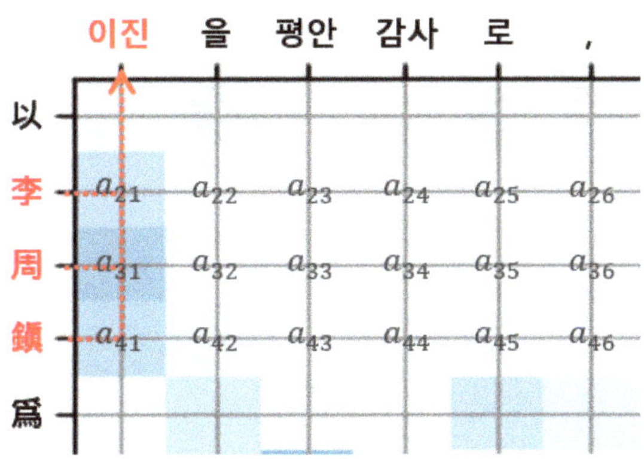

Figure 6. Korean tokens aligned with the Chinese named entities.

Table 5. Korean tokens aligned with the Chinese named entities. The underlined words are named entities. Among those words, red-colored ones are human names; blue-colored ones are place names; green-colored ones are book names.

Input	以李周鎭(1)爲平安監司, 元景淳(2)爲副校理, 尹敬周(3)爲正言。
Output	이진(1)을 평안 감사로, UNK(2)을 부교리로, 윤주(3)를 정언으로 삼았다.

We assumed that Korean token $y_{\hat{j}}$ was mistranslated. Finally, the aligned Korean tokens were replaced with a direct translation of the corresponding Chinese named entities from the look-up table. If the look-up table does not have the named entity, an identity copy of the Chinese named entity is an appropriate alternative. Figure 7 shows correction of the named entities in the translation results using look-up table. The corrections are: "이진(Lee Jin)" ⇒ "이주진(Lee Joo Jin)", "UNK" ⇒ "원경순(Won Kyung Soon)", and "윤주(Yoon Joo)" ⇒ "윤경주(Yoon Kyung Joo)". The subscripted, parenthesized numbers are found by the proposed method.

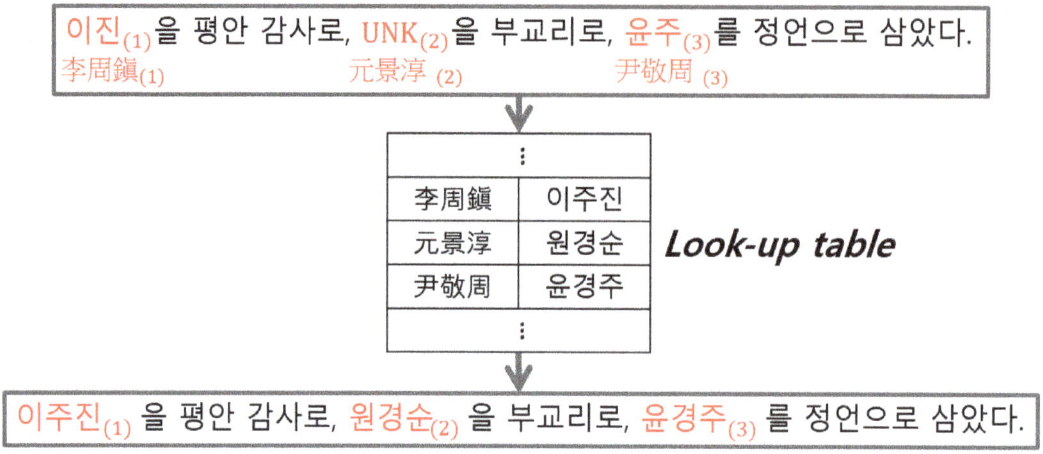

Figure 7. Named entity correction using the look-up table. The named entity, "이진(Lee Jin)" is corrected to "이주진(Lee Joo Jin)", "UNK" is to "원경순(Won Kyung Soon)", and "윤주(Yoon Joo)" to "윤경주(Yoon Kyung Joo)". All the named entities in this example are person names.

4. Experiments

We evaluated our approach on the Chinese-to-Korean translation task. The *Annals of the Joseon Dynasty* were used for our experiments as a parallel corpus. We compared the results for two cases: when the postprocessing was applied and when it was not applied.

4.1. Dataset: The Annals of the Joseon Dynasty

The *Annals* were written by the Joseon Dynasty of Korea in 1413–1865 and are listed in UNESCO's Memory of the World Registry. The *Annals* have been digitalized by the government of Korea since 2006 and are available on the website (http://sillok.history.go.kr/main/main.do, last access date: 1 July 2021) with the Korean translations and the original texts in Chinese. We used this parallel corpus to train our NMT models. To simulate real-world situations, we split the records according to the time they were written. Records from 1413 to 1623 were the training corpus, and records from 1623 to 1865 were the evaluation corpus. The training and evaluation corpus contained 230 K and 148 K parallel articles, respectively. We only used articles with Chinese and Korean tokens less than 200 in length, because the articles have an extremely variable length of letters. Figure 8 shows histograms for the sequence length of the Korean–Chinese parallel corpus. The Chinese–Korean pair sequences with the top 5% length were ignored in histogram Figure 8. For all Chinese sentences, the mean sequence length was 112.87 and the median was 54. For all Korean sentences, the mean was 124.56 and the median was 56. In Chinese (input), no tokenization was used. We simply split each Chinese sentence into a sequence of characters, because each Chinese character has its own meaning. In Korean (output), meanwhile, we used an explicit segmentation method [49] to split each Korean sentence into a sequence of tokens. Thus, the number of articles for training was 168 K and for evaluation was 113 K.

For the NER model, we also used the same corpus: the *Annals of the Joseon Dynasty*. The annotation of the Chinese named entities for this corpus is publicly available (https://www.data.go.kr/dataset/3071310/fileData.do, last access date: 1 July 2021). Additionally, Table 6 shows an analysis of the Chinese NER corpus. Approximately 7.5% of the characters belong to named entities, and the most frequently named entity type is *Person*.

Table 6. Analysis of the Chinese NER corpus.

# total characters			66M
# characters within the named entities			5M
# types of named entities			140K
Ratio of the named entity types			
Person	Location	Book	Era
73.3%	24.0%	2.4%	0.3%

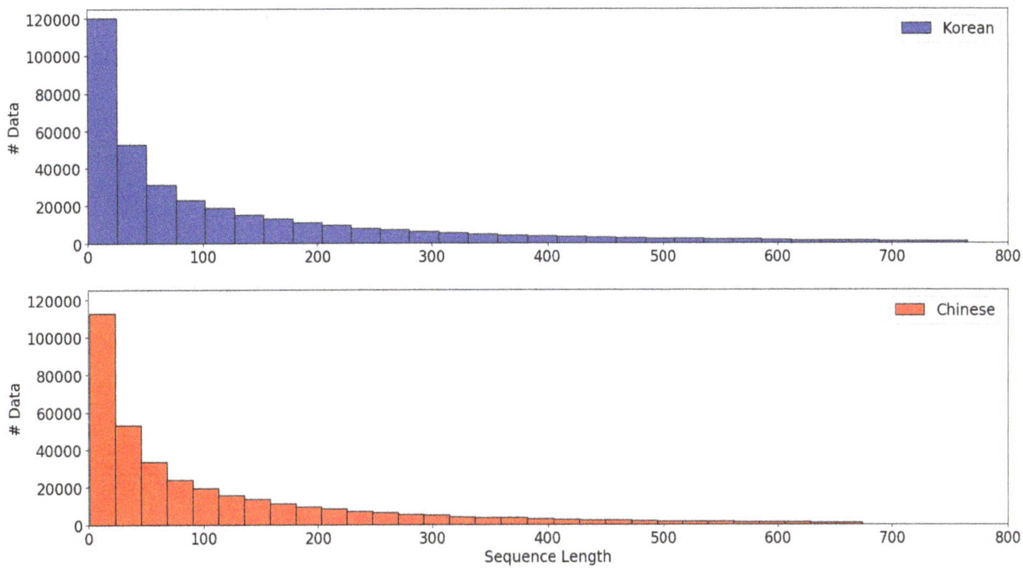

Figure 8. Histograms for the sequence length of the Korean–Chinese parallel corpus.

4.2. Models

For the seq2seq model, *Description* in Table 7 describes the model architecture used in our experiments. For the seq2seq model, *Description* means (embedding size, hidden units of encoder cells, # stack of encoder cells, hidden units of decoder cells, # stack of decoder cells). Embedding matrices for the source and target tokens were both pretrained by the word2vec algorithm [50], using only the training parallel corpus. The encoder is a stacked BiLSTM, and the decoder is a stacked unidirectional LSTM. During the learning process, the dropout approach [51,52] was applied to the output and states of the LSTM cells. Once training of the model was complete, beam-search decoding was used with a beam width of four to generate a translation that maximized the sum of the conditional probabilities of the sequence.

For the transformer model, *Description* in Table 7 represents (hidden size, # hidden layers, # heads, FFN filter size). To avoid overfitting of the model, the dropout [52] method was used among the layers in the training process. As the seq2seq model, beam-search decoding was implemented with a beam width of four. The NER model used in this study was the BiLSTM-CRF model. Specifically, the following model was used in our experiments. The embedding size was 500, and the embedding matrix was pretrained by the word2vec algorithm [50] using only the training dataset. Each cell had five-hundred twelve hidden units, and two cells are stacked. Here, we also used the dropout approach [51,52] in the learning phase.

Table 7. Performance improvements in the BLEU score. *Description*: model details. *Vocab*: the number of vocabularies. *Params*: the number of model parameters. *Original*: BLEU score without the proposed method. *Modified*: BLEU score for the results corrected by the proposed method. The best scores for both "Original" and "Modified" are achieved by using the second configuration of Seq2seq model, and those numbers are emphasized in bold face.

Model	Description	Vocab	Params	Original	Modified	Δ
Seq2seq	(500, 512, 3, 1024, 2)	40K	58M	35.75	39.29	+3.54
Seq2seq	(500, 512, 3, 1024, 2)	42K	59M	**35.83**	**39.53**	**+3.70**
Seq2seq	(500, 512, 3, 1024, 2)	50K	63M	35.66	39.13	+3.47
Seq2seq	(500, 512, 3, 1024, 2)	87K	81M	35.29	37.89	+2.60
Seq2seq-Reduced	(300, 256, 3, 512, 2)	42K	22M	33.95	37.26	+3.31
Seq2seq-Reduced	(300, 256, 3, 512, 2)	87K	54M	33.73	36.59	+2.86
Transformer-Big	(512, 6, 8, 2048)	42K	65M	33.90	37.07	+3.17
Transformer-Big	(512, 6, 8, 2048)	87K	88M	32.66	35.62	+2.96
Transformer	(256, 3, 4, 1024)	42K	16M	34.68	37.79	+3.11
Transformer	(256, 3, 4, 1024)	87K	27M	34.95	37.98	+3.03
Transformer-Reduced	(128, 2, 2, 256)	42K	6M	30.61	33.52	+2.91

4.3. Experimental Results

To evaluate our NER models, we introduced two types of F1-score: entity form and surface form [53]. First, the entity form is a conventional measurement calculated from the entity level. Second, the surface form evaluates the ability of NER models to find rare entity words. In Table 8, the lexicon used in *Dictionary search* was extracted only from the training corpus. The NER model used in the experiment was a two-stack LSTM model. Table 7 shows how the performance improved in the proposed method depending on the type of NMT model (seq2seq or transformer), the number of trainable parameters of the model, and the output (Korean) vocabulary size. Our experiments showed that the proposed method was effective regardless of these types, and the BLEU scores improved from 2.60 to 3.70. In Table 9, experimental results show that the proposed approach successfully corrected mistranslated named entities in the output of the NMT model.

Table 8. NER accuracy in 2 types of F1-score. *Entity Form* and *Surface Form* mean how many entities the model finds and how many types of entities the model finds, respectively.

Model	Entity Form	Surface Form
Dictionary search	4.4%	35.0%
1-layer LSTM Stack	91.1%	88.0%
2-layers LSTM Stack	91.9%	88.5%
3-layers LSTM Stack	91.8%	88.2%

Table 9. Named entity correction using the proposed method. *Baseline*: outputs of the seq2seq model. *Proposed*: results of our approach. Named entities are underlined. Human names are in red color; place names in blue; book names in green.

Truth	이주진을 평안 감사로, 원경순을 부교리로, 윤경주를 정언으로 삼았다.
English Translation	Lee Joo Jin is assigned as the Pyeongan inspector, Won Kyung Soon as the vice dictator, Yun Gyeong Joo as the dictator.
Baseline	이진을 평안 감사로, UNK을 부교리로, 윤주를 정언으로 삼았다.
Proposed	이주진을 평안 감사로, 원경순을 부교리로, 윤경주를 정언으로 삼았다.
Truth	암행어사 이윤명 · 김몽신 · 이우겸 등을 나누어 파견하여 여러 도를 검찰하게 하였다.
English Translation	The secret royal inspectors Lee Yun Myeong, Kim Mong Shin, and Lee Woo Gyeom were dispatched to investigate various provinces.
Baseline	암행어사 UNK · UNK · UNK 등을 나누어 보내어 두루 제도를 살피게 하였다.
Proposed	암행어사 이윤명 · 김몽신 · 이우겸 등을 나누어 보내어 두루 제도를 살피게 하였다.
Truth	강원도 양구현의 민가 99호가 한꺼번에 불타 없어졌는데, 도신이 계문하니, 임금이 휼전을 시행하라고 명하였다.
English Translation	99 civil houses in Yanggu Gangwon province were burnt down all at once, Do Shin requested the king to distribute food tickets to civilians.
Baseline	강원도 UNK민가 99호가 한꺼번에 불에타버렸다. 도주가 아뢰니, 상이 휼전을 행하라고 명하였다.
Proposed	강원도 양구현민가 99호가 한꺼번에 불에타버렸다. 도신가 아뢰니, 상이 휼전을 행하라고 명하였다.
Truth	임금이 영희전에 나아가 전알하고, 이어서 저경궁 · 육상궁 · 연호궁 · 선희궁에 나아가 전배하였다.
English Translation	The king went to Yeonghuijeon and perform a rites, and then to Jeogyeonggung, Sokseonggung, Yeonhogung, and Seonhuigung and performed rites.
Baseline	임금이 영모전에 나아가서 전알하고, 이어서 경복궁 · UNK · UNK · 경희전에 나아가 참배하였다.
Proposed	임금이 영희전에 나아가서 전알하고, 이어서 저경궁 · 육상궁 · 연호궁 · 선희궁에 나아가 참배하였다.
Truth	소대를 행하고 ≪명신주의≫를 강론하였다.
English Translation	Conducted a So Dae and lectured on ≪Myungshinism≫.
Baseline	소대를 행하고 ≪UNK≫를 강하였다.
Proposed	소대를 행하고 ≪명신주의≫를 강하였다.

5. Discussion

We found that the proposed method had several strengths and weaknesses. As for the strengths, the proposed method does not require retraining of the existing NMT models, and it can be directly applied to the NMT models without modifying the model architecture. It is suitable for any language pair. Moreover, it has a low computational complexity because of the small-sized vocabulary. As for the weaknesses, the proposed method does not work when predictions of the NER model are wrong. Additionally, tokens that should not be changed may be corrected if the alignment is not proper. The proposed method needs a look-up table to work properly. Table 10 shows examples of these weaknesses. In the above example, the NER model cannot find a named entity in the source sentence, so the *UNK* for "거려청" was not corrected. The *UNK* for "≪심경≫" was also not corrected, although the NER model recognized a token "≪必經≫" as a named entity, because our look-up table did not have "≪必經≫." In the final example, token "하" in *Baseline* was changed because the attention alignment map was not accurate.

Table 10. Weaknesses of the proposed method. *Source*: input sentence. *Truth*: ground truth. *Baseline*: outputs of the NMT models. *Proposed*: results of our approach. *NER output*: outputs of the NER model. Underlined words: named entities. Green-colored tokens: named entities for the names of books. Blue-colored: named entities for the names of place names.

Source	上御居廬廳, 召對, 命儒臣, 讀≪必經≫。
Truth	임금이 거려청에 나아가 소대하였다. 임금이 유신들에게 명하여 ≪심경≫을 읽게하였다.
English Translation	The king went to Georyeocheong and conducted a So Dae. The king ordered the subjects to read ≪Shim Gyung≫.
Baseline	상이 UNK에 나아가 소대하였다. 유신에게 명하여 ≪UNK≫을 읽게하였다.
Proposed	상이 UNK에 나아가 소대하였다. 유신에게 명하여 ≪UNK≫을 읽게하였다.
NER output	[≪必經≫, Book]
Source	進講于熙政堂。
Truth	희정당에서 진강하였다.
English Translation	He lectured at Huijeongdang.
Baseline	UNK에서 진강하였다.
Proposed	UNK에서 진강희정당였다.
NER output	[熙政堂, Location]

6. Conclusions

Even the NMT models that show state-of-the-art performance on multiple machine translation tasks are still limited when dealing with OOV and rarely occurring words. We found that the problem is particularly relevant for the translation of historical documents with multiple named entities. In this paper, we proposed a postprocessing approach to address this limitation. The proposed method corrects the machine translation output using the NER model and the attention map. The NER model finds named entities in the source sentence, and the attention map aligns the located named entities with the tokens in the translated sentence. Next, we assumed that the tokens aligned with the source named entities were mistranslated, and we replaced them using the look-up table or an identity copy. Experiments with various target vocabulary sizes in Section 4 demonstrated that our method is effective in the task of translation of historical documents from Chinese to Korean. Using the proposed NER method, the machine translation performance was improved up to 3.70 in terms of the BLEU score (35.83 to 39.53) in seq2seq translation models and up to 3.17 (33.90 to 37.07) in transformer models. Moreover, there was no BLEU score degradation due to the proposed method. The proposed method can be applied to an existing NMT model that uses the attention mechanism without retraining the model, if an NER model exists for the source language. Our method can be successfully applied not only to Chinese-to-Korean translation, but also to other language pairs. In our future work, we plan to explore this direction.

Author Contributions: Conceptualization, M.L., J.L. (Jungi Lee) and G.-J.J.; methodology, J.L. (Jangwon Lee); software, J.L. (Jangwon Lee); validation, J.L. (Jangwon Lee) and G.-J.J.; formal analysis, J.L. (Jungi Lee) and M.L.; investigation, J.L. (Jungi Lee); resources, G.-J.J. and J.L. (Jungi Lee); data curation, J.L. (Jangwon Lee); writing—original draft preparation, J.L. (Jangwon Lee) and G.-J.J.; writing—review and editing, M.L. and J.L. (Jungi Lee); visualization, J.L. (Jangwon Lee); supervision, G.-J.J.; project administration, G.-J.J.; funding acquisition, G.-J.J. All authors read and agree to the published version of the manuscript.

Funding: This work was supported by the National Research Foundation of Korea (NRF) grant funded by the Korea government (MSIP) (No. NRF-2017M3C1B6071399), and by the Technology Innovation Program (20016180, Forecast of overseas inflow of new infectious diseases and development of intelligent blocking technology) funded By the Ministry of Trade, Industry & Energy (MOTIE, Korea).

Institutional Review Board Statement: Not applicable because this study is involved with neither humans nor animals.

Informed Consent Statement: Not applicable.

Data Availability Statement: Not applicable.

Conflicts of Interest: The authors declare no conflict of interest.

Abbreviations

The following abbreviations are used in this manuscript:

NMT	Neural machine translation
NER	Named entity recognition
OOV	Out of vocabulary
RNN	Recurrent neural network
LSTM	Long short-term memory
BLSTM	Bi-directional long short-term memory
GRU	Gated recurrent unit
AAM	Attention alignment map

References

1. Koehn, P.; Och, F.J.; Marcu, D. Statistical Phrase-based Translation. In *Proceedings of the 2003 Conference of the North American Chapter of the Association for Computational Linguistics on Human Language Technology—Volume 1*; Association for Computational Linguistics: Stroudsburg, PA, USA, 2003; pp. 48–54. [CrossRef]
2. Cho, K.; van Merrienboer, B.; Gülçehre, Ç.; Bougares, F.; Schwenk, H.; Bengio, Y. Learning Phrase Representations using RNN Encoder-Decoder for Statistical Machine Translation. *arXiv* **2014**, arXiv:1406.1078.
3. Chiang, D. A Hierarchical Phrase-based Model for Statistical Machine Translation. In *Proceedings of the 43rd Annual Meeting on Association for Computational Linguistics*; Association for Computational Linguistics: Stroudsburg, PA, USA, 2005; pp. 263–270. [CrossRef]
4. Chen, K.; Zhao, T.; Yang, M.; Liu, L.; Tamura, A.; Wang, R.; Utiyama, M.; Sumita, E. A Neural Approach to Source Dependence Based Context Model for Statistical Machine Translation. *IEEE/ACM Trans. Audio Speech Lang. Process.* **2018**, *26*, 266–280. [CrossRef]
5. Wang, X.; Tu, Z.; Zhang, M. Incorporating Statistical Machine Translation Word Knowledge Into Neural Machine Translation. *IEEE/ACM Trans. Audio Speech Lang. Process.* **2018**, *26*, 2255–2266. [CrossRef]
6. Sutskever, I.; Vinyals, O.; Le, Q.V. Sequence to Sequence Learning with Neural Networks. In *Advances in Neural Information Processing Systems 27*; Ghahramani, Z., Welling, M., Cortes, C., Lawrence, N.D., Weinberger, K.Q., Eds.; Curran Associates, Inc.: Red Hook, NY, USA, 2014; pp. 3104–3112.
7. Kalchbrenner, N.; Blunsom, P. Recurrent Continuous Translation Models. In *Proceedings of the 2013 Conference on Empirical Methods in Natural Language Processing*; Association for Computational Linguistics: Seattle, WA, USA, 2013; pp. 1700–1709.
8. Cho, K.; van Merriënboer, B.; Bahdanau, D.; Bengio, Y. On the Properties of Neural Machine Translation: Encoder–Decoder Approaches. In *Proceedings of the SSST-8, Eighth Workshop on Syntax, Semantics and Structure in Statistical Translation*; Association for Computational Linguistics: Doha, Qatar, 2014; pp. 103–111. [CrossRef]
9. Li, S.; Zhao, J.; Shi, G.; Tan, Y.; Xu, H.; Chen, G.; Lan, H.; Lin, Z. Chinese Grammatical Error Correction Based on Convolutional Sequence to Sequence Model. *IEEE Access* **2019**, *7*, 72905–72913. [CrossRef]
10. Zhang, X.; Yin, F.; Zhang, Y.; Liu, C.; Bengio, Y. Drawing and Recognizing Chinese Characters with Recurrent Neural Network. *IEEE Trans. Pattern Anal. Mach. Intell.* **2018**, *40*, 849–862. [CrossRef] [PubMed]
11. Hochreiter, S.; Schmidhuber, J. Long Short-Term Memory. *Neural Comput.* **1997**, *9*, 1735–1780. [CrossRef] [PubMed]
12. Chung, J.; Gülçehre, Ç.; Cho, K.; Bengio, Y. Empirical Evaluation of Gated Recurrent Neural Networks on Sequence Modeling. *arXiv* **2014**, arXiv:1412.3555.
13. Bahdanau, D.; Cho, K.; Bengio, Y. Neural Machine Translation by Jointly Learning to Align and Translate. In Proceedings of the 3rd International Conference on Learning Representations, (ICLR 2015), San Diego, CA, USA, 7–9 May 2015.
14. Luong, M.; Pham, H.; Manning, C.D. Effective Approaches to Attention-based Neural Machine Translation. *arXiv* **2015**, arXiv:1508.04025.
15. Xu, Y.; Liu, W.; Chen, G.; Ren, B.; Zhang, S.; Gao, S.; Guo, J. Enhancing Machine Reading Comprehension With Position Information. *IEEE Access* **2019**, *7*, 141602–141611. [CrossRef]
16. Vaswani, A.; Shazeer, N.; Parmar, N.; Uszkoreit, J.; Jones, L.; Gomez, A.N.; Kaiser, L.u.; Polosukhin, I. Attention is All you Need. In *Advances in Neural Information Processing Systems 30*; Guyon, I., Luxburg, U.V., Bengio, S., Wallach, H., Fergus, R., Vishwanathan, S., Garnett, R., Eds.; Curran Associates, Inc.: Red Hook, NY, USA, 2017; pp. 5998–6008.
17. Seljan, S.; Dunđer, I.; Pavlovski, M. Human Quality Evaluation of Machine-Translated Poetry. In Proceedings of the 43rd International Convention on Information, Communication and Electronic Technology (MIPRO), Opatija, Croatia, 18 May 2020.

18. Dunđer, I.; Seljan, S.; Pavlovski, M. Automatic machine translation of poetry and a low-resource language pair. In Proceedings of the 43rd International Convention on Information, Communication and Electronic Technology (MIPRO), Opatija, Croatia, 28 September–2 October 2020. [CrossRef]
19. Dunđer, I. Machine Translation System for the Industry Domain and Croatian Language. *J. Inf. Organ. Sci.* **2020**, *44*, 33–50. [CrossRef]
20. Brkić, M.; Seljan, S.; Vičić, T. Automatic and Human Evaluation on English-Croatian Legislative Test Set. *Lect. Notes Comput. Sci. LNCS* **2013**, *7817*, 311–317. [CrossRef]
21. Zoph, B.; Yuret, D.; May, J.; Knight, K. Transfer Learning for Low-Resource Neural Machine Translation. In *Proceedings of the 2016 Conference on Empirical Methods in Natural Language Processing*; Association for Computational Linguistics: Stroudsburg, PA, USA, 2016; pp. 1568–1575. [CrossRef]
22. Yang, Z.; Cheng, Y.; Liu, Y.; Sun, M. Reducing Word Omission Errors in Neural Machine Translation: A Contrastive Learning Approach. In *Proceedings of the 57th Annual Meeting on Association for Computational Linguistics*; Association for Computational Linguistics: Stroudsburg, PA, USA, 2019; pp. 6191–6196. [CrossRef]
23. Tan, Z.; Wang, S.; Yang, Z.; Chen, G.; Huang, X.; Sun, M.; Liu, Y. Neural machine translation: A review of methods, resources, and tools. *AI Open* **2020**, *1*, 5–21. [CrossRef]
24. Jean, S.; Cho, K.; Memisevic, R.; Bengio, Y. On Using Very Large Target Vocabulary for Neural Machine Translation. *arXiv* **2014**, arXiv:1412.2007.
25. Luong, M.; Manning, C.D. Achieving Open Vocabulary Neural Machine Translation with Hybrid Word-Character Models. *arXiv* **2016**, arXiv:1604.00788.
26. Luong, T.; Sutskever, I.; Le, Q.; Vinyals, O.; Zaremba, W. Addressing the Rare Word Problem in Neural Machine Translation. In *Proceedings of the 53rd Annual Meeting of the Association for Computational Linguistics and the 7th International Joint Conference on Natural Language Processing (Volume 1: Long Papers)*; Association for Computational Linguistics: Stroudsburg, PA, USA, 2015; pp. 11–19. [CrossRef]
27. Sennrich, R.; Haddow, B.; Birch, A. Neural Machine Translation of Rare Words with Subword Units. In *Proceedings of the 54th Annual Meeting on Association for Computational Linguistics*; Association for Computational Linguistics: Stroudsburg, PA, USA, 2016; pp. 1715–1725. [CrossRef]
28. Haddad, H.; Fadaei, H.; Faili, H. Handling OOV Words in NMT Using Unsupervised Bilingual Embedding. In Proceedings of the 2018 9th International Symposium on Telecommunications (IST), Tehran, Iran, 17–19 December 2018; pp. 569–574. [CrossRef]
29. Ling, W.; Trancoso, I.; Dyer, C.; Black, A.W. Character-based Neural Machine Translation. *arXiv* **2015**, arXiv:1511.04586.
30. Costa-jussà, M.R.; Fonollosa, J.A.R. Character-based Neural Machine Translation. *arXiv* **2016**, arXiv:1603.00810.
31. Lee, J.; Cho, K.; Hofmann, T. Fully Character-Level Neural Machine Translation without Explicit Segmentation. *Trans. Assoc. Comput. Linguist.* **2017**, *5*, 365–378. [CrossRef]
32. Papineni, K.; Roukos, S.; Ward, T.; Zhu, W.J. BLEU: A Method for Automatic Evaluation of Machine Translation. In *Proceedings of the 40th Annual Meeting on Association for Computational Linguistics*; Association for Computational Linguistics: Stroudsburg, PA, USA, 2002; pp. 311–318. [CrossRef]
33. Mikolov, T. Statistical Language Models Based on Neural Networks. Ph.D. Thesis, Brno University of Technology, Brno-střed, Czech Republic, 2012.
34. Collobert, R.; Weston, J. A Unified Architecture for Natural Language Processing: Deep Neural Networks with Multitask Learning. In *Proceedings of the 25th International Conference on Machine Learning*; ACM: New York, NY, USA, 2008; pp. 160–167. [CrossRef]
35. Socher, R.; Lin, C.C.Y.; Ng, A.Y.; Manning, C.D. Parsing Natural Scenes and Natural Language with Recursive Neural Networks. In *Proceedings of the 28th International Conference on International Conference on Machine Learning*; Omnipress: Washington, WA, USA, 2011; pp. 129–136.
36. Glorot, X.; Bordes, A.; Bengio, Y. Domain Adaptation for Large-scale Sentiment Classification: A Deep Learning Approach. In *Proceedings of the 28th International Conference on International Conference on Machine Learning*; Omnipress: Washington, WA, USA, 2011; pp. 513–520.
37. Nut Limsopatham, N.C. Bidirectional LSTM for Named Entity Recognition in Twitter Messages. In Proceedings of the 2nd Workshop on Noisy User-Generated Text, Osaka, Japan, 11 December 2016; pp. 145–152. [CrossRef]
38. Aguilar, G.; Maharjan, S.; López Monroy, A.P.; Solorio, T. A Multi-task Approach for Named Entity Recognition in Social Media Data. In *Proceedings of the 3rd Workshop on Noisy User-Generated Text*; Association for Computational Linguistics: Copenhagen, Denmark, 2017; pp. 148–153. [CrossRef]
39. Devlin, J.; Chang, M.; Lee, K.; Toutanova, K. BERT: Pre-training of Deep Bidirectional Transformers for Language Understanding. *arXiv* **2018**, arXiv:1810.04805.
40. Schuster, M.; Paliwal, K. Bidirectional Recurrent Neural Networks. *Trans. Sig. Proc.* **1997**, *45*, 2673–2681. [CrossRef]
41. Zhou, P.; Shi, W.; Tian, J.; Qi, Z.; Li, B.; Hao, H.; Xu, B. *Attention-Based Bidirectional Long Short-Term Memory Networks for Relation Classification*; ACL: Berlin, Germany, 2016.
42. Graves, A.; Schmidhuber, J. Framewise phoneme classification with bidirectional LSTM networks. In Proceedings of the 2005 IEEE International Joint Conference on Neural Networks, Montreal, QC, Canada, 31 July–4 August 2005; Volume 4, pp. 2047–2052
43. Liao, Y.; Xiong, P.; Min, W.; Min, W.; Lu, J. Dynamic Sign Language Recognition Based on Video Sequence with BLSTM-3D Residual Networks. *IEEE Access* **2019**, *7*, 38044–38054. [CrossRef]

44. Gehring, J.; Auli, M.; Grangier, D.; Yarats, D.; Dauphin, Y.N. Convolutional Sequence to Sequence Learning. In Proceedings of the 34th International Conference on Machine Learning—Volume 70, Sydney, Australia, 6–11 August 2017; pp. 1243–1252.
45. He, K.; Zhang, X.; Ren, S.; Sun, J. Deep Residual Learning for Image Recognition. In Proceedings of the IEEE Conference on Computer Vision and Pattern Recognition (CVPR), Las Vegas, NV, USA, 21–26 July 2016.
46. Lei Ba, J.; Kiros, J.R.; Hinton, G.E. Layer Normalization. *arXiv* **2016**, arXiv:1607.06450.
47. Lample, G.; Ballesteros, M.; Subramanian, S.; Kawakami, K.; Dyer, C. Neural Architectures for Named Entity Recognition. *arXiv* **2016**, arXiv:1603.01360.
48. Lafferty, J.D.; McCallum, A.; Pereira, F.C.N. Conditional Random Fields: Probabilistic Models for Segmenting and Labeling Sequence Data. In *Proceedings of the Eighteenth International Conference on Machine Learning*; Morgan Kaufmann Publishers Inc.: San Francisco, CA, USA, 2001; pp. 282–289.
49. Park, E.L.; Cho, S. KoNLPy: Korean natural language processing in Python. In Proceedings of the 26th Annual Conference on Human and Cognitive Language Technology, Chuncheon, Korea, 10 October 2014.
50. Mikolov, T.; Sutskever, I.; Chen, K.; Corrado, G.S.; Dean, J. Distributed Representations of Words and Phrases and their Compositionality. In *Advances in Neural Information Processing Systems 26*; Burges, C.J.C., Bottou, L., Welling, M., Ghahramani, Z., Weinberger, K.Q., Eds.; Curran Associates, Inc.: Red Hook, NY, USA, 2013; pp. 3111–3119.
51. Gal, Y.; Ghahramani, Z. A Theoretically Grounded Application of Dropout in Recurrent Neural Networks. In *Advances in Neural Information Processing Systems 29*; Lee, D.D., Sugiyama, M., Luxburg, U.V., Guyon, I., Garnett, R., Eds.; Curran Associates, Inc.: Red Hook, NY, USA, 2016; pp. 1019–1027.
52. Srivastava, N.; Hinton, G.; Krizhevsky, A.; Sutskever, I.; Salakhutdinov, R. Dropout: A Simple Way to Prevent Neural Networks from Overfitting. *J. Mach. Learn. Res.* **2014**, *15*, 1929–1958.
53. Derczynski, L.; Nichols, E.; van Erp, M.; Limsopatham, N. Results of the WNUT2017 Shared Task on Novel and Emerging Entity Recognition. In *Proceedings of the 3rd Workshop on Noisy User-Generated Text*; Association for Computational Linguistics: Copenhagen, Denmark, 2017; pp. 140–147. [CrossRef]

applied sciences

MDPI

Article

Comparative Analysis of Current Approaches to Quality Estimation for Neural Machine Translation

Sugyeong Eo †, Chanjun Park †, Hyeonseok Moon †, Jaehyung Seo † and Heuiseok Lim *

Department of Computer Science and Engineering, Korea University, 145, Anam-ro, Seongbuk-gu, Seoul 02841, Korea; djtnrud@korea.ac.kr (S.E.); bcj1210@korea.ac.kr (C.P.); glee889@korea.ac.kr (H.M.); seojae777@korea.ac.kr (J.S.)

* Correspondence: limhseok@korea.ac.kr

† These authors contributed equally to this work.

Abstract: Quality estimation (QE) has recently gained increasing interest as it can predict the quality of machine translation results without a reference translation. QE is an annual shared task at the Conference on Machine Translation (WMT), and most recent studies have applied the multilingual pretrained language model (mPLM) to address this task. Recent studies have focused on the performance improvement of this task using data augmentation with finetuning based on a large-scale mPLM. In this study, we eliminate the effects of data augmentation and conduct a pure performance comparison between various mPLMs. Separate from the recent performance-driven QE research involved in competitions addressing a shared task, we utilize the comparison for sub-tasks from WMT20 and identify an optimal mPLM. Moreover, we demonstrate QE using the multilingual BART model, which has not yet been utilized, and conduct comparative experiments and analyses with cross-lingual language models (XLMs), multilingual BERT, and XLM-RoBERTa.

Keywords: quality estimation; neural machine translation; pretrained language model; multilingual pre-trained language model; WMT

Citation: Eo, S.; Park, C.; Moon, H.; Seo, J.; Lim, H. Comparative Analysis of Current Approaches to Quality Estimation for Neural Machine Translation. *Appl. Sci.* **2021**, *11*, 6584. https://doi.org/10.3390/app11146584

Academic Editors: Arturo Montejo-Ráez and Salud María Jiménez-Zafra

Received: 6 May 2021
Accepted: 15 July 2021
Published: 17 July 2021

1. Introduction

Quality estimation (QE) refers to automatically predicting translation quality using only source sentence and machine translation (MT) output [1]. The goal of QE is to estimate translation quality scores or categories for MT outputs without reference sentences at various levels of granularity (i.e., sentence, phrase, word). It is necessary to compare the MT output with a reference sentence to determine the quality of the translation in general. However, it is not easy to obtain a reference sentence, and constructing such a sentence requires large costs and human labor. Based on these issues, the need for QE research is increasing, and a considerable number of studies are being conducted in this area.

In the QE process, the quality of the MT output is indicated using quality annotations, such as numerical values or error tags. This allows the user to select or rank the system that exhibits the best translation results [2]. In addition, for low-quality sentences, efficiency can be increased during automatic post editing [3] by modifying only the low-quality words or phrases using quality annotations. Therefore, QE is an important process that can be widely applied.

According to recent research trends, there are a number of cases in which the QE task is conducted based on multilingual pretrained language models (mPLMs) [4–6]. mPLM is a case where a multilingual representation is learned by extending pretrained language model to multiple languages. In QE, where two languages are concatenated and entered as input, such a representation is required, so mPLMs are mostly used in this task. However, most studies are focused on improving performance by simply applying data augmentation while finetuning the QE task based on a large-capacity mPLM such as multilingual BERT (mBERT) [7], cross-lingual language model (XLM) [8], or XLM-RoBERTa (XLM-R) [9]. In addition, there are many cases in which QE models are trained based on XLM-R, which is the latest model with a state-of-the-art (SOTA) performance for cross-lingual transfer tasks [10,11] achieved by pretraining using an extremely large dataset [5,12–14]. However, unlike evaluation benchmarks for cross-lingual understanding that deal with multiple languages, QE differs from these because it requires measuring translation quality while referencing two languages at the same time. Thus, performance comparisons with other models should be preceded, but many papers tend to overlook this and simply use the XLM-R model [15].

Zhou et al. [16] compare the performance difference between mBERT and XLM-MLM for sub-task 1, and Baek et al. [13] additionally compare the performance difference of XLM-CLM, Ranasinghe et al. [17] compare the performance of mBERT and XLM-R. However, XLM models including the English and German languages are quite diverse, and, in particular, there has been no comparison with XLM-TLM models that learn information between languages in addition to multiple languages.

Unlike other previous studies that mostly utilize the SOTA model, we remove the effects of data augmentation that are utilized to achieve performance improvement and perform a comparative study between representative mPLMs based on sub-tasks 1 and 2 from WMT20. Each mPLM has a different capacity, training data size, or pretraining objective, and even the same model has different performance depending on how many languages it contains. Therefore, comparative analysis of various mPLMs in QE can serve as a good indicator of which model performs well for each task in future studies. In addition, because we compare pure performance, we can expect high performance by using data augmentation and new methodologies based on the model with high performance.

This study addresses two questions:

- Which mPLM is best for QE sub-tasks?
- Does the input order of the source sentence and the MT output sentence affect the performance of the model?

Considering the first question, the finetuning performance of mPLMs for a QE task can be validated using a quantitative analysis. To achieve this, we apply multilingual BART (mBART) [18], which has not been used in previous QE studies, and compare it with the existing mBERT, XLM, and XLM-R models. For XLM, we conduct performance comparisons between the causal language model (CLM), mask language model (MLM), and translation language model (TLM). In the case of XLM-MLM, the performances are compared according to the number of languages used for learning.

Considering the second question, it is possible to determine the criteria indicating which input structure should be adopted for QE embedding. Previous studies have used the input structure of *[BOS] Source sentence [EOS] [EOS] MT output [EOS]* or *[BOS] MT output [EOS] [EOS] Source sentence [EOS]* without a clear standard. Therefore, we investigate this process through a quantitative analysis by utilizing different input structures for all mPLMs. The contributions of this study are as follows:

- We conduct comparative experiments on finetuning mPLMs for a QE task, which is different from research concerning the performance improvement of the WMT shared-task competition. This quantitative analysis allows us to revisit the pure performance of mPLMs for the QE task. To the best of our knowledge, we are the first to conduct such research;
- Through a comparative analysis concerning how to construct an appropriate input structure for QE, we reveal that the performance can be improved by simply changing the input order of the source sentence and the MT output;
- In the process of finetuning mPLMs, we only use data officially distributed in WMT20 (without external knowledge or data augmentation) and use the official test set to ensure objectivity for all experiments.

2. Related Work and Background

A quality estimation (QE) task is a branch of machine translation. Representative metrics of NMT such as BLEU [19], METEOR [20] require reference sentences to evaluate quality of MT output. QE does not require access to reference outputs, and quality is indicated by OK/BAD tokens, numerical values, or spans, etc. QE research can be divided into three categories: the use of statistical methods, the use of recurrent neural networks (RNN) and long short-term memory (LSTM) after the advent of deep learning, and the use of pre-training and finetuning approaches with the advent of pretrained language models.

Most conventional QE studies have been conducted by extracting or selecting features to evaluate the quality of MT. When selecting such features, machine learning algorithms, such as Gaussian processes [21,22], support vector machines [23,24], and regression trees [1,25] are used. In the case of feature extraction, some studies have extracted useful features, such as linguistic features [26] and pseudo-reference features [27], using external resources such as parsers, taggers, and named entity recognizers [23,28]. However, these studies are focused on determining the complex relationship between features and references, and the process of selecting and extracting optimized features requires heuristic processes and high costs.

With the advent of deep learning, research using RNN and LSTM was mainly conducted in QE, and it achieved much higher performance improvement than statistical methods [29,30]. Kim et al. [31] proposed a new structure referred to as predictor-estimator. Predictor is a bilingual and bidirectional RNN-based word prediction model, which randomly selects and masks a word in a target sentence from a parallel corpus and then generates feature vectors by predicting it. In estimator, the generated feature vector is used as transferred knowledge to learn the QE model. This structure was able to alleviate the issue of data shortage while allowing an additional parallel corpus to be utilized for a limited amount of QE data, and it led to a dramatic performance improvement. Similar to this architecture, Wang et al. [32] constructed a QE brain model with two phases. In the first phase, features were extracted with the transformer model to be used as prior knowledge, and in the QE phase, these features were combined with human-craft features and fed into the Bi-LSTM structure to train for QE. A superior performance was also obtained using this method.

Since the advent of pre-trained language models (PLMs), the research flow of QE is mostly done based on mPLM. By designing the QE model based on the large-scale pretrained model, the performance is greatly improved. Kepler et al. [33] replaced the predictor component with a pretrained BERT or XLM model while training using the structure of a predictor-estimator. Kim et al. [34] finetuned the QE task based on mBERT. Ranasinghe et al. [35] proposed two unique approaches: MonoTransquest and Siamese-Transquest. The former finetuned for a single XLM-R, while the latter used two separate XLM-R models for each of the source and target sentences, and the cosine similarity of both outputs was measured to predict the translation quality at the sentence level. Lee [12] performed data augmentation using a parallel corpus and pretrained pseudo data with XLM-R. After the process, finetuning was performed using QE data provided by WMT. Wang et al. [36] considered the pretrained transformer model as a predictor and the task-specific regressors or classifiers as an estimator instead of mPLM. In the learning process, a bottleneck adapter layer was newly added to improve the efficiency of transfer learning and prevent over-fitting.

3. Multilingual Pretrained Language Models for QE

In this section, we describe mPLMs for QE performance comparison. We used mBERT, XLM, XLM-R, and mBART, which are multilingual pretrained models that include English and German.

3.1. Multilingual BERT

BERT [37] is built on a transformer [38] architecture, which consists solely of an encoder structure.

BERT performs a self-supervised learning process for large-scale mono-lingual corpus. Because the self-supervised learning process performs supervision on raw text on its own, it does not require labeled data, so it can utilize large amounts of raw data. After performing user-defined problems such as masked language model (MLM) and next sentence prediction (NSP) on unlabeled raw data, transfer learning is performed for downstream tasks. More specifically, the user generates arbitrary tasks and labels for raw text to learn language information, and uses the representations obtained through this process as initialization values for downstream tasks. For the case of BERT, MLM, and NSP are used as pretraining schemes.

MLM is a procedure of randomly masking tokens in the original sentence with [MASK] tokens. The objectives is to correctly predict these masked tokens based on left and right context of the sentence. In particular, the last hidden vector corresponding to the mask token goes through softmax and returns as the word with the highest probability in the vocabulary. In the process of masking, 15% of the original sentences are randomly sampled, then among them, 80% of these selected tokens are replaced by [MASK], 10% are replaced by random tokens in the vocabulary, and 10% remain unchanged. Through this masking process, a defective sentence $\bar{X} = \{\bar{x}_1, \bar{x}_2, \ldots, \bar{x}_n\}$ is generated from an unlabeled monolingual sentence $X = \{x_1, x_2, \ldots, x_n\}$. In the training process, $\bar{X}$ is fed into a BERT model, which is parameterized by θ, and the model is then trained to return X. This task can be described by Equation (1).

$$\max_{\theta} \sum_{(X,\bar{X}) \in D} \sum_{i=1}^{n} \log P(x_i \mid \bar{x}_1, \ldots, \bar{x}_n, \theta) \tag{1}$$

This equation indicates that a model is trained to predict an original token x_i by considering a defective sentence $\bar{X}$. By referring to nearby context while restoring a [MASK] token, a model can be trained using bidirectional contextual representation.

NSP is a binary classification task that aims to train by understanding sentence relationships. In the training process, two sentences are concatenated to construct inputs, and these sentences are then selected from an unlabeled monolingual corpus based on a probability. Successive sentences are selected for half of the time, while randomly picked sentences are chosen otherwise. The main objective of NSP is to distinguish whether these input sentences are successive or not. Through this training process, a model can obtain an improved understanding of relationships between sentences.

Multilingual BERT (mBERT) [7] is a BERT-based multilingual model. The same pretraining schemes as BERT (MLM and NSP) are adopted for mBERT. However, unlike BERT, mBERT is trained with a multilingual unlabeled corpus, which is comprised of 104 languages.

The way we adapt mBERT to a QE task is as follows. For the assessment of an entire sentence, we leverage the first hidden representation obtained from the mBERT model. By applying a linear classification head without the activation function, we can obtain the final prediction score of the sentence. Therefore, the sentence assessment score $score_{sentence}$ is derived from an encoded representation of the input sentence, $H = \{h_1, h_2, \ldots, h_m\}$, as shown in Equation (2).

$$score_{sentence} = W \cdot h_1 + b \tag{2}$$

In Equation (2), $W \in \mathbb{R}^{1 \times hidden}$ and $b \in \mathbb{R}^{1 \times 1}$ are trainable parameters where *hidden* indicates the hidden layer size of pretrained mBERT. During the QE training process, the mean squared error (MSE) loss between $score_{sentence}$ and the label score is considered.

3.2. Cross-Lingual Language Model

XLM [8] is a transformer-based model that extends existing language model pretraining methods, which mainly focus on a monolingual language representation, to the multiple language representation. XLM is pretrained through MLM and CLM by leveraging a multilingual unlabeled corpus. To achieve a better multilingual language understanding, TLM, which is a pretraining scheme utilizing a parallel corpus, is applied. Unlike mBERT, NSP is not considered during pretraining.

CLM is a pretraining scheme in which the objective is to model the probability of a word given the previous words in a sentence. This can be described as in Equation (3).

$$\max_{\theta} \sum_{X \in D} \sum_{i=1}^{n} \log P(x_i \mid x_{t<i}, \theta) \quad (3)$$

It can be said that the goal of CLM is to maximize the probability of a token based on preceding tokens. Through this process, a model can obtain an improved language understanding.

TLM is an extension of MLM and improves cross-lingual understanding by utilizing parallel data in the pretraining phase. The source and target sentences of a parallel corpus are first connected, and then some tokens in these sentences are replaced with [MASK] tokens. The training objective of TLM predicts masked tokens the same as in mBERT. However, masked tokens can be predicted by referring to the surrounding context of the masked tokens, as well as sentences from other languages concatenated. It is characterized by TLM that by predicting masked tokens by referencing both languages simultaneously, a representation containing information between languages can be obtained.

This can be described as shown in Equation (4).

$$\max_{\theta} \sum_{(X,Y,\bar{X},\bar{Y}) \in D} \left[\sum_{i \in M_x} \log P(x_i \mid \bar{X} : \bar{Y}, \theta) + \sum_{j \in M_y} \log P(y_j \mid \bar{X} : \bar{Y}, \theta) \right] \quad (4)$$

In Equation (4), $\bar{X} : \bar{Y}$ indicates corrupted input data where $\bar{X}$ is a source sentence component and $\bar{Y}$ is a target sentence component. M_x and M_y are index sets that consist of the indices indicating masked tokens in the source and target sentences, respectively. When predicting a masked word in a source sentence during the training process, a model can refer to the nearby source language context, as well as target sentence. This can encourage the model to acquire a better understanding of multilingual representation. Additionally, to obtaining decent multilingual representation, distinct language embeddings, and respective position embeddings are applied to each language.

XLM utilizes Wikipedia data for the pretraining of various languages. As the amount of established Wikipedia data differs for each language, bias towards high-resource languages can be obtained if such data are utilized without any preprocessing. To alleviate the data imbalance problem, different sampling ratios are applied in the training process. The applied sampling ratios are determined using a multinomial distribution, which is denoted in Equation (5).

$$q_i = \frac{p_i^{\alpha}}{\sum_{j=1}^{N} p_j^{\alpha}} \text{ where } p_i = \frac{n_i}{\sum_{j=1}^{N} n_j} \quad (5)$$

Here, q_i indicates a sampling ratio for the ith language data, with amount n_i, among the total dataset that comprises N languages. α is a hyperparameter that is set to 0.7 for the pretraining of XLM, such that the sampling ratio is increased for low-resource languages and decreased for high-resource languages.

For the XLM-based QE model, the overall training process is similar to Section 3.1, except that positional embeddings that encode absolute positions and language embeddings that indicate the language of each token are applied.

3.3. XLM-RoBERTa

Because XLM learns using Wikipedia, there is a limitation in that data on low resource language is insufficient. In XLM-R [9], the data are expanded to a much larger scale. XLM-R is a multilingual masked language model that adopts large-scale pretraining by utilizing CommonCrawl data [39], which comprises 100 languages. XLM-R gains state-of-the-art performance for cross-lingual classification, question answering, and sequence labeling. Among the three pretraining schemes for XLM, only MLM is utilized for XLM-R training, and MLM proceeds in the same way as XLM. By expanding the model capacity and leveraging larger data sizes than permitted for XLM, XLM-R alleviates the performance degradation caused by the curse of multilinguality.

The curse of multilinguality represents a trade-off between the number of languages in the training data and the model performance at a fixed model capacity. Increasing the number of languages in training data can encourage an improved performance for monolingual and cross-lingual benchmarks to a certain extent because the understanding of low-resource languages is supported by similar high-resource languages. However, if the model capacity is fixed, an excessive number of languages will lead to the overall performance degradation of this method because of the decrease in the per-language capacity. XLM-R alleviates this problem by extending the number of model parameters.

XLM-R adopts a multinomial distribution (5) for applying different sampling ratios to each language. Unlike XLM, XLM-R sets α to 0.3 to strengthen the sampling ratio of low-resource languages. The training process for the XLM-R-based QE model is similar to that of Section 3.1.

3.4. Multilingual BART

BART [40] is a denoising autoencoder that corrupts the text by adding arbitrary noise and trains the model to restore it to the original text. mBART [18] is an extension of BART that has been applied to large monolingual corpora across multiple languages. mBART was trained using a 25-language corpus from CommonCrawl data (CC25).

BART utilizes 5 pretraining schemes leveraging a monolingual corpus: token masking, token deletion, text infilling, document rotation, and sentence permutation. Among these pretraining schemes, mBART adopts text infilling and sentence permutation. In the case of text filling, unlike MLM in which one token in the original sentence is replaced with one [MASK] token, spans of tokens are replaced with one masked token. The total number of selected tokens is 35% of the entire sentence, and the length of the masked token is determined based on the Poisson distribution, which is described in Equation (6).

$$f(n : \lambda) = \frac{\lambda^n e^{-\lambda}}{n!} \tag{6}$$

Here, $f(n : \lambda)$ indicates the probability of selecting n as the masking length. mBART sets λ to 3.5 for pretraining. By training to reconstruct masked sentences, which are generated by text infilling, a model can be trained for bidirectional contextual understanding, as well as to determine how many tokens should be restored from a single mask token.

In the case of sentence permutation, the text is corrupted by changing the order of the sentences within each instance. In the process of restoring the noise injected by sentence permutation to the original text, the model can understand information about the relationship between sentences.

Similar to XLM and XLM-R, mBART adopts an up-down sampling method to achieve improved training for low-resource languages. The sampling ratio λ_i applied to the ith language data is provided by Equation (7).

$$\lambda_i = \frac{1}{p_i} \frac{p_i^{\alpha}}{\sum_{j=1}^{N} p_j^{\alpha}} \tag{7}$$

Here, p_i is the percentage of each language in the total dataset. The amount of training data for each language are rebalanced according to Equation (7), and, therefore, sampling from high-resource languages is relatively suppressed while sampling from low-resource languages is encouraged. The training process of the QE model leveraging mBART is similar to that of Section 3.1, wherein the same input structure as in pretraining is utilized.

4. Brief Introduction of the WMT20 QE Sub-Tasks

4.1. Sub-Task 1

Sub-task 1 is a sentence-level direct assessment task. This task consists of scoring MT output according to a perceived quality score called direct assessment. A limitation of human translation error rate (HTER) [41] is that it does not capture the extent to which MT errors affect the overall quality of a sentence. The objective of sub-task 1 is to measure the overall quality of sentences through direct assessment (DA) by translation experts. One of the goals of QE in relation to this task is to investigate the relationship between a model for predicting DA scores and a model trained to predict post-editing tasks [15]. The DA score is a value obtained by evaluating the quality of the MT output from 0 to 100 by at least three professional translators. Using a total of 7K training data and 1K evaluation data, systems participating in this sub-task measure quality by predicting the mean z-standardized DA score of the MT output.

4.2. Sub-Task 2

Sub-task 2 is word- and sentence-level post-editing efforts. The objective of sub-task 2 is to improve post-editing by tagging which tokens have been mistranslated, along with the overall quality of the sentence. At the word level, this task consists of evaluating whether the translation was successful for each token in the MT output and source sentence based on the human post-edited sentences. The tokens of the source and target sides are tagged as OK or BAD. In the case of the target sentences, a gap tag is added considering the case of missing words between the tokens. If the number of tokens in the target sentence is N, the total number of tag tokens is 2N+1. Participating systems predict tags for MT output tokens and source sentence tokens.

Similar to sub-task 1, a sentence-level post-editing effort task is used to measure the quality score for the MT output based on the human translation error rate (HTER) [41]. HTER is similar to the translation error rate (TER), wherein the TER compares the MT output with a reference translation and counts how many edits (substitutions, deletions, and insertions) must be performed to obtain a correct sentence. This value divided by the reference length is the TER score. HTER differs from TER in that humans create new reference translations for the MT output. Using these new reference translations can lead to correct sentences with minimal modifications compared to the use of other reference translations. Referring to the source sentence and the MT output, the participating system predicts the quality of the MT output sentence based on the HTER.

5. Question 1: Which mPLM Is Best for QE Tasks?

5.1. Dataset Details

In this study, we conducted experiments concerning sub-tasks 1 and 2 at the sentence-level of WMT20 based on various mPLMs. We experimented using the English–German language pair and used train, dev and test data provided by WMT20 (http://www.statmt.org/wmt20/quality-estimation-task.html, accessed on 15 July 2021). Table 1 shows a summary of the data for each sub-task.

In the case of sub-task 1, there is a total of 7k training data, and the numbers of source and MT output tokens are 98,127 and 97,453, respectively. The average of the mean z-standardized DA score is −0.008 and the median is 0.162. The development and test data consist of a total of 1K data, and there are approximately 14K source and MT output tokens. The development and test data provide average scores of −0.049 and 0.040, and the respective median scores are slightly higher at 0.211 and 0.319.

In the case of sub-task 2 at the sentence-level, the number of sentences is 7K in the training data and 1K in each of the development and test data, as in sub-task 1. The average HTER score is distributed around 0.3, and the median value either does not significantly differ or is slightly lower than the average value. HTER is centered around values lower than the error rate of 0.5.

Table 1. Summary of the QE dataset. We denote the number of instances in each dataset as # Instance. # SRC Token and # MT Token refer to the number of tokens in source- and target-side sentences for each dataset, respectively.

	Sub-Task 1			Sub-Task 2		
	Train	**Dev**	**Test**	**Train**	**Dev**	**Test**
# Instance	7000	1000	1000	7000	1000	1000
# SRC Token	98,127	14,102	14,043	114,980	16,519	16,371
# MT Token	97,453	14,003	14,019	112,342	16,160	16,154
Average Score	−0.008	−0.049	0.040	0.318	0.312	0.312
Median Score	0.162	0.211	0.319	0.3	0.295	0.286

5.2. Model Details

We conducted a finetuning performance comparison using a total of 9 models including XLM-R base, XLM-R large, mBERT, mBART, XLM-CLM, XLM-MLM, XLM-MLM-17, XLM-MLM-100, and XLM-TLM. English–German was used as the language pair for this experiment, and performance comparisons were conducted for each mPLM at sub-task 1 and sub-task 2 sentence-levels. These models are described as follows:

- **XLM-R-base**: Pretraining was performed with 220M parameters, 12 layers, 8 heads, and 768 hidden states.
- **XLM-R-large**: Pretraining was performed using 550M parameters. The hidden states were expanded to 1024, and 24 layers, and 16 heads were used, which is twice the scale of the base model.
- **mBERT**: The model parameters of mBERT were 110M, 12 layers, 768 hidden states, and 12 heads.
- **mBART**: mBART was pretrained with 610M parameters, 24 layers, 1024 hidden states, and 16 heads.
- **XLM-CLM**: A pretrained CLM for English and German. In total, 6 layers, 1024 hidden states, and 8 heads were used.
- **XLM-MLM**: A pretrained MLM for English and German. In total, 6 layers, 1024 hidden states, and 8 heads were used.
- **XLM-MLM-17**: Pretraining was conducted by expanding the MLM into 17 languages. It was trained using 570M parameters, 16 layers, 1280 hidden states, and 16 heads.
- **XLM-MLM-100**: Pretraining was conducted by expanding the MLM into 100 languages. It was trained using 570M parameters, 16 layers, 1280 hidden states, and 16 heads.
- **XLM-TLM**: TLM was performed for 15 languages. In total, 12 layers, 1024 hidden states, and 8 heads were used.

We performed finetuning using the pretrained model released in HuggingFace's transformers library [42]. We did not proceed with additional pretraining and data augmentation so that the pure performances of the mPLMs could be objectively evaluated and compared in the QE task.

In preprocessing, we performed subword tokenization using the tokenizer provided for each model in HuggingFace. For the model input, we added segment embeddings for mBERT, listing tokens separated by 0 and 1 to give a distinction between sentence 1 and sentence 2. XLM has added a position embedding that gives a number corresponding to the token index for each source sentence and MT output, as well as a language embedding that is segmented by a unique number for each language.

As a training procedure for finetuning, we first load mPLMs to initialize the parameters. After that, additional embeddings for each model are put as input to the model along with the sentences concatenated with the source and target sentences. We put the output corresponding to the position of the [CLS] token among the last hidden states as an input to the linear classifier and measured the loss between the predicted value and the label. We use the mean squared error (MSE) loss as the loss function.

We found that the model has a diverse range of performance fluctuations depending on the seed value, and we attempted to reduce the effect of the seed value on the general performance of the model. To achieve this, we conduct five experiments using the same model and compare the average values, as well as the minimum and maximum performance values, thereby increasing the reliability of the experimental results.

5.3. Experimental Results for Question 1

5.3.1. Sub-Task 1

To check which model out of various mPLMs performs well for the QE task, we raise question 1, and proceed with finetuning using mPLMs. The experimental results for the QE of sub-task 1 (i.e., the direct assessment at the sentence-level) are shown in Table 2.

Table 2. mPLM finetuning results for the test set of the WMT20 sub-task 1.

	Pearson			MAE			RMSE		
	Max	Min	Average	Min	Max	Average	Min	Max	Average
XLM-R-base	0.380	0.280	0.328	0.459	0.479	0.473	0.648	0.679	0.665
XLM-R-large	0.338	0.242	0.298	0.480	0.520	0.495	0.685	0.713	0.698
mBERT	0.407	0.322	0.382	0.452	0.468	0.458	0.642	0.672	0.655
mBART	0.402	0.306	0.351	0.465	0.534	0.490	0.642	0.729	0.677
XLM-CLM	0.296	0.168	0.253	0.474	0.516	0.489	0.683	0.703	0.691
XLM-MLM	0.219	0.192	0.206	0.493	0.526	0.503	0.693	0.728	0.708
XLM-MLM-17	0.318	0.143	0.253	0.465	0.525	0.490	0.670	0.731	0.696
XLM-MLM-100	0.256	0.191	0.232	0.482	0.536	0.498	0.690	0.702	0.695
XLM-TLM	0.442	0.336	0.394	0.451	0.683	0.517	0.631	0.805	0.681

As a result of the experiment, XLM-TLM showed the highest performance for sub-task 1 with a Pearson correlation coefficient of 0.442. In terms of the minimum and average performances, this system consistently demonstrated the highest performance compared to the other models. To investigate the cause of this result, we need to focus on the input data of XLM-TLM in the pretraining process.

The XLM-TLM model utilizes parallel data during pretraining and can refer to the context of either side when predicting the source- and target-side masked words. Likewise, in the QE field, the concatenating sentences of the source and target language are provided as an input to the model. This is similar to the form of the input for the XLM-TLM model in that it provides sentences in both languages as the input, while the other models use the mono data of multiple languages. According to Lample and Conneau [8], when predicting a masked word during XLM-TLM learning, the model can be encouraged to align the source and target language representations by attending the translated sentence along with the surrounding masked word. Therefore, when using the aligned representation derived between the source and target languages in the XLM-TLM model for QE, it is possible to infer what part of the translated sentence is wrong. The model with the second highest average performance is the mBERT model. This model provided approximately 0.012 less than that of the first-ranked model and demonstrates a comparable performance. mBART did not show a strong performance in the regression task, but the maximum value only showed a difference of about 0.005 compared to the mBERT model. Both models apply various noising schemes during pretraining, and it can be predicted that this strategy will help improve their performance.

In the case of XLM-R-large, many research groups that participated in WMT20 used this model; however, for sub-task 1, it was not ranked high. When comparing the average Pearson correlation coefficients of the models based on XLM, XLM-MLM-17 was 0.021 higher than that of XLM-MLM-100, and XLM-MLM, which learned only English and German, showed the lowest performance. XLM-MLM-17 and XLM-MLM-100 are approximately twice the size of XLM-MLM considering the number of layers and hidden states, etc. and the languages were also expanded to 17 and 100 languages, respectively. It can be inferred that the number of languages and model capacity helped to improve the performance for QE.

To answer subtask 2, we refer back to the question we posed. Which mPLM is best for QE tasks? For the question, XLM-TLM model that learned cross-lingual understanding performed the best in sub-task 1.

5.3.2. Sub-Task 2

The finetuning results for sub-task 2 (sentence-level post editing effort) are shown in Table 3. High performances were achieved in the descending order of XLM-TLM, XLM-R-large, mBART, XLM-R-base, mBERT, XLM-MLM-17, XLM-MLM-100, XLM-MLM, and XLM-CLM based on the average Pearson correlation coefficient. As a result of this experiment, XLM-TLM showed the highest performance based on the average, minimum, and maximum Pearson correlation coefficients, similar to the previous experimental results for sub-task 1. As analyzed in sub-task 1, because XLM-TLM was induced to learn alignment information for language pairs using parallel corpus, it can be predicted that this process contributes significantly to its performance improvement for QE, which requires knowledge of relationships between languages. In sub-task 2, the XLM-R-large model showed the best performance after XLM-TLM. A fairly comparable performance was demonstrated with an average Pearson correlation coefficient of 0.498. XLM-R-large is the latest model among the mPLM models considered in this study. As mentioned in Section 3.3, a state-of-the-art performance among cross-lingual models was achieved by expanding the number of parameters considering the large amount of data and the curse of multilinguality. Nevertheless, XLM-R did not learn the relationship between the source and target sentences because it learned the mono corpus in an unsupervised manner. In QE, the source sentence and MT output are referenced together to determine which part has been incorrectly translated, and, therefore, this characteristic did not produce an optimal effect compared to XLM-TLM. Although mBART is a sequence-to-sequence model, it ranks third in the regression task with a higher performance than all XLM models. As an extension of MLM, mBART uses a pretraining scheme referred to as text infilling and sentence permutation, and an average Pearson correlation coefficient of 0.463 was obtained. This result was significantly higher than those of XLM-MLM (0.334), XLM-MLM-17 (0.415), and XLM-MLM-100 (0.409), which used only MLM. Therefore, it can be confirmed that the additional strategy of mBART had a positive effect on the improvement of QE performance during finetuning. mBERT showed an average Pearson correlation coefficient of 0.417 in sub-task 2 and did not demonstrate a very high performance when compared with the sub-task 1 results. Considering the comparison of the various XLMs, XLM-MLM-17 performed slightly better than XLM-MLM-100 (as in sub-task 1), while XLM-CLM ranked lower than XLM-MLM, which exhibited the lowest performance in sub-task 1.

Table 3. mPLM finetuning results for the test set of the WMT20 sub-task 2.

	Pearson			MAE			RMSE		
	Max	Min	Average	Min	Max	Average	Min	Max	Average
XLM-R-base	0.456	0.438	0.448	0.146	0.156	0.150	0.189	0.204	0.195
XLM-R-large	0.507	0.489	0.498	0.141	0.155	0.145	0.178	0.204	0.186
mBERT	0.435	0.389	0.417	0.149	0.182	0.160	0.189	0.230	0.204
mBART	0.475	0.452	0.463	0.142	0.148	0.144	0.179	0.195	0.184
XLM-CLM	0.309	0.275	0.298	0.158	0.161	0.159	0.196	0.200	0.198
XLM-MLM	0.358	0.303	0.334	0.156	0.160	0.158	0.194	0.199	0.197
XLM-MLM-17	0.433	0.408	0.415	0.149	0.157	0.154	0.188	0.192	0.190
XLM-MLM-100	0.421	0.381	0.409	0.152	0.164	0.158	0.190	0.207	0.198
XLM-TLM	0.522	0.498	0.510	0.152	0.222	0.177	0.199	0.273	0.227

To answer subtask 2, we refer back to the question we posed. Which mPLM is best for QE tasks? For the question, we can explain that the XLM-TLM model also performed best in sub-task 2.

6. Question 2: Does the Input Order of the Source Sentence and the MT Output Sentence Affect the Performance of the Model?

6.1. Revisiting the QE Input Structure

In this section, we investigate the differences between the input structures used for QE training. Existing QE studies generally construct an input using the following shapes: *[BOS] Source sentence [EOS] [EOS] MT output [EOS]* or *[BOS] MT output [EOS] [EOS] Source sentence [EOS]*, where the beginning of sentence (BOS) and end of sentence (EOS) tokens can be viewed as *[CLS]* and *[SEP]*, respectively, depending on the pretraining methods that were used. Previously, Baek et al. [13], Fomicheva et al. [14], Ranasinghe et al. [35] adopted a prior structure as an input, while Moura et al. [4], Kepler et al. [33] adopted a posterior structure. Although decent performances can be achieved by adopting these structures, sufficient investigations concerning the selection of an input structure have not been conducted. In other words, clear criteria for constructing an adequate input structure have not yet been presented. Here, we focus on the inconsistent input structures utilized in current QE studies and quantitatively analyze the differences derived from adopting different input structures.

6.2. Experimental Results for Question 2

In order to check whether the order of the input sentence affects the performance while performing QE finetuning, we raise question 2 and compare the sentence order with the reversed sentence order when constructing the input sequence. The experimental results for sub-task 1 are shown in Table 4. As a result of this experiment, it can be observed that the model performance changes by simply reversing the order of the input sentence. In this table, we denote Avg Diff as the difference between the average Pearson correlation coefficients of the original input and reverse orders. As can be seen from the Avg Diff values, when the input sentence order was reversed, the average Pearson correlation coefficient of XLM-R-large improved by +0.032, while that of XLM-MLM-100 improved by +0.008. However, for all other models, the performance deteriorated when the order of the input sentences was reversed. Likewise, in Figure 1, it was confirmed that the overall reversed order input sentences in sub-task 1 did not help to improve the performance of the model.

Table 4. Results of mPLM finetuning with inverted inputs for the test set of WMT20 sub-task 1.

	Pearson				MAE			RMSE		
	Max	**Min**	**Average**	**Avg Diff**	**Min**	**Max**	**Average**	**Min**	**Max**	**Average**
XLM-R-base	0.365	0.272	0.326	−0.002	0.462	0.495	0.481	0.653	0.698	0.670
XLM-R-large	0.394	0.260	0.330	+0.032	0.447	0.508	0.479	0.644	0.729	0.681
mBERT	0.402	0.106	0.278	−0.104	0.453	0.553	0.498	0.648	0.762	0.700
mBART	0.388	0.277	0.346	−0.005	0.436	0.543	0.478	0.664	0.693	0.674
XLM-CLM	0.268	0.147	0.197	−0.056	0.483	0.515	0.502	0.688	0.714	0.698
XLM-MLM	0.250	0.128	0.177	−0.029	0.517	0.557	0.540	0.694	0.751	0.727
XLM-MLM-17	0.267	0.172	0.230	−0.023	0.482	0.502	0.491	0.682	0.713	0.693
XLM-MLM-100	0.314	0.189	0.260	+0.028	0.503	0.587	0.544	0.666	0.748	0.709
XLM-TLM	0.234	0.141	0.193	−0.201	0.563	1.115	0.896	0.739	1.237	1.061

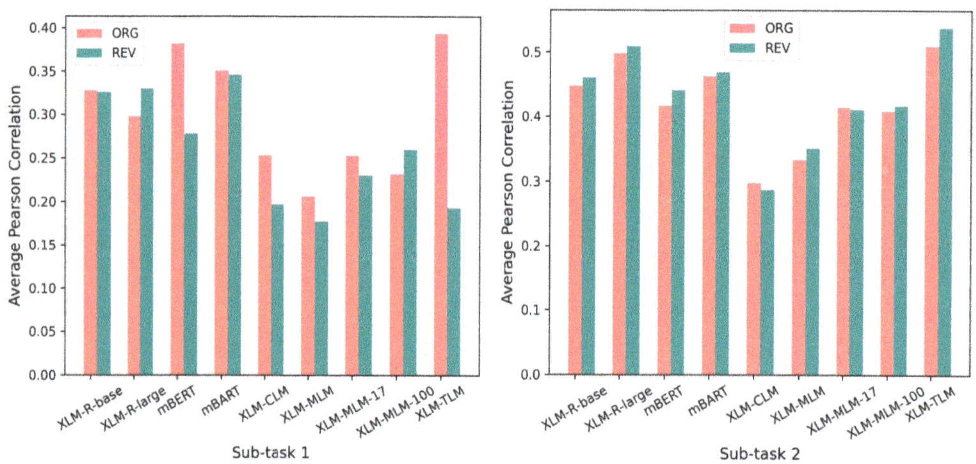

Figure 1. Comparison of the average Pearson correlation coefficients of original and reverse order inputs in sub-tasks 1 and 2.

Conversely, in the case of sub-task 2, the result of reversing the input sentences provided a better overall performance. As can be seen in Table 5 and Figure 1, only two models of XLM-CLM and XLM-MLM-17 declined in performance based on the average Pearson correlation coefficient, while all other models consistently exhibited improved performances. In particular, the range of performance fluctuations was high in both XLM-TLM and mBERT. These two models also showed the highest variation in sub-task 1, and it can, therefore, be said that these models respond most sensitively to the input sentence order. The models with the lowest performance fluctuations were XLM-MLM-17 and mBART. In the case of mBART, there was little change in performance even in sub-task 1, and there was no significant change in the performance in response to the varied input structure.

Table 5. Results of mPLM finetuning with inverted inputs for the test set of WMT20 sub-task 2.

	Pearson				MAE			RMSE		
	Max	Min	Average	Avg Diff	Min	Max	Average	Min	Max	Average
XLM-R-base	0.464	0.453	0.460	+0.012	0.144	0.153	0.148	0.184	0.199	0.191
XLM-R-large	0.523	0.501	0.509	+0.011	0.140	0.144	0.142	0.178	0.188	0.183
mBERT	0.449	0.434	0.441	+0.024	0.147	0.179	0.162	0.185	0.229	0.207
mBART	0.478	0.463	0.469	+0.006	0.141	0.151	0.145	0.179	0.196	0.187
XLM-CLM	0.297	0.283	0.287	−0.011	0.159	0.162	0.160	0.197	0.205	0.199
XLM-MLM	0.364	0.333	0.351	+0.017	0.153	0.159	0.156	0.193	0.200	0.196
XLM-MLM-17	0.420	0.405	0.411	−0.004	0.154	0.218	0.172	0.190	0.273	0.217
XLM-MLM-100	0.442	0.405	0.417	+0.008	0.151	0.183	0.161	0.187	0.220	0.196
XLM-TLM	0.552	0.526	0.538	+0.028	0.156	0.168	0.163	0.204	0.218	0.212

We refer again to the question we asked. Does the input order of the source sentence and the MT output sentence affect the performance of the model? Through these experiments, we determined that the performance fluctuation of the input order varies depending on the sub-task. To the question, we can answer that the structure of the input is a factor that affects the performance of the model, and it must, therefore, be considered before conducting such experiments.

7. Conclusions

Most recent studies of QE apply data augmentation with finetuning based on state-of-the-art large scale mPLM, such as XLM-R, to obtain a high performance for a WMT shared task. In this study, unlike typical QE research that focused on the competition involving a shared task, we conducted a pure performance comparison between various mPLMs. As a result of the experiments, we confirmed that the XLM-TLM model performed best on both sub-tasks, and that the induced learning of alignment between languages during pre-training had a positive impact. Additionally, we conducted experiments using mBART for the first time, and its additional noising schemes had a positive effect on QE research. Therefore, we confirmed the feasibility of using the mBART model in further QE research. We demonstrated that the order of the input sequence between the source sentence and its MT output can affect the model performance. In the future, we will further investigate data-centric issues that are not model-based [43,44]. By filtering data based on the HTER score, we will explore which score ranges contribute significantly to the performance of a model and provide a basis for future data-centric research on QE. In addition, we plan to conduct an in-depth study on low resource language QE. We plan to study a methodology that can automatically generate data based on a semi-supervised learning method.

Author Contributions: Conceptualization, C.P.; methodology/software, S.E.; validation, S.E. and H.M.; formal analysis, S.E. and C.P.; investigation, S.E. and H.M.; review and editing, H.M. and J.S.; supervision/project administration, C.P.; funding acquisition, H.L. All authors have read and agreed to the published version of the manuscript.

Funding: This research was supported by the MSIT (Ministry of Science and ICT), Korea, under the ITRC (Information Technology Research Center) support program (IITP-2018-0-01405) supervised by the IITP (Institute for Information & Communications Technology Planning & Evaluation) and the MSIT, Korea, under the ICT Creative Consilience program (IITP-2021-2020-0-01819) supervised by the IITP. Additionally, this work was supported by Institute for Information & communications Technology Planning & Evaluation (IITP) grant funded by the Korea government (MSIT) (No. 2020-0-00368, A Neural-Symbolic Model for Knowledge Acquisition and Inference Techniques).

Institutional Review Board Statement: Not applicable.

Informed Consent Statement: Not applicable.

Data Availability Statement: Publicly available datasets were analyzed in this study. The data can be found here: WMT20 English-German QE dataset: http://www.statmt.org/wmt20/quality-estimation-task.html (accessed on 15 July 2021).

Conflicts of Interest: The authors declare no conflicts of interest.

References

1. Specia, L.; Shah, K.; De Souza, J.G.; Cohn, T. QuEst-A translation quality estimation framework. In Proceedings of the 51st Annual Meeting of the Association for Computational Linguistics: System Demonstrations, Sofia, Bulgaria, 4–9 August 2013; pp. 79–84.
2. Specia, L.; Raj, D.; Turchi, M. Machine translation evaluation versus quality estimation. *Mach. Transl.* **2010**, *24*, 39–50. [CrossRef]
3. do Carmo, F.; Shterionov, D.; Moorkens, J.; Wagner, J.; Hossari, M.; Paquin, E.; Schmidtke, D.; Groves, D.; Way, A. A review of the state-of-the-art in automatic post-editing. *Mach. Transl.* **2020**, 1–43. [CrossRef]
4. Moura, J.; Vera, M.; van Stigt, D.; Kepler, F.; Martins, A.F. Ist-unbabel participation in the wmt20 quality estimation shared task. In Proceedings of the Fifth Conference on Machine Translation, Online, 19–20 November 2020; pp. 1029–1036.
5. Nakamachi, A.; Shimanaka, H.; Kajiwara, T.; Komachi, M. Tmuou submission for wmt20 quality estimation shared task. In Proceedings of the Fifth Conference on Machine Translation, Online, 19–20 November 2020; pp. 1037–1041.
6. Rubino, R. Nict kyoto submission for the wmt'20 quality estimation task: Intermediate training for domain and task adaptation. In Proceedings of the Fifth Conference on Machine Translation, Online, 19–20 November 2020; pp. 1042–1048.
7. Pires, T.; Schlinger, E.; Garrette, D. How multilingual is multilingual bert? *arXiv* **2019**, arXiv:1906.01502.
8. Lample, G.; Conneau, A. Cross-lingual language model pretraining. *arXiv* **2019**, arXiv:1901.07291.
9. Conneau, A.; Khandelwal, K.; Goyal, N.; Chaudhary, V.; Wenzek, G.; Guzmán, F.; Grave, E.; Ott, M.; Zettlemoyer, L.; Stoyanov, V. Unsupervised cross-lingual representation learning at scale. *arXiv* **2019**, arXiv:1911.02116.
10. Conneau, A.; Lample, G.; Rinott, R.; Williams, A.; Bowman, S.R.; Schwenk, H.; Stoyanov, V. XNLI: Evaluating cross-lingual sentence representations. *arXiv* **2018**, arXiv:1809.05053.
11. Lewis, P.; Oğuz, B.; Rinott, R.; Riedel, S.; Schwenk, H. Mlqa: Evaluating cross-lingual extractive question answering. *arXiv* **2019**, arXiv:1910.07475.
12. Lee, D. Two-phase cross-lingual language model fine-tuning for machine translation quality estimation. In Proceedings of the Fifth Conference on Machine Translation, Online, 19–20 November 2020; pp. 1024–1028.
13. Baek, Y.; Kim, Z.M.; Moon, J.; Kim, H.; Park, E. Patquest: Papago translation quality estimation. In Proceedings of the Fifth Conference on Machine Translation, Online, 19–20 November 2020; pp. 991–998.
14. Fomicheva, M.; Sun, S.; Yankovskaya, L.; Blain, F.; Chaudhary, V.; Fishel, M.; Guzmán, F.; Specia, L. Bergamot-latte submissions for the wmt20 quality estimation shared task. In Proceedings of the Fifth Conference on Machine Translation, Online, 19–20 November 2020.
15. Specia, L.; Blain, F.; Fomicheva, M.; Fonseca, E.; Chaudhary, V.; Guzmán, F.; Martins, A.F.T. Findings of the WMT 2020 Shared Task on Quality Estimation. In Proceedings of the Fifth Conference on Machine Translation, Online, 19–20 November 2020; pp. 743–764.
16. Zhou, L.; Ding, L.; Takeda, K. Zero-shot translation quality estimation with explicit cross-lingual patterns. *arXiv* **2020**, arXiv:2010.04989.
17. Ranasinghe, T.; Orasan, C.; Mitkov, R. TransQuest: Translation Quality Estimation with Cross-lingual Transformers. In Proceedings of the 28th International Conference on Computational Linguistics, Barcelona, Spain, 12 December 2020; pp. 5070–5081. [CrossRef]
18. Liu, Y.; Gu, J.; Goyal, N.; Li, X.; Edunov, S.; Ghazvininejad, M.; Lewis, M.; Zettlemoyer, L. Multilingual denoising pre-training for neural machine translation. *Trans. Assoc. Comput. Linguist.* **2020**, *8*, 726–742. [CrossRef]
19. Papineni, K.; Roukos, S.; Ward, T.; Zhu, W.J. BLEU: A method for automatic evaluation of machine translation. In Proceedings of the 40th Annual Meeting on Association for Computational Linguistics, Philadelphia, PA, USA, 7–12 July 2002; pp. 311–318.
20. Banerjee, S.; Lavie, A. METEOR: An automatic metric for MT evaluation with improved correlation with human judgments. In Proceedings of the Acl Workshop on Intrinsic and Extrinsic Evaluation Measures for Machine Translation and/or Summarization, Ann Arbor, MI, USA, 29 June 2005; pp. 65–72.
21. Shah, K.; Cohn, T.; Specia, L. A bayesian non-linear method for feature selection in machine translation quality estimation. *Mach. Transl.* **2015**, *29*, 101–125. [CrossRef]
22. Cohn, T.; Specia, L. Modelling annotator bias with multi-task gaussian processes: An application to machine translation quality estimation. In Proceedings of the 51st Annual Meeting of the Association for Computational Linguistics, Sofia, Bulgaria, 4–9 August 2013; Volume 1, pp. 32–42.
23. Hardmeier, C.; Nivre, J.; Tiedemann, J. Tree kernels for machine translation quality estimation. In Proceedings of the Seventh Workshop on Statistical Machine Translation, Montréal, QC, Canada, 7–8 June 2012; pp. 109–113.
24. Soricut, R.; Bach, N.; Wang, Z. The SDL language weaver systems in the WMT12 quality estimation shared task. In Proceedings of the Seventh Workshop on Statistical Machine Translation, Montreal, QC, Canada, 7–8 June 2012; pp. 145–151.

25. Moreau, E.; Vogel, C. Quality estimation: An experimental study using unsupervised similarity measures. In Proceedings of the Seventh Workshop on Statistical Machine Translation, Montréal, QC, Canada, 7–8 June 2012; pp. 120–126.
26. Felice, M.; Specia, L. Linguistic features for quality estimation. In Proceedings of the Seventh Workshop on Statistical Machine Translation, Montréal, QC, Canada, 7–8 June 2012; pp. 96–103.
27. Scarton, C.; Specia, L. Exploring consensus in machine translation for quality estimation. In Proceedings of the Ninth Workshop on Statistical Machine Translation, Baltimore, MD, USA, 26–27 June 2014; pp. 342–347.
28. Luong, N.Q.; Lecouteux, B.; Besacier, L. LIG system for WMT13 QE task: Investigating the usefulness of features in word confidence estimation for MT. In Proceedings of the 8th Workshop on Statistical Machine Translation, Sofia, Bulgaria, 8–9 August 2013; pp. 386–391.
29. Kim, H.; Lee, J.H. Recurrent neural network based translation quality estimation. In Proceedings of the First Conference on Machine Translation, Berlin, Germany, 11–12 August 2016; Volume 2, pp. 787–792.
30. Patel, R.N. Translation quality estimation using recurrent neural network. *arXiv* **2016**, arXiv:1610.04841.
31. Kim, H.; Lee, J.H.; Na, S.H. Predictor-estimator using multilevel task learning with stack propagation for neural quality estimation. In Proceedings of the Second Conference on Machine Translation, Copenhagen, Denmarkm, 7–8 September 2017; pp. 562–568.
32. Wang, J.; Fan, K.; Li, B.; Zhou, F.; Chen, B.; Shi, Y.; Si, L. Alibaba submission for WMT18 quality estimation task. In Proceedings of the Third Conference on Machine Translation: Shared Task Papers, Brussels, Belgium, 31 October–1 November 2018; pp. 809–815.
33. Kepler, F.; Trénous, J.; Treviso, M.; Vera, M.; Góis, A.; Farajian, M.A.; Lopes, A.V.; Martins, A.F. Unbabel's Participation in the WMT19 Translation Quality Estimation Shared Task. *arXiv* **2019**, arXiv:1907.10352.
34. Kim, H.; Lim, J.H.; Kim, H.K.; Na, S.H. QE BERT: bilingual BERT using multi-task learning for neural quality estimation. In Proceedings of the Fourth Conference on Machine Translation, Florence, Italy, 1–2 August 2019; Volume 3, pp. 85–89.
35. Ranasinghe, T.; Orasan, C.; Mitkov, R. TransQuest at WMT2020: Sentence-Level Direct Assessment. *arXiv* **2020**, arXiv:2010.05318.
36. Wang, M.; Yang, H.; Shang, H.; Wei, D.; Guo, J.; Lei, L.; Qin, Y.; Tao, S.; Sun, S.; Chen, Y.; et al. Hw-tsc's participation at wmt 2020 quality estimation shared task. In Proceedings of the Fifth Conference on Machine Translation, Online, 19–20 November 2020; pp. 1056–1061.
37. Devlin, J.; Chang, M.W.; Lee, K.; Toutanova, K. Bert: Pre-training of deep bidirectional transformers for language understanding. *arXiv* **2018**, arXiv:1810.04805.
38. Vaswani, A.; Shazeer, N.; Parmar, N.; Uszkoreit, J.; Jones, L.; Gomez, A.N.; Kaiser, Ł.; Polosukhin, I. Attention is all you need. In Proceedings of the Advances in Neural Information Processing Systems, Long Beach, CA, USA, 4–9 December 2017; pp. 5998–6008.
39. Wenzek, G.; Lachaux, M.A.; Conneau, A.; Chaudhary, V.; Guzman, F.; Joulin, A.; Grave, E. Ccnet: Extracting high quality monolingual datasets from web crawl data. *arXiv* **2019**, arXiv:1911.00359.
40. Lewis, M.; Liu, Y.; Goyal, N.; Ghazvininejad, M.; Mohamed, A.; Levy, O.; Stoyanov, V.; Zettlemoyer, L. Bart: Denoising sequence-to-sequence pre-training for natural language generation, translation, and comprehension. *arXiv* **2019**, arXiv:1910.13461.
41. Snover, M.; Dorr, B.; Schwartz, R.; Micciulla, L.; Makhoul, J. A study of translation edit rate with targeted human annotation. In Proceedings of the Association for Machine Translation in the Americas, Cambridge, MA, USA, 8–12 August 2006; Volume 200.
42. Wolf, T.; Debut, L.; Sanh, V.; Chaumond, J.; Delangue, C.; Moi, A.; Cistac, P.; Rault, T.; Louf, R.; Funtowicz, M.; et al. HuggingFace's Transformers: State-of-the-art natural language processing. *arXiv* **2019**, arXiv:1910.03771.
43. Park, C.; Yang, Y.; Park, K.; Lim, H. Decoding strategies for improving low-resource machine translation. *Electronics* **2020**, *9*, 1562. [CrossRef]
44. Lee, C.; Yang, K.; Whang, T.; Park, C.; Matteson, A.; Lim, H. Exploring the Data Efficiency of Cross-Lingual Post-Training in Pretrained Language Models. *Appl. Sci.* **2021**, *11*, 1974. [CrossRef]

applied sciences

MDPI

Article

English–Welsh Cross-Lingual Embeddings

Luis Espinosa-Anke [1,*,†], Geraint Palmer [2,†], Padraig Corcoran [1], Maxim Filimonov [1], Irena Spasić [1] and Dawn Knight [3]

[1] School of Computer Science and Informatics, Cardiff University, Cardiff CF24 3AA, UK; corcoranp@cardiff.ac.uk (P.C.); filimonovm@cardiff.ac.uk (M.F.); spasici@cardiff.ac.uk (I.S.)
[2] School of Mathematics, Cardiff University, Cardiff CF24 4AG, UK; palmergi1@cardiff.ac.uk
[3] School of English, Communication and Philosophy, Cardiff University, Cardiff CF10 3EU, UK; knightd5@cardiff.ac.uk
* Correspondence: espinosa-ankel@cardiff.ac.uk
† These authors contributed equally to this work.

Abstract: Cross-lingual embeddings are vector space representations where word translations tend to be co-located. These representations enable learning transfer across languages, thus bridging the gap between data-rich languages such as English and others. In this paper, we present and evaluate a suite of cross-lingual embeddings for the English–Welsh language pair. To train the bilingual embeddings, a Welsh corpus of approximately 145 M words was combined with an English Wikipedia corpus. We used a bilingual dictionary to frame the problem of learning bilingual mappings as a supervised machine learning task, where a word vector space is first learned independently on a monolingual corpus, after which a linear alignment strategy is applied to map the monolingual embeddings to a common bilingual vector space. Two approaches were used to learn monolingual embeddings, including word2vec and fastText. Three cross-language alignment strategies were explored, including cosine similarity, inverted softmax and cross-domain similarity local scaling (CSLS). We evaluated different combinations of these approaches using two tasks, bilingual dictionary induction, and cross-lingual sentiment analysis. The best results were achieved using monolingual fastText embeddings and the CSLS metric. We also demonstrated that by including a few automatically translated training documents, the performance of a cross-lingual text classifier for Welsh can increase by approximately 20 percent points.

Keywords: natural language processing; distributional semantics; machine learning; language model; word embeddings; machine translation; sentiment analysis

Citation: Espinosa-Anke, L.; Palmer, G.; Corcoran, P.; Filimonov, M.; Spasić, I.; Knight, D. English–Welsh Cross-Lingual Embeddings. *Appl. Sci.* **2021**, *11*, 6541. https://doi.org/10.3390/app11146541

Academic Editors: Arturo Montejo-Ráez and Salud María Jiménez-Zafra

Received: 18 May 2021
Accepted: 5 July 2021
Published: 16 July 2021

Publisher's Note: MDPI stays neutral with regard to jurisdictional claims in published maps and institutional affiliations.

1. Introduction

A popular research direction in current natural language processing (NLP) research consists of learning vector space representations of words for two or more languages, and then applying some kind of transformation to one of the spaces such that "cross-lingual synonyms", i.e., words with the same meaning across languages, are assigned similar vector space representations. The applications of these cross-lingual embeddings into downstream tasks is indisputable today, ranging from information retrieval [1], entity linking [2], text classification [3,4], as well as natural language inference or lexical semantics [5]. These cross-lingual embeddings are often learned and evaluated for language pairs, for which there is either a good availability of parallel or comparable text corpora, supervision signal, or, at the least, large enough raw but non-aligned corpora for each language (see, e.g., Mikolov et al. [6], Conneau et al. [7], Artetxe et al. [8,9]).

However, the availability of cross-lingual mappings between resource-rich and resource-poor languages still constitutes a challenge [10]. In this paper, we are particularly concerned with learning cross-lingual embeddings between the languages of English and Welsh. The last census indicated that there are currently 526,016 speakers of Welsh (https://gov.

wales/welsh-language-data-annual-population-survey-2019, accessed on 15 July 2021). Welsh is statistically a 'minority' language as there are more speakers of English than Welsh in Wales and the UK, but it is a healthy one, and Wales represents the largest bilingual community in the UK. As a minoritized language in the UK context (albeit with official status, alongside English, in the devolved nation of Wales), Welsh does not enjoy the same language technology resources as English or other major state languages, although there is an increasing interest in widening the availability of resources in this context. Welsh-language technologies that are currently available include POS (part of speech) taggers (including CyTag, [11]), WordNet Cymru (https://users.cs.cf.ac.uk/I.Spasic/wncy/index.html, accessed on 15 July 2021), and an extensive range of tools developed for the purposes of, for example, text-to-speech, speech recognition, machine translation, and terminology recognition, developed by Canolfan Bedwyr at Bangor University (see their online Welsh National Language Technologies Portal (http://techiaith.cymru/?lang=en, accessed on 15 July 2021). However, to the best of our knowledge, there has been no work on learning high-quality bilingual mappings between English and Welsh, which would drastically accelerate the current landscape for Welsh NLP technologies. In this paper, we thus propose to explore current state-of-the-art cross-lingual embeddings techniques for the Welsh language. We first train several monolingual models based on Skip-gram [12] and fastText [13], considering several configurations in terms of context window size, minimum frequency threshold, and vector dimensionality. Then, we apply VecMap [14], a method for learning cross-lingual mappings via orthogonality constraints to our monolingual embeddings. We also report results on a post-processing step based on applying an additional transformation obtained via a linear model trained on top of the bilingual synonym's mean vectors [5]. These cross-lingual representations are evaluated on the standard task of dictionary induction. Finally, as a further downstream task, we report the results of a sentiment analysis system for Welsh in zero-shot and few-shot settings, i.e., training it only with English data, or with limited instances of (automatically translated) task-specific Welsh data. Our results, while promising, also point to the challenges posed by under-represented, resource-poor languages in NLP development, and suggest that further research is needed to strengthen the landscape for Welsh language technologies. The contributions of this paper are as follows:

- Cross-lingual embeddings: we train, evaluate, and release a wealth of cross-lingual English–Welsh word embeddings.
- Train and test dictionary data: we release to the community a bilingual English–Welsh dictionary with a fixed train/test split, to foster reproducible research in Welsh NLP development.
- Sentiment analysis: we train, evaluate, and release a Welsh sentiment analysis system, fine-tuned on the domain of movie reviews.
- Qualitative analysis: we analyze some of the properties (in terms of nearest neighbors) of the cross-lingual spaces, and discuss them in the context of avenues for future work.

Our results suggest that gains in training English–Welsh bilingual embeddings can be obtained by carefully tuning the hyperparameters of the monolingual models, and that the distance metric chosen matters, with differences of up to 5% in accuracy. Overall, the best configuration across the board seems to always involve the fastText model (as opposed to skip-gram), and the CSLS distance metric (as opposed to cosine similarity and inverted softmax). Conversely, the cutoff threshold for minimum frequency and the context window seem to be less important for the final results, as there is not a clear pattern involving a consistent setting among the top-ranked results. In our external evaluation experiment, namely, zero and few-shot sentiment analysis, we verified that it is indeed possible to develop a competitive sentiment analysis system for Welsh only using cross-lingual embeddings and English training data, and that by adding synthetic Welsh training data (e.g., from a machine-translation engine), the performance of the model increases as well.

The remainder of this paper is organized as follows: Section 2 gives an account of research works in different areas relevant to the scope of this paper; Section 3 introduces resources we used for generating cross-lingual embeddings. Section 4 introduces the algorithm used for mapping monolingual embeddings into a shared space. Section 5 presents the results in two (intrinsic and extrinsic) experimental settings. Finally, Section 6 summarizes the main contributions of this work, and outline potential avenues for future work. Data, models, and software are publicly available at (https://github.com/cardiffnlp/en-cy-bilingual-embeddings, accessed on 15 July 2021).

2. Background

In this section, we present a review of related works with respect to NLP for the Welsh language and cross-lingual word embeddings.

2.1. Welsh Language NLP

Recently there has been much research in the space of applying NLP to non-English minority languages such as Welsh. The defining characteristic of a minority language is that the amount of corresponding data available for that language is significantly less than that available for the English language. Most state-of-the-art NLP models use deep learning where performance scales with the amount of available data. Given this, achieving performance on NLP tasks for minority languages is on par with that achieved for the same tasks for the English language represents a significant challenge. The Welsh Natural Language Toolkit (WNLT) is a Welsh-Government-funded project which focuses on the development of NLP tools for the Welsh language (https://hypermedia.research.southwales.ac.uk/kos/wnlt/, accessed on 15 July 2021). The tools in question are distributed under GNU Lesser General Public License (LGPL) and include tools for tokenization, lemmatization, POS tagging, and Named Entity Recognition (NER) (https://sourceforge.net/projects/wnlt-project/, accessed on 15 July 2021). Neale et al. [11] developed a rule-based part-of-speech (POS) tagger for the Welsh language entitled Cy-Tag. Although state-of-the-art POS taggers for the English language use deep learning, the authors argue there is insufficient Welsh language data to use such an approach for the Welsh language. The same authors later developed a rule-based semantic tagger, entitled CySemTagger [15]. Both of these tools are available under a free software (GPL version 3) licence (https://github.com/CorCenCC, accessed on 15 July 2021). Jones et al. [16] developed a statistical machine translation model for the English and Welsh language pair. Spasić et al. [17] developed a statistical method for multiple word term recognition in Welsh. This method builds on a previously proposed term-recognition method known as FlexiTerm [18].

2.2. Cross-Lingual Embeddings

Earlier attempts to train cross-lingual word embeddings required access to parallel (or, at least, comparable) corpora [19–25]. Finding such corpus especially for a minoritized language can prove challenging. Therefore, the research in this space gravitated towards using bilingual dictionaries instead of aligning the words in respective languages [6,26]. It was later shown that such cross-lingual supervision is not necessary to align word embedding [7]. Instead, adversarial training can be used to initialize a linear mapping between two vector spaces and produce a synthetic parallel dictionary. The success of this approach was based on the use of two metrices: one for unsupervised validation and the other one for similarity measure. Such combination reduces the hubness problem while improving the translation accuracy.

Hubness is a phenomenon that occurs in high-dimensional spaces, where some objects tend to concentrate around a centroid while others have few nearest neighbors [27]. Specifically, hubness associated with cross-lingual embeddings was explored in [28], who proposed incorporating a nearest-neighbor reciprocity as a way of managing hubness. Different measures were used to down-weigh similarities associated with hub words, including cross-domain similarity local scaling (CSLS) [7] and inverted softmax [29]. In addition, adding an orthogonality constraint, which conveniently has a closed-form solution, can improve performance further [30].

Alternatively, to align monolingual embedding spaces with no supervision, Zhang et al. [31] used adversarial training to exploit sudden drops in accuracy for model selection followed by minimizing the earth-mover distance [32]. Conversely, Conneau et al. [7] do not base model selection on its performance, which allows for hyper-parameters to be tuned specifically for a given language pair as they tend to vary significantly across languages. Similar approaches used to induce bilingual dictionaries from data [5,10,14,33] yielded state-of-the-art performance in many language pairs, although the experimental setup followed in the literature has also been closely scrutinized, and there exist studies that argue for experiments that account for different genres in source and target corpora, studying (dis)similarities between languages, etc. [34,35].

Although the advent of language models in the current NLP landscape (BERT, GPT, or RoBERTa) [36–38] has transformed the field, it is also true that even for languages where the availability of raw data is small, having access to pre-trained static word embeddings can make the difference between developing a language technology or not at all. Recent work has, for example, focused on dialectal Arabic, by combining BERT-based encodings with Arabic word embeddings for underrepresented domains and dialects [39].

3. Materials

This section describes the materials required for generating cross-lingual embeddings.

3.1. Corpora

While a number of Welsh corpora exist, there generally lacks extensive data sets of Welsh language that are freely/widely available. To undertake this study, we combined a number of existing Welsh corpora, sourced from different language contexts, including proceedings from the Welsh assembly (http://cymraeg.org.uk/kynulliad3/, accessed on 15 July 2021), scraped websites and blogs [40] and the National Corpus for Contemporary Welsh (CorCenCC, [41]), amongst others. The full list of corpora used are given in Table 1. We ensured that the collected corpus includes a diverse range of formats, genres and registers, including a balanced mix of formal and casual language, and general and specialized topics. For example, there are texts from the highly formal academic writing of academic journal papers and textbooks; the archaic writing of the bible; technical writing in the form of administrative documents and software documentation; journalistic writing from news and magazine articles; pieces of creative writing in prose, poetry and song; and everyday casual language including emails, tweets, text messages, and transcripts of spoken language.

In terms of the English corpus, we used a Wikipedia data dump for June 2018, which is a standard corpus in distributional semantics for learning word embeddings.

3.2. Text Corpus Creation

We developed Welsh and English corpora to train our bilingual embeddings, drawing on a range of pre-existing data sets. The full Welsh-language data set extended to 144,976,542 words after tokenization. The names and corresponding number of words in each individual text corpus are displayed in Table 1. We now provide a brief description of each individual text corpus.

Table 1. Names and corresponding number of words in each individual Welsh-language text corpus.

Corpus	Numb. Words
Welsh Wikipedia	21,233,177
Proceedings of the Welsh Assembly 1999–2006	11,527,963
Proceedings of the Welsh Assembly 2007–2011	8,883,870
The Bible	749,573
OPUS translated texts	1,224,956
Welsh Government translation memories	1,857,267
Proceedings of the Welsh Assembly 2016–2020	17,117,715
Cronfa Electroneg o Gymraeg	1,046,800
An Crúbadán	22,572,066
DECHE	2,126,153
BBC Cymru Fyw	14,791,835
Gwerddon	732,175
Welsh-medium websites	7,388,917
CorCenCC	10,630,657
S4C subtitles	26,931,013

Welsh Wikipedia—Wikipedia is a multilingual crowd-sourced online encyclopedia and one of the world's most popular websites. English Wikipedia was the first edition of Wikipedia and was founded in January 2001. As of 29 September 2019 (when these data were collected), there were 5,938,555 articles contained in this project. Given the large number of articles, English Wikipedia is a text corpus commonly used to train English language word embeddings. Welsh Wikipedia is the Welsh language edition of Wikipedia and was founded in July 2003. It is significantly smaller than English Wikipedia and as of 29 September 2019 it contains 106,128 articles. Web crawling of this was undertaken, specifically, using the Python library urllib and the Python library Beautiful Soup to extract all text within paragraph tags <p>. We subsequently removed all citations and mathematical equations.

National Assembly for Wales 1999–2006—The National Assembly for Wales is the devolved parliament of Wales, which has many powers, including those to make legislation and set taxes. By performing a web crawling of the Assembly website (http://xixona.dlsi.ua.es/corpora/UAGT-PNAW/, accessed on 15 July 2021), Jones et al. [16] created a bilingual aligned corpus of Welsh and English from the online version of the Proceedings of the Plenary Meetings of the Assembly between the years 1999 and 2006 inclusive. This is freely available as a plain text file. Only the Welsh part of this corpus was used for the purposed of the current project.

National Assembly for Wales 2007–2011—Donnelly [42] created the Kynulliad3 corpus, which is similar to the previous bilingual aligned corpus except that it covers the period between the years 2007 and 2011 inclusive. This corpus, which contains 350,000 aligned Welsh and English sentences, was extracted by querying an SQL database. Only the Welsh half of this corpus was used in the current project.

The Bible—Beibl.net (http://www.beibl.net, accessed on 15 July 2021) includes all books of the Bible in modern Welsh. Texts were scraped using urllib and Beautiful Soup in Python.

OPUS—OPUS is a collection of technical texts on the web, mainly including software documentation, in a number of languages. We extracted a range of en-cy (English–Welsh) texts from this resource in plain text format.

Welsh Government translations memories—The collection of translation memory files contains published bilingual documents and other materials from the Welsh Government (from August 2019 to May 2020). The data set comprises .tmx files, which were extracted using Python's translate toolkit package.

National Assembly for Wales 2016–2020—Records of the proceedings of the Welsh Assembly, including plenary information from the start of the Fifth Assembly (May 2016) and

Committee information from November 2017 to May 2020. The data set was downloaded as .xml, with text extracted using the Python library Beautiful Soup.
Cronfa Electroneg o Gymraeg—This corpus contains 500 articles of approximately 2000 words each, selected from a representative range of text types to illustrate modern (mainly post-1970) Welsh prose writing [43]. It includes articles from the fields of novels and short stories, religious writing, children's literature, non-fiction materials in the fields of education, science, business and leisure activities, public lectures, newspapers and magazines, reminiscences, academic writing, and general administrative materials (letters, reports, minutes of meetings).
An Crúbadán—This corpus was created by Scannell [40] by performing web crawling. It consists of a collection of Welsh Wikipedia articles, Welsh Tweets, Welsh Blogs, the Universal Declaration of Human Rights, and articles from a Jehovah's Witnesses website (JW.org) (https://www.jw.org/cy/, accessed on 15 July 2021). To prevent duplication of the previous Welsh Wikipedia corpus, we removed all Wikipedia articles.
DECHE—The Digitization, E-publishing, and Electronic Corpus (DECHE) project produces e-books out of Welsh language scholarly, academic books which are out of print and unlikely to be reprinted in traditional paper format [44]. Candidates for producing as e-books are nominated by lecturers working through the medium of Welsh and prioritized by the Coleg Cymraeg Cenedlaethol, who fund the project. We constructed a corpus from this project by manually downloading all books in epub format and extracting the plain text using the Python libraries epub_conversion and Beautiful Soup.
BBC Cymru Fyw—BBC Cymru Fyw is an online Welsh language service provided by BBC Wales containing news and magazine-style articles. Using the Corpus Crawler tool (https://github.com/google/corpuscrawler, accessed on 15 July 2021), we constructed a corpus containing all articles published on BBC Cymru Fyw between 1 January 2011 and 17 October 2019 inclusive.
Gwerddon—Gwerddon is a Welsh-medium academic e-journal which publishes research in the Arts, the Humanities, and the Sciences (http://www.gwerddon.cymru/, accessed on 15 July 2021). This corpus contains all text in articles contained in 29 editions of this journal. It was constructed by manually downloading the articles in question and extracting the corresponding text using the R programming language package pdftools. Some manual post-formatting was carried out to correct footnotes, etc.
Welsh-medium websites—Golwg360 (https://golwg360.cymru, accessed on 15 July 2021) and O'r Pedwar Gwynt (https://pedwargwynt.cymru, accessed on 15 July 2021) are Welsh-medium news websites. PoblCaerdydd (https://poblcaerdydd.com/, accessed on 15 July 2021) and Cylchgrawn Barn (https://barn.cymru/, accessed on 15 July 2021) are Welsh-medium online magazines. This corpus contains all text extracted from articles on these four websites. It was construed by performing web crawling using wget and extracting all relevant text using the Python library Beautiful Soup.
CorCenCC—CorCenCC (https://www.corcencc.org, accessed on 15 July 2021) [41] is the National Corpus of Contemporary Welsh (Corpws Cenedlaethol Cymraeg Cyfoes). This corpus contains over 11 million words of spoken, written, and electronic language data sampled from a range of genres, styles, registers, and dialect regions. The pre-processed version of the corpus was made available for use in this project.
S4C subtitles—Subtitles kindly received privately (i.e., not publicly available) from the Welsh-language TV channel S4C (https://www.s4c.cymru, accessed on 15 July 2021). Text manipulation was used to strip away the formatting and compile this corpus.

English corpora include the UMBC (https://ebiquity.umbc.edu/blogger/2013/05/01/umbc-webbase-corpus-of-3b-english-words/, accessed on 15 July 2021) web-based corpus and Wikipedia (www.wikipedia.org, accessed on 15 July 2021) corpus. UMBC contains over 3 billion words, including blog posts, news stories etc., that have been stripped from the web, cleaned, tokenized and pre-processed. The Wikipedia corpus includes all texts from the English Wikipedia site, with one sentence per line, tokenized, lemmatized, chunked, lower-cased and POS-tagged.

3.3. Word Embeddings

In our experiments, we compare two different word embeddings methods, namely, Skip–Gram with Negative Sampling (which we denote as *word2vec*) [12], and fastText [13], which is an improved *word2vec* architecture that accounts for subword information in order to capture morphological and subword information. For each of these two models, we experiment with different hyperparameters, namely, *vector size* (**DIM**), a word's *minimum frequency* threshold (**MF**), and *context window* (**CW**).

3.4. Bilingual Dictionary

Our initial bilingual dictionary was provided by Bangor University [45]. It contains over 100,000 bilingual entries, including named entities (e.g., "Alfred the Great"), multi-word terms (e.g., "acquired immunity"), or domain-specific terminology (e.g., for the chemical domain, "2,4-diisocyanato-1-methylbenzene"). For our purposes, we preprocessed this initial dictionary by removing all multi-word and ambiguous (i.e., words for which there was more than one entry—or *sense*— recorded in the dictionary) terms, and split it into training and test. The final size of this dictionary, which we used for mapping English to Welsh embeddings, and for evaluating these mappings, consisted of 9067 training pairs and 2268 test pairs.

4. Methods

Having a bilingual dictionary available makes it viable to cast the problem of learning bilingual mappings as a supervised machine learning task, where given two monolingual corpora, a word vector space is first learned independently for each language. This can be achieved with standard word embedding models such as Word2vec [6], GloVe [46], or fastText [13]. Second, a linear alignment strategy is used to map the monolingual embeddings to a common bilingual vector space. It is worth mentioning that we do not require parallel or comparable corpora to build these multilingual models [47,48], although it has also been shown that the higher the overlap in terms of domain, topic, genre, or linguistic typology, the better the alignments [35,49].

The learning model for these mappings is often a simple linear transformation trained on a bilingual dictionary. In the original paper by Mikolov et al. [6], a matrix **W** is trained, which minimizes the following objective:

$$\sum_{i=1}^{n} ||\mathbf{x_i}\mathbf{W} - \mathbf{z_i}||^2 \tag{1}$$

with $\mathbf{x_i}$ and $\mathbf{z_i}$ being the vector representations of cross-lingual synonyms (i.e., translations) of two words w_i and z_i, in two different languages, respectively. After training, the translation z' of any source word x' in the source language can be defined as $z' = \text{argmax}_{z'} d(\mathbf{Wx}, \mathbf{z}')$, with $d(\cdot)$ being a vector distance metric. In this paper, we consider as options for $d(\cdot)$ the following: (1) the well-known cosine similarity (NN); (2) inverted softmax (*invsoftmax*) [29]; and (3) cross-domain similarity local scaling (CSLS) [7]. This task, i.e., the retrieval of cross-lingual synonyms (or word translations) is known as *dictionary induction*, and is considered a good intrinsic testbed for assessing the quality of cross-lingual mappings. In this paper, we report experiments on the test split of the dictionary described in Section 3.4.

5. Results

We report results on the test set of our English–Welsh bilingual dictionary. We report these results in terms of *accuracy* (**ACC.**), i.e., we record a true positive only if the nearest neighbor in the mapped space is a translation of the source word. This is a strict measure (as we could have considered, for instance, *P@k*; $k \in \{1, 5, 10\}$), which serves as a strong baseline for upcoming research in English–Welsh crosslingual language technologies.

5.1. Quantitative Evaluation

The task of bilingual dictionary induction, a natural byproduct of learning bilingual mappings, and which we have introduced in Section 4, is a good proxy for evaluating the quality of cross-lingual mappings.

We thus report results of appplying the VecMap method. However, we also experimented with Meemi, but since the results were slightly lower across most configurations, we only report VecMap performance. Table 2 shows the top 20 configurations in terms of accuracy. As we can see, *fastText* consistently performs best when compared to *word2vec*, and CSLS clearly outperforms inverted softmax and cosine similarity in terms of retrieval metrics. On the other hand, the threshold for minimum frequency and context windows seem to be less relevant, as there is high variability among the best configurations. Regarding the overall scores, note that these are in line with what previous work has found when dealing with language pairs involving English and a low-resource language. For example, Doval et al. [49] report P@1 scores for their best models of 24.8 for English–Finnish, 21.5 for English–Farsi, or 19.3 for English–Russian, and Xu et al. [50] report roughly similar or worse results for dictionary induction experiments involving, e.g., Turkish (9.96) or Latvian (13.53). Note that theirs is an unsupervised approach.

Table 2. Top 20 configurations (ranked in descending order) in terms of accuracy (**ACC.**) for the bilingual dictionary induction task *when using VecMap*. We compare different monolingual embedding models (**MODEL**), vector size (**DIM.**), minimum frequency threshold (**MF**), context window (**CW**), and neighbor retrieval method (**RETRIEVAL**, cf. Section 3).

MODEL	DIM.	MF	CW	RETRIEVAL	ACC.
fastText	500	6	6	CSLS	22.92
fastText	500	6	4	CSLS	21.85
fastText	500	6	8	CSLS	21.75
word2vec	300	6	4	CSLS	21.75
word2vec	500	6	8	CSLS	21.46
word2vec	300	6	6	CSLS	21.46
word2vec	500	6	4	CSLS	21.46
word2vec	300	6	8	CSLS	21.36
word2vec	500	6	6	CSLS	21.36
fastText	500	3	4	CSLS	20.46
fastText	500	3	8	CSLS	19.75
fastText	500	3	6	CSLS	19.36
word2vec	300	3	8	CSLS	19.22
word2vec	500	3	6	CSLS	19.22
word2vec	500	3	8	CSLS	19.18
word2vec	300	3	6	CSLS	18.83
fastText	300	6	4	CSLS	18.57
word2vec	300	6	4	invsoftmax	18.48
word2vec	300	6	8	invsoftmax	18.48
word2vec	300	6	8	NN	18.43

5.2. Qualitative Evaluation

The cross-lingual vector space can be manually explored in order to evaluate how well both the monolingual embeddings capture semantic relationships within a language, and also how well the cross-lingual embeddings align. We start this by selecting a small set of prototype words in the first language, and inspect their nearest neighbors in the second language. We then compare this to the reverse procedure: selecting the same translated, words in the second language, and inspect their nearest neighbors in the first.

Table 3 lists a selection of ten words, and their translations, with their 10 nearest neighbors in their opposite languages. In general, the cross-lingual embeddings align well, with the common nouns, adjectives, and verbs mapping to very similar and very related words in both directions. We also attempted to find closely related words to *hiraeth*, a word often claimed to be untranslatable into English, which still gave accurate nearest neighbors, referring to feelings of longing and yearning for home.

More specialized vocabulary, such as foreign loadwords (*croissant*), and proper nouns (*French*, and place names such as *Cardiff* and *Tonypandy*) show some asymmetry in the alignment of the embeddings. Here, the Welsh nearest neighbors to English words are much more relevant and semantically related than the English nearest neighbors of Welsh words. For example, the Welsh nearest neighbors to *croissant* gives breakfast foods and pastries, while the English nearest neighbors are generic foodstuffs. Similarly, the Welsh nearest neighbors to *French* gives Euro-centric languages and adjectives, while the English nearest neighbors to *ffrangeg* (the French language) gives languages from further afield. It is also interesting to note that in Welsh, the words *ffrangeg* (the French language) is different to *ffrengig* (the French nationality), and all the English nearest neighbors to *ffrangeg* are languages or language-related terms, rather than words related to nationalities, while a mix of the two is seen in the Welsh nearest neighbors of *French*.

Geographic place names are also interesting, with the Welsh nearest neighbors of English place names giving more local and geographically closer place names than the English nearest neighbors of Welsh place names. This may be an effect of the English training corpus having a much more international and broader scope than the Welsh training corpus. For example, *Cardiff / caerdydd*, the capital of Wales and thus an important word in the Welsh language: its Welsh nearest neighbors are other major Welsh towns and cities, while it's English nearest neighbors are populated with Australian places, maybe referencing the much smaller Australian town of Cardiff.

Table 3. Table of a selected sample of cross-lingual nearest neighbors examples.

word_cy	Closest English Words to *word_cy*	*word_en*	Closest Welsh Words to *word_en*
nofio (*swim*)	swim, swimming, kayak, paddling, rowing, waterski, swam, watersport, iceskating, canoe	swim	nofio (*swim*), deifio (*diving*), cerdded (*walking*), blymio (*diving*), padlo (*paddling*), arnofio (*floating*), sblasio (*splashing*), troelli (*spinning*), neidio (*jumping*)
glaw (*rain*)	rain, snow, fog, heavyrain, downpour, rainstorm, heavyrains, snowfall, rainy, mist	rain	glaw (*rain*), eira (*snow*), cenllysg (*hail*), wlith (*dew*), cawodydd (*showers*), rhew (*frost*), taranau (*thunder*), barrug (*frost*), genllysg (*hail*), dafnau (*drops*)
hapus (*happy*)	happy, pleased, glad, grateful, delighted, thankful, anxious, eager, fortunate, confident	happy	hapus (*happy*), bodlon (*satisfied*), ffeind (*kind*), llawen (*joyful*), rhyfedd (*strange*), trist (*sad*), llon (*cheerful*), cysurus (*comfortable*), hoenus (*cheerful*), nerfus (*nervous*)
meddalwedd (*software*)	software, application, computer, system, tool, ibm, hardware, technology, database, device	software	meddalwedd (*software*), feddalwedd (*software*), caledwedd (*hardware*), dyfeisiau (*devices*), amgryptio (*encryption*), dyfeisiadau (*inventions*), cymwysiadau (*applications*), algorithm (*algorithm*), dyfais (*device*), ategyn (*plugin*)
ffrangeg (*French language*)	Arabic, Hebrew, Hindi, Arabiclanguage, language, urdu, sanskrit, haitiancreole, English	French	ffrengig (*french*), sbaenaidd (*Spanish*), almaeneg (*German language*), archentaidd (*Argentinian*), gwyddelig (*Irish*), twrcaidd (*Turkish*), llydewig (*Breton*), almaenaidd (*German*), danaidd (*Danish*), imperialaidd (*imperial*)
croissant (*croissant*)	frenchfries, yogurt, applesauce, currysauce, mulledwine, mozzarellacheese, noodlesoup, buñuelo, chilisauce, misosoup	croissant	bisgedi (*biscuits*), twmplenni (*dumplings*), byns (*buns*), teisennau (*cakes*), bacwn (*bacon*), caramel (*caramel*), melysfwyd (*confectionery*), cwstard (*custard*), marmaled (*marmalade*)
gwario (*spend money*)	expend, invest, reinvest, pay, allot, allocate, disburse, economize, retrench, accrue	spend	treulio (*spend time*), dreulio (*spend time*), gwario (*spend money*), aros (*wait*), threulio (*spend time*), gwastraffu (*wasting*), dychwelyd (*returning*), nychu (*languishing*), hala (*spend money*), byw (*live*)
hiraeth (*longing*)	longing, sadness, yearning, sorrow, anguish, loneliness, grief, feeling, ennui, heartache	longing	hiraeth (*longing*), galar (*grief*), anwyldeb (*dearness*), tristwch (*sadness*), nwyd (*passion*), gorfoledd (*exultation*), tosturi (*compassion*), tynerwch (*tenderness*), nwyf (*vivacity*), hyfrydwch (*loveliness*)
caerdydd (*Cardiff*)	Docklands, Southbank, Brisbane, Frankston, downtown, Thessaloniki, Coquitlam, Melbourne, Bayside, Glasgow	Cardiff	Abertawe (*Awansea*), nantporth (*Nantporth*), aberystwyth (*Aberystwyth*), llanelli (*Llanelli*), glynebwy (*Ebbw Vale*), caerdydd (*Cardiff*), porthcawl (*Porthcawl*), llandudno (*Llandudno*), awyr (*sky*), wrecsam (*Wrexham*)
Tonypandy (*Tonypandy*)	Edgewareroad, Blakelaw, Upperdicker, Ainleytop, Bilsthorpe, Romanby, Killay, Llanwonno, Penllergaer, Greenrigg	Tonypandy	aberdâr (*Aberdare*), Senghennydd (*Senghennydd*), aberpennar (*Mountain Ash*), brynaman (*Brynamman*), aberdar (*Aberdare*), coedpoeth (*Coedpoeth*), llanidloes (*Llanidloes*), brynbuga (*Usk*), penycae (*Penycae*), tymbl (*Tumble*)

5.3. Extrinsic Evaluation

The extrinsic evaluation assesses the performance of a language model in the context of a predefined task. In this study, this task was chosen to be that of sentiment analysis (SA), as it has been shown that cross-lingual systems can achieve high accuracy even in zero or few-shot settings [4]. Specifically, given the shortage of annotated Welsh corpora that can be used to train a Welsh SA model, we wanted to investigate to what extent cross-lingual embeddings can improve the performance of such a model by re-using a readily available annotated English data set.

To implement SA, we re-purposed an existing sentence classifier [51] based on a convolutional neural network for text classification [52], which has been extended by a bi-directional long–short-term memory (Bi-LSTM) [53] layer. This classification model is well equipped to capture both short- and long-range dependencies and extract general features of online reviews that would be useful for SA. The most important hyperparameters of the base model include 100 convolutional filters, a kernel of size 4 and strides of size 1, with a ReLu activation function. Further, the Bi-LSTM layer consisted of two 100-unit (forward and backward) LSTM layers. The model was trained using categorical cross-entropy with an Adam optimizer. In this model, each training instance is represented as a matrix, where each word is represented by the corresponding embedding. Such representation is suitable for cross-lingual training, as cross-lingual synonyms are expected to be represented by similar vectors in the joint vector space. Therefore, any abstractions learned by the model are also expected to be similar in the two languages.

All SA experiments were performed using a set of 50 K IMDB reviews, which represent a community standard for evaluating SA [54]. This data set is divided into two subsets of 25 K reviews, each to be used for training and testing, respectively. The original reviews were automatically translated from English to Welsh using Google Translate, a neural machine translation system [55] that proved mature enough to produce reliable data for training SA in languages other than English [56]. We used the best-performing bilingual English–Welsh embeddings as per their performance in the dictionary induction task (Section 5.1).

To perform cross-lingual training, we started by training an SA model using English data only and evaluated the results using Welsh data. We call this zero-short learning as no labeled data in Welsh were used at all. We then gradually added Welsh translations using increments of n reviews, where n = 100, 500, 1000, 2500, 5000, 7500, 10,000, 12,500, and 150,00. Given a fixed size n, a random subset was selected five times to check whether the evaluation results were reproducible. All experiments were evaluated against the Welsh test data.

Figure 1 shows the evaluation results in terms of accuracy (y axis) against exposure to labeled data in Welsh (x axis refers to the total number of reviews of Welsh that were combined with a total of 25 K reviews in English). The zero-shot model achieves an accuracy of 65%. The accuracy increased substantially by adding as little as a thousand reviews automatically translated to Welsh. Naturally, with increased exposure to Welsh during the training; the accuracy increased as well. Already at 5000 Welsh reviews, the average accuracy surged beyond 75%. In addition, the model stabilized as less variance was observed across the experiments using different subsets of a fixed size. The highest accuracy achieved fell just short of 80%. Further performance gains are expected to be obtained by tuning the hyperparameters or the neural network architecture itself to optimize its performance with Welsh. However, this is well beyond the scope of the current study. Nonetheless, our experiments confirmed that cross-lingual embeddings make zero-shot English-to-Welsh SA possible with few-shot settings contributing to considerable performance improvements. These results provide the evidence that existing NLP tools based on word embeddings can indeed be re-used to support NLP in Welsh.

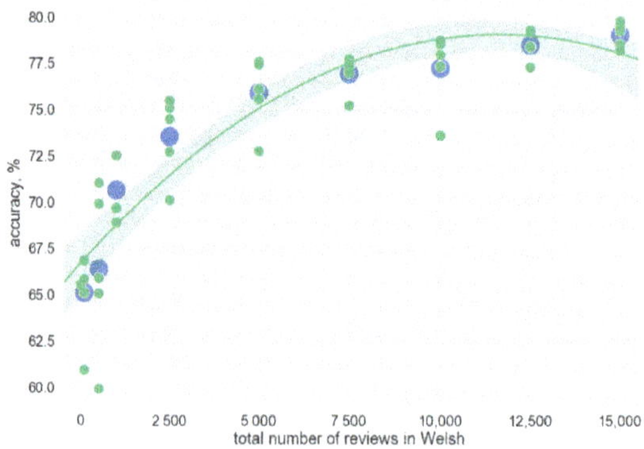

Figure 1. Accuracy results for the cross-lingual sentiment analysis experiment.

6. Conclusions

We have described the process of training bilingual English–Welsh embeddings. We start by discussing the corpora we used to train monolingual embeddings in both languages using both *word2vec* and *fastText*, and continue by explaining the curation of the supervision signal (the training bilingual dictionary), as well as the linear transformation method we use for mapping both monolingual spaces into a shared bilingual space.

We have evaluated this shared space both intrinsically and extrinsically. The intrinsic evaluation was based on *dictionary induction*, which was used to measure the alignment of two monolingual spaces directly by translating between the two languages and measuring the distance within a monolingual space. The best alignment was achieved by training the monolingual spaces using fastText and aligning them using the the CSLS metric. The true value of aligning two vector spaces lies in the ability to facilitate NLP applications in minoritized languages by taking advantage of readily available resources in a language such as English. To evaluate the cross-lingual embeddings extrinsically, we measured the effects of supplementing Welsh-language data with data in English on the accuracy of sentiment analysis in Welsh. We were able to use an existing neural network architecture based on CNNs and LSTMs originally developed for sentiment analysis in English. By training this neural network on cross-lingual embeddings and data from both languages, we managed to obtain highly competitive results in Welsh without having to modify the original method in any way. In particular, we demonstrated that a relatively small data set of 2 K documents in the target language seems to suffice. This opens exciting avenues for future work, where cross-lingual embeddings can be combined with neural architectures and data augmentation techniques to develop Welsh language technology at a negligible cost.

The Welsh language can be categorized, within the language resource landscape, as being a low-resource language, i.e., the availability of (raw and annotated) corpora, glossaries, thesauri, encyclopedias, etc. is limited when compared to other languages such as English, Chinese, Spanish, or Indo-Aryan languages. This study allows one to automatically compare the meaning of words not only within the Welsh language but also across the two languages, thus facilitating applications such as the creation of bilingual language resources, as well as the development of NLP systems for Welsh with limited Welsh training data, as we successfully demonstrated with sentiment analysis. Cross-lingual embedding we generated therefore unlocks access to a plethora of open-source NLP solutions developed originally for English. This in turn opens a possibility of supporting a wide range of applications, such as computer–assisted translation, cross-lingual information retrieval, and conversational artificial intelligence. These applications encourage the use of Welsh in activities of daily life, which contributes to maintaining and improving Welsh language skills.

Author Contributions: Conceptualization, D.K.; methodology, L.E.-A., G.P., P.C., M.F., and I.S.; software, L.E.-A., G.P., M.F., and I.S.; validation, I.S. and D.K.; formal analysis, L.E.-A., G.P., and I.S.; investigation, L.E.-A., G.P., and I.S.; resources, L.E.-A., G.P., P.C., I.S., and D.K.; data curation, L.E.-A., G.P., P.C., I.S., and D.K.; writing—L.E.-A.; writing—review and editing, L.E.-A., G.P., P.C., I.S., and D.K.; visualization, M.F.; supervision, L.E.-A., I.S., and D.K.; project administration, P.C., I.S., and D.K.; funding acquisition, I.S. and D.K. All authors have read and agreed to the published version of the manuscript.

Funding: This research was funded by the Welsh Government, under the Grant "Learning English-Welsh bilingual embeddings and applications in text categorisation".

Institutional Review Board Statement: Not applicable.

Data Availability Statement: Data and software to reproduce our results are available at: https://github.com/cardiffnlp/en-cy-bilingual-embeddings, accessed on 15 July 2021.

Acknowledgments: The research on which this article is based was funded by the Welsh Government as part of the "Learning English–Welsh bilingual embeddings and applications in text categorisation" project.

Conflicts of Interest: The funders had no role in the design of the study; in the collection, analyses, or interpretation of data; in the writing of the manuscript, or in the decision to publish the results.

References

1. Vulić, I.; Moens, M.F. Bilingual word embeddings from non-parallel document-aligned data applied to bilingual lexicon induction. In Proceedings of the 53rd Annual Meeting of the Association for Computational Linguistics and the 7th International Joint Conference on Natural Language Processing, Beijing, China, 26–31 July 2015; Short Papers; Volume 2, pp. 719–725.
2. Tsai, C.T.; Roth, D. Cross-lingual wikification using multilingual embeddings. In Proceedings of the 2016 Conference of the North American Chapter of the Association for Computational Linguistics: Human Language Technologies, San Diego, CA, USA, 12–17 June 2016; pp. 589–598.
3. Mogadala, A.; Rettinger, A. Bilingual word embeddings from parallel and non-parallel corpora for cross-language text classification. In Proceedings of the 2016 Conference of the North American Chapter of the Association for Computational Linguistics: Human Language Technologies, San Diego, CA, USA, 12–17 June 2016; pp. 692–702.
4. Camacho-Collados, J.; Doval, Y.; Martínez-Cámara, E.; Espinosa-Anke, L.; Barbieri, F.; Schockaert, S. Learning Cross-Lingual Word Embeddings from Twitter via Distant Supervision. In Proceedings of the International AAAI Conference on Web and Social Media, Atlanta, GA, USA, 8–11 June 2020; Volume 14, pp. 72–82.
5. Doval, Y.; Camacho-Collados, J.; Anke, L.E.; Schockaert, S. Improving Cross-Lingual Word Embeddings by Meeting in the Middle. In Proceedings of the 2018 Conference on Empirical Methods in Natural Language Processing, Brussels, Belgium, 31 October –4 November 2018; pp. 294–304.
6. Mikolov, T.; Le, Q.V.; Sutskever, I. Exploiting similarities among languages for machine translation. *arXiv* **2013**, arXiv:1309.4168.
7. Conneau, A.; Lample, G.; Ranzato, M.; Denoyer, L.; Jégou, H. Word Translation Without Parallel Data. In Proceedings of the ICLR, Vancouver, BC, Canada, 30 April–3 May 2018.
8. Artetxe, M.; Labaka, G.; Agirre, E. Learning principled bilingual mappings of word embeddings while preserving monolingual invariance. In Proceedings of the 2016 Conference on Empirical Methods in Natural Language Processing, Austin, TX, USA, 1–5 November 2016; pp. 2289–2294.
9. Artetxe, M.; Labaka, G.; Agirre, E. Generalizing and improving bilingual word embedding mappings with a multi-step framework of linear transformations. In Proceedings of the Thirty-Second AAAI Conference on Artificial Intelligence (AAAI-18), New Orleans, LA, USA, 2–7 February 2018.
10. Adams, O.; Makarucha, A.; Neubig, G.; Bird, S.; Cohn, T. Cross-lingual word embeddings for low-resource language modeling. In Proceedings of the 15th Conference of the European Chapter of the Association for Computational Linguistics, Valencia, Spain, 3–7 April 2017; Long Papers; Volume 1, pp. 937–947.
11. Neale, S.; Donnelly, K.; Watkins, G.; Knight, D. Leveraging lexical resources and constraint grammar for rule-based part-of-speech tagging in Welsh. In Proceedings of the Eleventh International Conference on Language Resources and Evaluation, Miyazaki, Japan, 7–12 May 2018.
12. Mikolov, T.; Sutskever, I.; Chen, K.; Corrado, G.S.; Dean, J. Distributed representations of words and phrases and their compositionality. In Proceedings of the Advances in Neural Information Processing Systems, Lake Tahoe, NV, USA, 5–10 December 2013; pp. 3111–3119.
13. Bojanowski, P.; Grave, E.; Joulin, A.; Mikolov, T. Enriching Word Vectors with Subword Information. *Trans. Assoc. Comput. Linguist.* **2017**, *5*, 135–146. [CrossRef]

14. Artetxe, M.; Labaka, G.; Agirre, E. A robust self-learning method for fully unsupervised cross-lingual mappings of word embeddings. In Proceedings of the 56th Annual Meeting of the Association for Computational Linguistics, Melbourne, Australia, 15–20 July 2018; Volume 1, pp. 789–798. [CrossRef]
15. Piao, S.; Rayson, P.; Knight, D.; Watkins, G. Towards a Welsh Semantic Annotation System. In Proceedings of the Eleventh International Conference on Language Resources and Evaluation, Miyazaki, Japan, 7–12 May 2018.
16. Jones, D.; Eisele, A. Phrase-based statistical machine translation between English and Welsh. In Proceedings of the 5th SALTMIL Workshop on Minority Languages at the 5th International Conference on Language Resources and Evaluation, Genoa, Italy, 24–26 May 2006.
17. Spasić, I.; Owen, D.; Knight, D.; Artemiou, A. Unsupervised Multi–Word Term Recognition in Welsh. In Proceedings of the Celtic Language Technology Workshop, Dublin, Ireland, 23 August 2014; pp. 1–6.
18. Spasić, I.; Greenwood, M.; Preece, A.; Francis, N.; Elwyn, G. FlexiTerm: A flexible term recognition method. *J. Biomed. Semant.* **2013**, *4*, 27. [CrossRef] [PubMed]
19. Klementiev, A.; Titov, I.; Bhattarai, B. Inducing crosslingual distributed representations of words. In Proceedings of the COLING 2012, Mumbai, India, 8–15 December 2012; pp. 1459–1474.
20. Zou, W.Y.; Socher, R.; Cer, D.M.; Manning, C.D. Bilingual Word Embeddings for Phrase-Based Machine Translation. In Proceedings of the EMNLP, Seattle, WA, USA, 18–21 October 2013; pp. 1393–1398.
21. Kneser, R.; Ney, H. Improved backing-off for m-gram language modeling. In Proceedings of the 1995 International Conference on Acoustics, Speech, and Signal Processing, New York, NY, USA, 9–12 May 1995; Volume 1, pp. 181–184.
22. Lauly, S.; Larochelle, H.; Khapra, M.M.; Ravindran, B.; Raykar, V.; Saha, A. An autoencoder approach to learning bilingual word representations. *arXiv* **2014**, arXiv:1402.1454.
23. Kočiský, T.; Hermann, K.M.; Blunsom, P. Learning bilingual word representations by marginalizing alignments. *arXiv* **2014**, arXiv:1405.0947.
24. Coulmance, J.; Marty, J.M.; Wenzek, G.; Benhalloum, A. Trans-gram, fast cross-lingual word-embeddings. *arXiv* **2016**, arXiv:1601.02502.
25. Wang, R.; Zhao, H.; Ploux, S.; Lu, B.L.; Utiyama, M.; Sumita, E. A novel bilingual word embedding method for lexical translation using bilingual sense clique. *arXiv* **2016**, arXiv:1607.08692.
26. Faruqui, M.; Dyer, C. Improving vector space word representations using multilingual correlation. In Proceedings of the 14th Conference of the European Chapter of the Association for Computational Linguistics, Gothenbourg, Sweden, 26–30 April 2014; pp. 462–471.
27. Radovanović, M.; Nanopoulos, A.; Ivanović, M. Hubs in space: Popular nearest neighbors in high-dimensional data. *J. Mach. Learn. Res.* **2010**, *11*, 2487–2531.
28. Dinu, G.; Lazaridou, A.; Baroni, M. Improving zero-shot learning by mitigating the hubness problem. In Proceedings of the ICLR Workshop Track, San Diego, CA, USA, 7–9 May 2015.
29. Smith, S.L.; Turban, D.H.; Hamblin, S.; Hammerla, N.Y. Offline bilingual word vectors, orthogonal transformations and the inverted softmax. In Proceedings of the ICLR, Toulon, France, 24–26 April 2017.
30. Xing, C.; Wang, D.; Liu, C.; Lin, Y. Normalized word embedding and orthogonal transform for bilingual word translation. In Proceedings of the 2015 Conference of the North American Chapter of the Association for Computational Linguistics: Human Language Technologies, Denver, CO, USA, 31 May–5 June 2015; pp. 1006–1011.
31. Zhang, M.; Liu, Y.; Luan, H.; Sun, M. Adversarial training for unsupervised bilingual lexicon induction. In Proceedings of the 53rd Annual Meeting of the Association for Computational Linguistics, Beijing, China, 26–31 July 2015 pp. 1959–1970.
32. Zhang, M.; Liu, Y.; Luan, H.; Sun, M. Earth mover's distance minimization for unsupervised bilingual lexicon induction. In Proceedings of the 2017 Conference on Empirical Methods in Natural Language Processing, Copenhagen, Denmark, 7–11 September 2017; pp. 1924—1935.
33. Artetxe, M.; Labaka, G.; Agirre, E. Learning bilingual word embeddings with (almost) no bilingual data. In Proceedings of the 55th Annual Meeting of the Association for Computational Linguistics, Vancouver, BC, Canada, 30 July–4 August, 2017; Volume 1, pp. 451–462.
34. Søgaard, A.; Ruder, S.; Vulić, I. On the Limitations of Unsupervised Bilingual Dictionary Induction. *arXiv* **2018**, arXiv:1805.03620.
35. Doval, Y.; Camacho-Collados, J.; Espinosa-Anke, L.; Schockaert, S. On the Robustness of Unsupervised and Semi-supervised Cross-lingual Word Embedding Learning. In Proceedings of the 12th Language Resources and Evaluation Conference, Marseille, France, 11–16 May 2020; pp. 4013–4023.
36. Devlin, J.; Chang, M.W.; Lee, K.; Toutanova, K. Bert: Pre-training of deep bidirectional transformers for language understanding. *arXiv* **2018**, arXiv:1810.04805.
37. Radford, A.; Wu, J.; Child, R.; Luan, D.; Amodei, D.; Sutskever, I. Language models are unsupervised multitask learners. *OpenAI Blog* **2019**, *1*, 9.
38. Liu, Y.; Ott, M.; Goyal, N.; Du, J.; Joshi, M.; Chen, D.; Levy, O.; Lewis, M.; Zettlemoyer, L.; Stoyanov, V. Roberta: A robustly optimized bert pretraining approach. *arXiv* **2019**, arXiv:1907.11692.

39. Alghanmi, I.; Anke, L.E.; Schockaert, S. Combining BERT with Static Word Embeddings for Categorizing Social Media. In Proceedings of the Sixth Workshop on Noisy User-generated Text (W-NUT 2020), Collocated with EMNLP 2020, 16 November 2020; pp. 28–33. Available online: https://www.aclweb.org/portal/content/6th-workshop-noisy-user-generated-text-wnut-2020 (accessed on 15 July 2021).
40. Scannell, K.P. The Crúbadán Project: Corpus building for under-resourced languages. In Proceedings of the 3rd Web as Corpus Workshop, Louvain-la-Neuve, Belgium, 1 January 2007; Volume 4, pp. 5–15.
41. Knight, D.; Loizides, F.; Neale, S.; Anthony, L.; Spasić, I. Developing computational infrastructure for the CorCenCC corpus: The National Corpus of Contemporary Welsh. *Lang. Resour. Eval.* **2020**, 1–28. [CrossRef]
42. Donnelly, K. Kynulliad3: A corpus of 350,000 Aligned Welsh-English Sentences from the Third Assembly (2007–2011) of the National Assembly for Wales. 2013. Available online: http://cymraeg.org.uk/kynulliad3 (accessed on 15 July 2021).
43. Ellis, N.C.; O'Dochartaigh, C.; Hicks, W.; Morgan, M.; Laporte., N. Cronfa Electroneg o Gymraeg (ceg): A 1 Million Word Lexical Database and Frequency Count for Welsh. 2001. Available online: https://www.bangor.ac.uk/canolfanbedwyr/ceg.php.en (accessed on 15 July 2021).
44. Prys, D.; Jones, D.; Roberts, M. DECHE and the Welsh National Corpus Portal. In Proceedings of the First Celtic Language Technology Workshop, Doublin, Ireland, 23 August 2014; pp. 71–75.
45. Uned Technolegau Iaith/Language Technologies Unit, Prifysgol Bangor University. Welsh-English Equivalents File. 2016. Available online: https://github.com/techiaith (accessed on 15 July 2021).
46. Pennington, J.; Socher, R.; Manning, C.D. GloVe: Global vectors for word representation. In Proceedings of the EMNLP, Doha, Quatar, 25–29 October 2014; pp. 1532–1543.
47. Zennaki, O.; Semmar, N.; Besacier, L. A neural approach for inducing multilingual resources and natural language processing tools for low-resource languages. *Nat. Lang. Eng.* **2019**, *25*, 43–67. [CrossRef]
48. Vulić, I.; Moens, M.F. Bilingual distributed word representations from document-aligned comparable data. *J. Artif. Intell. Res.* **2016**, *55*, 953–994. [CrossRef]
49. Doval, Y.; Camacho-Collados, J.; Espinosa-Anke, L.; Schockaert, S. Meemi: A simple method for post-processing cross-lingual word embeddings. *arXiv* **2019**, arXiv:1910.07221.
50. Xu, R.; Yang, Y.; Otani, N.; Wu, Y. Unsupervised Cross-lingual Transfer of Word Embedding Spaces. In Proceedings of the 2018 Conference on Empirical Methods in Natural Language Processing, Brussels, Belgium, 31 October–4 November 2018; pp. 2465–2474.
51. Espinosa-Anke, L.; Schockaert, S. Syntactically aware neural architectures for definition extraction. In Proceedings of the 2018 Conference of the North American Chapter of the Association for Computational Linguistics: Human Language Technologies, New Orleans, LA, USA, 1–6 June 2018; Volume 2, pp. 378–385.
52. Kim, Y. Convolutional Neural Networks for Sentence Classification. In Proceedings of the EMNLP, Doha, Qatar, 25–29 October 2014; pp. 1746–1751.
53. Hochreiter, S.; Schmidhuber, J. Long Short-Term Memory. *Neural Compasutation* **1997**, *9*, 1735–1780. [CrossRef] [PubMed]
54. Maas, A.; Daly, R.E.; Pham, P.T.; Huang, D.; Ng, A.Y.; Potts, C. Learning word vectors for sentiment analysis. In Proceedings of the 49th Annual Meeting of the Association for Computational Linguistics: Human Language Technologies, Prtland, OR, USA, 19–24 June 2011; pp. 142–150.
55. Wu, Y.; Schuster, M.; Chen, Z.; Le, Q.V.; Norouzi, M.; Macherey, W.; Krikun, M.; Cao, Y.; Gao, Q.; Macherey, K.; et al. Google's Neural Machine Translation System: Bridging the Gap between Human and Machine Translation. *arXiv* **2016**, arXiv:1609.08144.
56. Balahur, A.; Turchi, M. Multilingual sentiment analysis using machine translation? In Proceedings of the 3rd Workshop in Computational Approaches to Subjectivity and Sentiment Analysis, Jeju, Korea, 12 July 2012; pp. 52–60.

applied sciences

MDPI

Article

Conversational AI over Military Scenarios Using Intent Detection and Response Generation

Hsiu-Min Chuang * and Ding-Wei Cheng

Department of Computer Science and Information Engineering, Chung Cheng Institute of Technology, National Defense University, Taoyuan City 335, Taiwan; dinwei0108@gmail.com
* Correspondence: showmin1205@gmail.com

Abstract: With the rise of artificial intelligence, conversational agents (CA) have found use in various applications in the commerce and service industries. In recent years, many conversational datasets have becomes publicly available, most relating to open-domain social conversations. However, it is difficult to obtain domain-specific or language-specific conversational datasets. This work focused on developing conversational systems based on the Chinese corpus over military scenarios. The soldier will need information regarding their surroundings and orders to carry out their mission in an unfamiliar environment. Additionally, using a conversational military agent will help soldiers obtain immediate and relevant responses while reducing labor and cost requirements when performing repetitive tasks. This paper proposes a system architecture for conversational military agents based on natural language understanding (NLU) and natural language generation (NLG). The NLU phase comprises two tasks: intent detection and slot filling. Detecting intent and filling slots involves predicting the user's intent and extracting related entities. The goal of the NLG phase, in contrast, is to provide answers or ask questions to clarify the user's needs. In this study, the military training task was when soldiers sought information via a conversational agent during the mission. In summary, we provide a practical approach to enabling conversational agents over military scenarios. Additionally, the proposed conversational system can be trained by other datasets for future application domains.

Keywords: conversational AI; intent detection; slot filling; retrieval-based question answering; query generation

Citation: Chuang, H.-M.; Cheng, D.-W. Conversational AI over Military Scenarios Using Intent Detection and Response Generation. *Appl. Sci.* **2022**, *12*, 2494. https://doi.org/10.3390/app12052494

Academic Editors: Arturo Montejo-Ráez and Salud María Jiménez-Zafra

Received: 31 January 2022
Accepted: 24 February 2022
Published: 27 February 2022

1. Introduction

Breakthroughs in artificial intelligence and natural language processing (NLP) have made it possible for conversational agents to provide appropriate replies in various domains, helping to reduce labor costs [1–4]. Task-oriented conversational agents, in particular, are of great interest to many researchers. According to a 2018 VentureBeat article[5] over 300,000 chatbots are operating on Facebook. In addition, a 2021 Userlike survey showed that 68% of consumers liked that chatbots can provide fast answers or responses [6].As a result, text-based conversational systems or chatbots have become increasingly common in everyday life. Task-oriented conversational AI use NLP and NLU to perform intent detection and response generation based on domain-specific information, and are mainly used in entertainment [7], finance [8], medicine [9,10], law [11,12], education [13], etc.

Combat training emphasizes timeliness, coupled with the ever-changing battlefield. As a result, effectively predicting the combat information required by soldiers has become one of the key technologies on the frontline battlefield. Operators need to send and receive the type of data they want to enhance their situational awareness [14]. However, it is costly and practically difficult to provide a human assistant to every operator [15,16]. At the 2017 National Training and Simulation Association (NTSA) conference held in Florida, AI experts and military officials discussed valuable applications of AI in military training [15]. Considering that future battlefields and combat scenarios will be increasingly complex and difficult to navigate, the ability to use AI to design extremely realistic, intelligent entities that can be immersed in simulations will be an invaluable weapon for the Navy and Marine Corps. To reduce the risk to personnel in practice, since 2021, the U.S. The Navy has been planning to develop virtual assistants to assist in submarine hunting (https://voicebot.ai/2021/02/10/the-us-navy-wants-a-virtual-assistant-to-help-hunt-submarines/ (accessed on 31 January 2022)). For example, sonar operators on ships must manage the complexity of sonar technology and set settings based on weather, location, etc. Hence, the Navy wants to utilize artificial intelligence to enhance the operating system, improving sonar detection and reducing training costs. These information-processing AI systems can be tailored to specific industries.

In a recent study, the three main types of Human–Machine Interfaces (HMI) were text-based systems, voice-based systems, and interactive interface systems [17–19]. For example, Dr. Felix Gervits from the Army Research Laboratory worked with the U.S. Army Combat Capabilities Development Command and the University of Southern California's Institute for creative technologies to develop autonomous systems (https://eandt.theiet.org/content/articles/2021/04/military-bots-could-become-teammates-with-real-time-conversational-ai/ (accessed on 31 January 2022) [20] to derive intent from a soldier's speech via a statistical language classifier. By combining NLU with dialogue management and having the classifier learn the patterns between verbal commands, responses, and actions, they created a system that could respond appropriately to new commands and knew when to request extra information. In addition, Robb et al. [21] proposed a conversational multimodal interface by combining visual indicators with a conversational system, providing a natural way for users to gain information on vehicle status/faults and mission progress and to set reminders. The system can be used for operations in remote and hazardous environments.

During military training missions, soldiers must follow guidelines or personnel instructions. However, the overloading information may not be understood and completed effectively. In addition, traditional retrieval systems may delay user action. Hence, we constructed a conversational agent over a set of military scenarios that enables users to operate on constantly evolving battlefields and to obtain the information they need through conversation. Based on the survey of conversational systems in [22], we aim to design task-oriented dialogue systems for application in military scenarios, focusing on question answering with martial training intent and relevant entity information. Therefore, conversational goals for social and entertainment purposes, such as greetings, entertainment, and advertising, are outside the focus of our system. Therefore, the ability of a military conversational AI to correctly detect its user's intent and identify entities in a sentence will determine whether it can successfully reply to users.

One challenge for intent detection is that the questions of military users can be terse and ambiguous [23]. Furthermore, the answer often depends on the context of the conversations. To narrow down the range of possible intent types, we first defined the range of applications for the types of intent it was meant to detect, and then classified and annotated the conversational data. In general, although the users' queries are short, most will mention the important entities. The role of slot filling is to identify and annotate the entities in the sentence, e.g., persons, events, times, locations, and weapons. As for the response provided by the system to the user, the challenges are to choose the most appropriate answer and to generate questions that require explicit information when missing the primary entity from the user's query.

To enable the task-oriented conversational system, the architecture comprises four modules based on a pipeline strategy: (1) slot filling, (2) intent detection, (3) retrieval-based answering, and (4) query generation. For the (1) and (2) modules, we trained by our prepared dataset by named-entity recognition (NER) models and a support vector machine (SVM) classifier [24], respectively. The (3) module is used by the BM25 algorithm [25] to retrieve the Military List and then rank the most appropriate solution using the Learning to Rank (LTR) model [26]. For the final module, we adopted the template-based question generation for the database of the Army Joint Task List (AJTL) according to the user's intent.

The performance of our military conversational system was experimentally evaluated in terms of the performance of its intent detection, slot filling, LTR modeling, and question generation modules. The system performed well based on both quantitative and qualitative metrics. Therefore, this study established a new approach for the development of military conversational systems. The proposed architecture could also be trained using other domain-specific datasets to expand its scope of applicability. The contributions of this study may be summarized as follows.

- A task-oriented conversational system was designed based on the practical needs of military tasks. As its module functions and datasets are mutually independent, it is possible to use this architecture to accelerate the training of domain-specific conversational systems in other domains, as one simply has to replace the dataset.
- This study defined the four core tasks of a conversational system and used machine-learning technologies to enable the realization. They included using NER models for slot filling, a classifier for intent detection, answering by the retrieval-based and learning-to-rank (LTR) model, and generating new queries by the template-based method.
- The experimental results highlight the performance of the intention detection, slot filling, sentence ranking, and the overall user satisfaction for the conversational system. The result can serve as a promising direction for future studies.

The remainder of this paper is organized as follows. Section 2 describes related work and technologies. Section 3 introduces the proposed architecture and functions. Section 4 presents the experimental results evaluating. Section 5 summarizes the tasks and discusses future directions.

2. Related Work

This section reviews related work of conversational AI and military conversational systems. As well as tasks and models for NLU, emphasizing intent detection and slot filling, we conclude with a review of response generation methods.

2.1. Conversational AI

The rapid development of AI technologies has increased academic interest in human–computer interfaces, with applications ranging from domain-specific settings to open-domain conversations. In the business world, personalized AI assistants such as Siri (Apple), Assistant (Google), Cortana (Microsoft), Messenger (Facebook), and Alexa (Amazon) have become increasingly common. Owing to the extensive labeling of conversational databases and the application of deep-learning and NLP techniques, conversational systems have made considerable progress in understanding the semantics of natural language and contextual reasoning. We divide conversational AI into several categories according to their purposes as the following.

Conversational AI, which are also known as chatbots, may be divided into task-oriented or non-task-oriented dialogue systems [27]. Task-oriented dialogue systems are meant to help users perform a specific task, e.g., intelligent food ordering, legal queries [28], and smart customer service. In addition, they usually have domain-specific conversational dialogues and knowledge bases.

Non-task-oriented dialogue systems (e.g., chatbots for the elderly or children) are meant to provide reasonable responses to users and thus provide entertainment and have open-domain dialogues that are not specifically constrained in scope. The first chatbot in the world was the Eliza chatbot in 1996 [29], which used simple dialogue to mimic a psychologist conversing with a patient. In 2017, Fitzpatrick et al., developed Woebot [30], a cognitive-behavioral therapeutic (CBT) chatbot, which was able to converse with patients and provide CBT assistance. Zhang et al. [31] proposed a unified conversational search/recommendation framework called "System Ask—User Respond," which was trained using a large collection of user reviews in e-commerce. They then evaluated the performance of this framework using metrics such as the Normalized Discounted Cumulative Gain (NDCG).

Task-oriented dialogue systems are typically designed with a "pipeline" consisting of four modules: a NLU module, dialogue state tracker, dialogue policy learning module, and NLG module [27]. Recently, some workers have proposed end-to-end frameworks to expand the expressiveness of the state space and support dialogue beyond domain-specific corpora [32]. For example, Zhao and Eskenazi proposed an end-to-end framework that used reinforcement learning and policy learning to optimize the dialogue system for dialogue state tracking. They tested this framework using a 20-question game, where the conversational system asked the user a series of yes-or-no questions to find the answer to a specific question.

There are three main approaches for conversational response generation [4]: rule-based approaches, retrieval-based approaches, and generative approaches. A rule-based system often requires a large amount of manual design and labeling work and therefore has the highest costs. Retrieval-based conversational AI uses keyword matching with machine learning or deep learning to determine an optimal predefined response [33]. Finally, generative conversational AI can be trained in multiple stages using supervised/unsupervised learning, reinforcement learning, or adversarial learning. Recently, Zhang et al. [34] presented a graph-based self-adaptive conversational AI; it used a knowledge graph whose nodes and links represented key entities and semantic relationships, respectively, as a dynamic knowledge base. It allowed the system to gain knowledge through conversations with end-users. Based on the definition above and the category of conversational AI, the system presented in this work may be characterized as a domain-specific (military) task-oriented dialogue system. To ensure that the dialogues produced by the conversational AI are compatible with the expectations of military tasks, we used a pipeline design and retrieval-based response generation method.

Military-domain task-oriented conversational systems can be divided into three types according to their mode of interaction: voice-type, text-type, and interactive interface-type systems. A few successful examples are described below. The Siri chatbot developed by Apple in 2011 began as a part of the "Cognitive Assistant that Learns and Organizes" project funded by the Defense Advanced Research Projects Agency (DARPA); by using perceptual and experiential learning, Siri reduced the information overload faced by battlefield commanders. It has since become a virtual voice assistant of national importance. With the development of expert assistant systems and the application of text-based conversational AI in military applications, IBM's Watson system came to be used to provide occupational information to US military members and help them transition from active duty into civilian life. In 1998, DARPA presented a dialogue system based on conversational multimodal interfaces, which allowed its users to perform operational tasks more efficiently [28].

Due to the lack of relevant literature on military dialogue systems, this study summarizes relevant research on military dialogue systems in different periods in the past, as shown in Table 1. For example, Roque et al. [35] in 2006 proposed a spoken dialogue system that can engage in Call For Fire Radio dialogues (Radiobot-CFF) to help train soldiers in proper procedures for requesting artillery fire missions. They provided three modes: fully-automated, semi-automated, and passive mode, as the radio operator in a simulated Fire Direction Center (FDC) takes calls from a forward observer for artillery fire in training exercises.

The Hassan system [36] proposed by Gandhe in 2009 is a set of tactical question-answering dialogue systems, including a management interface for creating dialogue content and a dialogue manager, which can be used to build multiple virtual characters for tactical questions. The experiment consisted of 19 dialogues and 296 utterances. Furthermore, the experts expanded the range of possible responses provided by the virtual character by annotating other candidate utterances according to need. However, the system lacks the capabilities of question generation.

MIRIAM is a conversational multimodal interface developed for command-and-control systems proposed by Robb et al. [21]. The system improved situational awareness by providing information in multiple modalities (including audio, images, and text) to clarify the textual ambiguities that often arise in natural language and improve understanding. Therefore, multimodal conversational AI could become an important trend in the design of military conversational systems. The recent example of a successful military conversational AI would be the human-robot navigation system of Gervits et al. [20].

Table 1. Summary of military conversational systems.

	Interface	NLU	Dialogue State Tracking	NLG
Radiobot-CFF [35]	Spoken	✔	✔	✕
Hassan [36]	Text	✔	✔	✕
MIRIAM [21]	Multimodal	✔	✔	✔
Gervits [20]	Spoken	✔	✔	✔
Our study	Text	✔	✔	✔

2.2. Intent Detection and Slot Filling

In a conversation, the queries provided by the user are usually relatively short. Therefore, the dialogue system must first determine the aim or intent of the question. However, the information within the query may be incomplete or stated implicitly. Suppose the dialogue system does not possess the knowledge or context necessary to answer the question. In that case, it may require several rounds of dialogue to redress these issues and confirm the user's intentions.

In 2018, Zhang et al. [31] proposed a "circular" conversational architecture, where the dialogue system clarified a user's intentions through several rounds of dialogue. Intent detection pertains to the determination of intent by analyzing the structure of the question [37], e.g., by "5W2H" analysis (why, what, where, when, who, how, and how much) [38], to improve the accuracy of the retrieved answer. Furthermore, as the information contained by the query is critical for determining the correct answer, NER techniques can be used to identify key 12 persons, events, times, places, and objects and narrow down the query's scope.

Intent detection may be performed using statistical or rule-based methods. For example, Setyawan et al. [39] used Naïve Bayes and logistic regression machine-learning methods to perform intent detection, with the term frequency-inverse document frequency being the classification feature. In 2012, Wang et al. [40] converted short snippets into vectors to perform intent (sentiment) classification and compared the SVM, Naïve Bayes, and continuous bag-of-words methods. The typical representative one of the supervised learning methods is the SVM. Over the past 10 years, many scholars have uses SVM as a comparative method for intent classification or sentiment analysis, and the summary references are as shown in Table 2. In addition, deep-learning models have also become commonplace in intent detection. Nigam et al. [41] used a recurrent neural network to perform multi-staged named-entity learning and then used the named entities as classification features.

Table 2. Summary of SVM Models used in NLU Applications.

Reference	CA [1]	Domain	Language	Data Size	Models [2]	Optimal
Chen, 2012 [42]	✔	Community Question Answering	English	1.5 K	SVM, C4.5, RF, NB, KNN	SVM
Bhargava, 2013 [43]	✔	Audiovisual Media	English	27.5 K	SVM, HMM, CRF	SVM
Sarikaya, 2016 [44]	✔	Personal Assistant	English	400 K	SVM	SVM
Gaikwad, 2016 [45]	✕	Sentiment Analysis	English	8 K	SVM, NB, KNN	SVM
Sullivan, 2018 [46]	✕	Booking flights/ Accommodation	English	8 K	SVM, CNN	≈

Table 2. *Cont.*

Reference	CA [1]	Domain	Language	Data Size	Models [2]	Optimal
Troussas, 2020 [47]	×	Learning Styles	English	<1 K	SVM, NB, KNN, ensemble	ensemble
Rustamov, 2021 [48]	✓	Banking Services	Azerbaijani	161 K	LR, SVM, NN, DIET	DIET
Our study	✓	Military Training	Chinese	10 K	SVM	SVM

[1] The study is used to develop a conversational system (CA). [2] Model abbreviation: Random Forest (RF), Logistic Regression (LR), Naive Bayes (NB), *k* Neural Network (KNN), Convolutional Neural Network (VNN), Hidden Markov Model (HMM), Dual Intent and Entity Transformer (DIET).

Slot filling is another critical task in dialogue systems, as it provides semantic information and helps the conversation system determine which bits of information in a sentence should be searched for. In previous studies, slot filling has often been performed using generative models (such as the hidden Markov model) or discriminative models (such as the conditional random field (CRF) model) to estimate the conditional probabilities of slot labels in a sequence. However, with the emergence of deep-learning models, bidirectional recurrent neural network (RNN) models and long short-term memory (LSTM) models trained with contextually annotated sentences are now used for slot filling. In recent studies, CRF models have been combined with RNNs to train slot-filling models for unseen semantic labels and multi-domain tasks, enhancing their performance. Yang et al. [49] presented an intent-aware neural ranking model, which used the "Transformer" architecture to perform language representation learning and to analyze user intent patterns in information-seeking conversations.

Intent detection is usually viewed as a supervised classification problem, that is, mapping a sentence to some class within a finite set of classes. Slot filling, in contrast, is viewed as a token sequence labeling problem. Traditionally, intent detection and slot filling are performed separately or in a pipeline. Recently, some studies have investigated the use of joint models for simultaneously performing intent detection and slot filling [50], and have proven that these tasks are closely related to each other. Compared to the pipeline approach, joint models are less susceptible to error propagation between the intent detection and slot-filling models. In addition, they can be trained and tuned as a single model. However, joint models cannot be easily generalized to unseen data due to variations in natural language expressions for the same intent.

Furthermore, domains and label sets can change over time in real-world applications. There is a lack of publicly available task-oriented datasets among the conversational corpora used for training. Moreover, the available datasets are limited to a few specific domains. These corpora may be divided into two types: the first type comprises user–system conversations, such as the Air Travel Information System [51] and WOZ2.0 [52]; the second type includes simulated human–system dialogues subsequently and manually converted into natural language, such as the machine-to-machine dataset [53].

2.3. *Response Generation*

NLG is the phase in NLP where task-oriented dialogs are completed to meet user needs. Response generation by the system can be performed using retrieval-based or generation-based models. Retrieval-based answers generate dialogue by retrieving the best responses from the corpus through a ranking function and often have highly fluent and informative answers. However, it tends to be repetitive and cannot handle semantics outside its corpus. On the other hand, generation-based conversational systems use logic to infer spoken responses and are therefore not bound by response templates. Traditionally, NLG always involves sentence planning, where the input semantic symbol is mapped to an intermediary representation of the utterance (e.g., a tree-like or template structure). The intermediary process is then converted into the final response through surface realization.

In 2002, Sneiders [54] presented a template-based automated answering model. The model considered four entities (human names, locations, organizations, and times) and used NER to extract information (keywords). The keywords were then matched to question templates to create answers. However, answer generation alone may not produce an adequately correlated response with the original question. To address this problem, reference [55] used the co-occurrence of technical terms in a corpus to infer whether they were correlated. In 2003, Fiszman et al. [56] presented the SemRep system, which used manually-listed template rules for verbs to identify potential semantic relationships in a sentence and used identification criteria to select relevant technical terms and phrases.

Bhoir and Potey [57] proposed a heuristic retrieval-based conversational system for retrieving the most relevant answers from a predefined corpus based on a user's input and used complete sentences as candidate answers. It also considered the type of answer to select an appropriate response to the user. Choosing a proper reply is the most critical problem in question-answering systems, and the reaction must also be clear and concise. Therefore, selecting only the most critical information when formulating a reply is necessary. In another study [58], a similarity approach was used to predict whether a message was a reply to another message. This approach was validated by comparisons with a trained bidirectional encoder representation from the transformers model, with conversations generated by a bidirectional LSTM and RNN. The authors found that this approach helps to improve the understanding of the context.

Recently, there has been increasing interest in finding distributed vector representations (embeddings) for words, that is, by encoding the meanings of text into vectors. The text may range from words, phrases, and documents to human-to-human conversations. In 2017, Bartl and Spanakis [59] used a locality-sensitive hashing forest, an approximate nearest neighbor model, to generate context embeddings and find similar conversations in a corpus. The candidate answers were then ranked.

In 2018, Juraska et al. [60] presented a sequence-to-sequence natural language generator with an attentional mechanism and was able to produce accurate responses for a variety of conversational domains. In 2019, Song et al. [61] from Microsoft proposed the method based on the concept of pre-training and sequential neural networks. This method masked a segment of a random length and used an encoder–decoder attention mechanism to generate responses.

In 2020, Wang et al. [62] from Tencent proposed a deep-learning-based TransDG model for generating Chinese conversations. This model performed question–answer and semantics-named entity matching within its knowledge base and then selected the optimal strategy for response generation.

A recent study showed that template-based conversational AI faces two key challenges: (1) constructing a system grammar that balances the expressiveness necessary to conduct a task with the ability to infer parses from natural language correctly; and (2) dealing with parse ambiguities. Seungwhan et al. collected a new open-ended dialogue-KG parallel corpus called OpenDialKG [63], where each utterance from 15,000 human-to-human role-playing dialogues was manually annotated with a ground-truth reference to corresponding entities and paths from a large-scale knowledge graph with more than one million facts. They also proposed a DialKG Walker model for learning the symbolic transitions of dialogue contexts via structured traversals over the KG, and used an attention-based graph path decoder to predict the entities. Bockhorst et al. [64] addressed parse ambiguities by using a context-free grammar called episode grammar; the system constructed a semantic parse progressively for a multi-turn conversation, where the system's queries were derived from the parse uncertainty.

3. Methodology

This work proposes a conversational system architecture that uses machine-learning techniques through a Chinese corpus for military training missions. It includes the mission list of the joint training management system, the military dictionary, and the Army Joint Task List (AJTL). As the implementation of this system is independent of its domain and language, it can be used to enable conversational systems in other fields or languages by changing its corpus.

3.1. System Architecture

This study aims to develop a conversational AI for quickly answering soldiers' questions in the military training mission and supporting multiple conversation rounds. The architecture of our conversational system is shown in Figure 1. The user's query is first parsed by NLP, followed by a slot-filling module, which identifies important entities, and then the intent type is detected by the intent detection module. The system performs retrieval-based answer generation through the extracted entities and intents. The retrieval-based question-answering system ranked and selected the optimal responses. If the user confirms the answer is clear, take action; otherwise, the system will generate a new query to verify the user's intent. Slot filling and intent detection are the NLU stages for understanding, and retrieval-based answering and query generation are the NLG stages for responding. In the NLU stage, we use the CRF and SVM models to train slot-filling and intent-detection modules, respectively, which are practical and easy to implement due to the limited training information set. In addition, we make some summaries of the use of these models in related research. In the NLG stage, we use a learned ranking method to obtain retrieval-based answers. There are two basic types of generating sentences: extracted and abstract. This method is determined according to the greater probability of finding and querying within the existing corpus. The advantages of this method are that the grammar is relatively smooth and easy to understand and does not require a large number of training datasets—the main reason for responding to build mods. We use a template-based strategy in the query generation module. This template-based query generation method may propose a new query for the missing intent or entity and user to be confirmed with the user, that is, for the intent and entity to continue the dialogue, with the intention to avoid generating a new query and diverging context.

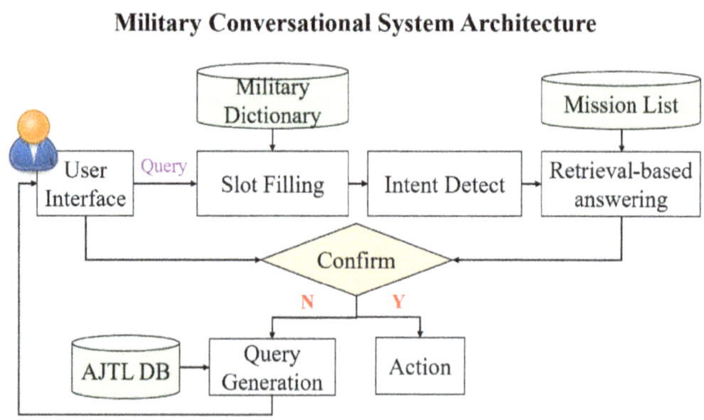

Figure 1. System architecture our conversation system.

3.2. Slot Filling and Intent Detection

Figure 2 illustrates the flow of a user's query. In the query "何時將完成後備部隊動員任務？(When will the reserve force complete the mobilization task?)," entities such as "何時 (when)" as B-time, "後備部隊 (reserve force)" as B-unit and I-unit, "動員任務 (mobilization task)" as B-event and I-event were annotated. Slot labels are labeled using the BIO format: B indicates the beginning of a slot span, I the middle of a span, and O indicates that the label does not belong to a slot. In addition, the query intent of this sentence is "when".

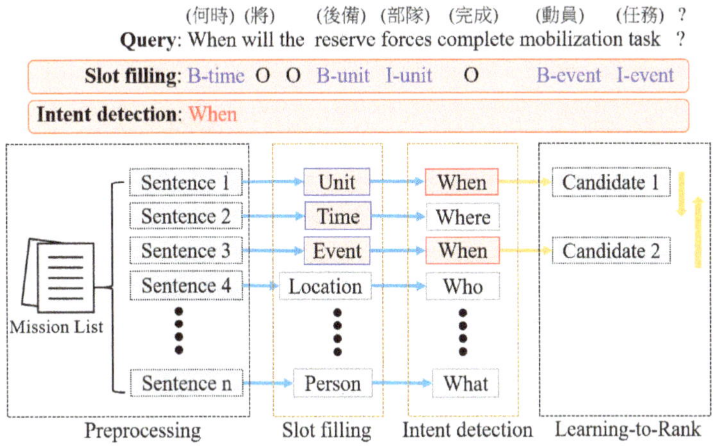

Figure 2. Flowchart of slot filling and intent detection.

To ensure the conversational system is able to deliver the correct intents and related entities, the system analyzes a soldier's utterances first to identify the entities mentioned and match them with intents stored in the mission list database, and then orders the appropriate response sentences. Because traditional retrieval systems rely on retrieving full-text search results for user queries, retrieval models are based on the similarity between the query and the text (e.g., vector space models). Therefore, the user may miss the correct answer because the intent of sentences with high similarity may not match the intent of the user's question. In other words, we prioritize intent and entity accuracy before evaluating similarity.

3.2.1. Slot Filling

The slot-filling task, in contrast, was defined as a sequence labeling problem, that is, an NER problem. We trained used five kinds of entities by the CRF model. Considering the amount of data and the implementation of integrating multiple modules, we choose the CRF-based method as the baseline for slot filling.

During the preprocessing phase, the CkipTagger (https://github.com/ckiplab/ckiptagger (accessed on 31 January 2022)) tool was used for Chinese word segmentation and part-of-speech (POS) tagging for the user's query. The CRF toolkit was performed to train five NER models. Five types of slots related to military missions were defined: the military unit and location, the name of military personnel (including job titles and ranks), the name of military event tasks, the name of the weapon, and time. As shown in Table 3, There are six types of features: POS tagging, vocabulary, specific terms, verbs, quantifiers, and punctuation. We match them to entities for vocabulary and specific terms, and the vocabulary source is the Military Dictionary. For verbs, quantifiers, and punctuation, we use them to determine the boundaries before and after entities. In summary, we trained five CRF models to predict five entities (i.e., location/unit, person, event, weapon, and time) for slot filling. The CRF model was then used to estimate the conditional probability of the sequence, as shown in Equation (1).

$$P(y|x) = \frac{1}{Z(x)} \prod\nolimits_{t=1}^{T} exp \sum\nolimits_{k=1}^{K} \theta_k f_k(y_t, y_{t-1}, x_t) \quad (1)$$

If it is assumed that x and y are random variables, given an observed sequence X, $P(y|x)$ is the conditional probability distribution of the hidden sequence Y, whose probability estimate in the state t depends on that in the state $t-1$. $Z(x)$ is a normalization function for normalizing the value of $P(y|x)$.

Table 3. Features of CRF models for slot filling.

	POS	Vocabulary	Specific Term	Verb	Quantifier	Punctuation
Location/unit	Y	Y	Address suffix	Y	N	Y
Person	Y	Y	Surname list	Y	Y	N
Event	Y	Y	Event suffix	Y	N	Y
Weapon	Y	Y	Digit/alphabet	Y	Y	Y
Time	Y	Y	Time suffix	Y	Y	Y

Note that the "Y" indicates that the feature is used, and the "N" indicates that the feature is not utilized.

3.2.2. Intent Detection

In this study, we adopted the SVM multi-class classification method [65] for intent detection. The algorithm constructs k SVM models, where k is the number of classes. All the examples in the mth class with positive labels are used to train the mth SVM and all the other examples with negative labels. Formally, given training data $(x_1, y_1), \ldots, (x_l, y_l)$, where $x_i \in R^n$, $i = 1, \ldots, l$, and $y \in 1, \ldots, k$ is the class of x_i, the mth SVM solves the following problem:

$$\begin{aligned} &\min_{w^m, b^m, \xi^m} \frac{1}{2}(w^m)^T w^m + C\sum_{i=1}^{l}\xi_i^m \\ &(w^m)^T\phi(x_i) + b_m \geq 1 - \xi_i^m \quad if\, y_i = m \\ &(w^m)^T\phi(x_i) + b_m \leq -1 + \xi_i^m \quad if\, y_i \neq m \\ &\xi_i^m \geq 0, i = 1, \ldots, l \end{aligned} \tag{2}$$

where the training data x_i are mapping to a higher dimension space by the function ϕ and C is the penalty parameter.

Here, intent detection is regarded as a multiclass classification problem, with the intent in a query consisting of four parts: who, where, when, and what. As we did not consider the possibility of multiple intents in one question, the hard classification performed the intent prediction with the highest probability of the query. SVM is one representative machine classifier for supervised learning methods. Many scholars use SVM as a comparison or combination method in recent studies, as discussed in Refs. [40,46,47,66,67]. However, despite these years of research, intent detection is still challenging. The classifier is used for intent detection by a SVM classifier, as SVMs are accurate for this task.

After extracting entities, there were 12 types of features for training user's intent, as shown in Table 4. Features 1–5 were the five types of entities extracted by NER models. Regarding Features 6–8, we used the Military Dictionary to match whether the query sentence contains military words and quantifiers. Features 9–12 were Common interrogative terms in Chinese. Formally, the multi-class classifier is used to predict an unseen sample x with labels 1 to k, which assigns the highest confidence score, as shown in Equation (4). We used a simple one-hot encoding to facilitate the classifier's training for nominal features, with matched features being one and unmatched features being 0.

$$y = argmax_{k \in \{1 \ldots K\}} f_k(x) \tag{3}$$

For a conversation system, there is difficulty remembering conversational intent from one sentence to the next. In other words, the procedure typically treats each query from the user as a new dialogue state. To alleviate this problem, our system stores intents and entities extracted from one round of conversations and intents and entities extracted from previous rounds of conversations. When the system confirms whether the response meets the user's information needs, if the user gives a negative reply, the intent and entity of the query are stored to avoid forgetting.

Table 4. Features of intent detection.

No.	Feature	Description
1	Location	Military location or organization
2	Person	Name, rank and title entities
3	Time	Time descriptor
4	Event	Military event
5	Weapon	Weapon or transport entity
6	Document	Military document name
7	Session	Phase name of combat missions
8	Unit	Commonly used quantifiers in military affairs
9	Who	An interrogative term about a person
10	Where	An interrogative term about a location
11	When	An interrogative term about time
12	What	An interrogative term about all other matters

3.3. Response Generation

After extracting the user intent and filling the slots, the second step is to answer the user using a retrieval-based response module. Suppose no intent or relevant entity was identified in the previous step. In that case, the system uses a template-based query generation module to ask the user for additional information to retrieve an appropriate answer. In practice, we use the Elasticsearch full-text search tool to build a retrieval model in the Chinese military domain. The model is built using the Okapi BM25 algorithm. The model determines the most appropriate response based on the correlation between query Q and database document D, as shown in Equation (4).

$$Score(D,Q) = \sum_{i=1}^{n} IDF(q_i)\frac{f(q_i,D)(K_1+1)}{f(q_i,D)+K_1(1-b+b\frac{|D|}{avgdl})} \tag{4}$$

where *IDF* denotes the inverse document frequency in this equation, and *avgdl* is the average length of all texts.

In the retrieval system, the entities extracted from the user query are used as the keywords to perform full-text retrieval. The top 10 most relevant results were then selected using the Learning-To-Rank Answering (LTRA) apporach, as shown in Algorithm 1. The input to the retrieval model is the searched sentences $S = \{s_1, \ldots, s_n\}$, query intent i_Q, and entities $E = \{e_1, \ldots, e_m\}$. First, intent prediction is performed for each sentence. If a sentence j intends i_j to be the same as i_Q, the candidate sentence j is kept; otherwise, it is discarded. The LTR model ranks the candidate sentences and considers the top k candidate sentences as the most suitable replies.

The LTR model was implemented using the LambdaMART algorithm [68]. LambdaMART is a listwise LTR that combines the LambdaRank and Multiple Additive Regression Tree (MART) algorithms, transforming the search candidate ranking problem into a regression tree problem. To train candidate sentences for ranking, we prepared 10 features, as shown in Table 5. Features 1–5 represent the five NER models for extracting entities from candidate sentences. Features 6 and 7 are used to determine if the candidate sentence matches the military document names and units in the Military Dictionary; the results are displayed as boolean values. Feature 8 represents the longest common subsequence (LCS) between the user query and candidate sentences, as shown in Equation (5). Feature 9 denotes the similarity between the user query and candidate sentence, as shown in Equation (6). Feature 10 denotes the term frequency-inverse document frequency (TFIDF), which is used to calcuate the word importance for user's query and sytem answers based on the corpus, as shown in Equation (7).

Algorithm 1 Learning-To-Rank Answering.

```
Input: search sentences S, query intent i_Q, entities E
Output: ranked sentences S'
Initialize candidate set C is empty
for sentence j = 1,...,n from S do
    if sentence intent i_j = i_Q then
        candidate set C ∪ sentence j
    end if
end for
while candidate set C ≠ {} do
    Rank sentence cs ∈ C by the LTR model
end while
Return top-k sentences S' = {cs_1,...,cs_k}
```

Table 5. Features of the learning to rank (LTR) model.

No.	Feature	Description	Value
1	Location	Military location or organization entity	yes/no
2	Person	Personnel name, rank and title entity	yes/no
3	Time	Time descriptor	yes/no
4	Event	Military event	yes/no
5	Weapon	Weapon or transport entity	yes/no
6	Document	Military document name	yes/no
7	Unit	Commonly used quantifiers in military affairs	yes/no
8	LCS	Common strings between user's query and system response	[0, 1]
9	Cosine	Similarity between user's query and system response	[0, 1]
10	TFIDF	Word importance for user's query and system response	[0, 1]

$$LCS(i,j) = \begin{cases} 0 \text{ if } i = 0 \text{ or } j = 0 \\ \text{LCS}(i-1, j-1) + 1 \text{ if } i, j > 0 \text{ and } \alpha_i = \beta_j \\ \max\{\text{LCS}(i-1, j), \text{LCS}(i, j-1)\} \text{ if } i, j > 0 \text{ and } \alpha_i \neq \beta_j \end{cases} \quad (5)$$

The above calculates the LCS for input sequences $A = \alpha_1, \alpha_2, \ldots, \alpha_m$ and $B = \beta_1, \beta_2, \ldots, \beta_n$, where $1 \leq i \leq m$ and $1 \leq j \leq n$.

$$Cosine(A,B) = \frac{(A \cdot B)}{(|A| \times |B|)} \quad (6)$$

$$TFIDF = \frac{tf_{i,j}}{\sum_k n_{k,j}} \times log \frac{|D|}{\{j : t_i \in d_j\}} \quad (7)$$

In the above, X and Y are discrete random variables, that is, the correlation between the entity sets of the user query and the candidate sentence.

The question generation module generated new queries based on the template-based representation of the intent–entity relations of queries. The structural composition was viewed as a set of intent–entity relationships within the state space, as shown in Figure 3. The intents of a question included who, where, when, and what, whereas named entities were in six types: person, unit, event, time, weapon, and document (Doc). Candidate sentences could be formed based on the occurrence probabilities of the intent–entity relationships. For example, the intent "who" and entity "event" was related by "be responsible for" in "the commander (who) is responsible for this combat readiness mission (event)." In general, as a response should also have a person as the intent (who) and a combat readiness mission (event) as the entity, the candidate sentences should be ranked according to the presence of the "event" entity and "who" intent in these sentences.

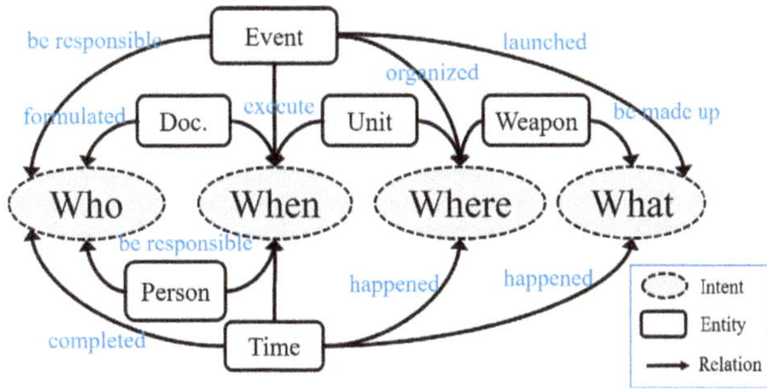

Figure 3. Intent–entity relations.

In the question generation module, the predicted intent types and the queried entities are combined according to the intent–entity relations shown in Figure 3. Table 6 shows some examples of such parsed sentences. Brackets correspond to slot-fill annotation entities, and bold corresponds to intent types. Templates can be applied in various combinations. To select the best new question from candidate sentences generated by multiple templates, we define Equation (8) to score each query–new-question pair. The higher the score, the stronger the semantic relevance between the newly generated question and the original query.

$$QQ_i = \alpha(LCS_i + Cosine_i) + (1 - \alpha)TFIDF_i \tag{8}$$

In this equation, i refers to the candidate sentence generated by the i-th rule for the same intent; α is a weighting parameter that ranges from 0 to 1; LCS is the longest common subsequence between the user query and generated query; Cosine is the cosine similarity between the user query and generated query; and TFIDF is the importance of words around user's query, which is calculated the product by term frequency and inverse document frequency.

Table 6. Exemplary template-based question generation via intents and corresponding entities.

No.	Intent	Entity	Templates
1	Who	Doc.	Who + is in charge of + [Doc.]?
	誰負責人員調度? Who is in charge of personnel management?		
2	Who	Person	Who + [Person] + reports to?
	作戰科科長需向誰提報旅部作戰計畫? Who does the Chief of operations need to present the brigade combat plan.		
3	Who	Event	Who + is in charge of + [Event]
	野戰照明工作是誰負責? Who is responsible for field lighting?		
4	Who	Unit	Who + are in the +[Location]?
	救災編組有哪些人? Who are in the disaster relief team?		
5	When	Event	When + the + [Event] + will happen
	什麼時候執行綜合演練? When will the joint drill happen?		
6	When	Unit	When + [Location] + will finish + [Event]
	想知道後備部隊在何時要完成動員整備任務? When the reserve forces will complete their mobilization and preparation?		

Table 6. *Cont.*

No.	Intent	Entity	Templates
7	Where	Unit	Where is the + [Location] + located?
	軍團指揮部位於哪裡? Where is the command of army located?		
8	Where	Weapon	Where did the + [Location] + discover the + [Weapon]?
	機旅在哪裡發現敵軍2部戰車? Where did the brigade discover two enemy tanks?		
9	What	Event	What + is the basis for performing this + [Event]?
	執行地面任務是依據什麼準則? What is the criteria for performing this ground mission?		
10	What	Weapon	What + is the range of this + [Weapon]?
	戰車砲攻擊距離有多遠? What is the range of this tank's main gun?		

4. Experiments

This section evaluates the performance of the proposed system, including describing the datasets and metrics used, the experimental evaluation of the intent detection and slot-filling modules, and the response generator's ranking performance evaluation. Finally, the overall performance of the dialogue system is discussed.

4.1. Datasets and Measures

A total of 1307 human-labeled sentences are included in the experimental dataset used for intent classification (who, where, when, and other). Table 7 shows the four types of intent quantitative sentences. As shown in Table 8, the experimental datasets were used to train the five NER models, including people, weapons, places, events, and time. The intent detection and slot-filling tasks in the NLU stage are evaluated using F1-score (Equation (9)) and accuracy (Equation (10)). In the following equations, true positive (TP) and true negative (TN) are the numbers of accurately predicted positives and negatives, respectively. Conversely, false positive (FP) and false negative (FN) are the numbers of wrongly predicted positives and negatives, respectively. Thus, Precision = $|TP| / |TP + FP|$ and Recall = $|TP| / |TP| + |FN|$.

Table 7. The number of datasets used for intent classification.

	Who	Where	When	What
# of sentence	335	320	329	323

Table 8. The number of datasets used for NER training.

	Person	Weapon	Location	Event	Time
# of sentence	1005	1007	1006	1005	1000
# of entity	3693	2465	1843	2399	1820

$$\text{F1-score} = \frac{2 \times \text{Precision} \times \text{Recall}}{\text{Precision} + \text{Recall}} \tag{9}$$

$$\text{Accuracy} = \frac{|\text{TP} + \text{TN}|}{|\text{TP} + \text{FP} + \text{TN} + \text{FN}|} \tag{10}$$

The LTR model was evaluated using the normalized discounted cumulative gain (NDCG), which gives the normalized relevance score of the files retrieved by the search engine at each rank position. Files closer to the top are given a higher weight (and therefore have a greater degree of influence on NDCG), as shown in Equation (11).

$$NDCG_p = \frac{DCG_p}{IDCG_p} \tag{11}$$

where *IDCG* is ideal discounted cumulative gain, and rel_p represents the list of relevant documents in the corpus up to position p.

$$DCG_p = \sum_{i=1}^{p} \frac{rel_i}{log_2(i+1)} \tag{12}$$

DCG is based on the principle that highly relevant documents appearing lower in a search result list will be penalized by having their relevance grade reduced logarithmically proportional to their position in a search result list. Finally, the query generation module is evaluated by quantifying user satisfaction.

4.2. Performance of the Intent Detection and Slot-Filling Modules

SVM models are used to perform multi-class classification. The dataset is randomly divided into training and test sets in three different proportions. According to the results shown in Figure 4, the 9:1 ratio outperforms the 7:3 ratio in terms of F1 score and accuracy, and achieves 90% accuracy. The performance for predicting the four intents is then performed based on the model trained with a data ratio of 9:1, as shown in Figure 5. The multi-class classifier had the highest F1 score for "where" intent (92%), followed by "when" (91.4%), "who" (91.1%), and finally "what" (88.8%). The average F1-score of the classifier was 90.1%. Due to the limited amount of training data (1307 sentences in four categories), the F1 performance of the trained intent detection model is 88 91%. However, from the perspective of the learning curve (dataset splitting rate), the performance improves as the amount of training data increases. Further comparing the performance of each category, we can see that the "What" category has the most errors, followed by "Who" and "When", and "Where" has the best performance. We analyzed the possible reasons and found two: one is language. For example, in Chinese grammar, the query for "What" is more complex than the other three categories, which may include "Why", "How", "which", etc. Another possible reason is that when the user's query has multiple intents, the classifier will only predict the class with the highest probability, so the number of false negatives for the "what" intent increases.

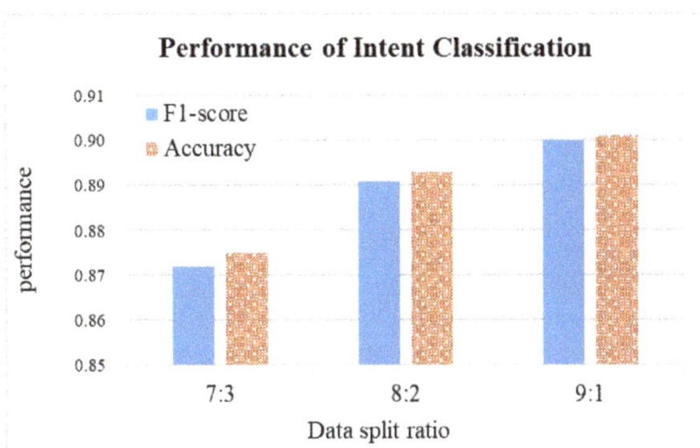

Figure 4. Performance of intent classification.

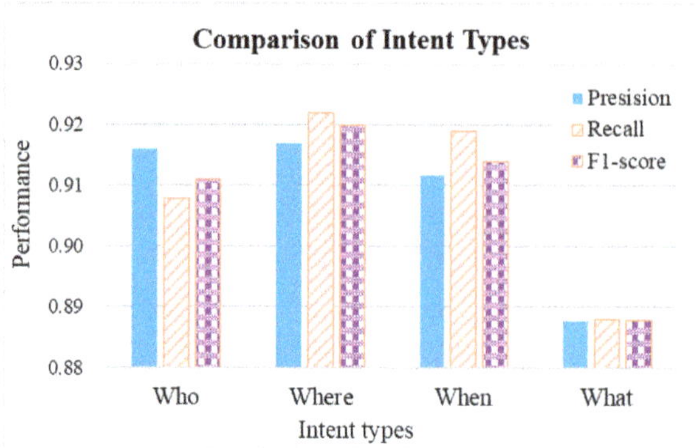

Figure 5. Performance of four types of intent classification.

The performance of the NER model is evaluated by performing five-fold cross-validation on the dataset. Figure 6 shows the performance of NER in recognizing military names, weapons, military locations, military events, and time entities. In terms of F1-score, the five models have the highest accuracy (0.943) for person names, followed by time. Conversely, it performed slightly worse at identifying military event, at 0.848.

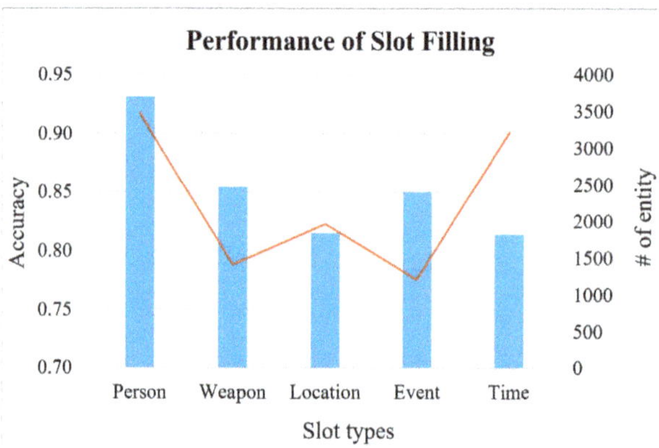

Figure 6. Performance of five types of named-entity recognition (NER).

4.3. *Performance in Response Generation*

Next, it evaluates the performance of retrieval-based answers. First, a question-and-answer dataset of 180 military joint tactical action lists is manually collected as standard answers. Three sentences are then randomly selected for each query, resulting in four candidates. Responses in the training dataset are ranked from 1 to 4, where 4 represents the highest score for the questions–answer pairing. We use the NDCG indicator as to the evaluation indicator for the LTR model. The training and testing datasets are divided into three different scales, and the experimental results for each scale are shown in Figure 7. Since the 9:1 ratio gives the highest NDCG score, this model is used as the LTR model for our conversational system.

Finally, eight military scenarios are randomly selected from 40 question-and-answer test data for performance evaluation. The Table 9 shows questions and answers for eight military scenarios. The intent is first identified for each question, and then responses are generated based on the extracted intents and entities and intent–entity relationships. Each intent may generate multiple sets of candidate sentences, and the first set of responses is the candidate sentence with the highest relevance to the query. The results of the dialogue were qualitatively assessed by 47 military-related personnel. The proportion of responses that meet user needs is shown in the Figure 8. If the user is not satisfied with the first response, the dialogue system generates a new query and then provides the second answer. Experimental results show that, on average, 66% or more of users are satisfied after the second round of conversations. Therefore, new queries generated by the system in the second round of dialogue (when the first answer is not satisfactory) significantly improved user satisfaction.

Table 9. Eight scenarios of question answering.

Session	Examples of Question Answering
S1	Q:通信單位用什麼裝備開設? What equipment will the communications group use to set up?
	A: 通信單位用多波道系統開設。 The communications group uses a multi-channel system to set up.
S2	Q:誰與火力協調機構協調建立安全空域? Who coordinates with the Fire Coordination Agency to establish safe airspace?
	A: 由聯絡官/防砲官協調火力協調機構。 Army liaison and anti-gunnery officers coordinate with the Fire Coordination Agency.
S3	Q: 誰決定勤務支援的優先順序? Who decides the prioritization of service support?
	A:指揮官決定勤務支援的優先順序。 The commander decides the prioritization of service support.
S4	Q:聯合空降將部隊投入哪裡? Where does Joint Airborne put troops?
	A:將作戰部隊及後勤支援部署到目標區域。 Deploy combat forces and their logistical support into target areas.
S5	Q:災害應變中心開設分幾級？ What are the levels of disaster response centers?
	A:共分兩級。情況輕者以二級開設，情況嚴重者一級開設。 There are two levels. Milder cases are opened at Level 2, and severe cases are offered at Level 1.
S6	Q:從駐地演習到動員需要多長時間? How long will it take from station maneuvers to mobilization?
	A:駐地機動至動員需2小時。 Station maneuvers to mobilization takes 2 h.
S7	Q: 火力支援組位於哪裡? Where is the fire support team?
	A:火力支援組編成位於作戰地區，建立聯合軍民防空情報傳遞系統。 The fire support team has been deployed in the combat zone to establish a joint military-civilian air-defense information transmission system.
S8	Q: 戰備整備是什麼? What is combat readiness?
	A:戰備整備依據地面防衛作戰構想，預想作戰推移與需求。 Combat readiness is based on the concept of ground defense operations, misunderstood combat progress, and needs.

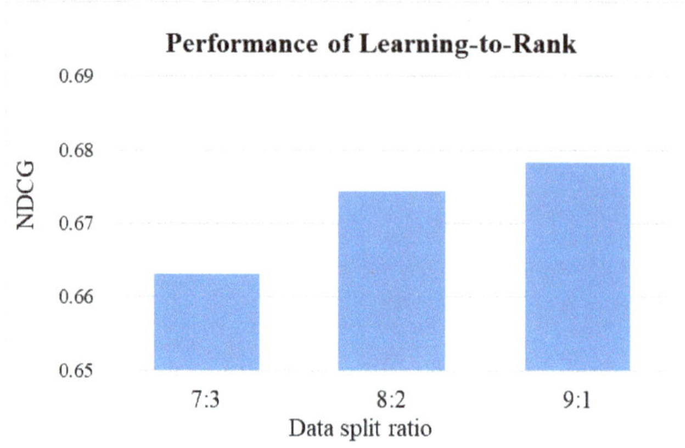

Figure 7. Performance of the learning-to-rank model.

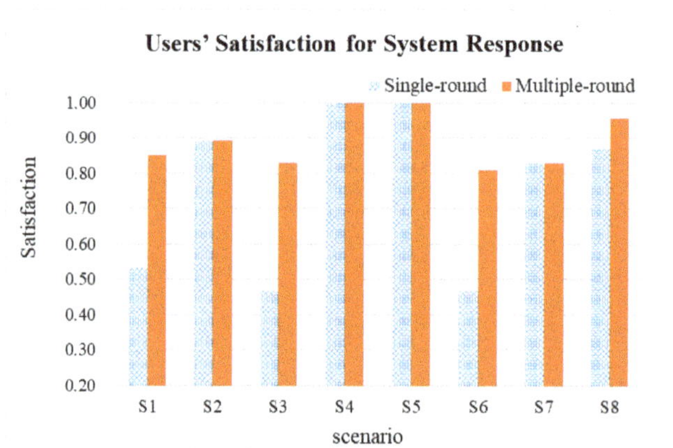

Figure 8. Users' satisfaction for system response.

4.4. Discussion

Here, we conduct an error analysis of the module performance and discuss the challenges of research and limitation. For the intent classification module, we found that the "what" category has the highest error rate (12%) because there are some queries with different interpretations, such as "why", "how much", "how to do", etc. For example, "how many days can each cavalry company unit fight independently?" "How many exchange centers and medium-sized communication centers can a communication force establish?" These false-negative examples of the "what" leads to lower recognition performance than other classes. It will be possible to further segment the user's intent in the future to guarantee that every feature of that intent is clearly defined.

For the slot-filling module, the accuracy for five types of entities is identified by the military personal name as the highest (0.917), followed by time (0.901), location (0.823), weapon (0.788), and military event (0.776) as the lowest. There are two main reasons for the poor attribution of military events: (1) Event names are longer than other entities, making it more challenging to identify the limits of the entity. Still, the system identifies part of the names of military activities. (2) The event name contains time or location, which is misjudged as another entity.

For the retrieval-based answering module, we use the learning-to-rank method to achieve results. From the learning curve point of view, with the increase in data, the efficiency of the system response improves (NDCG = 0.678). Candidate sentences are added only when the intent of the sentence is the same as the intent of the user's query. Then, we adopt these entities in the sentence as sorting features, and finally, sort them based on the LambdaMART algorithm. As we analyze why the correct sentence is not ranked first, we discover that pronouns may represent entities in sentences or omit them; thus, some candidate sentences do not identify entities related to the query, resulting in a low ranking score.

Based on the methods comparison in related literature, Sullivan [46] compared CNN and SVM, two ML algorithms with good performance records in the current NLP literature. However, the CNN model is not necessarily better than the SVM model based on a detailed statistical analysis of the experimental results. Under these experimental conditions, the SVM model using the radial basis function kernel produced statistically better results. However, SVM has its limitations. SVM is not suitable for large datasets because the complexity of algorithm training depends on the size of the dataset [69]; SVM is not ideal for training imbalanced datasets, which causes the hyperplane to be biased towards the minority class [70]. In terms of performance, choosing an "appropriate" kernel function is crucial. For example, using a linear kernel when the data are not linearly separable can lead to poor algorithm performance.

Two factors for the superior performance of the state-of-the-art are rich training datasets and high-speed hardware such as GPUs. We choose the CRF-based method, mainly considering the amount of data and training cost. Since obtaining a large amount of military training data in Chinese is a challenge, this study implements a dialogue AI system applied to military training scenarios using a limited military dialogue dataset. Using a CRF-based model is indeed a baseline approach. Nonetheless, this is an initial and fruitful result for the agency. For future work, we consider applying transfer learning (meta-learning) to extend multiple military domains with small datasets to improve the scalability of dialogue systems.

Another challenge in preparing military corpora is that for the Out-Of-Vocabulary (OOV) problem, the vocabulary of the user's question may not be included in the Mission list or Military Dictionary, so the retrieval system may not have a corresponding sentence for the entity. As a result, the question-generation module has to generate new queries to confirm the user's intent or generate further questions.

5. Conclusions

Conversational AI has found commercial applications in entertainment, food, and medicine. However, relatively little research has been conducted on AI applied to military dialogue. One of the challenges this study faces is that a large number of Chinese training datasets are not easy to obtain, and the existing research mainly uses English public social training datasets. Another challenge is to consider the practice of the whole system, which comprises several modules. In contrast, many studies have focused on improving several specific modules (NLU or NLG). The main contribution of this work is to combine multiple research topics into one framework, including intent detection, slot filling, and response generation. We applied various machine-learning techniques, including filling slots with NER models, intent detection with classifiers, answering with retrieval and learned-to-rank (LTR) models, and template-based methods to generate new queries. We design a task-oriented conversational system according to the actual needs of military missions. Since its module functions and datasets are independent of each other, this architecture can accelerate the training of problem-specific conversational systems in other service domains since only the datasets need to be replaced. Each method module can also be further considered to be replaced by methods with higher performance or efficiency in the future for comparison. From the evaluation results of the experiment, it is feasible to realize the application of dialogue AI in military scenarios based on intent detection and response generation technology. The experimental results show that the query satisfaction in eight scenarios is greater than 80% after two rounds of dialogue based on retrieval-based response generation. We integrated technologies such as natural language processing, information retrieval, and natural language generation, and used the limited military corpus to achieve the expected preliminary results of the plan. Through dialogue AI, we can help military trainers conduct multi-round question-and-answer sessions.

In future work, this research could improve in two directions: (1) Considering the amount of data and the feasibility of integrating multiple system modules, we choose the CRF-based method and SVM as the baseline for NLU tasks. This study has used the limited military conversational dataset to implement a conversational AI system for military training scenarios. In spite of this, using the CRF and SVM models are indeed preliminary approaches to implementation. During future research, we would like to apply transfer learning to expand multiple military domains with small datasets and apply few-shot learning to enhance performance. (2) Applying deep-learning architectures to replace template-based question-generation methods improves their accuracy and language expressiveness. Although current template-based queries have no obvious semantic problems, the generated sentence patterns are limited. Functional expansion based on the above two directions enables the system to take the proposal as a whole. In addition, in this military training scenario, we plan to use distant supervision to automatically label our data to expand the number of datasets and improve model training accuracy. Finally, we plan to study the feasibility of integrating Semantic Web technologies. Knowledge representation and reasoning and the construction of sentence generation using knowledge graphs architectures based on these techniques are refined to improve the NLG process of conversational AI systems.

Author Contributions: H.-M.C. conducted conceptualization, methodology, investigation, writing-original draft preparation, review & editing, supervision. D.-W.C. conducted data curation, software. All authors have read and agreed to the published version of the manuscript.

Funding: This research is sponsored by the Ministry of Science and Technology, Taiwan, under grant MOST 108-2221-E-606-013-MY2.

Institutional Review Board Statement: Not applicable.

Informed Consent Statement: Not applicable.

Data Availability Statement: The part of data that supports the findings of this study is available on request from the corresponding author. The data are not publicly available due to privacy and military restrictions.

Conflicts of Interest: The authors declare no conflict of interest.

References

1. Khari, J. Facebook Messenger Passes 300,000 bots. *VentureBeat*, 1 May 2018. Available online: https://venturebeat.com/2018/05/01/facebook-messenger-passes-300000-bots/ (accessed on 4 October 2021).
2. Leah. What Do Your Customers Actually Think About Chatbots? *Userlike*, 12 July 2021. Available online: https://userlike.com/en/blog/consumer-chatbot-perceptions (accessed on 4 October 2021).
3. Helena P. What Does the Future of Military Comms Look Like? STEM Awards 2020. Available online: https://www.telegraph.co.uk/education/stem-awards/defence-technology/military-communication-on-the-battlefield/ (accessed on 4 October 2021).
4. Shafquat, H.; Sianaki, O.A.; Ababneh, N. A Survey on Conversational Agents/Chatbots Classification and Design Techniques. In Proceedings of the Workshops of the 33rd International Conference on Advanced Information Networking and Applications (WAINA-2019), AINA Workshops, Matsue, Japan, 27–29 March 2019.
5. Singh, S.; Beniwal, H. A survey on near-human conversational agents. *J. King Saud Univ.-Comput. Inf. Sci.* 2021, *in press*. [CrossRef]
6. Goel, P.; Ganatra, A. A Survey on Chatbot: Futuristic Conversational Agent for User Interaction. In Proceedings of the 2021 3rd International Conference on Signal Processing and Communication (ICPSC), Coimbatore, India, 13–14 May 2021; pp. 736–740. [CrossRef]
7. Ramesh, K.; Ravishankaran, S.; Joshi, A.; Chandrasekaran, K. A Survey of Design Techniques for Conversational Agents. In Proceedings of the Second International Conference, ICICCT 2017, New Delhi, India, 13 May 2017.
8. Trieu, H.; Iida, H.; Bao, N.P.H.; Nguyen, L.M. Towards Developing Dialogue Systems with Entertaining Conversations. In Proceedings of the 9th International Conference on Agents and Artificial Intelligence (ICAART 2017), Porto, Portugal, 24–26 February 2017.

9. Altinok, D. An Ontology-Based Dialogue Management System for Banking and Finance Dialogue Systems. *arXiv* **2018**, arXiv:1804.04838.
10. Zeng, G.; Yang, W.; Ju, Z.; Yang, Y.; Wang, S.; Zhang, R.; Zhou, M.; Zeng, J.; Dong, X.; Zhang, R.; et al. MedDialog: Large-scale Medical Dialogue Datasets. In Proceedings of the 2020 Conference on Empirical Methods in Natural Language Processing (EMNLP), Online, 16–20 November 2020; pp. 9241–9250. [CrossRef]
11. Liu, W.; Tang, J.; Qin, J.; Xu, L.; Li, Z.; Liang, X. MedDG: A Large-scale Medical Consultation Dataset for Building Medical Dialogue System. *arXiv* **2020**, arXiv:2010.07497.
12. Sharma, M.; Russell-Rose, T.; Barakat, L.; Matsuo, A. Building a Legal Dialogue System: Development Process, Challenges and Opportunities. *arXiv* **2021**, arXiv:2109.00381.
13. Wang, C.; Chen, D.; Hu, Y.; Ceng, Y.; Chen, J.; Li, H. Automatic Dialogue System of Marriage Law Based on the Parallel C4.5 Decision Tree. *IEEE Access* **2020**, *8*, 36061–36069. [CrossRef]
14. Huang, C. The Intelligent Agent NLP-Based Customer Service System. In Proceedings of the 2021 2nd International Conference on Artificial Intelligence in Electronics Engineering, Phuket, Thailand, 15–17 January 2021; Association for Computing Machinery: New York, NY, USA, 2021; pp. 41–50. [CrossRef]
15. Heller, C.H. The Future Navy—Near-Term Applications of Artificial Intelligence. *Nav. War Coll. Rev.* **2019**, *72*, 7.
16. Chui, M.; Manyika, J.; Miremadi, M. *Where Machines Could Replace Humans—And Where They Can't (Yet)*; McKinsey & Company: Chicago, USA, 2016.
17. Kim, S.; Salter, D.; DeLuccia, L.; Tamrakar, A. Study on Text-Based and Voice-Based Dialogue Interfaces for Human-Computer Interactions in a Blocks World. In Proceedings of the 8th International Conference on Human-Agent Interaction, HAI'20, Virtual Event, 10–13 November 2020; Association for Computing Machinery: New York, NY, USA, 2020; pp. 227–229. [CrossRef]
18. Anwer, S.; Waris, A.; Sultan, H.; Butt, S.I.; Zafar, M.H.; Sarwar, M.; Niazi, I.K.; Shafique, M.; Pujari, A.N. Eye and Voice-Controlled Human Machine Interface System for Wheelchairs Using Image Gradient Approach. *Sensors* **2020**, *20*, 5510. [CrossRef]
19. Merdivan, E.; Singh, D.; Hanke, S.; Holzinger, A. Dialogue Systems for Intelligent Human Computer Interactions. *Electron. Notes Theor. Comput. Sci.* **2019**, *343*, 57–71. [CrossRef]
20. Gervits, F.; Leuski, A.; Bonial, C.; Gordon, C.; Traum, D., A Classification-Based Approach to Automating Human-Robot Dialogue. In *Increasing Naturalness and Flexibility in Spoken Dialogue Interaction, Proceedings of the 10th International Workshop on Spoken Dialogue Systems, Siracusa, Italy, 24–26 April 2019*; Marchi, E., Siniscalchi, S.M., Cumani, S., Salerno, V.M., Li, H., Eds.; Springer: Singapore, 2021; pp. 115–127. [CrossRef]
21. Robb, D.A.; Chiyah Garcia, F.J.; Laskov, A.; Liu, X.; Patron, P.; Hastie, H. Keep Me in the Loop: Increasing Operator Situation Awareness through a Conversational Multimodal Interface. In *ICMI '18, Proceedings of the 20th ACM International Conference on Multimodal Interaction, Boulder, CO, USA, 16–20 October 2018*; Association for Computing Machinery: New York, NY, USA, 2018; pp. 384–392. [CrossRef]
22. Allouch, M.; Azaria, A.; Azoulay, R. Conversational Agents: Goals, Technologies, Vision and Challenges. *Sensors* **2021**, *21*, 8448. [CrossRef]
23. He, T.; Xu, X.; Wu, Y.; Wang, H.; Chen, J. Multitask Learning with Knowledge Base for Joint Intent Detection and Slot Filling. *Appl. Sci.* **2021**, *11*, 4887. [CrossRef]
24. Cortes, C.; Vapnik, V. Support-Vector Networks. *Mach. Learn.* **1995**, *20*, 273–297. [CrossRef]
25. Manning, C.D.; Raghavan, P.; Schütze, H. *Introduction to Information Retrieval*; Cambridge University Press: Cambridge, MA, USA, 2008.
26. Liu, T.Y. Learning to Rank for Information Retrieval. *Found. Trends® Inf. Retr.* **2009**. *3*, 225–331. [CrossRef]
27. Hongshen Chen.; Liu, X.; Yin, D.; Tang, J. A Survey on Dialogue Systems: Recent Advances and New Frontiers. *arXiv* **2017**, arXiv:1711.01731.
28. Adebayo, K.J.; Caro, L.D.; Robaldo, L.; Boella, G. Legalbot: A Deep Learning-Based Conversational Agent in the Legal Domain. In Proceedings of the 22nd International Conference on Applications of Natural Language to Information Systems, NLDB 2017, Liège, Belgium, 21–23 June 2017.
29. Weizenbaum, J. ELIZA—A computer program for the study of natural language communication between man and machine. *Commun. ACM* **1966**, *9*, 36–45. [CrossRef]
30. Fitzpatrick, K.K.; Darcy, A.M.; Vierhile, M. Delivering Cognitive Behavior Therapy to Young Adults with Symptoms of Depression and Anxiety Using a Fully Automated Conversational Agent (Woebot): A Randomized Controlled Trial. *JMIR Ment. Health* **2017**, *4*, e7785. [CrossRef] [PubMed]
31. Zhang, Y.; Chen, X.; Ai, Q.; Yang, L.; Croft, W.B. Towards Conversational Search and Recommendation: System Ask, User Respond. In Proceedings of the 27th ACM International Conference on Information and Knowledge Management, CIKM'18, Torino, Italy, 22–26 October 2018; Association for Computing Machinery: New York, NY, USA, 2018; pp. 177–186. [CrossRef]
32. Zhao, T.; Eskénazi, M. Towards End-to-End Learning for Dialog State Tracking and Management using Deep Reinforcement Learning. *arXiv* **2016**, arXiv:1606.02560.
33. Goh, O.S.; Jaya Kumar, Y.; Sam, Y.H.; Leong, P. The Evaluation of User Experience Testing for Retrieval-based Model and Deep Learning Conversational Agent. *Int. J. Adv. Comput. Sci. Appl.* **2021**, *12*, 2021. [CrossRef]
34. Zhang, L.; Li, W.; Bai, Q.; Lai, E. Graph-Based Self-Adaptive Conversational Agent. In *AAMAS '21, Proceedings of the 20th International Conference on Autonomous Agents and MultiAgent Systems, Online, 3–7 May 2021*; International Foundation for Autonomous Agents and Multiagent Systems: Richland, SC, USA, 2021; pp. 1791–1793.

35. Roque, A.; Leuski, A.; Sridhar, V.K.R.; Robinson, S.; Vaswani, A.; Narayanan, S.S.; Traum, D.R. Radiobot-CFF: A spoken dialogue system for military training. In Proceedings of the INTERSPEECH, Pittsburgh, PA, USA, 17–21 September 2006.
36. Gandhe, S.; Whitman, N.; Traum, D.; Artstein, R. An Integrated Authoring Tool for Tactical Questioning Dialogue Systems. In Proceedings of the 6th IJCAI Workshop on Knowledge and Reasoning in Practical Dialogue Systems, Pasadena, CA, USA, 12 July 2009.
37. Malik, N.; Sharan, A.; Biswas, P. Domain knowledge enriched framework for restricted domain question answering system. In Proceedings of the 2013 IEEE International Conference on Computational Intelligence and Computing Research, Enathi, India, 26–28 December 2013; pp. 1–7.
38. Moldovan, D.; Paşca, M.; Harabagiu, S.; Surdeanu, M. Performance Issues and Error Analysis in an Open-Domain Question Answering System. In *ACL '02, Proceedings of the 40th Annual Meeting on Association for Computational Linguistics, Philadelphia, PA, USA, 6–12 July 2002*; Association for Computational Linguistics: Philadelphia, PA, USA, 2002; pp. 33–40. [CrossRef]
39. Setyawan, M.Y.H.; Awangga, R.M.; Efendi, S.R. Comparison Of Multinomial Naive Bayes Algorithm And Logistic Regression For Intent Classification In Chatbot. In Proceedings of the 2018 International Conference on Applied Engineering (ICAE), Batam, Indonesia, 3–4 October 2018; pp. 1–5.
40. Wang, S.; Manning, C. Baselines and Bigrams: Simple, Good Sentiment and Topic Classification. In Proceedings of the 50th Annual Meeting of the Association for Computational Linguistics (Volume 2: Short Papers), Jeju Island, Korea, 8–14 July 2012; Association for Computational Linguistics: Jeju Island, Korea, 2012; pp. 90–94.
41. Amber, N.; Sahare, P.; Pandya, K. Intent Detection and Slots Prompt in a Closed-Domain Chatbot. *arXiv* **2018**, arXiv:1812.10628.
42. Chen, L.; Zhang, D.; Mark, L. Understanding User Intent in Community Question Answering. In *WWW '12 Companion, Proceedings of the 21st International Conference on World Wide Web, Lyon, France, 16–20 April 2012*; Association for Computing Machinery: New York, NY, USA, 2012; pp. 823–828. [CrossRef]
43. Bhargava, A.; Celikyilmaz, A.; Hakkani-Tür, D.; Sarikaya, R. Easy contextual intent prediction and slot detection. In Proceedings of the 2013 IEEE International Conference on Acoustics, Speech and Signal Processing, Vancouver, BC, Canada, 26–31 May 2013; pp. 8337–8341. [CrossRef]
44. Draskovic, D.; Gencel, V.; Zitnik, S.; Bajec, M.; Nikolić, B. A software agent for social networks using natural language processing techniques. In Proceedings of the 2016 24th Telecommunications Forum (TELFOR), Belgrade, Serbia, 22–23 November 2016; pp. 1–4.
45. Gaikwad, G.; Joshi, D.J. Multiclass mood classification on Twitter using lexicon dictionary and machine learning algorithms. In Proceedings of the 2016 International Conference on Inventive Computation Technologies (ICICT), Coimbatore, India, 26–27 August 2016; Volume 1, pp. 1–6. [CrossRef]
46. Sullivan, K.O. Comparing the Effectiveness of Support Vector Machines and Convolutional Neural Networks for Determining User Intent in Conversational Agents. Master's Thesis, Technological University Dublin, Dublin, Ireland, 2018.
47. Troussas, C.; Krouska, A.; Sgouropoulou, C.; Voyiatzis, I. Ensemble Learning Using Fuzzy Weights to Improve Learning Style Identification for Adapted Instructional Routines. *Entropy* **2020**, *22*, 735. [CrossRef]
48. Rustamov, S.; Bayramova, A.; Alasgarov, E. Development of Dialogue Management System for Banking Services. *Appl. Sci.* **2021**, *11*, 10995. [CrossRef]
49. Liu, Y.; Qiu, M.; Qu, C.; Chen, C.; Guo, J.; Zhang, Y.; Croft, W.B.; Chen, H. IART: Intent-aware Response Ranking with Transformers in Information-seeking Conversation Systems. *arXiv* **2020**, arXiv:2002.00571.
50. Weld, H.; Huang, X.; Long, S.; Poon, J.; Han, S.C. A survey of joint intent detection and slot-filling models in natural language understanding. *arXiv* **2021**, arXiv:2101.08091.
51. Hemphill, C.T.; Godfrey, J.J.; Doddington, G.R. The ATIS Spoken Language Systems Pilot Corpus. In *Speech and Natural Language, Proceedings of the Workshop Held at Hidden Valley, Proceedings of the Workshop Held at Hidden Valley, PA, USA, 24–27 June 1990*; Texas Instruments Inc.: Dallas, TX, USA, 1990.
52. Mrksic, N.; Séaghdha, D.Ó.; Wen, T.; Thomson, B.; Young, S.J. Neural Belief Tracker: Data-Driven Dialogue State Tracking. *arXiv* **2016**, arXiv:1606.03777.
53. Shah, P.; Hakkani-Tür, D.; Tür, G.; Rastogi, A.; Bapna, A.; Nayak, N.; Heck, L.P. Building a Conversational Agent Overnight with Dialogue Self-Play. *arXiv* **2018**, arXiv:1801.04871.
54. Sneiders, E. Automated Question Answering: Template-Based Approach. Doctor's Thesis, Royal Institute of Technology and Stockholm University, Stockholm, Sweden, 2002.
55. Stapley, B.; Benoit, G. Biobibliometrics: Information retrieval and visualization from co-occurrences of gene names in MEDLINE abstracts. In Proceedings of the Pacific Symposium on Biocomputing, Honolulu, HI, USA, 5–9 January 2000; Volume 2000, pp. 529–540. [CrossRef]
56. Fiszman, M.; Rindflesch, T.; Kilicoglu, H. Integrating a Hypernymic Proposition Interpreter into a Semantic Processor for Biomedical Texts. In Proceedings of the AMIA...Annual Symposium Proceedings/AMIA Symposium, Washington, DC, USA, 8–12 November 2003; Volume 2003, pp. 239–243.
57. Bhoir, V.; Potey, M.A. Question answering system: A heuristic approach. In Proceedings of the Fifth International Conference on the Applications of Digital Information and Web Technologies (ICADIWT 2014), Bangalore, India, 17–19 February 2014; pp. 165–170.
58. Liu, Z.-X.; Chang, C.-H, Chatlog Disentanglement based on Similarity Evaluation Via Reply Message Pairs Prediction Task. *Int. J. Comput. Linguist. Chin. Lang. Process.* **2019**, *24*, 63–77.
59. Bartl, A.; Spanakis, G. A retrieval-based dialogue system utilizing utterance and context embeddings. *arXiv* **2017**, arXiv:1710.05780.

60. Juraska, J.; Karagiannis, P.; Bowden, K.; Walker, M. A Deep Ensemble Model with Slot Alignment for Sequence-to-Sequence Natural Language Generation. In Proceedings of the 2018 Conference of the North American Chapter of the Association for Computational Linguistics: Human Language Technologies, Volume 1 (Long Papers), New Orleans, LA, USA, 1–6 June 2018; Association for Computational Linguistics: New Orleans, LA, USA, 2018; pp. 152–162. [CrossRef]
61. Song, K.; Tan, X.; Qin, T.; Lu, J.; Liu, T. MASS: Masked Sequence to Sequence Pre-training for Language Generation. *arXiv* **2019**, arXiv:1905.02450.
62. Wang, J.; Liu, J.; Bi, W.; Liu, X.; He, K.; Xu, R.; Yang, M. Improving Knowledge-aware Dialogue Generation via Knowledge Base Question Answering. *arXiv* **2019**, arXiv:1912.07491.
63. Moon, S.; Shah, P.; Kumar, A.; Subba, R. OpenDialKG: Explainable Conversational Reasoning with Attention-based Walks over Knowledge Graphs. In Proceedings of the 57th Annual Meeting of the Association for Computational Linguistics, Florence, Italy, 28 July–2 August 2019; Association for Computational Linguistics: Florence, Italy, 2019; pp. 845–854. [CrossRef]
64. Bockhorst, J.; Conathan, D.; Fung, G.M. Knowledge Graph-Driven Conversational Agents. In Proceedings of the 33rd Conference on Neural Information Processing Systems (NeurIPS 2019), Vancouver, Canada, 8–14 December 2019.
65. Chih Wei, H.; Lin, C.J. A comparison of methods for multiclass support vector machines. *IEEE Trans. Neural Netw.* **2002**, *13*, 415–425. [CrossRef]
66. Yuan, W.; Ling-yu, Z.; Ya-xuan, Z.; Lu, H.; Ding-yi, F. Combining Support Vector Machines, Border Revised Rules and Transformation-based Error-driven Learning for Chinese Chunking. In Proceedings of the 2010 International Conference on Artificial Intelligence and Computational Intelligence, Sanya, China, 23–24 October 2010; Volume 1, pp. 383–387. [CrossRef]
67. Hamada, A.; Dafoulas, G.; Ismail, M. Intent Classification for a Management Conversational Assistant. In Proceedings of the 2020 15th International Conference on Computer Engineering and Systems (ICCES), Cairo, Egypt, 15–16 December 2020; pp. 1–6. [CrossRef]
68. Burges, C.J.C.; Svore, K.M.; Wu, Q.; Gao, J. *Ranking, Boosting, and Model Adaptation*; Technical Report MSR-TR-2008-109; Microsoft Research: Redmond, WA, USA, 2008.
69. Boser, B.E.; Guyon, I.M.; Vapnik, V.N. A Training Algorithm for Optimal Margin Classifiers. In *COLT '92, Proceedings of the Fifth Annual Workshop on Computational Learning Theory, Pittsburgh, PA, USA, 27–29 July 1992*; Association for Computing Machinery: New York, NY, USA, 1992; pp. 144–152. [CrossRef]
70. He, H.; Ma, Y. Class Imbalance Learning Methods for Support Vector Machines. In *Imbalanced Learning: Foundations, Algorithms, and Applications*; The Institute of Electrical and Electronics Engineers, Inc.: Piscataway, NJ, USA, 2013; pp. 83–99. [CrossRef]

applied sciences

Article

AraConv: Developing an Arabic Task-Oriented Dialogue System Using Multi-Lingual Transformer Model mT5

Ahlam Fuad * and Maha Al-Yahya

Department of Information Technology, College of Computer and Information Sciences, King Saud University, P.O. Box 145111, Riyadh 4545, Saudi Arabia; malyahya@ksu.edu.sa
* Correspondence: aabdulghni@ksu.edu.sa

Abstract: Task-oriented dialogue systems (DS) are designed to help users perform daily activities using natural language. Task-oriented DS for English language have demonstrated promising performance outcomes; however, developing such systems to support Arabic remains a challenge. This challenge is mainly due to the lack of Arabic dialogue datasets. This study introduces the first Arabic end-to-end generative model for task-oriented DS (AraConv), which uses the multi-lingual transformer model mT5 with different settings. We also present an Arabic dialogue dataset (Arabic-TOD) and used it to train and test the proposed AraConv model. The results obtained are reasonable compared to those reported in the studies of English and Chinese using the same mono-lingual settings. To avoid problems associated with a small training dataset and to improve the AraConv model's results, we suggest joint-training, in which the model is jointly trained on Arabic dialogue data and data from one or two high-resource languages such as English and Chinese. The findings indicate the AraConv model performed better in the joint-training setting than in the mono-lingual setting. The results obtained from AraConv on the Arabic dialogue dataset provide a baseline for other researchers to build robust end-to-end Arabic task-oriented DS that can engage with complex scenarios.

Keywords: task-oriented dialogue systems; Arabic; multi-lingual transformer model; mT5; natural language processing

Citation: Fuad, A.; Al-Yahya, M. AraConv: Developing an Arabic Task-Oriented Dialogue System Using Multi-Lingual Transformer Model mT5. *Appl. Sci.* **2022**, *12*, 1881. https://doi.org/10.3390/app12041881

Academic Editors: Arturo Montejo-Ráez and Salud María Jiménez-Zafra

Received: 7 January 2022
Accepted: 3 February 2022
Published: 11 February 2022

1. Introduction

Task-oriented dialogue systems (DS) are a type of conversational system designed to help users achieve pre-defined tasks. These systems are designed to help humans perform routine activities, such as make restaurant or hotel reservations, search for attractions, book flights, enquire about the weather forecast, and shop online. Task-oriented DS are considered the core modules of virtual assistants such as Google Assistant, Amazon Alexa, and Apple Siri, which utilize natural language interfaces for various online services [1]. Task-oriented DS allow users to ask questions using natural language and provide answers to those questions in the form of a conversation.

Despite the current progress of state-of-the-art English-based task-oriented DS, it remains a substantial challenge to build systems that can achieve coherent, sustained conversation on diverse topics [2]. Notably, task-oriented DS for Arabic lag behind [3], until now precluding the application of advanced data-intensive deep-learning models for the language [4], especially due to the shortage of Arabic dialogue datasets. Therefore, this study aimed to investigate the effectiveness of the multi-lingual pre-trained language model mT5 [5] for building end-to-end Arabic task-oriented DS. These end-to-end DS must be capable of handling both dialogue state tracking (DST) task and response generation task; in this context, DST is mainly responsible for helping to extract the goals and slot-value pairs from the conversation. As such, this work aimed to answer the following major research questions:

RQ1: To what extent can mT5, a multi-lingual pre-trained language model, produce satisfactory results for Arabic end-to-end task-oriented DS?

RQ2: To what extent can joint-training the mT5 model on Arabic dialogue data and data for one or two high-resource languages (namely, English or English and Chinese) improve the performance of Arabic task-oriented DS?

To answer these research questions, we conducted several experiments, leading this work to make the following contributions:

- Development of the first Arabic task-oriented dialogue dataset (Arabic-TOD) with 1500 dialogues. By translating the English BiToD dataset [1], we produced a valuable benchmark for further exploring Arabic task-oriented DS. Furthermore, Arabic-TOD is the first code-switching dialogue dataset for Arabic task-oriented DS.
- Introduction of the first Arabic end-to-end generative model, the AraConv model, short for Arabic Conversation, that achieves both DST and response generation tasks together in an end-to-end setting.

The paper comprises five sections. The next section explores related works in the area of task-oriented DS for both English and Arabic. The third section demonstrates the methodology used in this research, including the data collection process and the model architecture. Next, we detail our experiments, discussing the tasks and evaluation metrics, experimental setup, and findings. Finally, the fifth section summarizes our work and the significance of the AraConv before considering possible future research directions.

2. Related Works

There are two approaches in applying DS: traditional DS and end-to-end DS. Traditional DS use a pipeline that connects, trains, and evaluates each module separately. End-to-end DS are designed to train all modules as a single unit directly on both knowledge-based information and text transcripts [6]. This section discusses the evaluation of task-oriented DS for the English language before surveying the landscape of Arabic task-oriented DS.

2.1. English Task Oriented Dialogue Systems

Given the availability of multi-domain English task-oriented dialogue datasets, work on task-oriented DS in the language has progressed from modularized modeling to generative and end-to-end modeling. Given the fact that the traditional DS design complicates tracking the module responsible for interaction failure [6], some studies have built DS using the end-to-end paradigm [7–15]. However, building powerful task-oriented DS still engenders many challenges due to the system design complexity and the limited availability of human-annotated data. Therefore, the research community has focused on working with the pre-trained language models to reduce human supervision to the extent possible. This approach involves fine-tuning these models and helping to transfer the prior knowledge to improve various NLP tasks, including task-oriented DS. Large pre-trained language models, such as GPT2 and T5, have been used for various NLP tasks, especially language generation tasks. These new approaches model the dialogue pipeline in an end-to-end manner [6].

Given the high costs associated with data collection and annotation, researchers tend to train their models with the least number of samples using transfer learning. Transfer learning represents one of the most successful few-shot learning approaches for task-oriented DS. It refers to pre-training large language models on text or task-related data and then fine-tuning on a few samples. Such systems have proved their success in task-oriented DS such as the work presented in [12–22].

The task-oriented DS literature includes two study categories: studies targeting only DST and studies targeting both DST and response generation. Dialogue state tracking mainly helps to extract the goals (intents) and slot-value pairs from the conversation to maintain the dialogue belief state (BS) and the summary of the dialogue history. The BS contains information about the dialogue from the system perspective [6]. At each user turn during the conversation, the input to the DST comprises the previous BS, the outputs of the intent classification (the goal), and slot filling information; thus, the DST output is the new/updated BS. For end-to-end dialogue generation, the system indicates the correct required information and generates the appropriate response.

For the first category, studies targeting only DST, some studies focus on handling the DST task to guarantee building a good base for the whole dialogue system [23–29]. Meanwhile, other studies have targeted both DST and response generation in an end-to-end manner [11,12].

Table 1 summarizes the available models for task-oriented DS in English, including datasets and performance measures. Although the models have achieved promising results, they have been designed for English-language task-oriented DS, and, to the best of our knowledge, no research exists concerning Arabic-language task-oriented DS.

Nonetheless, the promising performance of pre-trained language models for English-language task-oriented DS has prompted efforts to produce multi-lingual models for task-oriented DS in other languages. Many of these languages are considered low-resource languages due to the absence of high-quality data in the language, and most existing task-oriented DS do not support low-resource languages, creating a gap between the performance of low-resource language systems and high-resource systems. Therefore, providing datasets for low-resource languages is critical to driving the development of efficient end-to-end task-oriented DS for these languages. Several existing studies have built task-oriented DS for low-resource languages using cross-lingual transfer learning [1,30,31]. This involves transferring knowledge from high-resource to low-resource languages, enabling the satisfactory performance of end-to-end task-oriented DS.

Table 1. Comparing the performances of the most common English-based task-oriented dialogue systems (DS). Bold numbers indicating the best system according to the column's metric value.

Model	Dataset	Back-Bone Models	Performance Metrics			
			BLEU	Inform Rate	Success Rate	JGA
DAMD [32]	MultiWOZ 2.1	multi-decoder seq2seq	16.6	76.4	60.4	51.45
Ham [10]	MultiWOZ 2.1	GPT-2	6.01	77.00	69.20	44.03
SimpleToD [11]	MultiWOZ 2.1	GPT-2	15.23	85.00	70.05	**56.45**
SC-GPT [16]	MultiWOZ	GPT-2	**30.76**	-	-	-
SOLOIST [12]	MultiWOZ 2.0	GPT-2	16.54	85.50	72.90	-
MARCO [33]	MultiWOZ 2.0	-	20.02	92.30	78.60	-
UBAR [13]	MultiWOZ 2.1	GPT-2	17.0	**95.4**	**80.7**	56.20
ToD-BERT [17]	MultiWOZ 2.1	BERT	-	-	-	48.00
MinTL [14]	MultiWOZ 2.0	T5-small T5-base BART-large	19.11 18.59 17.89	80.04 82.15 84.88	72.71 74.44 74.91	51.24 52.07 52.10
LABES-S2S [20]	MultiWOZ 2.1	A copy-augmented Seq2Seq	18.3	78.1	67.1	51.45
AuGPT [21]	MultiWOZ 2.1	GPT-2	17.2	91.4	72.9	-
GPT-CAN [15]	MultiWOZ 2.0	GPT-2	17.02	93.70	76.70	55.57
HyKnow [22]	MultiWOZ 2.1	multi-stage Seq2Seq	18.0	82.3	69.4	49.2

2.2. *Arabic Task-Oriented Dialogue Systems*

Considering the maturity of research concerning English-based task-oriented DS, we find that task-oriented DS research more broadly remains in its infancy for Arabic. This is due to a lack of fundamental NLP resources and a scarcity of datasets for Arabic task-oriented DS. Most of the research on Arabic task-oriented DS focuses on achieving specific tasks, such as intent classification [34–36] and entity classification [34]. However, there some attempts to build task-oriented DS have investigated specific domains, including home automation [34], flight bookings [37], education [38–40], hotel reservations [41], and Islamic knowledge enquires [42]. Some Arabic task-oriented DS have been designed to specifically serve the Arabic dialects (e.g., OlloBot [43] and Nabiha [44]). However, this review excludes some of these studies because they are categorized as chatbots rather than task-oriented DS because their system design does not follow a task-oriented DS structure [39,40,42–44].

Notably, Bashir et al. [34] used deep learning approaches to build a natural language understanding module for Arabic task-oriented DS for home automation. The module manages of both intent classification and entity extraction tasks. For intent classification, it uses LSTMs and CNNs; for entity extraction, BiLSTM and character-based word embeddings are used. The study used data collected via an online survey and the AQMAR dataset. The data were filtered and labeled according to the Conll-2003 NER format. The findings for the intent classification demonstrated that CNNs performed better than LSTMs (F-score = 94%). For entity extraction, the model obtained comparable results to the named entity recognition benchmarks in English (F-score = 94%).

Meanwhile, Elmadany et al. [35] used a multi-class hierarchical model to solve the dialogue acts classification issue associated with Arabic dialects. They used a manually collected and annotated dataset from multi-genre Egyptian call centers to evaluate their system performance. Using an SVM classifier produced an average F-score of 91.2%, indicating an improvement of 20% compared to the state-of-art approach. Elsewhere, Joukhadar et al. [36] examined different machine learning approaches to recognizing user acts in a text-based DS for the Levantine Arabic dialect. They manually produced 873 sentences for both restaurant orders and flight booking, reporting accuracy of 86% using the SVM model. However, their small dataset was insufficient to build an efficient dialogue system, suggesting an imperative to develop large multi-domain datasets or more efficient techniques.

For Arabic user-based DS, several studies [37,38,41] have applied either pattern matching, rule-based, or rule-based and data-driven hybrid approaches to task-oriented DS. Nonetheless, it is apparent that most Arabic task-oriented DS use either rule-based or pattern matching approaches, with very few using a hybrid approach. It is understandable that they use these approaches due to the challenges associated with building Arabic task-oriented DS in Arabic [3], among which is the lack of Arabic task-oriented dialogue datasets. Therefore, this study aimed to address this challenge by leveraging the pre-trained language models to build an Arabic task-oriented DS. Multi-lingual language models are among the most popular and common language models, observed to produce good performance on task-oriented DS for many languages. Accordingly, we explored the extent to which mT5 can be useful for building an Arabic task-oriented dialogue system. To the best of our knowledge, this work represents the first attempt at pre-training a large transformer-based language representation model on an Arabic task-oriented dialogue dataset (Arabic-TOD).

3. Method

A pre-trained language model is a deep learning model that has been trained on a large amount of data to perform particular NLP tasks [45]. Figure 1 shows a high-level view of the approach adopted. We began with the English BiToD dataset [1], translating the dialogues into Arabic to produce the Arabic-TOD dataset. The dataset was then preprocessed and prepared for the training step. Subsequently, we trained the models on the training Arabic-TOD dataset using different settings. Finally, we used the testing Arabic-TOD dataset to test the models and obtain the results for the AraConv model.

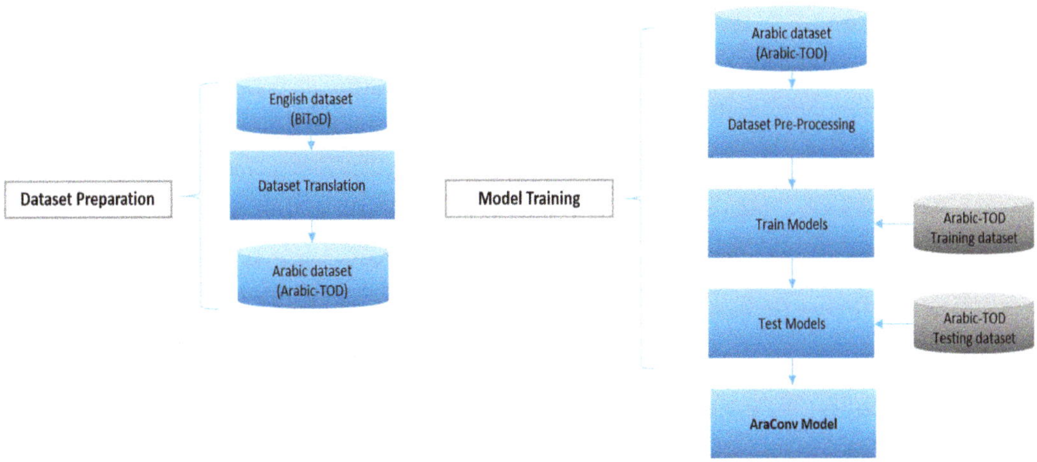

Figure 1. High-level view of our approach.

3.1. Arabic Task-Oriented DS Dataset

Because Arabic is a low-resource language, no human-annotated Arabic dataset for task-oriented DS has been produced (to the best of our knowledge). To obtain a good-quality dataset, we decided to use an existing dataset, translating a benchmark dataset for task-oriented DS (BiToD [1]) to develop a suitable training dataset for Arabic task-oriented DS.

Translating existing datasets is a practice frequently observed in the literature for low-resource languages, with examples including [46–48]. Recent translation techniques for crowd-sourced annotated datasets have produced reasonable results on training data for different languages, enabling many studies to address the lack of datasets by translating existing datasets for many downstream tasks in NLP. For example, for question answering (QA), the SQuAD dataset has been translated into Arabic [46] and Bengali [47], and for conversation generation, the EmpatheticDialogues dataset has been translated into Arabic [48].

Still, it is imperative for the research community to develop multi-lingual benchmarks to evaluate the cross-lingual transferability of end-to-end systems in general and task-oriented DS in particular [49]. For task-oriented DS, many multi-lingual datasets can be obtained by translating the English datasets. Table 2 presents some of these alongside their corresponding tasks and domains. Translation represents a good choice for low-resource languages to support the reuse of resources and save time spent creating and annotating long dialogues. Additionally, this enables the development of multi-lingual benchmarks for the research community to use.

Table 2. Datasets translated from English within the field of task-oriented DS. EN: English, ES: Spanish, DE: German, IT: Italian, TH: Thai, VI: Vietnamese, ZH: Chinese.

Dataset	Task	Language	Domains
Chinese ATIS [50]	Intent classification Slot extraction	ZH	Flight bookings
Multi-lingual WOZ 2.0 [51]	DST	EN, DE, IT	Restaurant bookings
SLU-IT [52]	Intent classification Slot extraction	IT	7 domains (Restaurant, Weather, Music, ...)
Almawave-SLU [53]	Intent classification Slot extraction	IT	7 domains (Restaurant, Weather, Music, ...)

Table 2. *Cont.*

Dataset	Task	Language	Domains
S. Schuster et al. [30]	Task-oriented DS	ES, TH	3 domains (Weather, Alarm, and Reminder)
Z. Liu et al. [54]	Task-oriented DS	ES, TH	3 domains (Weather, Alarm, and Reminder)
Z. Liu et al. [31]	DST Task-oriented DS	EN, DE, IT ES, TH	Restaurant booking 3 domains (Weather, Alarm, and Reminder)
Vietnamese ATIS [55]	Intent classification Slot extraction	VI	Flight bookings

3.2. Structure and Organization of Arabic-TOD Dataset

The Arabic-TOD dataset is based on the BiToD dataset, the first large bilingual task-oriented dialogue dataset created for training and evaluating end-to-end task-oriented DS. It contains annotated English and Chinese dialogues and features a total of 7232 dialogues with 144,798 utterances (3689 dialogues in English and 3543 dialogues in Chinese). The dialogues range between 10 and more than 50 turns with an average length of 19.98 turns. Each turn can be defined as one or more utterances from one speaker [56]. The BiToD dataset includes dialogues in five domains: Hotels, Restaurants, Weather, Attractions, and Metro.

Although there are many other common multi-domain task-oriented dialogue datasets, including MultiWOZ, we chose to translate the BiToD dataset to leverage certain useful features that distinguished it from other datasets [1]. Notably, the BiToD dataset supports mixed-language contexts, also known as code-switching. Some items in the knowledge base (and in daily life) feature mixed-language information, meaning English and Arabic texts appear in the same utterance. For example, there are some restaurant names in English that cannot be translated into Arabic, such as Chom Chom, which maintains the English name even if our conversation is in Arabic (i.e., "هل يمكنك أن تحجز لي مطعم Chom Chom"). Another advantageous feature of the BiToD dataset is its use of a deterministic API, which simplifies model evaluations. Deterministic API refers to the ability of the system to recommend the query-matched items on the basis of certain criteria (e.g., user rating). This differs from other API evaluation methods, which randomly return only one or two matched items with the API. Another important aspect of the BiToD dataset is the diversity of user tasks, meaning users might want to book hotels and restaurants within the same dialogue, as they might in a real human-based interactions. As such, we decided to contribute to enriching and augmenting the BiToD dataset by translating the English dialogues into Arabic, producing a multi-lingual dataset enabling the combined use of English, Chinese, and Arabic. Table 3 summarizes the different common multi-domain task-oriented dialogue datasets, indicating the features that we have tried to utilize.

Table 3. Summary of the characteristics of different common task-oriented dialogue datasets.

Dataset	Languages	Number of Dialogues	Avg. Turn Length	Number of Domains (Tasks)	Deterministic API	Mixed-Language Context
BiToD	EN, ZH	7232	19.98	5	Yes	Yes
MultiWoZ	EN	8438	13.46	7	No	No
Askmaster	EN	13,215	22.9	6	No	No
MetaLWOZ	EN	37,884	11.4	47	No	No

Table 3. *Cont.*

Dataset	Languages	Number of Dialogues	Avg. Turn Length	Number of Domains (Tasks)	Deterministic API	Mixed-Language Context
TM-1	EN	13,215	21.99	6	No	No
Schema	EN	22,825	20.3	17	No	No
SGD	EN	16,142	20.44	16	No	No
STAR	EN	5820	21.71	13	No	No
Frames	EN	1369	14.6	3	No	No
Multi-lingual WOZ 2.0	EN, DE, IT	3600	_	1	No	Yes
Arabic-TOD	AR	1500	19.98	4	Yes	Yes

For the translation task, three bilingual speakers of Arabic and English were paid to manually translate the English BiToD dataset into Arabic over 2.5 months, translating the utterances and slot-values in the dataset in the Hotels, Restaurants, Weather, and Attractions categories. We determined the strategy of translation and the used lexicons previously, and we gave them some examples of the target translated dialogues. Of the 3689 English dialogues, 1500 dialogues (30,000 utterances) were translated into Arabic. The translated utterances and slot-values were reviewed to verify the quality of translation and correctness of slot-value pairs on the basis of the English BiToD dataset.

Arabic-TOD dataset contains different lengths of dialogues, some of them with a single task and the others with multiple tasks varying between 2 and 4. For instance, some dialogues include multiple tasks in a single dialogue (e.g., a single dialogue can involve different tasks including enquiring about the weather, finding a restaurant to eat at, and an attraction to visit).

To the best of our knowledge, this Arabic-TOD is the first Arabic dataset supporting a mixed languages context for task-oriented DS that has been annotated following the BiToD dataset's structure [1].

3.3. Model Architecture

The AraConv model's generation process is based on a single multi-lingual Seq2Seq (mSeq2Seq) model that uses the pre-trained model mT5 [5], a multilingual variant of T5 [57], which can be formally defined as follows:

Assume the dialogue D represents a set of user utterances (Ut) and system utterances (St) at turn t, where $D = \{U_1, S_1, \ldots, U_t, S_t\}$.

The dialogue history (H) holds the previous user and system utterances of turn t, specified by the context window size (w), where $H_t = \{U_{t-w}, S_{t-w}, \ldots, S_{t-1}; U_t\}$. For turn t, the dialogue state is represented as B_t, and the knowledge state is represented as K_t.

Figure 2 illustrates the proposed workflow for response generation using the mSeq2Seq model based on the BiToD dataset [1].

Initially, we set the dialogue state and knowledge state to empty strings as B_0 and K_0. Then, we considered the current dialogue history (H_t), previous dialogue state (B_{t-1}), and previous knowledge state (K_{t-1}) as input at turn t. We added the prompt PB = "TrackDialogueState:" to indicate the generation task [57]. Therefore, the mSeq2Seq model produces Levenshtein Belief Spans at turn t (Lev_t), indicating a text span that contains the information for updating the dialogue state from (B_{t-1}) to B_t. Lev_t can be represented by the following equation:

$$Lev_t = \text{mSeq2Seq}(PB, H_t, B_{t-1}, K_{t-1}) \quad (1)$$

Then, the model generates an output (o/p) based on the new input as the updated dialogue state (B_t), and the response generation prompt—referred to as PR = "Response:" —at the current turn t. If there is a need for an API call, the model will generate an API name according to the following:

$$API = mSeq2Seq(PR,\ H_t,\ B_t,\ K_{t-1}) \tag{2}$$

In this case, the system queries the API with particular constraints in the dialogue state and updates the knowledge state form (K_{t-1}) to (K_t). The updated knowledge state (K_t) and API name (API) are subsequently combined to generate the next turn response. Otherwise, the model generates a textual response (R) that is returned directly to the user:

$$R = mSeq2Seq(PR,\ H_t,\ B_t,\ K_t,\ API\ ) \tag{3}$$

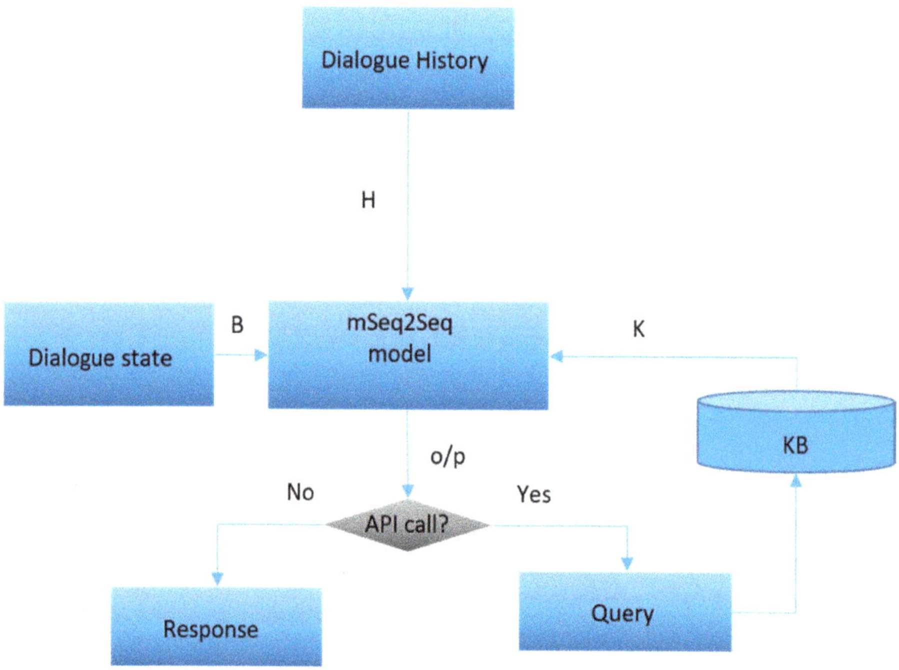

Figure 2. The multi-lingual Seq2Seq model workflow.

4. Experiments

This section first explains the evaluation metrics used to measure the performance of the AraConv model. Next, we describe the experimental setup and detail the experiments performed to test our hypothesis. Finally, we discuss the results of each experiment.

4.1. Evaluation Metrics

This study addresses two main tasks: DST and end-to-end dialogue generation, which includes both DST and response. To evaluate the DST performance of the AraConv model, we used the joint goal accuracy (JGA) metric to compare the predicted dialogue state to the ground truth for each dialogue turn. If all predicted slot values exactly match the ground-truth values, the model's output is considered correct. To evaluate the performance on the end-to-end generation task by the AraConv model, we used four metrics:

- the BLEU metric to assess the generated response fluency;
- the API call accuracy (API_{Acc}) metric to assess if the system generates the correct API call;
- the task success rate (TSR) metric to assess whether the system finds the correct entity and provides all of the requested information for a particular task. TSR can be defined as

$$TSR = \frac{\sum \text{success task}}{\text{total number of tasks}} \quad (4)$$

where the tasks involve searching task and booking task for hotel and restaurant domains, and search task for attraction and weather domains.

- the dialogue success rate (DSR) metric to evaluate whether the system accomplishes all of the dialogue tasks. DSR can be defined as

$$DSR = \frac{\sum \text{success dialogue}}{\text{total number } of \text{ dialogues}} \quad (5)$$

The evaluation method's main goal is to obtain an automated and repeatable evaluation procedure that enables efficient comparisons of the quality of different dialogue strategies. This involves focusing on the automatic evaluation metrics. However, further measurement of the quality of the generated responses also requires human review. Thus, following the literature [15], we evaluated the AraConv model's performance on end-to-end generation tasks according to two metrics:

- the language understanding score to indicate the extent to which the system understands user inputs; and
- the response appropriateness score to indicate whether the response is appropriate and human-like.

We performed a small-scale human review to measure these scores. The literature indicates two other common metrics used in human evaluation: TSR and DSR [56]. Given the costs and time-intensiveness of human evaluation, we measured these scores automatically (TSR and DSR).

4.2. Experimental Setup

Our framework uses the pre-trained multi-lingual model mT5-small. All of our experiments used the Transformers library [58] and the deep learning framework PyTorch [59]. We trained all of the models using an AdamW optimizer [60] (with an initial learning rate of 0.0005). We set the dialogue context window size (w) at 2 and the batch size at 128 in accordance with the approach observed to obtain the best results in the extant literature.

We split our Arabic-TOD dataset into 67%, 7%, and 26% for training, validation, and testing, resulting in 1000, 100, and 400 training, validation, and testing dialogues, respectively. For the mono-lingual setting, we trained the model for 20 epochs; for the bi-lingual and multi-lingual settings, we trained the models for 8 epochs. Training using Google Colab required approximately 22 hours.

4.3. Baseline

As this is the first work to build an Arabic end-to-end generative model for task-oriented DS, there is no directly comparable approach in the previous Arabic studies. Therefore, we experimented with several initial baselines (using the zero-shot setting, that is transferring the model, which is trained to solve task-oriented DS, in English to solve that specific task in Arabic). We trained the mT5 model on English using the English BiToD dataset then tested its performance directly on the Arabic-TOD dataset. This approach is a common practice similar to many downstream tasks such as QA [61,62] or task-oriented DS [63]. The performance of these initial baselines was very low; therefore, we set our baseline using the same concept of zero-shot setting where mT5 model is trained on mixed language training data by replacing the most task-related keyword entities in English BiToD language with their corresponding in Arabic language from a parallel dictionary.

4.4. Experiments

RQ1: To what extent can mT5, a multi-lingual pre-trained language model, produce satisfactory results for Arabic end-to-end task-oriented DS?

This experiment aimed to investigate the performance of an end-to-end Arabic task-oriented dialogue system using an mSeq2Seq model for Arabic. This mono-lingual setting only requires one language to train and test the model. Thus, we trained and tested the proposed mT5 model (AraConv) using the Arabic-TOD dataset. The AraConv model differs from the baseline with the training setting where AraConv trained on Arabic dialogues while the baseline did not (zero-shot learning).

Table 4 shows the results—-in terms of BLEU, APIACC, TSR, DSR, and JGA—-of the AraConv model in the mono-lingual setting in comparison to the English and Chinese experiments on the BiToD dataset [1]. The observed English results [1] outperformed the AraConv results. This is unsurprising because there are more data for English and Chinese. The model trained and tested on English or Chinese data still performed better than that tested on the Arabic-TOD dataset, which represented only 27% of the BiToD dataset [1]. Where the original mT5 model was trained using multiple languages, the English data represented 5.67% of the whole corpus, and Chinese and Arabic represented 1.67% and 1.66% of the total data, respectively [5], explaining the superior performance for English dialogue. Additionally, Arabic is a language with extensive grammatical case marking [5], which causes lower evaluation metrics compared to English. Meanwhile, despite the comparable sizes of the training data for Arabic and Chinese, the results of the mono-lingual model trained on Chinese BiToD dataset outperformed the AraConv model. This may have been due to the small size of the Arabic-TOD dataset compared to the Chinese BiToD dataset. Nonetheless, the AraConv model achieved a better BLEU value (by approximately 63%) than the Chinese model, meaning that the AraConv model can generate more fluent responses than the Chinese model.

Table 4. Mono-lingual experiment dialogue state tracking (DST) and end-to-end dialogue generation results for the AraConv model trained on Arabic-TOD dataset compared to the baseline and the mono-lingual BiToD experiments [1] using English (EN) and Chinese (ZH). Bold numbers indicating the best result according to the column's metric value.

	TSR	DSR	APIAcc	BLEU	JGA
			Arabic		
Baseline	3.95	1.16	4.30	3.37	8.21
AraConv	45.07	18.60	48.86	31.05	34.82
			Other languages		
EN [1]	**69.13**	**47.51**	**67.92**	**38.48**	**69.19**
ZH [1]	53.77	31.09	63.25	19.03	67.35

Still, the AraConv model did not achieve perfect results, potentially due to the complicated nature of the Arabic-TOD dataset, its complex ontology, and its diversity of user goals. Moreover, the DSR result was lower than the TSR result, likely because of the multiple tasks included in the dialogue (2–4 tasks). For instance, some dialogues included multiple tasks in a single dialogue (e.g., a single dialogue can involve the tasks of finding a hotel to stay at, a restaurant to eat at, an attraction to visit, and information about the weather).

RQ2: To what extent can joint-training the mT5 model on Arabic dialogue data and data for one or two high-resource languages (namely, English or English and Chinese) improve the performance of Arabic task-oriented DS?

Answering this research question requires performing two experiments to investigate the performance of building an end-to-end Arabic task-oriented dialogue system using an mSeq2Seq model in bi-lingual and multi-lingual settings. Because two languages are used to train and test the model in the bi-lingual setting, we trained the proposed model mT5 on both the Arabic-TOD and English-BiToD datasets [1].

In the experiments described in [1], the models were trained on almost the same number of English and Chinese dialogues (2952 and 2835). However, our Arabic-TOD dataset includes only 27% of the data included in the BiToD datasets. Accordingly, we investigated two cases:

- Non-equivalent (NQ): The size of the Arabic-TOD dataset is not equal to the English BiToD dataset. We trained the model with 1000 Arabic dialogues and 2952 English dialogues.
- Equivalent (Q): The size of the Arabic-TOD dataset and the English BiToD dataset are equal (1000 dialogues for training).

Because three languages were used to train and test the model for the multi-lingual experiment, the mT5 model was trained on the Arabic-TOD, the English BiToD, and the Chinese BiToD datasets [1]. As in the previous experiment, we investigated two cases:

- Non-equivalent (NQ): The size of the Arabic-TOD dataset is not equal to the English or Chinese BiToD dataset. We trained the model with 1000 Arabic dialogues, 2952 English dialogues, and 2835 Chinese dialogues.
- Equivalent (Q): The size of the Arabic-TOD dataset, the English BiToD dataset, and the Chinese BiToD are equal (1000 dialogues for training).

For the bi-lingual setting, Table 5 compares the AraConv results— in terms of BLEU, APIACC, TSR, DSR, and JGA—to the experiments reported in [1] regarding English and Chinese dialogues with the same settings. We observed that the non-equipollent bi-lingual AraConv model (AraConvBi-NQ) outperformed the equipollent bi-lingual AraConv model ($AraConv_{Bi\text{-}Q}$), demonstrating the impact of training dialogue dataset size on the final model given that the $AraConv_{Bi\text{-}NQ}$ model is trained on more data. Therefore, using more English data in training with Arabic helps to improve the result because of the semantics of the conversation, which is almost similar to Arabic, especially for the task-related words. However the model in [1], which was trained on both English and Chinese data and then tested on English, outperformed all models, assuming the dialogues in the two datasets were almost the same. As discussed, the distinguished performance of the English model could have been due to the amount of English data used to train the mT5 model. Nonetheless, we observed that the AraConv model performed better according to the BLEU metric than the Chinese model, despite training on the same English dataset (as a second dataset for joint-training), confirming the greater fluency of the AraConv model.

For the multi-lingual setting, Table 6 presents AraConv results calculated in terms of BLEU, APIACC, TSR, DSR, and JGA. Our findings emphasize the previous results of AraConv in the bi-lingual experiment, which saw the non-equipollent multi-lingual AraConv model ($AraConv_{M\text{-}NQ}$) perform better than the equipollent multi-lingual AraConv model ($AraConv_{M\text{-}Q}$). Accordingly, we recognize that joint-training on multiple languages including the target language (in this case, Arabic) improves the results in experiments on the target language, which aligns with the results reported in [30].

Table 5. Bi-lingual experiment DST and end-to-end dialogue generation results for the AraConv model trained on the Arabic-TOD dataset compared to the bi-lingual BiToD experiments [1] using English and Chinese BiToD datasets. The bold letters refer to the target language in the corresponding experiments (used to test the model). Bold numbers indicating the best result according to the column's metric value.

	TSR	DSR	APIAcc	BLEU	JGA
		Arabic			
AraConvBi-NQ (**AR**, EN)	45.57	21.90	56.23	30.41	37.35
AraConvBi-Q (**AR**, EN)	44.62	16.98	46.32	27.36	35.58
		Other languages			
ZH, **EN** [1]	**71.18**	**51.13**	**71.87**	**40.71**	**72.16**
ZH, EN [1]	57.24	34.78	65.54	22.45	68.70

Table 6. Multi-lingual experiment DST and end-to-end dialogue generation results for the AraConv model on the Arabic-TOD dataset. The bold letters refer to the target language. Bold numbers indicating the best result according to the column's metric value.

	TSR	DSR	APIAcc	BLEU	JGA
AraConvM-NQ (**AR**, EN, ZH)	**51.27**	**20.00**	**55.44**	**32.58**	**37.68**
AraConvM-Q (**AR**, EN, ZH)	47.17	16.98	53.07	31.05	36.13

Generally, for bi-lingual and multi-lingual experiments, the trained models can simultaneously handle dialogues in multiple languages (whether English, Chinese, or Arabic) without using any of the language identifiers supplied during testing.

For the human review, we aimed to rate dialogue or utterances on the basis of certain metrics identified in the literature [56]. Five expert researchers (who are independent from this paper author) were chosen for this task. We randomly selected 20 complete dialogue sessions from the generated dialogues of AraConv model. The researchers were asked to rate these dialogues by providing language understanding and response appropriateness scores. Their scores ranged from 0 (extremely bad) to 5 (extremely good), depending on the system's response. Subsequently, we evaluated the reliability of their rating using Fleiss' Kappa [64]. The overall Fleiss' kappa values for the language understanding and appropriateness scores were 0.253 and 0.229, respectively, indicating "fair agreement".

5. Conclusions and Future Work

To the best of our knowledge, this work represents to the first attempt to build an end-to-end Arabic task-oriented dialogue system (AraConv) using a pre-trained transformer-based multi-lingual language model. We utilized the highly regarded multi-lingual model mT5 to build an end-to-end Arabic task-oriented dialogue system with different settings and presented an Arabic-TOD dataset based on translating 27% of the BiToD dataset's English dialogue data into Arabic. The Arabic-TOD dataset is considered the first dialogue dataset for the Arabic task-oriented DS that supports code-switching. Although using the Arabic-TOD dataset to train and test the model in a mono-lingual setting demonstrates a reasonable performance for the AraConv model compared to the results observed for the English and Chinese BiToD datasets in the same settings, the performance is undermined by the small size of the Arabic TOD dataset. To overcome this problem, we considered joint-training the model on Arabic dialogue data and one or two high-resource languages (English or both English and Chinese). The findings reveal that the AraConv model in the multi-lingual setting outperformed the AraConv model in the mono-lingual setting, with multi-lingual training with English, Chinese, and Arabic observed to be better than bi-lingual training with only English and Arabic data. Thus, the AraConv model can be considered a good baseline for building robust end-to-end Arabic task-oriented DS that can engage with complex scenarios.

The main limitation of this work is the small size of the Arabic-TOD dataset. A related limitation concerns the Arabic-TOD dataset using non-Arabic entities, with the dataset code-switching due to entities in the original BiToD dataset. However, we leveraged this property to align the model with the routine usage of such entities in conversation. In the future, we aim to extend the Arabic-TOD dataset to equal the BiToD dataset in terms of the number of dialogues. Additionally, we plan to examine cross-lingual models, especially involving the Arabic-TOD dataset. Furthermore, we plan to develop Arabic task-oriented DS using other multilingual language models (e.g., mBART [65]). Another possible venue for future work is using a pre-trained Arabic model for Arabic task-oriented DS such as AraT5 [66], which was yet to be deployed at the time of working on this paper.

Author Contributions: Conceptualization, A.F. and M.A.-Y.; methodology, A.F.; software, A.F.; validation, A.F.; formal analysis, A.F. and M.A.-Y.; investigation, A.F. and M.A.-Y.; resources, A.F. and M.A.-Y.; data curation, A.F.; writing (original draft preparation), A.F.; writing (review and editing), A.F. and M.A.-Y.; visualization, A.F. and M.A.-Y.; supervision, M.A.-Y.; project administration, M.A.-Y.; funding acquisition, M.A.-Y. All authors have read and agreed to the published version of the manuscript.

Funding: This research is supported by a grant from the Researchers Supporting Project No. RSP-2021/286, King Saud University, Riyadh, Saudi Arabia.

Institutional Review Board Statement: Not applicable.

Informed Consent Statement: Not applicable.

Data Availability Statement: Upon request.

Acknowledgments: The authors extend their appreciation to the Researchers Supporting Project number RSP-2021/286, King Saud University, Riyadh, Saudi Arabia.

Conflicts of Interest: The authors declare no conflict of interest.

References

1. Lin, Z.; Madotto, A.; Winata, G.I.; Xu, P.; Jiang, F.; Hu, Y.; Shi, C.; Fung, P. BiToD: A Bilingual Multi-Domain Dataset For Task-Oriented Dialogue Modeling. *arXiv* **2021**, arXiv:2106.02787.
2. Huang, M.; Zhu, X.; Gao, J. Challenges in building intelligent open-domain dialog systems. *ACM Trans. Inf. Syst.* **2019**, *38*, 1–32. [CrossRef]
3. AlHagbani, E.S.; Khan, M.B. Challenges facing the development of the Arabic chatbot. *First Int. Work. Pattern Recognit. Int. Soc. Opt. Photonics* **2016**, *10011*, 7. [CrossRef]
4. Darwish, K.; Habash, N.; Abbas, M.; Al-Khalifa, H.; Al-Natsheh, H.T.; Bouamor, H.; Bouzoubaa, K.; Cavalli-Sforza, V.; El-Beltagy, S.R.; El-Hajj, W.; et al. A Panoramic Survey of Natural Language Processing in the Arab World. *Commun. ACM* **2021**, *64*, 72–81. [CrossRef]
5. Xue, L.; Constant, N.; Roberts, A.; Kale, M.; Al-Rfou', R.; Siddhant, A.; Barua, A.; Raffel, C. mT5: A Massively Multilingual Pre-trained Text-to-Text Transformer. In Proceedings of the 2021 Conference of the North American Chapter of the Association for Computational Linguistics: Human Language Technologies, online, 15–20 June 2020; Association for Computational Linguistics: Stroudsburg, PA, USA, 2020; pp. 483–498. [CrossRef]
6. McTear, M. *Conversational AI: Dialogue Systems, Conversational Agents, and Chatbots*; Morgan & Claypool Publishers LLC: San Rafael, CA, USA, 2020; Volume 13.
7. Qin, L.; Xu, X.; Che, W.; Zhang, Y.; Liu, T. Dynamic Fusion Network for Multi-Domain End-to-end Task-Oriented Dialog. In Proceedings of the 58th Annual Meeting of the Association for Computational Linguistics, online, 5–10 July 2020; pp. 6344–6354. [CrossRef]
8. Lei, W.; Jin, X.; Ren, Z.; He, X.; Kan, M.Y.; Yin, D. Sequicity: Simplifying task-oriented dialogue systems with single sequence-to-sequence architectures. In Proceedings of the 56th Annual Meeting of the Association for Computational Linguistics (Volume 1: Long Papers), Melbourne, Australia, 15–20 July 2018; pp. 1437–1447. [CrossRef]
9. Budzianowski, P.; Vulić, I. Hello, It's GPT-2-How Can I Help You? Towards the Use of Pretrained Language Models for Task-Oriented Dialogue Systems. In Proceedings of the 3rd Workshop on Neural Generation and Translation (WNGT 2019), Hong Kong, China, 4 November 2019; pp. 15–22. [CrossRef]

10. Ham, D.; Lee, J.-G.; Jang, Y.; Kim, K.-E. End-to-End Neural Pipeline for Goal-Oriented Dialogue Systems using GPT-2. In Proceedings of the 58th Annual Meeting of the Association for Computational Linguistics, online, 5–10 July 2020; Association for Computational Linguistics: Stroudsburg, PA, USA, 2020; Volume 2, pp. 583–592. [CrossRef]
11. Hosseini-Asl, E.; McCann, B.; Wu, C.S.; Yavuz, S.; Socher, R. A simple language model for task-oriented dialogue. *Adv. Neural Inf. Process. Syst.* **2020**, *33*, 20179–20191.
12. Peng, B.; Li, C.; Li, J.; Shayandeh, S.; Liden, L.; Gao, J. SOLOIST: Building Task Bots at Scale with Transfer Learning and Machine Teaching. *Trans. Assoc. Comput. Linguist.* **2021**, *9*, 807–824. [CrossRef]
13. Yang, Y.; Li, Y.; Quan, X. UBAR: Towards Fully End-to-End Task-Oriented Dialog Systems with GPT-2. *arXiv* **2020**, arXiv:2012.03539.
14. Lin, Z.; Madotto, A.; Winata, G.I.; Fung, P. MinTL: Minimalist Transfer Learning for Task-Oriented Dialogue Systems. In Proceedings of the 2020 Conference on Empirical Methods in Natural Language Processing (EMNLP), online, 16–20 November 2020; Association for Computational Linguistics: Stroudsburg, PA, USA, 2020; pp. 3391–3405. [CrossRef]
15. Wang, W.; Zhang, Z.; Guo, J.; Dai, Y.; Chen, B.; Luo, W. Task-Oriented Dialogue System as Natural Language Generation. *arXiv* **2021**, arXiv:2108.13679.
16. Peng, B.; Zhu, C.; Li, C.; Li, X.; Li, J.; Zeng, M.; Gao, J. Few-shot Natural Language Generation for Task-Oriented Dialog. *arXiv* **2020**, arXiv:2002.12328.
17. Wu, C.-S.; Hoi, S.C.H.; Socher, R.; Xiong, C. TOD-BERT: Pre-trained Natural Language Understanding for Task-Oriented Dialogue. In Proceedings of the 2020 Conference on Empirical Methods in Natural Language Processing (EMNLP), online, 16–20 November 2020; Association for Computational Linguistics: Stroudsburg, PA, USA, 2020; pp. 917–929. [CrossRef]
18. Madotto, A.; Liu, Z.; Lin, Z.; Fung, P. Language Models as Few-Shot Learner for Task-Oriented Dialogue Systems. *arXiv* **2020**, arXiv:2008.06239.
19. Campagna, G.; Foryciarz, A.; Moradshahi, M.; Lam, M. Zero-Shot Transfer Learning with Synthesized Data for Multi-Domain Dialogue State Tracking. *arXiv* **2020**, arXiv:2005.00891.
20. Zhang, Y.; Ou, Z.; Hu, M.; Feng, J. A Probabilistic End-To-End Task-Oriented Dialog Model with Latent Belief States towards Semi-Supervised Learning. In Proceedings of the 2020 Conference on Empirical Methods in Natural Language Processing (EMNLP), online, 16–20 November 2020; Association for Computational Linguistics: Stroudsburg, PA, USA, 2020; pp. 9207–9219. [CrossRef]
21. Kulhánek, J.; Hudeček, V.; Nekvinda, T.; Dušek, O. AuGPT: Dialogue with Pre-trained Language Models and Data Augmentation. *arXiv* **2021**, arXiv:2102.05126.
22. Gao, S.; Takanobu, R.; Peng, W.; Liu, Q.; Huang, M. HyKnow: End-to-End Task-Oriented Dialog Modeling with Hybrid Knowledge Management. In *Findings of the Association for Computational Linguistics: ACL-IJCNLP 2021*; Association for Computational Linguistics: Stroudsburg, PA, USA, 2021; pp. 1591–1602. [CrossRef]
23. Lee, H.; Lee, J.; Kim, T.Y. SUMBT: Slot-utterance matching for universal and scalable belief tracking. *arXiv* **2019**, arXiv:1907.07421.
24. Chao, G.L.; Lane, I. BERT-DST: Scalable end-to-end dialogue state tracking with bidirectional encoder representations from transformer. *Proc. Annu. Conf. Int. Speech Commun. Assoc. Interspeech* **2019**, *2019*, 1468–1472. [CrossRef]
25. Kim, S.; Yang, S.; Kim, G.; Lee, S.-W. Efficient Dialogue State Tracking by Selectively Overwriting Memory. *arXiv* **2020**, arXiv:1911.03906. [CrossRef]
26. Kumar, A.; Ku, P.; Goyal, A.; Metallinou, A.; Hakkani-Tur, D. MA-DST: Multi-Attention-Based Scalable Dialog State Tracking. *Proc. Conf. AAAI Artif. Intell.* **2020**, *34*, 8107–8114. [CrossRef]
27. Heck, M.; van Niekerk, C.; Lubis, N.; Geishauser, C.; Lin, H.-C.; Moresi, M.; Gašić, M. TripPy: A Triple Copy Strategy for Value Independent Neural Dialog State Tracking. *arXiv* **2020**, arXiv:2005.02877.
28. Li, S.; Yavuz, S.; Hashimoto, K.; Li, J.; Niu, T.; Rajani, N.; Yan, X.; Zhou, Y.; Xiong, C. CoCo: Controllable Counterfactuals for Evaluating Dialogue State Trackers. *arXiv* **2020**, arXiv:2010.12850.
29. Wang, D.; Lin, C.; Liu, Q.; Wong, K.-F. Fast and Scalable Dialogue State Tracking with Explicit Modular Decomposition. In Proceedings of the 2021 Conference of the North American Chapter of the Association for Computational Linguistics: Human Language Technologies, Mexico City, Mexico, 6–11 June 2021; Association for Computational Linguistics: Stroudsburg, PA, USA, 2021; pp. 289–295. [CrossRef]
30. Schuster, S.; Shah, R.; Gupta, S.; Lewis, M. Cross-lingual transfer learning for multilingual task oriented dialog. In Proceedings of the NAACL-HLT 2019, Minneapolis, MN, USA, 2–7 June 2019; Volume 1, pp. 3795–3805. [CrossRef]
31. Liu, Z.; Winata, G.I.; Lin, Z.; Xu, P.; Fung, P. Attention-informed mixed-language training for zero-shot cross-lingual task-oriented dialogue systems. *Proc. AAAI Conf. Artif. Intell.* **2020**, *34*, 8433–8440. [CrossRef]
32. Zhang, Y.; Ou, Z.; Yu, Z. Task-oriented dialog systems that consider multiple appropriate responses under the same context. *Proc. AAAI Conf. Artif. Intell.* **2020**, *34*, 9604–9611. [CrossRef]
33. Wang, K.; Tian, J.; Wang, R.; Quan, X.; Yu, J. Multi-Domain Dialogue Acts and Response Co-Generation. *arXiv* **2020**, arXiv:2004.12363.
34. Bashir, A.M.; Hassan, A.; Rosman, B.; Duma, D.; Ahmed, M. Implementation of A Neural Natural Language Understanding Component for Arabic Dialogue Systems. *Procedia Comput. Sci.* **2018**, *142*, 222–229. [CrossRef]

35. Elmadany, A.R.A.; Abdou, S.M.; Gheith, M. Improving dialogue act classification for spontaneous Arabic speech and instant messages at utterance level. In Proceedings of the Eleventh International Conference on Language Resources and Evaluation (LREC 2018), European Language Resources Association (ELRA), Miyazaki, Japan, 7–12 May 2018; pp. 128–134.
36. Joukhadar, A.; Saghergy, H.; Kweider, L.; Ghneim, N. Arabic Dialogue Act Recognition for Textual Chatbot Systems. In Proceedings of the First International Workshop on NLP Solutions for Under Resourced Languages (NSURL 2019), Trento, Italy, 11–12 September 2019; Association for Computational Linguistics: Stroudsburg, PA, USA, 2019; pp. 43–49.
37. Al-Ajmi, A.H.; Al-Twairesh, N. Building an Arabic Flight Booking Dialogue System Using a Hybrid Rule-Based and Data Driven Approach. *IEEE Access* **2021**, *9*, 7043–7053. [CrossRef]
38. Hijjawi, M.; Bandar, Z.; Crockett, K.; McLean, D. ArabChat: An arabic conversational agent. In Proceedings of the 2014 6th International Conference on Computer Science and Information Technology, CSIT 2014-Proceedings, Amman, Jordan, 26 March 2014; pp. 227–237. [CrossRef]
39. Almurtadha, Y. LABEEB: Intelligent Conversational Agent Approach to Enhance Course Teaching and Allied Learning Outcomes attainment. *J. Appl. Comput. Sci. Math.* **2019**, *13*, 9–12. [CrossRef]
40. Aljameel, S.; O'shea, J.; Crockett, K.; Latham, A.; Kaleem, M. LANA-I: An Arabic Conversational Intelligent Tutoring System for Children with ASD. *Adv. Intell. Syst. Comput.* **2019**, *997*, 498–516. [CrossRef]
41. Moubaiddin, A.; Shalbak, O.; Hammo, B.; Obeid, N. Arabic dialogue system for hotel reservation based on natural language processing techniques. *Comput. Sist.* **2015**, *19*, 119–134. [CrossRef]
42. Bendjamaa, F.; Nora, T. A Dialogue-System Using a Qur'anic Ontology. In Proceedings of the 2020 Second International Conference on Embedded & Distributed Systems (EDiS), Oran, Algeria, 3 November 2020; pp. 167–171.
43. Fadhil, A.; AbuRa'Ed, A. Ollobot-Towards a text-based Arabic health conversational agent: Evaluation and results. In Proceedings of the Recent Advances in Natural Language Processing (RANLP), Varna, Bulgaria, 2–4 September 2019; pp. 295–303. [CrossRef]
44. Al-Ghadhban, N.; Al-Twairesh, D. Nabiha: An Arabic dialect chatbot. *Int. J. Adv. Comput. Sci. Appl. Int. J. Adv. Comput. Sci. Appl.* **2020**, *11*, 452–459. [CrossRef]
45. Vaswani, A.; Shazeer, N.; Parmar, N.; Uszkoreit, J.; Jones, L.; Gomez, A.N. Attention is all you need. *Adv. Neural Inf. Process. Syst.* **2017**, *30*, 5999–6009.
46. Mozannar, H.; Maamary, E.; el Hajal, K.; Hajj, H. Neural Arabic Question Answering. *arXiv* **2019**, arXiv:1906.05394.
47. Mayeesha, T.T.; Sarwar, A.M.; Rahman, R.M. Deep learning based question answering system in Bengali. *J. Inf. Telecommun.* **2020**, *5*, 145–178. [CrossRef]
48. Naous, T.; Hokayem, C.; Hajj, H. Empathy-driven Arabic Conversational Chatbot. In Proceedings of the Fifth Arabic Natural Language Processing Workshop, Barcelona, Spain, 8 December 2020; Association for Computational Linguistics: Stroudsburg, PA, USA, 2020; pp. 58–68.
49. Razumovskaia, E.; Glavaš, G.; Majewska, O.; Ponti, E.M.; Korhonen, A.; Vulić, I. Crossing the Conversational Chasm: A Primer on Natural Language Processing for Multilingual Task-Oriented Dialogue Systems. *arXiv* **2021**, arXiv:2104.08570.
50. He, X.; Deng, L.; Hakkani-Tur, D.; Tur, G. Multi-style adaptive training for robust cross-lingual spoken language understanding. In Proceedings of the IEEE International Conference on Acoustics, Speech and Signal Processing, Vancouver, BC, Canada, 26–31 May 2013; pp. 8342–8346.
51. Mrkšić, N.; Vulić, I.; Séaghdha, D.; Leviant, I.; Reichart, R.; Gašić, M.; Korhonen, A.; Young, S. Semantic Specialization of Distributional Word Vector Spaces using Monolingual and Cross-Lingual Constraints. *Trans. Assoc. Comput. Linguist.* **2017**, *5*, 309–324. [CrossRef]
52. Castellucci, G.; Bellomaria, V.; Favalli, A.; Romagnoli, R. Multi-lingual Intent Detection and Slot Filling in a Joint BERT-based Model. *arXiv* **2019**, arXiv:1907.02884.
53. Bellomaria, V.; Castellucci, G.; Favalli, A.; Romagnoli, R. Almawave-SLU: A new dataset for SLU in Italian. *arXiv* **2019**, arXiv:1907.07526.
54. Liu, Z.; Shin, J.; Xu, Y.; Winata, G.I.; Xu, P.; Madotto, A.; Fung, P. Zero-shot cross-lingual dialogue systems with transferable latent variables. In Proceedings of the 2019 Conference on Empirical Methods in Natural Language Processing and the 9th International Joint Conference on Natural Language Processing (EMNLP-IJCNLP), Hong Kong, China, 11 November 2019; pp. 1297–1303. [CrossRef]
55. Dao, M.H.; Truong, T.H.; Nguyen, D.Q. Intent Detection and Slot Filling for Vietnamese. *arXiv* **2021**, arXiv:2104.02021.
56. Deriu, J.; Rodrigo, A.; Otegi, A.; Echegoyen, G.; Rosset, S.; Agirre, E.; Cieliebak, M. Survey on evaluation methods for dialogue systems. *Artif. Intell. Rev.* **2020**, *54*, 755–810. [CrossRef]
57. Raffel, C.; Shazeer, N.; Roberts, A.; Lee, K.; Narang, S.; Matena, M.; Zhou, Y.; Li, W.; Liu, P.J. Exploring the limits of transfer learning with a unified text-to-text transformer. *J. Mach. Learn. Res.* **2020**, *21*, 1–67.
58. Huggingface/Transformers? Transformers: State-of-the-Art Natural Language Processing for Pytorch, TensorFlow, and JAX. Available online: https://github.com/huggingface/transformers (accessed on 17 November 2021).
59. PyTorch. Available online: https://pytorch.org/ (accessed on 17 November 2021).
60. Loshchilov, I.; Hutter, F. Decoupled weight decay regularization. *arXiv* **2017**, arXiv:1711.05101.
61. Siblini, W.; Pasqual, C.; Lavielle, A.; Challal, M.; Cauchois, C. Multilingual Question Answering from Formatted Text applied to Conversational Agents. *arXiv* **2019**, arXiv:1910.04659.

62. Hsu, T.Y.; Liu, C.L.; Lee, H.Y. Zero-shot Reading Comprehension by Cross-lingual Transfer Learning with Multi-lingual Language Representation Model. In Proceedings of the 2019 Conference on Empirical Methods in Natural Language Processing and the 9th International Joint Conference on Natural Language Processing (EMNLP-IJCNLP), Hong Kong, China, 3–7 November 2019; pp. 5933–5940. [CrossRef]
63. Upadhyay, S.; Faruqui, M.; Tür, G.; Dilek, H.-T.; Heck, L. (Almost) Zero-shot cross-lingual spoken language understanding. In Proceedings of the 2018 IEEE International Conference on Acoustics, Speech and Signal Processing (ICASSP), Calgary, AB, Canada, 15–20 April 2018; pp. 6034–6038.
64. Fleiss, J.L. Measuring nominal scale agreement among many raters. *Psychol. Bull.* **1971**, *76*, 378–382. [CrossRef]
65. Liu, Y.; Gu, J.; Goyal, N.; Li, X.; Edunov, S.; Ghazvininejad, M.; Lewis, M.; Zettlemoyer, L. Multilingual Denoising Pre-training for Neural Machine Translation. *Trans. Assoc. Comput. Linguist.* **2020**, *8*, 726–742. [CrossRef]
66. Nagoudi, E.M.B.; Elmadany, A.; Abdul-Mageed, M. AraT5: Text-to-Text Transformers for Arabic Language Understanding and Generation. *arXiv* **2021**, arXiv:2109.12068.

applied sciences

MDPI

Article

Achieving Semantic Consistency for Multilingual Sentence Representation Using an Explainable Machine Natural Language *Parser* (*MParser*)

Peng Qin [1], Weiming Tan [1], Jingzhi Guo [1], Bingqing Shen [2] and Qian Tang [1,3,*]

[1] Faculty of Science and Technology, University of Macau, Macau 999078, China; yb77428@connect.um.edu.mo (P.Q.); wade.tan@connect.um.edu.mo (W.T.); jzguo@um.edu.mo (J.G.)
[2] School of Software, Shanghai Jiao Tong University, Shanghai 200240, China; sunniel@sjtu.edu.cn
[3] College of Business, Beijing Institute of Technology, Zhuhai 519088, China
* Correspondence: tang_qiansxy2022@126.com

Abstract: In multilingual semantic representation, the interaction between humans and computers faces the challenge of understanding meaning or semantics, which causes ambiguity and inconsistency in heterogeneous information. This paper proposes a Machine Natural Language Parser (MParser) to address the semantic interoperability problem between users and computers. By leveraging a semantic input method for sharing common atomic concepts, MParser represents any simple English sentence as a bag of unique and universal concepts via case grammar of an explainable machine natural language. In addition, it provides a human and computer-readable and -understandable interaction concept to resolve the semantic shift problems and guarantees consistent information understanding among heterogeneous sentence-level contexts. To evaluate the annotator agreement of MParser outputs that generates a list of English sentences under a common multilingual word sense, three expert participants manually and semantically annotated 75 sentences (505 words in total) in English. In addition, 154 non-expert participants evaluated the sentences' semantic expressiveness. The evaluation results demonstrate that the proposed MParser shows higher compatibility with human intuitions.

Keywords: document representation; semantic analysis; natural language processing; conceptual modeling; universal representation

Citation: Qin, P.; Tan, W.; Guo, J.; Shen, B.; Tang, Q. Achieving Semantic Consistency for Multilingual Sentence Representation Using an Explainable Machine Natural Language *Parser* (*MParser*). *Appl. Sci.* **2021**, *11*, 11699. https://doi.org/10.3390/app112411699

Academic Editors: Arturo Montejo-Ráez and Salud María Jiménez-Zafra

Received: 16 November 2021
Accepted: 8 December 2021
Published: 9 December 2021

1. Introduction

Multilingual semantic representation [1] presents words, phrases, texts, or documents in heterogeneous parties (e.g., English and Chinese) to achieve semantic consistency. It has been applied in several areas, such as machine translation [2], question answering [3], and document representation [4,5]. The process of parsing a natural language sentence to its semantic representation is called semantic parsing [6], which parses the sentences without representing the syntactic classification of the components of the sentence. Semantic parsing is an essential process and has attracted great attention in multilingual semantic representation and NLP research over the last few decades [6]. Typically, a semantic parser labels each word in the original sentence according to its semantic role or represents each compound component based on its meaning [7]. Several semantic approaches are proposed for parsing natural language sentences in semantic representation, such as Groningen Meaning Bank (GMB) [8] and abstract meaning representation (AMR) [9]. Still, their annotation schemes are designed for individual languages that have language-dependent features. Because many applications require multilingual capabilities, several efforts are underway to create more cross-lingual natural language resources such as universal conceptual cognitive annotation (UCCA) [10], universal networking language (UNL) [11], and universal dependencies (UD) [12]. They are the framework for cross-linguistically consistent grammatical annotation. Despite these efforts, some remaining interlanguage variations important for practical usage are not yet captured by the efforts. They create obstacles to a truly cross-lingual meaning representation that enables downstream applications written in one language to be applicable for other languages. Using cross-lingual language to perform cross-lingual semantic parsing for one language to improve the representation of another language remains a largely under-explored research question. This paper focuses on the problem of multilingual semantic interoperability in semantic representation.

In semantic analysis and labeling, texts and documents are generally very complex because of flexible structural and complex morphological grammars. The state-of-art semantic parser methods and applications have not achieved satisfying results. One technical challenge is the lack of consistent conversions across domains. The heterogeneous text may share heterogeneous meaning and cause semantic loss or misunderstanding between a computer and a user [13]. For example, Figure 1 shows an English inquiry sheet for illustrating the multilingual semantic interoperability problems. The table consists of 10 cells; cells 1–9 contain a single atomic concept, i.e., *"one cell one atomic concept"* (e.g., Date in cell 1). However, one atomic concept may have multiple meanings. For instance, the word *"company"* in cell 10 refers to several meanings such as *"a commercial business"* and *"the fact or condition of being with another or others, especially in a way that provides friendship and enjoyment"*. To achieve accurate atomic concept exchange and guarantee semantic consistency in cells 1–9, several document representation approaches [5,14] are proposed to solve the heterogeneous concept or meaning exchange problem. An effective solution is the collaboration mechanism that connects heterogeneous domains or contexts, allowing the exchange of heterogeneous semantic documents by a semantic input method (SIM) approach [15]. However, some sentences also contain sequences of atomic concepts for a free-text cell (e.g., cell 10), which makes it hard to ensure that the meaning (M_1) of an English sentence $Ei := List\ (w_1, w_2, \ldots, w_n)$ and the meaning (M_2) of a translated Chinese sentence $Cj := List\ (w_1, w_2, \ldots, w_n)$ will be semantically equivalent. The reasons for causing $M_1 \neq_s M_2$ ("$\neq_s$" refers to not semantically equal) include:

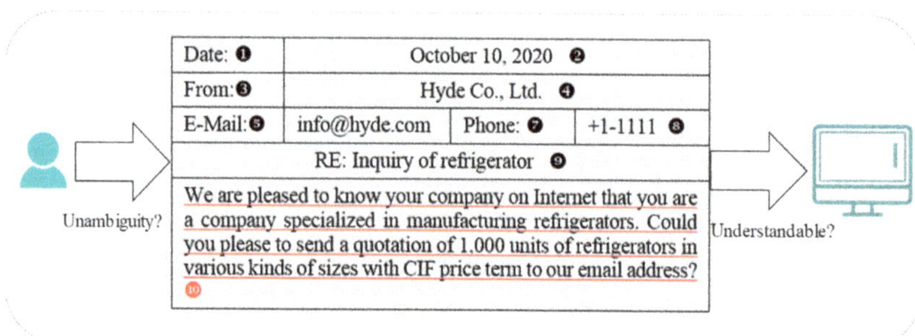

Figure 1. Complex document interaction between computers and users.

(1) Heterogeneous grammatical rules: The language grammars of the components in E_i and C_i have their own rules to generate a sentence and it is impossible to achieve a one-to-one mapping.
(2) Synonyms and homonyms: Each term in E_i may have several synonyms or homonyms. A wrong term in meaning may cause semantic ambiguity.
(3) Peculiar language phenomena: Some phenomena in E_i never appear in C_i, resulting in asymmetric mapping. For example, the particles of "を, に, で, へ, より" in Japanese do not have counterparts in Chinese.

Therefore, the same sentence will produce completely different scenarios in a heterogeneous text, and the original meaning in mind may be shifted to another meaning. The above problems are called semantic shift problems that change a sentence's original meaning in multilingual semantic representation. Moreover, in natural language texts, users cannot express their information needs in a computer-understandable way or interpret the representation correctly due to problems in representing complex semantics. Therefore, the development of a novel model has been motivated by the following aspects:

(1) Computer-human-understandable representation: providing information understandable by both computers and humans, realizing the accurate interpretation of sentences in the human-computer messaging cycle of humans and computers without ambiguity.
(2) Accurate semantic representation among computing applications: applying computer-human-understandable information in computing applications and enabling information to be semantically interoperable.
(3) Automated multilingual information processing by software agents: allowing multilingual information to be automatically processed across domains and contexts.

Thus, this research proposes a new multilingual semantic representation parser for sentence-based text or documents that enhances textual representation and reduces multilingual ambiguity. Based on our previous conceptual work [16], we propose a novel Machine Natural Language Parser (Mparser) to realize universal representations between computers and users unambiguously. The explainable MParser parses a simple English sentence, resolving complex concepts towards a bag of universal concepts sentence-readable and -understandable for any heterogeneous information, and mediates contextual human natural languages collaboratively, as shown in Figure 2. The universal concepts sentence shares a common concept at both the syntactic and semantic levels between users and computers.

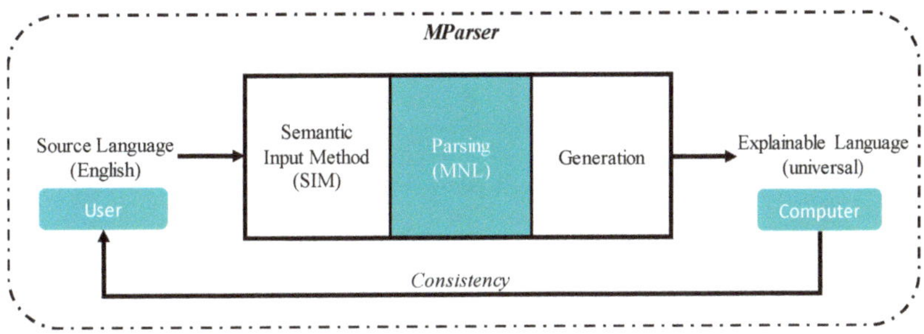

Figure 2. A general MParser process.

To achieve consistency and universal representation, *MParser* designs from human input and sentence generation:

(1) In the human input, each unique concept is collaboratively edited with SIM [15] based on a common dictionary (CoDic) [17] for eliminating atomic concept ambiguity and morphological features. Thus, a simple English sentence can be converted to a sequence of unique concepts across conversational contexts.
(2) To maintain complex semantic concept consistency between computers and users, an MParser for English sentences parses the semantic roles between English words and represents them for deriving a unique concept that can be accurately represented and understood by computers through case grammar [16]. The cases are used to label words, which are aligned from local language perspectives. The proposed parser utilizes powerful linguistic tools such as Stanford Parser and universal dependency relations.
(3) Evaluate the proposed MParser through annotator agreement between the expert's case labeling and MParser's outputs. Additionally, 154 non-expert participants investigated judgments of semantic expressiveness.

The rest of this paper is organized as follows. Section 2 compares the proposed approach with related work. Section 3 introduces the general process and methodology of

MParser. Section 4 introduces the activity of human semantic input. Section 5 introduces the activity of sentence computerization. Section 6 and 7 implement and evaluate MParser. Finally, a conclusion is given.

2. Related Work

Semantic representation presents the meaning of sentences, and the process should be reliable and computational [18]. The alternative approaches to semantic representation can be divided into two categories: document representation [1,19] and meaning representation.

Document Representation: Currently, document information exchange mainly has three approaches: (1) Standardization approaches define a semantic document by combining a set of standardized document compositions: for example, EDI-based (http://www.edibasics.com/ediresources/document-standards/), XML-based (ebXML. Available: http://www.ebxml.org) and Web service-based (http://www.edibasics.com/ediresources/document-standards/). The problem with this approach is that documents are only interoperable on representation syntax and templates, and these standards are heterogeneous and incompatible with each other. (2) Ontology modeling [20,21] approaches define a semantic document in a certain domain (e.g., RDF [22], RDFS [23] and OWL (http://www.edibasics.com/ediresources/document-standards/)). They are usually used to solve the problem of semantic interoperability and realize collaboration. Generally, an ontology clearly describes the relationships of entities [18] and can be employed for knowledge representation. However, if computers in different contexts participate in user-computer interaction, it will not be easy to achieve a consistent understanding, because an ontology is domain-dependent, preventing it from being understood between heterogeneous document descriptions. (3) Collaborative approaches [17,24] allow participants from different contexts to construct document terms and solve the cross-domain problem, but the document is constrained by a template and lacks flexibility. One issue is that the user still needs a user template to construct the document.

Current subjects of research on document representations are rule format [25,26], ontology [20,24], XML+Ontology [21], tree/graph [27], and collaborative approach [15,17,28]. First, it is not easy to embed and extract meanings to/from a document automatically. For example, it is not easy for a document written in natural languages to be automatically converted to a machine-processable format (e.g., RuleML [25,26]). Second, constructing semantic documents needs intensive work. For example, [5] proposes a semantic disambiguation solution by using a machine-readable semantic network (e.g., WordNet) as a common knowledge base. However, it is time-consuming and sometimes unnecessary because it also disambiguates unambiguous terms. To acquire accurate semantic concept representation for a document, [20] requires learning a concept border from a particular document collection based on a particular ontology in the same domain. However, there is a heavy workload and enormous data redundancy to construct and store concept borders for different domains. Third, it is not easy to maintain semantic consistency between heterogeneous document systems. For example, [24] claims accurate mapping between different ontologies' entities, and [20] requires the similarity computation between keywords in a received document and equivalent terms in a domain-wide ontology. Both approaches hardly reach a trade-off between low computational demand and semantic interoperability.

In short, these approaches rely on the homogeneity of concept in multilingual text or domain semantics, and sentence-based documents or complex concepts may cause semantic loss among different contexts through the above state-of-art approaches.

Semantic Representation: It defines the annotation to construct syntactic structure such as FrameNet [29] and Semlink [30], but focuses on argument out of other relations. In this context, there are several available semantic representation approaches. For instance, universal networking language (UNL) proposes independent language representation so that sentences inputted in any language can be translated into any other natural language. Abstract meaning representation (AMR) [9] proposes a relatively more straightforward sentence-level semantic parser to cover semantic role broad predictions. AMR manually annotates sentences and utilizes PropBank frames [31] to represent the semantic relationship between words. However, AMR faces difficulties across translation because the syntactical similarity is not suitable cross-linguistically [32]. Therefore, new multilayered solutions such as universal concept cognitive annotation (UCCA) [10] and universal decompositional semantics (UDS) [33] are applied in cross text for semantic annotation and word senses by BabelNet [34] and Open Multilingual Wordnet (http://compling.hss.ntu.edu.sg/omw/). They constructed substantial multilingual semantic nets to achieve universality by connecting resources such as WordNet and Wikipedia. The method adapts linguistic theory to build a manual and multilingual scheme. However, UCCA annotates short sentences (e.g., multiword expressions) where the same multiword or entity is annotated in many different sentences. Groningen Meaning Bank (GMB) is a new solution to integrate language phenomena into a single formalism instead of covering single phenomena in an isolated way. Additionally, universal dependencies (UD) [12] build cross text dependency-based annotations for multilingual sentences.

Most of the semantic representation methods use simple concepts such as UCCA, but some other methods adapt concepts such as WordNet synsets for UNL and PropBank frames for AMR. Furthermore, UNL has its relationships set while AMR uses PropBank relationships. UNL, UCCA, and AMR are fully manual annotated, but GMB produces meaning representations automatically and can be corrected by experts. However, such approaches (e.g., AMR, UCCA, GMB, and UNL) focus on lexical-semantic or multilingual words rather than on sentence semantics and cannot guarantee sentence-based semantic representation to be universal and unambiguous across languages. Most of the proposed semantic representation methods do not consider the morphological and syntactic characteristics of the language in the construction of sentence-level semantic labeling. Contributions made in the semantic representation of any language text will utilize the translated English resources, which may negatively affect the performance of other semantic representation methods. In our research work, MParser propose a universal semantic representation to extract semantic relationships from local language text using local language tools and resources, such as Stanford Parser. In addition, the proposed parser takes into account the syntactic and morphological features of a given sentence. It is worth noting that the proposed MParser model uses various tools, resources, and text features to reduce the negative impact of resource quality on semantic representation. Moreover, MParser achieves a universal representation and semantic consistency across languages.

3. MParser

3.1. Overview

MParser comprises two processes: (1) human semantic input (HSI) and (2) sentence computerization (SC), as shown in Figure 3. First, human semantic input is the process of converting human natural language (HNL_i) (here, i indicates English) through an editor typing from CoDic into a sequence of machine-readable sentences SiS_{ci}, which comprises sequentially converting sets of literals to a list of the symbolic signs. The editor (i.e., human user) inputs the HNL_i by SIM from CoDic to constrain sentence creation based on strict rules. Second, sentence computerization (f_c) is a process of converting a sentence SiS_{hi} to a sentence SiS_m that is universally readable and understandable by a computer in MParser, denoted by $f_c := SiS_m \leftarrow SiS_{hi}$. In particular, this involves a sequence of activities: sentence analysis (i.e., parsing a local language sentence based on the local grammatical rules through robust Stanford Parser and universal dependency), case generation (i.e., appending a case on each sign to represent its grammatical functions and properties), and machine representation (i.e., representing a sentence that is computer-readable and -understandable). Thus, sentence $SiS_m \subset MParser$ only readable and understandable by computers can be converted back to a human-readable and -understandable SiS_{hj} (here, j indicates other languages), such that $f_r := SiS_{hj} \leftarrow SiS_m$ to rebuild human-readable sentences based on SiS_m.

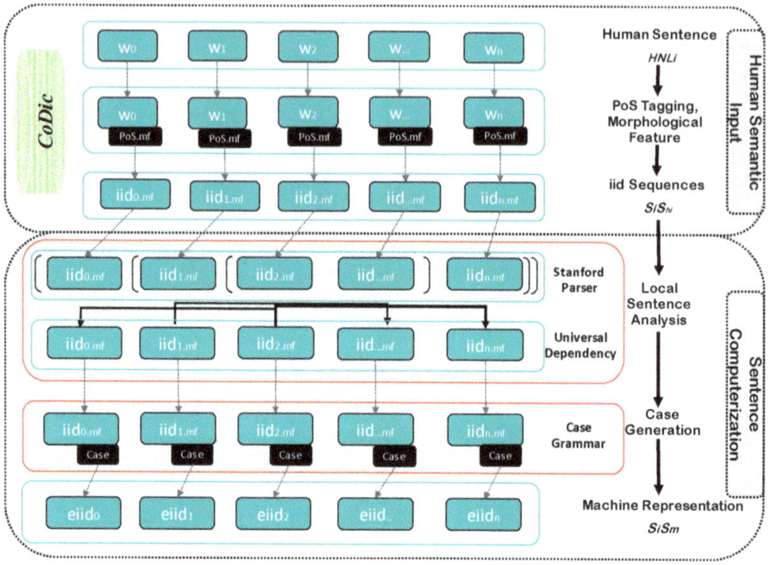

Figure 3. The process of MParser.

3.2. Methodology

The theoretical foundation of MParser comes from the sign description framework (SDF) [26], as shown in Figure 4. It is a language for representing signs in computing systems and is particularly intended to represent the interpreted meanings or ideas of all objects in reality, such as appearing in dictionaries, texts, software, and web pages. A sign: = (*sign*, *denoter*, *reifier*, *denotation*, *connotation*) is modeled by a bi-tree, consisting of three relationships of a *denotation*, a *connotation* and a *reification* between signs.

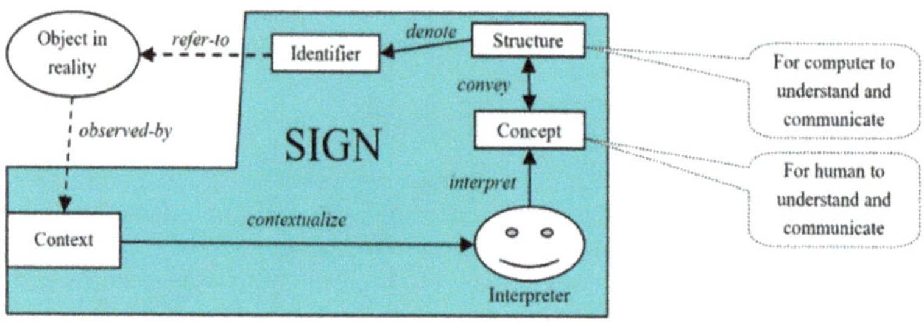

Figure 4. An SDF data model.

A *denotation* is an internal relationship between a sign and its denoter, such that the denoter denotes the properties of a sign. We can understand a denoter as a feature container, containing all features of a sign. For a natural language, these features consist of the form (e.g., iid, term, and pronunciation), sense (i.e., meaning), part of speech (e.g., noun), tense (e.g., past), aspect (e.g., perfective), gender (e.g., male), number (e.g., single), and context (e.g., English). In essence, denotation provides a way to define a sign in the context of a sentence by a set of properties provided by a denoter.

A *connotation* is an external relationship between signs, such that a sign is connoted by a set of signs, which builds a parse tree of a set of signs. For instance, when a set of signs constructs a sentence as a sign in language, it can be parsed through connotation in grammatical cases. For example, we replace the sign of a sentence, and connotation can then parse the sentence sign into many atomic signs.

A *reification* is an instantiation relationship between a reifier (often a particular sign) and a specific denoter (often an abstract sign). For instance, given a denoter denoting the sign of "color", then "white" is a reifier, and between "white" and "color", there is a reification relationship. Or the sign is INT datatype, and 1234 is the reifier.

By generalizing these represented concepts of objects into structured signs, SDF represents all objects in reality, such as objects of abstract and concrete, physical and virtual, and real and fictitious.

CoDic (CoDic http://www.cis.umac.mo/~jzguo/pages/codic/, accessed on 30 August 2021) [17] is a common dictionary and an application of the SDF consisting of 93,546 English words, 20,446 Chinese words, and 190,001 word senses. In CoDic, a concept is a basic element in a sentence and consists of words and phrases. Each concept has already been collaboratively edited without semantic ambiguity. Any dictionary term in CoDic (called a sign) is identified as a unique and internal identifier $iid \in IID$, which is neutral and independent of any natural language and can refer to any term of a natural language. PoS plays a very important role and includes 16 kinds of signs, which are: *Noun(n):= {Common (ncm), Pronoun(npr), Proper Organization (nop), Proper Geography(ngp), Pronoun(npr)}, Verb(v):= {Intransitive (vit), Transitive(vtr), Ditransitive(vdi), Copulative(cop)}, Adjective (adj), Adverb(adv), Preposition (prep), Conjunction(cnj), Interjection (int), Onomatopoeia (ono) and Particle (par)*. For the detailed description of PoS in CoDic, please see Appendix A. Given a simple sign *s = (t, iid) = (icebox, 5107df00b635) = (common noun, "An insulated chest or box into which ice is placed, used for cooling and preserving food.")* as shown in Figure 5. Specifically, the form of the sign is presented as follows:

- **IID**: = **POS**+Y+ID: indicates the universal sign representational form. For instance, iid = 5107df00b635, in which 1 after 5 refers to common noun, 7df refers to year 2015, and 00b635 is ID.
- **Term** indicates literal representational form for a sign, e.g., *"icebox"* is the literal representation of the sign 5107df00b635 in English context.
- **Meaning** is the sense of a sign, e.g., *"An insulated chest or box into which ice is placed, used for cooling and preserving food"* is the sense of 5107df00b635.

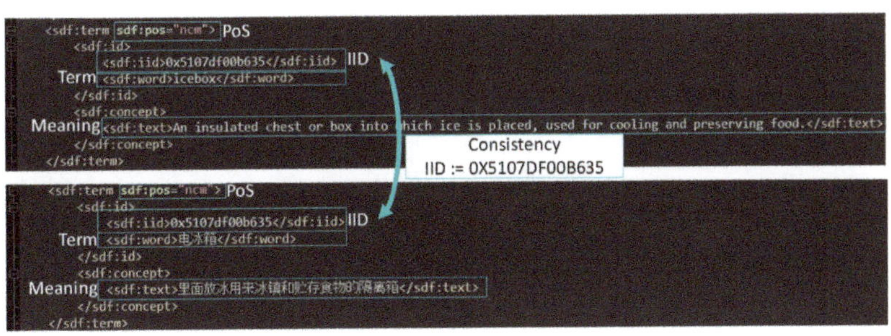

Figure 5. CoDic.

Thus, the meaning of *iid* is: *5107df00b635* = *"icebox"* = "アイスボックス" = "电冰箱" though they are in heterogeneous contexts.

4. Human Semantic Input (HSI)

In human semantic input, the user's initial intention is essential when they try to translate the transmitted concepts into unique semantic representations. If semantics are insufficient for a clear and accurate representation, in that case, the same literal words in users' minds may be different from different contexts between computers and users; it is possible to fail the information interaction because of ad hoc user input. Therefore, HSI tries to solve ad hoc input through a supervised sentence input that cannot casually input the words and phrases in users' minds.

In MParser, all written sentences are constrained by HSI, which is a supervised human-readable sentence via CoDic. We developed an editor to input any term by selecting PoS and the exact meaning, which has a unique identifier (*iid*), to point to the same meaning regardless of contexts. We use a simple English sentence *"I enjoy travel in summer."* to illustrate HSI. First of all, a user types words one by one by selecting terms as shown in Figure 6: the terms *"I"* (*ncm*,$0 \times 5107df00b5e2$), *"enjoy"* (*vtr, 0x5707df00184b*), *"travel"* (*ncm, 0x5107df01848b*), *"in"* (*prep, 0x5a07df000103*), and *"summer"* (*ncm, 0x5107df016d86*).

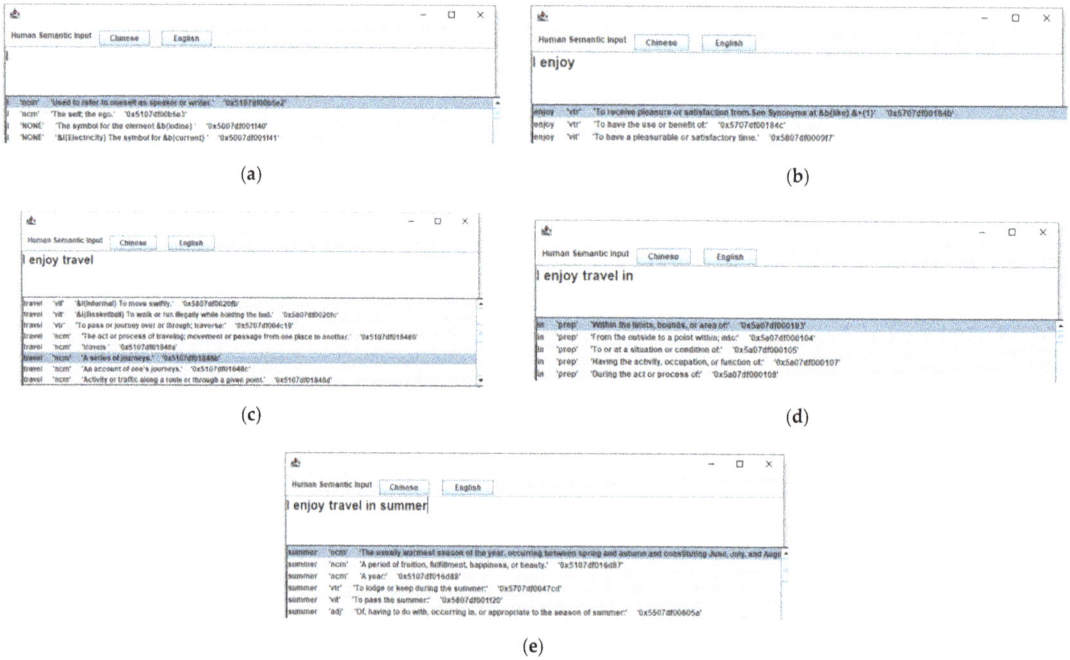

Figure 6. HSI for English sentence "I enjoy travel in summer." (**a**) term "I"; (**b**) term "enjoy"; (**c**) term "travel"; (**d**) term "in"; (**e**) term "summer".

CoDic resources are all on the level of lemmas, and the "term" can be seen as word senses in CoDic, which cannot realize different morphological forms for a word. For instance, in English, the lemma *"enjoy"* yields morphological features: *enjoys, enjoyed, enjoying*. Thus, the morphological feature (*mf*) for each lemma of CoDic is designed and lists the forms needed in each language. The morphologic feature (*mf*) has the *gender* (*G*) and *number* (*N*) features for nouns and the features of *tense* (*T*), *aspect* (*A*) and *voice* (*V*) for verbs. The morphological feature (*mf*) can be different in each language (for details of morphological features, please see Appendix B). The morphological features (*mf*) are parsed according to the local grammar rule because different languages have different morphological phenomena, which are language-dependent for each language. Actually, populating the morphological feature is an engineering effort of its own. In HSI, users manually select the correct feature for each term in the CoDic. Thus, when a user inputs nouns or verbs, he/she needs a second selection for words, including morphologic features (*mf*), as shown in Figure 7.

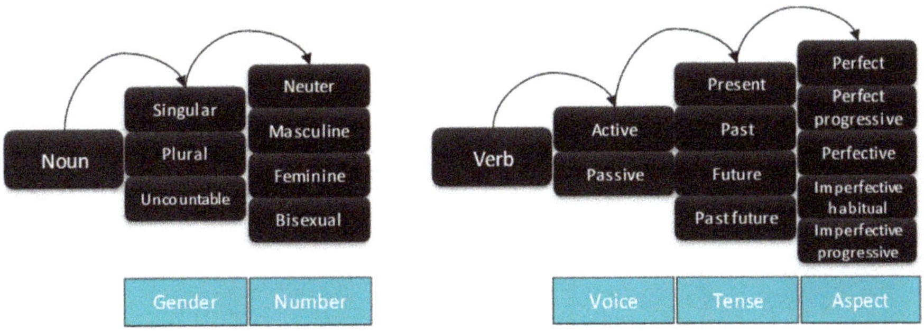

Figure 7. Morphological feature choice in MParser.

Thus, in the example sentence: [('I', 'ncm'), ('enjoy', 'vtr'), ('travel', 'ncm'), ('in', 'prep'), ('summer', 'ncm')], terms *"I"*, *"travel"* and *"summer"* choose *singular* and *neuter* (actually, no gender attribute in English, the default is *neuter*), and the hex is 0 for the noun. The term *"enjoy"* chooses *active present imperfective habitual*, and the hex is 03 for the verb. The morphologic feature identification algorithm is presented in Table 1. Table 2 shows the tenses of a sentence in English and *HSI* through a basic example (*"she go home"*). Following interesting observations from Table 2, it can be observed that helping verbs (***Bold font***) have been removed during the human sentence input for all tenses of verbs. MParser uses only the root form of the verb. These helping verbs, such as *"is, am, be, being, has, had"*, are represented by a hex of morphologic feature (*mf*). Thus, the human input sentence is universal for all languages.

Table 1. Morphologic feature identification algorithm.

```
Function (Input words)
Input
String ← Input word
if (String.pos= "ncm" or "npp" or "ntp") then
Gender(G): = n | m | f | b                        /* Select noun's gender */
Number(N): = s | p | u                            /* Select noun's number */
return ← noun morphological feature (mf)
if (String.pos= "vtr" or "vid" or "vit") then
Tense(T) = present | past | future | past future  /* Select a verb's tense */
Aspect(A) = f | g | w | h | p                     /* Select verb's aspect */
Voice(V): = active | passive                      /* Select verb's voice */
return ← verb morphological feature (mf)
```

Table 2. Human semantic input of tenses in English.

Tense of Sentence	English Sentence	HSI
Past perfect	She ***had gone*** home.	
Future perfect	She ***will have gone*** home.	
Present perfect continuous	She ***has been going*** home.	
Past perfect continuous	She ***had been going*** home.	
Future perfect continuous	She ***will have been going*** home.	She.$_{mf}$ *go*.$_{mf}$ home. (*mf* refers to defined Hex)
Simple present	She ***goes*** home.	
Simple past	She ***went*** home.	
Simple future	She ***will go*** home.	
Present continuous	She ***is going*** home.	
Past continuous	She ***was going*** home.	
Future continuous	She ***will be going*** home.	

HSI converted a sequence of human-readable literals HNL_i to a sequence of signs SiS_{ci} that a computer program can understand without semantic ambiguity. Formally, the concepts are defined below.

Definition 1. *(Human Simple Sentence "SiS_{hi}"): Given a well-formed sequence of literal words $(w_1, \ldots, w_k, \ldots, w_m)$ in HNL_i, input by a human user, in i context (i is English user), then:*

$$SiS_{hi} := (w_{0.mf}, w_{1.mf}, \ldots, w_{k.mf}, \ldots, w_{m.mf}) = \sum_{k=0}^{m} w_k \quad (1)$$

where w_0 is an automatically generated leading word signifying the beginning of a sentence, $0 < k \leq m$ is the word sequence number of the word w_k in SiS_{hi}.

Definition 2. *(Computer Sentence "SiS_{ci}"): Given SiS_{hi}: = $(w_{0.mf}, w_{1.mf}, \ldots, w_{k.mf}, \ldots, w_{m.mf})$ $=\sum_{k=0}^{m} w_k$, then SiS_{hi} is generated into iid sentence, called computer sentence "SiS_{ci}", such that:*

$$SiS_{ci} := (iid_{0.mf}, iid_{1.mf}, \ldots, iid_{k.mf}, \ldots, iid_{m.mf}) = \sum_{k=0}^{m} iid_k \quad (2)$$

where iid_0 is an automatically generated leading word signifying the beginning of a sentence, $0 < k \leq m$ is the iid sequence number of the iid_k in SiS_{ci}. For the result of human semantic input, SiS_{ci} is a supervised computer-readable sentence.

5. Sentence Computerization (SC)

Sentence computerization (SC) transforms a human sentence into a computer sentence. It consists of three main activities: (1) Analyze the constituency structure and universal dependency relationship from the outputting words in the HSI step. (2) Adapt the Stanford parser tool to extract potential relationships between outputting words. (3) Apply predefined case grammar rules to label semantic roles outputting words and generate a universal sentence. The activities involve local sentence analysis described in Section 5.1, case generation described in Section 5.2, and machine representation described in Section 5.3.

5.1. Local Sentence Analysis

Each word is tagged into the PoS and morphological feature (*mf*) in the sentence from the step of HSI. Local sentence analysis identifies the relationship between different words that constitute the English sentence. We adapted the Stanford parser tool [35], which provides full syntactic analysis, minimally a constituency (bracketed sentences) parse of local English sentences between different PoS. Constituency parse describes what the constituents are and how the words are put together. For instance, a sentence: *"the quick brown fox jumps over the lazy dog"* can transform into:

```
[('ROOT', [('NP', [('NP', [('DT', ['the']), ('JJ', ['quick']), ('JJ', ['brown']), ('NN', ['fox'])]), ('NP',
[('NP', [('NNS', ['jumps'])]), ('PP', [('IN', ['over']), ('NP', [('DT', ['the']), ('JJ', ['lazy']), ('NN',
['dog'])])])])])])]
```

The bracketed sentence represents grammatical functions, such as NP, VP, and PP based on English grammar. However, Stanford parser adapts the Penn PoS tagger rather than the CoDic PoS tagger, such that it is impossible to parse the sentence directly. Thus, we built a mapping between Penn PoS tagger and our CoDic PoS tagger, and the PoS mapping algorithm is shown in Table 3. In particular, particle words, which are only for the local English language, have no common *iid* to map other languages and will not appear in the final machine representation.

Table 3. PoS mapping algorithm.

```
if (isCoDicPos)
if (CoDicpos =par and iid= "xxx") {
Stanfordpos = "xxx";
} else if (CoDicpos = noun or verb and mf= "xxx") {
Stanfordpos = "xxx" or insert words and Stanfordpos = "xxx";
} else if (CoDicpos = other PoS) {
Stanfordpos = "xxx";
} else
print ="error"
end if; }
```

Stanford parser presents and parses a word's relationship by a pure constituency, but ignores their semantic role. For example, SVO (subject-verb-object) structure is presented as S → NP VP NP by the Stanford parser, and it is impossible to parse subject, object, and other semantic roles in a sentence. Nivre et al. [12] proposed a universal dependency (UD) that uses dependency labels and PoS tags to parse sentences for different languages. The UD annotation defines a classification of around 40 relations as the universal dependency label sets (https://universaldependencies.org/#language-tagset, accessed on 30 August 2021), such as *nsubj*: nominal subject, *amod*: adjectival modifier. Thus, when the UD appeared, it immediately became interesting to see its relationship with the Stanford parser. For instance, the sentence "*the quick brown fox jumps over the lazy dog*" can transform into:

[[((u'jumps', u'VBZ'), u'nsubj', (u'fox', u'NN')), ((u'fox', u'NN'), u'det', (u'The', u'DT')), ((u'fox', u'NN'), u'amod', (u'quick', u'JJ')), ((u'fox', u'NN'), u'amod', (u'brown', u'JJ')), ((u'jumps', u'VBZ'), u'nmod', (u'dog', u'NN')), ((u'dog', u'NN'), u'case', (u'over', u'IN')), ((u'dog', u'NN'), u'det', (u'the', u'DT')), ((u'dog', u'NN'), u'amod', (u'lazy', u'JJ'))]]

Finally, through Stanford Parser and UD, the local English sentence becomes a segmented sentence with dependency relationships for each word, as shown in Definition 3.

Definition 3. (*Segmented Simple Sentence "$SiS^q{}_{ci}$"*): *Given* SiS_{ci}: = $(iid_{0.mf}, iid_{1.mf}, \ldots, iid_{k.mf}, \ldots, iid_{m.mf}) = \sum_{k=0}^{m} iid_k$, *then* SiS_{ci} *is segmented into* $q + 1$ *subsequences, called q-subsequences* $SiS^q{}_{ci}$. *Each subsequence has p number of iid, such that:*

$$Segment: (iid_{0.mf}, (iid_{1.mf}, \ldots, iid_{p.mf})_1, \ldots, (iid_{1.mf}, \ldots, iid_{p.mf})_i, \ldots, (iid_{1.mf}, \ldots, iid_{p.mf})_q \leftarrow \sum_{k=0}^{m} iid_k \tag{3}$$

$$\begin{aligned} SiS^q{}_{ci} = Segment\ (iid_{0.mf}, iid_{1.mf}, \ldots, iid_{k.mf}, \ldots, iid_{m.mf}) = (iid_{0.mf}, (iid_{1.mf}, \ldots, iid_{p.mf})_1, \ldots, \\ (iid_{1.mf}, \ldots, iid_{p.mf})_i, \ldots, (iid_{1.mf}, \ldots, iid_{p.mf})_q) \end{aligned} \tag{4}$$

$$SiS^q{}_{ci} = iid \sum_{i=1}^{q} (\sum_{j=1}^{p} iid_j)_i \tag{5}$$

where the length of i-th subsequence $(iid_{1.mf}, \ldots, iid_{p.mf})_i = \sum_{j=1}^{p} iid_{j.mf}$ *is* p $(1 \leq p \in \mathbb{N})$.

5.2. Case Generation

MParser grammar is a set of machine natural language grammars such as universal grammar (UG) and case grammar (CG), originating from Fillmore's case study [36,37]. MParser grammar specifies various sequences of signs, forming a general natural language commonly read and understood both by humans and computer systems. It consists of morphological features (intrinsic) (discussed in HSI) and case grammar components (extrinsic). The morphological component varies from one language to another regarding the sets of morphological features, which are inflection forms themselves, but uses common naming conventions. Each case label either presents a syntactic, semantic, or computational function or marks a grammatical function in general and abstracts a particular grammatical phenomenon pertaining to a group of words, phrases, sentences, or others that appeared in natural languages.

In our previous work [16], we proposed a case grammar representing a universal and deep case (or semantic roles) that reflects in a sentence as the central means of explaining both the syntactic structure as well as the meaning of sentences. The case grammar component displays a common representation of syntactic structures and structural words and can be used as a resource for language processing tasks, such as translation, multilingual generation, and machine inference. The novel available cases are defined as follows:

- **Nominative Case** (NOM): denotes a semantic category of entities that initiate actions, trigger events, or give states. Nominative case often associates with the agentive properties of volition, sentience, instigation, and motion.
- **Predicative case** (PRE): denotes a semantic category of process in terms of action, event, or state. The process starts from a sign in the semantic category of the nominative.
- **Accusative case** (ACC): denotes a semantic class of patients who are the participants affected by the semantic class of agents marked by agentive case, which is the direct object of an agentive action.
- **Dative case** (DAT): denotes a semantic class of indirect participants relevant to an action or event. The objective participant marked by dative is called recipient or beneficiary of an action.
- **Genitive case** (GEN): denotes a semantic category of attributes that belong to things. It describes an attributive relationship of one thing to another thing.
- **Linking case** (LIN): denotes the thing that corresponds to the theme of thematic nominatives, such as attributes, classification, or identification of a theme.
- **Adverbial case** (ADV): denotes a semantic category of constraints belonging to predicative signs (i.e., a verb). It corresponds to the adverbial syntactic case.
- **Complementary case** (COM): denotes additional attributes of an entity, an action, an event, or a state, such as means, location, movement, time, causality, extent, and range. Under the PRE structure, COM is shown in COMv form. Under the NOM/ACC/DAT structure, COM is shown in COMn form. For other situations, it just shows COM form.

In this paper, cases are labels or tags that mark signs' syntactic, semantic, and computational functions in the marked forms such as marked words, phrases, and sentences within a natural language's text. For example, in the sentence *"earth moves around sun"*, the behavior *"move"* is performed by the entity *"earth"* and the behavioral method is *"around the sun"*. A case is used to label the functionality of a word or a phrase in the sentence, such as *"NOM.earth PRE.moves COMv.around NOM.sun"*. The universal case grammar provides a common grammar transformable to the grammar of any existing natural language.

Tree Generation

Case generation converts a sequence of single concepts (i.e., atomic signs) into complex concepts (i.e., a compound sign), that is self-described. MParser builds a sentence-based case concept associated with an *iid* defining how an *iid* grammatically functions and combines with other *iids* by the case grammar. It does not need to consider the order of the sentence, which is a bag of concepts. The key of case generation to a sign lies in two facts:

(1) There is a known PoS already associated with the term (HSI);
(2) The term has a clear grammatical relationship with other terms in a sentence (local sentence analysis).

A sentence is defined as a sequence of signs, each marked with a functionality label defined as a case. Each sign in a sentence can describe its case grammar relationship with other signs; that is the compound sign, called *SignX*, which is

Sentence :: = $\mathbf{SignX_1 \ldots SignX_i \ldots SignX_n}$

$\mathbf{SignX = IID.C_1 \ldots IID.C_i \ldots IID.C_n}$

For example:

[('**NOM**', [('**GEN**', ['the']), ('**GEN**', ['quick']), ('**GEN**', ['brown']), ('**NOM**', ['fox'])]), ('**PRE**',[('PRE', ['jump'])]), ('**COMv**', [('**COMv**', ['over']), ('**ACC**', [('**GEN**', ['the']), ('**GEN**', ['lazy']), ('**ACC**', ['dog'])])])])]

NOM and ACC cases are appending for nouns such as the words *"fox"* and *"dog"*, PRE case is appending for verbs such as the word *"jump"*. Thus, the case generation (fox_NOM (jump_PRE)) yielding the English *"fox jump"* can be turned into Chinese by just changing the lexical item: (狐狸 _NOM (跳_PRE)) yielding "狐狸跳". The case is appending NOM and PRE to form correct sentences in both languages. Meanwhile, the morphology feature (*mf*) builds inflection features for nouns and verbs in both languages.

Based on sign theory [28], every concept (e.g., *fox, jump)* is a meaning group, which appends a single case (e.g., NOM, PRE) to modify a larger meaning group in a tree of concepts. If each concept in a sequence is unique, then the sequence is also unique. The tree is defined as $T = (N, E)$, where N indicates a group of nodes, and E indicates a group of edges, where $E \subseteq N \times N$. The path in a tree is a sequence of nodes $n_1, n_2, \ldots, n_{k-1}, n_k$, where each pair (n_1, n_2) has $e(n_1, n_2) \in E$. A cycle is a path $n_1, n_2, \ldots, n_{k-1}, n_k$ $(k > 2)$ that consists of distinct nodes, except $n_1 = n_k$. In our tree generation, we present a sentence in a tree-based *SignX* representation as T_{SignX}. Nodes N contains two main types: *iid* node N_{iid} and case node N_c. Formally, the node-set is:

$$N = \{N_{iid}, N_c \mid iid \in IID, c \in C\}$$

where IID is a group of all words' *iids* in the sentence, and each *iid* in the local sentence is represented as a node in the T_{SignX}. C is a group of predefined case concepts, including NOM, PRE, and so on. Additionally, edges E link any two nodes in a tree, where:

$$E \subseteq \{n_f, n_c \mid f, c \in N\}$$

An important principle is designed in sentence construction, which is the *father–child* relationship. Each edge $e(n_f, n_c)$, where $n_f, n_c \in N$ is connected with a father–child relationship that represents the structure relationship between its two connected nodes n_f and n_c —whether a father code (*f*) is modified by another child node (*c*) or not, while the father node proceeds. A *father node* is a key sentence constituent. Differently, a *child node* is always dependent and belongs to a father node. This correspondence can be illustrated in Figure 8. Applying this principle, we can always construct a sequence of sentences in different order of atomic concepts but still ensure structural equivalence.

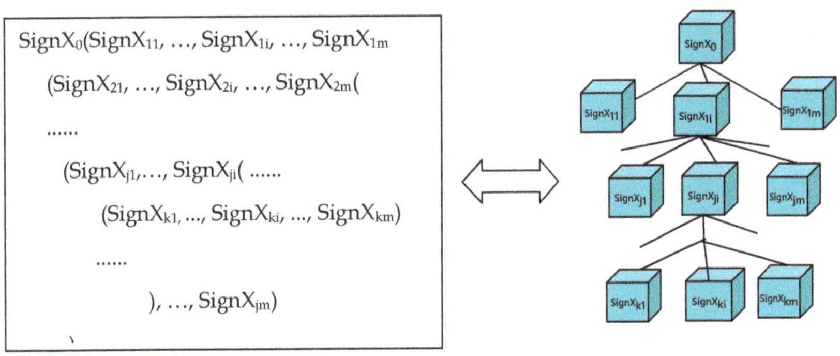

Figure 8. MParser SignX Tree T_{SignX}.

The case generation is converted into T_{SignX} using the tree generation algorithm. T_{SignX} provides a phrase-based structure, such as SVO, OVS sequences, and case labeling, and is a non-redundant representation. The T_{SignX} Tree algorithm is derived as follows:

1. *Linearize input to a term sequence S.*
2. *Connect each term in S to its smallest subtree in* T_{SignX}.

3. *Append one case in each node of T_{SignX} based on case grammar rules.*
4. *Parse the universal dependency labels at each branching node N of the T_{SignX}.*
5. *Find the dependency relationship in the node of each word:*
   a. *If exist corresponding dependency label, then replace the current case using dependency mapping rules;*
   b. *If no dependency relationship, keep the current case.*

The proposed T_{SignX} model can represent different sentences with the same tree if they have the same semantics. Because the order of the words does not affect its representation, it reduces the influence of language, which has the property of flexible order. A sentence becomes a case sentence through the case generation, which appends a case concept for each word, as shown in Definition 4.

Definition 4. (*Case Sentence "SiS_c"*): *Given* $SiS^q{}_{ci} = (iid_{0.mf}, (iid_{1.mf}, \ldots, iid_{p.mf})_1, \ldots, (iid_{1.mf}, \ldots, iid_{p.mf})_i, \ldots, (iid_{1.mf}, \ldots, iid_{p.mf})_q)$, *then* $SiS^q{}_{ci}$ *is appended cases for the sentence, called Case Sentence* SiS_c. *Each word has one case, such that:*

$$SiS_c = (iid_{0.mf.C}, (iid_{1.mf.C}, \ldots, iid_{p.mf.C})_1, \ldots, (iid_{1.mf.C}, \ldots, iid_{p.mf.C})_i, \ldots, (iid_{1.mf.\,.C}, \ldots, iid_{p.mf.\,.C})_q) \quad (6)$$

where the length of i-th subsequence $(iid_{1.mf.C}, \ldots, iid_{p.mf.C})_i = \sum_{j=1}^{p} iid_j$ *is* p $(1 \leq p \in \mathbb{N})$, *and C is appended case.*

5.3. Machine Representation

After attaching a case to a word, a machine universal language representation shows a computer-readable and -understandable sentence without huge extra data to process it.

Definition 5. (*Computer-Understandable Simple Sentence "SiS_m"*): *Given a sign-based sentence* $SiS_c = (iid_{0.mf.C}, (iid_{1.mf.C}, \ldots, iid_{p.mf.C})_1, \ldots, (iid_{1.mf.C}, \ldots, iid_{p.mf.C})_i, \ldots, (iid_{1.mf.\,C}, \ldots, iid_{p.mf.C})_q)$, SiS_m *is a set of extend iid, called eiid, such that:*

$$SiS_m = (S, eiid_1, \ldots, eiid_k, \ldots, eiid_n) \quad (7)$$

where an extended iid (eiid):

eiid : = Term.iid.mf.Case.F.C

(***term*** and "***iid***" refers to a sense in CoDic, "*PoS*" is already defined in *iid*, ***mf*** refers to morphological feature, ***F*** is the index of the higher level father sign in MParser tree, and index of the lower level child node "***C***" in MParser tree). Additionally, the machine representation referring to PoS is defined:

(1) If *PoS* is noun, *eiid* = Term.IID.mf.Case.F.C, in which mf refers to the morphological feature of the noun.
(2) If *PoS* is verb, *eiid* = Term.IID.mf.Case.F.C, in which mf refers to the morphological feature of the verb.
(3) If *PoS* is adjective | adverb | prep | conjunction | … , *eiid* = Term. IID.Case.F.C;
(4) If *PoS* is a particle, delete the node. (Unlike a noun or a verb, a particle is localized and meaningless in a sense for other languages, only confers a local grammatical meaning, and it is not possible to map it to other languages.)

Finally, through the machine representation activity, a sentence becomes a bag of semantic concepts without considering the sequence of the sentence through term index and can be self-described for understanding by computers.

6. Implementation

The MParser is implemented in Python and Java under macOS version 11.0.1 system, and runs under python 3.7 and JDK 1.8. CoDic is represented in XML format for

English and Chinese. In addition, Stanford Parser and universal dependency APIs are called by MParser. In the implementation, several sentences are processed and analyzed to describe how to represent a sentence and maintain semantic consistency from English sentences. In MParser, the user first types words one by one by selecting terms and additional morphological features such as "*I enjoy travel in summer*" in the HSI step. By calling the constructInfo function in MParser, the sentence is generated into:

constructInfo [('I', 'ncm', '0x5107df00b5e2', '0'), ('enjoy', 'vtr', '0x5707df00184b', '03'), ('travel', 'ncm', '0x5107df032b53', '0'), ('in', 'prep', '0x5a07df000103', ''), ('summer', 'ncm', '0x5107df016d86', '0')]

The step of ***constructInfo*** constructs the information for each typed word in HSI, such as term, PoS, iid and morphological features for nouns and verbs. Next, the sentence goes to the *Sentence Computerization* step, which is an automated analysis without user participation. ***ParserList*** function of MParser calls the Stanford Parser API to construct a phrase-based structure sentence based on predefined PoS tagger mapping rules between Stanford Parser and CoDic:

paserList ['(ROOT', ' (S', ' ((ncm I))', ' ((vtr enjoy)', ' ((ncm travel))', ' ((prep in)', ' ((ncm summer))))', ' (. .)))']

Meanwhile, a universal dependency is parsed by calling the ***dependency_parse*** function in MParser, and finding each word dependency relationship by the ***everyWordDep*** function in the sentence:

dependency_parse [('ROOT', 0, 2), ('nsubj', 2, 1), ('dobj', 2, 3), ('case', 5, 4), ('nmod', 2, 5), ('punct', 2, 6)]
everyWordDep {'I': 'nsubj', 'travel': 'dobj', 'in': 'case', 'summer': 'nmod', '.': 'punct'}

After the local sentence analysis, the English sentence includes phrase-based structure and dependency semantic roles. Then, the sentence is analyzed based on case rules:

(1) This sentence begins from an S, which is a *declarative* sentence.
(2) The noun (*ncm*) *we* is case *NOM* [*I-ncm-NOM*] if it is before a verb such at *ncm-NOM* $\leftarrow$ *vtr-PRE* (except GEN, ADV and others).
(3) The verb (*v*) *enjoy* is case *PRE* [*enjoy-vtr-P*] where *transitive verb* (*vtr*) follows only one *noun* structure, such that *vtr-PRE* $\rightarrow$ *vtr-PRE noun* [*supplementary*: *vit-PRE*; *vdi-PRE* $\rightarrow$ *vdi-PRE* $noun_1$ $noun_2$.]
(4) The noun (*ncm*) *travel* is case *ACC* [*travel-ncm-ACC*] if it is before a vtr verb such at *vtr-PRE* $\leftarrow$ *ncm-ACC*.
(5) The preposition (*prep*) *in* is case *COMv* [*in-prep-COMv*] under PRE structure.
(6) The noun (*ncm*) *summer* is case *NOM* [*summer-ncm-NOM*], such that *in-prep-COMv*$\leftarrow$ *summer-ncm-NOM*.

We applied our case grammar rules to generate the MParser tree. The tree visualizations are presented by NLTK API (NLTK API: http://nltk.org). Figure 9 shows the structure and tree screenshot from MParser.

Finally, the machine representation generated a universal sentence:

S.0.0(I.0x5107df00b5e2.0.NOM.0.1(enjoy.0x5707df00184b.03.PRE.0.1(travel.0x5107df01848b.0.ACC.1.2(in.0x5a07df000103.COMv.1.2(summer.0x5107df016d86.0.NOM.2.3)))))

The universal sentence presents a sequence of extracted meaningful concepts related to each other using cases and syntactical relationships. The sentence also can map into Chinese words for Chinese CoDic via unique *iid*. An illustration shows a transformation from local English HNL (*i*) to a universal sentence, then Chinese HNL (*j*) in Table 4.

```
(ROOT
  (S
    (NOM (ncm I))
    (PRE (vtr enjoy)
      (ACC (ncm travel))
      (COMv (prep in)
        (NOM (ncm summer))))
  (. .)))
```

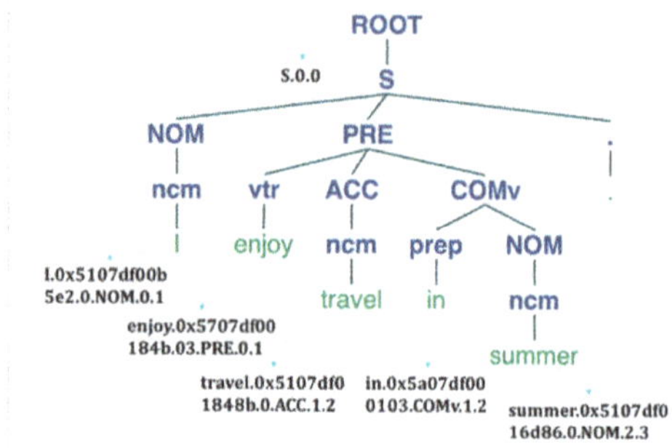

Figure 9. Structure and MParser tree for English sentence "I enjoy travel in summer".

Table 4. Transformation from English to Chinese in MParser.

I	Enjoy	Travel	In	Summer	English(HNL$_i$)
0x5107df00b5e2	*0x5707df00184b*	*0x5107df01848b*	*0x5a07df000103*	*0x5107df016d86*	*iid*
0x5107df00b5e 2.0.NOM.0.1	***0x5707df00184 b.03.PRE.0.1***	***0x5107df01848 b.0.ACC.1.2***	***0x5a07df00010 3.COMv.1.2***	***0x5107df016d 86.0.NOM.2.3***	*eiid*
我	享受	旅程	在	夏天	*Chinese (HNL$_j$)*

First, the English sentence is converted to machine-readable *iid* sequences from English CoDic. Then, through case generation and machine representation steps, the English computer-understandable sentence is converted into a universal computer-readable and -understandable *eiid* sentence that is a bag of unique concepts. Finally, the *eiid* sentence can be translated into another language such as Chinese based on local rules. MParser ensures that any sentence in an HNL$_i$ can be transformed into HNL$_j$ without any semantic loss.

We also tested a passive sentence in English to illustrate the difference between NOM and ACC from the semantic role, which is "*dog is hit by man heavily.*", as shown in Figure 10.

From the example, we found that "*dog*" is ACC, and "*man*" is NOM in a passive sentence, and they meet the standard semantic role for a passive sentence. The PoS of the word "*is*" is null since it is inserted during local sentence analysis, not from CoDic, and it does not appear in final machine representation. We illustrate from tenses of three English sentences, shown in Table 5.

Table 5. Tense test of MParser in English.

HSI	English Sentence Analysis	Machine Representation
I.0 go.00 home.0 (*I **have gone** home.*)	**I**/NN (have/VBP) **go**/VBN **home**/NN.	S.0.0(I.0x5107df00b5e2.0. NOM.0.1(go.0x5707df00203d. 00.PRE.1.2(home.0x5107df00afcc.0.ACC.2.3)))
I.0 go.40 home.0. (*I **have been going** home.*)	**I**/NN (have/VBP been/VBN) **go**/VBN **home**/NN.	S.0.0(I.0x5107df00b5e2.0.NOM.0.1(go.0x5707df00203 d.40.PRE.1.2(home.0x5107df00afcc.0.ACC.2.3)))
I.0 go.04 home.0 (*I **am going** home.*)	**I**/NN (is/VBP) **go**/VBG **home**/NN.	S.0.0(I.0x5107df00b5e2.0. NOM.0.1(go.0x5707df00203 d.04.PRE.1.2(home.0x5107df00afcc.0.ACC.2.3)))

constructInfo [('dog', 'ncm', '0x5107df006e14', '0'), ('hit', 'vtr', '0x5707df0022b5', '42'), ('by', 'prep', '0x5a07df0000b3', ''), ('man', 'ncm', '0x5107df00dbf4', '0'), ('heavily', 'adv', '0x5607df00031b', '')]
paserList ['(ROOT', ' (S', ' ((ncm dog))', ' ((null is)', ' ((vtr hit)', ' ((prep by)', ' ((ncm man)))', ' ((adv heavily))))', ' (. .)))']
dependency_parse [('ROOT', 0, 3), ('nsubjpass', 3, 1), ('auxpass', 3, 2), ('case', 5, 4), ('nmod', 3, 5), ('advmod', 5, 6), ('punct', 3, 7)]
everyWordDep {'dog': 'nsubjpass', 'is': 'auxpass', 'by': 'case', 'man': 'nmod', 'heavily': 'advmod', '.': 'punct'}

```
(ROOT
  (S
    (ACC (ncm dog))
    (PRE (null is)
      (PRE (vtr hit)
        (COMv (prep by)
          (NOM (ncm man)))
      (ADV (adv heavily))))
  (. .)))
```

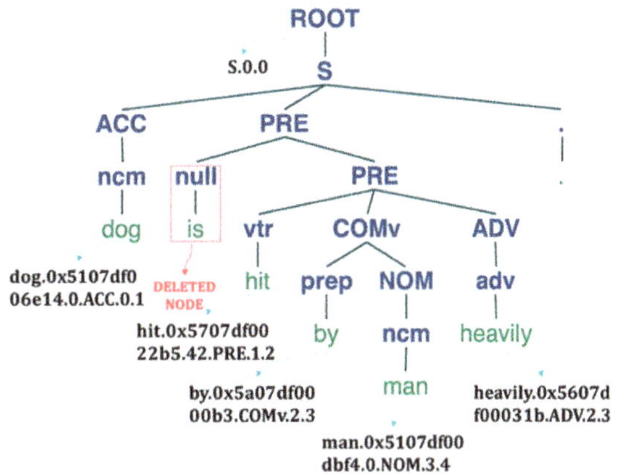

S.0.0(dog.0x5107df006e14.0.ACC.0.1(hit.0x5707df0022b5.42.PRE.1.2(by.0x5a07df0000b3.COMv.2.3(man.0x5107df00dbf4.0.NOM.3.4(heavily.0x5607df00031b.ADV.2.3)))))

Figure 10. MParser for English sentence "dog is hit by man heavily".

7. Evaluation

Human manual evaluation is the crucial and ultimate criterion for validating semantic case labeling given our definition of semantics as a meaning as it is understood by a language speaker [38]. In this research, MParser was evaluated using intrinsic and extrinsic evaluation. Intrinsic evaluation (reader-focused) aimed to evaluate the properties of MParser output by asking participants about the degree of semantic expressiveness of the output in a questionnaire. The extrinsic (expert-focused) evaluation aimed to evaluate the agreement rate of case labeling between MParser outputs and experts.

7.1. Dataset

In our experiment, we randomly selected 100 sentences from a dataset (https://www.kaggle.com/c/billion-word-imputation/data, accessed on 30 August 2021)[39],which is a large corpus of English language sentences, to manually input each word for each sentence in MParser, and finally output 75 retained sentences ($N = 75$) (please see Appendix C for 75 automatic sentence outputs from MParser) because we removed some unrecognizable words from CoDic and unparseable sentences. Taking into account the validity of the questionnaire, we divided the 75 sentences ($N = 75$) into 5 groups (each with 15 sentences ($N = 15$)), which were Group A, B, C, D, and E. Table 6 shows our test dataset, which were 50 short sentences with less than 8 words and 25 long sentences with more than 8 words.

Table 6. Number of MParser outputs.

Sentence Type	Number of Words					Total
	Group A (N = 15)	*Group B (N = 15)*	*Group C (N = 15)*	*Group D (N = 15)*	*Group E (N = 15)*	
***Short Sentence** (length <= 8, N = 50)*	*46 (N = 10)*	*59 (N = 10)*	*50 (N = 10)*	*54 (N = 10)*	*60 (N = 10)*	**269**
***Long Sentence** (length > 8, N = 25)*	*45 (N = 5)*	*44 (N = 5)*	*45 (N = 5)*	*51 (N = 5)*	*51 (N = 5)*	**236**
Total	***91***	***103***	***95***	***105***	***111***	***505***

7.2. Experiment Settings

Intrinsic: An intrinsic (reader-focused) design usually requires a larger sample of (non-expert) participants. In order to investigate judgments of the semantic expressiveness of MParser outputs, we used 154 valid participants to judge the degree of semantic expressiveness for 75 generated sentences through a questionnaire [40]. The semantic expressiveness criterion was: *"how clear is it to understand what is being described"* or *"how clear it would be to identify the case label from the description"*. We adapted the 5-point Likert scale of semantic expressiveness, as follows:

1. Very unclear 2. Unclear 3. Acceptable 4. Clear 5. Perfectly clear

Readers were from cohorts of undergraduate and graduate students pursuing English-related degrees. Before completing the questionnaire, they were expected to understand the attributes of each MNL case label; each group required at least 25 readers to complete.

Extrinsic: In the semantic case labeling evaluation, ideally, by asking the annotator to make some semantic prediction or annotation based on pre-specified criteria and comparing it with the case extracted from the proposed method, the degree of agreement between the proposed method and the expert's annotation could be determined. Thus, a small number of expert annotators were recruited to label cases of the MParser [41]. We used three experts, two Ph.D. students majoring in an English linguistics-related research area, and one university English lecturer to label the 75 sentences. Before labeling, they were required to fully understand the description of attributes of each MNL case through learning case grammars. Additionally, five groups of sentences (each with 15 sentences) required three experts to be completed. This meant that every expert needed to label 75 sentences. To facilitate labeling by the experts and compare it to test data of MParser, we split each word of each sentence, and the experts only needed to select the case for each word. We measured pairwise agreement of extrinsic evaluation among experts and MParser outputs using the kappa coefficient (κ), which is widely used in computational linguistics for measuring agreement in category judgments [42]. It is defined as

$$K = \frac{P(A) - P(E)}{1 - P(E)} \tag{8}$$

where *P(A)* is the observed agreement rate of case labeling for one annotator such as expert 1, and *P(E)* is the expected agreement rate for another expert 2. The simple Kappa coefficient adapts binary classification. Thus, case labeling was achieved by a binary classification where each case has *Yes* (1) or *No* (0). For example, a NOM case label might be NOM case (1) or non-NOM case (0) in one word for annotators. We calculated κ from two aspects: inter-annotator agreement and intra-annotator agreement. Inter-annotator agreement was calculated for 75 sentences, which were annotated by two experts. Intra-annotator agreement followed a similar process but was calculated for 75 sentences that were annotated between expert and MParser outputs. The interpretation standard of Kappa varied (−1 to 1) according to Landis and Koch [43]: ***<0 Poor** | **0−0.2 Slight** | **0.2−0.4 Fair** | **0.4−0.6 Moderate** | **0.6−0.8 Substantial** | **0.8–1 Perfect.***

7.3. Results

From Table 7 and Figure 11, the judgments of semantic expressiveness indicated that MParser had better results since *Clear* and *Perfectly clear* had the largest percentage overall. Additionally, the *Perfectly clear* percentage between short sentences ($N = 50$) and long sentences ($N = 25$), at 44% and 23%, respectively, indicated that performance with short sentences was more significantly clear in semantic expressiveness.

Table 7. The judgements of semantic expressiveness in intrinsic evaluation.

	Perfectly Clear	*Clear*	*Acceptable*	*Unclear*	*Very Unclear*	*Total*
Group A	119	152	128	21	15	435
Group B	130	192	98	9	6	435
Group C	143	223	109	4	1	480
Group D	127	166	90	3	4	390
Group E	139	226	101	11	3	480
	30%	**43%**	**24%**	**2%**	**1%**	**2220**

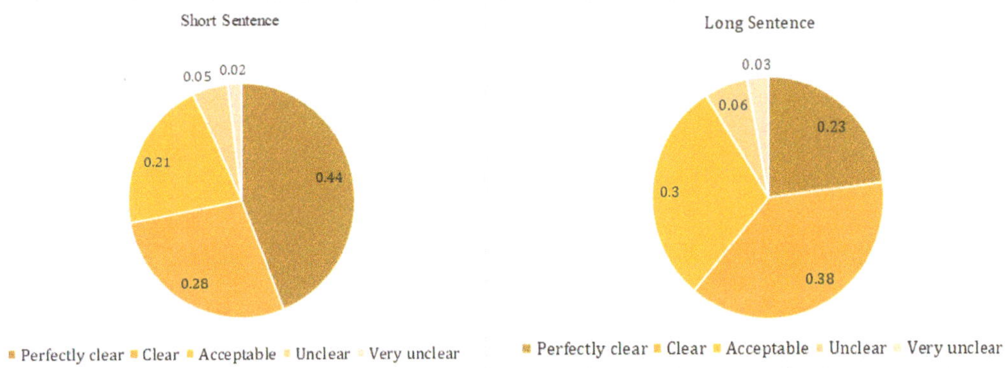

Figure 11. The percentage of semantic expressiveness for short and long sentences in intrinsic evaluation.

Table 8 shows the experimental results using MParser and human expert labeling. The average κ values were 0.693 for inter-annotator agreement and 0.717 for intra-annotator agreement. As $0.6 < \kappa < 0.8$ indicates substantial agreement, the empirical results showed good consistency between the predictions generated by our approach and those of experts. The analysis of the κ values between three experts found that the agreement κ values for experts 2 and 3 were relatively higher. Experts 1 and 2, 3 had a slight gap, but the κ values were still within the range $0.6 < \kappa < 0.8$. Table 8 found that experts 2 and 3 had higher average κ values than expert 1 in intra-annotator agreement. In addition, we calculated average κ values for intra-annotator agreement between short sentences and long sentences, as shown in Table 9. The average κ value for long sentences was significantly lower than that for short sentences. This result is consistent with the trend for our intrinsic evaluation, which showed that the higher complexity of a sentence was more likely to cause disagreement in case grammar labeling. In summary, comparing expert and MParser outputs, inter-annotator and intra-annotator agreement presented substantial results, and there was no major disagreement between our MParser results and those of the experts.

Table 8. Kappa agreement between experts and Mparser.

Group	Inter-Annotator (Expert, Expert)	κ *	$\kappa_{avg.}$	Intra-Annotator (Expert, MParser)	κ *	$\kappa_{avg.}$
Group **A** (*N* = 15)	*(Expert 1, Expert 2)*	*0.688*	*0.741*	*(Expert 1, MParser)*	*0.637*	*0.753*
	(Expert 2, Expert 3)	***0.831***		*(Expert 2, MParser)*	***0.853***	
	(Expert 1, Expert 3)	*0.703*		*(Expert 3, MParser)*	*0.768*	
Group **B** (*N* = 15)	*(Expert 1, Expert 2)*	*0.597*	*0.663*	*(Expert 1, MParser)*	*0.537*	*0.668*
	(Expert 2, Expert 3)	***0.766***		*(Expert 2, MParser)*	*0.685*	
	(Expert 1, Expert 3)	*0.627*		*(Expert 3, MParser)*	***0.781***	
Group **C** (*N* = 15)	*(Expert 1, Expert 2)*	*0.648*	*0.642*	*(Expert 1, MParser)*	*0.603*	*0.734*
	(Expert 2, Expert 3)	***0.673***		*(Expert 2, MParser)*	*0.779*	
	(Expert 1, Expert 3)	*0.605*		*(Expert 3, MParser)*	***0.821***	
Group **D** (*N* = 15)	*(Expert 1, Expert 2)*	*0.694*	*0.687*	*(Expert 1, MParser)*	*0.613*	*0.724*
	(Expert 2, Expert 3)	***0.775***		*(Expert 2, MParser)*	***0.835***	
	(Expert 1, Expert 3)	*0.593*		*(Expert 3, MParser)*	*0.724*	
Group **E** (*N* = 15)	*(Expert 1, Expert 2)*	*0.686*	*0.730*	*(Expert 1, MParser)*	*0.616*	*0.706*
	(Expert 2, Expert 3)	***0.837***		*(Expert 2, MParser)*	*0.749*	
	(Expert 1, Expert 3)	*0.668*		*(Expert 3, MParser)*	***0.753***	
Avg.	***Substantial***		***0.693***	***Substantial***		***0.717***

* *p* value < 0.001.

Table 9. Kappa intra-annotator agreement between short and long sentences.

Sentence Type	Inter-Annotator (Expert, MParser)	$\kappa_{avg.}$
All Sentences *(N = 75)*	*(Expert 1, MParser)*	*0.601*
	(Expert 2, MParser)	***0.780***
	(Expert 3, MParser)	*0.769*
Short Sentence *(length <= 8) (N = 50)*	*(Expert 1, MParser)*	*0.728*
	(Expert 2, MParser)	***0.834***
	(Expert 3, MParser)	*0.819*
Long Sentence *(length > 8) (N = 25)*	*(Expert 1, MParser)*	*0.505*
	(Expert 2, MParser)	***0.726***
	(Expert 3, MParser)	*0.719*

7.4. Discussion

7.4.1. Case Labeling

From the experimental results in 7.3, we can see that our MParser had better results. We also calculated each case match rate (MR) for all words (*N* = *505*) between experts and MParser outputs as the ratio of *MatchedCase* to *TotalCase*.

From the results shown in Table 10, we found that PRE and GEN cases had extremely high MRs, which were 0.986 and 0.959, respectively. ADV, ACC, and LIN cases came next. The MR of DAT was relatively low because of the differences in the judgment of the infinitive. To our surprise, the MR of the NOM case was relatively low. Through one-to-one analysis of sentences, we found that when nouns were under the COM (COM_n/COM_v) structure, some experts still labeled the COM case for nouns, and our MParser identified the nouns as NOM case. For COM, COM_n, and COM_v cases, the MR was not very high because the experts had different labels on which COM case to use for prepositions. However, if the COM case was considered a general COM case, COM_{all}, the average of the MR achieved a very high score, which was 0.920, indicating a consensus on the COM case.

Table 10. Case Match Rate (MR) between Experts and MParser outputs.

Intra-Annotator (Expert, MParser)	MR (*N* = 505)											
	NOM	*PRE*	*ACC*	*DAT*	*GEN*	*LIN*	*ADV*	*COM*	*COMv*	*COMn*	COM_{all}	*Avg.*
(Expert 1, MParser)	0.684	0.979	0.804	0.647	0.958	0.756	0.840	0.682	0.649	0.690	0.916	**0.782**
(Expert 2, MParser)	0.706	1	0.847	0.684	0.973	0.807	0.891	0.639	0.711	0.687	0.907	**0.805**
(Expert 3, MParser)	0.715	0.979	0.828	0.749	0.947	0.784	0.874	0.662	0.684	0.648	0.938	**0.801**
Avg.	**0.702**	**0.986**	**0.826**	**0.693**	**0.959**	**0.782**	**0.868**	**0.661**	**0.681**	**0.675**	**0.920**	

7.4.2. Semantic Consistency

Here, we discuss the multilingual semantic consistency of MParser between English and Chinese. In MParser, a sentence is a concept tree, consisting of simple sentences defined by a sequential list *SiS*, where each atomic concept *iid* is a *low-level concept llc*∈LLC in the step of human semantic input (HSI), and compound concept *eiid* ∈EIID is a *high-level concept hlc*∈HLC generated in MParser, acting as a sentence constituent in the step of sentence computerization (SC). Given two sentences, SiS_i, which is an English sentence, and SiS_j, which is a Chinese sentence, if low-level concept equivalence and high-level concept equivalence are equal such that $SiS_i =_m SiS_j$ ($=_m$ indicates semantic equivalence), then they are semantically consistent. As low-level concept equivalence is semantic consistency of terms, or word-based, high-level concept equivalence is sentence-based semantic consistency.

1. *Low-level concept equivalence:* *SiS_i and SiS_j are equivalent if and only if:*

(1) $\forall\, LLC_i \subset IID_i \subset CoDic$
(2) $\forall\, LLC_j \subset IID_j \subset CoDic$
(3) *Mapping relationship:* $LLC_i \leftrightarrow LLC_j$

This guarantees that two heterogeneous single concepts are semantically consistent, as two sentences share a common *iid* ∈ *CoDic*.

2. *High-level concept equivalence:* *SiS_i and SiS_j are equivalent if and only if:*

(1) $\forall\, HLC_i \subset EIID_i$
(2) $\forall\, HLC_j \subset EIID_j$
(3) *Mapping relationship:* $HLC_i \leftrightarrow HLC_j$

HLC achieves complex concept consistency by converging all heterogeneous structures onto an isomorphic grammatical structure through MParser.

3. *LLC ⇔ HLC:* *LLC and HLC are equivalent if and only if:*

(1) *Mapping relationship: IID ↔ EIID, which is iid in Def. 4 mapped to eiid in Def. 5*

Thus, if and only if the following mapping path exists for semantic equivalence:

SiS_i (local concept) ↔ LLC_i (local concept, IID_i,) ↔ Map (IID_i, Common concept) ↔ HLC_i (Common concept, $EIID_i$) ↔ HLC_j ($EIID_j$, Common concept) ↔ Map (Common concept, IID_j) ↔ LLC_j (IID_j, local concept) ↔ SiS_j (local concept)

It is obvious that if all three conditions are met, then $SiS_i =_m SiS_j$. Figure 12 illustrates that languages *i* and *j* are semantically consistent as they share common tree concepts in cross languages through the unique *iid* and *eiid* in MParser.

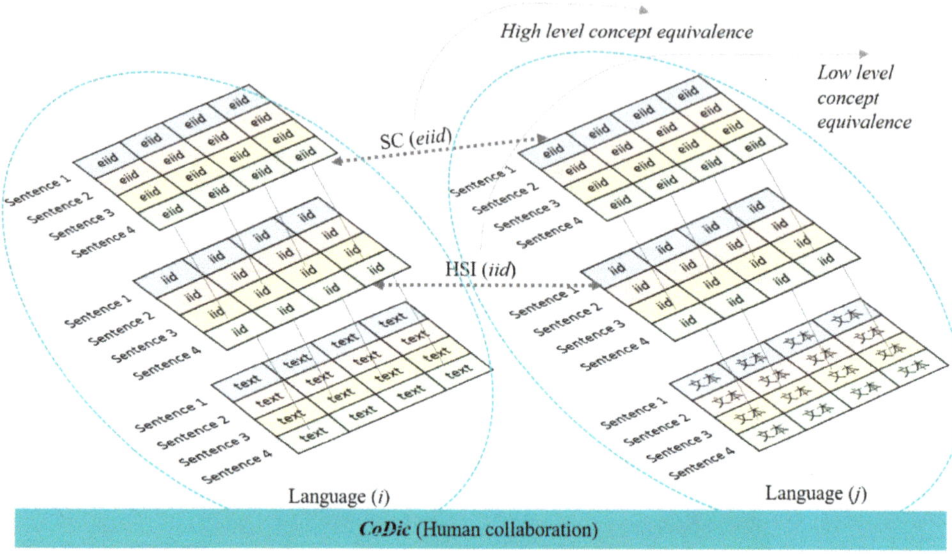

Figure 12. An illustration on semantic consistency.

8. Conclusions and Future Work

Creating a common semantic representation for multilingual languages is an essential goal of the NLP community. To facilitate multilingual sentence representation and semantic interoperability, this research presented an MParser for parsing local language sentences and providing a common understanding across the heterogeneous sentence. MParser converts complex concepts into a computer-readable and -understandable universal sentence for any simple multilingual sentence. This approach has provided a universal grammatical feature such that any sentence can be processed as a bag of concepts and refer to any term of a natural language. Additionally, it has laid a theoretical foundation for enabling humans and computers to understand sentences semantically through unique *iid* and *eiid*.

In the future, we plan to apply the approach to more real-world applications. For example, we will conduct research on how to achieve content persistence during construction of the Metaverse [44] by proposing a content-level persistence maintenance model since the ambiguity of the language, the use of synonyms to express a single idea, creates problems. In the blockchain, we will explore the question of how to achieve semantic interoperability between IoT devices and users [45]. In the field of smart contracts, we will study the cross-context issues of smart contracts between unknown business partners such as developers or anybody who even comes from different backgrounds or languages. Since language barriers prevent cross-language searches, most users do not have easy access to most of this [46]. Moreover, it also will be necessary to extend the research, including semantic inference on extracted meaning. We hope that our novel method will inspire the community to integrate various functions into our work.

Author Contributions: Conceptualization, P.Q. and J.G.; methodology, P.Q., J.G., W.T. and B.S.; software, P.Q.; writing—original draft preparation, P.Q.; writing—review and editing, W.T., B.S. and Q.T.; supervision, J.G.; project administration, J.G.; funding acquisition, J.G., Q.T. All authors have read and agreed to the published version of the manuscript.

Funding: This research is partially supported by the University of Macau Research Grant No. MYRG2017-00091-FST, MYRG2019-00024-FST and FDCT-NSFC Grant No. 0004/2019/AFJ.

Institutional Review Board Statement: Not applicable.

Informed Consent Statement: Not applicable.

Data Availability Statement: Not applicable.

Acknowledgments: Not applicable.

Conflicts of Interest: The authors declare no conflict of interest.

Appendix A. Parts of Speech (PoS)

Table 1. PoS in CoDic.

PoS		Abbr.	Definition
Noun (n)	Common	*Ncm*	A term class denoting a common entity.
	Proper Person	*npp*	A term class denoting a proper person entity.
	Proper Organization	*nop*	A term class denoting a proper organizational entity.
	Proper Geography	*ngp*	A term class denoting a proper geographical entity.
	Pronoun	*npr*	A term class substituting a noun or a noun phrase.
Verb (v)	Intransitive	*vit*	A term class denoting an action, an event, or a state without following any entity.
	Transitive	*vtr*	A term class denoting an action, an event, or a state following only one entity.
	Ditransitive	*vdi*	A term class denoting an action, an event, or a state without following only two entities.
	Copulative	*cop*	A term class denoting a linkage between an entity and a *copulated component* (coc) that expresses a state of being. Adopting "coc" is to avoid the confusion of current use of "predicative expression".
	Adjective	*adj*	A term class describing the attributes of an entity.
	Adverb	*adv*	A term class describing the attributes of an action, an event, or a state.
	Preposition	*prep*	A term class denoting a relation to other noun-formed term(s) before, in the middle, or after.
	Conjunction	*conj*	A term class connecting terms, phrases and clauses, such as *and*, *or*, and *if*.
	Interjection	*int*	A term class expressing a spontaneous feeling or reaction.
	Onomatopoeia	*ono*	A term class imitating, resembling, or suggesting a sound.
	Particle	*par*	A term class indicating a case encompassed by it.

B. Grammatical Features

In MParser, the gender and number features are only attributed to nouns. The features of tense, aspect, and voice are only attributed to verbs. For the grammatical aspects, we have the following definitions:

- *Perfect* (*prf*): a verb form that indicates that an action or circumstance occurred earlier than the time under consideration, often focusing attention on the resulting state rather than on the occurrence itself. E.g., "I have made dinner".
- *Perfect Progressive* (*pfg*): a verb form that indicates that an action was progressive and finished at a time. E.g., "I had been doing homework until 6 PM yesterday".
- *Perfective* (*pfv*): a grammatical aspect that describes an action viewed as a simple whole, i.e., a unit without interior composition. Sometimes called the aoristic aspect, which is a verb form to usually refer to past events. For example, "I came".
- *Imperfective* (*ipfv*): a grammatical aspect used to describe a situation viewed with interior composition. The imperfective is used to describe ongoing, habitual, repeated, or similar semantic roles, whether that situation occurs in the past, present, or future. Although many languages have a general imperfective, others have distinct aspects for one or more of its various roles, such as progressive, habitual, and iterative aspects.

1. *Imperfective habitual* (*iph*): describes habitual and repeated actions. For example, "I read". "The rain beat down continuously through the night".
2. *Imperfective progressive* (*ipp*): describes ongoing actions or events. For example, "The rain was beating down".

Thus, we now have the feature combinations for noun and verb as shown in Tables 2 and 3.

Table 2. Grammatical features of noun on morphological change.

Number	Gender	Binary Postfix	Hex Postfix
Countable singular	Neuter	0000	0
	Masculine	0001	1
	Feminine	0010	2
	Bisexual	0011	3
Countable plural	Neuter	0100	4
	Masculine	0101	5
	Feminine	0110	6
	Bisexual	0111	7
Uncountable	Neuter	1000	8
	Masculine	1001	9
	Feminine	1010	A
	Bisexual	1011	B

Table 3. Grammatical features of verb on morphological change.

Voice	Tense	Aspect	Binary Postfix	Hex Postfix
active	Present	Perfect	0000 0000	00
		Perfect progressive	0000 0001	01
		Perfective	0000 0010	02
		Imperfective habitual	0000 0011	03
		Imperfective progressive	0000 0100	04
	Past	Perfect	0001 0000	10
		Perfect progressive	0001 0001	11
		Perfective	0001 0010	12
		Imperfective habitual	0001 0011	13
		Imperfective progressive	0001 0100	14
	Future	Perfect	0010 0000	20
		Perfect progressive	0010 0001	21
		Perfective	0010 0010	22
		Imperfective habitual	0010 0011	23
		Imperfective progressive	0010 0100	24
	Past future	Perfect	0011 0000	30
		Perfect progressive	0011 0001	31
		Perfective	0011 0010	32
		Imperfective habitual	0011 0011	33

C. MParser Output—75 Sentences

In MParser, we manually input 75 valid sentences and automatically output parsed results for each sentence, as shown in Table 4.

Table 4. 75 sentences from Mparser output.

1.	*I like apples.* *(I.NOM like.PRE apple.ACC)*
2.	*I miss those times and cherish them often.* *(I.NOM miss.PRE those.GEN time.ACC cherish.PRE them.ACC often.ADV)*
3.	*She has been found.* *(She.NOM find.PRE)*
4.	*Nobody can understand.* *(Nobody.NOM can.PRE understand.PRE)*
5.	*His method was strange but impressive.* *(His.GEN method.NOM was.PRE strange.LIN impressive.LIN)*
6.	*She said she is waiting until night.* *(she.NOM said.PRE she.NOM. wait.PRE. until.COMv night.NOM)*

Table 4. *Cont.*

7.	*We need to speed into perspective.* *(we.NOM. need.PRE speed.PRE into.COMv perspective.NOM)*
8.	*The size of sample will change user behavior.* *(size.NOM of sample.NOM change.PRE user.ACC behavior.ACC)*
9.	*The car was sold with a three warranty.* *(Car.ACC sell.PRE with.COMv three.NOM warranty.NOM)*
10.	*The crash occurred in our province.* *(Crash.NOM occurr.PRE in.COMv our.GEN province.NOM)*
11.	*Russia remains hostage oil and gas prices.* *(Russia.NOM remain.PRE hostage.ACC oil.ACC gas.ACC price.ACC)*
12.	*Previous appointees stayed the role until their deaths.* *(Previous.GEN appointee.NOM stay.PRE role.ACC until.COMv their.GEN death.NOM)*
13.	*Everyone has been for their particular skill.* *(Everyone.NOM is.PRE for.LIN their.GEN particular.GEN skill.NOM)*
14.	*They have their cake and eat it too.* *(They.NOM have.PRE their.GEN cake.ACC eat.PRE it.ACC too.ADV)*
15.	*It was experiencing some hard moments.* *(It.NOM experience.PRE some.GEN hard.GEN moment.ACC)*
16.	*I 'm going to join the club.* *(I.NOM go.PRE join.PRE club.ACC)*
17.	*This dispute with the legal is just beginning.* *(This.GEN dispute.NOM with.COMn legal.NOM is.LIN just.ADV beginning.COMn)*
18.	*She said the outage started in the afternoon.* *(She.NOM said.PRE outage.NOM started.PRE in.COMv afternoon.NOM)*
19.	*Our teacher's appearance looks bad and dirty.* *(Our.GEN teacher.NOM appearance.NOM look.LIN bad.COM dirty.COM)*
20.	*The quick brown fox jumped over the lazy dog.* *(Quick.GEN brown.GEN fox.NOM jump.PRE over.COMv lazy.GEN dog.NOM)*
21.	*I wish you are lucky too.* *(I.NOM wish.PRE you.NOM are.PRE lucky.LIN too.ADV)*
22.	*I spoke to my mum at last night.* *(I.NOM spoke.PRE my.GEN mum.ACC at.COMv last.GEN night.NOM)*
23.	*Everybody wants to their mark.* *(Everybody.NOM want.PRE their.GEN mark.ACC)*
24.	*The dog is hit by the man heavily.* *(Dog.ACC hit.PRE by.COMv man.NOM heavily.ADV)*
25.	*The day finally dawned.* *(Day.NOM finally.ADV dawn.PRE)*
26.	*They are just excited about the honor.* *(They.NOM are.PRE just.ADV excited.LIN about.COMv honor.NOM)*
27.	*She detailed the highs and lows.* *(She.NOM detail.PRE high.ACC low.ACC)*
28.	*Two of the soldiers were catching ride.* *(Two.NOM soldier.NOM catch.PRE ride.ACC)*
29.	*The students also track the men's progress.* *(Student.NOM also.ADV track.PRE man. ACC progress.ACC)*
30.	*He is popular in all of the House.* *(He.NOM is.PRE popular.LIN in.COMv all.GEN House.NOM)*
31.	*Fame released in UK cinemas.* *(Fame.NOM release.PRE in.COMv UK.NOM cinema.NOM)*
32.	*I enjoy travel in summer.* *(I.NOM enjoy.PRE travel.ACC in.COMv summer.NOM)*
33.	*We relied on the integrity of truth.* *(We.NOM rely.PRE integrity.ACC truth.ACC)*
34.	*His sense of taste is returning.* *(His.GEN sense.NOM taste.NOM return.PRE)*

Table 4. *Cont.*

35.	*Home builders also jumped most financials.* *(Home.NOM builder.NOM also.ADV jump.PRE most.GEN financial.ACC)*
36.	*They were taxed income when we earned them.* *(They.NOM tax.PRE income.ACC we.NOM earn.PRE them.ACC)*
37.	*She joined a sport during primary school.* *(She.NOM join.PRE sport.ACC during.COMv primary.NOM school.NOM)*
38.	*Your friends are good men.* *(Your.GEN friend.NOM are.LIN good.GEN man.ACC)*
39.	*You will find links to this news.* *(You.NOM find.PRE link.ACC to.COMv this.GEN news.NOM)*
40.	*Some radio channels will move new position.* *(Some.GEN radio.NOM channel.NOM move.PRE new.GEN position.ACC)*
41.	*She has also worked with battery hens.* *(She.NOM also.ADV work.PRE with.COMv battery.NOM hen.NOM)*
42.	*The group now owns venues across the country.* *(Group.NOM now.NOM own.PRE venue.ACC across.COMv country.NOM)*
43.	*The student finished their season in one hour.* *(Student.NOM finish.PRE their.GEN season.ACC in.COMv one.NOM hour.NOM)*
44.	*It sets the two on collision courses.* *(It.NOM set.PRE two.ACC on.COMv collision.NOM course.NOM)*
45.	*The two people were taking in the class.* *(Two.GEN people.NOM talk.PRE in.COMv class.NOM)*
46.	*The financial crisis has many of those bets.* *(Financial.GEN crisis.NOM has.PRE many.GEN those.GEN bet.ACC)*
47.	*The party is at a new location.* *(Party.NOM is.PRE at.LIN new.GEN location.NOM)*
48.	*This is great place to start the trip.* *(This.NOM is.PRE great.COM place.LIN start.PRE trip.ACC)*
49.	*I want to pick something else really.* *(I.NOM want.PRE pick.PRE something.ACC else.GEN really.ADV)*
50.	*You should find a similar thing like sport.* *(You.NOM should.ADV find.PRE similar.GEN thing.ACC like.COMv sport.NOM)*
51.	*The violence was some of the worst ethnic in China for decades.* *(Violence.NOM is.PRE some.GEN worst.GEN ethnic.ACC in.COMn China.NOM for.COMv decade.NOM)*
52.	*The market is mired in scandals and has not recovered good.* *(Market.NOM mired.PRE in.COMv scandals.NOM not.ADV recover.PRE good.COM)*
53.	*The insurgents often attack police and sometimes city officials at night.* *(Insurgent.NOM often.ADV attack.PRE police.ACC sometimes.ADV city.ACC official.ACC at.COMv night.NOM.)*
54.	*The cake is made by the shop after months slowly.* *(Cake.ACC made.PRE by.COMv shop.NOM after.COMv month.NOM slowly.ADV)*
55.	*His detention began in this week when he was trying to leave the city on a false passport.* *(His.GEN detention.NOM begin.PRE in.COMv this.GEN week.NOM he.NOM try.PRE leave.PRE city.ACC on.COMv false.GEN passport.NOM)*
56.	*I want to thank every member of congress who stood tonight with courage.* *(I.NOM want.PRE thank.PRE every.GEN member.ACC congress.ACC stand.PRE tonight.ADV with.COMv courage.NOM)*
57.	*It was his job to fight the war and make an assessment when the time came.* *(It.NOM is.PRE his.GEN job.ACC fight.PRE war.ACC make.PRE assessment.ACC time.NOM come.PRE)*

Table 4. *Cont.*

58.	*The Justice Department scheduled a news conference Tuesday afternoon to announce the indictment.* *(Justice.NOM Department.NOM schedule.PRE news.ACC conference.ACC in.COMv afternoon.NOM announce.PRE indictment.ACC)*
59.	*The president had been scheduled to leave for the trip on Sunday.* *(President.NOM schedule.PRE leave.PRE for.COMv trip.NOM on.COM Sunday.NOM)*
60.	*A sale has been hit after a robbery in a store.* *(Sale.ACC hit.PRE after.COMv robbery.NOM in.COMv store.NOM)*
61.	*I have won this race twice and it would be great to win it again.* *(I.NOM win.PRE this.GEN race.ACC twice.ADV it.NOM is.LIN great.COM win.PRE it.ACC again.ADV)*
62.	*We 've got great commanders on the ground in leadership.* *(We.NOM get.PRE great.GEN commander.ACC on.COMv ground.NOM in.COMv leadership.NOM)*
63.	*He intends to return to the company within next year.* *(He.NOM intend.PRE return.PRE company.ACC within.COMv next.GEN year.NOM)*
64.	*Providing sensitive information to strangers by phone is dangerous.* *(Providing.PRE sensitive.GEN information.ACC to.COMv stranger.NOM by.COMn phone.NOM is.LIN dangerous.COM)*
65.	*She heard the noise and thought someone must have been making it for the event.* *(She.NOM hear.PRE noise.ACC think.PRE someone.NOM must.ADV make.PRE it.ACC for.COMv event.NOM)*
66.	*He had been banned over fears that raised the chances of contamination.* *(He.ACC ban.PRE over.COMv fear.NOM raise.PRE chance.NOM contamination.NOM)*
67.	*Readers who want local color in their mysteries usually seek exotic foreign.* *(Reader.NOM want.PRE local.GEN color.ACC in.COMv their.GEN mystery.NOM usually.ADV seek.PRE exotic.GEN foreign.ACC)*
68.	*He said he will develop a new investment strategy for several months.* *(He.NOM said.PRE he.NOM develop.PRE new.GEN investment.NOM strategy.NOM for.COMv several.GEN month.NOM)*
69.	*The emerging legislation is at his economic recovery program for further years.* *(Emerging.GEN legislation.NOM is.PRE at.LIN his.GEN economic.NOM recovery.NOM program.NOM for.COMv further.GEN year.NOM)*
70.	*All the records were always at hand if we must call about something.* *(All.GEN record.NOM are.LIN always.ADV at.COM hand.NOM we.NOM must.ADV call.PRE about.COMv something.NOM)*
71.	*The TV series has become a big hit among viewers who find empathy with characters in the drama.* *(TV.NOM series.NOM become.PRE big.GEN hit.NOM among.COMv viewer.NOM find.PRE empathy.ACC with.COMv character.NOM in.COMv drama.NOM)*
72.	*The chain of workers involved in real estate deals has grown over the years.* *(Chain.NOM worker.NOM involved.PRE in.COMv real.GEN estate.NOM deal.NOM grow.PRE over.COMv year.NOM)*
73.	*Rival studios have come together to push consumers to rent more movies on their cable boxes.* *(Rival.NOM studio.NOM come.PRE together.ADV push.PRE consumer.ACC rent.PRE more.GEN movie.ACC on.COMv their.GEN cable.NOM boxe.NOM)*
74.	*He fled to a neighboring town where he took a family hostage.* *(he.NOM fled.PRE neighbour.GEN town.ACC he.NOM take.PRE family.NOM hostage.NOM)*
75.	*Everyone was expecting France teams to make the finals competition.* *(Everyone.NOM expect.PRE France.ACC team.ACC make.PRE final.GEN competition.ACC)*

References

1. Zou, Y.; Lu, W. Learning Cross-lingual Distributed Logical Representations for Semantic Parsing. In Proceedings of the 56th Annual Meeting of the Association for Computational Linguistics (Volume 2: Short Papers), Melbourne, Australia, 15–20 July 2018.
2. Balahur, A.; Perea-Ortega, J.M. Sentiment analysis system adaptation for multilingual processing: The case of tweets. *Inf. Process. Manag.* **2015**, *51*, 547–556. [CrossRef]
3. Noraset, T.; Lowphansirikul, L.; Tuarob, S. WabiQA: A Wikipedia-Based Thai Question-Answering System. *Inf. Process. Manag.* **2021**, *58*, 102431. [CrossRef]
4. Zheng, J.; Li, Q.; Liao, J. Heterogeneous type-specific entity representation learning for recommendations in e-commerce network. *Inf. Process. Manag.* **2021**, *58*, 102629. [CrossRef]
5. Etaiwi, W.; Awajan, A. Graph-based Arabic text semantic representation. *Inf. Process. Manag.* **2020**, *57*, 102183. [CrossRef]
6. Liang, P. Learning executable semantic parsers for natural language understanding. *Commun. ACM* **2016**, *59*, 68–76. [CrossRef]
7. Liang, P.; Potts, C. Bringing Machine Learning and Compositional Semantics Together. *Annu. Rev. Linguistics* **2015**, *1*, 355–376. [CrossRef]
8. Bos, J.; Basile, V.; Evang, K.; Venhuizen, N.J.; Bjerva, J. The Groningen Meaning Bank. In *Handbook of Linguistic Annotation*; Springer: Berlin/Heidelberg, Germany, 2017; pp. 463–496.
9. Banarescu, L.; Bonial, C.; Cai, S.; Georgescu, M.; Griffitt, K.; Hermjakob, U.; Knight, K.; Koehn, P.; Palmer, M.; Schneider, N. Abstract meaning representation for sembanking. In Proceedings of the LAW, Sofia, Bulgaria, 8–9 August 2013; pp. 178–186.
10. Abend, O.; Dvir, D.; Hershcovich, D.; Prange, J.; Schneider, N. Cross-lingual Semantic Representation for NLP with UCCA. In Proceedings of the 28th International Conference on Computational Linguistics: Tutorial Abstracts, Barcelona, Spain, 8–13 December 2020; pp. 1–9.
11. Boguslavsky, I.; Frid, N.; Iomdin, L.; Kreidlin, L.; Sagalova, I.; Sizov, V. Creating a Universal Networking Language module within an advanced NLP system. In Proceedings of the 18th Conference on Computational Linguistics, Saarbrücken, Germany, 31 July–4 August 2000; Volume 1, pp. 83–89.
12. Nivre, J.; Marneffe, M.-C.D.; Ginter, F.; Goldberg, Y.; Hajic, J.; Manning, C.D.; McDonald, R.; Petrov, S.; Pyysalo, S.; Silveira, N.; et al. Universal dependencies v1: A multi-lingual treebank collection. In Proceedings of the of LREC, Portorož, Slovenia, 23–28 May 2016; pp. 1659–1666.
13. Xiao, G.; Guo, J.; Da Xu, L.; Gong, Z. User Interoperability with Heterogeneous IoT Devices Through Transformation. *IEEE Trans. Ind. Inform.* **2014**, *10*, 1486–1496. [CrossRef]
14. Nikiforov, D.; Korchagin, A.B.; Sivakov, R.L. An Ontology-Driven Approach to Electronic Document Structure Design. In *Communications in Computer and Information Science*; Springer: Berlin/Heidelberg, Germany, 2017; pp. 3–16.
15. Xiao, G.; Guo, J.; Gong, Z.; Li, R. Semantic input method of Chinese word senses for semantic document exchange in e-business. *J. Ind. Inf. Integr.* **2016**, *3*, 31–36. [CrossRef]
16. Qin, P.; Guo, J. A novel machine natural language mediation for semantic document exchange in smart city. *Futur. Gener. Comput. Syst.* **2020**, *102*, 810–826. [CrossRef]
17. Guo, J. Collaborative conceptualisation: Towards a conceptual foundation of interoperable electronic product catalogue system design. *Enterp. Inf. Syst.* **2009**, *3*, 59–94. [CrossRef]
18. Li, W.; Suzuki, E. Adaptive and hybrid context-aware fine-grained word sense disambiguation in topic modeling based document representation. *Inf. Process. Manag.* **2021**, *58*, 102592. [CrossRef]
19. Medjahed, B.; Benatallah, B.; Bouguettaya, A.; Ngu, A.H.; Elmagarmid, A.K. Busi-ness-to-business interactions: Issues and enabling technologies. *VLDB J.* **2003**, *12*, 59–85. [CrossRef]
20. Bing, L.; Jiang, S.; Lam, W.; Zhang, Y.; Jameel, S. Adaptive concept resolution for document repre-sentation and its applications in text mining. *Knowl.-Based Syst.* **2015**, *74*, 1–13. [CrossRef]
21. Tekli, J. An Overview on XML Semantic Disambiguation from Unstructured Text to Semi-Structured Data: Background, Applications, and Ongoing Challenges. *IEEE Trans. Knowl. Data Eng.* **2016**, *28*, 1383–1407. [CrossRef]
22. Decker, S.; Melnik, S.; Van Harmelen, F.; Fensel, D.; Klein, M.; Broekstra, J.; Erdmann, M.; Horrocks, I. The Semantic Web: The roles of XML and RDF. *IEEE Internet Comput.* **2000**, *4*, 63–73. [CrossRef]
23. Wang, T.D.; Parsia, B.; Hendler, J. A survey of the web ontology landscape. In Proceedings of the International Semantic Web Conference, Athens, GA, USA, 5–9 November 2006.
24. Rico, M.; Taverna, M.L.; Caliusco, M.L.; Chiotti, O.; Galli, M.R. Adding Semantics to Electronic Business Documents Exchanged in Collaborative Commerce Relations. *J. Theor. Appl. Electron. Commer. Res.* **2009**, *4*, 72–90. [CrossRef]
25. Governatori, G. REPRESENTING BUSINESS CONTRACTS IN RuleML. *Int. J. Cooperative Inf. Syst.* **2005**, *14*, 181–216. [CrossRef]
26. Tsadiras, A.; Bassiliades, N. RuleML representation and simulation of Fuzzy Cognitive Maps. *Expert Syst. Appl.* **2013**, *40*, 1413–1426. [CrossRef]
27. Marneffe, M.; Maccartney, B.; Manning, C. Generating Typed Dependency Parses from Phrase Structure Parses. In Proceedings of the LREC'06, Genoa, Italy, 22–28 May 2006; pp. 449–454.
28. Guo, J. SDF: A Sign Description Framework for Cross-context Information Resource Representation and Inter-change. In Proceedings of the 2nd Int'l Conference on Enterprise Systems (ICES 2014), Shanghai, China, 2–3 August 2014.
29. Ruppenhofer, J.; Ellsworth, M.; Schwarzer-Petruck, M.; Johnson, C.R.; Baker, C.F.; Scheffczyk, J. *FrameNet II: Extended Theory and Practice*; International Computer Science Institute: Berkeley, CA, USA, 2006.

30. Loper, E.; Yi, S.-T.; Palmer, M. Combining lexical resources: Mapping between PropBank and VerbNet. In Proceedings of the 7th International Workshop on Computational Linguistics, Syktyvkar, Russia, 23–25 September 2007.
31. Palmer, M.; Gildea, D.; Kingsbury, P. The Proposition Bank: An Annotated Corpus of Semantic Roles. *Comput. Linguist.* **2005**, *31*, 71–106. [CrossRef]
32. Xue, N.; Bojar, O.; Hajic, J.; Palmer, M.; Uresova, Z.; Zhang, X. Not an intelingua, but close: Comparison of English AMRs to Chinese and Czech. In Proceedings of the Ninth International Conference on Language Resources and Evaluation (LREC'14), Reykjavik, Iceland, 26–31 May 2014; pp. 1765–1772.
33. White, A.S.; Reisinger, D.; Sakaguchi, K.; Vieira, T.; Zhang, S.; Rudinger, R.; Rawlins, K.; Van Durme, B. Universal Decompositional Semantics on Universal Dependencies. In Proceedings of the 2016 Conference on Empirical Methods in Natural Language Processing (EMNLP), Austin, TX, USA, 1–5 November 2016; pp. 1713–1723.
34. Ehrmann, M.; Cecconi, F.; Vannella, D.; McCrae, J.P.; Cimiano, P.; Navigli, R. Representing multilingual data as linked data: The case of babelnet 2.0. In Proceedings of the LREC, Reykjavik, Iceland, 26–31 May 2014; pp. 401–408.
35. Klein, D.; Manning, C.D. Accurate unlexicalized parsing. In Proceedings of the Proceedings of the 41st Annual Meeting on Association for Computational Linguistics—ACL '03, Sapporo, Japan, 7–12 July 2003; pp. 423–430.
36. Cook, V.J. Chomsky's universal grammar and second language learning. *Appl. Linguist.* **1985**, *6*, 2–18. [CrossRef]
37. Starosta, S.; Anderson, J.M. *On Case Grammar: Prolegomena to a Theory of Grammatical Relations*; Routledge: Abingdon, UK, 2018.
38. Gkatzia, D.; Mahamood, S. A Snapshot of NLG Evaluation Practices 2005–2014. In Proceedings of the Proceedings of the 15th European Workshop on Natural Language Generation (ENLG), Brighton, UK, 10–11 September 2015; pp. 57–60.
39. Chelba, C.; Mikolov, T.; Schuster, M.; Ge, Q.; Brants, T.; Koehn, P.; Robinson, T. One billion word benchmark for measuring progress in statistical language modeling. *arXiv* **2013**, arXiv:1312.3005.
40. Brysbaert, M. How Many Participants Do We Have to Include in Properly Powered Experiments? A Tutorial of Power Analysis with Reference Tables. *J. Cogn.* **2019**, *2*, 16. [CrossRef] [PubMed]
41. Shrotryia, V.K.; Dhanda, U. Content Validity of Assessment Instrument for Employee Engagement. *SAGE Open* **2019**, *9*, 2158244018821751. [CrossRef]
42. Carletta, J. Assessing agreement on classification tasks: The kappa statistic. *Comput. Linguist.* **1996**, *22*, 249–254.
43. Landis, J.R.; Koch, G.G. The Measurement of Observer Agreement for Categorical Data. *Biometrics* **1977**, *33*, 159–174. [CrossRef]
44. Shen, B.; Tan, W.; Guo, J.; Zhao, L.; Qin, P. How to Promote User Purchase in Metaverse? A Systematic Literature Review on Consumer Behavior Research and Virtual Commerce Application Design. *Appl. Sci.* **2021**, *11*, 11087. [CrossRef]
45. Shen, B.; Guo, J.; Yang, Y. MedChain: Efficient Healthcare Data Sharing via Blockchain. *Appl. Sci.* **2019**, *9*, 1207. [CrossRef]
46. Qin, P.; Tan, W.; Guo, J.; Shen, B. Intelligible Description Language Contract (IDLC)—A Novel Smart Contract Model. *Inf. Syst. Front.* **2021**, 1–18. [CrossRef]

applied sciences

MDPI

Article

Causal Pathway Extraction from Web-Board Documents

Chaveevan Pechsiri [1,*] and Rapepun Piriyakul [2]

1 College of Innovative Technology and Engineering, Dhurakij Pundit University, Bangkok 10210, Thailand
2 Department of Computer Science, Ramkhamhaeng University, Bangkok 10240, Thailand; rapepunnight@yahoo.com

Abstract: This research aim is to extract causal pathways, particularly disease causal pathways, through cause-effect relation (CErel) extraction from web-board documents. The causal pathways benefit people with a comprehensible representation approach to disease complication. A causative/effect-concept expression is based on a verb phrase of an elementary discourse unit (EDU) or a simple sentence. The research has three main problems; how to determine CErel on an EDU-concept pair containing both causative and effect concepts in one EDU, how to extract causal pathways from EDU-concept pairs having CErel and how to indicate and represent implicit effect/causative-concept EDUs as implicit mediators with comprehension on extracted causal pathways. Therefore, we apply EDU's word co-occurrence concept (wrdCoc) as an EDU-concept and the self-Cartesian product of a wrdCoc set from the documents for extracting wrdCoc pairs having CErel into a wrdCoc-pair set from the documents after learning CErel on wrdCoc pairs by supervised-machine learning. The wrdCoc-pair set is used for extracting the causal pathways by wrdCoc-pair matching through the documents. We then propose transitive closure and a dynamic template to indicate and represent the implicit mediators with the explicit ones. In contrast to previous works, the proposed approach enables causal-pathway extraction with high accuracy from the documents.

Keywords: cause-effect relation; transitive closure; word co-occurrence

Citation: Pechsiri, C.; Piriyakul, R. Causal Pathway Extraction from Web-Board Documents. *Appl. Sci.* **2021**, *11*, 10342. https://doi.org/10.3390/app112110342

Academic Editors: Arturo Montejo-Ráez and Salud María Jiménez-Zafra

Received: 14 September 2021
Accepted: 28 October 2021
Published: 3 November 2021

1. Introduction

The objective of this research is to extract causal pathways, particularly disease causal pathways, from downloaded disease documents from several Thai hospital web-boards. The causal pathway extraction of the research is based on determining a sequence of Cause-Effect pairs having a cause-effect relation (called 'CErel') from the documents where a Cause-Effect pair (called 'CEpair') is an ordered pair; Cause is a causative event/state concept; Effect is an effect event/state concept. According to Khoo [1], CErel is a semantic relation which is a directional link between concepts as entities that participate in the relation. Where the concepts connected by a relation are often represented as follow:

<Concept1>—(Relation)—<Concept2>

where the '<... >' and '(...)' symbols represent a concept and a relation type respectively. A dash line is a directional link between Concept1 and Concept2 and is labeled to indicate the type or meaning of the relation. With regard to our research, CErel as a cause-effect relation type is represented as follow:

<CausativeConcept>—(CErel)—><EffectConcept>

where CausativeConcept and EffectConcept are a causative concept and an effect concept respectively of either event or state occurrences on the documents. Moreover, Khoo [1] stated that "concepts and relations are the foundation of knowledge and thought while the concepts are the building blocks of knowledge and the relations are the cement linking up the concepts into the knowledge structures," e.g., a causal chain or a causal pathway contains CausativeConcept, EffectConcept and CErel to become the knowledge structure). With regards to Staplin et al. [2], the causal pathway in epidemiologic studies is a path starting at the exposure and ending at the disease that follows the direction of the arrows. All arrows of the causal pathway point in the same direction from the exposure toward the outcome, e.g., the disease occurrence is the outcome, [3]. According to Gaskell and Sleigh [3], the causal pathway as A $\rightarrow$ M_j $\rightarrow$ B contains the exposure (A) which might cause the outcome (B) or through an intermediate process or variable called a mediator (M_j) where M_j is a single mediator if $j = 1$; and M_j is either sequential-mediators or multilevel-mediators if $j = 2,3, \ldots$,num which is an integer. However, our research concerns the extraction of the causal pathway with either a single mediator or sequential mediators mostly occurring on the documents. With regards to our disease document, A, M_j, and B are either a causative event/state concept or an effect event/state concept which is mostly based on a verb phase of an elementary discourse unit (EDU) where an EDU is a simple sentence or a clause [4]. The EDU expression of the research is based on the general linguistic expression in Figure 1 after stemming words and the stop word removal. Where NP1 and NP2 are noun phrases; VP is a verb phrase; Noun is a noun concept set; $Verb_{strong}$ is a strong verb concept set consisting of causative-verb concepts and effect-verb concepts; $Verb_{weak}$ is a weak verb concept set requiring more information, i.e., Noun, to become the causative/effect concept, e.g., 'มี/*have*+ไขมัน/*fat*+สะสม/*accumulate*' ('*have accumulated fat*') and 'เป็น/*be*+โรค/*disease*' ('*get disease*'); Adv is an adverb concept set; Adj is the adjective concept set; and Adjphrase is an adjective phrase.

EDU → NP1 VP | VP
VP → Verb NP2 | Verb adv | Verb
Verb → $Verb_{weak}$ Noun | $Verb_{strong}$
NP1 → pronoun | Noun | Noun Adj | Noun Adjphrase
NP2 → Noun | Noun Adj | Noun Adjphrase | Noun Prepphrase
$Verb_{weak}$ → {'เป็น/be','ไม่เป็น/*notBe*','มี/*have*','ไม่มี/*notHave*','ใช้,นำ/*use, Take*'}
$Verb_{strong}$ → { 'เกิด/*occur*', 'ตีบ/*constrict*', 'ตัน/*block up*', 'ไม่ตอบสนอง/*not_respond*', 'เสื่อม/*deteriorate*', 'ขับ/*excrete*', 'เพิ่ม,ขยาย/*increase, enlarge*', 'เปลี่ยนแปลง/*change*', 'ขาด/ *lack*', 'ตอบสนอง/ *Response*', 'ทำลาย/*damage*', 'อักเสบ/ *becomeInflamed*', 'ตาย/*die*', 'แข็ง/ *beStiff*', 'หนา/*beThick*', 'ฉีกขาด/*Tear*', 'เลี้ยง/*supply*', 'เลือดออก/*bleed*', 'วาย,ล้มเหลว/*beFailure*', 'เกาะ,สะสม/*Deposit*, 'ไหล,ผ่าน/*flow,pass*', 'กระตุ้น/*stimulus*',.. }
Adj → {'สูง/*high*', ...}
Adv → {'ยาก/ *difficultly*', 'เหลว/ *liquidly*', ...}
Noun → {' ', 'แผล/*scar*', 'ผู้ป่วย/*patient*', 'หลอดเลือด/*bloddVessel*', 'หัวใจ/*heart*', 'ตับ/*liver*', 'ไต/*kidney*', 'กล้ามเนื้อหัวใจ/ *myocardium*', 'สมอง/*brain*','อวัยวะ/*humanOrgan*', 'เลือด/*blood*', 'ปัสสาวะ/*urine*', 'ความดัน/*pressure*', 'น้ำตาล/*sugar*', 'ไขมัน/*fat*', 'โปรตีน/*protein*', 'อาหาร/*food*', 'การหดตัว/*contraction*', 'สี../..*color*', 'ตัวเร่ง/*catalyst*',...}

Figure 1. The general Thai linguistic expression of EDU after stemming words and eliminating stop words.

The example of a causal pathway is expressed on the document by a sequence of Cause-Effect pairs (CEpairs) having CErel on the document is shown in Example 1.

Example 1. *Topic Name:* *ฮอร์โมนอินซูลินมีความสำคัญต่อร่างกายอย่างไร/**How important is the hormone insulin for the body?***
(From Bangkok Hospital Phuket) ...

EDU1. *"เมื่อผู้ป่วยขาดอินซูลิน/**When a patient lacks insulin,"***
*เมื่อ/**When** (ผู้ป่วย/**a patient**)/NP1 ((ขาด/**la cks**)/$Verb_{strong}$ (อินซูลิน/**insulin**)/NP2)/VP*

EDU2. *"<u>ทำให้</u> ร่างกายไม่สามารถนำน้ำตาลไปใช้เป็นพลังงานให้ส่วนต่างๆ ของร่างกายได้/*
<u>causing</u> the body to be unable to use sugar as energy in various parts of the body".
*<u>ทำให้/**causing**</u> ((ร่างกาย/**the body**)/NP1*
*(ไม่สามารถ/**is unable**) (นำ/**to take**)/$Verb_{strong}$ (น้ำตาล/**sugar**ไปใช้เป็นพลังงาน/**to use as energy** ให้ส่วนต่างๆของร่างกายได้/**in various part of body**)/NP2)/VP*

EDU3. *"<u>ทำให้</u> [ผู้ป่วย] มีระดับน้ำตาลในเลือดสูง/**<u>Causing</u> [the patient] to have hyperglycaemia".***
*<u>ทำให้/**causing**</u> ([ผู้ป่วย/**the patient**])/NP1*
*((มี/**has**)/$Verb_{weak}$ (ระดับน้ำตาลในเลือด สูง/**hyperglyc aemia**)/NP2)/VP*

*EDU4. "และส่งผลให้ [ผู้ป่วย] เป็นโรคเบาหวาน/**And causing** **[the patient] to be diabetes**". . . . .*
*และส่งผลให้/**and causing** ([ผู้ป่วย/**the patient**])/NP1*
*((เป็น/**gets**)/Verb$_{weak}$ (โรคเบาหวาน/**diabet es**)/NP2)/VP*

where the [..] symbol means a term/terms inside the symbol being ellipsis; and a 'ทำให้/causing' term is a causal verb with the part of speech as a causal conjunction between a causative-concept EDU and an effect-concept EDU.

Example 1 shows a sequence of CEpairs having CErel on an EDU-pair sequence as shown in the following expression.

EDU1-EDU2 Pair as the first CEpair (CEpair$_1$): EDU1<Cause>—(CErel)—> <Effect>EDU2

EDU2-EDU3 Pair as the second CEpair (CEpair$_2$): EDU2<Cause>—CErel)—> <Effect>EDU3

EDU3-EDU4 Pair as the third CEpair (CEpair$_3$): EDU3 <Cause>—(CErel)—> <Effect>EDU4

The sequence of CEpair$_1$ through CEpair$_3$ is the causal pathway as EDU1→EDU2→ EDU3→EDU4.

According to Example 1, EDU1 (a causative-concept EDU), and EDU4 (an effect-con cept EDU) are the exposure and the outcome respectively whilst EDU2 and EDU3 (which are effect/causative-concept EDUs) are sequential-mediators.

The extracted causal pathways would support the problem-solving system by supporting unprofessional persons to have a more comprehensible approach to the disease complication through social media which results in compliance to the physician's suggestion of the appropriate treatment. Therefore, the research concerns extracting the causal pathways represented by the CEpair$_i$ sequences (where i is an index of a CEpair in a sequence) from the document.

There are several techniques [5–13] having been applied for determining or extracting the causal pathways or the causal chains through the cause-effect/causal relation determination between two entities/events from the documents (see Section 2). The features used for the CErel determination from the previous research are mainly version a noun variable pair or a verb variable pair from a noun phrase pair or a verb phrase pair respectively within one simple sentence or a simple sentence pair. Whilst the causative/effect concepts of the events/states on the Thai documents are mostly based on the EDUs' verb phrase expressions where the same verb phrase expressions with the different NP1 concepts have the different causative/effect concepts of the events/states, e.g., EDU1: "(ลิ่มเลือด/*Blood clots*) NP1 (ไหลในเลือด/*flow in the artery*)/VP" and EDU2:"(ไขมัน/Fat)/NP1 (ไหลในเลือด/*flow in the artery*)/VP" have the *flow*(*BloodClot, artery*) and *flow*(*Fat, artery*) concepts respectively after stemming words and stop words removal. Therefore, the features used for the CErel determination of our research are based on composite variables relied on a predicate-argument term set for obtaining a causative/effect concept. Where a composite variable is a variable made up of two or more individual variables, called indicators, into a single variable [14]. Each indicator alone doesn't provide sufficient information, but altogether they can represent the more complex concept. In addition, the entailment classification of the previous research [15] is based on the similarity scores which cannot apply to our disease documents, e.g., the relation between EDU1: "ผู้ป่วยเป็นโรคหลอดเลือดแข็ง/*The patient get arteriosclerosis*" and EDU2: " เพราะไขมันไปเกาะที่ผนังหลอดเลือดแดง/*because fat deposits on the artery wall*". is CErel with the similarity score approaching zero. Moreover, the actual causal pathway determination from texts of the previous research are mostly based on two steps of the cause-effect relation type, e.g., A causes B and B causes C, without concerning the implicit mediator whereas our disease documents contain several steps or more than two steps of the cause-effect relation type including the implicit mediators. With regard to the causal pathway for the problem-solving system, the implicit mediators on the causal pathway should be represented by the explicit mediators to have the complete causal pathway for understanding the mechanism through which the composite variable affects the outcome.

However, the Thai documents have several specific characteristics, such as zero ana phora or the implicit noun phrase, without word and sentence delimiters, etc. All of these characteristics are involved in three main problems (see Section 3): (1) how to determine CErel on each EDU-concept pair from the documents where there are some EDU occurrences with both the causative concept in one CErel and an effect concept in another CErel; (2) how to extract causal pathways from several EDU-concept pairs having CErel; and (3) how to indicate the implicit mediators or the implicit effect/causative-concept EDUs on the correct extracted causal pathways from the documents for representing the implicit mediators in the form of the explicit effect/causative-concept EDUs or the explicit mediators for clear comprehension. Regarding these three main problems, we need to develop a framework which combines machine learning and the linguistic phenomena to represent each EDU occurrence by an EDU's word co-occurrence (called 'wrdCo') based on wrdCoPattern on Equation (1) relying on a predicate argument pattern of an EDU occurrence (see Figure 1) after stemming words and eliminating stop words. In addition, each EDU concept of an event/state is represented by an EDU's wrdCo concept (called 'wrdCoc ') as a feature or an element of a wrdCo concept set or a wrdCoc set (WC).

$$\text{wrdCoPattern} = \text{V} + \text{W1} + \text{W2} \tag{1}$$

where V is a predicate verb set; V = $\text{Verb}_{\text{strong}} \cup \text{V}_{\text{inf}}$; $v_a \in \text{V}$; Each element of V_{inf} consists of $v_{\text{weak},b} + w_{\text{inf},c}$ ($v_{weak,b} \in \text{Verb}_{\text{weak}}$, $w_{\text{inf,c}} \in \text{Noun}$, and $w_{\text{inf,c}}$ is a word right after $v_{\text{weak},b}$; *a*, *b*, *c*, *d*, and *e* are an integer.;

W1 is an agent argument set; $w_{1,d} \in \text{W1}$; $w_{1,d}$ is a head noun or a Noun element of NP1 and $w_{1,d}$ is a Noun element of the previous EDU's NP1 if the current EDU's NP1 is ellipsis;

W2 is a linguistic patient/information set; $w_{2,\text{e}} \in \text{W2}$; W2 = $\text{Noun} \cup \text{Adv} \cup \text{Adj}$ and $w_{2,\text{e}}$ is also a word sequence right after v_a; $w_{2,\text{e}}$ has a null value if $w_{2,\text{e}}$ doesn't exist;

And all $\text{Verb}_{\text{strong}}$, $\text{Verb}_{\text{weak}}$, Noun, Adv, Adj sets are based on Figure 1)

Likewise, three contributions of this paper are statistically-based approaches involved with linguistic phenomena and machine learning. The first one is that each wrdCoc feature used for the CErel determination by machine learning is the composite-variable consisting of the elements of V, W1, and W2 for the causative concept/effect concept representation. The second one is that our extracted causal pathways contain more than two steps of the cause-effect relation type and are the actual causal pathways from the documents. And the third one is that some extracted causal pathways contain the implicit mediators (or the implicit effect/causative-concept EDUs) as the implicit wrdCoc features which have to be represented by the explicit wedCoc features for clear comprehensible pathways. Moreover, our implicit wrdCoc features are the qualitative data whereas the previous research [16] discovered the hidden semantics or the latent semantics as the implicit features by the graph regularization where the latent semantics of [16] is the quantitative data.

We then apply the self-Cartesian product of WC × WC [17] (the first WC as the causative-concept set, the second WC as the effect-concept set) to a test corpus for extracting wrdCoc pairs having CErel into WCP (WCP is a set of wrdCoc pairs having CErel) after learning CErel on wrdCoc pairs by naïve Bayes (NB) [18], support vector machine (SVM) [19], and logistic regression (LR) [20] from a learning corpus. According to the test corpus, all WC elements are determined by wrdCo-expression matching between wrdCo expressions of the test corpus and wrdCo expressions of the semi-automatic annotated corpus having annotated wrdCoc features. WCP is used for extracting the causal pathways through wrdCoc-pair matching on the documents (see Sections 3.1 and 3.2). We then propose using transitive closure of a binary relation over a causative concept set and an effect concept set [21] to indicate the implicit mediator occurrences on the correct extracted causal pathways and also using a dynamic template to collect the correct extracted causal pathways with the explicit mediators used for representing those implicit mediators (where the explicit mediators are the explicit effect/causative-concept EDUs represented by EDUs'wrdCoc features) (see Section 3.3).

Our research is organized into six sections. In Section 2, related work is summarized. Problems in the causal-pathway extraction from the documents are described in Sections 3 and 4 shows our framework for the causal pathway extraction from the documents. In Section 5, we evaluate our proposed model including discussion and then present a conclusion in Section 6.

2. Related Works

Several strategies [5–13] have been proposed to determine/extract a causal pathway, a causal chain, or causal path of a graph/network through the cause-effect/causal relation determination except [13] working on the implicit knowledge completion where [5,6] working on only the causal/cause-effect relation determination from texts. Girju [5] proposed decision-tree learning the causal relation from a sentence based on the lexico-syntactic pattern (NP1 causal-verb NP2) where NP1 is a cause and NP2 is an effect or vice versa. Cao et al. [6] also used syntactic patterns by manually annotating one sentence or between two sentences having a cue (a word or a phrase) as a cause-effect link to express the cause-effect relation which is the core of scientific papers. Their cause-effect links were extracted by a syntactic pattern-based algorithm from scientific papers with 47% and 70% on average precision and recall respectively. Chang and Choi [7] extracted causality/causal relation with an F-score of 77.37% based on one complex sentence or two simple sentences by using a cue-phrase set to connect two noun phrases (or an NP pair) as a cause and an effect including probabilities. The extracted causal relations were used for constructing the causal paths of the causal network for the term protein having two relations; the causal relation and the hypernym relation. Pechsiri and Piriyakul [8] applied verb-pair rules resulting from machine learning techniques to extract the causality or the cause-effect relation from several simple sentences to construct one cause with several effect paths on an explanation knowledge graph. The cause-effect paths of [8] were emphasized on the consequence or concurrent occurrence of the extracted effect events. Whilst a causal chain [9] was generated by connecting the extracted causal relations with sentence's word similarity and topic matching between a causative sentence of one causal relation and an effect sentence of another causal relation where the causal relation extraction was based on clue words. Kang et al. [10] applied the Granger causality model with features, i.e., N-words, topics, sentiments, etc., to detect cause-effect relationships from texts for a time series. And [10] also applied a neural reasoning algorithm based on human annotation along with BLEU (bilingual evaluation understudy) scores used for measuring the connection of two cause-effect relationships to construct a causal chain with 57% accuracy based on expert judgments. However, the cause/effect events or entities [10] are mainly expressed by noun phrases based on day-by-day time series. Izumi and Sakaji [11] applied a causal verb set as the edge/relation to construct causal paths by connecting between a cause node and an effect node expressed by noun phrases within one sentence. The causal chain was constructed by manually selecting word vector similarities between effect nodes and cause nodes from different causal relations. Nordon et al. [12] extracted several causal relations based on the lexico-syntactic pattern [5], and then applied the text analysis, i.e., word co-occurrence and Word2vec, to determine the edge weights for solving each causal path of the causal graph from the extracted causal relations. Moreover, Ref. [13] applied the similarity score between two word-pairs as an event pair including the notified event location to calculate the event relevant for automatically discovering implicit event knowledge occurring among the sequential event chain of actions from a Japanese web blog corpus without the CErel consideration between the event pair, e.g., "roll on the floor", "sitting on a sofa", and "drinking tea" were the event chain of actions with the Living room location added. [13] evaluated the knowledge completion for the chain of actions (including the notified event location) by the graduate students scoring as 3.0 based on a five-point Likert scale.

Therefore, the causal relation determination of the previous research [5–12] is mostly based on noun phrases within one or two simple sentences except [8] using only verb pairs to extract the causal relation from several simple sentences. However, CErel of our research is based on wrdCoPattern on Equation (1) included an NP1 head noun and an EDU's verb phrase because the different agents (NP1) with the same predicate verb provide the different semantics of causative/effect concepts. The causal pathways, the causal chain, or the causal-graph paths of the previous research [7–12] are determined/extracted from documents without concerning the implicit mediator on the certain path/chain whilst [13] emphasizes the implicit knowledge completion on the event chain of actions without the CErel consideration. However, there are a few works on extracting the causal pathway from texts with little concerning in the implicit mediator.

3. Problems of Causal-Pathway Extraction

3.1. How to Determine CErel on an EDU-Concept Pair/a wrdCoc Pair

According to the corpus behavior study of the medical care domain, most of the causative/effect-concept EDUs are the events or states expressed by verb phrases. There are some verb phrase expressions with both the causative concepts and the effect concepts on the documents as shown in Example 1, e.g., EDU2 is an effect-event concept and a causative-event concept for $CEpair_1$ and $CEpair_2$ respectively where $CEpair_1$ and $CEpair_2$ are consecutive. Moreover, lack of the sentence delimiter in the Thai documents causes a problem of determining EDU's concept pairs (e.g., an EDU1-EDU2 pair or an EDU2-EDU3 pair) having CErel from three consecutive EDUs if the second EDU contains a discourse-marker cue set, {*'เพราะ/because'*, *'เนื่องจาก/since'*, … }, as shown in Example 2. Where each EDU concept is represented by wrdCoc, i.e., an EDU_j concept is represented by EDU_j's wrdCoc called $wrdCoc_{EDUj}$, j is an integer.

Example 2. *Topic Name:* โรคเบาหวาน/**Diabetes** …

EDU1. *"ผู้ป่วยเบาหวานอาจเป็นโรคหัวใจ/A diabetic patient might get heart disease".*
*(ผู้ป่วยเบาหวาน/A **diabetic patient**)/NP1*
*((อาจเป็น/**might get**)/$Verb_{weak}$(โรคหัวใจ/**the heart disease**)/NP2)/VP*
$wrdCoc_{EDU1}$ = getHeartDisease(person)

EDU2. *"<u>เนื่องจาก</u> [ผู้ป่วย]มีภาวะน้ำตาลในเลือดสูง/**Since [the patient] has hyperglycaemia,**"*
*<u>เนื่องจาก/**Since**</u> ([ผู้ป่วย/**the patient**])/NP1*
*((มี/**has**)/$Verb_{weak}$ ภาวะน้ำตาลในเลือด/**hyperglycaemia**)/VP*
$wrdCoc_{EDU2}$ = haveHyperGlycaemia(person)

EDU3. *ทำให้สารเคมีบางชนิดเพิ่มสูงขึ้นในเลือด/**causing some chemicals increase in the blood.** "* …
*ทำให้/**causing** (สารเคมีบางชนิด/**some chemicals**)/NP1*
*((เพิ่มสูงขึ้น/**increase**)/$Verb_{strong}$ ใน/**in** เลือด/**the blood.**)/VP*
$wrdCoc_{EDU3}$ = increase(chemical,blood)

Example 2 contains a $CEpair_i$ with CErel as shown in the following:
$wrdCoc_{EDU2}$-$wrdCoc_{EDU3}$ Pair as$CEpair_1$:$wrdCoc_{EDU2}$<Cause>—(CErel)—> <Effect> $wrdCoc_{EDU3}$.

Therefore, we apply the self-Cartesian product of WC × WC having the first WC as a causative concept set and the second WC as an effect concept set to the test corpus for extracting the wrdCoc pairs of an EDU pairs having CErel into WCP after learning CErel on each wrdCoc pair by NB, SVM, and LR from the learning corpus (see Section 4.2). The WC elements are collected by the wrdCo-expression matching between the wrdCo expressions of the test corpus and the wrdCo expressions of the learning corpus with the semi-automatic annotated wrdCoc features (see Section 4.1).

3.2. How to Extract the Causal Pathways

During the causal pathway extraction, some causal pathways mingle with non-causative/effect concept EDU(s) and remain a challenge, e.g., Example 1 mingles with a non-causative/effect concept EDU(s) such as EDU2: " อินซูลินมีหน้าที่ส่งสัญญาณให้เซลน้ำน้ำตาลไปใช/***Insulin has***

a function of signaling cells to take sugar for use". which intervenes after EDU1 of Example 1 as shown in the following:

EDU1. "เมื่อผู้ป่วยขาดอินซูลิน/***When a patient lacks insulin,***"
$wrdCoc_{EDU1}$ = *lack(person,insulin)*

EDU2. " อินซูลินมีหน้าที่ส่งสัญญาณให้เซลนำน้ำตาลไปใช/***Insulin has a function of signaling cells to take sugar for use***".
(อินซูลิน/***Insulin***)/*NP1*
((มี/***has***)/$Verb_{weak}$ หน้าที่/***a function of*** ส่งสัญญาณให้เซล/***signaling cells***
นำน้ำตาล/***to take sugar*** ไปใช้/***for use***)/*VP*
$wrdCoc_{EDU3}$ = *hasFunction(insulin,signaling)*

EDU3. "ทำให้ ร่างกายไม่สามารถนำน้ำตาลไปใช้เป็นพลังงานให้ส่วนต่างๆ ของร่างกายได้/***causing the body to be unable to use sugar as energy in various parts of the body***".
$wrdCoc_{EDU1}$ = *beUnableToUseSugar(person)*

EDU4. " ทำให้ [ผู้ป่วย] มีระดับน้ำตาลในเลือดสูง/***Causing*** **[*the patient*]** ***to have hyperglycaemia***".
$wrdCoc_{EDU1}$ = *haveHyperglycaemia(person)*

EDU5. "และส่งผลให้ [ผู้ป่วย] เป็นโรคเบาหวาน/***And causing*** **[*the patient*]** ***to be diabetes***"......... ...
$wrdCoc_{EDU1}$ = *getDiabetes(person)*

Each wrdCoc feature having a predicate verb $v_a \in \text{Verb}_{\text{strong}} \cup \text{V}_{\text{inf}}$ on the test-corpus document is sequentially collected into an array of wrdCoc features for the causal pathway extraction. Where all wrdCoc features of the test-corpus document are obtained by the wrdCo-expression matching between the wrdCo expressions of the test-corpus document and the wrdCo expressions of the annotated corpus having the annotated wrdCoc features.

Therefore, we apply WCP to extract each causal pathway by the wrdCoc-pair matching on sliding window size of two consecutive wrdCoc features (or a wrdCoc pair) on the array of wrdCoc features to match among WCP elements with one wrdCoc distance through the array. If there is no match on wrdCoc-pair matching, we will stop sliding the window and then obtain a causal pathway (Section 4.4).

3.3. How to Indicate Implicit Mediators for Explicit Mediator Representation

Some determined causal pathways contain the implicit mediators (the implicit effect/causative-concept EDUs) as in Examples 3–4 of the same disease group.

Example 3. *Topic Name:* **โรคไตจากเบาหวาน (รพวิภาวดี)/*Diabetic Nephropathy*** *(from Vibhavadi Hospital)*

EDU1. "ผู้ป่วยเป็นเบาหวานมานานหลายปี/*A Patient gets a diabetic disease for several years".)*
(ผู้ป่วย/A patient)/NP1
((เป็น/get)/$Verb_{weak}$โรคเบาหวาน/a diabetesมานานหลายปี/for several years)/VP
$wrdCoc_{EDU1}$ = *getDiabetes(person)*

EDU2. "เนื่องจาก[ผู้ป่วย]มีระดับน้ำตาลในเลือดสูงอยู่เป็นระยะเวลานาน/
Since [the patient] have hyperglycaemia for a long period of time, "
เนื่องจาก/*Since ([ผู้ป่วย/the patient])/NP1*
((มี/has)/$Verb_{weak}$ระดับน้ำตาลในเลือดสูง/hyperglycaemia
อยู่เป็นระยะเวลานาน/for a long period of time)/NP2)/VP
$wrdCoc_{EDU2}$ = *haveHy perglycaemia(person,long-time)*

EDU3. "ทำให้หลอดเลือดทั่วร่างกายจะแข็ง และหนา/*causing blood vessels of whole body to be stiff and thick*".
ทำให้/*causing (หลอดเลือดทั่วร่างกาย/blood vessels of whole body)/NP1*
(จะ/will(แข็ง/be stiff)/$Verb_{strong}$และ/and(หนา/thick)/$Verb_{strong}$)/VP
$wrdCoc_{EDU3}$ = *beStiff&Thick(bloodVessel)*

EDU4. "ทำให้เลือดไปเลี้ยงได้น้อยในส่วนต่าง ๆ ของร่างกาย/
Causing the blood supply less to the parts of the body".
ทำให้/causing (เลือด/*blood)/NP1*

((ไปเลี้ยง/supplies)/Verb$_{strong}$ น้อย/lessในส่วนต่างๆของร่างกาย/to the parts of the body)/VP
$wrdCoc_{EDU4}$ = beSupplied(blood,less)

EDU5. "ส่งผลให้ [ผู้ป่วย]เป็นโรคไต/Causing [the patient] gets a chronic kidney disease". . . . .
ส่งผลให้/causing ([ผู้ป่วย/the patient])/NP1
(เป็น/gets)/Verb$_{weak}$ (โรคไต/kidney disease)/NP2)/VP
$wrdCoc_{EDU5}$ = g etKidneyDisease(person)

Example 3 shows a sequence of CEpairs having CErel as follow:

$wrdCoc_{EDU2}$-$wrdCoc_{EDU1}$ Pair as $CEpair_1$: $wrdCoc_{EDU2}$<Cause>—(CErel)—> <Effect> $wrdCoc_{EDU1}$

$wrdCoc_{EDU2}$-$wrdCoc_{EDU3}$ Pair as $CEpair_2$: $wrdCoc_{EDU2}$<Cause>—(CErel)—> <Effect> $wrdCoc_{EDU3}$

$wrdCoc_{EDU3}$-$wrdCoc_{EDU4}$ Pair as $CEpair_3$: $wrdCoc_{EDU3}$<Cause>—(CErel)—> <Effect> $wrdCoc_{EDU4}$

$wrdCoc_{EDU4}$-$wrdCoc_{EDU5}$ Pair as $CEpair_4$: $wrdCoc_{EDU4}$<Cause>—(CErel)—> <Effect> $wrdCoc_{EDU5}$

$CEpair_1$ (EDU1–EDU2) is the first causal pathway expression and $CEpair_2$–$CEpair_4$ (EDU2–EDU5) are the second causal pathway expression as show in Figure 2.

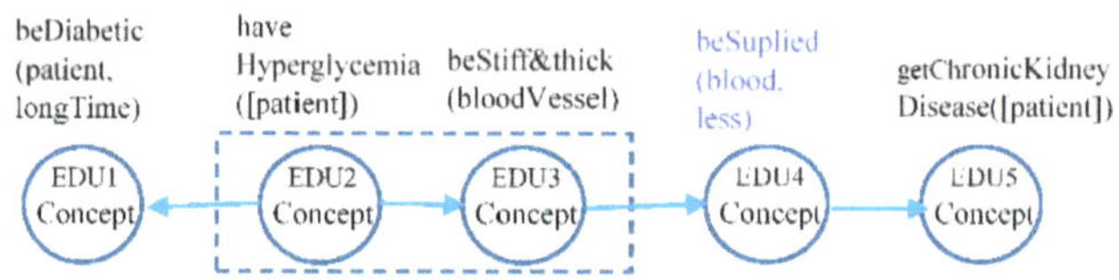

Figure 2. Show causal pathway expressions of Example 3.

Example 4. *TopicName:* โรคเบาหวาน กับโรคไต/***Diabetes and kidney disease*** *(from Sukumvit Hospital) . . .*

*EDU1. "ถ้าระดับน้ำตาลในเลือดสูงเกิดขึ้นเป็นระยะเวลานาน/**If hyperglycaemia occurs for a long-term,**"*
*ถ้า/**If** (ระดับน้ำตาลในเลือดสูง/**Hyperglycaemia**)/NP1*
*((เกิดขึ้น/**occurs**)/Verb$_{strong}$เป็นระยะเวลานาน/**for a long-term**)/VP*
$wrdCoc_{EDU1}$ = occur(hyperglycaemia,long-term)

*EDU2. "ผนังหลอดเลือดจะอักเสบ/**the vascular wall will become inflamed**".*
*(ผนังหลอดเลือด/**The vascular wall**)/NP1*
*(จะ/**will** (อักเสบ/**become inflamed**)/Verb$_{strong}$)/VP*
$wrdCoc_{EDU2}$ = becomeInflamed(bloodVesselWall)

*EDU3. "ทำให้หลอดเลือดแข็งและตีบ/**Causing the arteries to be stiff and narrow**".*
*ทำให้/**causing** (หลอดเลือด/**the arteries**)/NP1*
*((แข็ง/**be stiff**)/Verb$_{strong}$และ/**and**(ตีบ/**be narrow**)/Verb$_{strong}$)/VP*
$wrdCoc_{EDU3}$ = beStiff&Narrow(bloodVessel)

EDU4. "ดังนั้นหลอดเลือดเล็ก ๆ เช่น หลอดเลือดไตมักจะได้รับผลกระทบก่อน/
***Then, small blood vessels such as renal arteries are often affected first**".*
*ดังนั้น/**Then** (หลอดเลือดเล็กๆ/**small blood vessels***
*เช่น หลอดเลือดไต/**such as renal arteries**)/NP1*
*(มักจะ/**often** (ได้รับ/**gets**)Verb$_{weak}$ ผลกระทบก่อน/**affected first**)/VP*
$wrdCoc_{EDU4}$ = getAffect(bloodVessel)

*EDU5. "ทำให้ผู้ป่วยเกิดภาวะไตวาย/**Causing the patient to have kidney failure**". . . .*
*ทำให้/**causing** (ผู้ป่วย/**the patient**)/NP1*
*((เกิด/**have**)/Verb$_{weak}$ภาวะไตวาย/**kidneyfailure**)/VP*
$wrdCoc_{EDU5}$ = haveKidneyFailure(person)

Example 4 shows a sequence of CEpairs having CErel as follow:

wrdCoc$_{\text{EDU1}}$-wrdCoc$_{\text{EDU2}}$ Pair as CEpair$_1$: wrdCoc$_{\text{EDU1}}$<Cause>—(CErel)—> <Effect> wrdCoc$_{\text{EDU2}}$

wrdCoc$_{\text{EDU2}}$-wrdCoc$_{\text{EDU3}}$ Pair as CEpair$_2$: wrdCoc$_{\text{EDU2}}$<Cause>—(CErel)—> <Effect> wrdCoc$_{\text{EDU3}}$

wrdCoc$_{\text{EDU3}}$-wrdCoc$_{\text{EDU5}}$ Pair as CEpair$_3$: wrdCoc$_{\text{EDU3}}$<Cause>—(CErel)—> <Effect> wrdCoc$_{\text{EDU5}}$

According to Example 4, the causal pathway of Figure 3, particularly in a dash-line square, contains EDU2 as an implicit mediator between EDU2 and EDU3 in another dash-line square of the causal pathway in Figure 2. Whilst EDU4 of the causal pathway in Figure 2 is another implicit mediator between EDU3 and EDU5 of the causal pathway in Figure 3. Where the chronic kidney disease and the kidney failure have the same concept of the kidney deterioration. Therefore, we propose using TransCEPair (which is a set of CEpairs having CErel to be transitive (which is equivalent to an implicit mediator) from Transitive Closure of the binary relation over all correct extracted causal pathways) to indicate the implicit mediators on each correct extracted causal pathway and using the following ExplicitCEpairWithCErelPathways template as the dynamic template to collect the extracted causal pathways with the explicit mediators represented by EDUs' wrdCoc features used for representing the implicit mediators (see Section 4.5).

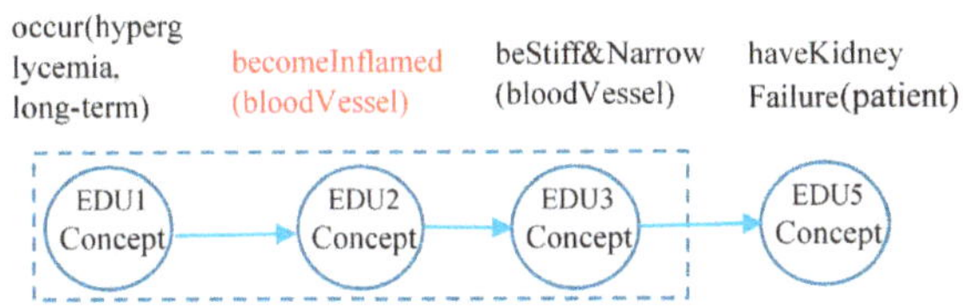

Figure 3. Show a causal pathway expression (EDU1, EDU2, EDU3 and EDU5) of Example 4.

Dynamic ExplicitCEpairWithCErelPathways Template:

wrdCoc$_{\text{EDU}j}$ -wrdCoc$_{\text{EDU}j+1}$ Pair as CEpair$_{p1}$:wrdCoc$_{\text{EDU}j}$<Cause>—(CErel)—><Effect>wrdCoc$_{\text{EDU}j+1}$

wrdCoc$_{\text{EDU}j+1}$-wrdCoc$_{\text{EDU}j+2}$Pair as CEpair$_{p2}$:wrdCoc$_{\text{EDU}j+1}$<Cause>—(CErel)—><Effect>wrdCoc$_{\text{EDU}j+2}$

... ...

wrdCoc$_{\text{EDU}j+n-1}$-wrdCoc$_{\text{EDU}j+n}$ Pair as CEpair$_{pn}$:wrdCoc$_{\text{EDU}j+n-1}$<Cause>—(CErel)—><Effect>wrdCoc$_{\text{EDU}j+n}$

(CEpair$_{p1}$ CEpair$_{p2}$ ... CEpair$_{pn}$)$_p$: **ExplicitCEpairWithCErelPathways**

where p is a causal-pathway number; $p = 1, 2, \ldots m$; m, j, n, are an integer; $EDU_{j+t} <> EDU_{j+t+1}$; $t = 0, 1, 2, \ldots, n-1$.

In addition to TransCEPair, the TransCEPair elements are collected by calculating the transitive closure of the binary relation [21,22] (see Figure 4) linking each node (c_{j+t}) having a Cause/causative-concept (where c_{j+t} is represented by wrdCoc$_{\text{EDU}j+t}$; $j = 1; t = 0, 1, 2, \ldots, n-1$; n is the number of nodes) to an e_{j+t+s} node having Effect/effect-concept (where e_{j+t+s} represented by wrdCoc$_{\text{EDU}j+t+s}$; $s = 2, 3, \ldots n-1$) on the correct extracted causal pathways (Equation (2)).

$$TransCEPair = \cup_{k=1}^{numberCP}\left\{c_1e_3, c_1e_4, \ldots, c_1e_{n\_k}, c_2e_4, \ldots, c_2e_{n\_k}, \ldots, c_{(n\_k)-2}\ e_{n\_k}\right\} \quad (2)$$

where numberCP is the number of correct extracted causal pathways.

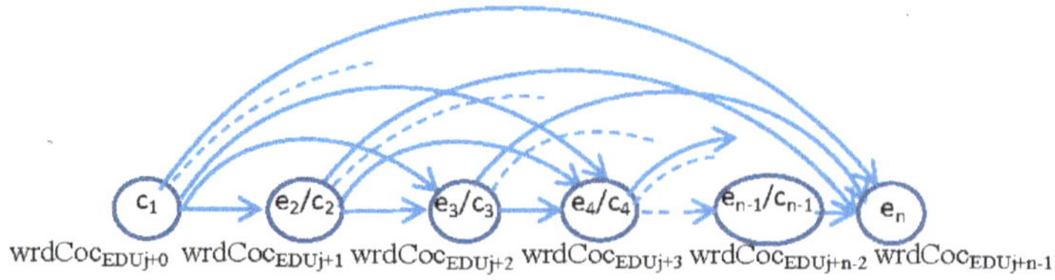

Figure 4. Apply the transitive closure on a correct extracted causal pathway started with $c_{j+t} = c_1$ where $j = 1$; $t = 0, 1, 2, \ldots, n-1$.

4. System Overview

There are five steps in our framework, Corpus Preparation, CErel Learning on Each wrdCoc Pair, Determination of wrdCoc Pairs Having CErel, Causal Pathway Extraction, and Implicit-Mediator Indication and Representation with Explicit-Mediator from Dynamic Template as shown in Figure 5.

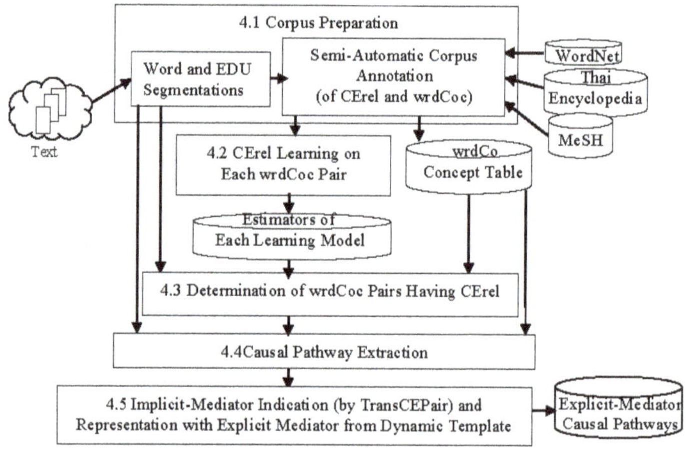

Figure 5. System overview.

4.1. Corpus Preparation

4.1.1. Word and EDU Segmentations

This step is to prepare an EDU corpus from disease-explanation documents downloaded from several hospital web-boards (http://haamor.com/;http://www.bangkokhealth.com; http://www.si.mahidol.ac.th/sidoctor/e-pl/; https://www.bumrungrad.com; etc. accessed on 10 August 2021). The step involves using Thai word-segmentation tools [23] and Named-Entity recognition [24,25]. After the word segmentation is achieved, EDU Segmentation [26,27] is then operated to provide an 8000 EDU corpus (consists of 4000 EDUs from a diabetes and kidney disease group and 4000 EDUs from a heart and artery disease group). This 8000 EDUs' corpus is separated into 2 parts after stemming words and the stop word removal. The first part (which consists of 2000 EDUs from the diabetes and kidney disease group and 2000 EDUs from the heart and artery disease group) is the corpus for semi-automatic annotations of the wrdCo concepts (as the wrdCoc features) and the relation-class of each wrdCoc pair by the experts on the next step of Section 4.1.2 where this annotated corpus is used as a learning corpus in Section 4.2. The second part is a test corpus which consists of 2000 EDUs from the diabetes and kidney disease group and 2000 EDUs from the heart and artery disease group. The test corpus of each disease group is used for (1) determining and collecting wrdCoc pairs as CEpairs having CErel into WCP in Section 4.3 and (2) extracting the causal pathways in Section 4.4.

4.1.2. Semi-Automatic Corpus Annotation

The semi-automatic corpus annotation of each disease group consists of the wrdCoc feature annotation on the wrdCo expressions and the CErel annotation on each wrdCo–expression pairs. We semi-automatically annotate the corpus by using an element of a discourse-marker cue set, {'ทำให้/*causing*', 'เพราะ/*because*', 'เนื่องจาก/*since*'}, to anchor on the corpus documents for obtaining predicate verbs, v_a, ($v_a \in$ z$\cup$V$_{\text{inf}}$; $a = 1, 2, \ldots,$ *numofPredicateVerbs*) from all EDU occurrences right before and right after the anchored causal-cue set elements. Then we obtain a V-pair set = the result of the self-Cartesian product (V × V). We use all V-pair set elements to search two adjacent predicate verbs (v_{a1} v_{a2} where v_{a1},v_{a2} $\in$ V; v_{a1} $\diamond v_{a2}$; $a1 \diamond a2$) of two adjacent EDU occurrences along with automatically annotating the v_a, $w_{1,d}$, and $w_{2,e}$ terms of two adjacent EDUs' wrdCo expressions for the wrdCo concept annotation as the wrdCoc features by the experts selecting the concepts from Lexitron Dictionary after the Thai-to-English translation. Where the concepts from Lexitron Dictionary are referred to Thai Encyclopedia (https://www.saranukromthai.or.th/index2.php), MeSH (https://www.ncbi.nlm.nih.gov/mesh accessed on 10 August 2021), and Wordnet [28] (http://word-net.princeton.edu/obtain accessed on 10 August 2021). Additionally, the relation class (CErel/nonCErel) between two annotated wrdCoc features as a wrdCoc pair (or CEpair) on the annotated corpus is also annotated by the expert as shown in Figure 6 for learning the relation-class in Section 4.2. Both the wrdCo expressions and the wrdCoc features from both disease groups are collected into wrdCo-Concept Table (see Table 1) containing several wrdCo expressions with the same wrdCoc feature where the duplicate entries are eliminated.

4.2. *CErel Learning on Each wrdCoc Pair*

The objective of this step is CErel learning on each wrdCoc pair (which is a wrdCoc$_{\text{EDU}}$ pair as CEpair) with the CErel/nonCErel class from the annotated corpus used as the learning corpus to obtain WCP of each disease group in the next section. Regarding the annotated corpus of each disease group from Section 4.1, each annotated corpus contains several EDUs with the wrdCoc-pair class annotations by the wrdCocPair tag. The wrdCoc features of each disease group, e.g., a CwrdCoc feature and a EwrdCoc feature (where CwrdCoc is a causative wrdCo concept; EwrdCoc is an effect wrdCo concept), are obtained by the wrdCo tag containing 'Concept' and 'type' (Figure 6). All annotated wrdCoc pairs as CwrdCoc, EwrdCoc pairs with CErel/nonCErel by the wrdCocPair tag of each disease group are used for learning CErel by NB, SVM, and LR based on ten-fold cross validation.

(a) NB [18]. The NB learning results of each disease group by this step based on using Weka (http://www.cs.wakato.ac.nz/ml/weak/ accessed on 10 August 2021) are the probabilities of CErel and nonCErel of CwrdCoc features and EwrdCoc features in wrdCoc pairs as shown in Table 2. Where CwrdCoc $\in$ CWC which is a causative-wrdCo-concept set; EwrdCoc $\in$ EWC which is an effect-wrdCo-concept set; and CWC$\cap$EWC$\neq\emptyset$.

(b) SVM [19]. The SVM learning is a linear binary classification applied to classify the CErel and nonCErel of each wrdCoc pairs from the annotated corpus by using Weka. This linear function, $f(\mathbf{x})$, of the input $\mathbf{x} = (x_1, x_2, \ldots, x_n)$ assigned to the Cerel class if $f(\text{x}) > 0$, and otherwise to the nonCErel class, is as Equation (3).

$$\begin{aligned} f(x) &= \langle w \cdot x \rangle + b \\ &= \textstyle\sum_{j=1}^{n} w_j x_j + b \end{aligned} \quad (3)$$

where x is a dichotomous vector number, w is weight vector, b is bias, and (w, b) $\in \text{R}^n \times \text{R}$ are the parameters that control the function. The SVM learning is to determine w_j and b for each wrdCoc feature (x_j) which is either a CwrdCoc feature or a EwrdCoc feature in each wrdCoc pair with CErel or nonCErel from the annotated corpus of each disease-group.

<Topic_name Entity-concept=Diabetes/disease>โรคเบาหวาน</Topic_name>**…………**

"..ผู้ป่วยเป็นโรคเบาหวาน" EDU1 เนื่องจากร่างกายไม่สามารถนำน้ำตาลในร่างกายไปใช้ได้อย่างเต็มที่" EDU2 เพราะ [ร่างกาย] ขาดฮอร์โมนอินซูลิน EDU3 หากผู้ป่วย มีระดับน้ำตาลในเลือดสูงเป็นเวลานาน EDU4 ……."

"..A patient gets a diabetes disease EDU1 *since the body cannot fully use sugar inside the body* EDU2 *because [the body] lacks insulin* EDU3 *If the patient to have a high blood sugar level (hyperglycaemia) for long* EDU4 *………."*

ผู้ป่วย<**เป็นโรคเบาหวาน**>/**Vinf** EDU1 เนื่องจากร่างกาย<**ไม่สามารถนำน้ำตาล**>/**Vinf** ในร่างกายไปใช้ได้อย่างเต็มที่ EDU2

A patient< ***gets a diabetes disease***>/**Vinf** EDU1 *since the body* <***cannot fully use sugar***>/**Vinf** *inside the body* EDU2

<wrdCocPair#1 Class= **CErel**>

<EDU1><wrdCo: type=**Effect**; **Concept= 'getDiabetes(person)'**>

<NP1-HeadNounWord: w1 concept='person'>ผู้ป่วย/*patient* </NP1>

<VP: v, w2 ><v: Type='Vinf' concept='getDiabetes' >

<v-inf: Type='Vweak' >เป็น/*be* </v-inf>

<w-inf: Type='Noun'>โรคเบาหวาน/*diabetes* </w-inf></v>

<w2: Type=null >''</w2></VP></wrdCo></EDU1>

<EDU2><wrdCo: type=**Cause** ; **Concept='notTakeSugar(body)'**>

<NP1-HeadNounWord: w1 concept='body'>ร่างกาย/*body*</NP1>

<VP: v, w2 ><v: Type='Vinf' concept='notTakeSugar'>

<v-inf: Type='Vweak' >ไม่สามารถนำ /*notTake* </v-inf>

<w-inf: Type='Noun'>น้ำตาล/*sugar* </w-inf></v>

<w2: Type=null >''</w2></VP></wrdCo></EDU2></wrdCocPair #1>

เนื่องจากร่างกาย<**ไม่สามารถนำน้ำตาล**>/**Vinf** ในร่างกายไปใช้ได้อย่างเต็มที่ EDU2 เพราะ [ร่างกาย] <ขาด>/**Vstrong** ฮอร์โมนอินซูลิน EDU3

since the body <***cannot fully use sugar***>/**Vinf** *inside the body* EDU2 *because [the body]* <*lacks*>/**Vstrong** *insulin* EDU3

<wrdCocPair #2 Class= **CErel**>

<EDU2><wrdCo: type=**Effect**; **Concept='notTakeSugar(body)'**>

<NP1-HeadNounWord: w1 concept='body'>ร่างกาย/*body*</NP1>

<VP: v, w2 ><v: Type='Vinf' concept='notTakeSugar'>

<v-inf: Type='Vweak' >ไม่สามารถนำ /*notTake* </v-inf>

<w-inf: Type='Noun'>น้ำตาล/*sugar* </w-inf></v>

<w2: Type=null >''</w2></VP></wrdCo></EDU2>

<EDU3><wrdCo: type=**Cause** ; **Concept='lack([body],insulin)'**>

<NP1-HeadNounWord: w1 concept='body'>ϕ</NP1>

<VP: v, w2 ><v: Type='Vstrong' concept='lack'> ขาด/*lack*</v>

<w2: Type='Noun' concept='insulin'>ฮอร์โมน อินซูลิน/*insulin*</w2></VP></wrdCo></EDU3></wrdCocPair#2>

เพราะ [ร่างกาย] <ขาด>/**Vstrong** ฮอร์โมนอินซูลิน EDU3 หาก ผู้ป่วย <**มีระดับน้ำตาลในเลือดสูง**>/**Vinf** เป็นเวลานาน EDU4

because [the body]<***lacks***>/**Vstrong** *insulin* EDU3 *If the patient* <***to have a high blood sugar level***>/**Vinf** *for long* EDU4

<wrdCocPair #3 Class= non**CErel**>

<EDU3><wrdCo: type=**Cause/Effect** ; **Concept='lack([body],insulin)'**>

<NP1-HeadNounWord: w1 concept='body'>ϕ</NP1>

<VP: v, w2 ><v: Type='Vstrong' concept='lack'> ขาด/*lack*</v>

<w2: Type='Noun' concept='insulin'>ฮอร์โมนอินซูลิน/*insulin*</w2></VP></wrdCo></EDU3>

<EDU4><wrdCo: type= **Cause/Effect**; **Concept= 'haveHyperglycaemia([person])'**>

< NP1-HeadNounWord: w1 concept='person'>ผู้ป่วย/*patient* </NP1>

< VP: v, w2 ><v: Type='Vinf' concept='haveHyperglycaemia'>

<v-inf: Type='Vweak'>มี/*have*</v-inf>

<w-inf: Type='Noun' >ระดับน้ำตาลในเลือดสูง/*Hyperglycaemia*</w-inf></v>

<w2: Type=null>''</w2></VP></wrdCo></EDU4></wrdCocPair #3>

……………………

Figure 6. Annotation of wrdCo concepts or wrdCoc features including CErel/nonCErel Class between wrdCoc pair where v, w1, and w2 symbols in a wrdCo tag is v_a, $w_{1,d}$, and $w_{2,e}$ terms/elements respectively of wrdCoPattern.

Table 1. wrdCo-Concept Table from the annotated corpus.

wrdCo Expression Based on wrdCoPattern			wrdCoc Feature
V	W1	W2	
สะสม/deposit <deposit>	*ไขมัน/fat* <fat>	*ผนังหลอดเลือด/blood vessel wall* <wall>	**beDeposited(fat,bloodVessel)**
เกาะ/deposit <deposit>	*ไขมัน/fat* <fat>	*เส้นเลือดแดง/artery* <bloodVessel>	**beDeposited(fat,bloodVessel)**
จับ/deposit <deposit>	*การมีไขมัน/having fat* <fat>	*ผนังหลอดเลือด/blood vessel wall* <wall>	**beDeposited(fat,bloodVessel)**
มีไขมัน/have fat <haveFat>	*หลอดเลือดแดง/artery* <bloodVessel>	*สะสม/deposit* <deposit>	**beDeposited(fat,bloodVessel)**
เกาะ/deposit <deposit>	*ไขมัน/fat* <fat>	*ตะกรัน/plaque* <bePlaque>	**beDeposited(fat,bePlaque)**
คือตะกรันไขมัน/is fatty-plaque <bePlaque>	*สิ่งอุดตันในหลอดเลือด/embolism* <embolism>	null	**bePlaque(embolism)**
ก่อตัว/form <form>	*ตะกรันไขมัน/ fatty-plaque* <plaque>	*หนา/thick* <thick>	**Form(plaque,thick)**
ตีบแคบ/be narrow <beNarrow>	*หลอดเลือด/blood vessel* <bloodVessel>	null	**beNarrow(bloodVessel)**
ตีบแคบ/be narrow <beNarrow>	*หลอดเลือด/blood vessel* <bloodVessel>	*ลง/more* <more>	**beNarrow(bloodVessel)**
หล่อเลี้ยง/supply <supply>	*เลือด/blood* <blood>	*กล้ามเนื้อหัวใจ/ myocardium* < myocardium>	**beSuppliedInsufficiently (blood, myocardium)**
ขาด/lack <lack	*กล้ามเนื้อหัวใจ/ myocardium* < myocardium>	*เลือด/blood* <blood>	**beSuppliedInsufficiently (blood, myocardium)**
ไปเลี้ยง/supply <supply>	*เลือด/blood* <blood>	*เนื้อเยื่อ/tissue* <tissue>	**beSuppliedInsufficiently (blood, tissue)**
ขาด/lack <lack	*สมอง/brain* <brain>	*เลือด/blood* <blood>	**beSuppliedInsufficiently (blood, brain)**
เกิด/occur <occur>	*การสร้างสารเคมี/chemical forming* < chemicalForming>	null	**Occur(chemicalForming)**
เกิด/occur <occur>	*หลอดเลือด/blood vessel* <bloodVessel>	*การอักเสบ/inflammation* < inflammation>	**Occur (bloodVessel,inflammation)**
เกิดขึ้น/occur <occur>	*การอักเสบ/inflammation* < inflammation>	*หลอดเลือด/blood vessel* <bloodVessel>	**Occur (bloodVessel,inflammation)**
แข็งตัว/be stiff <beStiff>	*หลอดเลือด/blood vessel* <bloodVessel>	null	**beStiff(bloodVessel)**
อุดตัน/get clogged <beClogged>	*หลอดเลือด/blood vessel* <bloodVessel>	null	**beClogged(bloodVessel)**
มีระดับน้ำตาลในเลือดสูง/have hyperglycaemia <haveHyperglycaemia>	*ผู้ป่วย/patient* <person>	null	**haveHyperglycaemia(person)**
แตก/break <break>	*หลอดเลือด/blood vessel* <bloodVessel>	null	**beBroken(bloodVessel)**
ถูกทำลาย/damage <beDamaged>	*เนื้อเยื่อ/tissue* <tissue>	null	**beDamaged(tissue)**
… … … … … … … ..	… … … … … … ..	… … … … … …	… … … … … … … ..

Table 2. Show the CErel and nonCErel probabilities of CwrdCoc features and EwrdCoc features in wrdCoc pairs of the diabetes and kidney disease group and the heart and artery disease group.

Disease Group	CwrdCoc	CErel	Noncerel	EwrdCoc	CErel	Noncerel
Diabetes & Kidney Disease Group	<beLost(protein,urine)>	0.0698	0.0784	<beLost(protein,urine)>	0.0074	0.0066
	<beFailure(kidney)>	0.1581	0.0784	<beFailure(kidney)>	0.0355	0.0333
	<haveHyperglycaemia (person)>	0.0452	0.0294	<haveHyperglycaemia (person)>	0.0411	0.0200
	<notTakeSugar(body)>	0.0266	0.0392	<notTakeSugar(body)>	0.0112	0.0133
	<beNarrow(bloodVessel)>	0.0185	0.0098	<beNarrow(bloodVessel)>	0.0374	0.0533
	... ...	...	...	... ...	...	...
Heart & Artery Disease Group	<beDeposited(fat, bloodVessel)>	0.0540	0.0288	<beDeposited(fat, bloodVessel)>	0.0011	0.0193
	<beThick(bloodVessel)>	0.0149	0.0288	<beThick(bloodVessel)>	0.0169	0.0038
	<beDamages(bloodVesselWall)>	0.0218	0.0041	<beDamages(bloodVesselWall)>	0.0101	0.0038
	<becomeInflamed(bloodVessel)>	0.1323	0.1028	<becomeInflamed(bloodVessel)>	0.0079	0.0115
	<beClogged(bloodVessel,organ)>	0.0448	0.0288	<beClogged(bloodVessel,organ)>	0.0418	0.0424
	... ...	...	...	... ...	...	...

(c) LR [20]. The logistic regression model of the research is based on the linear logistic regression with binary vector data. The distinguishing feature of the logistic regression model is that the variable is binary or dichotomous. Usually, the input data with any value from negative to positive infinity would be used to establish which attributions are influential in predicting the given outcome with values between 0 and 1, and hence is interpretable as a probability. The logistic function can be written as:

$$F(x) = \frac{1}{1 + e^{-(\beta_0 + \beta_1 x_1 + \beta_2 x_2)}} \quad (4)$$

$F(x)$ is interpreted as the probability of the given outcome to be predicted where x_1 and x_2 are attribute variables; β_0 is bias; and β_1, and β_2 are the model estimators which play the role of momentum for each attribute. The LR learning is to determine β_0, β_1, and β_2 for each CwrdCoc feature and each EwrdCoc feature as x_1 and x_2 features respectively in each wrdCoc pair (CwrdCoc, EwrdCoc) with either the positive/CErel class or the negative/nonCErel class formed by supervised learning on the learning corpus of each disease-group.

The learning results by NB, SVM, and LR models are the estimators which are used for determining wrdCoc pairs having CErel from the test corpus of each disease group in the next step of Section 4.3. Moreover, all precisions of learning by NB, SVM, and LR from the learning corpus of each disease group are greater than 0.8.

4.3. Determination of wrdCoc Pairs Having CErel

The WC elements are determined from all wrdCo expressions on the test corpus of each disease group by the wrdCo-expression matching between the wrdCo expressions on this test corpus and the wrdCo expressions on wrdCo-Concept Table (Table 1) to obtain the wrdCoc features or the WC elements. The result of the self-Cartesian product (WC × WC) is a wrdCo-concept ordered pair set which is used for determining and collecting the wrdCoc pairs having CErel into WCP of each disease group by the following NB, SVM, and LR.

(a) NB. Regarding Equation (5) and the CErel and nonCErel probabilities of CwrdCoc and EwrdCoc (Table 2), the CwrdCoc EwrdCoc pairs as the wrdCoc pairs having Cause-EffectRelationClass as CErel is determined from the self-Cartesian product (WC × WC) result and then collected into WCP of each disease group on which CwrdCoc and EwrdCoc are independent.

$$\begin{aligned}\text{Cause} - \text{EffectRelationClass} &= \underset{class \in Class}{\text{argmax}}\ P(class|CwrdCoc, EwrdCoc) \\ &= \underset{class \in Class}{\text{argmax}}\ P(CwrdCoc|class)P(EwrdCoc|class)P(class)\end{aligned} \tag{5}$$

where $CwrdCoc$ is a causative − wrdCo concept;

$EwrdCoc$ is an effect − wrdCo concept;

Class = {"CErel", "nonCErel"}.

(b) SVM. The bias, b, and the weight vector, w, of the CWC elements and the EWC elements in the wrdCoc pairs from the SVM learning (Section 4.2 (b)) are used to determine and collect the CwrdCoc EwrdCoc pairs as the wrdCoc pairs having the CErel class into WCP of each disease group from the self-Cartesian product (WC × WC) result with Equation (3).

(c) LR the research applies Equation (4) along with Equation (6) to determine the CErel class between the CWC elements and the EWC elements in the wrdCoc pairs from both the positive/CErel class determination and the negative/nonCErel class determination by using the estimators from the LR learning (Section 4.2 (c)).

$$\text{CErelClass} = \text{Max}(F(x)_{\text{CErelClass}}, F(x)_{\text{nonCErelClass}}) \tag{6}$$

According to (6), x_1 and x_2 as CwrdCoc and x_2 as EwrdCoc are the attribute variable pair of each wrdCoc pair from the test corpus of each disease group where β_0, β_1, and β_2 of CwrdCoc and EwrdCoc are obtained by the supervised learning with LR on the learning corpus of each disease group. The wrdCoc pair (or the CwrdCoc and EwrdCoc pair) with the CErel class is determined and collected into WCP of each disease group from the self-Cartesian product (WC × WC) result.

4.4. Causal Pathway Extraction

All wrdCoc features per the test-corpus document of each disease group are sequentially collected in an array of wrdCoc features (wcc [ ]) after the wrdCo-expression matching between the wrdCo expressions of this test-corpus document and the wrdCo expressions on wrdCo-Concept Table (Table 1) to obtain the wrdCoc features. The causal pathways are then extracted by the wrdCoc-pair matching between wcp_k ($wcp_k \in$ WCP; k = 1, 2, . . . , *numberOfWCPelements*) and each wrdCoc pair in wcc [ ] as a $wcc_{[ct]}$ $wcc_{[ct+1]}$ pair (ct = 1, 2, . . . , *numberOFwrdCocFromTestCorpusDocument*) by sliding a window size of two consecutive wrdCoc features ($wcc_{[ct]}$ $wcc_{[ct+1]}$) with one wrdCoc distance ($wcc_{[ct++]}$) on wcc [ ]. We stop sliding the window if there is no match on wrdCoc-pair matching. We then obtain a causal pathway as shown in Algorithm 1 where CEpair$_i$ is a wrdCoc pair ($wcc_{[ct]}$ $wcc_{[ct+1]}$) in wcc [ ]; and allPathways (which is an array of arrayList with an 'a' variable of an array size) contains several causal pathways.

4.5. Implicit-Mediator Indication and Representation with Explicit-Mediators

The correct extracted causal pathways of the allPathways result for each disease group by the CausalPathwayExtraction algorithm on Section 4.4 consists of the explicit mediator causal pathways and the implicit mediator causal pathways. The allPathways result of each disease group also contains some duplicate causal pathways. Therefore, allPathways is sorted and then is eliminated the duplicate causal pathways to become PathWays (which is an array of arrayList with an updated 'a' variable) before indicating the implicit mediators on the correct extracted causal pathways. With regard to PathWays of each disease group, the causal pathways containing the explicit mediators represented by EDUs' wrdCoc features are collected into the dynamic template as the ExplicitCEpairWithCErelPathways template (see Section 3.3) which is an ExplicitPath variable in an ExplicitCausalPathwayRepresentation algorithm (Algorithm 2) whilst the causal pathways containing the implicit mediators are collected into an ImplicitPath variable as a temporary template.

Algorithm 1 Causal Pathway Extraction

```
CAUSAL_PATHWAY_EXTRACTION
/* (Extraction of several CEpairi sequences as causal pathways.)
/* Assume that each EDU is represented by (NP1 VP).
/* L is a list of EDUs from one test-corpus document after stemming words and the stop word removal.
/* CEpairi is a wrdCoc pair with index i of the causal pathway.
/* wcc[ ] is an array of wrdCoc and is collected from this test corpus.
/* WCP is a set of wrdCoc pairs having CErel.
ct = 1;j = 1;ct = 1; a = 0; string wcc[];
ArrayList<string> []allPathways = new ArrayList[a];
/* array of arrayList.
while j≤ Length[L] do
{1 wcexp_j = getWrdCo(EDU_j);
/* Get wrdCo expression of EDUj from the test corpus.
    If wcexp_j.v∈Vstrong ∪ Vinf then
/* wcexp_j.v is a predicate verb v_a on a wrdCo expression
with index j.
    { wcc[ct] = getWrdCoConcept; ct++};
/* getWrdCoConcept by the wrdCo-expression matching between wrdCo
expressions of the test-corpus document and the wrdCo expressions
with the wrdCoc features on wrdCo-Concept Table (Table 1).
  j++ }1;
count = ct; ct = 1; i = 1; flagce = 0; flagec = 0; fl = 0;
while ct ≤ count-1 do
{1 while (wcc[ct] + wcc[ct+1]∈ WCP) ·∧· (ct ≤ count-1) do
/* a causal pathway extraction by wrdCoc-pair matching
wcc[ct]+wcc[ct+1]) among wcp_k (where wcp_k∈WCP).
    {2 CEpair_i = wcc[ct]+ wcc[ct+1];
/* CErel occurs on Text as EDUCauseEDUEffect.
    If flagce = 0 then {a++; flagce = 1};
    allPathways[a].AddNewCause¬EffectPair(CEpair_i); i++;ct++ }2;
    If flagce = 1 then { flagce = 0; fl = 1; i = 1};
    while (wcc[ct+1] + wcc[ct] ∈ wcpk) ∧ (ct ≤ count-1) do
/* another causal pathway extraction by wrdCoc-pair
matching (wcc[ct+1]+wcc[ct])among wcp_k (where wcp_k∈WCP).
    {3 CEpair_i = wcc[ct+1] + wcc[ct];
/* CErel occurs on Text as EDU_Effect EDU_Cause.
/* wcc[ct+1] is Cause and wcc[ct] is Effect.
        If flagec = 0 then {a++; flagec = 1};
        allPathways[a].AddNewCause¬EffectConceptPair(CEpair_i);
        i++; ct++}3;
    If flagec = 1 then { flagec = 0; fl = 1; I = 1};
    If fl = 0 then ct++;
      else fl = 0; }1;
}Return allPathways /* Return causal pathways.
```

Algorithm 2 Explicit Causal Pathway Representation

```
EXPLICIT_CAUSAL_PATHWAY_REPRESENTATION (ArrayList<string> []Pathways; a)
/*  Assume Pathways is allPathways (by Algorithm 1) with eliminating the duplicate causal pathways.
/*  trsvSet is TransCEPair which is a set of CEpairs with CErel to be transitive.
/*  ExplicitPath is a dynamic ExplicitCEpairWithCErelPathways template.
trsv = 0; check1 = 0; trsvSet←∅;
ArrayList<string> []ExplicitPath = new ArrayList[a];
ArrayList<string> []ImplicitPath = new ArrayList[a];
ArrayList<String> fill = new ArrayList<>();
num1 = a
/* where a is the number of Pathways elements(the Pathways array size).
For (α = 1 to num1; α++) /*determine trsvSet from Transitive Closure
{trsvSet←trsvSet ∪ Pathways[α].transitiveClosureDetermination };
For (α = 1 to num1; α++)
{1 i = 1; /*collect Explicit-Mediator Causal Pathways to ExplicitPath.
      while (i ≤ Pathways[α].numberOfCauseEffectConceptPairs) ∧
(Pathways[α].Get(CEpair_i)∉TrsvSet) do
/*add explicitCEpair_i to ExplicitPath.
        { ExplicitPath[α].Add(Pathways[α].Get(CEpair_i)); i++ }
      while (i ≤ Pathways[α].numberOfCauseEffectConceptPairs) do
    {If (Pathways[α].Get(CEpair_i)∈TrsvSet) then
/* Identify and mark CEpairi having implicit Mediator as
causalTransitivity with '*'; and then
add '*'CEpair_i&all subsequent CEpair_i,...to ImplicitPath.
              ImplicitPath[α].Add('*'+Pathways[α].Get(CEpairi))
else ImplicitPath[α].Add(Pathways[α].Get(CEpairi)); i++ } }1
/* replace'*'CEpairi on ImplicitPath with explicit mediator from
ExplicitPath.
  while check1 = 0 do
  {1For (α =1 to num1; α++)
    {2If ImplicitPath[α]isNotEmpty then
      {3idi = 1; id = 1; check2 = 0; fill.clear()
        while idi ≤ImplicitPath[α].numberOfCEpairs do
/* find'*'CEpairidion ImplicitPath.
        {4 If ImplicitPath[α].Get(CEpairidi).contain('*') = true then
          {5 ImplicitPath[α].Get(CEpairidi).replace('*',"");
            C = CEpair_idi.wcc_idi; E = CEpair_idi.wcc_idi+1;
/* get C/cause & E/effect.
b = 1; idd = 1; f1 = 0; f2 = 0; f3 = 0;
            while b ≤ num1 ∧ f1 = 0 do
/*find an explicit mediator: C+ mediator(s)(CEpairidi...)+E
from ExplicitPath.
  {6 while idd ≤ExplicitPath[b].numberOfCauseEffectConceptPairs
∧f3=0 do
    {7 If C = ExplicitPath[b].Get(CEpairidd).wccidd then f2 = 1
      elseIf(E=ExplicitPath[b].Get(CEpair_idd).wcc_idd+1)∧f2 = 1 then
f3=1;
      If f2 = 1∨f3 = 1 then {fill[id]=ExplicitPath[b].Get(CEpair_idd);
id++};
        idd++ }7; /* fill contains C+ mediator(s)(CEpair_idi...)+E.
    If f3 = 0 then {idd = 1; f2 = 0; fill.clear()} /* no E in fill.
      else {f1 = 1; ExplicitPath[α].addAll(fill); check2 = 1};
/* add fill to ExplicitPath.
Id = 1; b++ }6 }5
      elseIf5 ImplicitPath[α].Get(CEpairidi).contain('*') = false
∧ check2 = 1 then
{ExplicitPath[α].add(ImplicitPath[α].Get(CEpair_idi)}5
idi++ }4
```

30: If check2 = 1 then {ImplicitPath$_{[\alpha]}$.clear(); check2 = 0 } }3 }2
31: For (α = 1 to num1; α++) /* **check ImplicitPath being empty.**
32: {If ImplicitPath$_{[\alpha]}$ isNotEmpty then check2 = 1 };
33: If check2 = 0 then check1 = 1;
34: }$_1$; ExplicitPath.sortRowOfArrayOfArrayList;
ExplicitPath.removeDuplicateRow;
35: }Return **ExplicitPath**

The implicit mediator indication and representation on PathWays with the explicit mediators from ExplicitPath/the ExplicitCEpairWithCErelPathways template is based on the following steps of the ExplicitCausalPathwayRepresentation algorithm (Algorithm 2):

(1) Determine TransCEPair from all correct extracted causal pathways (Pathways$_{[\alpha]}$; $\alpha = 1, 2, \ldots, a$).
(2) Use TransCEPair to indicate CEpair$_i$ having the implicit mediator as the causal transitivity on Pathways$_{[\alpha]}$; if Pathways$_{[\alpha]}$ contains CEpair$_i$ $\notin$ TransCEPair (where $i = 1, 2, \ldots, numOfCauseEffectConceptPairs$), Pathways$_{[\alpha]}$ is the explicit- mediator causal pathway; and is added to ExplicitPath$_{[\alpha]}$.
(3) If Pathways$_{[\alpha]}$.CEpair$_i$ $\in$TransCEPair, we mark '*'on CEpair$_i$ having causal transitivity or the implicit mediator, add *CEpair$_i$ and subsequent CEpair$_{i+1}$, CEpair$_{i+2}$, ..., CEpair$_{numOfCauseEffectConceptPairs}$ of Pathways$_{[\alpha]}$ to ImplicitjPath$_{[\alpha]}$, and add CEpair$_1$, CEpair$_2$ ... CEpair$_{i-1}$ of Pathways$_{[\alpha]}$ to ExplicitPath$_{[\alpha]}$.
(4) Replace *CEpair$_i$ with the explicit mediator(s) of CEpair$_{idd}$ from ExplicitPath as shown in Algorithm 2.
(5) Check the ExplicitPath result from the ExplicitCausalPathwayRepresentation algorithm does not contain the implicit-mediator(s)/the causal transitivity by comparing trsvSet (line no. 5 of Algorithm 2) to TransCEPair determined from the ExplicitPath result on line no. 35 of Algorithm 2. If the TransCEPair determination from ExplicitPath is the same as trsvSet, ExplicitPath contains the explicit- mediator causal pathways without the causal transitivity or the implicit mediator, otherwise the ExplicitCausalPathwayRepresentation algorithm is re-executed after copying ExplicitPath to Pathways of Algorithm 2 and then setting ExplicitPath to empty.

Therefore, the representation of the correct extracted causal pathways with the explicit mediators by Algorithm 2 is shown in Figure 7.

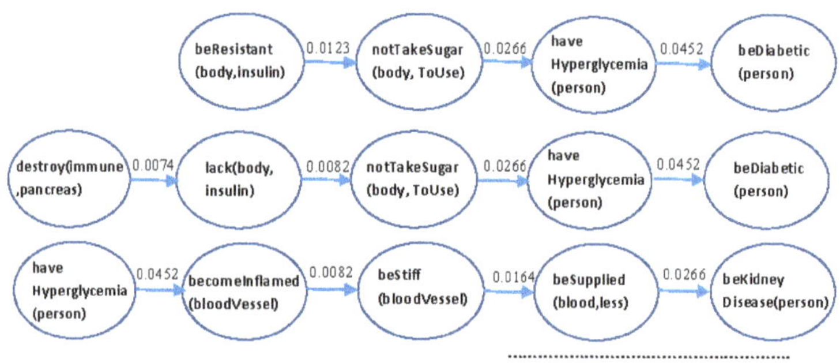

Figure 7. The representation of the correct extracted causal pathways with the explicit mediators of the diabetes and kidney disease group where the numeric label on each arrow represents the CErel probability of each CEpair by NB.

5. Evaluation and Discussion

The test corpus of 4000 EDUs employed to evaluate the proposed methodology for the causal pathway extraction through determining wrdCoc pairs having CErel is collected from the downloaded disease documents on Thai hospital web-boards. The test corpus consists of 2000 EDUs from the diabetes and kidney disease group documents and the 2000 EDUs from the heart and artery disease group documents. There are three evaluations, (1) the determination of wrdCoc pairs having CErel, (2) the causal pathway extraction, and (3) the implicit-mediator indication and representation with the explicit mediators from the dynamic template.

5.1. Determination of wrdCoc Pairs Having CErel

The evaluation results of extracting the EDU-concept pairs/wrdCoc pairs having CErel from the documents of the diabetes and kidney disease group and the heart and artery disease group are the precisions and the recalls based on three experts with max win voting as shown in Table 3 including the number of different wrdCoc features which results in the frequencies of wrdCoc features as shown in Figure 8.

Table 3. The accuracy of determining wrdCoc pairs having CErel.

Disease Group (2000 EDUs/Group)	#of Different wrdCoc Features		Extraction of wrdCoc Pairs Having CErel					
			NB		SVM		LR	
	Cause	Effect	Precision	Recall	Precision	Recall	Precision	Recall
Diabetes and kidney disease group	40	92	0.844	0.777	0.893	0.803	0.877	0.795
Heart and artery disease group	93	110	0.826	0.746	0.841	0.760	0.831	0.754

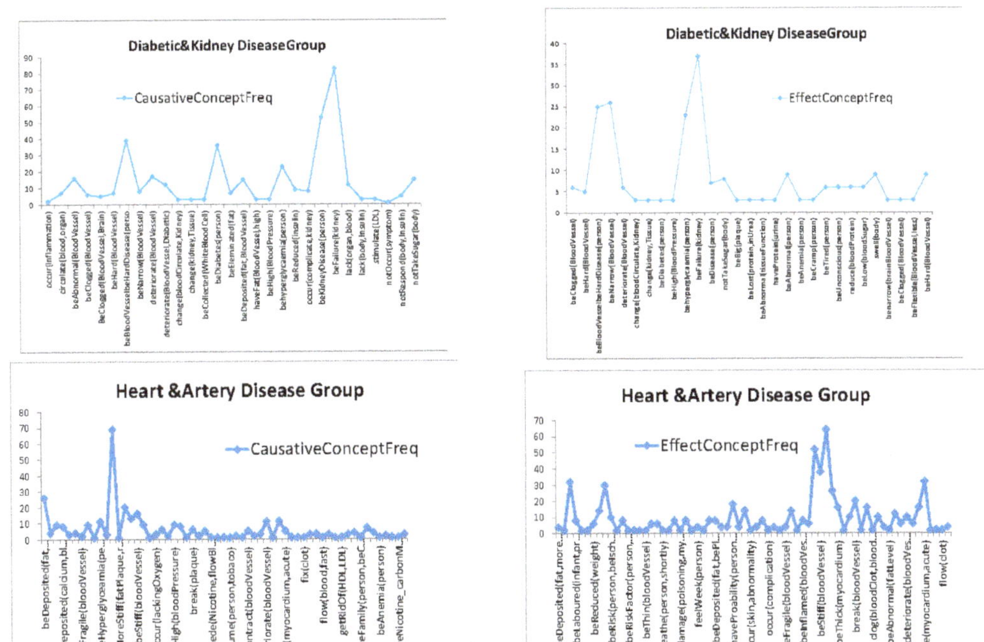

Figure 8. Show the wrdCoc frequencies with the causative concepts and the effect concepts from the diabetes and kidney disease group and the heart and artery disease group.

From Table 3, the average precisions of extracting wrdCoc pairs having CErel from the documents of the diabetes and kidney disease group and the heart and artery disease group are 0.871 and 0.833 respectively, with the average recalls of 0.791 and 0.753 and the average F-score of 0.830 and 0.791 respectively. Whereas the causality or cause-effect relation extraction from the previous research [7] based on the probabilities of words on NP pair and the cue phrase probability from the complex sentence or two simple sentences from the medical domain has an F-score of 0.774. With regard to our research results on Table 3, the reason for the diabetes and kidney disease group having the higher precision and recall of extracting wrdCoc pairs having CErel than the heart and artery disease group is that the heart and artery disease group have more diversity of the wrdCoc features in both the causative concepts and the effect concepts than the diabetes and kidney disease group. The high diversity of wrdCoc features results in low frequencies of most wrdCoc features. In addition, there are some dependency occurrences among wrdCoc features, e.g.,

EDU1: 'haveHyperglycaemia(person)' → EDU2:'becomeInflamed(BloodVessel)', where EDU1′s wrdCoc and EDU2′s wrdCoc mostly occur as a cause-effect relation or a dependency occurrence on documents but some documents contain EDU1′s wrdCoc followed by EDU2′s wrdCoc without the cause-effect relation. Thus, the wrdCo diversity and the wrdCoc dependency result in SVM having highest precision in both the diabetes and kidney disease group and the heart and artery disease group. However, both the diabetes and kidney disease group and the heart and artery disease group have low recalls because the diversity of wrdCo expressions occurs on the downloaded documents of both disease groups.

5.2. *Causal Pathway Extraction*

The causal pathway extraction from the test corpus is evaluated by the precision and recall based on three experts with max wins voting as shown in Table 4.

Table 4. The accuracy of extracting causal pathways.

Disease Group (2000 EDUs/Group)	Causal Pathway Extraction	
	Precision	**Recall**
Diabetes and Kidney Disease Group	0.840	0.724
Heart and artery Disease Group	0.828	0.706

The causal pathways determination from the documents of two disease group as shown in Table 4 have an average precision of 0.834 with the average recall of 0.715. The reason for having low recall of extracting causal pathways from the documents is that some causal pathways start with EDUs containing the causative/effect concept expressed by either NP1 or NP2 as shown in the following EDU1 of Example 5 instead of the predicate verb or Verb on the general linguistic expression in Figure 1.

Example 5 .

EDU1. *"โรคไตเสื่อมเกิดจากการเป็นเบาหวานมาเป็น/*
***Kidney disease** is caused by **being diabetic for a long time**".*
*(โรคไตเสื่อม/**Kidney disease**)/**NP1***
*(เกิดจาก/is caused by (การเป็นเบาหวานมาเป็นเวลานาน /**being diabetic for a long time**)/**NP2**)/VP.*

EDU2. *"ทำให้ผนังหลอดเลือดถูกทำลาย/Causing artery wall to **be destroyed**".*
(ทำให้/Causing)/conj(ผนังหลอดเลือด/artery wall)/NP2
*((ถูกทำลาย/**is destroyed**)/$Verb_{strong}$)/**VP***

EDU3. *"แล้วการทำหน้าที่กรองของไตจะเสื่อม/Then the filtration function of the kidneys **will deteriorate**".*
แล้ว/Then(การทำหน้าที่กรองของไต/the filtration function of the kidneys)/NP1
*((จะเสื่อม/**will deteriorate**)/$Verb_{strong}$)/**VP**.*

EDU4. *"ทำให้โปรตีนรั่วออกมาในปัสสาวะ/Causing protein **to leak out** in the urine".*

(ทำให้/Causing)/conj (โปรตีน/protein)/NP1
*((รั่วออกมา/**leaks out**)/Verb$_{strong}$ใน/**in**(ปัสสาวะ/the urine)/NP2)/**VP**.*

However, the evaluation of the previous work [10] on extracting and constructing the causal chain/pathways from a large corpus on an on-line social media (tweets, news articles, and blogs) relied on time series through prediction of noun phrases as the next effect is 57% accuracy based on expert judgments whereas our causal pathways relied on the actual events/states with the causative/effect concepts.

5.3. Implicit-Mediator Indication and Representation with Explicit-Mediators

We evaluate the implicit mediator indication and representation with the explicit mediators (from the dynamic template) in term of a Likert scale (1 to 5) for concise and comprehensible representations of the correct extracted causal pathways. The evaluation results with the average scores (based on the Likert scale) of the concise and comprehensible representations of Doc (which is the causal pathway representation by explanation on the documents) and Graph (which is the causal pathway representation by the correct extracted causal pathway with the explicit mediators from the documents) by the 30 end-users (who are non-professional persons) are presented on Table 5 and Figure 9 of both disease groups.

From Table 5 and Figure 9, Graph Representations of both disease groups have higher concise and higher comprehensible representations than Doc Representations of both disease groups. Moreover, from Table 5, the average scores of the concise representation and the comprehensible representation by Graph Representations from both disease groups are 4.4 and 4.25 respectively whereas the evaluation of the implicit knowledge completion on the event chain of actions from the Japanese web-blog corpus of the previous work [13] based on the similarity scores of event pairs on the chain with the higher than thresholds is 3.0 (based on the Likert scale 1–5) without the CErel consideration between event-pairs on the chain.

According to the evaluation of the comprehensible representation of our research on both disease group, there are a few causal pathways requiring more explicit mediators for more clear representation as shown in Example 6.

Example 6 . *A causal pathway representation of arteriosclerosis from the downloaded 'Heart Disease and Blood Vessel' documents on the hospital web-board:*

bedeposited(cholesterol,BloodVessel)→ beInflamed(bloodVesselWall)→occur(sclerosis)

where "beInflamed(bloodVesselWall)" is an explicit mediator. Whilst the extracted causal pathway of the "Heart Disease and Blood Vessel" document from Thai Encyclopedia contain several explicit mediators as shown in the following

beDeposited(cholesterol,BloodVessel)→oxidize(cholesterol,OxygenDirivative)→
beInflame(bloodVesselWall)→consume(whiteBloodCell,fatParticle)→
accumulate(whiteBloodCell, bloodVesselWall) →
(beThickas(bloodVesselWall,plaque)

where (beThickas(vascularWall,plaque) is occur(sclerosis).

Table 5. The evaluation of the concise and comprehensible representations of Doc and Graph is based on scoring with the Likert scale (1 to 5).

Disease Group	Concise Representation		Comprehensible Representation	
	Average Score by Doc Representation	Average Score by Graph Representation	Average Score by Doc Representation	Average Score by Graph Representation
Diabetes and Kidney	3	4.5	3	4.3
Heart and Artery	2.2	4.3	2.7	4.2

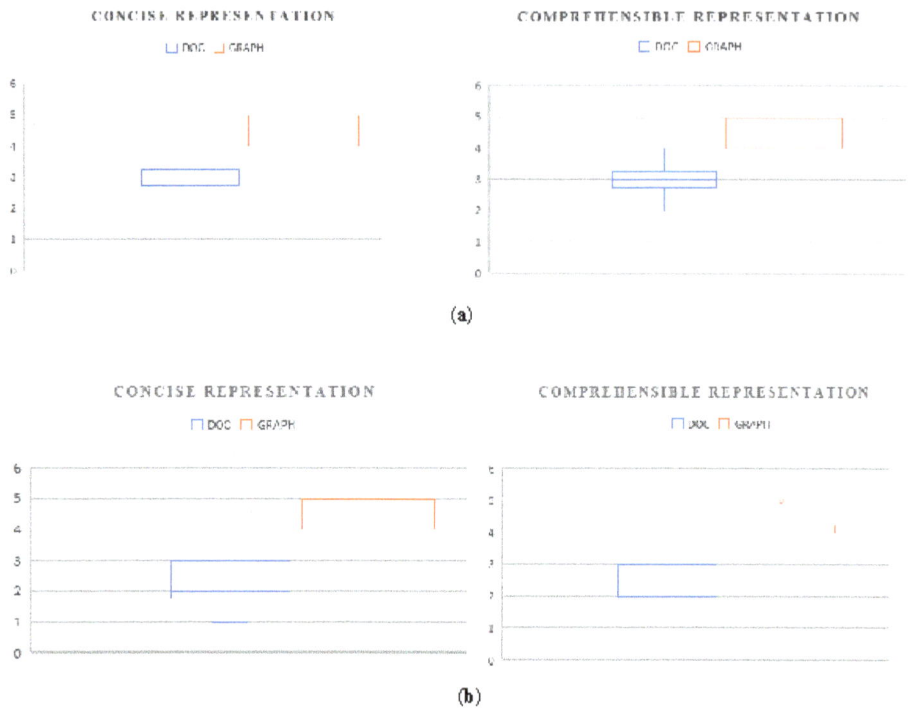

Figure 9. Show Box plot of the concise and comprehensible representations by Doc and Graph with the Likert scale 1–5. (**a**) Diabetes and Kidney Disease Group. (**b**) Heart and Artery Disease Group.

6. Conclusions

In this paper, we presented the extraction of the causal pathways containing the explicit and/or implicit mediators through learning and determining the wrdCoc pairs or CE-pairs having CErel from the downloaded documents of the diabetes and kidney disease group and the heart and artery disease group on the Thai hospital web-boards. Where each explicit mediator is expressed on the document by an effect/causative-concept EDU represented by an EDU's wrdCoc feature. We also represent the implicit mediators by the explicit mediators within the correct extracted causal pathways. With regard to the limited literation of the causal pathway extraction from texts, the extracted causal pathways including the explicit mediator representation of our research supports the preliminary causal inference and also makes non-professionals understand an etiological pathway including disease complication through the social media for the compliance to the preventive treatments. Our proposed method of extracting and representing the causal pathways in terms of the explicit mediators even the implicit mediator occurrences from the documents is based on (1) the wrdCoc-pair matching between the wrdCoc pairs on the test corpus and the WCP elements through the sliding window on the test corpus for the causal pathway extraction where each wrdCoc feature is obtained by the wrdCo-expression matching between the wrdCo expressions on the test corpus and the wrdCo expressions on the wrdCo-Concept Table. In addition, the wrdCoc features from the wrdCo-expression matching are based on the v_a, $w_{1,d}$, and $w_{2,e}$ terms with complete matching as in [29]. Since the precisions of determining wrdCoc pairs having CErel from the learning corpus and the test corpus are consistent, the causal pathways extracted by the wrdCoc-pair matching are strengthened (where the WCP elements obtained by the correct determination of wrdCoc pairs having Cerel) And (2) applying the transitive closure to obtain TransCEPair for indicating the implicit mediators on the correct extracted causal pathways to represent these implicit mediators with the explicit ones from the dynamic template. To evaluate the proposed method, the accuracy of determining the wrdCoc pairs having CErel depends on both the diversity of the wrdCoc features (including the diversity of wrdCo expressions) and the dependency between wrdCoc features; which later affect to the causal pathway extraction and representation. In contrast to the previous researches, our proposed method provides three contributions: (1) the CErel or cause-effect relation determination with high F-scores of our research is based on a wrdCoc pair for representing an event concept pair expressed by two EDU's verb phrases with the NP1 head noun consideration whereas the cause-effect relation determinations of the previous researches are based on either the NP1-NP2 pairs [5–7,9–12] within one/two simple sentences or the Verb pairs within two EDUs [8]. The event/state occurrences with the causative/effect concepts on our corpus contain the verb phrases expressions (which relate to the NP1s' head nouns) more than only the noun phrase expressions in the literature. (2) the causal pathway extraction with high precisions of our research is based on the actual event/state occurrences with the causative/effect concepts and also emphasizes on the boundary of the sequent wrdCoc pairs through the wrdCoc-pair matching between a wrdCoc pair of each slided-window on the test corpus and the WCP elements. Whereas the causal pathway/chain of the previous works are relied on either the prediction of the next effect from the previous noun phrase events based on the time series [10] or two steps of the cause-effect relation based on noun terms/phrases connected by either the similarity score [11] or the edge weights [12], e.g., 'A causes B' and 'B causes C' are connected by B similarity score without considering a boundary of a sequence event pairs having CErel. (3) our research applies the transitive closure and the dynamic template to indicate and represent the implicit mediator with the explicit mediators respectively to the correct extracted causal pathways with the high concise and clear comprehensible representations whereas the previous work [13] on the implicit knowledge completion of the event chains is based on the similarity scores between each event pairs from the corpus without the CErel consideration whilst our method of using the transitive closure and the dynamic template can apply to present the implicit knowledge completion of [13].

In the future, the temporal feature and the condition feature should be considered to increase the accuracy of the causal pathway extraction by reducing the wrdCoc diversity in terms of conditional cases. Moreover, the proposed method can also be applied in other languages, and the causal pathway extraction (Figure 7) can provide health literacy for non-professional persons to have clear comprehension of disease complications in order to follow the preventive treatments suggested by the physician.

Author Contributions: The following is a list of contributions by each author: Conceptualization, C.P.; methodology, C.P.; software, C.P.; validation, R.P.; formal analysis, R.P.; investigation, C. Pechsiri; resources, C.P.; data curation, C.P.; writing—original draft preparation, C.P.; writing—review and editing, C.P. and R.P.; visualization, C.P. and R.P.; supervision, C.P.; project administration, C.P.; funding acquisition, C.P. All authors have read and agreed to the published version of the manuscript.

Funding: This research received no external funding.

Institutional Review Board Statement: Not applicable.

Informed Consent Statement: Informed consent was obtained from all subjects involved in the study.

Data Availability Statement: The data for corpus preparation was obtained from hospitals website http://haamor.com/ (accessed on 10 August 2021); http://www.bangkokhealth.com (accessed on 10 August 2021); http://www.si.mahidol.ac.th/sidoctor/e-pl/ (accessed on 10 August 2021); https://www.bumrungrad.com (accessed on 10 August 2021); etc.

Conflicts of Interest: The authors declare no conflict of interest.

References

1. Khoo, C.; Na, J.C. Semantic relations in information science. *Annu. Rev. Inf. Sci. Technol.* **2006**, *40*, 157–228. [CrossRef]
2. Staplin, N.; Herrington, W.G.; Judge, P.K.; Reith, C.A.; Haynes, R.; Landray, M.J.; Baigent, C.; Emberson, J. Use of causal diagrams to inform the design and interpretation of observational studies: An example from the study of heart and renal protection (SHARP). *Clin. J. Am. Soc. Nephrol.* **2017**, *12*, 546–552. [CrossRef] [PubMed]
3. Gaskell, A.L.; Sleigh, J.W. An Introduction to causal diagrams for anesthesiology research. *Anesthesiology* **2020**, *132*, 951–967. [CrossRef] [PubMed]
4. Carlson, L.; Marcu, D.; Okurowski, M.E. Building a discourse-tagged corpus in the framework of rhetorical structure theory. *Curr. New Dir. Discourse Dialogue* **2003**, *22*, 85–112.
5. Girju, R. Automatic detection of causal relations for question answering. In Proceedings of the 41st annual meeting of the association for computational linguistics, workshop on multilingual summarization and question answering-Machine learning and beyond, Sapporo, Japan, 11–12 July 2003; pp. 76–83.
6. Cao, M.; Sun, X.; Zhuge, H. The contribution of cause-effect link to representing the core of scientific paper-The role of Semantic Link Network. *PLoS ONE* **2018**, *13*, 0199303. [CrossRef] [PubMed]
7. Chang, D.-S.; Choi, K.-S. Incremental cue phrase learning and bootstrapping method for causality extraction using cue phrase and word pair probabilities. *Inf. Process. Manag.* **2006**, *42*, 662–678. [CrossRef]
8. Pechsiri, C.; Piriyakul, R. Explanation knowledge graph construction through causality extraction from texts. *J. Comput. Sci. Technol.* **2010**, *25*, 1055–1070. [CrossRef]
9. Sawamaru, H.; Kobayashi, I. An Approach to Extraction of Causal Chain among Events in Multiple Documents. SCIS-ISIS. In Proceedings of the 6th International Conference on Soft Computing and Intelligent Systems, and the 13th International Symposium on Advanced Intelligence Systems, Kobe, Japan, 20–24 November 2012; IEEE: Piscataway, NJ, USA, 2012; pp. 1104–1108.
10. Kang, D.; Gangal, V.; Lu, A.; Chen, Z.; Hovy, E. Detecting and explaining causes from text for a time series event. In Proceedings of the Conference on Empirical Methods in Natural Language Processing, Copenhagen, Denmark, 9–11 September 2017; pp. 2758–2767.
11. Izumi, K.; Sakaji, H. Economic Causal-Chain Search using Text Mining Technology. In Proceedings of the 1st Workshop on Financial Technology and Natural Language Processing, Macao, China, 12 August 2019; pp. 61–65.
12. Nordon, G.; Koren, G.; Shalev, V.; Kimelfeld, B.; Shalit, U.; Radinsky, K. Building causal graphs from medical literature and electronic medical records. In Proceedings of the 33rd AAAI Conference on Artificial Intelligence, Honolulu, HI, USA, 27 January–1 February 2019; pp. 1102–1109.
13. Takishita, S.; Rzepka, R.; Araki, K. Implicit Knowledge Completion Using Relevance Calculation of Distributed Word Representations. In Proceedings of the IJCAI Workshop on Bridging the Gap between Human and Automated Reasoning, Macao, China, 12 August 2019; pp. 60–64.
14. Song, M.-K.; Lin, F.-C.; Ward, S.E.; Fine, J.P. Composite Variables. *Nurs. Res.* **2013**, *62*, 45–49. [CrossRef] [PubMed]
15. Ayinde, B.O.; Inanc, T.; Zurada, J.M. Regularizing deep neural networks by enhancing diversity in feature extraction. *IEEE Trans. Neural Netw. Learn. Syst.* **2019**, *30*, 2650–2661. [CrossRef] [PubMed]
16. Leng, C.; Zhang, H.; Cai, G.; Cheng, I.; Basu, A. Graph regularized Lp smooth non-negative matrix factorization for data representation. *IEEE/CAA J. Autom. Sin.* **2019**, *6*, 584–595. [CrossRef]
17. Weisstein, E.W. "Cartesian Product". Available online: www.mathworld.wolfram.com (accessed on 5 September 2020).
18. Mitchell, T.M. *Machine Learning*; The McGraw-Hill Co., Inc.; MIT Press: Singapore, 1997.
19. Cristianini, N.; Shawe-Taylor, J. *An Introduction to Support Vector Machines*; Cambridge University Press: Cambridge, UK, 2000.
20. Freedman, D.A. *Statistical Models: Theory and Practice*; Cambridge University Press: Cambridge, UK, 2009.
21. Weisstein, E.W. "Transitive Closure". from MathWorld—A Wolfram Web Resource. Available online: https://mathworld.wolfram.com/TransitiveClosure.html (accessed on 30 August 2021).
22. Eve, J.; Kurki-Suonio, R. On computing the transitive closure of a relation. *Acta Inform.* **1977**, *8*, 303–314. [CrossRef]
23. Sudprasert, S.; Kawtrakul, A. Thai word segmentation based on global and local unsupervised earning. In Proceedings of the NCSEC 2003, Chonburi, Thailand, 28–30 October 2003; pp. 1–8.
24. Chanlekha, H.; Kawtrakul, A. Thai named entity extraction by incorporating maximum entropy model with simple heuristic information. In Proceedings of the IJCNLP 2004, Hainan Island, China, 22–24 March 2004; pp. 1–7.
25. Tongtep, N.; Theeramunkong, T. Pattern-based Extraction of Named Entities in Thai News Documents. *Thammasat Int. J. Sci. Technol.* **2010**, *15*, 70–81.
26. Chareonsuk, J.; Sukvakree, T.; Kawtrakul, A. Elementary discourse unit segmentation for Thai using discourse cue and syntactic information. In Proceedings of the NCSEC 2005, Bangkok, Thailand, 27–28 October 2005; pp. 85–90.
27. Ketui, N.; Theeramunkong, T.; Onsuwan, C. Thai elementary discourse unit analysis and syntactic-based segmentation. *Information* **2013**, *16*, 7423–7436.
28. Miller, G.A. WordNet: A lexical database. *Commun. ACM* **1995**, *38*, 39–41. [CrossRef]
29. Adhikari, B.K.; Zuo, W.; Maharjan, R.; Han, X.; Liang, S. Detection of Sensitive Data to Counter Global Terrorism. *Appl. Sci.* **2020**, *10*, 182. [CrossRef]

applied sciences

MDPI

Article

A Query Expansion Method Using Multinomial Naive Bayes

Sergio Silva [1,2,3,*], Adrián Seara Vieira [1,2,3], Pedro Celard [1,2,3], Eva Lorenzo Iglesias [1,2,3] and Lourdes Borrajo [1,2,3]

[1] Computer Science Department, Escuela Superior de Ingeniería Informática, Universidade de Vigo, 32004 Ourense, Spain; adrseara@uvigo.es (A.S.V.); pedro.celard.perez@uvigo.es (P.C.); eva@uvigo.es (E.L.I.); lborrajo@uvigo.es (L.B.)

[2] CINBIO-Biomedical Research Centre, Universidade de Vigo, 36310 Vigo, Spain

[3] SING Research Group, Galicia Sur Health Research Institute (IIS Galicia Sur), SERGAS-UVIGO, 36312 Vigo, Spain

* Correspondence: smachado@alumnos.uvigo.es

Abstract: Information retrieval (IR) aims to obtain relevant information according to a certain user need and involves a great diversity of data such as texts, images, or videos. Query expansion techniques, as part of information retrieval (IR), are used to obtain more items, particularly documents, that are relevant to the user requirements. The user initial query is reformulated, adding meaningful terms with similar significance. In this study, a supervised query expansion technique based on an innovative use of the Multinomial Naive Bayes to extract relevant terms from the first documents retrieved by the initial query is presented. The proposed method was evaluated using MAP and R-prec on the first 5, 10, 15, and 100 retrieved documents. The improved performance of the expanded queries increased the number of relevant retrieved documents in comparison to the baseline method. We achieved more accurate document retrieval results (MAP 0.335, R-prec 0.369, P5 0.579, P10 0.469, P15 0.393, P100 0.175) as compared to the top performers in TREC2017 Precision Medicine Track.

Keywords: query expansion; information retrieval; multinomial naive bayes; relevance feedback

Citation: Silva, S.; Seara Vieira, A.; Celard, P.; Iglesias, E.L.; Borrajo, L. A Query Expansion Method Using Multinomial Naive Bayes. *Appl. Sci.* **2021**, *11*, 10284. https://doi.org/10.3390/app112110284

Academic Editor: Arturo Montejo-Ráez

Received: 15 September 2021
Accepted: 28 October 2021
Published: 2 November 2021

1. Introduction

Information Retrieval (IR) is a field of computer science that processes text documents and retrieves those that are more similar to a user query based on the resemblance of the contents of the documents and the keywords of the query. In a particular way, the task of information retrieval is gaining importance in the field of biomedicine.

The exponentially growing amount of clinical data makes it remarkably difficult to extract relevant information that meets the needs of each individual user [1]. The most well-known techniques make use of keywords to search for specific items that contain them, but language semantics, polysemy, synonymy, and hyponymy make keywords useless in many cases [2]. Therefore, the retrieval process in information retrieval systems must be improved in order to deal with all this complexity and deliver appropriate results that meet what the user is looking for.

One of the most widely-used techniques to improve the retrieval process is query expansion (QE). QE is the process of reformulating a given query in order to retrieve the more suitable documents that meet a user's needs. Over the years, several query expansion techniques have been analyzed, but even recent elaborate architectures are having problems surpassing the performance of classic techniques [3]. Because of this, our work is focused on the extraction of terms that expand the original query in order to improve the relevance of the retrieved documents.

Related Works

So far, several authors have worked on diverse research lines related to query expansion, the improvement of efficiency and performance being the common aim, in order to offer more relevant information to the user and better fulfill their needs.

Zhu et al. [2] evaluated the use of auxiliary collections to address polysemy, synonymy, and hyponymy in clinical text retrieval. These semantic relations complicate the retrieval process as different words can relate to the same topic. In order to deal with this problem, they proposed a pseudo-relevance feedback method that looks for new terms in the auxiliary collections in order to expand the initial query. The authors concluded that the use of all available data, in some cases, is inadequate and may not lead to improvements in the recovery system. In these cases, the authors suggested additional resources and a selection of the collection that is suitable for the query.

Ehman et al. [4] proposed the Normalized Difference Measure metric, a measure that takes into account the relative frequency of documents and terms in order to improve text classification. This metric analyzes all the terms found in the documents and benefits from the inclusion of new relevant terms in a query when used as a classifier in information retrieval systems.

Araújo et al. [5] implemented a pseudo-feedback query expansion method that allows the user to select expansion words from a list of possible relevant terms. The authors used the top three retrieved documents to extract terms based on document and word frequencies, word length, and query length. The obtained results showed an overall improvement in the number of relevant retrieved documents, despite the fact that the results in some cases were not better than the base case due to the low number of relevant documents.

Afun et al. [6] suggested a combination of several query expansion methods such as Ontologies, Association Rules, WordNet, Methathesaurus, Synonym Mapping, Local Co-occurrence, and Latent Semantic Indexing. The authors noted several limitations to the previous techniques (e.g., performance reduction, term relationship loss), emphasizing the importance of choosing the right technique for each specific case.

Agosti et al. [7] reviewed multiple query expansion techniques that had been applied to information retrieval systems used in clinical trials. The authors concluded that it is not possible to build an expansion technique pattern that correctly applies to a huge text corpus. They reported that the use of weighted keyword expansion and query reduction (removal of words that are not relevant) improved the performance of information retrieval in clinical trials.

Xu et al. [8] proposed a supervised query expansion model that could be applied to highly diverse biomedical datasets. The authors performed a term extraction for each query, proceeded to assign labels to each term, and then ranked them to know which were the most relevant. Owing to these three steps, and with the use of rank weights, the authors were able to enhance the queries and improve the performance of biomedical information retrieval.

Azad and Deepak [1] surveyed multiple query expansion techniques, weighting methods, and ranking methodologies for information retrieval. They found that the most frequent queries are composed of one, two, or three words, which increases its ambiguity and makes the retrieval process difficult. This exposed the increasing need for query expansion techniques to enhance the original queries with the use of relevant terms, making it easier for information retrieval systems to obtain more suitable elements.

McDonald et al.[3] proposed a technique that extends deep learning architectures where queries and documents are analyzed together in order to obtain their similarities. The authors claimed that even state-of-the-art complex architectures do not improve the performance of classic algorithms such as BM25 and BM25+extra. Their proposed method helps to achieve better performance, offering an improvement of BM25, although unable to surpass BM25+extra in some cases.

Wang et al. [9] implemented a pseudo-relevance feedback technique to expand the queries using terms extracted from the top-ranked documents returned in a first search. The authors used Rochio+BM25 to extract the expansion terms, outperforming baseline models in terms of MAP and precision at different positions.

This paper shows the work developed to expand an initial query from a set of first-retrieved documents using the Multinomial Naive Bayes technique as an autonomous selector of terms. The proposed technique improves performance and helps information retrieval systems to obtain a higher number of relevant documents for specific queries.

It is organized according to the following structure: Section 2 presents a detailed overview of the information retrieval process, techniques used in text preprocessing and representation, the selection of attributes, and the measures used in the evaluation of the information retrieval system. Section 2.3 describes the procedures carried out and the proposed query expansion technique. Section 3 describes the evaluation methodology followed and the results obtained. Finally, in Section 4, the conclusions of the study are presented and future perspectives are discussed.

2. Materials and Methods

The general information retrieval process obtains documents that are relevant to a given query. This involves tasks related to text preprocessing, document indexing, and the execution of an initial query that could then be expanded to obtain better results. Figure 1 shows an overview of the process.

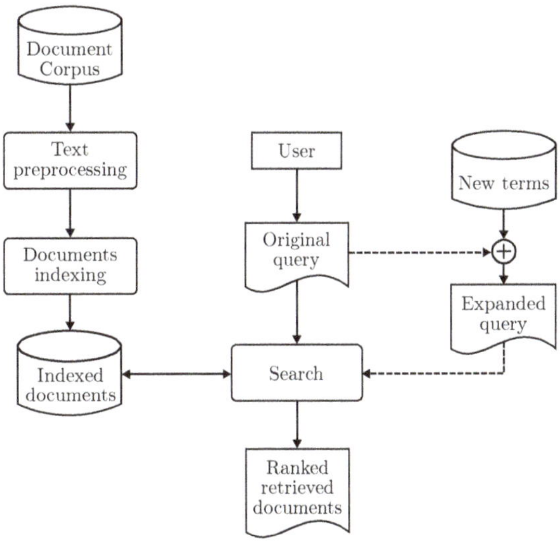

Figure 1. Information retrieval—general process.

2.1. Text Preprocessing and Matching

The information retrieval process involves data preprocessing and matching. The *pre-processing* step includes tasks related to tokenization, stopword removal, sentence detection, stemming, lemmatization, and term weighting.

Tokenization allows for the transformation of a document into words using white spaces, commas, periods, and tab delimiters as separators during the token-building process. According to [10], there are different text delimiters that can lead to a complex process of tokenization. Since most scientific documents are written in English, the recognition and extraction of tokens is carried out considering a specific set of characters. A whole set of special characters is disregarded as they contribute nothing to the knowledge and only function as token separators. Among them are: (,), /, {, }, [,], :, ;.

There are some issues that need to be taken into account, such as the identification of abbreviations, dates, acronyms, and letter capitalization. Case transformation allows for the standardization of the words contained in documents, thus dropping different versions of the same word. Given that the stopword list is presented in lowercase, all the letters are transformed to its lowercase variant.

Stopword removal is based on the elimination and non-consideration of words that are very frequent and offer little significance. The main advantage of this procedure is the reduction of data size, and thus the decrease of computational cost and the improvement of accuracy. There are lists of stopwords available for the English language. However, new terms may be added to these lists depending on the structure of the documents and the needs of particular circumstances.

Stemming is a process of reducing words to their word stem, preserving only the morphological root. Suffixes of words such as plurals and gerunds, among others, are removed. According to the literature, *Porter Stemmer* and *Krovetz Stemmer* are the most frequently used stemmers in information retrieval systems in the English corpora. Porter Stemmer was developed by Martin Porter at Cambridge University in 1980 and was first published in Porter, M.F [11]. It is a process of removing word suffixes, such as gerunds and plurals, and replacing inflectional endings. It consists of rules dealing with a specific suffixes and according to certain conditions. Lemmatization uses dictionaries and a morphological analysis of words in order to reflect the base form of a word, consequently collapsing the inflectional forms.

Document indexing is based on the frequency of the words that each document contains. Words with a high number of repetitions have a higher frequency, while the others have a lower frequency [12]. This is not always desirable behavior as some words such as and, the, and or appear frequently in documents but do not offer relevant information. One of the most widely applied algorithms is Term Frequency-Inverse Document Frequency (TF-IDF) [13], a statistical measure that assesses the importance of a word for a document in a collection or corpus. The TF-IDF value increases proportionally to the number of times a word appears in the document, but is offset by the frequency of the word in the document collection, which recognizes the fact that some words are generally more common than others.

The matching step allows for the calculation of the similarity between documents and queries, with an associated weighting of terms. In general, a retrieval system returns a list of ordered documents where the first is the document most similar to the query. Taking this into account, it is possible to reformulate the query and expand it to be more representative of the need of the user; this technique is known as query expansion. According to the literature, query expansion techniques can be classified as: query-specific, corpus-specific, or language-specific [14].

Query-specific terms are based on the extraction of new terms from a subset of documents retrieved by a specific query. It is an approach of relevance feedback systems in which the new terms are obtained from a set of relevant documents. Although this technique is widely used and very effective, it requires users to indicate which documents are relevant.

In the corpus-specific technique, the entire content of a specific full-text database is analyzed to identify terms that are used in similar ways. This can be performed manually (although this requires a lengthy and ad hoc process) or automatically. Traditional automatic thesaurus construction techniques group words based on their patterns of occurrence at the document level [15,16]; that is, words that often occur together in documents are considered similar. These thesauri can be used for automatic expansion or manual consultation.

Language-specific is a technique present in online thesaurus that is not adapted for any specific text collection. Liddy and Myaeng [17] used Longman's Dictionary of Contemporary English, a semantically encoded dictionary. Others such as Voorhees [18] turned to WordNet [19], a network of lexical relationships built by hand. Borrajo et al. [20] studied the use of dictionaries in the classification of biomedical texts with three different dictionaries (BioCreative [21], NLPBA [22], and an ad hoc subset of the UniProt database called Protein [23]).

In this work, Indri [24] is used as the search engine to perform the matching between a given query and a set of documents. Indri uses a combination of language modeling and inference networks for the information retrieval procedure. It is able to evaluate a query against a previously indexed corpus, returning a collection of the most relevant documents.

Indri uses a Dirichlet likelihood function for query evaluation prior to term weight smoothing. This function takes into account the frequency of words in a document and in the document collection, and a parameter μ, which takes the value of 2500 by default [25]. The score returned by the Dirichlet probability function is given by:

$$\log\left([C(W,D) + \mu * C(W,C)/|C|)/(|D| + \mu)\right] \tag{1}$$

- $C(W, D)$ represents the word count in the document D;
- $C(W, C)$ represents the word count in the document collection;
- $\mu = 2500$ default.

2.2. Corpus

In this work, the Clinical Trial corpus is used, which is composed of a set of clinical documents, topics (descriptions of the user needs), and relevance judgments performed by specialists in the field [26]. Roberts et al. [27] discussed in greater detail how the corpus was created and showed multiple works using it as an experimental corpus. Clinical Trials are available on the TREC official web page http://www.trec-cds.org/2017.html, accesed on 22 July 2021. The database contains 241,006 documents in txt and xml format. For this work, the *txt* format is selected.

Given that the main objective of this work is to present a new technique for query expansion, all the topics available are used. The topics consist of *disease*, *genetic variants*, *demographic*, and potentially other information about the patients.

The relevance judgments file corresponding to the Clinical Trial collection contains the relevant documents to each query, except for the query related to topic 10. In this case, there is no relevant information, and the query associated with this topic is disregarded.

In general, the documents contain a title, a detailed description of what is carried out in the study, information on the patient condition, intervention, and eligibility factors (these may include gender, age, or the respective criteria for inclusion or exclusion from the study). All documents are indexed with all its content, which means that no specific field in the document is selected.

In order to index the corpus, the documents were preprocessed using Porter stemming, and a list of stopwords for the English language were removed https://www.ranks.nl/stopwords, accessed on 15 May 2021. The terms *age*, *condition*, *detailed*, *eligibility*, *exclusion*, *inclusion*, *intervention*, *title*, *criteria*, *description*, *gender*, and *summary* were added to the stopword list because they were terms related to the names of the field labels in the documents; therefore, they were not relevant. All 241,006 documents were indexed for retrieval. Figures 2 and 3 show examples of document and topic structures used in the Corpus Clinical Trial.

TITLE:
Information Presentation Formats
CONDITION:
Meningioma
INTERVENTION:
Check Symptoms
SUMMARY:
Prevention and early detection of medical problems
can greatly reduce health care costs ...
DETAILED DESCRIPTION:
We will present individuals with medically accurate
information about a medical condition and measure ...
ELIGIBILITY:
Gender: All
Age: 18 Years to N/A
...

Figure 2. Clinical Trial—sample document.

<topic number="1">
<disease>Liposarcoma</disease>
<gene>CDK4 Amplification</gene>
<demographic>38-year-old male</demographic>
<other>GERD</other>
</topic>

Figure 3. Clinical Trial—sample topic.

2.3. System Architecture

The new query expansion technique presented in this study is based on relevance feedback and is presented in Figure 4.

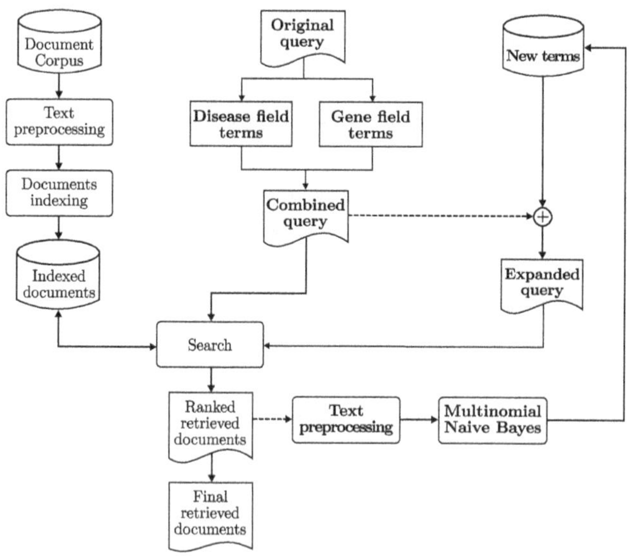

Figure 4. The proposed method.

The main elements of the proposed technique are the Original Query (OQ) found in the corpus, a combination of the terms found in the data fields of the OQ called Combined Query (CQ), and an Expanded Query (EQ) obtained from a combination of the words of the CQ and new terms from the relevant documents retrieved by the CQ in a first search. From this point, new searches could be performed to further improve the query.

2.3.1. Combined Query (CQ)

The CQ was obtained by using a combination of terms referring to the fields *disease* and *gene*. In general, documents containing terms related to these fields were retrieved. More specifically, it was expected that all documents containing the terms related to the disease and with each of the genes (and their variants, if any) would be retrieved.

In this study, the language modeling tool Lemur (Lemur Project) was used for indexing and query execution. Lemur is a software tool designed to facilitate research in language modeling and IR, using weighting algorithms that provide query analysis methods, document indexing, and query-related document retrieval. This tool was developed by the Center for Intelligent Information Retrieval (CIIR) at the University of Massachusetts, and by the Institute of Language Technologies (LTI) at Carnegie Mellon University. It is an open source and freely accessible that incorporates Indri as its query language,https://www.lemurproject.org/ software accessed on 12 April 2021.

The Indri query language is based on the Inquery language. It allows for the building of complex queries, and its grammar provides options for term detection, proximity, synonyms, wildcard operations, field restriction, combined beliefs (operators), filter operators, numeric and date field operators, document priors, etc.

Some operations used in this work are:

- *#band(w1 w2 ... wn)* returns documents containing all the terms *w1, w2, ..., wn*;
- *#combine (w1 w2 ... wn)* returns a scored list of documents that contains at least one of the terms;
- *#syn(w1 w2 ... wn)* returns the score of documents containing one of the terms *w1, w2,..., wn*, but considering these as synonyms.

The combined query is written as *#band(Disease GENE variant)*, but if the topic has more than one variant, it is changed to *#band(Disease GENE variant1)*, *# band(Disease GENE variant2)*, etc. For example, for topic 1, we get the combined query *#band(liposarcoma cdk4 amplification)*.

2.3.2. Extraction of New Terms

A set of 29 CQ was executed, saving the recovered documents as a training base consisting of 29 categories, one for each query. Each category is composed of the documents retrieved by the respective query (under category 1 we have the documents retrieved by query 1, and so on).

The training database built upon the retrieved documents was preprocessed the same way as the original documents (stemming, tokenization, stopword removal, case converter, and weighting). To simultaneously carry out the aforementioned operations on the documents, the free WEKA (Waikato Environment for Knowledge Analysis) software was used. It is available on Waikato official web page, https://www.cs.waikato.ac.nz/ml/weka, accessed on 13 May 2021. Weka includes data analysis tools such as textual data preprocessing, filtering, Naive Bayes Multinomial algorithm execution, and data visualization.

2.3.3. Attribute Selection

Attribute selection aims to reduce the number of attributes present in the data. More specifically, in text documents, these attributes refer to words that contain irrelevant information. The application of a classifier is performed on a smaller number of attributes considered the most relevant, which leads to the acquisition of more relevant terms or attributes for each category. The *GainRatio* technique selects attributes that maximize the information gain while minimizing the number of values of an attribute. After calculating the relevance for each attribute, a ranking is generated and the attributes of that ranking are selected, according to a *threshold* value. In this case, this value was 0.

2.3.4. Multinomial Naive Bayes

The expansion of queries is the process of reformulating a given query (Combined Query) in order to improve the performance of the information retrieval system. An evaluation of the initial consultation is carried out, which is expanded with new additional terms in order to be able to retrieve more relevant documents. In general, the expansion of queries may involve the search for synonyms or semantically related words. Moreover, it may employ associated procedures to correct spelling errors, reduce terms to a morphological form, or reweight the terms of the initial consultation, among others. In this study, a Query-specific term approach was adopted using relevant feedback.

A small set of documents was retrieved from an initial consultation, and all of them were considered relevant without any intervention from the user [28]. The content of the retrieved documents was used to obtain the new terms for the CQ expansion. The new query (Expanded Query) was obtained by combining the new terms and the CQ.

The extraction of the new terms is based on the probability that a word belongs to a given category. Once the training base (list of retrieved documents for a query) is organized by categories and the attribute selection is performed, the Multinomial Naive Bayes algorithm is applied.

The Naive Bayes algorithm is widely used in works involving text classification. It is based on probabilistic techniques, assuming the independence of variables. It is assumed that the presence or absence of a given characteristic of a category is not associated with the presence or absence of any other characteristic that is given that category.

The Multinomial Naive Bayes model considers how often the word occurs in documents x_t instead of the binary occurrence. It is calculated as follows, where $|V|$ represents the length of the vocabulary, and $n(C_i)$ is the total number of words in the category C_i:

$$P(d_j|C_i) = \prod_{t=1}^{|V|} P(w_t|C_i)^{x_t} \tag{2}$$

$P(w_t|C_i)^{x_t}$ is the probability of a term w_t occurring in a category C_i, and $n(w_t, C_i)$ is the number of occurences of w_t in the category C_i, as given by:

$$P(w_t|C_i) = \frac{1 + n(w_t, C_i)}{|V| + n(C_i)} \tag{3}$$

Finally, the classification is given by the maximizing function:

$$c^*(d) = argmax_{C_i} P(C_i) \prod_{t=1}^{|V|} P(w_t|C_i)^{x_t} \tag{4}$$

Therefore, the Multinomial Naive Bayes model is a reliable alternative for categorizing documents. In this case, instead of relying on binary values, it uses the frequency of the term. That is, it takes into account the number of times a word or *token* occurs in a document (also called gross frequency) [29]. In particular, the Multinomial Naive Bayes algorithm calculates the probability of a word belonging to a given category.

In this study, for each category (topic), the words w_t that verify the condition $P(w_t|C_i)^{x_t} > 0$ were considered as new terms for the query C_i expansion.

2.4. Expanded Query

In the first stage, the (CQ) was generated. It was from here that the expansion of the queries was processed. After the training documents for each query (category) were established, the Naive Bayes Multinomial algorithm was applied.

The CQ was obtained by the terms referring to the fields *disease* and *gene*: *#band(Disease GENE variant)*. Documents containing terms referring to these fields were retrieved at the same time. This initial consultation was performed on the indexed corpus. When the gene had more than one variant, the CQ was written as *#band(Disease GENE variant1)*

#band(Disease GENE variant2). The *band* method uses an *AND* boolean operator, so all documents containing all the terms related to the disease and the genes (and their variants, if any) were retrieved.

Finally, an expanded query (EQ) was built as *#combine(*t_1 t_2 ... t_n n_1 n_2 ... n_n*)*, employing the boolean operation *OR*. The terms $t_1, t_2, \ldots, t_n$ are the words contained in the *disease* and *gene* fields of the combined query, and $n_1, n_2, \ldots, n_n$ are the new terms extracted by the described process. The EQ was, again, performed over the full indexed corpus.

3. Results and Discussion

After the execution of the queries, the measures were extracted using the *trec_eval* tool. It receives the recovered documents and the qrels file as parameters. This tool has been officially developed for its use in many of the tasks organized by the Text REtrieval Conference (TREC). For each query, the values of MAP, R-prec, and P@n were recorded for $n \in \{5, 10, 15, 100\}$. This procedure was exactly the same for both CQ and EQ.

Among the most frequently used measures in information retrieval are MAP, R-prec, and P@n. The Mean Average Precision (MAP) is the mean of the average precision scores for each query:

$$MAP = \frac{\sum_{q=1}^{Q} Ave(P)}{|Q|} \tag{5}$$

The average precision ($Ave(P)$) emphasizes the assignment of a higher ranking to relevant documents. It is the average of the precision of each of the relevant documents in the ranked sequence:

$$Ave(P) = \frac{\sum_{k=1}^{n} (P(k) \times rel(k))}{\text{number of relevant documents}} \tag{6}$$

- k is the rank in the sequence of retrieved documents;
- n is the number of retrieved documents;
- $rel(k)$ is a binary function that assumes the value of 1 if the item at rank k is a relevant document, and zero if otherwise;
- $P(k)$ is the precision at cut-off k on the list.

R-prec is the precision after R documents have been retrieved, where R is the number of relevant documents for the topic. P@n is the accuracy of the first n documents recovered.

After the evaluation of the results in terms of the average values of the aforementioned measures, there is a clear improvement obtained by the expanded query. As can be seen in Figure 5, there is an increase of approximately 30% in the value of the MAP measure, from 0.261 to 0.335, with the use of query expansion. Regarding the R-prec measure, there is a general improvement of 12%. In relation to P@5 and P@10, the improvement is still significant at about 12% and 13%, which means that even with an increase in the considered number of the first retrieved documents, the system remains robust. In the measure P@15, the improvement was about 12%, while in P@100, it was about 24%.

In Table 1, the *MAP*, *R-prec*, *P@5*, *P@10*, *P@15*, and *P@100* values obtained for each combined and expanded query are recorded. This shows a general improvement in all queries resulting from the expansion of the initial consultation.

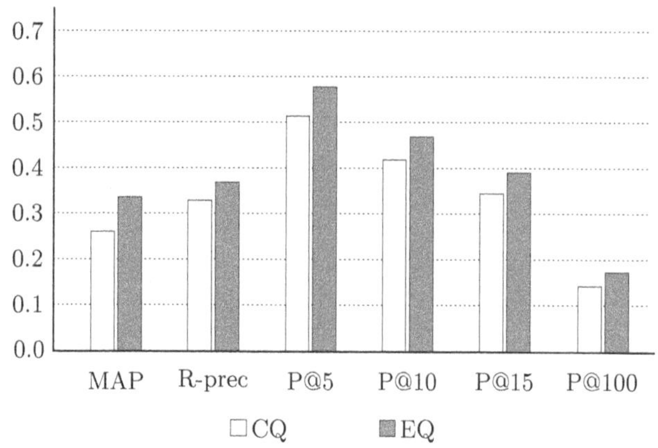

Figure 5. Mean values of measures.

Table 1. Measures of the evaluation of the combined and expanded queries.

	MAP		**R-prec**		**P@5**		**P@10**		**P@15**		**P@100**	
Query	**CQ**	**EQ**	**CQ**	**EQ**	**CQ**	**EQ**	**CQ**	**EQ**	**CQ**	**EQ**	**CQ**	**EQ**
1	0.293	**0.408**	0.353	0.353	1.000	1.000	0.500	**0.600**	0.333	**0.400**	0.050	**0.080**
2	0.243	**0.345**	0.394	**0.402**	**0.800**	0.600	0.700	0.700	0.733	0.733	0.440	**0.450**
3	0.405	**0.615**	0.417	**0.625**	1.000	1.000	**0.900**	0.800	0.667	**0.733**	0.100	**0.200**
4	0.410	**0.454**	0.491	**0.509**	0.600	**0.800**	0.600	**0.700**	**0.600**	0.533	0.360	**0.380**
5	**0.205**	0.173	0.194	0.194	**0.200**	0.000	0.100	**0.200**	0.133	**0.200**	**0.210**	0.170
6	0.404	**0.411**	**0.444**	0.370	1.000	1.000	**0.700**	0.500	**0.600**	0.533	0.160	**0.170**
7	0.369	**0.571**	0.390	**0.546**	0.600	**1.000**	0.600	**0.900**	0.667	**0.867**	0.480	**0.680**
8	0.450	**0.504**	**0.541**	0.525	0.600	**0.800**	0.600	**0.700**	0.667	0.667	0.420	0.420
9	0.314	**0.520**	0.339	**0.532**	0.600	**0.800**	0.600	**0.800**	0.467	**0.600**	0.300	**0.460**
11	0.319	**0.402**	0.316	**0.368**	0.600	**0.800**	0.400	**0.600**	0.333	**0.400**	**0.150**	0.140
12	0.118	**0.266**	0.231	**0.256**	0.600	**0.800**	0.300	**0.700**	0.200	**0.467**	0.100	**0.190**
13	0.090	**0.161**	**0.324**	0.206	0.200	0.200	0.200	0.200	**0.267**	0.200	0.110	**0.120**
14	**0.579**	0.563	0.714	0.714	0.800	0.800	0.500	0.500	0.333	0.333	0.050	**0.060**
15	0.250	**0.253**	0.250	0.250	0.200	0.200	0.100	0.100	0.067	0.067	0.010	0.010
16	0.300	**0.327**	0.400	0.400	0.400	0.400	0.200	0.200	0.133	0.133	0.020	**0.040**
17	0.150	**0.191**	0.303	0.303	0.400	0.400	0.400	0.400	0.400	**0.533**	0.090	0.090
18	0.000	**0.044**	0.000	**0.079**	0.000	**0.200**	0.000	**0.200**	0.000	**0.133**	0.020	**0.050**
19	0.044	**0.307**	0.044	**0.304**	0.200	**0.400**	0.100	**0.400**	0.067	**0.267**	0.010	**0.130**
20	0.200	**0.234**	0.200	0.200	0.200	0.200	0.100	0.100	0.067	0.067	0.040	0.040
21	0.088	**0.246**	0.209	**0.269**	0.400	**0.600**	**0.500**	0.300	**0.400**	0.333	0.190	**0.240**
22	0.059	**0.087**	0.118	**0.147**	**0.600**	0.400	**0.600**	0.400	**0.400**	0.333	0.120	**0.160**
23	**0.243**	0.236	**0.367**	0.333	**0.400**	0.200	**0.500**	0.400	**0.400**	0.333	**0.190**	0.170
24	0.333	**0.580**	0.333	**0.556**	1.000	1.000	0.600	**0.900**	0.400	**0.667**	0.060	**0.110**
25	0.265	**0.459**	0.375	**0.475**	0.600	**0.800**	0.600	**0.900**	0.467	**0.733**	0.210	**0.250**
26	0.213	**0.225**	0.200	0.200	0.200	0.200	0.100	0.100	0.067	0.067	0.010	0.010
27	0.250	**0.313**	0.393	**0.429**	0.400	**0.800**	0.500	0.500	0.400	0.400	0.080	**0.110**
28	0.250	0.250	0.500	0.500	0.200	0.200	0.100	0.100	0.067	0.067	0.010	0.010
29	0.277	**0.385**	0.250	**0.375**	0.400	**0.600**	0.200	**0.300**	0.133	**0.200**	0.010	**0.020**
30	**0.263**	0.198	**0.600**	0.290	0.600	0.600	0.400	0.400	0.400	0.400	0.110	**0.120**
ALL	0.261	**0.335**	0.330	**0.369**	0.514	**0.579**	0.418	**0.469**	0.347	**0.393**	0.142	**0.175**

The bold font refers to the highest value for each measure.

We can draw a comparison between the obtained results and the outcome of the participants of the TREC2017 Precision Medicine Track. As reported in [27], Table 8 shows the best, median, and worst results per topic from over 133 runs at P@5, P@10, and P@15 for the TREC 2017 Precision Medicine Track using clinical trials. Our method clearly outperforms most of median results in all precision ranges. More specifically, it achieves better precision than the median in 25 out of 29 topics for P@5, 26 out of 29 topics for P@10, and 26 out of 29 topics for P@15. In order to better analyze the results, Figures 6–8 show a comparison of the mean precision of the participants of the TREC2017 Precision Medicine Track and that of the proposed method in the top 5, 10, and 15 retrieved documents.

Firstly, Figure 6 shows how our proposed method obtained better or equal results as the mean participants of the TREC2017 Precision Medicine Track, except for the fifth query, which means that none of the first five retrieved documents were relevant due to the high difficulty of the query.

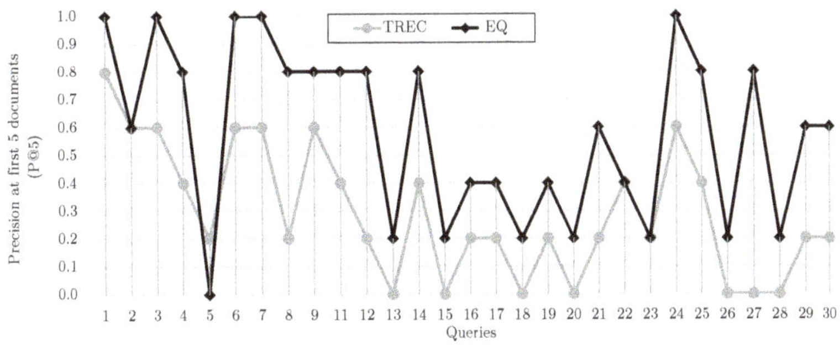

Figure 6. Comparison of the mean precision of the participants of the TREC2017 Precision Medicine Track (TREC) and the proposed method (EQ) in the top five retrieved documents.

Secondly, it can be seen in Figure 7 how the proposed method obtained the same results as the other teams at query number 5. Unlike the precision in five documents, the proposed method always offers better or equal results at P@10.

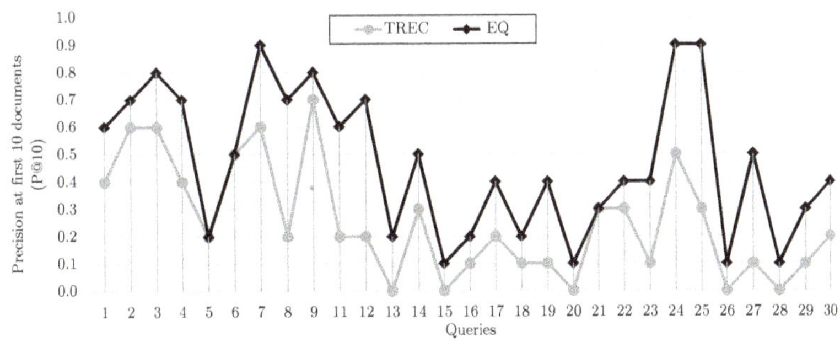

Figure 7. Comparison of the mean precision of the participants of the TREC2017 Precision Medicine Track (TREC) and that of the proposed method (EQ) in the top ten retrieved documents.

Lastly, Figure 8 proves how the results keep getting better than the mean results of the other teams in most cases, only being unable to reach it at query number 9. This query only has two definitely relevant documents and 60 partially relevant ones, making the retrieval process deeply difficult as the number of first-analyzed documents is increased. Even in this case, the obtained precision (0.60) is very close to the mean precision of the teams (0.667).

Furthermore, of the top overall systems in Table 6 [27], the proposed method surpassed the best team run [30] at P@5 (0.5448), P@10 (0.4448), and P@15 (0.3885), where we attained 0.579, 0.469, and 0.393, respectively. If we analyze the standard deviation of the best team results excluding the best team and duplicates, the values we get are 0.0175 (P@5), 0.0178 (P@10), and 0.0195 (P@15), while the proposed method improves them to 0.034 (P@5), 0.024 (P@10), and 0.005 (P@15). This means that our method obtains a significant improvement when analyzing the first five and ten documents, only managing to match the best team results at P@15. To better visualize this analysis, Figure 9 shows a comparison of the mean improvement of the teams and their respective previous team to the improvement of the proposed method and the best team results.

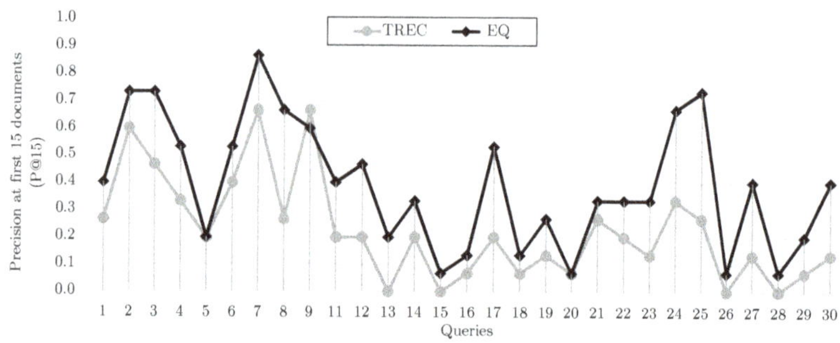

Figure 8. Comparison of the mean precision of the participants of the TREC2017 Precision Medicine Track (TREC) and the proposed method (EQ) in the top 15 retrieved documents.

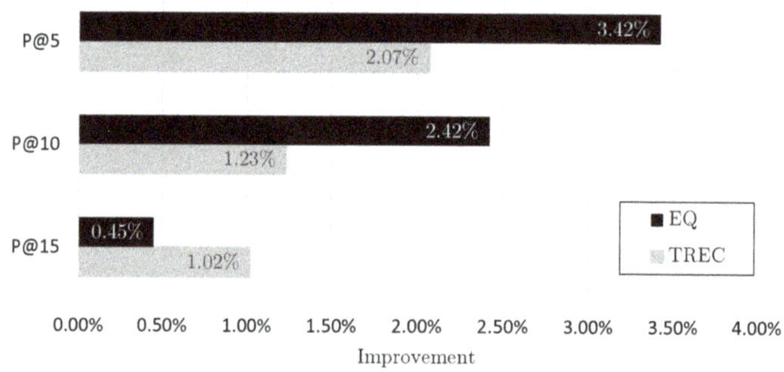

Figure 9. Mean improvement of the best teams at the TREC2017 Precision Medicine Track (TREC) over their respective previous team as compared to the improvement of the proposed method (EQ) over the best team results.

4. Conclusions

In this study, a new unsupervised query expansion technique using Multinomial Naive Bayes was presented. An expanded query was obtained by the combination of terms found in the original query and the new terms retrieved by the Multinomial Naive Bayes method. The extraction of the vocabulary from the documents retrieved by the combined query, and the selection of terms that were likely to belong to a category both proved to be effective in recovering more relevant documents.

More specifically, the application of this Pseudo-Feedback technique proved to be satisfactory considering the MAP and Precision results. The first 5, 10, 15, and 100 documents retrieved were considered in the evaluation process. Even when the first 100 documents were taken into account, the results improved, which shows that it is possible to increase their quantity and continue to improve the retrieval process quality.

The proposed query expansion technique allows for the improvement of a lightly defined query made by the user in order to obtain better results. This will help users to find relevant documents that fulfill their needs, and easily filter documents found in large and specialized collections of documents, such as the medical corpora, where technical lexicon vocabulary make it difficult to find relevant content through a short query composed of keywords.

Inspired by the success of this new method, more techniques could be researched. We plan to review some topic modeling techniques that could offer more terms inspired in the topic covered by the first retrieved documents, which would allow for a more complex and wider query expansion. In addition, new ways of combining the terms found in the original query and the expansion terms are being studied.

Author Contributions: Conceptualization, S.S.; methodology, S.S.; software, S.S.; validation, L.B., E.L.I. and A.S.V.; formal analysis, S.S.; investigation, S.S. and A.S.V.; resources, S.S.; data curation, S.S. and A.S.V.; writing—original draft preparation, S.S. and P.C.; writing—review and editing, S.S., P.C. and L.B.; visualization, L.B.; supervision, E.L.I. and L.B.; project administration, E.L.I. and L.B. All authors have read and agreed to the published version of the manuscript.

Funding: This research received no external funding.

Institutional Review Board Statement: Not applicable.

Informed Consent Statement: Not applicable.

Data Availability Statement: The Clinical Trial corpus used in this work is available on the TREC official web page http://www.trec-cds.org/2017.html.

Acknowledgments: The SING group thanks CITI (Centro de Investigación, Transferencia e Innovación) from the University of Vigo for hosting its IT infrastructure. We also appreciate the support provided by Consellería de Educación, Universidades e Formación Profesional (Xunta de Galicia) under the scope of the strategic funding of ED431C2018/55-GRC Competitive Reference Group. Pedro Celard is supported by a pre-doctoral fellowship from Xunta de Galicia (ED481A 2021/286). The funders had no role in study design, data collection and analysis, the decision to publish, or the preparation of the manuscript.

Conflicts of Interest: The authors declare no conflicts of interest.

References

1. Azad, H.K.; Deepak, A. Query expansion techniques for information retrieval: A survey. *Inf. Process. Manag.* **2019**, *56*, 1698–1735. [CrossRef]
2. Zhu, D.; Wu, S.; Carterette, B.; Liu, H. Using large clinical corpora for query expansion in text-based cohort identification. *J. Biomed. Inform.* **2014**, *49*, 275–281. [CrossRef] [PubMed]
3. McDonald, R.; Brokos, G.I.; Androutsopoulos, I. Deep relevance ranking using enhanced document-query interactions. In Proceedings of the 2018 Conference on Empirical Methods in Natural Language Processing, EMNLP 2018, Brussels, Belgium, 31 October–4 November 2018; pp. 1849–1860.
4. Rehman, A.; Javed, K.; Babri, H.A. Feature selection based on a normalized difference measure for text classification. *Inf. Process. Manag.* **2017**, *53*, 473–489. [CrossRef]
5. Araújo, G.; Mourão, A.; Magalhães, J. NOVASearch at Precision Medicine 2017. In Proceedings of the Twenty-Sixth Text REtrieval Conference (TREC 2017) Proceedings, Gaithersburg, MD, USA, 15–17 November 2017.
6. Afuan, L.; Ashari, A.; Suyanto, Y. A Study: Query Expansion Methods in Information Retrieval. *J. Phys. Conf. Ser.* **2019**, *1367*, 012001.
7. Agosti, M.; Di Nunzio, G.M.; Marchesin, S. An analysis of query reformulation techniques for precision medicine. In Proceedings of the 42nd International ACM SIGIR Conference on Research and Development in Information Retrieval, Paris, France, 21–25 July 2019; pp. 973–976.
8. Xu, B.; Lin, H.; Yang, L.; Xu, K.; Zhang, Y.; Zhang, D.; Yang, Z.; Wang, J.; Lin, Y.; Yin, F. A supervised term ranking model for diversity enhanced biomedical information retrieval. *BMC Bioinform.* **2019**, *20*, 1–11. [CrossRef] [PubMed]
9. Wang, J.; Pan, M.; He, T.; Huang, X.; Wang, X.; Tu, X. A Pseudo-relevance feedback framework combining relevance matching and semantic matching for information retrieval. *Inf. Process. Manag.* **2020**, *57*. [CrossRef]

10. Junior, J.R.C. *Desenvolvimento de uma Metodologia para Mineração de Textos*; Pontificia Universidad Catolica de Rio de Janeiro: Rio de janeiro, Brasil, 2007.
11. Porter, M.F. An algorithm for suffix stripping. *Program* **1980**, *14*, 130–137. [CrossRef]
12. Zipf, G.K. *Human Behaviour and the Principle of Least-Effort: An Introduction to Human Ecology*; Martino Fine Books: Eastford, CT, USA, 1949.
13. Baeza-Yates, R.A.; Ribeiro-Neto, B. *Modern Information Retrieval*; Addison-Wesley Longman: Reading, MA, USA, 1999.
14. Gauch, S.; Wang, J.; Rachakonda, S.M. A corpus analysis approach for automatic query expansion and its extension to multiple databases. *ACM Trans. Inf. Syst. (TOIS)* **1999**, *17*, 250–269. [CrossRef]
15. Crouch, C.J.; Yang, B. Experiments in automatic statistical thesaurus construction. In Proceedings of the 15th Annual International ACM SIGIR Conference on Research and Development in Information Retrieval, Copenhagen, Denmark, 21–24 June 1992; pp. 77–88.
16. Qiu, Y.; Frei, H.P. Concept based query expansion. In Proceedings of the 16th Annual International ACM SIGIR Conference on Research and Development in Information Retrieval, Pittsburgh, PA, USA, 27 June–1 July 1993; pp. 160–169.
17. Liddy, E.D.; Myaeng, S.H. DR-LINK's linguistic-conceptual approach to document detection. In Proceedings of the 1st Text Retrieval Conf. (TREC-1), Gaithersburg, MD, USA, 4–6 November 1992. [CrossRef]
18. Voorhees, E.M. *Query Expansion Using Lexical-Semantic Relations*; SIGIR '94; Springer: London, UK, 1994; pp. 61–69.
19. Miller, G.A.; Beckwith, R.; Fellbaum, C.; Gross, D.; Miller, K.J. Introduction to WordNet: An on-line lexical database. *Int. J. Lexicogr.* **1990**, *3*, 235–244. [CrossRef]
20. Borrajo, L.; Romero, R.; Iglesias, E.L.; Marey, C.R. Improving imbalanced scientific text classification using sampling strategies and dictionaries. *J. Integr. Bioinform.* **2011**, *8*, 90–104. [CrossRef]
21. Hirschman, L.; Yeh, A.; Blaschke, C.; Valencia, A. Overview of BioCreAtIvE: critical assessment of information extraction for biology. *BMC Bioinform.* **2005**, *6*, S1. [CrossRef] [PubMed]
22. Zhou, G. Recognizing names in biomedical texts using hidden markov model and SVM plus sigmoid. In Proceedings of the International Joint Workshop on Natural Language Processing in Biomedicine and Its Applications (NLPBA/BioNLP), Geneva, Switzerland, 28–29 August 2004; pp. 1–7.
23. Apweiler, R.; Bairoch, A.; Wu, C.H.; Barker, W.C.; Boeckmann, B.; Ferro, S.; Gasteiger, E.; Huang, H.; Lopez, R.; Magrane, M.; et al. UniProt: the universal protein knowledgebase. *Nucleic Acids Res.* **2004**, *32*, D115–D119. [CrossRef]
24. Strohman, T.; Metzler, D.; Turtle, H.; Croft, W.B. Indri: A language model-based search engine for complex queries. In Proceedings of the International Conference on Intelligent Analysis, Atlanta, GA, USA, 19–20 May 2005; Volume 2, pp. 2–6.
25. Turtle, H.; Flood, J. Query evaluation: strategies and optimizations. *Inf. Process. Manag.* **1995**, *31*, 831–850. [CrossRef]
26. Hiemstra, D.; van Leeuwen, D. Creating a Dutch information retrieval test corpus. In *Computational Linguistics in the Netherlands 2001*; Brill Rodopi: Leiden, The Netherlands, 2002; pp. 133–147.
27. Roberts, K.; Demner-Fushman, D.; Voorhees, E.M.; Hersh, W.R.; Bedrick, S.; Lazar, A.J.; Pant, S. Overview of the TREC 2017 precision medicine track. In Proceedings of the Text Retrieval Conference (TREC) NIH Public Access, Gaithersburg, MD, USA, 15–17 November 2017; Volume 26.
28. Mitra, M.; Singhal, A.; Buckley, C. Improving automatic query expansion. In Proceedings of the 21st Annual International ACM SIGIR Conference on Research and Development in Information Retrieval, Melbourne, Australia, 24–28 August 1998; pp. 206–214.
29. Raschka, S. Naive Bayes and Text Classification I-Introduction and Theory. *arXiv* **2014**, arXiv:1410.5329.
30. Mahmood, A.A.; Li, G.; Rao, S.; McGarvey, P.B.; Wu, C.H.; Madhavan, S.; Vijay-Shanker, K. *UD_GU_BioTM at TREC 2017: Precision Medicine Track*; TREC: Gaithersburg, MD, USA, 2017.

applied sciences

MDPI

Article

NASCA and NASES: Two Monolingual Pre-Trained Models for Abstractive Summarization in Catalan and Spanish

Vicent Ahuir *, Lluís-F. Hurtado, José Ángel González * and Encarna Segarra

VRAIN: Valencian Research Institute for Artificial Intelligence, Universitat Politècnica de València, Camí de Vera sn, 46022 València, Spain; lhurtado@dsic.upv.es (L.-F.H.); esegarra@dsic.upv.es (E.S.)
* Correspondence: viahes@eui.upv.es (V.A.); jogonba2@dsic.upv.es (J.Á.G.)

Abstract: Most of the models proposed in the literature for abstractive summarization are generally suitable for the English language but not for other languages. Multilingual models were introduced to address that language constraint, but despite their applicability being broader than that of the monolingual models, their performance is typically lower, especially for minority languages like Catalan. In this paper, we present a monolingual model for abstractive summarization of textual content in the Catalan language. The model is a Transformer encoder-decoder which is pretrained and fine-tuned specifically for the Catalan language using a corpus of newspaper articles. In the pretraining phase, we introduced several self-supervised tasks to specialize the model on the summarization task and to increase the abstractivity of the generated summaries. To study the performance of our proposal in languages with higher resources than Catalan, we replicate the model and the experimentation for the Spanish language. The usual evaluation metrics, not only the most used ROUGE measure but also other more semantic ones such as BertScore, do not allow to correctly evaluate the abstractivity of the generated summaries. In this work, we also present a new metric, called *content reordering*, to evaluate one of the most common characteristics of abstractive summaries, the rearrangement of the original content. We carried out an exhaustive experimentation to compare the performance of the monolingual models proposed in this work with two of the most widely used multilingual models in text summarization, mBART and mT5. The experimentation results support the quality of our monolingual models, especially considering that the multilingual models were pretrained with many more resources than those used in our models. Likewise, it is shown that the pretraining tasks helped to increase the degree of abstractivity of the generated summaries. To our knowledge, this is the first work that explores a monolingual approach for abstractive summarization both in Catalan and Spanish.

Keywords: abstractive summarization; monolingual models; multilingual models; transformer models; transfer learning

Citation: Ahuir, V.; Hurtado, L.-F.; González, J.Á.; Segarra, E. NASCA and NASES: Two Monolingual Pre-Trained Models for Abstractive Summarization in Catalan and Spanish. *Appl. Sci.* **2021**, *11*, 9872. https://doi.org/10.3390/app11219872

Academic Editors: Arturo Montejo-Ráez and Salud María Jiménez-Zafra

Received: 17 September 2021
Accepted: 20 October 2021
Published: 22 October 2021

1. Introduction

The purpose of the summarization process is to condense the most relevant information from a document or a set of documents into a small number of sentences. This process can be performed in an extractive or an abstractive way. While extractive summarization consists of identifying and copying those sentences in the original document that contain the most remarkable and useful information, abstractive summaries require abstractive actions that must be mastered. In this way, summaries are not mere clippings of the original documents; rather, abstractive summarizations are created by choosing the most important phrases of the documents and paraphrasing that content, creating a combination of some phrases, introducing new words, searching for synonyms, creating generalizations or specifications of some words or reordering content. All these actions must be done while preserving the linguistic cohesion and the coherence of the information [1–5].

Nowadays, Transformer-based language models excel in text generation, especially due to the transfer learning paradigm, by means of self-supervised pretraining on large text corpora, and later fine-tuning on downstream tasks. The generation capabilities achieved by these models boosted the state of the art in automatic summarization. However, most of the models proposed in the literature, such as BART [6], PEGASUS [7], or T5 [8] are intended to the English language and are not directly applicable to other languages. Multilingual models such as mBART [9] or mT5 [10] were also studied in the literature to address that language constraint, but despite their applicability being broader than that of the monolingual models, their performance is typically lower, especially on languages that are underrepresented in the pretraining corpora, or differ so much in linguistic terms from the most represented languages [11–14]

For minority languages like Catalan, the data resources available are much lower than other languages like English, Chinese, or Spanish. Additionally, the multilingual models typically either do not include data of minority languages, or if they do, its proportion in the pretraining sets is much lower than those of the majority languages. In this work, we hypothesize that monolingual models are a better choice for those minority languages, such as the Catalan language, which are underrepresented in the pretraining datasets of the multilingual models, but for which reasonable amounts of data are available.

In this work, a BART-like summarization model for the Catalan language is pretrained from scratch, and then fine-tuned on the summarization task. During the pretraining step, we include several self-supervised tasks to enhance the of the degree of abstractivity of the generated summaries. Furthermore, to test our hypothesis about monolingual models, we compare the performance of our proposal against well-known pretrained multilingual models such as mBART and mT5. It is also interesting to study the performance of our proposal in languages with higher resources than Catalan. For this reason, we replicate the model and the experimentation for the Spanish language to extract conclusions about abstractivity and monolingual models in two different languages.

We performed experimentation on the Dataset for Automatic summarization of Catalan and Spanish newspaper Articles (DACSA) corpus [15] This corpus provides pairs of news article and its summary from different journals in the Catalan and the Spanish languages. The experimental results show that the monolingual models generalize better than the multilingual ones, obtaining a more stable summarization performance on the test partitions of the DACSA dataset. The provided experimentation also illustrate the improvements in abstractivity as a result of the addition of the pretraining tasks. We analyze the abstractivity of the models through the use of abstractivity indicators [2]. Following some of these indicators, which correspond to actions done by professional summary writers, we quantify the degree of abstractivity of the generated summaries as the summaries generated by the models. One of the common actions when a person writes an abstractive summary is to rearrange the information from the original document. To our knowledge, no metrics were proposed for this specific action. For this reason, in this work the *content reordering* metric, which aims to quantify the rearrangement degree of the information in the summary with respect to the document, is proposed.

The contributions of this work are the following:

- A monolingual abstractive text summarization model, News Abstract Summarization for Catalan (NASCA), is proposed. This model, based on the BART architecture [6], is pretrained with several self-supervised tasks to improve the abstractivity of the generated summaries. For fine-tuning the model, a corpus of online newspapers is used (DACSA).
- An evaluation of the performance of the model on the summarization task and an evaluation of the degree of abstractivity of its generated summaries are presented. We compare the results of each NAS model with the results obtained by the summarization models based on well-known multilingual language models (mBART [9] and mT5 [10]) fine-tuned for the summarization task for each language using the DACSA corpus.

- A text summarization model with the same pretraining process than NASCA is also trained and evaluated for Spanish, News Abstract Summarization for Spanish (NASES).
- The *content reordering* metric is proposed, which helps to quantify if the extractive content within the abstractive summary is written in a different order than in the document.

The monolingual models, NASCA (https://huggingface.co/ELiRF/NASCA, accessed on 19 October 2021) and NASES (https://huggingface.co/ELiRF/NASES, accessed on 19 October 2021), proposed in this work were publicly release through HuggingFace model hub [16].

2. Related Work

Abstractive summarization works normally focused on the creation of models using approaches different to those used for extractive summarization [17–22]. Recently, abstractive summarizers became ubiquitous due to their powerful generation capabilities, achieved by using encoder-decoder architectures with Transformers [23] as backbone, and by pretraining them with self-supervised language modeling tasks on massive text corpora. This kind of models, especially PEGASUS [7], BART [6], T5 [8] and ProphetNet [24], fine-tuned for summarization tasks, are the state of the art in abstractive summarization benchmarks.

While all these models are nearly identical regarding their architecture, they mainly differ in the self-supervised tasks used in the pretraining stage. In some cases, such as BART, T5, and ProphetNet, these tasks aims the models to learn general aspects of the language, e.g., by masking tokens or reordering sentences. More specifically, BART is pretrained to reconstruct masked spans (text infilling) and to arrange sentences in the original order after being permuted (sentence permutation). Similarly, T5 is pretrained on encoder-decoder masked language modeling, in order to address universally all text-based language problems in a text-to-text format. Regarding ProphetNet, it is pretrained on future n-gram prediction to encourage the model to plan for future tokens instead of the next token, which prevents overfitting on strong local correlations. However, in other cases such as PEGASUS, the self-supervised tasks intentionally resemble the summarization task to encourage whole-document understanding and summary-like generation. In contrast to the previous models, PEGASUS is trained with Gap Sentences Generation (GSG), which consists of reconstructing the sentences that maximize the ROUGE with respect to the whole document. In this way, the authors of PEGASUS hypothesize that GSG is more suitable for abstractive summarization than other pretraining strategies, as it closely resembles the downstream task.

Other works are also based on strategies that involve pretraining to improve the abstractivity of the generated summaries. For instance, in [25], domain transfer and data synthesis techniques by using pretrained models are explored to improve the performance of abstractive summarization models in low-resource scenarios. Also, the authors of [26] propose to use pretrained language models to incorporate prior knowledge about language generation, which provides results comparable to state-of-the-art models in terms of ROUGE, while increasing the level of abstraction of the generated summaries, measured in terms of n-gram overlapping. Finally, in [27] a combination of several pretraining tasks is introduced to tailor the models to abstractive summarization, improving performance upon other Transformer-based models with significantly less pretraining data. Specifically, three tasks were proposed for pretraining: sentence reordering, next segment generation and masked document generation. While sentence reordering and masked document generation are identical to the text infilling and sentence permutation tasks used in BART, next segment generation aims to complete a document given a prefix of that document. Therefore, our work is similar to [27] in the sense that we combine the pretraining tasks of BART and PEGASUS to improve the abstractive skills of monolingual models trained for Catalan and Spanish.

All the models and proposals discussed in this section are intended for the English language, however, there are many other languages that deserve attention. Some efforts were done to consider other languages along with the English language by means of multilingual models such as mBART [9] or mT5 [10]. Although these efforts are very convenient and useful in many cases, the performance of the multilingual models is typically lower on languages that are underrepresented in the pretraining data or differ so much, in linguistic terms, from the most represented languages [13,14]. Learning monolingual models from scratch was extensively explored for language understanding by means of pretraining monolingual BERT models, with excellent results in many languages such as French [12,28], Dutch [29], or Spanish [11,30]. However, monolingual pretraining in languages other than English is still unexplored for language generation tasks such as abstractive summarization. To our knowledge, this is the first work that explores a monolingual approach for abstractive summarization both in Catalan and Spanish.

3. Newspapers Summarization Corpus

As stated above, the models proposed in this work are focused on the specific domain of newspaper articles. To train the models, the Dataset for Automatic summarization of Catalan and Spanish newspaper Articles (DACSA) [15] corpus was used. This corpus provides pairs of news article and its summary from different newspapers for both, the Catalan and the Spanish languages.

Regarding the Catalan set, there are 725,184 sample pairs from 9 newspapers, and their distribution is shown in the Table 1:

Table 1. Statistics of Catalan set. Sources marked with * were not used for training the models.

		Article				Summary		
Source	Docs	Tokens	|V|	Sents Per Doc	Words Per Sent	|V|	Sents Per Doc	Words Per Sent
#1	238,233	114,500,016	614,146	17.68	27.19	115,954	1.14	20.16
#2	194,697	105,119,526	621,612	19.99	27.01	112,904	1.28	19.14
#3	137,447	63,683,416	485,286	14.99	30.92	91,975	1.05	22.65
#4	56,827	24,891,291	276,720	14.84	29.52	58,071	1.21	17.52
#5	44,381	26,977,332	277,225	18.04	33.69	55,216	1.15	23.86
#6	35,763	17,181,460	202,931	11.31	42.49	42,289	1.05	22.79
#7 *	7104	3,800,842	83,942	18.04	29.66	19,267	1.02	26.51
#8 *	5882	9,414,192	185,977	66.04	24.24	31,006	2.54	24.84
#9 *	4850	2,667,185	102,024	23.61	23.29	19,584	1.16	28.05
Set	725,184	368,235,260	1,326,343	17.71	28.67	223,978	1.17	20.59

Regarding the Spanish set, the corpus provides 2,120,649 sample pairs from 21 newspapers, distributed as it is detailed in the Table 2:

When the distributions of the samples on both subsets are analyzed, the amount of samples by source is far from being homogeneous. If these distributions preserve over the partitions (training, validation, and test set), the models will focus their learning on the newspapers that are predominant. To avoid this bias and achieve more general models, the test and validation sets were created in a way that ensured that all newspapers had roughly the same number of samples on those sets. To achieve this balance in the validation and test sets, the sources with less samples were discarded. In this way, it is guaranteed that all sources represent at least 5% of samples in each one of these two sets. The sources that were excluded are marked with an asterisk in the Tables 1 and 2.

The three sets for Catalan include 6 of the 9 newspapers, creating a training set that contains 636,596 samples and 35,376 samples for validation and test sets. In the case of Spanish, the three sets are composed of 13 of the 21 newspapers provided in the Spanish set of DACSA: the training set contains 1,802,919 samples, and the validation and test sets contain 104,052 samples each.

Table 2. Statistics of Spanish set. Sources marked with * were not used for training the models.

			Article			Summary		
Source	Docs	Tokens	|V|	Sents Per Doc	Words Per Sent	|V|	Sents Per Doc	Words Per Sent
#1	550,148	420,786,144	1,473,628	31.36	24.39	210,079	1.40	19.02
#2	342,045	174,411,220	907,312	16.66	30.61	148,271	1.06	22.34
#3	196,410	93,755,039	622,073	15.40	31.00	110,728	1.02	20.59
#4	168,065	105,628,806	659,054	23.35	26.92	112,908	1.09	22.30
#5	148,053	105,453,102	626,058	28.35	25.13	109,546	1.47	20.46
#6	116,561	93,956,373	524,177	26.16	30.81	169,025	1.27	43.20
#7	107,162	70,944,634	470,244	19.90	33.26	87,901	1.29	25.27
#8	99,098	65,352,628	495,495148148	25.03	26.35	81,654	1.25	18.38
#9	81,947	42,825,867	363,075	15.54	33.63	71,913	1.03	22.41
#10	74,024	57,782,514	470,826	30.28	25.78	81,793	1.31	20.23
#11 *	70,193	29,692,261	272,248	11.06	38.26	84,898	1.22	44.48
#12	57,235	28,198,002	294,175	16.06	30.68	58,580	1.21	19.49
#13	35,163	20,156,337	260,690	19.22	29.83	50,556	1.15	21.20
#14	35,112	28,408,974	309,194	30.48	26.55	78,751	1.18	28.35
#15 *	17,379	10,099,958	153,598	16.82	34.54	41,512	1.85	26.89
#16 *	16,965	13,791,564	166,446	28.26	28.77	29,955	1.07	25.18
#17 *	2450	4,545,924	135,761	74.97	24.75	23,588	3.16	26.72
#18 *	1374	641,752	39,094	17.08	27.34	12,365	1.98	29.43
#19 *	643	398,834	26,797	17.73	34.99	2495	1.04	16.02
#20 *	467	233,873	22,699	18.70	26.78	3857	1.22	24.23
#21 *	155	199,140	19,750	39.06	32.89	2098	1.91	21.79
Set	2,120,649	1,367,262,946	3,189,783	23.44	27.50	516,307	1.24	22.95

All the sources excluded were used as a separate test set. This partition allows to evaluate the generalization capabilities of the models. In this work, we refer to the test set with newspapers included in the training set as TESTI and to the test set that contains newspapers not included in the training set as TESTNI. The statistics of all the sets are shown in Tables 3 and 4.

Table 3. Statistics of partitions for Catalan language.

			Article			Summary		
Partition	Docs	Tokens	|V|	Sents Per Doc	Words Per Sent	|V|	Sents Per Doc	Words Per Sent
Training	636,596	316,817,625	1,206,292	17.39	28.62	206,616	1.17	20.36
Validation	35,376	17,831,029	258,999	16.17	31.17	51,940	1.15	20.93
TESTI	35,376	17,704,387	262,148	16.13	31.03	51,958	1.15	20.89
TESTNI	17,836	15,882,219	247,154	35.38	25.17	45,997	1.56	25.93

Table 4. Statistics of partitions for Spanish language.

			Article			Summary		
Partition	Docs	Tokens	|V|	Sents Per Doc	Words Per Sent	|V|	Sents Per Doc	Words Per Sent
Training	1,802,919	1,172,626,265	2,920,894	23.94	27.17	454,179	1.24	21.99
Validation	104,052	67,669,381	550,213	23.01	28.27	109,460	1.21	23.36
TESTI	104,052	67,363,994	550,910	22.93	28.23	109,706	1.21	23.34
TESTNI	109,626	59,603,306	447,679	16.25	33.46	116,201	1.35	36.84

4. Summarization Models

In this work, a monolingual news summarization model is proposed: News Abstractive Summarization for Catalan (NASCA). It is a Transformer encoder-decoder model with the same architecture and hyper-parameters as BART [6]. Inspired by the work of Zou et al. [27], we decided to combine several pretraining tasks to inject linguistic knowledge during the pretraining stage with the aim of increasing the abstractivity of the summaries generated by the model. Specifically, four tasks were combined: sentence permutation, text infilling [6], Gap Sentence Generation (GSG) [7], and Next Segment Generation (NSG) [27]. NASCA is pretrained simultaneously with the four tasks, which are randomly selected at each batch following a uniform distribution.

We hypothesize that the combination of these four pretraining tasks leads to improvements in the summarization task, especially concerning the abstractivity of the generated summaries. Firstly, with sentence permutation and text infilling, the model should acquire capabilities of content reordering and phrase replacements. Secondly, GSG should tailor the model to whole-document understanding, summary-like generation and paraphrasing. Finally, with NSG, the model could increase the cohesion of the whole summary, as the task consists of generating continuations of documents given a prefix.

NASCA was pretrained with the documents of the Catalan training set of the DACSA corpus (including some documents discarded in the corpora creation process [15]), the Catalan subset of the OSCAR corpus [31], and the dump from 20 April 2021 of the Catalan version of the Wikipedia. In total, 9.3 GB of raw text (2.5 millions of documents) were used to pretrain it.

Additionally, we replicated NASCA for the Spanish language. We refer to this model as News Abstractive Summarization for Spanish (NASES). NASES is identical to NASCA in terms of architecture and pretraining tasks, but they differ in the pretraining dataset. To pretrain NASES, we only used the Spanish documents of the DACSA corpus and the dump from 20 April 2021 of the Spanish version of the Wikipedia. We did not consider for NASES the Spanish subset of OSCAR corpus so as to not increase excessively the difference in the amount of data available for the Spanish model regarding the Catalan one. In total, 21 GB (8.5 million documents) were used to pretrain NASES. Note that even though we did not use the OSCAR corpus, the size of the pretraining dataset for Spanish is twice the size of the Catalan pretraining dataset.

In addition to the monolingual models, two multilingual models were used for the experimental comparison in the summarization task. We worked with two of the most widely used multilingual models in text summarization, mBART and mT5. Regarding the mBART model, we used the *mbart-large-cc25* version, released by Facebook and available online through HuggingFace (https://huggingface.co/facebook/mbart-large-cc25, accessed on 19 October 2021) [16]. For the mT5 model, we used the *mt5-base* version, published by Google, that is also available online (https://huggingface.co/google/mt5-base, accessed on 19 October 2021)).

All the monolingual and multilingual models were fine-tuned and evaluated for the summarization task using the DACSA corpus. The monolingual models proposed in this work were publicly released (https://huggingface.co/ELiRF/NASCA, accessed on 19 October 2021), (https://huggingface.co/ELiRF/NASES, accessed on 19 October 2021).

5. Metrics

To evaluate the performance of the summarization models we used the usual evaluation metrics, the most used ROUGE measure [32] which is based on n-grams, and a more semantic such as BertScore [33], which is based on contextual embeddings provided by a BERT language model. However, these metrics do not allow to correctly evaluate the abstractivity of the generated summaries.

Measuring the abstractivity of the summaries generated by the models is, except counting the introduced new words, not trivial. In some studies, abstractivity was measured as the absence of n-gram overlap [34,35], however, creating abstractive summaries is not just about solely of using different vocabulary [2]. In this work, we used a set of metrics as abstractivity indicators to asses the level of abstractivity. In particular, the following metrics were selected: *extractive fragment coverage* [34], *abstractivity*$_p$ [35], *novel 1-grams*, *novel 4-grams* [26]. Also in this work, we present a new metric, called *content reordering*, to evaluate one of the most common characteristics of abstractive summaries, the rearrangement of the original content.

The *content reordering* metric was defined to quantify the percentage of reordering that the information in the summary suffered with respect to its original order in the document. This metric correlates positively with the abstractivity, and thus, by reordering the information, the summary increases its abstractivity.

The measure is based on the inversion concept. The inversion operation extracts all pairs of items that are out of order: $INV(\pi) = \{(a_i, a_j) | i < j \wedge a_i > a_j\}$, where π is a list of comparable elements [36]. For instance, with the list $[1, 5, 4, 2]$, the inverse operation results in $[(5, 4), (5, 2), (4, 2)]$.

Given a list of pairs (u, v), where u is the position of a maximum length segment in the original document, and v is the position in which such segment is placed in the summary, this list is sorted by u and the number of inversions that must be made to order the list of pairs by v is calculated. Thus, this allows us to quantify the disorder established in the list of the second component of the pairs when we take into account the order of the first component.

Let $\mathcal{F}(T, S)$[34] be the operation that returns the longest common extractive segments between a text T and its summary S, let $|S|$ be the number of words of the summary, and let $Reordered(T, S)$ be the operation that counts the number of extractive reordered segments; *content reordering* is defined as follows:

$$ContentReordering(T, S) = \begin{cases} \frac{\sum_{f \in \mathcal{F}(T,S)} |f|}{|S|} \cdot \frac{Reordered(T,S)}{|\mathcal{F}(T,S)| - 1}, & |\mathcal{F}(T,S)| > 1. \\ 0, & otherwise. \end{cases}$$

The output value range of the function is $[0, 1]$, where 1 is the highest degree of information rearrangement.

To illustrate this metric, we provide a full example with the following text (T):

> $_1$*Content reordering is a metric that* $_7$*quantifies how the extracted information from the original document is rearranged in the summary.* $_{21}$*Reorder the content* $_{24}$*is a common action used* $_{28}$*in abstractive summarization.*

and the following summary (S):

> $_1$*In abstractive summarization,* $_4$*reorder the content* $_7$*is a common action,* $_{11}$*content reordering* $_{13}$*quantifies it.*

The highlighted text are fragments in common between the original text and its summary. The subindex before the fragment indicates the starting position in words of the fragment. Thus, the list of the pairs (u, v) of the extractive fragments is the following one when it is ordered by u:

$$[(1, 11), (7, 13), (21, 4), (24, 7), (28, 1)]$$

The resulting list of the INV operation applied on the list made up with the second components of the pairs of the previous list is:

$$INV([11, 13, 4, 7, 1]) = [(\underline{11}, 4), (11, 7), (11, 1), (\underline{13}, 4), (13, 7), (13, 1), (\underline{4}, 1), (\underline{7}, 1)]$$

The $Reorder(T, S)$ operation is 4 since there are 4 extractive reordered segments. This value is computed as the unique values in the first components of the pairs in the previous list (11, 13, 4, 7). Additionally, the length (in words) of the summary is 14, there are 5 extractive fragments, and the sum of their length is 13. With all this information, the *content reordering* metric is calculated as follows:

$$ContentReordering(T, S) = \frac{13}{14} \cdot \frac{4}{5-1} = 0.93$$

With this result, we conclude that there is a certain degree of abstractivity in the summary introduced by a high degree of rearrangement of the information. This fact can be verified in the summary of the example. This abstractivity was introduced by the rearrangement of the extractive segments, and not due to the absence of text overlapping between the summary and the original text.

6. Results

In this section, we present the conducted experimentation with the summarization models. Firstly, we present the results of the performance obtained by the three models for Catalan in the summarization task: the NASCA model, the mBART model, and the mT5 model. Secondly, we show the results regarding the abstractivity of these models for Catalan. Additionally, we show the results for the three models for Spanish, the NASES model and the two multilingual ones. All the models were evaluated on the two test partitions, TESTI and TESTNI.

6.1. Summarization Performance of the Models for CATALAN

The performance of the models was evaluated using the ROUGE metrics [32] and BERTScore metric [33]. For each metric, we calculated the average F1 score and its 95% confidence interval by using bootstrapping. Results are shown in Table 5.

The average F1 scores are shown in a normal font size and their confidence intervals in a smaller font size, placed at the right-side of the score. The best average score for each metric within a test partition is remarked in bold style. The confidence intervals are shown in blue color if their range intersects with the confidence interval of the best score value of the metric within the same test partition; in other case, the confidence intervals are presented in black color.

Table 5. Average F1 scores and confidence intervals of models in summarization task in Catalan.

Partition	Model	ROUGE-1	ROUGE-2	ROUGE-L	ROUGE-Ls	BERTScore
TESTI	NASCA	**28.84** (28.68, 29.01)	11.68 (11.51, 11.85)	22.78 (22.61, 22.94)	23.30 (23.13, 23.46)	71.85 (71.78, 71.92)
	mBART	28.59 (28.42, 28.77)	**11.89** (11.73, 12.06)	**23.00** (22.82, 23.16)	**23.39** (23.22, 23.56)	**72.03** (71.96, 72.10)
	mT5	27.01 (26.84, 27.18)	10.70 (10.54, 10.87)	21.81 (21.65, 21.97)	22.12 (21.98, 22.29)	71.55 (71.49, 71.61)
TESTNI	NASCA	**28.19** (27.97, 28.42)	11.20 (10.99, 11.43)	**21.45** (21.20, 21.65)	**22.44** (22.21, 22.67)	70.14 (70.05, 70.22)
	mBART	27.46 (27.24, 27.69)	11.04 (10.81, 11.29)	21.13 (20.93, 21.37)	22.01 (21.78, 22.24)	70.33 (70.25, 70.43)
	mT5	27.00 (26.77, 27.23)	**11.28** (11.04, 11.52)	21.27 (21.03, 21.51)	22.01 (21.78, 22.23)	**70.56** (70.47, 70.65)

The Table 5 shows, regarding the TESTI partition, that the NASCA model performs similarly compared to the multilingual mBART model. mBART presents significantly better BERTScore result than NASCA while there are overlappings in the confidence intervals in the ROUGE measures. The mT5 model has obtained a significant lower performance than the other two models, despite the fact that mT5 contains the Catalan language in its pretraining phase unlike the mBART model. We hypothesize that the pretraining dataset could influence the results. It could be that the data considered for Catalan to pretrain mT5 differs so much from our domain. Also, the proportion of languages similar to Catalan in the pretraining corpus could be related to this effect.

In the case of the TESTNI partition, there is a significant overall reduction of the performance in most of the metrics of the three models in comparison to the TESTI partition. Generally speaking, the NASCA model has significantly better performance in almost all ROUGE metrics compared to the multilingual models, although there is an overlapping between the confidence interval of NASCA and that of mT5 in ROUGE-2. According to BERTScore, the mT5 model obtains significant differences in comparison to the scores of the NASCA and mBART models.

Taking into account the higher scores and the generalization capabilities, the results of the monolingual model are significant better than the multilingual ones. In one side, mBART has similar performance than NASCA model in the TESTI partition, however, the performance reduction in the second test partition indicates that the model generalizes worse than the other two models. On the other side, the mT5 model generalizes better than mBART, since the drop of the perfomance between the TESTI and the TESTNI is lower in mT5 than mBART, however, mT5 presents significantly lower performance than that of the NASCA model.

6.2. Abstractivity of the Summaries Generated by the Models for Catalan

To evaluate the abstractivity, 4 metrics were used: *extractive fragment coverage* [34] (henceforth, we refer to it simply as *coverage*), *abstractivity*$_p$ [35], *novel n-grams* [26]and *content reordering*. From now on, we refer those metrics as indicators, since each indicator complements, in some way, the other indicators to obtain a global perception of the level of abstractivity. The Table 6 shows the average scores and their confidence intervals. The scores are calculated by comparing the generated summaries against to their respective article text. The scores remarked in bold styles indicates the highest abstractivity. In this experimentation, the lowest value is emphasized in the *extractive fragment coverage* indicator since it correlates negatively with the abstractivity and the highest value is remarked in the remaining abstractivity indicators, since they correlate positively.

Table 6. Abstractivity indicators and confidence intervals for Catalan. Values are shown as percentages.

Partition	Model	Extractive Fragment Coverage	Content Reordering	Abstractivity$_p$ (p = 2)	Novel 1-Grams	Novel 4-Grams
TESTI	NASCA	**96.99** (96.94, 97.04)	46.17 (45.79, 46.55)	**47.19** (46.90, 47.46)	**03.21** (03.15, 03.26)	**28.65** (28.41, 28.92)
	mBART	97.73 (97.68, 97.77)	**47.85** (47.44, 48.23)	37.70 (37.42, 37.97)	02.40 (02.36, 02.45)	23.80 (23.55, 24.02)
	mT5	98.59 (98.55, 98.62)	41.25 (40.84, 41.67)	38.04 (37.78, 38.28)	01.51 (01.48, 01.55)	21.89 (21.71, 22.08)
TESTNI	NASCA	**96.66** (96.55, 96.77)	42.37 (41.84, 42.88)	**41.89** (41.44, 42.37)	**03.52** (03.40, 03.63)	**26.32** (25.91, 26.68)
	mBART	97.08 (96.99, 97.16)	**42.96** (42.40, 43.56)	36.98 (36.55, 37.41)	03.01 (02.92, 03.09)	24.32 (23.95, 24.70)
	mT5	98.31 (98.26, 98.36)	38.82 (38.24, 39.41)	39.18 (38.83, 39.54)	01.80 (01.74, 01.85)	23.20 (22.92, 23.48)

As it is shown in Table 6, all the models show a predominant extractivity behavior in the same way as the most abstractive models in the literature. All the scores of the abstractivity indicators denote low abstractivity. For instance, the *coverage* and *novel 1-grams* indicators show that the models reuses a lot of words from the original documents. Although all the models present high-extractivity in their generated summaries, there are significant differences among the models that can be analyzed.

Regarding the TESTI partition, the scores of most of the abstractivity indicators of the NASCA model reflect significantly better abstractivity than that of the multilingual models. Also, we can observe that the multilingual models have relatively similar scores in most of the indicators, although, the indicators of the mBART model show slightly more abstractivity than the mT5 model.

In the case of the TESTNI partition, the NASCA model indicators reflect better abstractivity than in the multilingual models. However, compared to the values in TESTI, NASCA reduced most of their abstractivity indicators scores except the *coverage* indicator, which is slightly better. In this partition, the differences in the values between the NASCA model and the multilingual models are lower than in the TESTI partition.

Overall, it is noticeable that the NASCA model reuses a lot of content from the original text. The model uses a lot of words from the original text which is reflected in the low value of the *novel 1-grams* indicator. However, despite the fact that the model reuses a lot of words, the extractive fragments tend to be shorter than in the multilingual models, since the *novel 4-grams* indicator shows a significantly higher value than in the multilingual models;

this fact is also exposed by the *abstractivity*$_p$ indicator, which presents a difference between the 5% and the 10% depending on the partition and the multilingual model. For all these observations in the indicators, we conclude that the NASCA model generates summaries with higher degree of abstractivity than the multilingual models.

With the aim of better analyzing the behavior of the models, we computed the cumulative distributions of the abstractivity indicators for each model and test partition. The results are presented in the Figure 1.

The plots show in the x-axis the indicator measured, and in the y-axis, the percentage of generated summaries that present less or equal score to the value in the x-axis. These plots are helpful to evaluate the abstractivity of the generated summaries by taking into account how they are distributed based on certain score. If a metric correlates negatively with the abstractivity, it is desired that the scores be lower; that is, the model accumulates the samples fast. In contrast, if the metric correlates positively, it is desired that the scores be higher. In this case, we say that the model accumulates the samples slowly.

TESTI

TESTNI

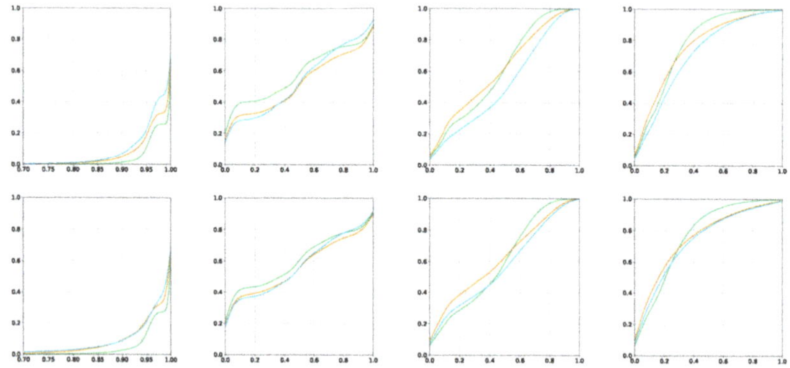

(a) *Ext. Frag. Coverage* (b) *Content Reordering* (c) Abstractivity$_p$ ($p = 2$) (d) *Novel 4-grams*

Figure 1. Cumulative distribution of 4 abstractivity indicators for models NASCA, mBART, mT5 for Catalan.

In Figure 1, regarding the *coverage* indicator, which correlates negatively with abstractivity, we observe that the NASCA model stays always on top of the multilingual models, so this indicates that the samples are accumulated faster, which is a positive indication for the abstractivity. In the remaining indicators, which correlate positively with the abstractivity, the NASCA model tends to accumulate the samples slower than the multilingual models, which is also positive concerning the abstractivity, except the *content reordering* indicator. Regarding this indicator, although NASCA present a lower value than the mBART model in the Section 6.2, the NASCA model's distribution stays below the mBART until 40%, and later reaches and surpasses the multilingual models. This means that the *NASca* model, overall, introduces less *content reordering* on their summaries; however, the amount of summaries with rearrangement of the information is higher than in the ones generated by the multilingual models.

The results presented in the Table 6 and the Figure 1 show enough evidences to conclude that the NASCA model presents better abstractivity than the rest of the trained models. Additionally, to verify if the improvement in the abstractivity indicators is due to the pretraining tasks, we pretrained a BART model specifically for Catalan using only the pretraining tasks proposed in the original work [6]. The results show that both models, NASCA and BART, have a similar performance in the summarization task, however, the NASCA model presents significant higher abstractivity indicators. For instance, in the *coverage* indicator of the TESTNI partition, the NASCA model scores 96.99 (96.94, 97.04) and

BART 97.29 (97.24, 98.41). In the case of novel 4-grams, and also for TESTNI, the NASCA model scores 26.65 (25.91, 26.68) and BART 25.48 (25.12, 25.82).

An example of an article and the summaries generated by the three models is shown in Appendix A.

6.3. Summarization Performance and Abstractivity of the Summaries Generated by the Models for Spanish

It is also interesting to study the performance of our proposal in languages with higher resources than Catalan. For this reason, we replicated the model and the experimentation for the Spanish language. The summarization performance results and the results related to the abstractivity indicators are shown in Tables 7 and 8, respectively. In addition, the cumulative distributions of the abstractivity indicators are presented in Figure 2.

Table 7. Average F1 scores and confidence intervals of models in summarization task in Spanish.

Partition	Model	ROUGE-1	ROUGE-2	ROUGE-L	ROUGE-Ls	BERTScore
TESTI	NASES	**33.24** (33.12, 33.38)	**15.79** (15.63, 15.93)	**26.76** (26.63, 26.89)	**27.56** (27.43, 27.69)	**73.11** (73.05, 73.16)
	mBART	31.09 (30.98, 31.20)	13.56 (13.44, 13.68)	24.67 (24.56, 24.78)	25.48 (25.37, 25.58)	72.25 (72.21, 72.30)
	mT5	31.72 (31.60, 31.85)	14.54 (14.39, 14.67)	25.76 (25.63, 25.89)	26.31 (26.18, 26.44)	72.86 (72.82, 72.91)
TESTNI	NASES	30.60 (30.52, 30.68)	10.75 (10.66, 10.83)	22.29 (22.21, 22.37)	23.06 (22.99, 23.15)	70.66 (70.62, 70.69)
	mBART	**30.66** (30.58, 30.74)	12.08 (11.98, 12.18)	23.13 (23.06, 23.22)	23.89 (23.81, 23.98)	71.07 (71.04, 71.10)
	mT5	30.61 (30.51, 30.70)	**12.36** (12.25, 12.47)	**23.53** (23.43, 23.62)	**24.05** (23.95, 24.14)	**71.26** (71.22, 71.30)

Table 7 shows that the NASES model presents the best performance of the three models in the TESTI partition. All the scores obtained by the NASES model are significantly better compared to those of the multilingual models. Specifically, the NASES model achieve, on average, 8.2% higher performance than mBART and 4.5% higher than mT5. Regarding the TESTNI partition, the NASES model reduces its performance in average, while mT5 achieves the best results in almost all the metrics.

The results show that the NASES excelled in the TESTI partition, which contains newspapers included in the training partition. However, NASES presents lower generalization capabilities than the multilingual models due to the noticeable performance reduction in the TESTNI partition, which contains newspapers not included in the training partition.

Table 8. Abstractivity indicators and confidence intervals for Spanish. Values are shown as percentages.

Partition	Model	Extractive Fragment Coverage	Content Reordering	Abstractivity$_p$ (p = 2)	Novel 1-Grams	Novel 4-Grams
TESTI	NASES	**97.65** (97.62, 97.68)	**45.27** (45.04, 45.50)	**38.15** (37.97, 38.31)	**02.55** (02.52, 02.58)	**21.17** (21.04, 21.31)
	mBART	98.14 (98.10, 98.18)	37.70 (37.45, 37.92)	35.17 (35.00, 35.32)	01.85 (01.81, 01.89)	17.58 (17.47, 17.70)
	mT5	98.74 (98.72, 98.76)	38.67 (38.42, 38.92)	32.41 (32.25, 32.58)	01.36 (01.34, 01.38)	17.39 (17.29, 17.49)
TESTNI	NASES	**98.16** (98.13, 98.19)	**46.58** (46.33, 46.82)	29.76 (29.60, 29.92)	**02.00** (01.97, 02.03)	**15.76** (15.65, 15.88)
	mBART	98.92 (98.90, 98.94)	39.38 (39.13, 39.61)	**30.48** (30.33, 30.64)	01.03 (01.01, 01.05)	14.68 (14.59, 14.78)
	mT5	99.24 (99.23, 99.26)	37.17 (36.91, 37.43)	24.19 (24.06, 24.32)	00.83 (00.81, 00.84)	12.08 (12.00, 12.16)

Regarding the abstractivity indicators on the TESTI partition, presented in Table 8, all the scores of the NASES model are significantly better than those of the multilingual models. In the TESTNI partition, the models present less abstractivity in comparison to the TESTI partition. Also in TESTNI, the NASES model shows significant differences compared to the multilingual models in all the indicators, excluding *abstractivity*$_p$ where mBART obtains better scores than NASES and the mT5 models. We also computed the cumulative distributions of the abstractivity indicators for each model and test partition. The results are presented in the Figure 2.

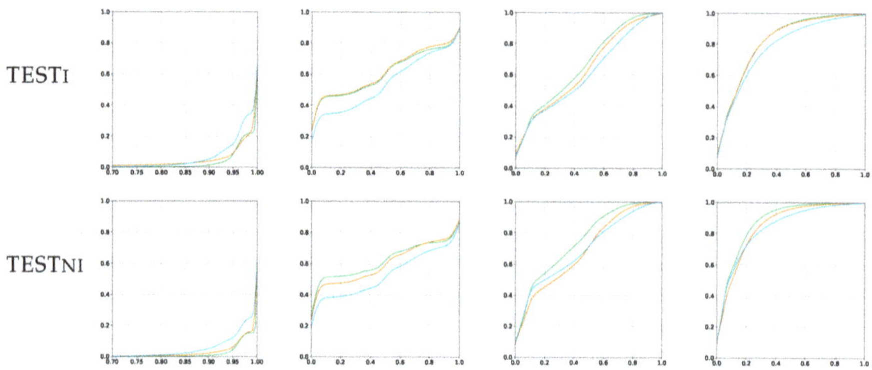

(**a**) *Ext. Frag. Coverage* (**b**) *Content Reordering* (**c**) Abstractivity$_p$ ($p = 2$) (**d**) *Novel 4-grams*

Figure 2. Cumulative distribution of 4 abstractivity indicators for models NASES, mBART, mT5 for Spanish.

The plots presented in Figure 2 help us to reinforce the observations extracted from the numerical results showed in Table 8. The NASES model tends to accumulate slightly higher percentage of samples in the *coverage* indicator after the 90% of *coverage* is achieved. Regarding the remaining indicators, the accumulation tends to occur slower than in the other two models.

The abstractivity indicators analysis shows that the summaries generated by NASES have a significant higher abstractivity than those generated by the multilingual models, something that complements the observations made in the Sections 6.1 and 6.2 about the models for Catalan.

7. Conclusions

In this work, a monolingual model for abstractive summarization in Catalan, NASCA, was presented. The model was pretrained from scratch based on the BART architecture and using four self-supervised tasks with the aim of increasing the abstractivity of the generated summaries. The fine-tuning phase was carried out using the DACSA dataset, a corpus of articles obtained from online newspapers. The experimentation conducted supports the correctness of our proposal considering the three evaluated aspects: the performance of the model, the abstractivity of the generated summaries, and the generalization capabilities of the model.

Following the same architecture and the same training strategy, a model for abstractive summarization in Spanish, NASES, was also trained and evaluated, and it also provided very good results. To our knowledge, this is the first work that explores a monolingual approach for abstractive summarization both in Catalan and Spanish.

Additionally, in this work, we also proposed a new metric, *content reordering*, with the aim of helping to quantify the rearrangement of the original content within an abstractive summary. This characteristic is common in abstractive summaries, but it is not considered by the metrics in the literature.

Author Contributions: V.A. conceptualization, software, formal analysis, resources, data curation, writing—original draft preparation, and visualization. L.-F.H.: methodology, validation, formal analysis, resources, writing—review and editing, supervision, project administration, and funding acquisition. J.Á.G.: methodology, software, investigation, data curation, writing—original draft preparation, and visualization. E.S.: conceptualization, validation, investigation, writing—review and editing, supervision, project administration, and funding acquisition. All authors have read and agreed to the published version of the manuscript.

Funding: This work was partially supported by the Spanish Ministerio de Ciencia, Innovación y Universidades and FEDER founds under the project AMIC (TIN2017-85854-C4-2-R), and by the Agencia Valenciana de la Innovació (AVI) of the Generalitat Valenciana under the GUAITA (INNVA1/2020/61) project.

Institutional Review Board Statement: Not applicable.

Informed Consent Statement: Not applicable.

Conflicts of Interest: The authors declare no conflict of interest.

Abbreviations

The following abbreviations are used in this manuscript:

DACSA	Dataset for Automatic summarization of Catalan and Spanish newspaper Articles
GSG	Gap Sentences Generation
MDG	Masked Document Generation
NASCA	News Abstractive Summarization for Catalan
NASES	News Abstractive Summarization for Spanish
NSG	Next Segment Generation
SR	Sentence Reordering

Appendix A. Summarization Example

An example of an article, its reference summary, and the summaries generated by the three models are shown in Figure A1. It also shows the different metrics achieved by each summary. All the generated summaries are syntactically and semantically correct. Based on the low values of the ROUGE scores, we can affirm that all the generated summaries are very different from the reference one. Regarding the coverage indicator, although the three summaries are quite extractive, since they use several segments from the article, mT5 is by far the most extractive. Considering all the abstractive indicators, NASCA and mBART are better than mT5, and NASCA outperforms mBART especially in terms of novel n-grams and abstractivity$_p$.

Article: La clau va ser el ritme. El ritme amb què Marc Márquez va arrencar al Gran Premi de l'Argentina i amb què el va acabar. El pilot de Cervera, que sempre assegura que li agraden les curses en grup, va fer avançaments, va buscar els forats i va passar-s'ho bé dalt de la moto: a l'Argentina va decidir ser, per un dia, infidel al seu estil. Sabia que tenia ritme, ho havia demostrat durant totes les sessions d'entrenaments lliures i també als oficials (havia dominat cinc de les sis sessions), i a la cursa no va tenir rival. Va sortir, va posar el "mode creuer", com va dir, i va perdre de vista la resta de rivals. En una volta, un segon d'avantatge, i ja s'escapava de 12 segons dels perseguidors quan va decidir passar a controlar la cursa, sense prendre més riscos dels necessaris. "No és el meu estil, però després del que va passar l'any passat tenia ganes de fer una cursa així. Va passar el que va passar i volia demostrar el meu ritme", va assegurar després de baixar de la moto. Márquez va marcar la pole i la volta ràpida, i va ser líder des que es van apagar els semàfors fins al final. Va aconseguir el que es coneix com un Grand Chelem: el de Cervera, de fet, tan sols n'ha aconseguit cinc des que va debutar a MotoGP; tres a Austin (2014, 2016 i 2018), un a Jerez (2014) i el de diumenge a l'Argentina. "Pocs dies a l'any et trobes amb aquestes sensacions dalt de la moto. Calia aprofitar-ho, ha sigut perfecte", reconeixia. La manera més dolça de marcar el ritme. La victòria es va començar a coure molt abans de la sortida, al box, amb el seu equip, llegint els temps de les sessions d'entrenaments. "Els papers deien que era qui tenia més ritme. He intentat marcar les diferències en les set primeres voltes i, després, mantenir l'avantatge", explicava el català. Com si fos un rellotge, clavava volta a volta un 1:39. Al final, els 12 segons d'avantatge es van reduir a 9.816, que, si bé no és la distància més gran amb què Márquez ha guanyat una cursa (a Brno el 2017 va acabar primer amb 12.438 respecte a Pedrosa), sí que és la més gran que ha aconseguit el de Cervera en una cursa en sec: tant a Brno fa dos anys com a Sachsenring en fa tres, en què va acabar a 9.857 de Crutchlow, la pluja va marcar les curses. Lluny també queden els més de 37 segons d'avantatge amb què Dani Pedrosa va guanyar a Xest el 2012 sobre Nakasuga, també sota la pluja, després de la caiguda de Lorenzo. "Com que hem guanyat per deu segons, sembla que som en un altre món, però no, la distància és només de quatre punts respecte a Dovizioso", afegia Márquez. Just abans del podi es va veure segurament una de les imatges de l'any: Valentino Rossi, que va acabar segon, va encaixar la mà amb Márquez, un gest que no es veia des de feia un any, quan el de Cervera, precisament a Termas de Río Hondo, va tocar l'italià i el va fer caure, cosa que va comportar l'inici d'un terratrèmol. Diumenge, ja al podi, els dos campions van fer xocar les ampolles de xampany, però sense dirigir-se la paraula.

Reference: El triomf de Márquez a l'Argentina, el més ampli en sec del de Cervera a MotoGP.

NASCA: El de Cervera va marcar la 'pole' a l'Argentina i va ser líder del Mundial en una volta.
(ROUGE-1: 5.97; ROUGE-2: 4.42; ROUGE-L: 4.72; BertScore: 67.08)
(Coverage: 85.00; Reordering: 85.00; Abstractivity$_p$: 87.75; Novel 1-grams: 15.79; Novel 4-grams: 94.12)

mBART: El de Cervera marca la 'pole' a l'Argentina i és líder des que es van apagar els semàfors.
(ROUGE-1: 6.28; ROUGE-2: 4.72; ROUGE-L: 5.97; BertScore: 69.17)
(Coverage: 85.00; Reordering: 85.00; Abstractivity$_p$: 79.75; Novel 1-grams: 15.00; Novel 4-grams: 70.59)

mT5: El pilot de Cervera, que sempre assegura que li agraden les curses en grup, va fer avançaments, va buscar els forats i va passar-se bé dalt de la moto.
(ROUGE-1: 9.58; ROUGE-2: 8.68; ROUGE-L: 9.27; BertScore: 72.96)
(Coverage: 96.97; Reordering: 48.48; Abstractivity$_p$: 35.54; Novel 1-grams: 3.70; Novel 4-grams: 13.33)

Figure A1. Text of the article, the reference summary, and the summaries generated by the models.

References

1. Rane, N.; Govilkar, S. Recent Trends in Deep Learning Based Abstractive Text Summarization. *Int. J. Recent Technol. Eng.* **2019**, *8*, 3108–3115. [CrossRef]
2. Jing, H. Using Hidden Markov Modeling to Decompose Human-Written Summaries. *Comput. Linguist.* **2002**, *28*, 527–543. [CrossRef]
3. Verma, P.; Pal, S.; Om, H. A Comparative Analysis on Hindi and English Extractive Text Summarization. *ACM Trans. Asian Low-Resour. Lang. Inf. Process.* **2019**, *18*, 1–39. [CrossRef]
4. Widyassari, A.P.; Rustad, S.; Shidik, G.F.; Noersasongko, E.; Syukur, A.; Affandy, A.; Setiadi, D.R.I.M. Review of automatic text summarization techniques & methods. *J. King Saud Univ. Comput. Inf. Sci.* **2020**. [CrossRef]
5. National Information Standards Organization. *Guidelines for Abstracts*; Standard, American National Standards Institute: Gaithersburg, MD, USA, 1997.
6. Lewis, M.; Liu, Y.; Goyal, N.; Ghazvininejad, M.; Mohamed, A.; Levy, O.; Stoyanov, V.; Zettlemoyer, L. BART: Denoising Sequence-to-Sequence Pre-training for Natural Language Generation, Translation, and Comprehension. In Proceedings of the 58th Annual Meeting of the Association for Computational Linguistics, Association for Computational Linguistics, Online, 5–10 July 2020; pp. 7871–7880. [CrossRef]
7. Zhang, J.; Zhao, Y.; Saleh, M.; Liu, P. PEGASUS: Pre-training with Extracted Gap-sentences for Abstractive Summarization. In Proceedings of the 37th International Conference on Machine Learning, Vienna, Austria, 13–18 July 2020; pp. 11328–11339.
8. Raffel, C.; Shazeer, N.; Roberts, A.; Lee, K.; Narang, S.; Matena, M.; Zhou, Y.; Li, W.; Liu, P.J. Exploring the Limits of Transfer Learning with a Unified Text-to-Text Transformer. *J. Mach. Learn. Res.* **2020**, *21*, 1–67.

9. Liu, Y.; Gu, J.; Goyal, N.; Li, X.; Edunov, S.; Ghazvininejad, M.; Lewis, M.; Zettlemoyer, L. Multilingual Denoising Pre-training for Neural Machine Translation. *Trans. Assoc. Comput. Linguist.* **2020**, *8*, 726–742. [CrossRef]
10. Xue, L.; Constant, N.; Roberts, A.; Kale, M.; Al-Rfou, R.; Siddhant, A.; Barua, A.; Raffel, C. mT5: A Massively Multilingual Pre-trained Text-to-Text Transformer. In Proceedings of the 2021 Conference of the North American Chapter of the Association for Computational Linguistics: Human Language Technologies, Mexico City, Mexico, 6–11 June 2021; pp. 483–498. [CrossRef]
11. Cañete, J.; Chaperon, G.; Fuentes, R.; Ho, J.H.; Kang, H.; Pérez, J. Spanish Pre-Trained BERT Model and Evaluation Data. 2020. Available online: https://users.dcc.uchile.cl/~jperez/papers/pml4dc2020.pdf (accessed on 19 October 2021).
12. Martin, L.; Muller, B.; Ortiz Suárez, P.J.; Dupont, Y.; Romary, L.; de la Clergerie, É.V.; Seddah, D.; Sagot, B. CamemBERT: A Tasty French Language Model. In Proceedings of the 58th Annual Meeting of the Association for Computational Linguistics, Online, 5–10 July 2020.
13. Virtanen, A.; Kanerva, J.; Ilo, R.; Luoma, J.; Luotolahti, J.; Salakoski, T.; Ginter, F.; Pyysalo, S. Multilingual is not enough: BERT for Finnish. *arXiv* **2019**, arXiv:1912.07076 [CrossRef]
14. Pires, T.; Schlinger, E.; Garrette, D. How Multilingual is Multilingual BERT? In P*roceedings of the 57th Annual Meeting of the Association for Computational Linguistics*; Association for Computational Linguistics: Florence, Italy, 2019; pp. 4996–5001. [CrossRef]
15. DACSA: A Dataset for Automatic summarization of Catalan and Spanish newspaper Articles. Unsubmitted.
16. Wolf, T.; Debut, L.; Sanh, V.; Chaumond, J.; Delangue, C.; Moi, A.; Cistac, P.; Rault, T.; Louf, R.; Funtowicz, M.; et al. Transformers: State-of-the-Art Natural Language Processing. In Proceedings of the 2020 Conference on Empirical Methods in Natural Language Processing: System Demonstrations, Online, 16–20 November 2020; pp. 38–45.
17. Zhong, M.; Liu, P.; Chen, Y.; Wang, D.; Qiu, X.; Huang, X. Extractive Summarization as Text Matching. In Proceedings of the 58th Annual Meeting of the Association for Computational Linguistics, Online, 6–10 July 2020; pp. 6197–6208. [CrossRef]
18. Liu, Y.; Lapata, M. Text Summarization with Pretrained Encoders. In Proceedings of the 2019 Conference on Empirical Methods in Natural Language Processing and the 9th International Joint Conference on Natural Language Processing (EMNLP-IJCNLP), Hong Kong, China, 3–7 November 2019; pp. 3730–3740. [CrossRef]
19. Nallapati, R.; Zhai, F.; Zhou, B. SummaRuNNer: A Recurrent Neural Network Based Sequence Model for Extractive Summarization of Documents. In Proceedings of the Thirty-First AAAI Conference on Artificial Intelligence, San Francisco, CA, USA, 4–9 February 2017; AAAI'17; p. 3075–3081.
20. Rush, A.M.; Chopra, S.; Weston, J. A Neural Attention Model for Abstractive Sentence Summarization. In Proceedings of the 2015 Conference on Empirical Methods in Natural Language Processing, Lisbon, Portugal, 17–21 September 2015; pp. 379–389. [CrossRef]
21. Nallapati, R.; Zhou, B.; dos Santos, C.; Gúllçehre, Ç.; Xiang, B. Abstractive Text Summarization using Sequence-to-sequence RNNs and Beyond. In Proceedings of the 20th SIGNLL Conference on Computational Natural Language Learning, Berlin, Germany, 11–12 August 2016; pp. 280–290. [CrossRef]
22. See, A.; Liu, P.J.; Manning, C.D. Get To The Point: Summarization with Pointer-Generator Networks. In Proceedings of the 55th Annual Meeting of the Association for Computational Linguistics, Vancouver, BC, Canada, 30 July–4 August 2017; pp. 1073–1083. [CrossRef]
23. Vaswani, A.; Shazeer, N.; Parmar, N.; Uszkoreit, J.; Jones, L.; Gomez, A.N.; Kaiser, L.; Polosukhin, I. Attention is All You Need. In Proceedings of the 31st International Conference on Neural Information Processing Systems, Red Hook, NY, USA, 20–22 June 2017; p. 6000–6010.
24. Qi, W.; Yan, Y.; Gong, Y.; Liu, D.; Duan, N.; Chen, J.; Zhang, R.; Zhou, M. Prophetnet: Predicting future n-gram for sequence-to-sequence pre-training. In Proceedings of the 2020 Conference on Empirical Methods in Natural Language Processing: Findings, Online, 16–20 November 2020; pp. 2401–2410.
25. Magooda, A.; Litman, D.J. Abstractive Summarization for Low Resource Data using Domain Transfer and Data Synthesis. In Proceedings of the The Thirty-Third International Flairs Conference, North Miami Beach, FL, USA , 17–20 May 2020.
26. Kryściński, W.; Paulus, R.; Xiong, C.; Socher, R. Improving Abstraction in Text Summarization. In Proceedings of the 2018 Conference on Empirical Methods in Natural Language Processing, Brussels, Belgium, 31 October 31–4 November 2018; pp. 1808–1817. [CrossRef]
27. Zou, Y.; Zhang, X.; Lu, W.; Wei, F.; Zhou, M. Pre-training for Abstractive Document Summarization by Reinstating Source Text. In Proceedings of the 2020 Conference on Empirical Methods in Natural Language Processing (EMNLP), Online, 16–20 Novermber 2020; pp. 3646–3660. [CrossRef]
28. Le, H.; Vial, L.; Frej, J.; Segonne, V.; Coavoux, M.; Lecouteux, B.; Allauzen, A.; Crabbé, B.; Besacier, L.; Schwab, D. FlauBERT: Unsupervised Language Model Pre-training for French. In Proceedings of the 12th Language Resources and Evaluation Conference, Marseille, France, 11–16 May 2020; pp. 2479–2490.
29. de Vries, W.; van Cranenburgh, A.; Bisazza, A.; Caselli, T.; van Noord, G.; Nissim, M. BERTje: A Dutch BERT Model. *arXiv* **2019**, arXiv:1912.09582.
30. Ángel González, J.; Hurtado, L.F.; Pla, F. TWilBert: Pre-trained deep bidirectional transformers for Spanish Twitter. *Neurocomputing* **2021**, *426*, 58–69. [CrossRef]
31. Ortiz Suárez, P.J.; Romary, L.; Sagot, B. A Monolingual Approach to Contextualized Word Embeddings for Mid-Resource Languages. In Proceedings of the 58th Annual Meeting of the Association for Computational Linguistics, Online, 5–10 July 2020; pp. 1703–1714.

32. Lin, C.Y. ROUGE: A Package for Automatic Evaluation of Summaries. In *Text Summarization Branches Out*; Association for Computational Linguistics: Barcelona, Spain, 2004; pp. 74–81.
33. Zhang, T.; Kishore, V.; Wu, F.; Weinberger, K.Q.; Artzi, Y. BERTScore: Evaluating Text Generation with BERT. In Proceedings of the 8th International Conference on Learning Representations, ICLR 2020, Addis Ababa, Ethiopia, 26–30 April 2020.
34. Grusky, M.; Naaman, M.; Artzi, Y. Newsroom: A Dataset of 1.3 Million Summaries with Diverse Extractive Strategies. In Proceedings of the 2018 Conference of the North American Chapter of the Association for Computational Linguistics: Human Language Technologies, Volume 1 (Long Papers), New Orleans, LA, USA, 1–6 June 2018; pp. 708–719. [CrossRef]
35. Bommasani, R.; Cardie, C. Intrinsic Evaluation of Summarization Datasets. In Proceedings of the 2020 Conference on Empirical Methods in Natural Language Processing (EMNLP), Association for Computational Linguistics, Online, 16–20 November 2020; pp. 8075–8096. [CrossRef]
36. Barth, W.; Mutzel, P.; Jünger, M. Simple and Efficient Bilayer Cross Counting. *J. Graph Algorithms Appl.* **2004**, *8*, 179–194. [CrossRef]

MDPI
St. Alban-Anlage 66
4052 Basel
Switzerland
Tel. +41 61 683 77 34
Fax +41 61 302 89 18
www.mdpi.com

Applied Sciences Editorial Office
E-mail: applsci@mdpi.com
www.mdpi.com/journal/applsci

www.ingramcontent.com/pod-product-compliance
Lightning Source LLC
Chambersburg PA
CBHW041122250926
R18908400001B/R189084PG51850CBX00001B/1
9783036544397